Red Shephard

Andr

CW01636888

# A Treatise on Gyrostatics

and

# Rotational Motion

Elibron Classics
www.elibron.com

Elibron Classics series.



ISBN 1-4212-5592-8 (paperback)
ISBN 0-543-86951-2 (hardcover)

This Elibron Classics Replica Edition is an unabridged facsimile
of the edition published in 1918 by Macmillan and Co., Ltd., London.

By PROFESSOR ANDREW GRAY

*8vo. 14s. net*

A TREATISE ON

MAGNETISM AND ELECTRICITY

---

*2 Vols. Cr. 8vo*

THE THEORY AND PRACTICE OF

ABSOLUTE MEASUREMENTS IN ELECTRICITY AND MAGNETISM

VOL. I. 12*s*. 6*d*.
VOL. II. In Two Parts. 25*s*.

---

By PROF. ANDREW GRAY AND DR. J. G. GRAY

*Cr. 8vo. 10s. net*

A TREATISE ON DYNAMICS

---

By PROF. ANDREW GRAY AND PROF. G. B. MATHEWS

*8vo. 14s. net*

A TREATISE ON

BESSEL FUNCTIONS AND THEIR APPLICATIONS TO PHYSICS

---

LONDON: MACMILLAN & CO., LIMITED

MACMILLAN'S

# SCIENCE MONOGRAPHS

# A TREATISE ON GYROSTATICS AND ROTATIONAL MOTION

MACMILLAN AND CO., LIMITED
LONDON · BOMBAY · CALCUTTA · MADRAS
MELBOURNE

THE MACMILLAN COMPANY
NEW YORK · BOSTON · CHICAGO
DALLAS · SAN FRANCISCO

THE MACMILLAN CO. OF CANADA, LTD.
TORONTO

# A TREATISE ON
# GYROSTATICS
AND
# ROTATIONAL MOTION


BY

ANDREW GRAY, F.R.S.

PROFESSOR OF NATURAL PHILOSOPHY IN THE UNIVERSITY OF GLASGOW


THEORY AND APPLICATIONS


MACMILLAN AND CO., LIMITED

ST. MARTIN'S STREET, LONDON

1918

GLASGOW: PRINTED AT THE UNIVERSITY PRESS
BY ROBERT MACLEHOSE AND CO. LTD

# PREFACE

This book is an attempt to supply a systematic discussion of gyrostatic action and rotational motion which may be of use both to students of dynamics and to practical men. To many engineers the gyrostatic action of machinery is more or less a mystery, while to the student gyrostatics is an affair of certain formal equations by means of which, as by a kind of "hocus pocus," certain well-known elementary results can be obtained. These results seem to have been erected into a code of *dogmata* to be quoted and applied in all manner of circumstances. This has been especially notable in the history of invention during the war. In attempts to construct devices which should operate by gyrostatic action, inventors have often tried to combine mutually exclusive conditions, a fact which shows that the criteria of fulfilment of any specified conditions have not been clearly understood. And if inventors have been deficient in clear perception of the conditions of success in the construction of gyrostatic devices, it cannot be said that judges and critics of such appliances have been conspicuously more successful.

All this is no doubt to a considerable extent the result of the fact that as a rule the discussion of dynamical and especially of gyrostatic problems in the lecture-room and in text-books has either been restricted to one or two simple special cases, or, when more comprehensive, been far too much an affair of Euler's or Lagrange's equations, which, by providing a walled in path along which the mind can travel, withdraw its attention from the incidents, often exceedingly instructive, which attend its progress. This process may "pay" as a preparation for the academic puzzles of the examination room, but, as a training for dealing at once and from first principles with the practical scientific questions which arise in the course of a great European war, it is sadly deficient. What is required is a training in the analytical and numerical discussion of problems of actual apparatus, accompanied by the practical study of gyrostatic devices. This training can be best obtained in the lecture-room, laboratory, and workshop of a physical institute where real dynamics is made an important subject.

In the present work my aim has been to refer, as far as possible, each gyrostatic problem directly to first principles, and to derive the solutions by steps which could be interpreted at every stage of progress. In this way light is thrown on the formal processes, which, when their results are seen to be the outcome of a few simple primary considerations, acquire a freshness and life which as mere rules they do not possess. A large number of problems of all kinds have been thus dealt with, and in a final chapter of the present volume a considerable number of others have been collected, and solved either fully in detail or in a summary of steps with their results. Some of these problems were rather famous in the early history of the subject and were solved sixty or seventy years ago by Bour, Lottner, and others, using the formal equations; the remainder are those, many of them of great physical interest, which have presented themselves in the later history of the subject.

As a rule the method employed has been one of calculating rates of growth of angular momentum for different axes, which amounts to a reduction to practice of vector ideas, and which I devised nearly twenty years ago for use in my own teaching and dynamical work. At the foundation it is really only the simple notion that the *velocity* (in its fullest sense) of the outer extremity of the vector of resultant angular momentum represents in direction and magnitude the moment of the resultant applied couple, as given by the couple-axis. This notion, *mutatis mutandis*, can of course be applied to any vector quantity and its rate of change.

It has been stated that all such methods merely amount in substance to the formal equations given in all the usual treatises, and no doubt that is true in a certain sense. Nevertheless the criticism is pointless. There is a very important sense in which every process of a vector nature (whether it follow the notation of a vector analysis or not) differs from the usual equations: the former declares interpretations at every step to those who can read them, the latter does not. Thus, insight into mechanical action is fostered; a continual appeal to first principles accustoms the mind of the user to the application of fundamental ideas to all kinds of contrivances and in all circumstances. The perception of the action of more or less complicated gyrostatic or other dynamical devices becomes, in time and with the continual exercise of thought, intuitive when the processes are fundamental and interpretative; on the other hand, the traditional and scholastic method fosters the habit of reliance on the operation of a kind of machine of which the solver of the problem merely turns the handle. How to arrange the machine, what to feed into it and what to leave out, puzzles him at every change of application and environment.

Free use has been made of mathematical analysis, though only as a means of obtaining and explaining results of physical interest, and of preparing these for numerical computation, and, further, of placing students in a position to make progress in the discussion of new problems or of carrying old problems to a further stage of development.

It is impossible to enumerate here all the cases in which I am indebted to original researches or to dynamical literature, but in the proper place I have always tried to make due acknowledgment. As was to be expected, I am under special obligation to the writings of Sir George Greenhill. His *Report on Gyrostatic Theory* will long form the chief digest of learning and research on the applications of Elliptic Functions to Dynamics, but besides my indebtedness to that important work I owe much stimulation and help to our correspondence from time to time on matters gyrostatic.

As regards the reading of proofs, I have been helped in the earlier part of the book by my colleagues Professor G. A. Gibson, who read the first 250 pages with much care and gave me much valuable advice, and Dr. R. A. Houstoun. But the exigencies of teaching and work in great University departments, with their staffs depleted by the demands of the war, rendered the continuance of such help impossible, and latterly I have had mainly to depend on my own unaided scrutiny of the proofs. Hence many slips may have escaped observation. A few sheets were, however, very kindly looked over by Sir George Greenhill, while my old colleague Dr. G. B. Mathews has come also to my assistance from time to time.

I must also acknowledge the great care which the compositors and readers of the Glasgow University Press have bestowed on every detail of type-setting and printing, and their patience in meeting all the troubles which arose in carrying through a very difficult piece of work.

It was originally intended to include in the volume an account of a variety of gyrostatic devices which are of use in naval and military affairs and in engineering. Any such account, written on lines permissible at present, would however be more or less fragmentary, and therefore unsatisfactory; and it is proposed to issue, if that is possible and convenient, a supplementary Second Part after the war has come to an end. That will also include, besides a series of diagrams of the motion of tops, various numerical results illustrative of some interesting parts of the elliptic function theory.

In the discussion of practical applications I hope to have the assistance of my son Dr. J. G. Gray, who for several years at Glasgow, where the subject is traditional, has made gyrostatics his main scientific work.

ANDREW GRAY.

" For as whipped tops or bandied balls,
The learned hold are animals ;
So horses they affirm to be
Mere engines made by geometry."

*Hudibras.*

". . . stupet inscia supra
Impubesque manus, mirata volubile buxum."

Verg., *Æn.* VII, 381-2.

## CONTENTS

### CHAPTER I

### INTRODUCTORY

## CHAPTER II

### DYNAMICAL PRINCIPLES



## CHAPTER III

### ELEMENTARY DISCUSSION OF GYROSTATIC ACTION



## CHAPTER IV

### SYSTEMS OF COORDINATES AND THEIR RELATIONS. SPACE-CONE AND BODY-CONE

## CHAPTER V

### THE SIMPLER THEORY OF TOPS AND GYROSTATS



## CHAPTER VI

### FURTHER DISCUSSION OF THE RISE AND FALL OF A TOP WHEN THE INITIAL PRECESSION IS NOT ZERO

## CHAPTER VII

### GYROSTATS AND VARIOUS PHYSICAL APPLICATIONS OF GYROSTATS



## CHAPTER VIII

### VIBRATING SYSTEMS OF GYROSTATS. SUGGESTIONS OF GYROSTATIC EXPLANATION OF PROPERTIES OF MATTER



## CHAPTER IX

### THE MOTION OF CHAINS OF GYROSTATIC LINKS. MAGNETO-OPTIC ROTATION

## CHAPTER X

### THE EARTH AS A TOP. PRECESSION AND NUTATION. GYROSTATIC THEORY OF THE MOTION OF THE NODES OF THE MOON'S ORBIT



## CHAPTER XI

### THE FREE PRECESSION OF THE EARTH. FURTHER DISCUSSION

## CHAPTER XII

### CALCULATION OF THE PATH OF THE AXIS OF A TOP BY ELLIPTIC INTEGRALS



## CHAPTER XIII

### LIQUID GYROSTAT. MISCELLANEOUS INVESTIGATIONS

## CHAPTER XIV

### EFFECTS OF AIR FRICTION AND PRESSURE. BOOMERANGS



## CHAPTER XV

### THE SPHERICAL PENDULUM. MOTION OF A PARTICLE ON A SURFACE OF REVOLUTION

## APPENDIX TO CHAPTER XV



## CHAPTER XVI

### DYNAMICS OF A MOVING FRAME CONTAINING A FLYWHEEL



## CHAPTER XVII

### MOTION OF AN UNSYMMETRICAL TOP

## CHAPTER XVIII

### THE RISING OF A SYMMETRICAL TOP SUPPORTED ON A HORIZONTAL SURFACE



## CHAPTER XIX

### GENERAL DYNAMICS OF GYROSTATIC AND CYCLIC SYSTEMS

## CHAPTER XX

### THEORY OF GYROSTATIC DOMINATION



## CHAPTER XXI

### GEOMETRICAL REPRESENTATION OF THE MOTION OF A TOP

## CHAPTER XXII

### ANALOGY BETWEEN A BENT ROD AND THE MOTION OF A TOP. WHIRLING OF SHAFTS, CHAINS, ETC.

## CHAPTER XXIII

### EXAMPLES OF GYROSTATIC ACTION AND ROTATIONAL MOTION

## CORRIGENDA

p. 426, line 16, *for* $-\kappa\theta\sin\theta$ *read* $-\kappa\dot{\theta}\sin\theta$
*and for* $-\psi\kappa\partial(\cos\theta)/\partial\theta$ *read* $-\dot{\psi}\kappa\partial(\cos\theta)/\partial\theta$

p. 428, line 2, *for* $(A+ma^2)\theta$ *read* $(A+ma^2)\theta$
(The dots have failed to print.)

p. 520, line 9 from foot, *for* united *read* initial

p. 524, line 38, *after* and *insert* , except for the couple mentioned in the next paragraph,

# CHAPTER I

## INTRODUCTORY*

**1.** *Gyroscopes and gyrostats.* A gyroscope is generally regarded as a toy, the behaviour of which is mysterious and unnatural. It consists of a flywheel, generally in the form of a disk with a massive rim, mounted in an open frame, which may be supported in various ways, for example as shown in the adjoining diagram. But the flywheel may be enclosed in a

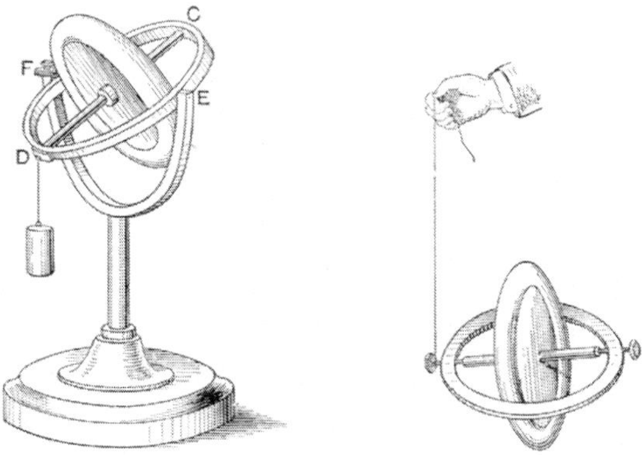


(a) (b)

case which completely conceals it, as when, in the music-hall entertainment, a cheese-shaped body is set up on edge and successfully resists the efforts of a strong man to turn it down flat. The concealment of the rapidly rotating flywheel in this case gives an additional element of mystery. When the flywheel is thus concealed we have what Lord Kelvin called a gyrostat, because, in virtue of the rotation of the flywheel, the arrangement stands with any of the edges of the case resting on a hard smooth table.

The arrangements of the apparatus shown in the diagram above allow

* In this chapter a general account of gyrostatic phenomena and of a number of experiments is given; full explanations of all these will be found in later chapters of the book.

some remarkable facts to be verified. In diagram (*a*) of the figure the flywheel is shown with its axis CD held by a ring, which can turn about bearings E, F at right angles to CD. These bearings are at the extremities of a vertical fork carried by an upright stem, which we shall suppose is free to turn in a vertical socket carried by a massive base-piece resting on the table. The bearings E, F are so arranged that the ring and flywheel can be readily removed from the fork and securely replaced when desired. For the present we suppose that there is no weight applied at D, and that the axis EF passes through the centre of gravity of the wheel and frame, which coincides with the centre of the wheel.

2. *Experiment showing permanence of direction of axis of rotation. Action of torpedo.* Now let the following experiment be tried. The flywheel being at rest and the arrangement in equilibrium, the base-piece is

turned round on the table. The upright stem with its fork is carried round by friction in the socket, and so the flywheel is also carried round, and its axis changes direction in space. The ring with the flywheel is now removed from the fork EF, and placed so that the axle of the wheel rests on the rubber edge of a disk, carried on the overhanging end of the spindle of an electric motor [see figure]. By means of a starting resistance the motor is run slowly at first, and its speed gradually increased to its utmost value. The operator presses the spindle on the disk, the flywheel gets into motion, and runs faster and faster, as the speed of rotation of the spinner increases. Slip between the spindle and disk is avoided. The ring is now replaced on the bearings EF with the axle CD in any convenient inclined position. It is found now that the behaviour of the apparatus is entirely changed.

First it is seen that when the base-piece is turned round on the table the vertical stem is no longer carried round, but remains apparently fixed, so that the axis CD points in a constant direction. No matter how the base-piece is turned round or shifted on the table, the direction of the axis of the wheel remains practically unchanged. The wheel has now acquired the property of resisting any action tending to produce alteration of the direction of its axis of rotation.

The reader will now be able to divine the explanation of the music-hall experiment. A massive flywheel revolves within the cheese-shaped body, with its axis along the axis of symmetry of the outward shape of the body. When an attempt to throw the body down flat is made, the flywheel powerfully resists any change in the direction of its spin axis, and the attempt is unavailing. For the frontal attack on the body must be substituted what is literally a "flank movement," which may perhaps be inferred from what follows.

This permanence of direction of axis of spin explains the stability and precision of a rifle bullet. It is also utilised in the torpedo to maintain the projectile in the direction in which it started. The hollow body of the torpedo contains a gyrostat placed with its axis of spin along the axis of

the "cigar," and this gyrostat is started by the release of a powerful spring at the moment of firing. If the "cigar" swerves a little from the initial direction the two axes diverge at a small angle, since the axis of spin preserves its direction. In consequence of this angle a motor is called into play to actuate a rudder in such a way as to annul the deflection. Thus the gyrostat acts as if it had both a nervous and a muscular system. It detects the swerve and calls into play the forces required to correct it.

3. *Motor-driven gyrostats.* For various purposes a motor-driven gyrostat is very convenient, for example a combination of gyrostats can be much more readily adjusted and controlled if the gyrostatic elements are motor driven. Dr. J. G. Gray and Mr. George Burnside have constructed a very compact arrangement in which the flywheel of the gyrostat is the rotor of a continuous-current electric motor. The parts are shown in the diagram. The armature of the rotor is a Gramme ring wound on a ring of malleable cast-iron, with radial iron projections between the segments of the winding. The field is supplied by a small electro-magnet, of which the cone and pole pieces are shown on the right at the bottom of the diagram. This magnet forms the stator: it carries two windings, one on each side of the shaft.

The armature ring is supported centrally from the shaft, which is stationary, by means of two magnalium disks perforated to allow the armature and field magnet to be cooled by circulation of air. The disks are recessed near their outer edges to fit tightly into the inner periphery of the armature ring, and have gun-metal bushes. The magnalium is very light, and so practically the whole mass of the rotating system is contained in the ring. The commutator is mounted on one of the disks: the brushes make contact with phosphor-bronze studs projecting inward from the magnalium disk which carries the armature ring.

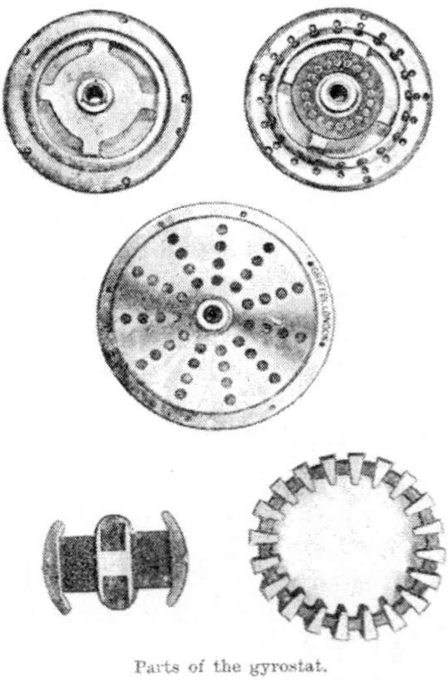

Parts of the gyrostat.

The case is made in halves, provided with flanges to enable them to be screwed together after they are placed over the motor.

The form of gyrostat with open frame, shown in the preceding diagram as being spun on the motor-driven disk apparatus, has however been improved so much by Dr. Gray and Messrs. Griffin as regards balance, management of bearings, and lubrication that it can be used for all kinds of purposes without danger of derangement, spun many times with one oiling, and will continue to spin for from 10 to 40 minutes according to circumstances. Thus for a large number of ordinary experiments this type is perfectly convenient, and its cost is necessarily considerably less than that of the motor instrument.

4. *Effect of couple applied by weight hung on one side.* Let the weight be hung on at D, as shown in the diagram ($a$), in 1. If the flywheel were not turning on its axis the effect would be simply to turn the ring and wheel about EF, and the weight would descend until it came into contact with the supporting stem. But with the wheel rotating rapidly the result is widely different. A very slight downward motion of the point D can be detected as the immediate effect. but as soon as the least such deflection has taken place the whole system begins to turn round, so that the point D moves sideways, or, with reference to the diagram ($a$), p. 1, at right angles to the paper. This motion goes on, and presently, as far as the eye can detect, a steady motion of the stem, fork, ring and flywheel (all as if they composed a rigid body) about the vertical has been set up. The point D of the axis does not appear to descend further, and

the axis CD remains at a constant inclination to the horizontal while its azimuth changes, that is the vertical plane containing CD turns steadily round the vertical. If there were no friction between the moving parts, what would be seen would be that the axis CD alternately rose and fell through a small range, so that on the average the apparatus turned steadily round as just described.

5. *Experiment of gyrostat supported by cord attached at point in line of axis.* As shown in diagram (*b*), in 1, let the ring and flywheel be removed from the fork and hung by a string attached to a knob projecting from the ring. The wheel is steadied until the axis of rotation is horizontal and the arrangement is left to itself. It is supposed that the wheel is rotating rapidly about its axis Again a downward deflection, so slight as to be hardly observable, takes place, and a sideways motion is set up, so that the axis of spin of the flywheel turns round in a horizontal plane. Strictly speaking, the gyroscope does not turn about the string as a vertical axis, for the string is not exactly vertical. If the point at which it is gripped by the hand remains fixed, the string is inclined so that its lower end is a little displaced from the vertical towards the supported gyroscope, for upon the gyroscope moving as a whole round the vertical through the point of support—the upper end of the string—the string must exert an inward component of pull. But this displacement is slight, and we may say, with a considerable approach to accuracy, that the axle of the flywheel turns steadily round in a horizontal plane while the string remains vertical. In reality the string describes a narrow cone about a vertical axis.

6. *Behaviour of gyroscope dynamical but apparently unnatural.* The ordinary undynamical observer who knows a little of the facts, but practically nothing of the reasons, of the behaviour of ordinary non-rotating bodies, immediately asks the question, "Why does the gyroscope not fall down?" The question is a natural one, for the behaviour of the gyroscope appears to him to be most unnatural. He does not pause to consider that if it did fall down the planes of rotation of the matter of a massive wheel would have to be quickly changed, and that such a change may possibly be difficult or impossible to effect by a direct overturning action applied to the wheel. His attitude is just that of the ordinary music-hall spectator of the unoverturnable "cheese."

The idea that suggests itself to those who, without any dynamical knowledge, try to think out causes, is the erroneous one that gravity is neutralised in some way by the rotation. On this supposed neutralisation of gravity has been based a proposal (perhaps several proposals) for the construction of a flying machine. Such observers fail to note the significant fact that the gyroscope, when its axis remains horizontal in this apparently paradoxical manner, is turning round in azimuth, and that if this turning is checked by a resistance applied to the turning axle, the wheel at once begins to descend in the "natural" way. But here another puzzling result is obtained. If, instead of trying to impede the motion of the axle in azimuth (the *precessional motion*, as it is called), the observer tries to assist it, the outer end of the axle *rises*, that is to say the centre of gravity of the gyroscope is raised.

7. *Elementary phenomena and their dynamical explanation.* The dynamical theory of rotational motion and of gyrostatic action is fully discussed in the chapters which follow; but at the risk of some repetition an indication of the principles of gyrostatics is given here for the sake of general readers.

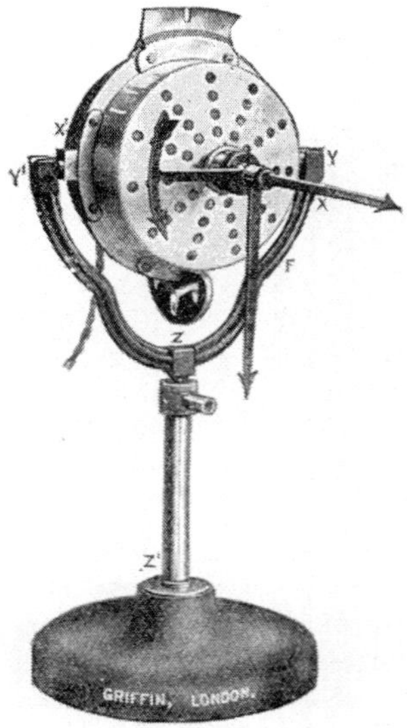


We consider the case of steady precessional motion. It is supposed to be mysterious, but it is really very simple. It is illustrated by the pedestal gyrostat shown in the first figure, or by the motor-driven pedestal gyrostat shown in the adjoining figure. The curved arrowhead shows the direction of rotation, the projecting arrow the axis of spin, the arrow pointing down can be turned so as to show the direction of the axis of any applied couple, that is moment about YY′ of any applied forces. It is maintained by a spring washer in any position in which it may be placed. At the other end of the axis a weight balancing all these rods is hung, so that the centre of gravity may remain at the centre of the wheel.

Let then the arrow last specified be turned so that it points horizontally to the right. It represents now a couple which would turn the spin axis down, if the flywheel were not rotating. It is at right angles to the parallel planes in which, in obedience to the couple, particles of the body would move, if there were no rotation; and it is drawn towards the observer if the turning due to the couple, looked at from a point beyond the arrow-head, is in the counter-clock direction. The length is made equal to the moment, and thus the line (which is called the axis of the couple) represents the couple in all respects. [The phrase "spin axis" means the axis of the flywheel.]

A couple may be applied by pushing down the front of the case by hand or otherwise: it is found that the spin axis turns, as far as can well be observed, in a horizontal plane, in the direction in which it follows the couple axis. The two axes turn together since they are rigidly connected, but the spin axis is at each instant turning towards the position which the couple axis has at that instant, that is, towards the right.

If the back of the case is pushed down, the horizontal turning of the spin axis is towards the left. But as the couple is reversed in sign from its former value, the pointer, to represent it, should be turned towards the left, so that the general rule that the spin axis follows the couple axis again holds.

This turning is due to the fact that as the spin axis, with the angular momentum $Cn$ about it, turns towards the couple axis with angular speed $\omega$, there is a rate of production of angular momentum measured by $Cn\omega$ about the latter axis. Thus, when there is steady turning, the rate of production of angular momentum about the couple axis, equal numerically to the moment of the couple, is supplied by the motion of the body.

If it should happen that there is turning of the spin axis (OX in the diagram) towards a direction, OY say, at right angles, about which no couple acts, then the body will begin turning about the second axis OY in the direction to neutralise the rate of production $Cn\omega$ of angular momentum due to the motion of the spin axis. [This in general will cause turning about OZ, which we do not now discuss.] If the consequent displacement brings into play a couple about OY sufficient to account for this $Cn\omega$, we may have steady motion set up. The configuration then remains unchanged, the spin axis chases, so to speak, the couple axis round, and we have a state of relative equilibrium with constant precession (as it is called) of angular speed $\omega$.

If however the neutralising motion about OY referred to above is, as it may be, in the direction to call into play a moment of forces, or couple, in the opposite direction about OY to that required to account for $Cn\omega$, there will be instability, the deflection will go on indefinitely increasing. We have an example in the gyrostat with axis vertical and carried round on a tray in azimuth in a direction opposed to that of the rotation of the flywheel, the experiment described in 20, below.

Next we observe that when it is attempted to retard the precessional motion, by applying force in the "natural" way to effect this result, the axis descends, if it is in like manner attempted to accelerate the precession the axis rises. This experiment shows that the horizontality of the axis depends on the freedom of the gyrostat to precess at a certain definite rate. This rate, as we shall see presently, depends, in the case of the hanging gyrostat described above, on the couple applied by the weight of the gyrostat acting downward in one vertical line, and the string pulling upward in another line nearly vertical, and on the angular momentum of the flywheel.

We look at the thing now in this way. The axis of rotation round which the flywheel has angular momentum is turning towards the axis A of the couple, with angular speed, $\omega$ say. Now, and this is the point not recognised as a rule, *this motion itself creates a rate of production of* *angular momentum about the axis A of the couple.* For when an axis O$l$ with which is associated a directed quantity, L say, is turning towards a fixed direction O$m$ at right angles to it with angular speed $\omega$, there is a time-rate of production of the quantity of the same kind associated with the latter direction measured by the product L$\omega$.*

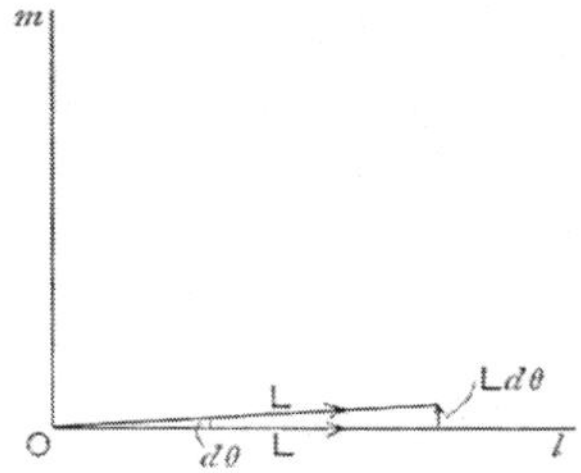


Now the flywheel is revolving with angular speed $n$, so that if its moment of inertia is C, it has angular momentum C$n$ about the axis of rotation O$l$; but with angular speed $\omega$ the axis O$l$ is turning towards the instantaneous position of the axis O$m$, a fixed direction to which O$l$ is at the moment perpendicular, and, in consequence of this turning, a rate of production of angular momentum C$n\omega$ exists about O$m$. [See 12, II below.]

Now for the steady motion of the gyrostat, that is steady turning in azimuth without rising or falling of the axis, it is only necessary that this rate should be equal to the moment of the couple about O$m$, G let us say. Thus we get $Cn\omega = G$, which gives $\omega = G/Cn$.

If the precession is hurried by a little impulse applied to the wheel, and the gyrostat is then left to itself, the hurried motion, if it continued afterwards in the horizontal plane, would result in a more rapid generation of angular momentum about O$m$ than there is moment of couple to account for, and the gyrostat would begin to turn about O$m$, in the direction to cause the angular momentum to be produced at the proper rate, that is the axis would begin to rise. In the same way an impulse towards delaying the precession would cause the axis to begin to descend. In each case the result would be a succession of alternate rises and descents; but the subject of vibrations about steady motion is too difficult to discuss here, and will be found treated later in the book.

**8.** *Two possible precessions in general case.* In any case of steady motion there are two possible precessional motions for the same spin and the same inclination of the axis of spin to the vertical, which are given in the theory as the roots of a certain quadratic equation [see 18, II]. One is great, the other small. The former to the first approximation does not

*For example, a particle of mass $m$ is moving at any point P along a curve, that is along the tangent to the curve at P, and therefore the direction of motion is changing at P with angular speed $v/r$, where $r$ is the radius of curvature at P. We may regard the tangent as an axis with which is associated the momentum $mv$, and which turns round with angular speed $v/r$, as the point of contact advances along the curve. Thus along the direction towards the centre of curvature at P, which direction is fixed for P, and towards which the tangent at P is turning, there is a rate of growth of momentum measured by $mv\omega = mv \cdot v/r = mv^2/r$, a very well-known result. The same process holds for any directed quantity (momentum, angular velocity, angular momentum, etc., associated with an axis).

depend on applied forces, the other does. Lord Kelvin called the former "adynamic," the other "precessional." But in strictness both involve the forces, and they appear as the roots of a certain equation. One of these is at once approximately realised when the wheel is spun fast, the gyrostat set on the plate at rest, and left to itself. The motion is one of small oscillation about the steady motion, which is characterised by slow precession, given very nearly, but not quite exactly, by the same formula as before. The other motion of the axis in the same cone is one of much greater precessional angular speed, which is given by a slightly more complicated formula [see 18, II].

In strictness we must regard this second precessional motion as characteristic also of the gyrostat when its axis is horizontal, but in that case the precessional angular speed is infinite, and only the slow motion is realisable.

The rule often stated that hurrying a gyrostat in its precession causes tilting up of the axis, and delaying the precession causes tilting downward, is true only of the slower more usual precession. For the faster precession exactly the reverse rule holds good. This fact does not seem to be generally known, as the rule is often stated absolutely.

It is important to notice that if the centre of gravity of the gyrostat is above the point of support, supposed on the line of the axis, the two precessional motions are in the same direction; if on the other hand the centre of gravity be below the point of support, the precessional motions are in opposite directions. The faster motion changes sign in passing through an infinite value, when the axis is horizontal.

By the effect of hurrying or retarding the precession is explained the rising and falling of a top spinning on a rounded peg in contact with a rough floor along which the top can move. At first the spin is fast and the slipping is such as to produce a hurrying friction couple which causes the erection of the top. After the spin has fallen off the slipping is the other way and a couple which produces the reverse effect results, and the top falls. But the action is complex, and requires mathematical discussion: it will be dealt with in a later chapter.

**9.** ***Bessemer's gyrostatically controlled saloon.*** The idea occurred to Bessemer of taking advantage of the resistance which a rapidly spinning flywheel offers to change of its plane of motion for the construction of a ship's saloon, the floor of which should be kept, by gyrostatic control, fixed in direction as the ship rolled or pitched. The idea was quite reasonable, and justified an attempt to realise the saloon. So a cabin weighing 180 tons was suspended from a fore and aft axis [see *The Engineer*, 1875], and a gyroscope with axis of spin vertical, for the ship at rest on even keel, was provided to control it. But in doing so the designers and constructors ignored the fact that, as in the experiment described above, gyrostatic resistance to change of direction of the axis of rotation is bound up with freedom of the gyroscope to turn as a whole with precessional motion. This freedom of precession (about an axis athwart the ship) was not provided for in the construction of the saloon, and so the experiment was unsuccessful so far as mitigation of rolling was concerned. The arrangement however was such as to modify the pitching motion, though that fact does not seem to have been perceived by the designers or their critics. The same idea has been revived in recent years and realised with fair success in the Schlick apparatus, described in Chapter VIII below, for controlling the rolling of a ship.

10. *Gyroscope or gyrostat is merely a spinning top. The earth is a spinning top. Spinning of an ordinary top.* The gyroscope is however merely a glorified spinning top, and the person who asks why the gyroscope hung by a cord, and precessing with the axis of rotation of the flywheel horizontal, does not fall down, seldom seems to notice anything remarkable in the fact that a top spinning upright on its peg stands up stably, and when its spin has died away lies down on its side on the ground, and will not stand up without spin. Top spinning is practised by all small boys, and the behaviour of a top is taken as a matter of course. "Familiarity breeds contempt," says the proverb; that familiarity with scientific devices which consists merely in their daily use, or in seeing them frequently, produces indifference.

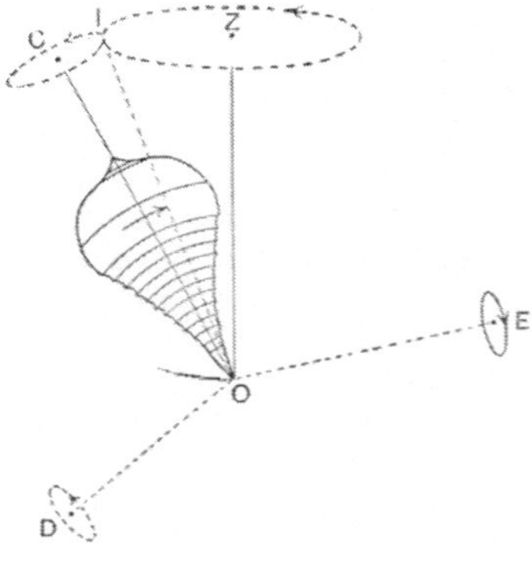


When however the top is elaborated a little into a gyroscope, and new experiments are made with it, such as that described above of making it hang on one side of a nearly vertical string, many people, who accept without question the ordinary top, and its quite as wonderful properties, ask at once the question discussed in 6. Only a very few ordinary observers grasp all the essential facts of what they see. And comparatively few realise that the earth is merely an ordinary top, which spins about its axis of figure at the rate of one turn in about four minutes less than a solar day. If an ordinary top is spun by the usual process of throwing it from the hand, so that it alights on its peg just after a string has been quickly unwound from the body of the top, its axis is usually at first inclined at a somewhat large angle to the vertical, and sways round with a conical motion. Gradually, by the action of frictional forces which has been referred to in 8, the angle of inclination of the axis to the vertical becomes smaller and smaller, and the centre of gravity rises until the axis is vertical and the top "sleeps."

This conical motion is indicated in the last diagram. The circles, round the vertical and the axis of the top, there shown, indicate the conical motion as produced by the rolling of a cone, axis OC, fixed in the body, round a cone, axis OZ, fixed in space, in the manner further described below in connection with astronomical precession.

After a little as the spin falls off, mainly in consequence of air friction, the rotation becomes too slow for maintenance of the vertical position, and the axis of figure becomes again inclined to the vertical: the conical motion is resumed and goes on with increasing inclination until the top has fallen.

As will be shown later, a certain limiting angular speed of spin is necessary that the top may be in stable equilibrium with its axis in the upright position. Also it will be proved that, when the axis has fallen away from the vertical, the conical motion of the axis—the precession of the top—is due to the action of gravity on the inclined over body, that same action which in the case of the gyroscope makes the instrument turn round in azimuth, and which, according to the undynamical observer, ought to bring the centre of gravity down at once to the lowest possible position. But if the falling off of rapidity of spin could be stopped by withdrawing all resistance to the spinning motion, while the axis is left inclined over and moving properly sideways, the conical motion would continue indefinitely and the inclination of the axis to the vertical would remain unchanged.

**11.** *The earth's precessional motion.* The earth has, besides the rotation about its axis of figure, a precisely analogous conical motion of its axis, or, to be more exact, its axis would move in a cone if its translational motion in space were annulled. This conical motion is due to the gravitational action of the sun and of the moon on that belt of matter round the earth's equator, which may be regarded as constituting the deviation of the earth's mass from a spherical distribution. The earth's centre moves, with very slight deviations, in a certain plane round the sun, the plane of the ecliptic, as it is called, to which the plane of the earth's equator is inclined at an angle of 23° 27′ 8″ (the so-called "obliquity of the ecliptic"), so that the axis of spin of the earth, that is the line joining the poles, is inclined to the ecliptic at the complementary angle 66° 32′ 52″. The "falling down" of the earth which this gravitational action would by analogy "naturally" produce, is a turning of the plane of the equator into coincidence with the ecliptic, so that the earth's axis should set itself at right angles to the ecliptic. If this were to happen there would be a disappearance of the succession of seasons; only the slight variation caused by the eccentricity of the earth's orbit would be left, and a practically unvarying climate would prevail in each zone of the earth's surface. Instead of undergoing any such catastrophic change in position, the earth's axis moves round slowly in the conical path referred to above, and the angles of inclination of the equator and of the axis of spin to the ecliptic remain unaltered. Thus the direction of the earth's axis in space undergoes a great, though slowly proceeding, periodic change.

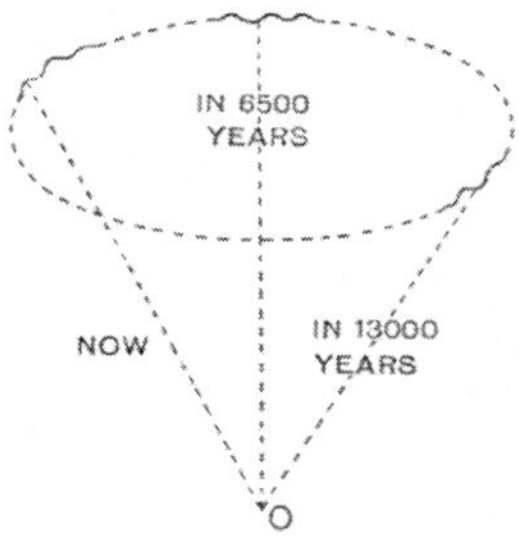


The position of the north or south pole on the celestial sphere moves slowly round in a circle, the angular diameter of which subtended at the earth's centre is twice the obliquity, or 46° 54′ 16″, and completes a circuit in 25,800 years. The fixed stars are so distant—the nearest is about 275,000 times as far from the earth as the sun is—that the translational motion of the earth in its nearly circular orbit of about 186,000,000 miles in diameter has hardly an observable effect on the apparent path of the pole among the fixed stars. About 12,000 years hence the pole will point nearly to the star α Lyrae; about 4,000 years ago, when the Pyramids of Egypt were built, the pole star was the star α Draconis. The sloping passages in the Pyramids seem to have been constructed so as to be used for the observation of that star.

**12.** *Conical motion of axis produced by rolling of body cone on space cone.* This conical motion may be regarded as due to the rolling of a narrow cone fixed in the earth (and therefore moving with the earth), having its axis coincident with the earth's axis, upon a cone fixed in space, the axis of which is at right angles to the ecliptic, and the semi-angle of which is 23° 27′ 8″. This is illustrated by the diagram, which shows a model of the earth mounted so as to show the precessional motion. The former is called the body cone, the latter the space cone. The upper surface of the flat ring round the stand is the plane of the ecliptic, the dark line NE is part of the equator, PN is a meridian drawn from the pole P to the intersection N of the equator with the ecliptic. The point N represents one of the equinoxes or "nodal points" at which the sun in its apparent motion among the stars passes from one side of the plane of the equator to the other. Lines drawn from the centre (not shown) of the globe to points on the ring shown as situated parallel to the ecliptic at the top of the model, give the space cone. The

body cone is represented by its upper end P, which is a rod of brass resting on the inner surface of the ring. This rod is a very narrow cone with its apex also at the centre of the globe, and its axis of figure is that of the globe. With the rod resting on the ring the globe is set spinning about P in the counter-clock direction as seen from above. It will be seen that, while for the ordinary top as shown in the diagram in 10, the body cone rolls on the outside of the space cone, the body cone here rolls on the inside of the space cone.

The cone P then rolls round the ring, of which it makes one circuit in the time $T.D/d$, where T is the time of one turn of the globe on its axis and $D/d$ the ratio of the diameter of the ring to the diameter of the cone at P. [Strictly speaking, the ratio should be $(D-d)/d$, but $d$ is very small in comparison with D.] As the cone rolls round the meridian NP travels with it, and the equinoctial point N moves round along the ecliptic. Thus we have the precession of the equinoxes.

**13.** "*Diameter of earth's axis.*" We can now solve an interesting problem which was sometimes proposed to the students of the Natural Philosophy Class of the University of Glasgow. Supposing the polar radius of the earth to be 4000 miles and the precessional period to be 25,800 years, find the diameter of the earth's axis! This problem puzzled most students at first sight, as its statement seemed to imply the possession by the earth of an actual material axle, but an inspection of the model globe, of which the diagram above is a picture, made the question into an interesting exercise, both on the geometry of motion and on an important question of physical astronomy. The time T is one sidereal day, and the time $TD/d$ [strictly, $T(D-d)/d$] is 25,800 equinoctial years of $366\frac{1}{4}$ sidereal days each. Thus we have approximately $25800 \times 366\frac{1}{4} = D/d$. But

$$D = 2 \times 4000 \times \sin(23° \, 27' \, 8'').$$

Hence we get, in feet,

$$d = \frac{2 \times 4000 \times 5280 \times \sin 23° \, 27' \, 8''}{25800 \times 366\frac{1}{4}} = 1{\cdot}78.$$

The answer to the question is therefore 21 inches.

**14.** *Free vibrational motion of a top. Free and forced vibrations.* The motion of the terrestrial top in the gravitational field of force, due mainly to the sun and moon acting on the equatorial belt of matter, is so far as we have gone analogous to that of a child's top spinning in the field of force due to the earth's attraction on bodies at its surface. But the analogy can be traced into further details of the motion. The child's top, in the early part of its spin, rises till its axis is vertical and then sleeps. If there were no frictional forces acting to raise the axis to the vertical, in the manner referred to in 8, it would continue spinning steadily about its axis while the conical motion, due to the deviation of the axis from the vertical and the consequent couple about a horizontal axis, would continue without alteration of rate or of the inclination of the axis to the vertical. But as it is any little inequality of the pavement, encountered by the peg, produces a little disturbance, and the top "wobbles." This wobble is soon extinguished by the frictional forces acting, but if these did not exist there would be a periodically repeated disturbance of the motion. This would be seen by a small vibrational motion of the axis of the top about the motion which it had before the disturbance arose.

The vibration has a period which depends on the system of forces under which the ordinary undisturbed motion takes place, on the distribution of matter in the top, and on the inclination of the axis to the vertical in that motion. This period is called a "natural" or free period of vibration of the top

The period arising from a single small disturbance must not be confused with the periodic motion which would be produced by a periodically occurring disturbance. In this case what is called a "forced vibration" would be produced in the period of the disturbance, not in any natural period of the body, unless a natural period and the forced period happened to agree, in which case a large disturbance might quickly grow up.

A good illustration of this is found in the rolling of a ship. If the ship is inclined over, in otherwise still water, and left to herself, she will then roll in a period which depends on the ship alone, and is a natural or free period. If, however, she steams in a seaway, in such a direction that waves pass under her transversely, she rolls in the period in which she encounters the successive waves, that is she performs transverse oscillations in a *forced* period.

In the same way the earth-top is subject to periodic alterations of the forces applied by the sun and moon, as the system of bodies periodically change their relative position, and so we have forced vibrational variations of precession, and those changes of inclination of the axis to the ecliptic which are called *nutation*. But, comparatively recently, the question has been asked, Is the motion of the earth-top affected by any free period oscillation? We can calculate the period of a free oscillation of the axis about the mean position, and the calculation is given in X below. It comes out 306 days approximately.

15. *Periodic changes of latitude.* The next question is how such a free oscillation of the earth-top would disclose itself. It is clear from what has been stated above that the axis of figure, remaining fixed in the body, would vibrate about its mean direction in space. This mean direction would be, as we shall see later, that of the axis OK, say, of resultant angular momentum of the earth, due to its spin and the precessional motion

combined. This axis, nearly coinciding with the axis of figure, we may, for the consideration of the motion of the axis of figure, take as at rest, though as a matter of fact it is displaced in consequence of the forced precession and nutation. Hence, as the earth turns, OK, thus fixed directionally in space, describes a cone in the earth, and the pole of the earth moves round among the stars in a circular path about the intersection of OK with the celestial sphere, in the period of free vibration. But the usual way of determining the latitude of a place is by observing the

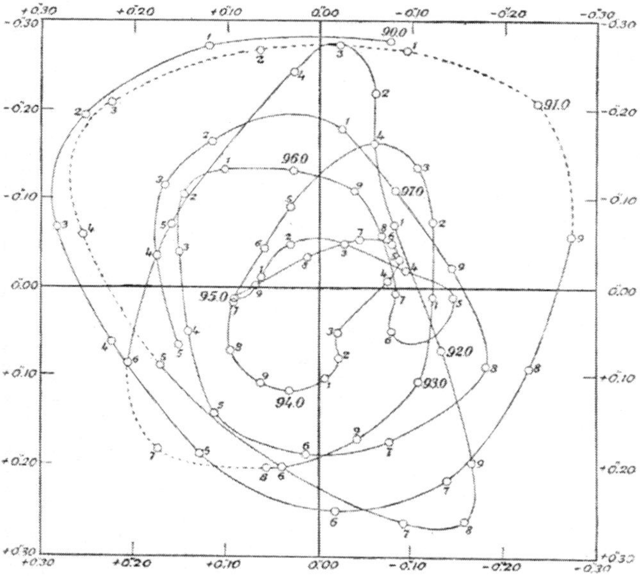


Albrecht, *Astr. Nachr.* No. 3489; *Nature*, May 12, 1898.

apparent altitude of a star at an interval of 12 hours, and therefore, since the pole moves round a fixed point on the celestial sphere, a periodic alteration of latitude will be noticed if the observational apparatus is sufficiently delicate.

It is found (see *loc. cit.* below) that if the moment of inertia of the earth about the axis of figure be C, and the mean value of that about an equatorial diameter be A, the period of a free oscillational variation is A/(C−A) times the period of rotation. Now A/(C−A) is about 304, so that the natural period for a rigid earth is about 304 sidereal days.

A careful scrutiny of observations of latitude, which was made by Mr. S. C. Chandler of Cambridge (U.S.), has disclosed two periodic disturbances, one in a yearly period and another in a period of about 428 days. The former, a motion of the invariable line OK with reference to the earth, is a narrow ellipse, about 30 feet long at the North Pole, and is no doubt due to the periodic deposit and melting of snow and ice in the polar regions, which must alter the action producing precession. Accordingly this is a forced oscillation.

The latter however is a motion in a circle at the pole of about 26 feet in diameter in the fourteen months' period stated. In the chart on p 15, the path of the north pole of the earth for five years from 1890 to 1895 is pictured. It will be seen that the curve is apparently very irregular: this is due to the variation in position of the ellipse giving the yearly motion. This is confirmed by the results of later observations, notably by the chart for the seven years 1906-1912, which is given in Chap. XI below.

The discrepancy between 428 days and 304 days was at first puzzling, but it was pointed out by the American astronomer Simon Newcomb that any elastic yielding of the earth, or change of distribution of matter due to mobility of surface water, that might accompany the vibrational motion, would increase the period. For, clearly, such yielding would diminish the forces called into play by the vibrational displacements, and the period would be increased. The exact amount of the increase of period due to this cause cannot well be estimated [see however Chap. XI].

**16.** *Egg-shaped solid body stable with long axis vertical when spinning. "Liquid gyrostats."* The question of the behaviour of a spheroidal shell filled with a liquid and made to spin is an interesting one. We take a

spheroidal piece of wood, a nearly egg-shaped piece, as shown in the diagram, and laying it with its long axis horizontal, apply two horizontal forces in opposite directions at its ends with the fingers so as to make it spin rapidly, say in the direction of the circular arrow on the left. The wood does not continue to spin with the long axis horizontal, but raises its centre of gravity until the long axis is vertical, as shown on the right, and spins in stable equilibrium in this position, like an ordinary top.

If now the same experiment be tried with a metal shell filled with a liquid, or with a fresh egg, it will not succeed. The arrangement will spin in a feeble way, but it will not behave as an ordinary top and stand up on end.

The rotational motion set up is not stable, and dies away fairly quickly. But the shell can be stopped by placing the finger on the egg for a moment. When the finger is removed the shell moves on again, being dragged round by the still rotating liquid. But if the egg is boiled hard it will behave like the wooden spheroid. This is one way of solving the problem of Columbus—to make an egg stand on end!

17. *Stability of rotational motion in a spheroidal shell.* The stability of the rotation of the liquid contents of a spheroidal shell depends on the form of the shell. If the spheroid is oblate, that is if the axis of figure is, like the axis of the earth, the shortest diameter, the rotation is stable, and will endure so that the shell filled with liquid may be used as the flywheel of a gyrostat. If however the shell is prolate, that is if the axis of figure is the long axis, as in the case of the piece of wood, the motion is unstable, unless indeed the shell be sufficiently prolate. We may place the gyrostat, with prolate liquid spheroid as wheel, on the spinner and get up a great speed; as soon however as the gyrostat is removed from the machine it is found that the spin has disappeared.

On p. 18 is a diagram showing an oblate shell and a prolate one filled with water, both mounted in frames like a gyrostat flywheel, as shown in the third diagram. The deviation from sphericity is 5 per cent. in each case, but in opposite directions. The oblate one admits of stable motion of the liquid, the other does not.

Oblateness however is not absolutely essential for steady rotational motion of a liquid round the axis of figure in a spheroidal case turning with the liquid. It was shown by Sir George Greenhill in 1880 that steady motion is possible in a prolate spheroid, if it be sufficiently prolate. The axial diameter, in fact, must either be shorter than the equatorial diameter, or be more than three times as long.* As Sir George Greenhill points out, a modern elongated projectile if filled with a liquid would not rotate steadily about its axis of figure, and therefore would not have a definite trajectory as a rifle bullet has; it would (unless abnormally long) turn broadside on to the direction of motion.

The possibility of spinning an oblate ellipsoidal mass of liquid was discovered mathematically by Colin MacLaurin, Professor of Mathematics at Edinburgh in the first half of the eighteenth century. He showed that, provided the angular speed was kept under a certain limit depending on the density of the fluid, there were two revolution ellipsoids of different

* *Proceedings of the Cambridge Philosophical Society*, 1880; *Encyclopædia Britannica*, article, "Hydromechanics."

eccentricities, which were figures of equilibrium for a mass of liquid spinning about the axis of figure, with its surface free. With one or other of the eccentricities proper for the speed, the case may be supposed removed without affecting the equilibrium. Of course it is understood that there is no terrestrial gravity to produce disturbance: the spinning ellipsoid of liquid, without enclosing case, could not be realised except in a laboratory at the centre of the earth, and perhaps for various reasons not even there.

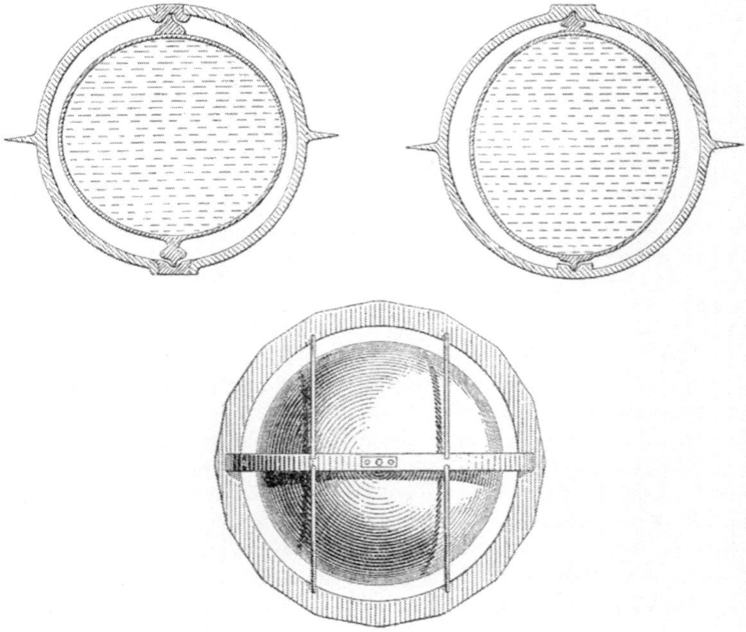

All this is connected with a subject of very great interest, the equilibrium of spinning masses of liquid. It is of much importance in its bearing on the genesis of the earth and moon; but for further information the reader is referred to Thomson and Tait's *Natural Philosophy*, and to memoirs by Poincaré and Sir George Darwin.

**18.** *Gyrostatic experiments: gyrostat supported above universal flexure joint. Double instability stabilised.* Some of the usual gyrostatic experiments have already been described in 4 . . . 7 above. We shall now describe a few others of a somewhat more complicated kind, and indicate some speculations which have been put forward, principally by Lord Kelvin, as to the explanation of certain physical properties of matter by gyrostatic structure.

A very instructive gyrostatic experiment is that with a gyrostat supported on a universal gimbal joint as shown in the diagram. This drawing shows one gimbal axis clearly: the other axis is at right angles to this and is partly hidden by the upper ring to which the knife-edges are seen attached. The gyrostat thus mounted forms an inverted pendulum, which has two freedoms of motion, and when the wheel is without spin is unstable in both. With sufficient spin the gyrostat is stable in both freedoms.

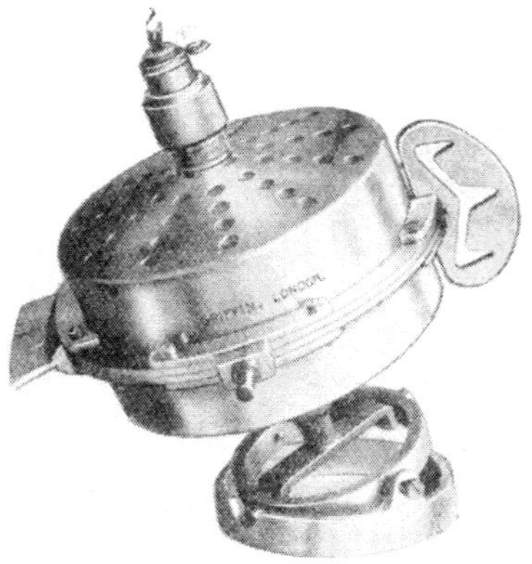

Motor-gyrostat on gimbals.

This is an example of a doubly unstable arrangement rendered stable by spin.* In Thomson and Tait's *Natural Philosophy* the interesting theorem is proved (at least it is implied in the discussion in § 345$^{viii}$ *et seq.*) that in a gyrostatic system an even number, but not an odd number, of freedoms can be stabilised by rotation of flywheels [see also XX below]. Of this we have here a particular case.

An arrangement of double instability, of which the idea is said to be due to the late Professor Blackburn, is shown on p. 20. A trapeze is attached at its ends to two vertical chains by two rings attached to swivels, so that the trapeze can turn about its own longitudinal axis. The trapeze is made of two slips of metal between which the rim of the gyrostat can be slipped and secured by a pin. One of the chains has inserted in it a large ring, so placed that by unhooking one of the chains, passing it through the ring, and then rehooking it, and turning the trapeze end for end, we have it suspended by crossed chains. Such a bifilar suspension is of course unstable, as the trapeze tends to turn round towards assuming the arrangement of two parallel or uncrossed chains, in which of course the centre of gravity is lower than in the other case, and which so far as the bifilar is concerned is the arrangement of stable equilibrium without spin.

The gyrostat when without spin and hanging below the trapeze, with the chains uncrossed, has two freedoms for both of which it is stable: (1)

* It is understood here and elsewhere that there is no frictional resistance involved. The existence of such resistance must be reckoned with in contrivances for practical purposes. The exact meaning also of stability cannot be discussed here.

the system can swing as a pendulum about the swivels at the end of the trapeze; (2) the trapeze can turn in azimuth about a vertical axis through its middle point, in vibrations in which the chains are carried in opposite

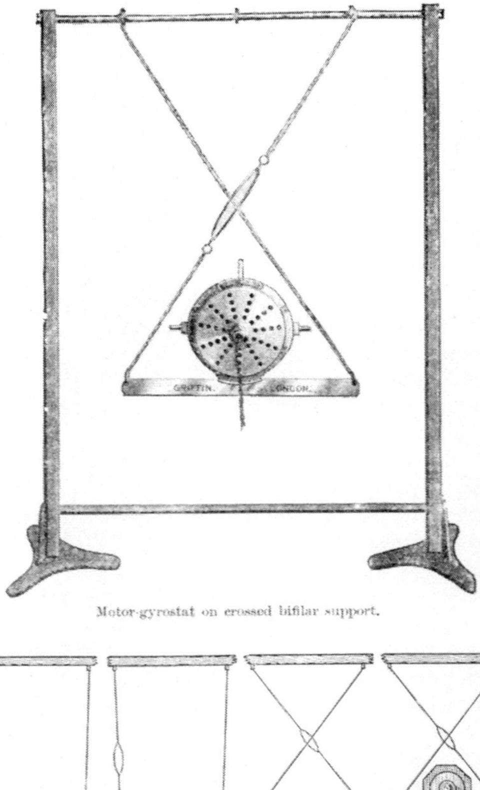

Motor-gyrostat on crossed bifilar support.

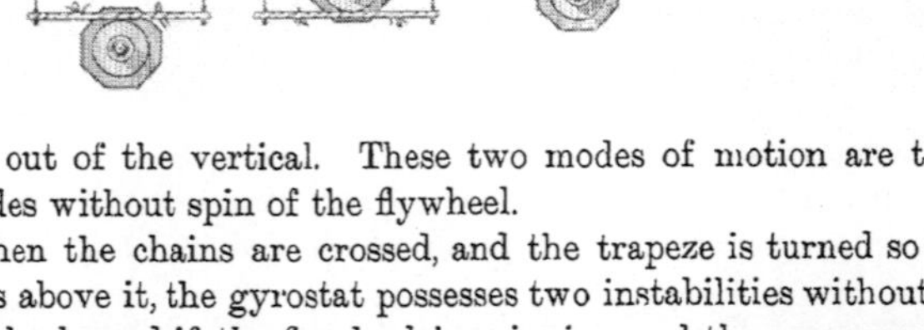

directions out of the vertical. These two modes of motion are thus both stable modes without spin of the flywheel.

Now when the chains are crossed, and the trapeze is turned so that the gyrostat is above it, the gyrostat possesses two instabilities without rotation of its flywheel; and if the flywheel is spinning and the arrangement is set up so that the chains are in one plane with the casing of the gyrostat vertical, the contrivance exhibits remarkable balancing power. The effect of friction at the supports however causes the arrangement to leave the vertical. Gyrostatic oscillations come into existence. The gyrostat oscillates about the horizontal swivels, and the chains oscillate about their mean positions. The amplitudes of these oscillations continually increase, until finally the gyrostat falls over. The balancing in any oscillation can be traced out in detail as an effect of "hurrying" precession.

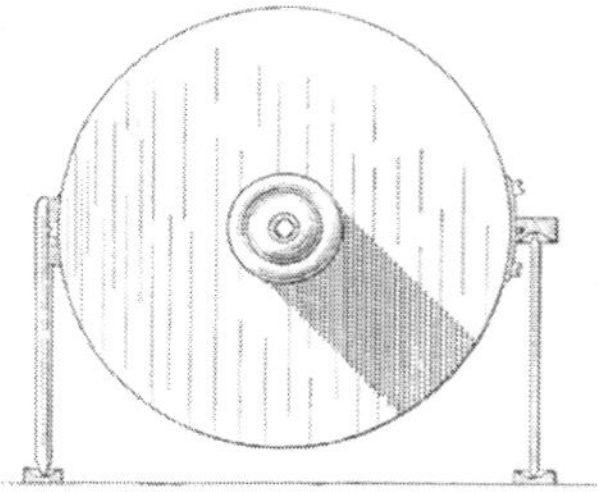

A gyrostat may also be supported on two stilts as shown. One of these is rigidly attached to the case, parallel to the plane of the flywheel; the other is merely a stiff wire with rounded points, the upper of which rests loosely in an inverted cup on the case. The lower ends of the stilts rest in cups on the table, or they may be merely set on a roughened metal plate. There is, with the freedom to turn about the line of the lower points and at the same time to swing about the vertical centre line, which the arrangement possesses, considerable stability

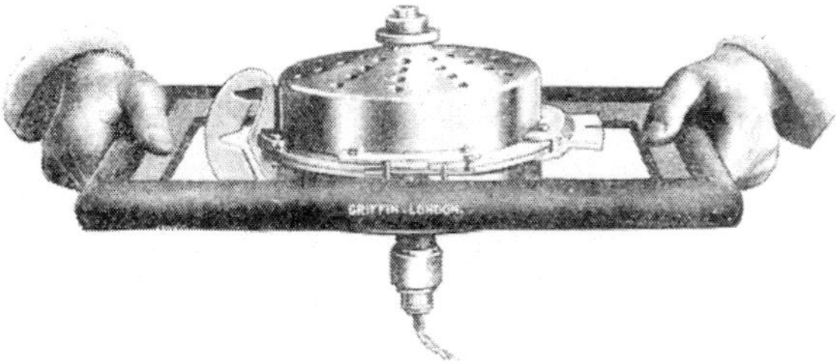


Motor-gyrostat mounted to demonstrate the principle of the gyrostatic compass.

(quickly undermined however by increasing amplitude of vibration) with rapid rotation of the flywheel. There is of course no stability without rotation.

**19**. *Gyrostatic experiments: gyrostat on tray carried round in azimuth.* The gyrostat of the pedestal instrument carries two trunnions in line with the centre of the wheel. The gyrostat is now mounted by placing these trunnions on bearings attached to the square wooden frame shown in the figure, so that when the tray is held in a horizontal position, the gyrostat rests with its axis vertical or nearly so. The direction in which the wheel is spinning is shown by the arrow on the upper side. The frame is carried round in azimuth in the direction of spin: nothing happens; the gyrostat spins on placidly. If however the frame be carried slowly round the other way, the gyrostat immediately turns upside down on the trunnions; and now, as the motion round in that direction is continued, the gyrostat is quiescent as at first; but the spin, by the inversion of the gyrostat, has been brought into the same direction as the azimuthal motion.

The gyrostat behaves as if it possessed volition—a very decided will of its own. It cannot bear to be carried round in the direction opposed to the rotation, and, as it cannot help the carrying round, it accommodates itself to circumstances by turning upside down so that the two turning motions are made to agree in direction. Again the azimuthal motion is reversed, and once more the gyrostat inverts itself, so that the wheel turns in the same direction in space as at first.

It will be noticed that when this curious one-sided stability and instability is displayed, the gyrostat is affected by a precession impressed upon it from without. The system was not left to itself, it was carried round. The gyrostat has little or no gravitational stability—the centre of gravity is nearly on a level with the trunnions; but even if it were gravitationally unstable, sufficiently rapid azimuthal motion would keep it upright if that motion agreed with the spin, while the least motion the other way round would cause it to capsize.

If the gyrostat be placed on the trunnions so that the axis of the wheel is in the plane of the frame, azimuthal turning causes one end or the other of the axis to rise, according to the direction of turning.

Better than anything else, this experiment affords an example of the two forms of solution of a certain differential equation, which, when the inclination $\theta$ of the spin axis to the vertical is small, may be written

$$\mathrm{A}\frac{d^2\theta}{dt^2}+\omega\mathrm{N}\theta=0,$$

where N is the angular momentum of the wheel, and $\omega$ the angular speed with which the frame was carried round. When the turnings were in the same direction, $\omega$ and N had the same sign, but when the turnings were in opposite directions the product $\omega$N had a negative value. When the product is positive we have a solution giving oscillations about the vertical, in the period $2\pi(\mathrm{A}/\omega\mathrm{N})^{\frac{1}{2}}$: the equilibrium is stable. When however $\omega$ is reversed the product must be given the opposite sign, and we get a solution in real exponentials, giving the beginning of continued falling away from the upright position until the opposite position, which is stable, is attained. N has now also been reversed in space, and the product $\omega$N in the differential equation is again positive.

A turning moment about the vertical is required by the gyrostat from the frame constraining it to move round in azimuth. The average amount of this moment, multiplied by the time occupied in the inversion, is 2N. There is thus at each instant of the turning in azimuth, before the inversion has been completed, a couple required from the frame, and this couple is greater the greater the angular speed $n$ of spin The operator must apply an equal couple to the frame.

The couple arises thus Let the gyrostat axis have been displaced from the vertical through an angle $\theta$ about the trunnion axis. In consequence of the azimuthal *turning*, at rate $\omega$ say, the outer extremity of the axis of angular momentum is being moved parallel to the instantaneous position of the line of trunnions, and thus there is rate of production R of angular momentum about that line; but there being no applied couple about the trunnions, the gyrostat must begin to turn about them in order to neutralise R. This turning tends to erect or to capsize the gyrostat according as the spin and azimuthal motions agree or are opposed in direction. The consequent motion involves production of angular momentum about the vertical for which a couple must be applied by the frame, and of course to the frame by the operator. This couple is greater the greater N, and therefore if the operator cannot apply so great a couple an azimuthal turning at rate $\omega$ cannot take place. With sufficiently great angular momentum the resistance to azimuthal turning could be made for any stated values of $\theta$ and $\omega$ greater than any specified amount.

The magnitude of this couple, which measures the resistance to turning at a given rate, is greatest when the angle $\theta$ is 90°, that is when the axis of the flywheel is in the plane of the frame.

**20.** *Gyrostatic structure of a rigid solid.* We come now to an interesting application of these ideas. Lord Kelvin endeavoured to frame something like a kinetic theory of elasticity, that is he conceived the idea that, for example, the rigidity of bodies, their elasticity of shape, depends on motions of the parts of the bodies, hidden from our ordinary senses as the flywheel of a gyrostat is hidden from our sight and touch by the case.* Consider this diagram of a web. It represents two sets of squares, one shown by full the other by fine lines; the former are supposed to be rigid squares, the latter flexible. Unlike ordinary fabrics, which are almost unstretchable except in a direction at 45° to the warp and woof, this web is equally stretchable in all directions. If the web is strained slightly by a small change of each flexible square into a rhombus, or into a not rectangular parallelogram, the areas are to the first order of small quantities unaltered.

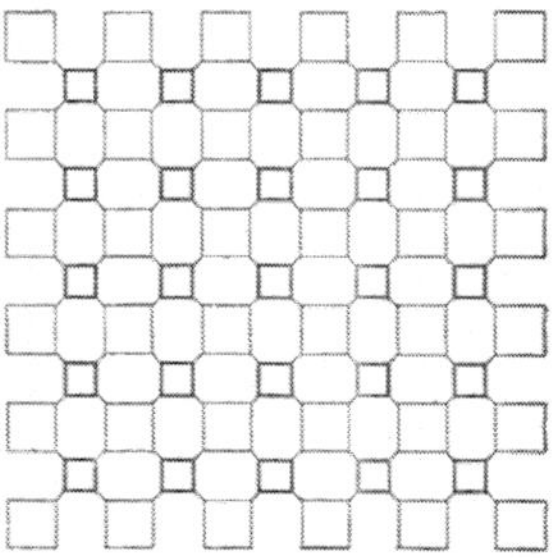

* "Ether, Electricity, and Ponderable Matter." Presidential Address to Inst. Elect. Engineers, Jan. 10, 1889. *Collected Papers*, III, p. 484.

Now imagine that a gyrostat is mounted in each of the rigid squares, so that the axis of the trunnions and the axis of rotation are in the plane of the square shown in the figure. If the angular speeds of the flywheels are sufficiently great, it is impossible to turn the squares in

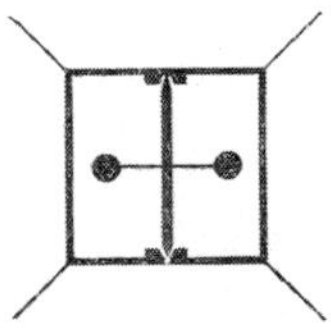

azimuth at any given small angular speed, for the gyrostat would very strongly resist the change of direction of its axis. Thus any strain involving turning of the small squares is resisted, and we have azimuthal rigidity conferred on the web by the gyrostats. There is however no resistance to non-rotational displacement of the squares as wholes.

To get a model in three dimensions Lord Kelvin imagined an analogous structure made up of cubes, each composed of a rigid framework to play the part of the squares, and connected by flexible cords joining adjacent corners of the cubes. In each cube he supposed mounted three gyrostats with their trunnions at right angles to the three pairs of sides. This arrangement would, like the web of squares, resist rotation, but now about any axis whatever; and there would be no resistance to mere translation of the cubes as wholes. Thus the body so constituted would be undistinguishable from an ordinary elastic solid as regards translatory motion, but would resist turning.

21. *Gyrostatic spring balance.* In this connection we may refer to another arrangement suggested by Lord Kelvin—a gyrostatic imitation of a spiral spring—in which a constant displacement is produced and maintained by the action of a constant force in a fixed direction, involving the application of a couple of constant moment, though not of constant direction of axis. This gyrostatic spring balance is indicated in a paper entitled "On a Gyrostatic Adynamic Constitution for Ether," published partly in the *Comptes rendus,** and partly in the *Proceedings of the Royal Society. of Edinburgh.*† It is described in some detail in his *Popular Lectures and Addresses.*‡ The arrangement of gyrostats is shown in the diagram, and is fairly simple in conception. It does not however, except under

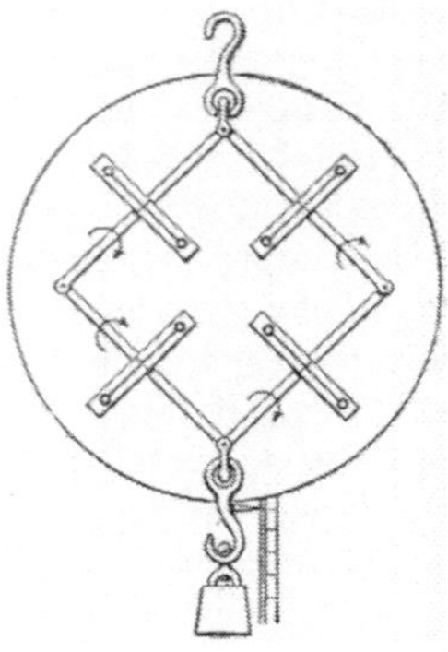

certain conditions not easily realisable even approximately, possess the peculiar property of a spiral spring of being drawn out a distance proportional to the weight hung on the lower hook. The gyrostatic arrangement

* *Comptes rendus,* vol. 109, p. 453, 1889; *Math. and Phys. Papers,* vol. 3, p. 466.

† *Proceedings of the Royal Society of Edinburgh,* vol. 11, 1890. ‡ Vol. 1, p. 237, *et seq.*

is very difficult to realise with ordinary gyrostats, but presents no difficulty with properly constructed motor driven instruments. It consists of a frame of four equal bars, constructed by jointing the bars freely together at their extremities in the manner shown by the diagram, and hung from a vertical swivelling pin at one corner, so that one diagonal of the frame is vertical, and another vertical swivelling pin at the lowest corner carries a hook. Four equal gyrostats are inserted, one in each bar as shown, with its axis along the bar, and are given equal rotations in the directions shown by the circular arrows. Under the couples tending to change the directions of the axes of the flywheels, and applied by the weights of the gyrostats and bars, the system precesses round the two swivels, and so preserves a constant configuration. If now a weight is hung on the hook at the lower end, the frame is elongated a little, and a new precessional motion gives again a constant configuration of the frame, different of course from the former one. Two gyrostats, the upper or lower pair, would serve quite well to give the effect.

Lord Kelvin suggested that if the frame were surrounded by a case, leaving only the swivel-pins at top and bottom protruding, it would be impossible, apart from special knowledge of the construction of the interior, to discern the difference between the system and an enclosed spiral or coach spring, surrounded by a case and fitted with hooks for suspension and attachment of weights. It is found however when the steady motion of the system under gravity is worked out, that, in strictness, the distance through which the frame is lengthened can be regarded as simply proportional to the load applied only in special circumstances. The ratio of the extension of the vertical diagonal to the addition of load on the hook is a function of the inclination of the arms of the frame to the vertical, and therefore the spiral spring law only holds for very small deviations from a steady motion position. The system acts certainly as a spring, but, constructed with actual practical gyrostats, it has not the properties possessed, though only approximately even in their case, by ordinary springs.

The theory of the combination, and an account of an easily constructed substitute, will be given later [see 1, VIII].

The idea however underlying the contrivance is very suggestive, and goes a long way towards enabling us to obtain a definite idea of how the elastic properties of bodies may be explained by motion.

**22.** *Gyrostatic pendulum.* Next we consider a pendulum consisting of a rigid suspension rod, and a bob rigidly attached to it, which contains a gyrostat with axis of rotation directed along the suspension rod as shown in the figure. Without rotation, the two freedoms of this system are stable, and if the bob be made to describe a circle about the vertical through the point of support, the period of revolution is the same for both directions of the circular motion. When the gyrostat is spun the behaviour is very different. Circular motion may take place in either direction, but the periods are different, that of the circular motion in the same direction as the rotation being the smaller. The ratio of the periods depends on the arrangement; the theory is set forth below. [See 4, VIII.]

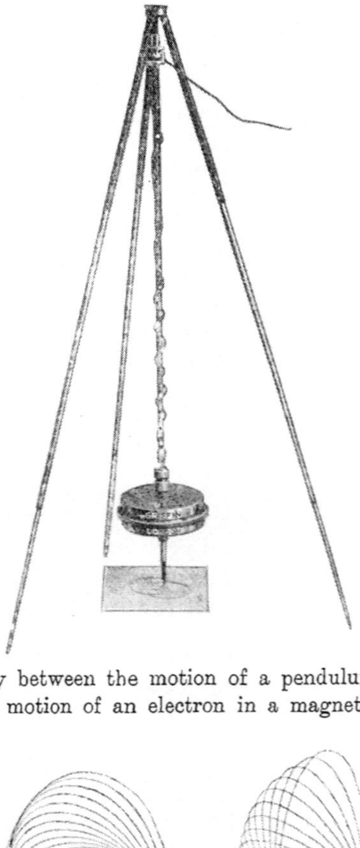

A combination of the two circular motions in different periods and in opposite directions gives a star figure, which in the diagram the pen attached below the bob is shown describing. The peculiar appearance of the graphs here pictured is due to a very rapid falling off of amplitude, and therefore shortening of the rays, due to friction.

There is a remarkable analogy between the motion of a pendulum with a gyrostat in its bob and the motion of an electron in a magnetic field, which is dealt with later.

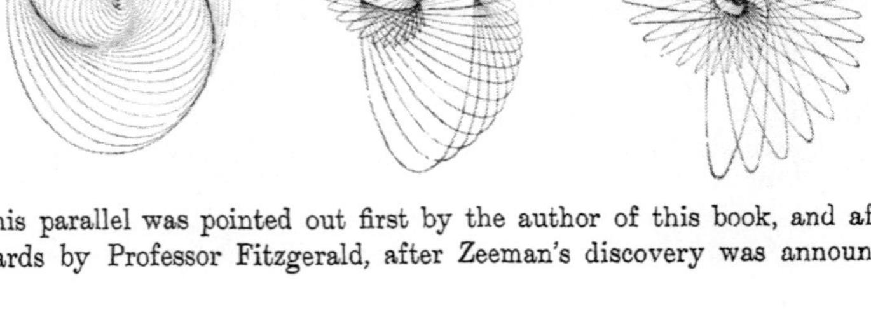

This parallel was pointed out first by the author of this book, and afterwards by Professor Fitzgerald, after Zeeman's discovery was announced.

The gyrostatic explanation of the Zeeman effect is attended with difficulties on account of the complex of spectral lines observed in the Zeeman phenomenon.

23. *Experiments: Gyrostat weighted on one side. Motor gyrostat on chain-suspension.* We shall now describe a number of gyrostatic experiments, most of them new and made with novel apparatus. For the most part the experiments and apparatus are due to Dr. J. G. Gray The descriptions given below are to a great extent taken from discussions written by Dr. Gray and the author.

The direction of rotation of the rotor of the gyrostat is shown by the circular arrow head, and the axis OX, drawn from the centre (all on the remote side of the instrument in the diagram), represents the angular momentum N, or $Cn$ ($C =$ moment of inertia of rotor, $n =$ angular speed of rotor).

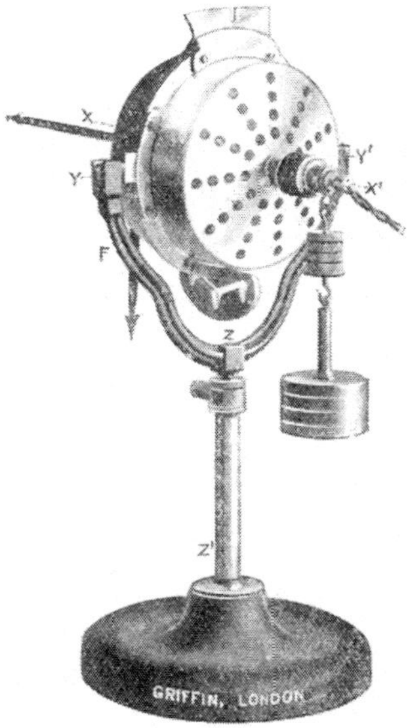

Motor-gyrostat in pedestal, with weight attached.

One of the weights supplied with the instrument is hung from the counterpoise of the rods and arrows; or, if counterpoise and rods are removed, it is hung from the loop shown at X'. If the flywheel is without rotation, the gyrostat simply turns about YY'. The weight is now removed, the flywheel is spun, and the weight is replaced. It is seen now that the wheel and case with the supporting fork turn as a whole about ZZ', and do so more quickly the larger the weight attached. Again, with a given attached weight it is found that the greater the angular speed of the rotor, the slower is the speed of turning about ZZ'. We have here illustrations of the gyrostatic equation which holds in all these cases, $Cn\omega = L$, where $\omega$ is the angular speed about the axis Z'Z, and L the moment of the couple about Y'Y. By the turning about Z'Z with speed $\omega$ angular momentum is being produced about Y'Y at rate $Cn\omega$, and this must be equal to L, the moment of the couple produced about Y'Y by the attached weight.

Again, when the weight is attached let a couple be applied by hand in the direction which seems the natural one to hurry the turning motion about Z'Z. It is found that the gyrostat now turns about Y'Y, so as to *raise* the weight. On the other hand, if the couple is applied so as

to endeavour to delay the precession, the gyrostat will turn about Y′Y so as to allow the weight to descend.

The gyrostat may be hung by the chain suspension from one of the notches in the side piece shown in the adjoining diagram, and made to give the couple by its own weight. If it is hung in the right-hand notch, as shown, and the spin axis, indicated by the arrow index, be drawn to the right, the axis of the couple due to the inclined position of the gyrostat would be at right angles to the paper and drawn towards the reader. The gyrostat would thus turn in azimuth so that the extremity of the spin axis would follow the couple axis, that is would turn with angular speed $\omega$ towards the observer.

If the chain were attached at the left-hand notch, the moment of the couple and the angular velocity $\omega$ would be reversed in direction.

An attempt to hurry or retard precession has an exactly similar result to that already described.

When the chain is hooked into the middle notch, the couple is zero, and there is no azimuthal turning.

**24.** *Experiment: Gyrostat on skate.* The adjoining diagram shows a motor-gyrostat resting on a rounded convex skate screwed to the flanges which unite the two halves of the case. If the gyrostat leans over at an angle $\theta$ from the vertical through the skate edge,

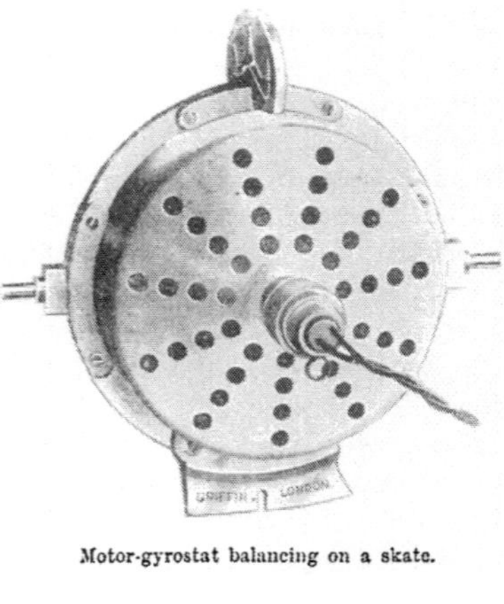

Motor-gyrostat balancing on a skate.

a couple of moment equal to the force of gravity on the gyrostat multiplied into $h \sin \theta$, where $h$ is the distance of the edge of the skate from the centre of gravity, is applied, and the gyrostat without alteration of its inclination moves round in azimuth with an angular speed $\omega =$ *moment of couple/angular momentum of wheel*, at least this will be the angular speed if the gyrostat is simply set down on the skate (preferably on a supporting horizontal tray of thick glass). But there is another and faster angular speed of azimuthal motion which can be realised by properly starting the motion on the glass plate.

An explanation of how these two possible speeds of steady motion arise will be given in a later chapter.

All the experiments on hurrying and retarding precession can also be made with this arrangement.

**25.** *Experiment: Gyrostatic bicycle rider.* The rider of a bicycle is here replaced by a gyrostat. The apparatus may be so contrived that the balancing is entirely due to gyrostatic action, or so that it is effected as a human rider does it. In the former case the bicycle is stable both when at rest and when in motion, in the latter it is stable when driven in the forward direction. We give here the second case, in which an old type of bicycle is used.

Gyrostatic bicycle rider.

The gyrostat is spun in the direction of rotation of the wheels in the forward motion of the cycle. Then tilting of the machine to the left, say, and the consequent alteration of direction of the axis of spin of the fly-wheel, causes, as can easily be seen, precession of the gyrostat, which turns the wheel to the left. For the tilting gives a rate of production of angular momentum about a downward axis, and there being initially no couple about that axis, the gyrostat and wheel begin to turn round in azimuth in the direction to neutralise the angular momentum produced. The forward motion of the rider and machine then gives the upright making action in the usual way.

When perfectly balanced the bicycle has a straight path. When a weight is hung on one side the path becomes curved.

There are two or three other ways in which a gyrostatic rider can keep upright, but that just described and the following must suffice for the present.

The next figure shows a gyrostatic rider on a modern form of bicycle.

The saddle is replaced by a sleeve-joint on which the rider is supported, and two arms are applied as shown to the handle bar. When the wheels of the bicycle are in one vertical plane the sleeve-joint is inclined backward from the vertical, and so, if the gyrostat turns on the sleeve, the brass ball carried by the gyrostat moves in a circle, the highest point of which is in the plane of the back wheel. Thus, when the wheels are in one plane the rider is unstably mounted. But the whole machine is unstable on the wheels. Thus there are two instabilities without rotation of the gyrostat flywheel, and these, in accordance with gyrostatic theory [to be given later], are both stabilised by spin of the gyrostat.

With rapid spin in the direction of forward motion of the wheels, the gyrostat, as already described, imitates the human rider, and tilting action causes the machine to turn to the right or left as the case may be

In all these bicycle tops the gyrostat not only detects but sets about correcting any tendency of the bicycle to capsize.

**26.** *Experiment: " Walking gyrostat."* Another entirely novel experiment is shown in the figure. A box is suspended by two arms of equal length from two horizontally stretched wires. The wires are carried by a frame mounted on two trunnions mounted on wooden uprights as shown in the diagram, which however display only one end of the arrangement of wires and supports. The wires are conveniently made about 12 or 14 feet long, and strung upon them are two rings, to which the arms attached to the box are hooked. Fitted within the box is a gyrostat with its axis horizontal and in the plane of the arms.

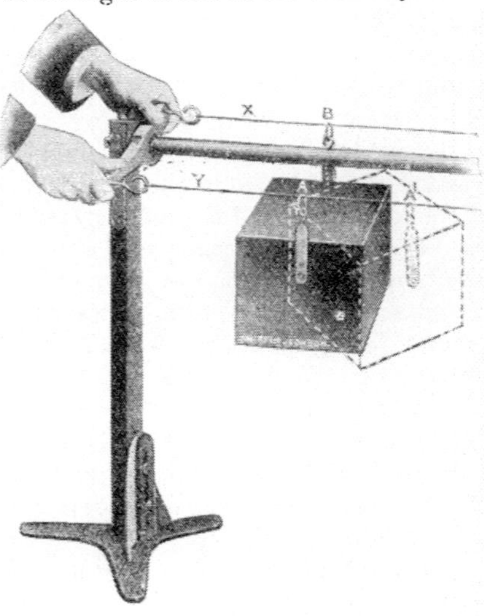

Walking gyrostat.

For the experiment the gyrostat is taken out and spun rapidly, and replaced in the box. When the frame is oscillated to and fro on the trunnions, the box proceeds hand over hand along the wires, as if it were endowed with life.

It will be seen that the tilting of the frame to and fro throws the weight of the gyrostat and box alternately on each of the arms The illustration shows the weight of the box thrown on the arm B: since Y is lower than X. The resulting precessional motion of the gyrostat (which is supposed to be spinning counter-clockwise as viewed by the reader) causes the arm A to move forward to A′. At this instant the wire frame is tilted so that the weight is thrown on the arm A′, when the arm B swings forward, and so on At the start of the motion the spin is great and the precession small, and the box has a slow and stately motion. As the spin falls off the precession, and consequently the rate of walking, increases; until finally the box literally runs along the wires.

When the box has walked from one end of the frame to the other it must be brought back to the starting end to repeat the experiment. The direction of walking depends on the direction of the spin: if it is desired to cause the box to walk in the reverse direction it must be reversed on the wires.

**27.** *Experiment: A top on a top.* This figure illustrates a form of acrobatic top. Fitted to the frame of the large gyrostat, in a position in line with the axis of the flywheel, is a tube, into which is fitted a peg as shown. If the gyrostat is spun rapidly and placed vertically with the peg resting on a table, it will balance. In consequence of the fact that the centre of gravity of the gyrostat lies above the point of support it possesses two instabilities without rotation of the flywheel, and the result, as in 18, is complete stability with rotation. The action is in fact identical with that involved in the "gyrostat on gimbals" experiment there described. The friction between the table and peg being small, the friction at the pivots is sufficient to cause the gyrostat frame to rotate in the direction in which the flywheel is rotating.

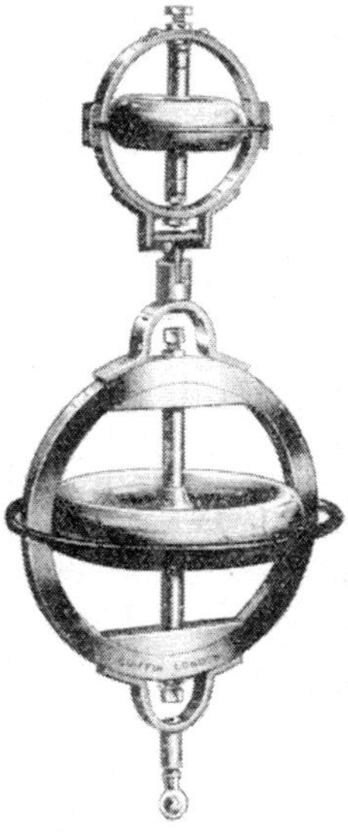

Acrobatic top.

Into a tube fitted to the top of the frame and in line with the axis of the flywheel can be fitted a second tube attached to a horizontal bar, and to this is hinged the frame of a smaller gyrostat.

To perform the experiment the small gyrostat is spun rapidly and fitted above the large one as above described. The system should now be held in the vertical position (this may be accomplished by holding the frame of the small gyrostat in the hand) and given an impulse in the direction in which the flywheel of the large gyrostat is rotating. Providing that the flywheels of the two gyrostats are rotating in the same direction as viewed from above, the small gyrostat will balance on the horizontal bar.

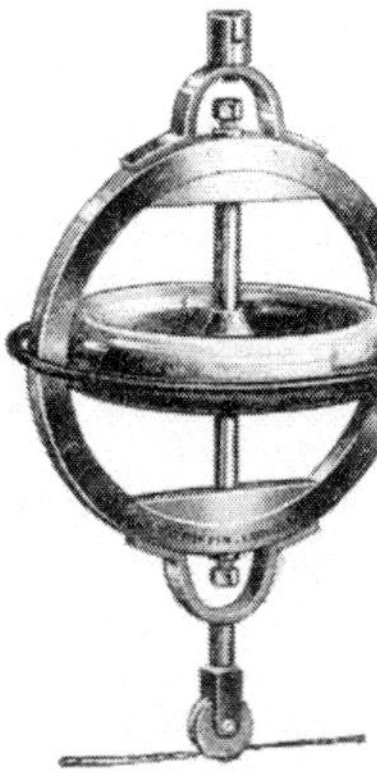

A variation of the "gyrostat on gimbals" experiment may be performed by means of the large gyrostat of the last experiment. The arrangement of the apparatus is shown in the figure. The tube carrying the peg is removed from its socket and replaced by a tube terminating in a grooved wheel. The gyrostat is spun rapidly and placed in the vertical position with the wheel engaging on a tight or slack wire. If the wire is stretched horizontally the gyrostat may be pushed to and fro by hand. The forces should be applied to the tube in the neighbourhood of the grooved wheel, and not to the frame of the gyrostat. A preferable method of carrying out the experiment consists in starting the gyrostat from one end of a slack wire, when it will run to and fro, and finally come to rest.

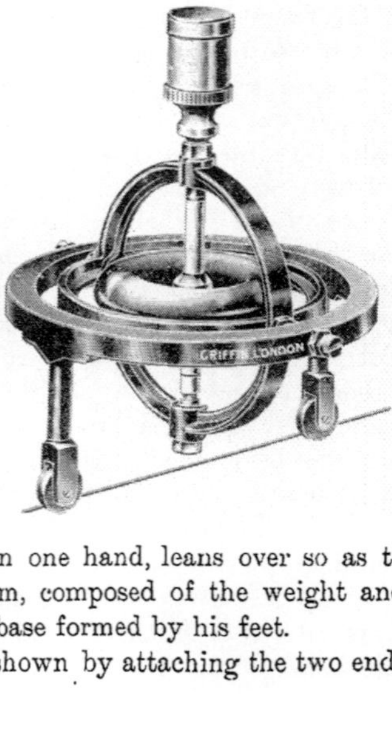


This gives an idea of gyrostatic balancing of a monorail carriage. This is however better shown in this monorail top which runs along a wire on two small wheels. With the weight above there is double instability: (1) for turning about the wire, (2) for turning about an axis transverse to and above the wire. These two instabilities are stabilised by rotation of the flywheel.

If a weight is suspended from one side of the top it will lean over to the other side of the wire, thus imitating the action of a man who, holding a heavy weight in one hand, leans over so as to bring the centre of gravity of the system, composed of the weight and himself, vertically above the centre of the base formed by his feet.

The balancing action of this top is best shown by attaching the two ends of a long wire to two fixed points, one at each side of a lecture theatre. The top is placed on the wire at one end, when it runs to and fro on the wire, coming to rest in the middle position. As an alternative the wire may be tightly stretched between two points not on the same level. If the slope of the wire is small and the top is started from the highest point of the wire, it will run slowly down the incline. The balancing action in this case is striking. The top may be pushed to and fro on the wire without any danger of capsizing.

28. *Gyrostatic action of rotating machinery. Examples: Paddles of a steamer, wheels of a carriage.* The gyrostatic action of the rotating parts of machinery is, if the speed of rotation is great or the dimensions large, of importance in connection with the running of ships or vehicles propelled by such machinery. We take a paddle-steamer as the first example, though in consequence of the slow speed of rotation the gyrostatic action of the wheels is not of large amount. The spin axis for the wheels must be drawn out to port if the steamer is going ahead. Now let the steamer roll to starboard: angular momentum about a vertically upward axis is produced by the motion, and in the direction about this axis round to port. Hence the bow of the steamer must begin to turn to starboard to neutralise this. As she turns a couple begins to act, resisting her turning, and produces the angular momentum due to the change of position of the spin axis. The precessional motion carries as usual the spin axis towards the instantaneous position of the couple axis. Of course the reverse action takes place if the ship rolls to port. The gyrostatic action of the wheels thus results to a certain extent in changes in the direction of the ship's head taking the place of rolling, and the ship is so far steadied. It will be noticed that when the ship rolls to starboard the starboard wheel becomes more deeply immersed in the water than the port wheel, and so tends to turn the bow to the port side. If it were not for this fact it would be a matter of more difficulty to steer a straight course with a paddle-steamer in a cross-sea.

Similar considerations show that if the helm is put over to starboard the steamer will, in consequence of the gyrostatic action of the paddle-wheels, heel over to starboard. This causes the starboard paddle-wheel to dip further into the water than the port wheel, and turning of the steamer to port is assisted. Similarly if the helm is put over to port the steamer tilts over to port, and again a couple turning the steamer in the desired direction is brought to bear on her.

The gyrostatic action of the paddle-wheels may be readily demonstrated by means of the motor gyrostat. For this purpose it is fitted with the curved rod to show the direction of rotation and of the armature, the rods to represent the spin and couple axes, and a fourth rod which serves to show the direction in which the steamer is supposed to be proceeding. The gyrostat, so fitted, is placed in the "fork and pedestal" mounting as described in 7. The direction of rotation of the armature being counter-clockwise as viewed by an observer looking at the gyrostat from the side to which the rods are attached, the arrow showing the direction of motion of the steamer must be put over to the left. The effect of seas striking the port side of the steamer is imitated by the experimenter pushing the top of the gyrostat from him (the experimenter is supposed to be in the position of the observer referred to above). In consequence of gyrostatic action the fork will turn about the axle $ZZ'$, and the arrow representing the bow of the steamer will turn to starboard. To imitate the effect of a sea on the starboard side the lower side of the gyrostat is pushed from the experimenter; here the arrow turns to port.

If the fork is grasped by the hands and turned about the axis $ZZ'$ the gyrostat will turn about the axis $YY'$. If the turning about $ZZ'$ turns the bow arrow to port the gyrostat will heel to starboard, and *vice versâ*.

The gyrostatic action of the wheels of a carriage is similar to that of the paddles of a steamer. It produces when the carriage is passing round a curve a gyrostatic couple turning the carriage over towards the outside of the curve. The gyrostatic couple is $NCv^2/rR$, where C is the moment of inertia and $r$ the radius of a wheel, N the number of wheels, $v$ the speed of the carriage and R the radius of the curve.

**29.** *Gyrostatic action of flywheel of motor-car.* In the motor-car a massive flywheel, placed with its plane across the car, revolves with considerable angular velocity in the clockwise direction as viewed from behind the car. The gyrostatic action of this flywheel has important effects upon the running of the car. If the car turns a corner it is easy to see that it will be subjected to a gyrostatic couple in a fore and aft vertical plane. Here the spin axis is represented by a straight line drawn from the car in the direction of the front of the car. If a corner is turned to the left a couple is brought to bear on the gyrostat, represented by a line drawn vertically upwards. The spin axis turns to the left, and hence for this direction of turning a corner the effect of the gyrostatic couple is to diminish the forces applied by the ground to the front wheels and to increase those applied by the ground to the back wheels. The magnitude of the gyrostatic couple is proportional, at any instant, to the rate of turning of the car in azimuth, and if this is very great, that is if the corner is turned very quickly, the diminution in the forces between the front wheels and the ground may be sufficiently great to endanger the steering power of the car. Turning a corner to the right, it will be seen, results in the forces between the front wheels and the ground being increased, and those between the back wheels and the ground being diminished.

Again, in the case of a motor-car, it will be seen that if the car encounters a dip or brow in the road the effect of the flywheel will be to apply a couple in a horizontal plane to the car. This couple tends to produce skidding of the wheels on the ground.

**30.** *Gyrostatic action of turbines, aeroplane propellers, etc.* The gyrostatic action in steamers in which the main engines are of the steam turbine type is similar to that of the flywheel of the motor-car. The turbines are placed fore and aft, and rolling of the ship brings no gyrostatic action into play. If the ship be propelled by two main turbines, and if the rotors be equal in all respects and run at the same speed, but in opposite directions, the total couple exerted on the ship will be zero, since equal and opposite couples will be exerted by the two rotors. Internal stresses will be exerted in consequence of the opposite couples, and the stresses will form a self-balanced system within the ship

When the ship's head turns the gyrostatic action of each rotor results in a couple lying in a vertical plane being applied to the ship. Here, as before, if the ship is propelled by two turbines as described, the resultant couple is zero.

A very interesting, and at the same time very important, example of gyrostatic action is afforded by the aeroplane. The rotor of the high-speed engine and the propeller form a powerful gyrostat. The gyrostatic action of a two-blade propeller is partly of an alternating character. In consequence of gyrostatic action the effect of turning the aeroplane in azimuth in one direction is to cause the aeroplane to dive; turning it in azimuth in the opposite direction causes the front of the aeroplane to rear and the tail to be depressed. The aeroplane may be maintained more or less nearly level by means of the tilting planes; but it is to be remembered that if it is so maintained large couples are brought to bear on the machine, and great stresses fall to be borne by the framework. Dangers from the gyrostatic action of the propeller and rotor of the engine would be avoided by balancing the gyrostatic action of the rotor and propeller by means of a second flywheel rotating in the opposite direction, or, if that were possible, by doubling the propelling system, and running the rotors in opposite ways.

The gyrostatic action of the flywheel of a motor-car, of the turbines on board ship, and of the rotor and propeller of an aeroplane may be readily demonstrated by means of the motor-gyrostat when fitted up in the "fork and pedestal." The mode of doing so will easily occur to the reader.

**31.** *Schlick controller of rolling of a ship.* In the Schlick device for steadying a ship at sea a gyrostat is carried on bearings placed athwart the ship, and in line with the centre of gravity of the flywheel. A weight is attached to the frame of the gyrostat in a position in line with the axis of the flywheel. It will thus be seen that when the ship is on an even keel the gyrostat rests with its axis vertical and with the weight directly below the centre of gravity of the flywheel. Heeling of the ship in one direction causes the gyrostat to precess in one direction on the bearings in which it is mounted; heeling in the other direction causes precession opposite to the former, and couples resisting the rolling motion are brought to bear on the ship. The device may be employed in two ways. In the first place, if the bearings on which the frame of the gyrostat is carried within the ship are smooth, the effect of the gyrostat is to resist the rolling force of the waves and to bring about a lengthening of the free period of rolling of the ship; for the introduction of the gyrostat results in a large virtual increase of moment of inertia, giving a long period. Excessive rolling of a ship is due to the cumulative action of the waves, and such cumulative action is only possible where the period of the ship and that of the waves are nearly the same. A ship may have a long period, with the result that the motion of the ship does not synchronise with that of the waves. The free period of an ordinary ship, however, is often of the same order as that of the waves usually encountered, and in consequence the ship rolls badly.

One effect of introducing a gyrostatic control, operated in the manner above described, is to endow the small ship with the period of a very large one. With this long period however coexists, as will be shown in 6, VIII, below, a short period. But the short period motion is damped out by the friction brake described below.

As the ship rolls the gyrostat and its frame oscillate to and fro on the bearings. There is a phase difference of 90° between the motions. When the ship is passing through the upright position the axis of the gyrostat is at one extremity of its swing, and when the ship is at one extremity of the roll the gyrostat is passing through its mean position; and so on. Now, if C is the moment of inertia of the flywheel, $n$ its angular velocity, and $\omega$ the instantaneous value of the rate of turning of the frame on its bearings, the instantaneous value of the gyrostatic couple resisting the waves is $Cn\omega$. Hence, when the ship is in the upright position the gyrostatic couple is zero, since at that instant the frame of the gyrostat is at rest, so that $\omega$ is zero. At the instant at which the ship is at one extremity of its swing the gyrostatic couple has its greatest value, since at that instant the rate of turning of the gyrostat on its bearings is a maximum.

Again, consider the condition of affairs which holds for an instant at which the axis of the gyrostat makes an angle $\theta$ with the vertical. The gyrostatic couple at the instant is $Cn\omega$, where $\omega$ is the instantaneous rate of turning of the frame on the bearings which connect it with the ship. This couple acts in a plane perpendicular to the axis of the gyrostat, and hence resolves into the two component couples, $Cn\omega \sin\theta$ in a vertical plane athwart the ship, and $Cn\omega \cos\theta$ in a horizontal plane. The former couple alone affects the period of the vessel.

Friction is applied to resist the motion of the frame of the gyrostat in its bearings, $\theta$ never becomes greater than 45°, and, further, the phase difference is made less than 90°. The oscillations of the ship, and especially any rapid vibrations that may grow up, are now damped; the energy of the rolling motion is converted into heat at the bearings.

In applying this second mode of using a gyrostat to steady a ship in a cross-sea, Schlick arranges a brake pulley in the line of the bearings on which the frame of the gyrostat turns, and friction of a graded amount is applied by means of a special device. With this second mode of operating the gyrostat the ship is forcibly prevented from rolling. In the trials of the device it was found that, with the control in operation, the angle of roll of the ship did not exceed 1° in a cross-sea which produced a total swing of 35° when the control was out of action. It is interesting to notice that,

Motor-gyrostat fitted up to demonstrate Schlick's method of steadying a ship in a cross-sea.

contrary to the opinions which were expressed when the device was first suggested, the preventing of the rolling of a ship does not result in the waves breaking over her; a ship thus controlled is a dry ship.

In the figure is shown a motor-gyrostat fitted up in a skeleton frame representing a ship The frame is mounted on two bearings arranged in two wooden uprights, and may be oscillated on these bearings so as to imitate the rolling of a ship in a cross-sea. The frame of the gyrostat is mounted on two bearings placed athwart the ship frame, and a weight is attached to the outside of the case in a position in line with the axis of the flywheel. The centre of gravity of the gyrostat is in line with the bearings. A clip device is provided which allows the gyrostat to be rapidly clamped to the skeleton frame, and provision is made whereby a graded amount of friction may be applied at one of the bearings.

The model so arranged serves to demonstrate both of Schlick's devices. In using it the ship should be set oscillating with the flywheel of the gyrostat at rest. The current should then be turned on and the frame oscillated. If the gyrostat is clamped to the frame no effect is produced by it upon the rolling, and thus the necessity for allowing the gyrostat freedom to precess is demonstrated. If, however, the gyrostat is free, and there is no friction at the bearings, the frame oscillates with a vastly increased period. Finally, the friction collar provided at one of the bearings may be screwed up, when it will be seen that if the frame is made to roll the rolling motion is quickly damped out. Impulses may be applied to the frame so as to imitate the action of the waves of a cross-sea, and the resisting power of the gyrostat demonstrated. [A more detailed account of the action of this controller will be given later.]

# CHAPTER II

## DYNAMICAL PRINCIPLES

1. *Kinematics of a body turning about a fixed point.* We shall not discuss in detail the elementary kinematics of a rigid body, and shall assume the usual theorems regarding moments and products of inertia of a system of particles. A rigid body here means a configuration of particles such that if any plane whatever be drawn in it, the particles found in that plane remain in a plane, and do not change their distances apart, as the body moves.

If such a body turn round a point of itself, or a point rigidly connected with it, which is fixed, any displacement of the body is equivalent to a turning about a line drawn through the fixed point in a determinate direction. For let a sphere be described from the fixed point as centre with any convenient radius. The particles on that spherical surface remain on it as the body moves. Take any two which are at the points A, B (Fig. 1) at a given instant, and at the points A′, B′ at a subsequent instant. This change of position is, we can suppose, produced by a displacement of the rigid body. For join A to A′ along an arc AA′ of a great circle, and B to B′ along another great circle of the sphere. Bisecting AA′, BB′ in C, D, draw through these points great circles on the sphere at right angles to AA′, BB′ respectively. These in general will not be coincident, and will meet in two diametrically opposite points K, K′ on the sphere. Clearly the displacement may be produced by turning the body about the line KK′, or OK as axis; and the same thing is true for the displacements of any other two points on the spherical surface.

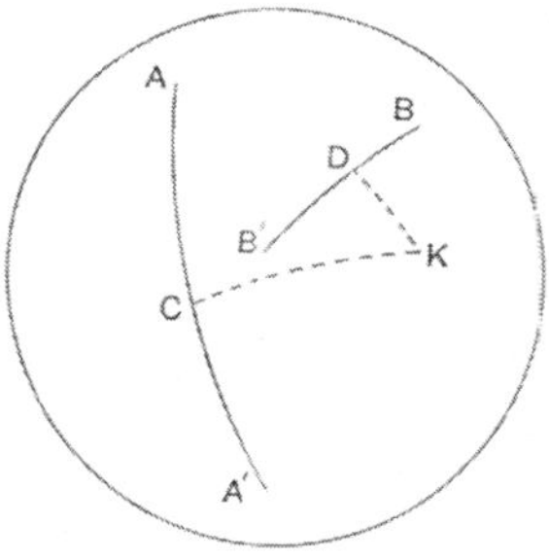

Fig. 1.

The most general displacement of a rigid body in space can be produced by turning the body through a definite angle about a determinate axis, and displacing it without rotation through a certain distance parallel to that axis. For let the body be moved without rotation until a point in it is displaced from its initial position A to its final position A′ along the straight line AA′. Then, by the proposition just established, the body can be turned about an axis through A′ until the final configuration is reached.

The points of the body which lie in planes perpendicular to that axis turn in these planes. But in the translation parallel to AA′ the successive positions of these planes were parallel to one another, and the direction of the translatory motion of each point was inclined to these planes at the same angle. We might therefore have first displaced the body in the direction perpendicular to these planes, and then (still without rotation) have displaced it parallel to these planes so as to bring the point A to A′. But the latter displacement, with the subsequent turning about A′, could be compounded into a turning about a parallel axis. Thus the whole displacement can, as stated above, be produced by a translation parallel to a certain axis and a rotation about that axis.

These displacements—the translation and the rotation—may be regarded as taking place simultaneously; and the body will then have the motion of a nut along a screw. The motion of a rigid body has been very fully dealt with from this point of view by Sir Robert Ball in his *Treatise on the Theory of Screws.*

**2.** *Angular momentum* [A.M.]. The angular momentum (A.M.) of a body, or of a system of particles, about any axis OA is found in the following manner. Let $l, m, n$ be the direction cosines of OA with reference to rectangular axes OX, OY, OZ and $\dot{x}, \dot{y}, \dot{z}$ the speeds with reference to these axes of a small element P of the body (a particle) of mass $\mu$ situated at the point $x, y, z$. Then, if $v$ be the total speed of P, the cosines of the direction of motion of P at the instant are $(\dot{x}, \dot{y}, \dot{z})/v$.* We resolve $v$ into two components, $v_1$ in the direction $l, m, n$ and $v_2$ at right angles to the plane AOP. The cosines of a normal to this plane are

$$(mz - ny,\ nx - lz,\ ly - mx)/r \sin\theta,$$

where $\theta$ is the angle AOP, and $r = (x^2 + y^2 + z^2)^{\frac{1}{2}}$. [The choice of cyclical order here made between $l, m, n, x, y, z$ is explained below.] The component of momentum of P at right angles to the plane AOP is therefore

$$\mu v_2 = \mu \frac{\dot{x}(mz - ny) + \dot{y}(nx - lz) + \dot{z}(ly - mx)}{r \sin\theta}. \quad \text{..............(1)}$$

The A.M. of P about the axis OA is the product of this by the length of the perpendicular from P on the line OA, that is by $r \sin\theta$. Hence this A.M. is

$$\mu\{\dot{x}(mz - ny) + \dot{y}(nx - lz) + \dot{z}(ly - mx)\},$$

or, arranged according to the cosines $l, m, n$,

$$l\mu(\dot{z}y - \dot{y}z) + m\mu(\dot{x}z - \dot{z}x) + n\mu(\dot{y}x - \dot{x}y).$$

* This expression is an abbreviation for $\dot{x}/v$, $\dot{y}/v$, $\dot{z}/v$. Similar contractions, which will usually explain themselves, are frequently used in what follows.

The usual system of rectangular axes is one in which any axis (*e.g.* OX) is changed into the next in cyclical order, OY, by a turning about the third axis OZ, in the counter-clock direction to an observer looking towards O from a point on OZ. An ordinary right-handed screw (the usual cork-screw) placed so as to point along OZ, and with half of the handle along OX, would, if that part of the handle were turned in the direction from OX to OY, advance along OZ.

To agree with this specification of axes the angular momentum should, for positive values of $x$, $y$, $z$, $\dot{x}$, $\dot{y}$, $\dot{z}$, correspond to motion of the particle round the axis OA, in the counter-clock direction as seen by an observer looking towards O from a point beyond the plane through P at right angles to OA. If, for example, the axis OA coincide with OX so that $l=1$, $m=n=0$, while $\dot{y}$ is also zero, so that the motion is parallel to the plane ZOX, we get $\mu\dot{z}y$ for the A.M. about OA, that is the counter-clock turning is taken as positive. The direction cosines of the normal to the plane AOP have been taken above so as to fulfil this condition.

Denoting the A.M. of the whole system about OA by $H_{OA}$, we get

$$H_{OA}=l\Sigma\{\mu(\dot{z}y-\dot{y}z)\}+m\Sigma\{\mu(\dot{x}z-\dot{z}x)\}+n\Sigma\{\mu(\dot{y}x-\dot{x}y)\}\,; \quad\ldots\ldots\ldots\ldots(2)$$

or, if we write

$$H_x=\Sigma\{\mu(\dot{z}y-\dot{y}z)\},\quad H_y=\Sigma\{\mu(\dot{x}z-\dot{z}x)\},\quad H_z=\Sigma\{\mu(\dot{y}x-\dot{x}y)\},\ \ldots(3)$$

$$H_{OA}=lH_x+mH_y+nH_z. \quad\ldots\ldots\ldots\ldots(4)$$

$H_x$, $H_y$, $H_z$ are the components of A.M. about the axes OX, OY, OZ.

Thus $\mathbf{H}_{OA}$, regarded as a vector associated with the axis OA, is compounded of the three vectors $\mathbf{H}_x$, $\mathbf{H}_y$, $\mathbf{H}_z$ associated with the axes of coordinates. When it is desired to insist on the vector quality of a quantity, it is usual to employ a special symbol for the quantity. We shall use the same letter as that ordinarily employed, but taken from the Clarendon fount. Thus, what we have referred to as the vector $H_{OA}$ would be denoted by $\mathbf{H}_{OA}$. Thus H means the scalar, or numerical, value of the vector. The component vectors are $\mathbf{H}_x$, $\mathbf{H}_y$, $\mathbf{H}_z$.

These components give a vector of which the scalar value is

$$H=(H_x^2+H_y^2+H_z^2)^{\frac{1}{2}}, \quad\ldots\ldots\ldots\ldots(5)$$

the direction cosines of which are $(H_x, H_y, H_z)/H$. Thus $\mathbf{H}$ is not to be confounded with $\mathbf{H}_{OA}$. Denoting the angle between the two vectors by $(H, H_{OA})$, we have

$$\cos(H, H_{OA})=\frac{1}{H}(lH_x+mH_y+nH_z)=\frac{H_{OA}}{H}. \quad\ldots\ldots\ldots\ldots(6)$$

**3.** *Relations of components of* A.M. *to momental ellipsoid.* Now let the system revolve as a rigid body about an axis OI. OI is called the instantaneous axis, or axis of resultant angular velocity, and every particle of the system has at the same instant the same angular speed, $\omega$ say, about that axis. In some of the cases of motion which are considered below, the system is a rigid body, and turns about a point fixed both in space and in the body, and this point may then be conveniently taken as the origin of coordinates. In the more general case, however, of the motion of a rigid body, the instantaneous axis changes in position both in space and in the body, and is not subject to the condition of always passing through a definite point either in the body or in space.

We shall denote the direction cosines of the instantaneous axis by $\alpha, \beta, \gamma$, reserving $l, m, n$ for *any* axis OA. The angular speeds $\omega_x, \omega_y, \omega_z$ are the components of angular velocity about the axes of coordinates. It is easy to see that they give

$$\dot{x} = \omega_y z - \omega_z y, \quad \dot{y} = \omega_z x - \omega_x z, \quad \dot{z} = \omega_x y - \omega_y x. \qquad (1)$$

By (3), 2, we obtain

$$\begin{aligned} H_x &= \Sigma[\mu\{(\omega_x y - \omega_y x)y - (\omega_z x - \omega_x z)z\}] \\ &= \omega_x \Sigma\{\mu(y^2+z^2)\} - \omega_y \Sigma(\mu xy) - \omega_z \Sigma(\mu zx) \qquad (2) \end{aligned}$$

In the same way similar expressions are obtained for $H_y$, $H_z$.

But for a rigid body we have the three moments of inertia,

$$A = \Sigma\mu(y^2+z^2), \quad B = \Sigma\mu(z^2+x^2), \quad C = \Sigma\mu(x^2+y^2), \qquad (3)$$

and the three products of inertia,

$$D = \Sigma(\mu yz), \quad E = \Sigma(\mu zx), \quad F = \Sigma(\mu xy). \qquad (4)$$

Hence for the components of A.M. we have

$$\left.\begin{aligned} H_x &= A\omega_x - F\omega_y - E\omega_z, \\ H_y &= -F\omega_x + B\omega_y - D\omega_z, \\ H_z &= -E\omega_x - D\omega_y + C\omega_z. \end{aligned}\right\} \qquad (5)$$

It may be noticed that these component angular momenta give, since now $\omega_x = \alpha\omega$, etc., for the A.M. about OI the value

$$\begin{aligned} &\alpha(A\omega_x - F\omega_y - E\omega_z) + \beta(-F\omega_x + B\omega_y - D\omega_z) + \gamma(-E\omega_x - D\omega_y + C\omega_z) \\ &\qquad = \omega(A\alpha^2 + B\beta^2 + C\gamma^2 - 2D\beta\gamma - 2E\gamma\alpha - 2F\alpha\beta). \qquad (6) \end{aligned}$$

The moment of inertia I about OI is therefore given by

$$I = A\alpha^2 + B\beta^2 + C\gamma^2 - 2D\beta\gamma - 2E\gamma\alpha - 2F\alpha\beta, \qquad (7)$$

the well-known expression on which Poinsot's theory of the momental ellipsoid is founded. For by the definition of A.M. stated in 2 the A.M. of the system about OI is $\omega\Sigma(\mu p^2)$ where $p$ is the length of the perpendicular on OI from the particle considered.

The direction cosines of OI are $(\omega_x, \omega_y, \omega_z)/\omega$, those of the axis of resultant A.M. (OH say) are $(H_x, H_y, H_z)/H$. Hence we get for the angle (H, I) between OH and OI,

$$\cos(H, I) = \frac{1}{\omega H}(\omega_x H_x + \omega_y H_y + \omega_z H_z). \qquad (8)$$

The A.M., $H_{OA}$, about an axis OA, of which the direction cosines are $l, m, n$, is given by

$$\begin{aligned} H_{OA} &= l(A\omega_x - F\omega_y - E\omega_z) + m(-F\omega_x + B\omega_y - D\omega_z) + n(-E\omega_x - D\omega_y + C\omega_z) \\ &= \omega_x(Al - Fm - En) + \omega_y(-Fl + Bm - Dn) + \omega_z(-El - Dm + Cn). \qquad (9) \end{aligned}$$

4. Equations (1), 3, give for the kinetic energy of a rigid body turning about a fixed point, the expression

$$T=\frac{1}{2}\Sigma\mu\{(\omega_y z-\omega_z y)^2+(\omega_z x-\omega_x z)^2+(\omega_x y-\omega_y x)^2\}, \qquad (1)$$

that is, with the values of A, B, C, D, E, F inserted,

$$T=\frac{1}{2}(A\omega_x^2+B\omega_y^2+C\omega_z^2-2D\omega_y\omega_z-2E\omega_z\omega_x-2F\omega_x\omega_y). \qquad (2)$$

We can write this in the form

$$T=\frac{1}{2}\omega_x(A\omega_x-F\omega_y-E\omega_z)+\frac{1}{2}\omega_y(-F\omega_x+B\omega_y-D\omega_z)$$
$$+\frac{1}{2}\omega_z(-E\omega_x-D\omega_y+C\omega_z), \qquad (3)$$

that is,

$$T=\frac{1}{2}(\omega_x H_x+\omega_y H_y+\omega_z H_z). \qquad (4)$$

Thus the kinetic energy is half the sum of the products obtained by multiplying each component of A.M. by the corresponding component of angular velocity.

It is to be observed that if the motion be impulsively generated from rest by impulsive couples about the axes, $H_x$, $H_y$, $H_z$ are the time-integrals of these couples for the infinitely short interval in which the motion is generated.

Equation (4) expresses in ordinary notation the theorem that

$$T=-\frac{1}{2}S.\Omega\mathbf{H}, \qquad (5)$$

in quaternion notation, or, in words, that the kinetic energy is *minus* half the scalar product of the two vectors, $\Omega$ representing the angular velocity, and **H** representing the vector of angular momentum.

In (3) T is expressed as a homogeneous quadratic function of $\omega_x$, $\omega_y$, $\omega_z$, and we see that we have then

$$\frac{\partial T}{\partial\omega_x}=H_x,\quad \frac{\partial T}{\partial\omega_y}=H_y,\quad \frac{\partial T}{\partial\omega_z}=H_z. \qquad (6)$$

We infer from the form of (4) that if by solution of the equations (5), 3, for $H_x$, $H_y$, $H_z$ in terms of $\omega_x$, $\omega_y$, $\omega_z$ we found $\omega_x$, $\omega_y$, $\omega_z$ in terms of $H_x$, $H_y$, $H_z$, and substituted in (4), we should obtain T expressed as a homogeneous quadratic function of the latter variables, and should then have

$$\frac{\partial T}{\partial H_x}=\omega_x,\quad \frac{\partial T}{\partial H_y}=\omega_y,\quad \frac{\partial T}{\partial H_z}=\omega_z. \qquad (7)$$

This transformation is easily carried out. Writing $\Delta$ for the determinant

$$\begin{vmatrix} A, & -F, & -E \\ -F, & B, & -D \\ -E, & -D, & C \end{vmatrix}=ABC-AD^2-BE^2-CF^2-2DEF, \qquad (8)$$

we get

$$\left.\begin{aligned}\omega_x&=\frac{1}{\Delta}\{H_x(BC-D^2)+H_y(DE+CF)+H_z(BE+FD)\},\\ \omega_y&=\frac{1}{\Delta}\{H_x(DE+CF)+H_y(CA-E^2)+H_z(AD+EF)\},\\ \omega_z&=\frac{1}{\Delta}\{H_x(BE+FD)+H_y(AD+EF)+H_z(AB-F^2)\}.\end{aligned}\right\} \qquad (9)$$

The kinetic energy is now given by

$$T=\frac{1}{2\Delta}\{H_x^2(BC-D^2)+H_y^2(CA-E^2)+H_z^2(AB-F^2)$$
$$+2H_yH_z(AD+EF)+2H_zH_x(BE+FD)+2H_xH_y(DE+CF)\}. \quad\text{...(10)}$$

It will be observed that if the axes of reference be the principal axes of moment of inertia, this reduces to

$$T=\frac{1}{2}\left(\frac{1}{A}H_x^2+\frac{1}{B}H_y^2+\frac{1}{C}H_z^2\right); \quad\text{...(11)}$$

and it may be noticed that

$$\frac{\partial T}{\partial A}=-\frac{1}{2}\frac{H_x^2}{A^2}, \quad \frac{\partial T}{\partial B}=-\frac{1}{2}\frac{H_y^2}{B^2}, \quad \frac{\partial T}{\partial C}=-\frac{1}{2}\frac{H_z^2}{C^2}. \quad\text{...(12)}$$

By (2) above we have, for the same choice of the axes of reference,

$$T=\tfrac{1}{2}(A\omega_x^2+B\omega_y^2+C\omega_z^2), \quad\text{...(13)}$$

and this of course is the same as (11). But we have now

$$\frac{\partial T}{\partial A}\left(=\tfrac{1}{2}\omega_x^2\right)=\frac{1}{2}\frac{H_x^2}{A^2}, \quad \frac{\partial T}{\partial B}=\frac{1}{2}\frac{H_y^2}{B^2}, \quad \frac{\partial T}{\partial C}=\frac{1}{2}\frac{H_z^2}{C^2}; \quad\text{...(14)}$$

so that here the partial differential coefficients of T with reference to A, B, C have the same absolute values, but positive or negative signs according as T is expressed in terms of the component angular velocities, or in terms of the components of angular momentum.

5. A.M. *and kinetic energy.* As we have seen, if the body turns about an axis OI, drawn in the direction $\alpha$, $\beta$, $\gamma$, with angular speed $\omega$, $(\omega_x, \omega_y, \omega_z)$,

$$2T=A\omega_x^2+B\omega_y^2+C\omega_z^2-2D\omega_y\omega_z-2E\omega_z\omega_x-2F\omega_x\omega_y.$$

Consider now the expression

$$A\xi^2+B\eta^2+C\zeta^2-2D\eta\zeta-2E\zeta\xi-2F\xi\eta,$$

where $\xi$, $\eta$, $\zeta$ are the coordinates of a point on the line OI. If $\rho$ be the distance of the point from O, we have

$$\alpha=\frac{\xi}{\rho}=\frac{\omega_x}{\omega}, \quad \beta=\frac{\eta}{\rho}=\frac{\omega_y}{\omega}, \quad \gamma=\frac{\zeta}{\rho}=\frac{\omega_z}{\omega};$$

so that if we take $\rho=\omega$, we have $(\xi, \eta, \zeta)=(\omega_x, \omega_y, \omega_z)$. If now we suppose the value of T to be the same for different directions of the axis of resultant angular velocity, the surface given by $(\xi, \eta, \zeta)=(\omega_x, \omega_y, \omega_z)$ is an ellipsoid (centre O) and $\omega^2$ is numerically equal to $\xi^2+\eta^2+\zeta^2$, the square of the length of the radius vector in the direction chosen. Thus we have the theorem that if the body rotate about any axis through the fixed point with an angular speed proportional to the length of the radius vector of this ellipsoid which coincides with the axis, the kinetic energy of rotation is the same for every axis. The ellipsoid is similar and similarly situated to any Poinsot's momental ellipsoid for the given body and the fixed point as centre. This result is of course obvious from the fact that the moment of inertia about any radius vector of the momental ellipsoid is inversely proportional to the square of the length of that radius vector.

The direction cosines of the perpendicular to the plane which touches this ellipsoid at the outer extremity of the radius vector in the direction $\alpha$, $\beta$, $\gamma$ are proportional to $\partial T/\partial\omega_x$, $\partial T/\partial\omega_y$, $\partial T/\partial\omega_z$, that is to $H_x$, $H_y$, $H_z$. Hence these direction cosines are $(H_x, H_y, H_z)/H$. Since $(\xi, \eta, \zeta) = (\omega_x, \omega_y, \omega_z)$ the length $\varpi$ of the perpendicular is given by

$$\varpi = (\omega_x H_x + \omega_y H_y + \omega_z H_z)/H = \frac{2T}{H}.$$

The direction of the resultant A.M. is therefore perpendicular to the diametral plane which is conjugate to the axis of rotation. This result is useful as telling at once how the axis of rotation (the radius vector) and the axis of A.M. (the perpendicular to the tangent plane), which are lines passing through the origin, are relatively situated.

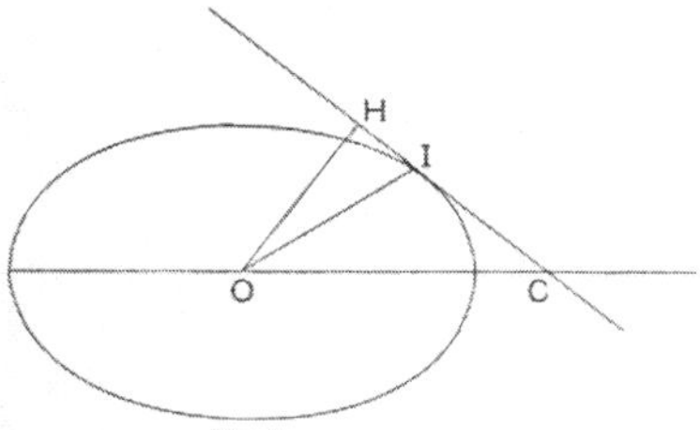

FIG. 2.

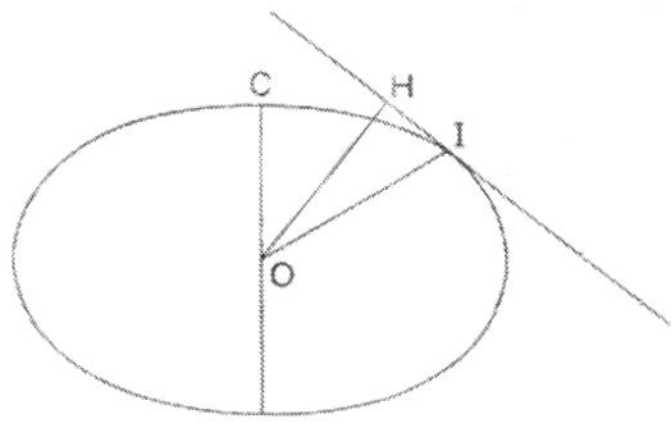

FIG. 3.

For example, consider a body of which two principal moments of inertia, A and B, are equal. According as $A > \text{or} < C$ the momental ellipsoid is prolate or oblate, and in each case the moment of inertia about the axis of figure is C. In the case of the oblate ellipsoid the axis OH of A.M. is nearer to the axis of figure OC than is the axis of rotation OI: the reverse is true in the other case (see Figs. 2 and 3).

6. *Motion of rotation combined with motion of translation.* If the body is not in motion about a fixed point, we consider separately the motion of the centroid of the body, and the turning about an axis, or about coordinate axes, passing through the centroid. Thus, if $u$, $v$, $w$ be the component velocities of the centroid, and $H_x$, $H_y$, $H_z$ the angular momenta and $\omega_x$, $\omega_y$, $\omega_z$ the angular speeds about axes drawn from the centroid as origin, and M denote the whole mass of the body, the kinetic energy is given by the equation

$$T = \tfrac{1}{2}\{M(u^2 + v^2 + w^2) + \omega_x H_x + \omega_y H_y + \omega_z H_z\}.$$

Examples will be found below in connection with the motion of a top on a horizontal plane.

7. *Time-rate of change of* A.M. *Equations of motion.* The time-rate of change of A.M. about OX is

$$\dot{H}_x = A\dot{\omega}_x - F\dot{\omega}_y - E\dot{\omega}_z + \dot{A}\omega_x - \dot{F}\omega_y - \dot{E}\omega_z, \quad \ldots\ldots\ldots\ldots\ldots\ldots(1)$$

for as the body is in motion relatively to the axes, the quantities A, B, C, D, E, F are subject to variation.

If at the instant considered the principal axes of the body coincide with the axes of coordinates, A, B, C are the principal moments of inertia, and $D = E = F = 0$. Equation (1) becomes then

$$\dot{H}_x = A\dot{\omega}_x + \dot{A}\omega_x - \dot{F}\omega_y - \dot{E}\omega_z. \quad \ldots\ldots\ldots\ldots\ldots\ldots\ldots\ldots\ldots(2)$$

Similar equations hold for $\dot{H}_y$, $\dot{H}_z$.

If the material system be a rigid body turning about O as a fixed point, the values of $\dot{A}$, $\dot{E}$, $\dot{F}$ are to be calculated subject to the conditions expressed in (1), 3. Since

$$A = \Sigma\mu(y^2+z^2), \quad E = \Sigma\mu zx, \quad F = \Sigma\mu xy, \quad \ldots\ldots\ldots\ldots\ldots\ldots(3)$$

we have $\dot{A} = 2\Sigma\mu(y\dot{y}+z\dot{z})$, $\dot{E} = \Sigma\mu(\dot{z}x+\dot{x}z)$, $\dot{F} = \Sigma\mu(\dot{x}y+\dot{y}x)$; ........(4)

and (1), 3, give

$$y\dot{y}+z\dot{z} = \omega_z xy - \omega_y xz, \quad \dot{x}y+\dot{y}x = \omega_z(z^2+x^2) - \omega_z(y^2+z^2) + \omega_y yz - \omega_x xz.$$

Multiplying the last-found expressions by $\mu$ and summing for all the particles, we find (since at the instant $D = E = F = 0$, $\dot{A} = 0$) $\dot{F} = \omega_z(B-A)$. Similarly we obtain $\dot{E} = \omega_y(A-C)$. The equation of motion, obtained by equating $\dot{H}_x$ to the moment L of the applied forces about OX, is therefore

$$A\dot{\omega}_x - (B-C)\omega_y\omega_z = L. \quad \ldots\ldots\ldots\ldots\ldots\ldots\ldots\ldots\ldots\ldots(5)$$

Similar equations of motion hold for the other two axes.

8. *Equations of motion for moving axes. Euler's equations.* In these equations of motion the angular speeds $\omega_x$, $\omega_y$, $\omega_z$ are those about the fixed axes of coordinates with which the principal axes of moment of inertia of the body coincide at the instant considered, that is at time $t$. But since, obviously, there is no difference between the angular speed about a fixed axis and that about a moving axis coincident with it, we identify $\omega_x$, $\omega_y$, $\omega_z$ with the angular speeds $\omega_1$, $\omega_2$, $\omega_3$, say, about the principal axes of the momental ellipsoid, which, of course. moves with the body. It requires examination, however, to decide whether $\dot{\omega}_x$, say, the time-rate of variation of the angular speed $\omega_x$ about the fixed axis OX, may be identified with the time-rate of variation $\dot{\omega}_1$ of the angular speed about the moving axis: for after the lapse of an interval $dt$ the moving axis has separated from the fixed axis.

To decide this point, and at the same time obtain some useful results, we consider the general case in which the trihedral system of axes OA, OB, OC (not necessarily principal axes of the body) turns about the axes OA, OB, OC themselves with the angular speeds $\theta_1$, $\theta_2$, $\theta_3$. Let these axes coincide at time $t$ with the fixed axes OX, OY, OZ: then $\omega_1=\omega_x$, $\omega_2=\omega_y$, $\omega_3=\omega_z$, that is the angular speeds of the body about OA, OB, OC and about OX, OY, OZ are the same at the instant. At time $t+dt$, however, $\omega_x$, $\omega_y$, $\omega_z$ have become $\omega_x+\dot{\omega}_x dt$, $\omega_y+\dot{\omega}_y dt$, $\omega_z+\dot{\omega}_z dt$, while $\omega_1$, $\omega_2$, $\omega_3$ have become $\omega_1+\dot{\omega}_1 dt$, $\omega_2+\dot{\omega}_2 dt$, $\omega_3+\dot{\omega}_3 dt$. But, in the interval $dt$, OA, OB, OC have moved away from coincidence with the fixed axes. OA has moved through the angle $\theta_2 dt$ about OB (or OY), and through the angle $\theta_3 dt$ about OC (or OZ). This, of course, is not exactly true; but it can easily be shown that the amount of inexactitude becomes vanishingly small in comparison with $\theta_2 dt$ and $\theta_3 dt$, as $dt$ is diminished towards zero. Thus in the time interval $dt$, taken very small, OA has turned round O in such a way that a point $a$ on it, at unit distance from O, has at time $t+dt$ coordinates the values of which, measured along OX, OY, OZ, are 1, $\theta_3 dt$, $-\theta_2 dt$. Since the length O$a$ is unity, these coordinates may be taken as the direction cosines of the new position of O$a$.

The instantaneous axis OI still passes at time $t+dt$ through O, but it may have changed its direction in space. Let its direction cosines with respect to OX, OY, OZ be $\alpha+d\alpha$, $\beta+d\beta$, $\gamma+d\gamma$ at time $t+dt$, and the angular speed about it be $\omega+d\omega$. The angle between the new positions of OI and OA is

$$\cos^{-1}\{\alpha+d\alpha+(\beta+d\beta)\theta_3 dt-(\gamma+d\gamma)\theta_2 dt\},$$

or, to small quantities of the first order,

$$\cos^{-1}\{\alpha+d\alpha+(\beta\theta_3-\gamma\theta_2)dt\}.$$

The angular speed $\omega_1+\dot{\omega}_1 dt$ about OA in the new position is therefore given by

$$\omega_1+\dot{\omega}_1 dt=(\omega+d\omega)\{\alpha+d\alpha+(\beta\theta_3-\gamma\theta_2)dt\},$$

which, since $\omega_1=\omega\alpha$, $\omega_2=\omega\beta$, $\omega_3=\omega\gamma$, can be written

$$\dot{\omega}_1=\frac{d(\omega\alpha)}{dt}+\omega_2\theta_3-\omega_3\theta_2. \quad \text{.............................(1)}$$

Thus, although $\omega\alpha=\omega_1$, $d(\omega\alpha)/dt$ is not $d\omega_1/dt$, but $d\omega_x/dt$, for it is the time-rate of change of the component of $\omega$ with reference to the fixed axis OX, while $d\omega_1/dt$ (or $\dot{\omega}_1$) is the time-rate of change of the component of $\omega$ with reference to the moving axis OA, and following that axis. Similar results can be obtained in the same way for the other axes. Thus we have the important equations

$$\left.\begin{aligned}\dot{\omega}_1&=\dot{\omega}_x+\omega_2\theta_3-\omega_3\theta_2,\\ \dot{\omega}_2&=\dot{\omega}_y+\omega_3\theta_1-\omega_1\theta_3,\\ \dot{\omega}_3&=\dot{\omega}_z+\omega_1\theta_2-\omega_2\theta_1.\end{aligned}\right\} \quad \text{.............................(2)}$$

It is convenient to write equations (2) in the form

$$\dot{\omega}_x=\dot{\omega}_1-\omega_2\theta_3+\omega_3\theta_2,\ \text{etc, etc.} \quad \text{.........................(2')}$$

Then $\dot{\omega}_x$ can be built up, in the manner indicated in 12 and 13 below, of $\dot{\omega}_1$ arising from the time-rate of variation, of $\omega_1$, $\omega_3\theta$ arising from the turning about OB, and $-\omega_2\theta_3$ arising from the turning about OC.

If the axes OA, OB, OC are fixed in the body, we have

$$(\theta_1, \theta_2, \theta_3) = (\omega_1 . \omega_2, \omega_3),$$

and therefore $$\dot\omega_1 = \dot\omega_x, \quad \omega_2 = \dot\omega_y, \quad \dot\omega_3 = \omega_z. \qquad \text{.........................(3)}$$

Thus, whether OA, OB, OC are principal axes or not, provided only they are fixed in the body, the two angular accelerations are the same.

Returning now to (5), 7, we see that we can write that equation, and the two similar equations which hold for the other axes,

$$\left.\begin{array}{l} A\dot\omega_1 - (B - C)\omega_2\omega_3 = L, \\ B\dot\omega_2 - (C - A)\omega_3\omega_1 = M, \\ C\dot\omega_3 - (A - B)\omega_1\omega_2 = N, \end{array}\right\} \qquad \text{.........................(4)}$$

for OA, OB, OC are here principal axes of moment of inertia, and therefore fulfil the condition of being fixed in the body if that be rigid. L, M, N are the moments of the applied forces about the instantaneous positions of OA, OB, OC. These are Euler's equations for a rigid body turning about a fixed point.

9. *Example. A symmetrical top turning about a fixed point* O. To make the discussion of moving axes clear, we take a case in which three axes OC, OD, OE, fixed neither in the body nor in space, are chosen instead of OA, OB, OC, which we reserve for the representation of the principal axes. The body (Fig. 4) is supposed to be symmetrical about the axis OC. Let OC at the instant considered be inclined at an angle $\theta$ to the vertical OZ, and further let the plane ZOC be turning about the vertical with angular speed $\dot\psi$, while a plane fixed in the body, and containing the axis of figure, is turning relatively to the plane ZOC with angular speed $\omega'$, in the directions shown by the circular arrows in the figure. We refer the motion to OC, and to other axes OD, OE at right angles to one another and to OC. OD is at right angles to the plane ZOC, OE lies in that plane, so that, while OC is fixed in the body, OD, OE are not, but move with the plane ZOC. The angular speeds of the body about OC, OD, OE at time $t$ are respectively $\omega' + \dot\psi\cos\theta$, $\dot\theta$, $\dot\psi\sin\theta$. According to the notation used above, these represent $\omega_1$, $\omega_2$, $\omega_3$. The angular speeds of this frame of moving axes about OC, OD, OE respectively are then $\dot\psi\cos\theta$, $\dot\theta$, $\dot\psi\sin\theta$, and these represent the $\theta_1$, $\theta_2$, $\theta_3$ of 8. By (2), 8, we get for the angular accelerations with respect to fixed axes OX, OY, OZ, with which OC, OD, OE coincide at the instant $t$,

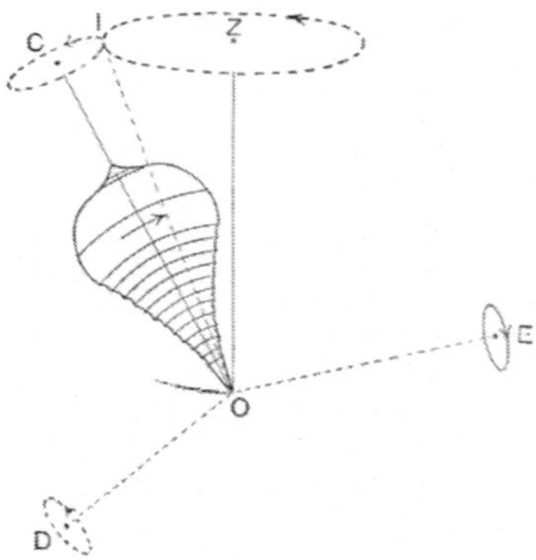

FIG. 4.

$$\left.\begin{array}{ll} \dot\omega_x = \dot\omega_1 - \omega_2\dot\theta_3 + \omega_3\theta_2 = \dot\omega_1 - \dot\theta\dot\psi\sin\theta + \dot\psi\sin\theta\,.\,\dot\theta & = \dot\omega_1, \\ \dot\omega_y = \dot\omega_2 - \omega_3\theta_1 + \omega_1\theta_3 = \dot\omega_2 - \dot\psi^2\sin\theta\cos\theta + (\omega' + \dot\psi\cos\theta)\dot\psi\sin\theta & = \dot\omega_2 + \omega'\psi\sin\theta, \\ \dot\omega_z = \dot\omega_3 - \omega_1\theta_2 + \omega_2\theta_1 = \dot\omega_3 - (\omega' + \dot\psi\cos\theta)\dot\theta + \dot\theta\dot\psi\cos\theta & = \dot\omega_3 - \omega'\dot\theta. \end{array}\right\} \text{.....(1)}$$

**10.** *Body turning about a fixed point* O. *More general equations of motion for moving axes.* Exactly the same process can be applied to the comparison of the rates of change of the components of any vector whatever, with reference to fixed axes, with the rates of change of the same vector with reference to a system of axes turning with angular speeds $\theta_1$, $\theta_2$, $\theta_3$ about the individual axes of the same system, and coinciding with the fixed axes at time $t$. Denoting the components in the two cases by $H_x$, $H_y$, $H_z$, $H_1$, $H_2$, $H_3$, we have only to use H instead of $\omega$ in (2), 8, to obtain

$$\left.\begin{aligned} \dot{H}_x &= \dot{H}_1 - H_2\theta_3 + H_3\theta_2, \\ \dot{H}_y &= \dot{H}_2 - H_3\theta_1 + H_1\theta_3, \\ \dot{H}_z &= \dot{H}_3 - H_1\theta_2 + H_2\theta_1. \end{aligned}\right\} \qquad (1)$$

Since the axes coincide at time $t$, we have of course

$$(H_x, H_y, H_z) = (H_1, H_2, H_3).$$

If $H_x$, $H_y$, $H_z$ be the components of angular momentum about the fixed axes with which the moving axes coincide at the instant, we obtain the equations of motion

$$\left.\begin{aligned} \dot{H}_1 - H_2\theta_3 + H_3\theta_2 &= L, \\ \dot{H}_2 - H_3\theta_1 + H_1\theta_3 &= M, \\ \dot{H}_3 - H_1\theta_2 + H_2\theta_1 &= N. \end{aligned}\right\} \qquad (2)$$

These are more general than (4), 8, inasmuch as they hold for *any* system of moving axes about which the moments of applied forces are L, M, N at the instant considered.

As has been noticed, the values of $H_1$, $H_2$, $H_3$ are identical with $H_x$, $H_y$, $H_z$, and this being the case, the non-identity of the two sets of time-rates of variation is sometimes regarded as mysterious. But while $H_1$, $H_2$, $H_3$, like $H_x$, $H_y$, $H_z$, refer to a set of axes in a definite position, $\dot{H}_1$, $\dot{H}_2$, $\dot{H}_3$ refer to the change from the values for one definite position of the axes to those for *another* (but near) definite position, while $\dot{H}_x$, $\dot{H}_y$, $\dot{H}_z$ refer always to the same position of the axes as that with which $H_x$, $H_y$, $H_z$ $(= H_1, H_2, H_3)$ are associated.

**11.** *Body turning about a fixed point* O. *Relations between moving and fixed axes.* If the moving axes OA, OB, OC are not coincident with the fixed axes at the moment under consideration, we may define their positions by the three sets of direction cosines which specify their directions with respect to OX, OY, OZ. We denote these by $l_1, m_1, n_1$, $l_2, m_2, n_2$, $l_3, m_3, n_3$. Thus we have

$$\left.\begin{aligned} H_x &= H_1 l_1 + H_2 l_2 + H_3 l_3, \\ H_y &= H_1 m_1 + H_2 m_2 + H_3 m_3, \\ H_z &= H_1 n_1 + H_2 n_2 + H_3 n_3. \end{aligned}\right\} \qquad (1)$$

These give
$$\left.\begin{aligned}\dot{H}_x &= \dot{H}_1 l_1 + \dot{H}_2 l_2 + \dot{H}_3 l_3 + H_1 \dot{l}_1 + H_2 \dot{l}_2 + H_3 \dot{l}_3,\\ \dot{H}_y &= \dot{H}_1 m_1 + \dot{H}_2 m_2 + \dot{H}_3 m_3 + H_1 \dot{m}_1 + H_2 \dot{m}_2 + H_3 \dot{m}_3,\\ \dot{H}_z &= \dot{H}_1 n_1 + \dot{H}_2 n_2 + \dot{H}_3 n_3 + H_1 \dot{n}_1 + H_2 \dot{n}_2 + H_3 \dot{n}_3.\end{aligned}\right\} \quad\text{..........(2)}$$

Equations (2) are quite general, and may be specialised in various ways. For example, we may obtain from them the results already established, but before making any applications it is convenient to notice the vector significance of the expressions. This may be exhibited as follows. We are considering a vector **H** which is changing both in direction and in magnitude, and the components of which, with reference to OA, OB, OC, are $H_1$, $H_2$, $H_3$. The component $H_l$, say, of the vector in a direction $Ol$ making with these axes angles $\alpha$, $\beta$, $\gamma$ at time $t$ (and moving in any way that may be specified) is given by

$$H_l = H_1 \cos\alpha + H_2 \cos\beta + H_3 \cos\gamma. \quad\text{........................(3)}$$

The time-rate of variation of this *for the fixed direction with which $Ol$ coincides at the instant*, is

$$\frac{dH_l}{dt} = \dot{H}_1 \cos\alpha + \dot{H}_2 \cos\beta + \dot{H}_3 \cos\gamma - H_1 \dot{\alpha} \sin\alpha - H_2 \dot{\beta} \sin\beta - H_3 \dot{\gamma} \sin\gamma. \quad\text{...(4)}$$

Here $\dot{H}_1$, $\dot{H}_2$, $\dot{H}_3$ are the rates of variation of $H_1$, $H_2$, $H_3$ on the supposition that $\alpha$, $\beta$, $\gamma$ are constant, and are therefore rates of variation following a system of axes OA, OB, OC, moving with $Ol$. But for the moving $Ol$ we denote the rate of growth of the vector ($\alpha$, $\beta$, $\gamma$ constant) by $\partial H/\partial t$, and have

$$\frac{\partial H}{\partial t} = \dot{H}_1 \cos\alpha + \dot{H}_2 \cos\beta + \dot{H}_3 \cos\gamma.$$

Thus
$$\frac{dH}{dt} = \frac{\partial H}{\partial t} - H_1 \dot{\alpha} \sin\alpha - H_2 \dot{\beta} \sin\beta - H_3 \dot{\gamma} \sin\gamma. \quad\text{..............(5)}$$

Now it is easy to prove that

$$\left(\frac{d\cos\alpha}{dt}\right)^2 + \left(\frac{d\cos\beta}{dt}\right)^2 + \left(\frac{d\cos\gamma}{dt}\right)^2, \quad\text{or}\quad \dot{\alpha}^2\sin^2\alpha + \dot{\beta}^2\sin^2\beta + \dot{\gamma}^2\sin^2\gamma,$$

is the square of the angular speed with which the direction $Ol$ is turning round. If we take a vector $\Theta$, of length equal to this angular speed, in the direction $Ok$, say, about which this turning takes place, this vector will represent the angular velocity with which the axes OA, OB, OC are turning. If then, dropping the suffix $l$, we take **H** as the vector considered, we see that $d\mathbf{H}/dt$ exceeds $\partial\mathbf{H}/\partial t$ by the vector product $V(\mathbf{H}\Theta)$, that is, the equation

$$\frac{d\mathbf{H}}{dt} = \frac{\partial\mathbf{H}}{\partial t} + V(\mathbf{H}\Theta) \quad\text{..............................(6)}$$

holds. This merely states the obvious fact that *the rate of growth of any vector drawn from a fixed point is represented in magnitude and direction by the velocity of the outer extremity of the vector.*

If **i**, **j**, **k** denote unit vectors along the instantaneous positions of the axes OA, OB, OC, with respect to which $H_1$, $H_2$, $H_3$ are taken, and $\theta_1$, $\theta_2$, $\theta_3$ be the numerical values of the components of $\Theta$ with reference to the same axes, we can write the last equation in the form

$$\frac{d\mathbf{H}}{dt}-\frac{\partial\mathbf{H}}{\partial t}=(\mathbf{i}H_1+\mathbf{j}H_2+\mathbf{k}H_3)(\mathbf{i}\theta_1+\mathbf{j}\theta_2+\mathbf{k}\theta_3)+H_1\theta_1+H_2\theta_2+H_3\theta_3$$
$$=\mathbf{i}(H_2\theta_3-H_3\theta_2)+\mathbf{j}(H_3\theta_1-H_1\theta_3)+\mathbf{k}(H_1\theta_2-H_2\theta_1). \quad\text{..........(7)}$$

Thus we infer that the components of $dH/dt-\partial H/\partial t$ are

$$H_2\theta_3-H_3\theta_2,\quad H_3\theta_1-H_1\theta_3,\quad H_1\theta_2-H_2\theta_1.$$

The vector product on the right of (6) is a vector at right angles to the plane of the two vectors $\mathbf{H}$, $\Theta$. The equation is indeed obvious, when it is considered that the turning about O$k$ moves the outer extremity of the vector $\mathbf{H}$ at right angles to this plane, so that the total time-rate of change of $\mathbf{H}$, $d\mathbf{H}/dt$, exceeds $\partial\mathbf{H}/\partial t$ by the vector representing the rate of growth thus produced. Thus, from the vector equation, which really requires no demonstration, equations (1) and (2), 10, might be inferred. As a matter of fact, the simple theorem stated at the beginning of next article, and then proved and illustrated in various ways, is clearly true from the most simple commonsense considerations, and is a practical application of the vector theorem (6), in the simple but exceedingly important case in which the vectors $\Theta$, $\mathbf{H}$ are represented by their components $\omega_1$, $H_2$ along the axes with which OA and OB coincide at the instant. The turning about OA with angular speed $\omega_1$ gives a rate of production of the vector along OC equal to $\omega_1 H_2$, and similarly for other components of the two vectors.

The equations of motion, (2) of § 12, are now summed up by the single vector-equation

$$\frac{d\mathbf{H}}{dt}=\frac{\partial\mathbf{H}}{\partial t}+V(\mathbf{H}\Theta)=\mathbf{L}, \quad\text{..............................(8)}$$

where $\mathbf{H}$ is the resultant angular momentum and $\mathbf{L}$ the resultant moment of applied forces, both taken as vectors.

We may also in this equation interpret $\mathbf{H}$ as the vector A.M. about *any* specified axis, $\Theta$ as the vector angular velocity with which that axis is changing in direction, and $\mathbf{L}$ as the vector applied couple, which of course has the direction of $d\mathbf{H}/dt$.

In the case in which $\mathbf{H}$, the value of the A.M. for the moving axis, is constant in numerical amount, the variation $d\mathbf{H}/dt$ is wholly due to the turning; and we have then the *direction* of $\mathbf{H}$ constantly changing so as to produce exactly the time-rate of change of the vector due to the action applied to the body. Thus, as we shall see, in the case of a gyrostat in steady motion, the axis of A.M. always "follows," that is, turns towards, the position of the 90° distant axis of the resultant couple.

**12.** *Simple rules for forming equations of motion.* The following simple theorem, which admits of direct and elementary proof, gives a convenient working rule for the use of the foregoing analysis in practical cases. Though formally derived for a particular case, it gives an intuitive, and therefore easily remembered, method of dealing with even the most complicated problems of the motion of tops and gyrostats.

Consider (Fig. 5) two axes, $Op$, $Oq$, at right angles to one another, which are turning in their own plane with angular speed $\omega$ about the point O, and let $Op'$, $Oq'$ be two fixed axes with which at the instant $Op$, $Oq$ coincide. Let P, Q associated with $Op$, $Oq$ be components of a given vector A in that plane. Along the direction of A measure a length to represent the magnitude of the vector, and on the same scale measure distances along $Op$, $Oq$ respectively to represent P, Q. We shall suppose $Op$, $Oq$ themselves to be these lengths. The motion of the extremity of the line representing the vector A will have components along $Op'$, $Oq'$ representing the component rates of change of the vector in these directions. But since P, Q, taken along the moving axes, continue to represent the vector, the sum of the displacements of the two points $p$, $q$ (in any interval of time), parallel to $Op'$ and $Oq'$ respectively, must represent the components of the total changes which the vector has undergone in those directions in the interval.

In an interval of time $dt$ we have the displacements

$$\dot{P}\,dt\,.\cos(\omega\,dt)-Q\sin(\omega\,dt),\ \text{parallel to } Op',$$
$$\dot{Q}\,dt\,.\cos(\omega\,dt)+P\sin(\omega\,dt),\ \text{parallel to } Oq'.$$

Since $dt$ is supposed to be very small, these become

$$(\dot{P}-\omega Q)dt,\quad (\dot{Q}+\omega P)dt,$$

and the component rates of change along $Op'$, $Oq'$ are

$$\dot{P}-\omega Q,\quad \dot{Q}+\omega P.$$

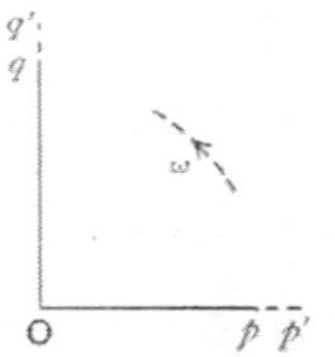


FIG. 5.

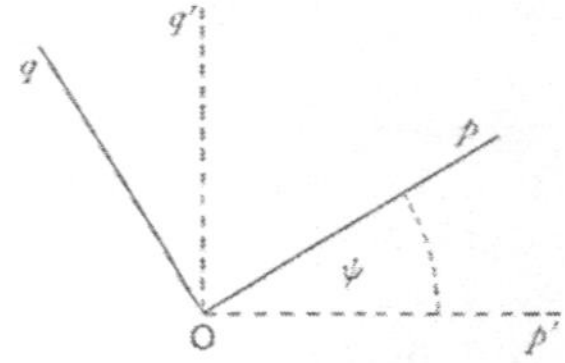


FIG. 6.

These results may also be obtained in the following manner. Let P′, Q′ be components of a vector with reference to the fixed rectangular axes $Op'$, $Oq'$, and P, Q those of the same vector with reference to axes $Op$, $Oq$ turning in the plane $p'Oq'$. Let, however, at the instant the axis $Op$ make with $Op'$ the angle $\psi$ as in the diagram (Fig. 6). Then we have

$$P'=P\cos\psi-Q\sin\psi,\quad Q'=P\sin\psi+Q\cos\psi. \qquad (1)$$

These equations can be united in one as

$$P'+iQ'=(P+iQ)e^{i\psi}. \qquad (2)$$

Hence we obtain by differentiation

$$\dot{P}'+i\dot{Q}'=\{\dot{P}-Q\dot{\psi}+i(\dot{Q}+P\dot{\psi})\}e^{i\psi}. \qquad (3)$$

Equating real and imaginary parts on the two sides of the last equation, we find

$$\left.\begin{aligned}\dot{P}'&=(\dot{P}-Q\dot{\psi})\cos\psi-(\dot{Q}+P\dot{\psi})\sin\psi,\\ \dot{Q}'&=(\dot{P}-Q\dot{\psi})\sin\psi+(\dot{Q}+P\dot{\psi})\cos\psi.\end{aligned}\right\} \qquad (4)$$

Thus, if at the instant considered $\psi=0$, and we write $\omega$ for $\dot{\psi}$, we get

$$\dot{P}'=\dot{P}-\omega Q, \quad Q'=\dot{Q}+\omega P. \qquad (5)$$

Equations (4) [as well as equations (5)] show that $\dot{P}-\omega Q$, $\dot{Q}+\omega P$ are the component rates of change of the vector for fixed axes with which the revolving axes at the instant coincide. The reader may give a vectorial interpretation to the steps of this process

The theorem stated in the earlier part of this article is extended for practical working as follows: Let there be three moving axes, O$p$, O$q$, O$r$ (Fig. 7), which remain at right angles to one another as they move, and at time $t$ are coincident with three fixed axes, O$p'$, O$q'$, O$r'$. The motion of the axes may be specified in the following manner: O$q$, O$r$ turn together with angular speed $\omega_1$ about O$p$; O$r$, O$p$ turn with angular speed $\omega_2$ about O$q$; O$p$, O$q$ turn with angular speed $\omega_3$ about O$r$. These rates of turning, existing simultaneously, maintain the axes at right angles to one another, whatever their (finite) time-rates of change may be. For, by considering angular displacements about the axes, effected separately in successive intervals of time, we could show that, to small quantities of the first order of magnitude inclusive, the same result is obtained when the displacements take place in any order or combination of orders.

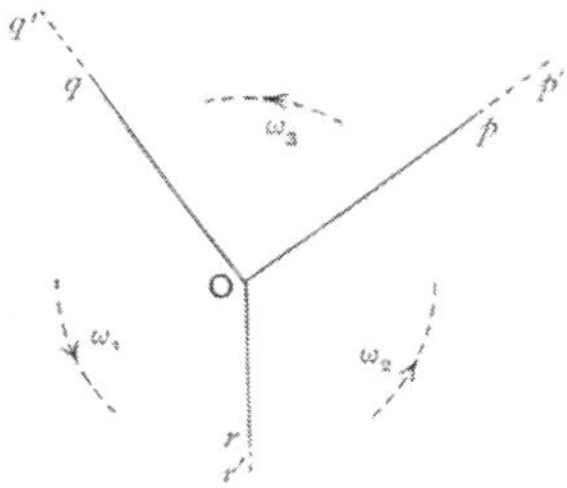

FIG. 7.

That the axes remain at right angles in the simultaneous turnings may be seen by noticing, for example, that while, by the turning about O$p$, O$q$ and O$r$ are moving in their own plane, the turning about O$q$ is moving O$r$, and the turning about O$r$ is moving O$q$, in each case at right angles to that plane, and that each of these motions is followed by O$p$. These displacements of O$p$ no doubt change in $dt$ in a slight degree the plane of motion of O$q$, O$r$, but the effects of this are infinitesimals of a higher order than the first.

Let now the components P, Q, R of a vector A for the axes O$p$, O$q$, O$r$ be represented by the segments O$p$, O$q$, O$r$ themselves, which are therefore of lengths numerically equal to P, Q, R. Then the sum of the displacements of the extremities $p$, $q$, $r$ of these lines, taken parallel to any direction that may be specified, gives the magnitude of the vectorial change in that direction, and corresponding to these displacements.

In a short interval of time $dt$ we have the displacements:

$$\dot{P}\,dt \,.\, \cos\alpha - Q\sin(\omega_3 dt) + R\sin(\omega_2 dt), \text{ parallel to O}p',$$
$$\dot{Q}\,dt \,.\, \cos\beta - R\sin(\omega_1 dt) + P\sin(\omega_3 dt), \text{ parallel to O}q',$$
$$\dot{R}\,dt \,.\, \cos\gamma - P\sin(\omega_2 dt) + Q\sin(\omega_1 dt), \text{ parallel to O}r',$$

where $\alpha$, $\beta$, $\gamma$ are the angles which the vector A in its new position makes with the fixed axes O($p'$, $q'$, $r'$).

These angles are evanescent in the limit when $dt$ is taken indefinitely small. These displacements therefore give, for the time-rates of change of the components along $Op'$, $Oq'$, $Or'$,

$$\dot{P}-Q\omega_3+R\omega_2, \quad \dot{Q}-R\omega_1+P\omega_3, \quad \dot{R}-P\omega_2+Q\omega_1,$$

which, when $H_1$, $H_2$, $H_3$ are used instead of P, Q, R, and $\theta_1$, $\theta_2$, $\theta_3$ for $\omega_1$, $\omega_2$, $\omega_3$, are the values of the component time-rates of change exhibited in (1), 10.

**13.** *Illustration.* We may illustrate the foregoing article by using its method to establish Euler's equations, which have been obtained otherwise in 7 and 8. As we have seen in 7, the moving axes chosen are the axes of principal moments of inertia of the rigid body. These move with the body: the moments of inertia are denoted by A, B, C, and the angular speeds at the instant are $\omega_1$, $\omega_2$, $\omega_3$, so that the angular momenta about the axes are $A\omega_1$, $B\omega_2$, $C\omega_3$. These latter quantities we identify with P, Q, R. Hence the three component rates of change exhibited at the end of 12 are

$$A\dot{\omega}_1-(B-C)\omega_2\omega_3, \quad B\dot{\omega}_2-(C-A)\omega_3\omega_1, \quad C\dot{\omega}_3-(A-B)\omega_1\omega_2,$$

which, equated to the moments of the applied forces about the axes in the positions $Op'$, $Oq'$, $Or'$, give the equations

$$\left.\begin{aligned} A\dot{\omega}_1-(B-C)\omega_2\omega_3&=L,\\ B\dot{\omega}_2-(C-A)\omega_3\omega_1&=M,\\ C\dot{\omega}_3-(A-B)\omega_1\omega_2&=N.\end{aligned}\right\} \qquad (1)$$

The reader will find the following way of building up such expressions easy to remember and apply. Take the rate of change of A.M. about $Op'$. We have first the time-rate of change of $A\omega_1$, the component of A.M. about $Op$. That gives $A_1\dot{\omega}$. Then we notice that the turning with angular speed $\omega_2$ about $Oq$ is causing the axis $Or$ to approach $Op'$, and as the A.M. about $Or$ is $C\omega_3$, this gives for $Op'$ a rate of change of A.M. $C\omega_3\omega_2$. Again, the turning of the axes with angular speed $\omega_3$ is causing the axis $Oq$ to recede from $Op'$, and as the A.M. about $Oq$ is $B\omega_2$, this brings about a rate of change of A.M. of amount $-B\omega_2\omega_3$ for the axis $Op'$. Hence the total rate of change is $A\dot{\omega}_1-(B-C)\omega_2\omega_3$. The same process may of course be followed for the axes $Oq'$, $Or'$.

**14.** *Simple rules for passage from one set of axes to another set.* The method of 12 enables the equations for passage from the principal axes of moment of inertia OA, OB, OC to any other system of axes $Ox$, $Oy$, $Oz$, to be written down at once. The body turns about an instantaneous axis OI with angular speed $\omega=(\omega_1^2+\omega_2^2+\omega_3^2)^{\frac{1}{2}}$, where the positive value of the square root is taken. The direction cosines of OI are $(\omega_1, \omega_2, \omega_3)/\omega$. Take now, for example, the axis $Ox$, which has direction cosines $\lambda_1, \lambda_2, \lambda_3$ with reference to the axes OA, OB, OC, and let $\vartheta$ denote the angle $x$OI. The angular momenta give a component E about $Ox$, a component F about an axis $Of$ at right angles to $Ox$ and in the plane $x$OI, and a component G about an axis $Og$ at right angles to the plane $x$OI. The angular speed $\omega$ resolves into two components, $\omega\cos\vartheta$ about $Ox$, and $\omega\sin\vartheta$ about $Of$. Hence, if $Ox$, $Og$ be of lengths numerically equal to E and G, the total rate of change of A.M. for the axis $Ox$ is $\dot{E}-\omega G \sin \vartheta$, where in $\dot{E}$ the cosines are supposed to be constant. Now we have

$$\dot{E}=\frac{d}{dt}(\lambda_1 A\omega_1+\lambda_2 B\omega_2+\lambda_3 C\omega_3)=\lambda_1 A\dot{\omega}_1+\lambda_2 B\dot{\omega}_2+\lambda_3 C\dot{\omega}_3, \quad \ldots\ldots\ldots\ldots\ldots(1)$$

and $$G=\frac{1}{\omega \sin \vartheta}\{A\omega_1(\lambda_3\omega_2-\lambda_2\omega_3)+B\omega_2(\lambda_1\omega_3-\lambda_3\omega_1)+C\omega_3(\lambda_2\omega_1-\lambda_1\omega_2)\}. \quad (2)$$

Similarly we can deal with the axes $Oy$, $Oz$, the cosines of which are $\mu_1, \mu_2, \mu_3, \nu_1, \nu_2, \nu_3$. Taking the total rate of change for each axis, and equating to the moment of forces about that axis, we get the equations

$$\left.\begin{aligned}
&A\lambda_1\dot{\omega}_1+B\lambda_2\dot{\omega}_2+C\lambda_3\dot{\omega}_3+A\omega_1(\lambda_2\omega_3-\lambda_3\omega_2)\\
&\qquad+B\omega_2(\lambda_3\omega_1-\lambda_1\omega_3)+C\omega_3(\lambda_1\omega_2-\lambda_2\omega_1)=L',\\
&A\mu_1\dot{\omega}_1+B\mu_2\dot{\omega}_2+C\mu_3\dot{\omega}_3+A\omega_1(\mu_2\omega_3-\mu_3\omega_2)\\
&\qquad+B\omega_2(\mu_3\omega_1-\mu_1\omega_3)+C\omega_3(\mu_1\omega_2-\mu_2\omega_1)=M',\\
&A\nu_1\dot{\omega}_1+B\nu_2\dot{\omega}_2+C\nu_3\dot{\omega}_3+A\omega_1(\nu_2\omega_3-\nu_3\omega_2)\\
&\qquad+B\omega_2(\nu_3\omega_1-\nu_1\omega_3)+C\omega_3(\nu_1\omega_2-\nu_2\omega_1)\quad=N'.
\end{aligned}\right\}\ldots\ldots\ldots\ldots(3)$$

Here $L'$, $M'$, $N'$ are the moments of forces about $Ox$, $Oy$, $Oz$. They may be derived from the moments L, M, N about OA, OB, OC by the equations

$$L'=\lambda_1 L+\lambda_2 M+\lambda_3 N,\quad M'=\mu_1 L+\mu_2 M+\mu_3 N,\quad N'=\nu_1 L+\nu_2 M+\nu_3 N, \ldots(4)$$

but it is usually more convenient to calculate them directly.

Equations (3) may be derived from Euler's equations by multiplying the first by $\lambda_1$, the second by $\lambda_2$, the third by $\lambda_3$, and adding. The result is the first of (3). The others are derived by means of $\mu_1, \mu_2, \mu_3, \nu_1, \nu_2, \nu_3$ in the same way.

The axes $Ox$, $Oy$, $Oz$ may be the instantaneous positions of a set of moving axes, but it is to be remembered that equations (3) apply to a set of fixed axes in these positions.

**15.** ***Example: A top spinning under gravity forces.*** As another example of 12 we take the top spinning about a fixed point on its axis under gravity forces, and choose the system of axes OD, OE, OC specified in 9, and shown in Fig. 4. These have directions exactly corresponding to those of an ordinary system of axes $Ox$, $Oy$, $Oz$. The angular speeds about these axes are, according to the notation explained (*loc. cit. supra*), $\dot{\theta}$, $\dot{\psi}\sin\theta$, $\omega'+\dot{\psi}\cos\theta$. If A be the moment of inertia of the top about the axis OD, A is also the moment of inertia about OE. We denote the moment of inertia about OC by C.

The components of A.M. associated with the three axes are therefore $A\dot{\theta}$, $A\dot{\psi}\sin\theta$, $C(\omega'+\dot{\psi}\cos\theta)$, or $Cn$. Taking distances along the chosen axes from O equal numerically to these quantities, and considering the motions of the extremities due to time-rates of change of the lengths of the axes, and to the turnings, we get for the rates of growth of A.M about the fixed axes $OD_1$, $OE_1$, $OC_1$, with which OD, OE, OC coincide,

$$\begin{aligned} &A\ddot{\theta}+(Cn-A\dot{\psi}\cos\theta)\dot{\psi}\sin\theta, &&\text{for } OD_1,\\ &A\ddot{\psi}\sin\theta+2A\dot{\psi}\dot{\theta}\cos\theta-Cn\dot{\theta}, &&\text{for } OE_1,\\ &C\dot{n}+A\dot{\psi}\sin\theta\,.\,\dot{\theta}-A\dot{\theta}\dot{\psi}\sin\theta, &&\text{for } OC_1.\end{aligned}$$

The only moment of forces of gravity is that about $OD_1$, which amounts to $Mgh\sin\theta$. In strictness there is a couple about $OC_1$ due partly to the action of the air, and partly in some tops, to friction at the pivots of a rotating flywheel, the axis of which is coincident with the axis of symmetry. We denote the moment of this couple by $\mathcal{N}$. There may also, in some arrangements, be a couple about $OE_1$ and an additional couple about $OD_1$: if these couples exist we denote their moments by $\mathcal{L}$ and $\mathcal{M}$. In most cases $\mathcal{L}$, $\mathcal{M}$, $\mathcal{N}$ are taken as zero. Hence we have

$$\left.\begin{aligned} A\ddot{\theta}+(Cn-A\dot{\psi}\cos\theta)\dot{\psi}\sin\theta &= Mgh\sin\theta+\mathcal{L},\\ A\ddot{\psi}\sin\theta+(2A\dot{\psi}\cos\theta-Cn)\dot{\theta} &= \mathcal{M},\\ C\dot{n} &= \mathcal{N}.\end{aligned}\right\}\quad\ldots\ldots\ldots\ldots(1)$$

If $\mathcal{N}$ may be taken as zero, we see that $n$, and also $Cn$, remain unaltered as the top moves. In some gyrostatic tops, made recently by Dr. J. G. Gray, the spin has been found to be still rapid after the top has been left to itself for nearly an hour, showing that the retarding couple in these instruments, which have a carefully designed form of ball bearings, has been brought down to a very small value.

The A.M. about the vertical OZ is $Cn\cos\theta+A\dot{\psi}\sin^2\theta$, and this, if $\mathcal{M}$ is zero, is constant, that is, we have

$$Cn\cos\theta+A\dot{\psi}\sin^2\theta=G,\quad\ldots\ldots\ldots\ldots(2)$$

where G is a constant. The reader may verify by differentiation that (2) is equivalent to the second of (1) with $\mathcal{M}$ put equal to zero.

**16.** *Euler's equations for rigid body under gravity forces.* Returning to Euler's equations, take the case of any rigid body whatever, under the action of gravity. Let the mass of the body be M, the coordinates of the centroid G be $\xi$, $\eta$, $\zeta$, and the direction cosines of the upward vertical be $\alpha$, $\beta$, $\gamma$, in both cases with reference to the principal axes of moment of inertia OA, OB, OC, drawn through the fixed point O. The components of gravity forces parallel to OA, OB, OC, respectively, are $-Mg(\alpha, \beta, \gamma)$. The second and third of these have moments about OA, the third and first about OB, and the first and second about OC. The total moments are easily found to be $Mg(\beta\zeta-\gamma\eta)$, $Mg(\gamma\xi-\alpha\zeta)$, $Mg(\alpha\eta-\beta\xi)$. Hence Euler's equations (1), 14, become

$$\left.\begin{aligned} A\dot{\omega}_1-(B-C)\omega_2\omega_3 &= Mg(\beta\zeta-\gamma\eta),\\ B\dot{\omega}_2-(C-A)\omega_3\omega_1 &= Mg(\gamma\xi-\alpha\zeta),\\ C\dot{\omega}_3-(A-B)\omega_1\omega_2 &= Mg(\alpha\eta-\beta\xi).\end{aligned}\right\}\quad\ldots\ldots\ldots\ldots(1)$$

But since the vertical is fixed, $\alpha$, $\beta$, $\gamma$ fulfil a certain set of relations. For $\alpha$, $\beta$, $\gamma$ are the coordinates of a point on OZ at unit distance from O, and the component velocities of that point referred to fixed axes, with which OA, OB, OC coincide at the instant, are therefore zero. But by (1), 10, identifying the $H_1$, $H_2$, $H_3$ of these equations with $\alpha, \beta, \gamma$, and $\theta_1, \theta_2, \theta_3$ with $\omega_1, \omega_2, \omega_3$, we obtain

$$\dot{\alpha}-\beta\omega_3+\gamma\omega_2=0, \quad \dot{\beta}-\gamma\omega_1+\alpha\omega_3=0, \quad \dot{\gamma}-\alpha\omega_2+\beta\omega_1=0. \quad \text{.........(2)}$$

It will be observed that this set of relations gives, as it ought,

$$\alpha\dot{\alpha}+\beta\dot{\beta}+\gamma\dot{\gamma}=0.$$

The kinetic energy of the motion of the rigid body is $\frac{1}{2}(A\omega_1^2+B\omega_2^2+C\omega_3^2)$, and the potential energy is $Mg(\alpha\xi+\beta\eta+\gamma\zeta)$. Hence we have the energy equation

$$\tfrac{1}{2}(A\omega_1^2+B\omega_2^2+C\omega_3^2)+Mg(\alpha\xi+\beta\eta+\gamma\zeta)=E, \quad \text{................(3)}$$

where E is the total energy.

Again, the angular momentum about the vertical OZ does not vary, and therefore

$$A\alpha\omega_1+B\beta\omega_2+C\gamma\omega_3=G, \quad \text{.............................(4)}$$

where G is a constant.

These equations are easily obtained from (1) and (2). If we multiply equations (1) respectively by $\omega_1$, $\omega_2$, $\omega_3$, add and integrate, taking account of (2), we obtain (3); if we multiply (1) by $\alpha$, $\beta$, $\gamma$, respectively, add and integrate, we obtain (4).

17. *Steady motion of symmetrical top under gravity.* It is possible to have a motion of unvarying value of $\theta$, and unvarying precession. It will be seen that if $\mathcal{M}$ is zero in (1), 15, and $\dot{\theta}$ be zero, then $\ddot{\psi}$ is zero, at least if $\theta$ is not equal to 0. If then $\dot{\theta}=0$, and also $\ddot{\theta}=0$, we get from the first of these equations, if $\theta$ is not zero,

$$(Cn-A\dot{\psi}\cos\theta)\dot{\psi}=Mgh. \quad \text{...........................(1)}$$

The value of $\dot{\psi}$ is now constant, and the motion is said to be *steady*. Denoting for distinction the steady value of $\dot{\psi}$ by $\mu$, we have the quadratic equation

$$A\cos\theta\,.\,\mu^2-Cn\mu+Mgh=0, \quad \text{.........................(2)}$$

which gives two values of $\mu$, which are possible when the inequality

$$C^2n^2 > 4AMgh\cos\theta$$

is fulfilled. Thus, provided $Cn$ exceeds a limit here indicated, there are, for any given inclination $\theta$ of the axis to the vertical, two values of $\mu$, and if $Cn$ is very great, that is if the angular speed of spin is high enough, one of these values of $\mu$ is very great in comparison with the other. A first approximation to the small root is given by the equation

$$-Cn\mu+Mgh=0, \quad \text{................................(3)}$$

which is obtained by neglecting the term in $\mu^2$ in (2). Thus

$$\mu=\frac{Mgh}{Cn}. \quad \text{....................................(4)}$$

A first approximation to the large root is obtained by neglecting the constant term in (2). Thus we find

$$\mu = \frac{Cn}{A\cos\theta}. \qquad\qquad (5)$$

It is remarkable that, if these first approximations are taken, the large root does not depend on the applied forces at all, while the small root is directly proportional to the product of $Mg$ by $h$, that is to what has been called the "preponderance" of the top. Thus Lord Kelvin proposed (Thomson and Tait's *Natural Philosophy*, § 345$^{xxiv}$) that the azimuthal motion specified by (5) should be called "adynamic" The precessional motion of the earth is that given by the small root of the corresponding equation to (2), and therefore he proposed also that the azimuthal motion specified by (4) should be called *precessional*.

But either azimuthal motion can be realised, except in some particular cases, by starting the top properly. When, however, the top is spun by the rapid unwinding of a string, while the axis is held at rest at some inclination $\theta$ to the vertical, the subsequent motion, when the top is left to itself spinning rapidly, will be one of alternate falling slightly below, and returning to, the initial value of $\theta$, accompanied by a varying value of $\dot\psi$, oscillatory like that of $\theta$, and in the same period, but of mean value given by (4), to a first approximation.

The roots can be calculated to any degree of approximation by writing the quadratic equation in the form

$$\mu^2 - b\mu + a = 0, \qquad\qquad (6)$$

that is by putting $Cn/A\cos\theta = b$, and $Mgh/A\cos\theta = a$. Then if we write the quadratic as

$$\mu = b - \frac{a}{\mu},$$

we get for $\mu$ the continued fraction

$$\mu = b - \frac{a}{b-}\,\frac{a}{b-}\,\frac{a}{b-}\cdots, \qquad\qquad (7)$$

which gives the large root. On the other hand, if we write the quadratic as

$$\mu = \frac{a}{b-\mu},$$

we obtain the continued fraction $\mu = \frac{a}{b-}\,\frac{a}{b-}\,\frac{a}{b-}\cdots, \qquad (8)$

which gives the small root.

A closer approximation to the large root than that expressed by (5) is given by (7), namely,

$$\mu = \frac{Cn}{A\cos\theta} - \frac{Mgh}{Cn}\left(1 + \frac{Mgh}{C^2n^2}A\cos\theta\right), \qquad\qquad (9)$$

while the corresponding approximation to the small root is

$$\mu = \frac{Mgh}{Cn}\left(1 + \frac{Mgh}{C^2n^2}A\cos\theta\right). \qquad\qquad (10)$$

Thus both roots involve $Mgh$, and neither is, strictly speaking, adynamic. They are conjugate roots of the quadratic equation, and it seems desirable to designate them in some way that takes account of their common origin. One may be called the *fast precession*, the other the *slow precession*.

It will be observed that the fast precession becomes infinite when $\theta=\frac{1}{2}\pi$, that is when the axis of the top is horizontal. Adopting an understood notation, we see that when $\theta=\frac{1}{2}\pi-0$, $\mu=+\infty$, and when $\theta=\frac{1}{2}\pi+0$, $\mu=-\infty$. The value of $\mu$ changes sign in passing through $\frac{1}{2}\pi$, but it does so in passing from positive infinity to negative infinity.

It is important to notice that a steady motion is possible when $h=0$, and therefore no couple acts. The roots of (2) are then $\mu=0$, $\mu=Cn/A\cos\theta$. Thus, if the top is supported at its centroid, the axis may either preserve a constant direction in space, or, *if the top is properly started,* revolve about the vertical with angular speed $Cn/A\cos\theta$. Here the motion is truly adynamic.

**18.** *Different cases of steady motion. Direct and retrograde precession. Stability of steady motion.* From the discussion given above it will be seen that if $\theta=\frac{1}{2}\pi$, the larger root of the steady motion equation (2) is infinite, so that there is only one realisable value of $\mu$, that is $\mu=Mgh/Cn$. If $\theta$ be greater than $\frac{1}{2}\pi$, the value of the larger root $Cn/A\cos\theta$ is negative. Thus take the three positions shown in the diagram (Fig. 8), and suppose that the couple about an axis through O is due to the gravity of the gyrostat situated as shown, and that we are not concerned with resultant couples in the contrary direction due to the presence of non-rotating back-weights. If the direction of rotation of the flywheel be the same in all three (that is counter-clockwise as viewed by an eye looking towards O along the axis of rotation from the side of the gyrostat remote from O), the direction of the precession measured by the root which is approximately $Mgh/Cn$ is counter-clockwise, to an eye looking towards O from Z, for all three cases. The precession is said to be *direct.* The turning motion given by the other approximate root is, when looked at in the same way, in the counter-clock direction in case (*a*), and is reckoned positive; in case (*b*) it is $+\infty$ when the angle ZOC is infinitely little less than $\frac{1}{2}\pi$, and is $-\infty$ when that angle is infinitely little greater than $\frac{1}{2}\pi$, and in case (*c*), for which the angle ZOC is between $\frac{1}{2}\pi$ and $\pi$, it is negative, that is clockwise, and of numerical value $Cn/A\cos\theta_2$. The precession is said to be *retrograde.* A more detailed discussion of direct and retrograde precession is given in 16 and 17, V, below.

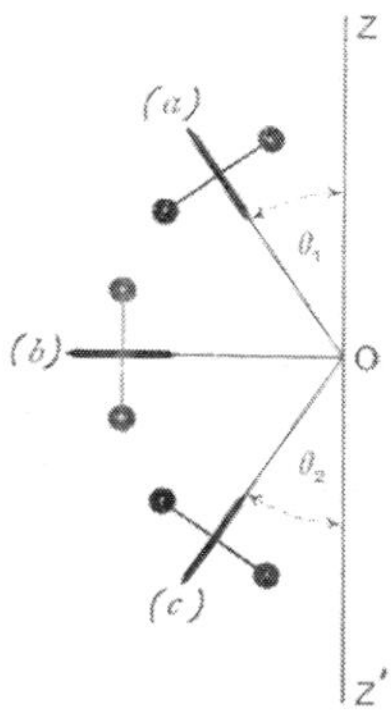

FIG. 8.

It is the smaller root of an equation corresponding to (2) that gives the precession of the equinoxes in the case of the terrestrial top. But both roots are derived from the quadratic equation in the same way: they are conjugate values of the conical motion of the axis. When expressed exactly both involve the applied forces, though one depends on these much more than does the other. Also both are possible motions, if the top is properly started. If the top is simply spun rapidly, and left with the axis at rest inclined at an angle $\theta$ to the vertical, the axis will oscillate between the value of $\theta$ and a slightly larger value (for which of course the axis is lower). The mean precessional angular speed will be approximately that given by the smaller root of the steady motion equation ($Mgh/Cn$). The actual precessional speed will vary from zero at the smaller value of $\theta$ to $2Mgh/Cn$ at the larger value, where $\dot{\theta}$ is again zero. If a sphere of unit radius be described about the point of support as centre, the axis will move between two close horizontal circles corresponding to the turning values of $\theta$. The path of the intersection of the axis with this sphere will be a succession of curves, each terminating in two cusps at the upper circle, and at its middle point tangential to the lower circle. That there are cusps at the upper circle follows from the fact that the axis is instantaneously at rest at the upper circle, with $\ddot{\theta}$ a maximum, and $\dot{\psi}$ zero, while only $\dot{\theta}$ is zero at the lower, so that the excess of kinetic energy at the lower circle above that at the upper is represented by the kinetic energy of the azimuthal motion.

When the spin is very fast, the oscillations from limiting circle to limiting circle are very small, and the mean angular speed is as stated above. When the speed of spin is not very great the limiting circles are wider apart, and then the motion can be worked out completely by the theory of elliptic functions. The results here indicated for a fast spinning top will be found worked out in 14, V, below.

The general theory shows, as we shall see later, that an even number $2m$ of freedoms of a gyrostatic system, can, if unstable before spin, be completely stabilised by gyrostatic domination, and if stable before spin remain so after spin. If the total number of freedoms of the system be $2n$, there are $n$ small roots approximately given by an equation of the $n$th degree, and $n$ large roots approximately given by another equation of the $n$th degree. The latter, if the spins are great enough, are practically "adynamic."

A so-called proof is sometimes given, by what appears to be a simple process of composition of vectors, that the angular speed of the azimuthal motion is $Mgh/Cn$ for any finite value of $\theta$. The other value of $\mu$ is ignored, and the process is quite fallacious. When $\theta$ is constant, equation (2) is obtained as follows by the process set forth in § 15. The rate of production of A.M. about OD is then $(Cn - A\mu\cos\theta)\mu\sin\theta$. The term $Cn\mu\sin\theta$ is due to the motion of the extremity of the vector $Cn$ produced by the turning with angular speed $\mu\sin\theta$ about OE. This turning is towards $OD_1$, the instantaneous position of OD. The other term, of amount $A\mu^2\cos\theta\sin\theta$, is due to the motion which the extremity of the vector of length $A\mu\sin\theta$ has parallel to $D_1O$, in consequence of the turning at rate $\mu\cos\theta$ about OC caused by the turning of the plane $zOC$ about $Oz$ with angular speed $\mu$. OE is being made to recede from $OD_1$ to a greater angle than $\frac{1}{2}\pi$, and hence the term is subtractive.

When the azimuthal motion is so fast that the rate of increase of A.M. $Cn\mu\sin\theta$ is nearly balanced by the rate of diminution $A\mu^2\cos\theta\sin\theta$, we have the larger root.

We shall return to the steady motion of a top or gyrostat in 9, V, and shall there show that the steady motion is stable, that is, that a slight disturbance of the motion will result in a periodic rising and falling of the axis about a steady motion inclination to the vertical, with a corresponding variation of the azimuthal turning; but without continually increasing departure from the steady motion. There is thus variation of $\theta$, with, in a different phase, an accompanying variation of precession. This periodic change of $\theta$ is called *nutation*.

Like *precession*, the term *nutation* is derived from the motion of the earth, which is exactly that of a spinning top. [See Chap. X, in which the earth's precession and nutation are discussed.]

## CHAPTER III

### ELEMENTARY DISCUSSION OF GYROSTATIC ACTION

1. *Elementary explanation of motion of a top.* Although the behaviour of a top can be deduced from the equations of motion of a rigid body, which themselves are but the expression of fundamental dynamical principles, it is paradoxical and perplexing to many people. The stability of the upright position when the speed of rotation is considerable, and still more the precessional (or sideways) motion of the axis of the top, produced by the action of gravity when the axis is inclined to the vertical, appears to be inexplicable. Hence, before carrying the general theoretical discussion further, we shall endeavour to give in the present chapter an elementary explanation of these phenomena.

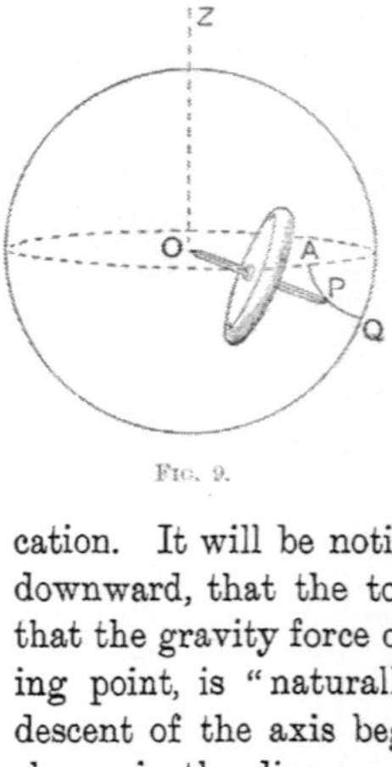

FIG. 9.

Let the top (supposed of the construction shown in Fig. 4) be set in rotation about its axis, by the rapid withdrawal of a string wound on the top, or in any other way, and be then left to itself supported on a fixed point O of the axis OC. *If the axis be at rest when the top is freed from constraint,* the course of events will be as follows; but if the initial condition be not as here stated, the present description will require modification. It will be noticed that the axis at first has an accelerated motion downward, that the top begins to fall. This is a beginning of the effect that the gravity force on the top, together with the reaction of the supporting point, is "naturally" expected to produce. But no sooner has the descent of the axis begun than the top also begins to move sideways, as shown in the diagram (Fig. 9) by the curve AP. This curve is the path of the intersection of the axis with a spherical surface described round the point, O, of support of the top as centre. The difficulty, to a person untrained in higher dynamics, is to understand this sideways motion, or *precession.*

It will be seen, moreover, that the axis gradually ceases to tilt further downward, so that at a certain instant, when it has the position Q, it has a sideways motion only, and thereafter begins to have an upward tilting motion, and continues to ascend, still with the same direction of sideways progressive displacement, until it has attained the same inclination to the vertical as it had at starting, when it will be observed that precession has for the moment ceased. The axis then begins again a tilting motion downward, and the former behaviour is repeated, and so on, until the exact repetition has been seriously interfered with by diminution of the spin by the action of friction. But until the spin has fallen off perceptibly the axis moves backward and forward between two limiting circles on the sphere, and each to and fro passage carries it a certain way round the vertical OZ. Thus the curve showing the path of intersection of the axis with the sphere is a series of repeated parts, each concave upward, touching the lower circle tangentially, and meeting the upper circle at right angles.

We shall explain this motion at first by reference to a special arrangement, and then go on to a quantitative discussion of an elementary kind.

[The difference of behaviour owing to difference in starting conditions is explained in 17, V, below ]

2. *Top made with hollow rim containing balls. Analysis of action of balls.* Let the top be made in the form shown in Fig. 10, that is, consist of a spindle OC, carrying rigidly fixed to it a disk provided with a massive rim. Let the rim be, as Figure 11 indicates, a hollow tube containing a number of equal balls in contact, fitting the tube and arranged so as to form a ring of particles, which is carried round with the disk when that rotates about its axis of figure OC. Suppose the top, after being set in rotation with the axis OC inclined to the vertical OZ, to be left to itself under the action of gravity. As we have seen, it begins to turn about a horizontal axis OD at right angles to the plane COZ. This turning may be regarded as consisting of a motion of the centre of the disk at right angles to OC and in the plane COZ, and a turning of the whole about an axis AB through the centre of the disk and parallel to OD. It is this latter motion that causes precession, as we shall now show.

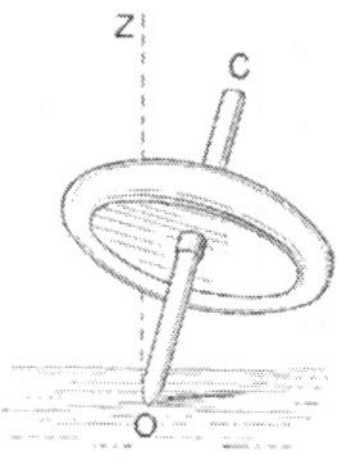

FIG. 10.

Consider then a disk or flywheel (Fig. 11), loaded with balls in the manner just described, to have its centre at rest, and to be rotating, or "spinning," in the plane of the paper about an axle through its centre O′, while it is turning about a horizontal axis AO′B at a definite speed. In consequence of the spin the ball $a$ in the quadrant BE′ of the ring is, at any instant while it is there, being carried further away from the axis AB.

In virtue of its inertia the ball tends to go on without alteration of its speed and direction of motion at the instant. But as it increases its distance from AB in the tube, the ball is caused by the action of the tube to move faster in the turning motion about AB. The turning about AB has the direction, say, which brings E′ nearer the reader, and therefore the ball presses against the back of the tube.

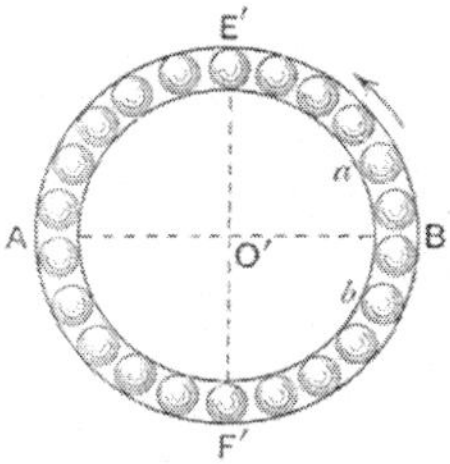

FIG. 11.

Now consider a ball $b$ in the quadrant F′B; it is approaching the axis AB, and therefore its speed at right angles to the plane of the ring must diminish towards zero. This diminution is brought about by the action of the tube on the ball, and the ball therefore reacts on the tube. It will be seen that this reaction is in the same direction as that exerted by the ball $a$ at the same instant, and that these two reactions conspire in tending to turn the ring about the axis E′F′, in the plane of the instantaneous position of the ring and at right angles to AB.

Consideration of the balls, as $c$, $d$, which are at the same moment in the quadrants E′A and AF′, shows that, while the actions of $c$ and $d$ on the tube are in the same direction, that direction is opposite to the direction of the actions of $a$ and $b$. Hence the assemblages of balls occupying at a given instant the halves of the ring separated by the diameter E′F′, exert then equal and opposite actions on the tube, forming a couple tending to produce turning of the ring about E′F′; and unless a couple about this axis is applied from without to balance that which we have seen to be due to the inertia of the balls, the horizontal axis AB will turn round in azimuth: in other words, the vertical plane COZ (Fig. 10) will turn round OZ, and we have the sideways motion of the top. If at any instant the angular speed of turning about OZ be $\dot{\psi}$, and $\theta$ be the angle COZ, the turning about OZ is equivalent to two turnings with angular speed, $\dot{\psi}\cos\theta$, $\dot{\psi}\sin\theta$, respectively, about OC and OE, where OE is at right angles to OC in the plane COE.

3. *Mode of production of precessional motion. Calculation of forces. Gyrostatic couple and "gyrostatic resistance."* It will now be seen how the precessional motion of the top about OZ, or the component of that turning about OE, to which we have been giving more special attention in what precedes, arises. The loaded disk which forms the main part of the mass is made up of particles having OC (Fig. 10) as common axis, and each ring of particles plays the part of the balls in the ring-groove. The turning about AB, required in the turning of the system about OD (see Fig. 4), involves a turning motion of the whole about E′F′ (Fig. 11), which is parallel

to the axis OE; and the amount of this at any instant depends on the angle through which the axis OC has descended from its elevation at the beginning of the motion.

Following out these ideas, we can find the force-systems concerned. For this purpose we consider, in the first place, a flywheel symmetrical about an axis of figure OC, and revolving about that axis with angular speed $n$. Let AB be horizontal and pass through a point O′ (*e.g.* O of Fig. 10, so that AB is in that case coincident with OD in the plane BO′E′) anywhere on the axis of the flywheel. Consider a particle P of mass $m$ in the plane BO′E′, let $O'P = r$, and $\angle PO'B = \phi$. If the turning about AB have angular speed $\omega$, the particle P has a speed $\omega r \sin\phi$ at right angles to the plane in which it is at the instant being carried round by the spin at speed $n$, and it has also the speed $r\cos\phi \,.\, \dot\phi$ in the direction OE. A section of the wheel through P at right angles to OC is in general a ring bounded by circles which have OC as common axis, and the speeds which have just been specified are in addition to the motion which P has in virtue of that of the common centre of these circles. They are components of the motion of P; the latter is due to the turning about the instantaneous position of OC, the former to the turning of the circle about its horizontal diameter.

Now, after time $dt$ the plane of the section in which P is situated has turned through the angle $\omega\, dt$ about the horizontal diameter, and the two speeds specified above have become for the new orientation of the plane,

$$\omega r \sin\phi + d(\omega r \sin\phi), \quad r\cos\phi \,.\, \dot\phi + d(r\cos\phi \,.\, \dot\phi),$$

respectively. These give components perpendicular to the former position of the plane, which are, respectively,

$$\{\omega r \sin\phi + d(\omega r\sin\phi)\}\cos(\omega\, dt),$$

or in the limit,
$$\omega r \sin\phi + d(\omega r \sin\phi),$$

and
$$\{r\cos\phi \,.\, \dot\phi + d(r\cos\phi \,.\, \dot\phi)\}\sin(\omega\, dt),$$

or in the limit,
$$\omega r\cos\phi \,.\, \dot\phi\, dt = \omega n r\cos\phi \,.\, dt,$$

since the angular speed $\dot\phi = n$. The total growth in $dt$ of speed of P perpendicular to the position of the plane at the initial instant of $dt$ is therefore $d(\omega r\sin\phi) + \omega r\cos\phi \,.\, n\, dt$, or

$$2\omega n r\cos\phi \,.\, dt + \dot\omega r \sin\phi\, dt.$$

To produce this acceleration a force

$$m(2\omega n r\cos\phi + \dot\omega r\sin\phi)$$

is required, and is applied to the particle through its connections with the other particles of the flywheel. An equal and opposite force is applied to the wheel by the particle as reaction. If there is no system of forces applied from outside the wheel to balance these reactions, the motion about E′F′ will take place, so as on the whole to render the effect of these accelerations zero.

If the wheel be supposed placed so that its mean plane coincides with the paper, and its top move towards the reader, then while P is on the right of the diameter E′F′ (Fig. 11), the part of the force represented by the first term, $2m\omega nr\cos\phi$, is towards an observer looking at the paper, and the corresponding reaction is in the opposite direction. For a particle on the left of E′F′ these directions are reversed. The second part of the force has one sign for matter on one side of the horizontal diameter, and the opposite sign for matter on the other side. In taking moments, therefore, about E′F′ of the forces on the particles of a thin uniform slice of the flywheel, taken at right angles to OC, we may disregard the force $m\dot{\omega}r\sin\phi$ on each particle, as these forces contribute nothing to the result.

Summing the moments for a ring of matter of linear density $\sigma$, every particle of which is at a distance $r$ from the centre of the disk, we get, taking $m=\sigma r\,d\phi$,

$$\left.\begin{array}{c}\text{Sum of moments}\\ \text{about E}'\text{F}'\end{array}\right\}=2\omega n\sigma r^3\int_0^{2\pi}\cos^2\phi\,d\phi=2\pi\sigma r^3\omega n=\mu\omega n,$$

if $\mu$ denote the moment of inertia about OC of a complete ring of particles. This is also the moment of the forces about any line parallel to E′F′ through a point of the axis.

The same thing holds for every ring of which the disk is composed, and for every disk comprised in the flywheel; and therefore if C be the moment of inertia of the flywheel about OC, we have the following result:

*The couple about the axis E′F′* (Fig. 11) [*or about any axis (e.g. OE) intersecting OC and parallel to E′F′*], *which must be applied to balance the reactions which arise when the body rotating about OC with angular speed $n$, is made to turn about a fixed axis OD (parallel to AB) with angular speed $\omega$, is $Cn\omega$.* The reaction couple thus balanced is often called the *gyrostatic couple.* Also it is sometimes called the *gyrostatic resistance*, since to change the direction of the axis of the rotation $n$ at rate $\omega$ requires the application of a couple of moment $Cn\omega$.

In other words, there is a rate of production $Cn\omega$ of A.M. about the axis OE in the direction opposite to that of the gyrostatic reaction couple. This, in the absence of an applied couple, is nullified by acceleration of the whole body about OE in the direction of action of that couple. The directions are as follows: The reacting couple is in the counter-clock direction as seen by an observer looking from beyond E (or E′) towards O (or O′). The direction of the rate of production of A.M. about OE (or the parallel axis E′F′), which gives rise to the reacting couple—to be balanced by one externally applied if OD is to remain fixed, that is, if the rate of growth of A.M. about OE is not to be counteracted—is clockwise.

We may put the matter in symbolical form as follows: Let A and $\dot{\omega}'$ be the moment of inertia and angular acceleration about OE. The total rate of production of A.M. about OE is $A\dot{\omega}'+Cn\omega$. But if there be no applied couple about OE, we have

$$A\dot{\omega}'+Cn\omega=0, \quad \text{or} \quad A\dot{\omega}'=-Cn\omega.$$

Thus $-Cn\omega$ may be regarded as a couple producing A.M. $A\dot{\omega}'$ about the moving axis OE, in the opposite direction to $Cn\omega$.

We obtain, of course, a similar result for the case of turning about the axis OE with angular speed $\omega'$, say, in the counter-clockwise direction. If there is to be no turning about OD, the reacting couple of amount $Cn\omega'$ must be balanced by a couple applied from without about the axis OD, and is clockwise as seen by an observer looking from beyond D towards O.

It follows from this that a couple of magnitude $Cn\omega'$ applied about OD from without in the counter-clock direction, will be consistent with turning at angular speed $\omega'$ in the counter-clock direction about OE, without any turning about OD. The axis OD will, however, move with angular speed $\omega'$ in a plane perpendicular to OE, while OE remains fixed.

4. *Deduction of principal equation of motion.* We now obtain an important result, already established in another way. Consider a top spinning about its axis of figure with angular speed $n$, while that axis (OC) turns round the vertical OZ (or line answering to the vertical if the top is not in a gravitational field) in consequence of the angular speed $\dot{\psi}$ of the plane ZOC about OZ. The inclination of OC to OZ is $\theta$, and the axes OD, OE (specified in 9, II) turn with the plane ZOC.

Apply, as explained in 3, a couple $Cn\omega'$ about OD in the counter-clock direction. This, as we have seen, is consistent with a turning at speed $\omega'$, counter-clockwise about the axis OE, while that axis remains at rest. Thus the axis OD turns about OE, counter-clockwise.

Again, let the axis O′E′ be identified with OE, and the angular speed there denoted by $\omega'$ with $\dot{\psi}\sin\theta$, and imagine turning at speed $\dot{\psi}\cos\theta$ about OC in the counter-clock direction. The turning with this speed about OC as a fixed axis involves a reacting couple which must be balanced by the application of a couple of moment $A\dot{\psi}^2\sin\theta\cos\theta$, clockwise about OD.* Thus we see that the total couple $Cn\dot{\psi}\sin\theta-A\dot{\psi}^2\sin\theta\cos\theta$, in the counter-clock direction about OD, is consistent with a motion of the top compounded of a turning (counter-clockwise) about OE with angular speed $\dot{\psi}\sin\theta$, and a turning (counter-clockwise) about OC with angular speed $\dot{\psi}\cos\theta$. But these turnings about OC and OE are clearly together equivalent to turning with angular speed $\dot{\psi}$ about OZ. Thus the combination of couples just found, equated to the total applied couple about OD, gives an equation of motion which holds either for steady motion about the line OZ,

* We here take OE, and presently OD, as an axis of spin. This may be justified by an extension of the reasoning in 3.

or for an instant at which the acceleration $\ddot{\theta}$ of turning about the axis OD is zero. This equation is

$$(Cn - A\dot{\psi}\cos\theta)\dot{\psi}\sin\theta = L, \quad \text{.........................(1)}$$

if L be the moment of the couple. If there is angular acceleration $\ddot{\theta}$, the couple L must furnish the rate of production of A.M., $A\ddot{\theta}$, in addition to the combination of couples specified. Thus, if the field of force is gravitational, we have $L = Mgh\sin\theta$, and the complete equation of motion is

$$A\ddot{\theta} + (Cn - A\dot{\psi}\cos\theta)\dot{\psi}\sin\theta = Mgh\sin\theta. \quad \text{.................(2)}$$

This is the equation found in 15, II, in an entirely different manner.

In the same way we can find the equation of motion for the axis OE. Referring back to the general result of 3, consider the couple $Cn\omega$ about OE there referred to, and identify $\omega$, about OD, with $\dot{\theta}$. The turning about OD is counter-clockwise, the couple of moment $Cn\dot{\theta}$ about OE which would balance the reacting couple is clockwise.

Now take for $\omega$ the angular speed $\dot{\psi}\cos\theta$ about OC, referred to above, and consider a couple about OE (instead of about OD) which would balance the reaction couple due to the turning $\dot{\theta}$ about AB (or OD). This is a couple $A\dot{\psi}\cos\theta\,.\,\dot{\theta}$ in the counter-clock direction about OE. Thus we get a total couple $(A\dot{\psi}\cos\theta - Cn)\dot{\theta}$. But if there is no time-rate of variation of the angular speed $\dot{\psi}\sin\theta$ about OE, and there be no applied couple about OE, this total couple is zero. The complete equation in the absence of externally applied couple about OE is therefore

$$A\frac{d}{dt}(\dot{\psi}\sin\theta) + (A\dot{\psi}\cos\theta - Cn)\dot{\theta} = 0, \quad \text{.....................(3)}$$

or

$$A\ddot{\psi}\sin\theta + (2A\dot{\psi}\cos\theta - Cn)\dot{\theta} = 0. \quad \text{.....................(3')}$$

To put the matter most simply, the total reaction couple is $-(A\dot{\psi}\cos\theta - Cn)\dot{\theta}$, and this must be equal to $Ad(\dot{\psi}\sin\theta)/dt$.

These results have been obtained in the most elementary manner from first principles. Any question that may arise as to the legitimacy of supposing that the motions giving the different reacting couples, taking place as they do simultaneously, can be taken account of by adding the separate couples of reaction together, may be answered in a general manner as follows: If we were to suppose the different changes to be effected in successive intervals of time $dt$, the effect of the existence of the first change on the amount of the second would, as regards direction, be accounted for by multiplication by the cosine of an indefinitely small angle; and as regards magnitude, introduce quantities involving $dt$ as a factor in addition to the finite quantities specified above.

5. ***Elementary quantitative analysis of motion of top.*** We now endeavour to predict, from the elementary considerations adduced above, the behaviour of a top which is set into rapid rotation about the axis OC,

when in the position OA, say, shown in Fig. 9, and is then left to itself. The axis begins at once to acquire a descending motion of turning about O, and each of the disks at right angles to OC, of which the top is composed, begins to turn about a horizontal diameter, while the whole top turns about the parallel horizontal axis OD. If the turnings about OC and OD have the directions in Fig. 4, the top is, as we have seen, acted on by a couple $(Cn - A\dot{\psi}\cos\theta)\theta$ about OE in the direction of the circular arrow. Thus the axis of the top begins to have a component of turning in the direction at right angles to the first motion, and towards the instantaneous position of OD.

The intersection of the axis with a sphere described round O as a centre is therefore a curve like AP (Fig. 9). Here OD is to be drawn horizontally back from O, and OE in the angle ZOP, both at right angles to OP When the axis is at P the top is turning about OD, now of course in a new position, counter-clockwise, and in consequence is still gaining counter-clock rotation about OE, which is also in a new position, corresponding to that of OD. This will be clear from consideration of (3), 4, which shows that the rate of increase of $A\dot{\psi}\sin\theta$ is $(Cn - A\dot{\psi}\cos\theta)\dot{\theta}$, which (if $Cn > A\dot{\psi}\cos\theta$) is positive, since of course we suppose that $\dot{\theta}$ is still positive. The acceleration about OE is increasing in amount because the couple about OD applied by gravity is increasing $\dot{\theta}$. By the formula just quoted, this acceleration about OE is proportional to the accumulated rate of turning about OD, except for the factor $Cn - A\dot{\psi}\cos\theta$, and so OC turns faster and faster towards the instantaneous position of OD.

But the counter-clock turning about OE produces A.M. about the instantaneous position of OD, represented by the term $Cn\dot{\psi}\sin\theta$ of (2), 4, giving clockwise acceleration about that axis, though at the same time, in consequence of the turning with angular speed $\dot{\psi}\cos\theta$, this (if $\theta < \frac{1}{2}\pi$) is counteracted to some extent by the term $A\dot{\psi}^2\cos\theta\sin\theta$. For we can write (2), 4, in the form

$$A\ddot{\theta} = Mgh\sin\theta - (Cn - A\dot{\psi}\cos\theta)\dot{\psi}\sin\theta. \quad \text{.................(1)}$$

An instant arrives at which the right-hand side of this equation is zero, and so $\dot{\theta}$ becomes for the instant unvarying. Thereafter counter-clock turning about OD diminishes, while, the spin being great and $(Cn - A\dot{\psi}\cos\theta)\dot{\theta}$ being positive, the turning about OE continues to increase, until the turning speed $\dot{\theta}$ about OD is reduced to zero. The axis OC has now ceased to descend. At that instant the counter-clock turning about OE is at its greatest, and the clockwise *acceleration* about OD a maximum, so that angular speed in that direction begins to grow up; the axis begins to rise. It continues to rise with increasing speed, for so long time as the counter-clock rotation about OE (which is now diminishing on account of the reversal of the turning about OD) is sufficient to produce (see the equation (1) last written) a couple, $(Cn - A\dot{\psi}\cos\theta)\dot{\psi}\sin\theta$, greater than that due to gravity. After a certain elevation has been attained, equal to that at which the acceleration about OD ceased in the downward progress, the increase of the speed of rise ceases, and the upward motion slows off to zero at an elevation equal to that of the starting position.

It is easy to see (since, if at any point all the motions were exactly reversed, the path would be traversed in the reverse direction) that as the axis mounts higher it continues to swing round the vertical, and does so in such a manner that the path from the turning point to the highest elevation is exactly the image of the descending path in a mirror coinciding with the meridian of the sphere which passes through the lowest point reached by the axis. Once the original elevation has been regained the axis begins once more to descend, and a path on the sphere is described, which is exactly that described before, only shifted round the vertical through an angle equal to that swung round through by the axis in descending and ascending.

It will be observed that, since

$$\frac{d}{dt}(\mathrm{A}\dot{\psi}\sin\theta)=(\mathrm{C}n-\mathrm{A}\dot{\psi}\cos\theta)\dot{\theta} \qquad (2)$$

is zero at the upper circle limiting the motion on the sphere, the value of $\ddot{\psi}$ is zero as well as that of $\dot{\psi}$, while $\mathrm{A}\ddot{\theta}$ has the value given by the gravitation couple without any diminution. There is thus neither azimuthal turning nor azimuthal acceleration at that circle, while the inclinational acceleration has its maximum value. The curve of intersection of the axis with the spherical surface therefore meets the upper circle at right angles, that is the points of contact are a series of cusps on the path, when the mode of starting is as specified above.

At the lower turning points $\ddot{\psi}$ is zero, while $\dot{\psi}$ has its maximum value, and as $\dot{\theta}$ is zero the path of the axis on the sphere is tangential to the lower circle at each turning point.

# CHAPTER IV

## SYSTEMS OF COORDINATES AND THEIR RELATIONS. SPACE-CONE AND BODY-CONE

1. *Relations of systems of coordinates.* It is convenient to give here for reference most of the relations between different sets of coordinates that are likely to be required. In Fig. 12 are shown three sets of axes drawn from a common origin O, namely, O($x$, $y$, $z$), O(A, B, C), O(D, E, C). Of these O$z$ may be identified, when convenient, with the upward vertical, and OC with the axis of a top; O(A, B, C) may be a set moving in any manner about themselves, for example, they may move with the body, while OD and OE are respectively perpendicular and parallel to the plane $z$OC. It will be convenient also to refer to OE′ drawn from O in the opposite direction to OE.

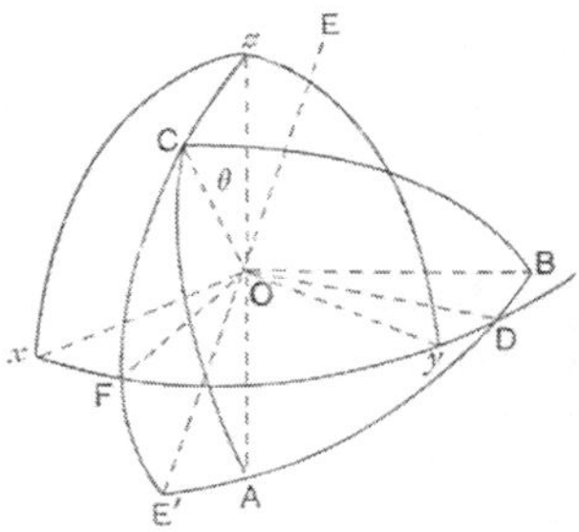


Fig. 12.

The point O may be taken as the centre of a sphere of unit radius, on the surface of which quadrantal arcs $xy$, $yz$, $zx$, AB, BC, CA, DE, EC, CD, and DC, CE′, E′D may be drawn. The arc $z$C produced gives the quadrant CE′, and intersects the quadrant $xy$ in F, so that $z$F is also a quadrant. The angle $z$OC is denoted by $\theta$, that between the planes OCA and OCE′ by $\phi$, and that between the planes OCE′ and O$zx$ by $\psi$. (Thus, for a top, $\psi$ is the angle through which the plane $z$OC has turned about the vertical from its position at a chosen instant, that is, the integral precessional displacement.) Since OD is at right angles to O$z$ and to OC, and all the lines are, as we suppose, of unit length, the point D lies at the intersection of the quadrant AB with the quadrant $xy$ produced, and the angle D of the spherical triangle FDE′ is $\theta$. Clearly also arc DB $= \phi$, arc $y$D $= \psi$. The angles $\theta$, $\phi$, $\psi$ are called the *Eulerian angles.*

The direction cosines of O(A, B, C) with reference to the fixed axes

$O(x, y, z)$, or *vice versa*, can be expressed in terms of the three angles $\theta$, $\phi$, $\psi$ by the formulae of spherical trigonometry or otherwise. For example, the cosine of the angle $AOx$ is the cosine of the side $Ax$ of the spherical triangle $ACx$; the side $Cx$ and the angle $FCx$ can be found from the spherical triangle $Czx$, and hence the angle $ACx$, so that $Ax$ can be found.

But the simplest method is that of projection. We shall find the direction cosines of OA, OB, OC with reference to $Ox$ by this process, and leave the reader to verify the values of the others as given in the scheme which follows. We resolve OA into two components, one, OG, along OE′, and the other at right angles to OE′. These are (since $OA = 1$) $\cos\phi$, $\sin\phi$. The former projected on OF is $\cos\theta\cos\phi$, and this projected on $Ox$ gives $\cos\theta\cos\phi\cos\psi$. The other component ($\sin\phi$) is perpendicular to the plane OCE′, and therefore to OF. Its projection on $Ox$ is therefore $-\sin\phi\sin\psi$. Hence the total projection of OA on $Ox$ is $\cos\theta\cos\phi\cos\psi - \sin\phi\sin\psi$. The direction cosine of OA with respect to $Ox$ is therefore

$$\cos\theta\cos\phi\cos\psi - \sin\phi\sin\psi.$$

Again, resolving OB into two components, along and at right angles to OD, we see that these components are $\cos\phi$, $\sin\phi$. Along $Ox$ the former has the component $-\cos\phi\sin\psi$, and the latter $-\cos\theta\sin\phi\cos\psi$. Thus the direction cosine of OB with respect to $Ox$ is

$$-\cos\theta\sin\phi\cos\psi - \cos\phi\sin\psi.$$

Finally, OC resolves into two components, $\sin\theta$ along OF, and $\cos\theta$ at right angles to OF and to the plane $xOy$. The latter has no component along $Ox$, the former has the component $\sin\theta\cos\psi$.

The following scheme gives in a convenient manner the direction cosines of the axes of either system with reference to the axes of the other system, and also the coordinates of any point taken with reference to one set of axes, in terms of the coordinates of the same point with reference to the other set. Thus, for example, the $Ox$ direction cosine of OB, or the OB direction cosine of $Ox$, is $-\cos\theta\sin\phi\cos\psi - \cos\phi\sin\psi$; and if $x_1$, $y_1$, $z_1$ be the coordinates of a point with respect to the axes O(A, B, C), and $x$, $y$, $z$ those of the same point with respect to $O(x, y, z)$, we have

$$x_1 = (\cos\theta\cos\phi\cos\psi - \sin\phi\sin\psi)x + (\cos\theta\cos\phi\sin\psi + \sin\phi\cos\psi)y$$
$$-\sin\theta\cos\phi \, . \, z,$$

$$x = (\cos\theta\cos\phi\cos\psi - \sin\phi\sin\psi)x_1 - (\cos\theta\sin\phi\cos\psi + \cos\phi\sin\psi)y_1$$
$$+\sin\theta\cos\psi \, . \, z_1.$$

The lines in the following scheme give the direction cosines and coordinates for the axes O(A, B, C) in terms of those for the axes $O(x, y, z)$: the columns give the cosines and coordinates for the axes $O(x, y, z)$ in terms of those for the axes O(A, B, C).

| | $Ox, x$ | $Oy, y$ | $Oz, z$ |
|---|---|---|---|
| $OA, x_1$ | $\cos\theta\cos\phi\cos\psi-\sin\phi\sin\psi$ | $\cos\theta\cos\phi\sin\psi+\sin\phi\cos\psi$ | $-\sin\theta\cos\phi$ |
| $OB, y_1$ | $-\cos\theta\sin\phi\cos\psi-\cos\phi\sin\psi$ | $-\cos\theta\sin\phi\sin\psi+\cos\phi\cos\psi$ | $\sin\theta\sin\phi$ |
| $OC, z_1$ | $\sin\theta\cos\psi$ | $\sin\theta\sin\psi$ | $\cos\theta$ |

The angles $\theta$, $\phi$, $\psi$ were first used by Euler for the specification of the positions of the axes of one set with reference to the other set, and are usually referred to as the Eulerian angles.

A similar scheme for the axes O(E′, D, C), $O(x, y, z)$ can be deduced from the above scheme by simply putting $\phi = 0$. If we denote coordinates of a point with reference to the first set of axes by $\xi$, $\eta$, $\zeta$, and with respect to $O(x, y, z)$ by $x, y, z$ as before, we get

| | $Ox, x$ | $Oy, y$ | $Oz, z$ |
|---|---|---|---|
| $OE', \xi$ | $\cos\theta\cos\psi$ | $\cos\theta\sin\psi$ | $-\sin\theta$ |
| $OD, \eta$ | $-\sin\psi$ | $\cos\psi$ | $0$ |
| $OC, \zeta$ | $\sin\theta\cos\psi$ | $\sin\theta\sin\psi$ | $\cos\theta$ |

It will be observed that if we take any two columns in either one of these schemes, and multiply the direction cosines in the same line together and add, we obtain a zero sum, and that the same result is obtained if we take any two lines and multiply the direction cosines in the same column together and add. This is a condition imposed by the fact that any pair of axes belonging to either system are at right angles to one another.

2. *The Eulerian angles and their rates of variation.* Certain special systems of coordinates are useful in the detailed treatment of the motion of the axes of a top by the methods of elliptic functions, and these will be set forth where this subject is dealt with. The following illustration of the Eulerian angles bears directly on the elliptic function theory.

The axes O(A, B, C) in their positions at time $t$ may be regarded as turned from coincidence with $O(x, y, z)$ through an angle $\vartheta$, say, about a determinate axis OK, where K is a point on the sphere of unit radius whose centre is O. Beginning with the axes coincident as stated, let the system O(A, B, C) be first turned through the angle $\psi$ about $Oz$; the plane $z$OK [supposing OK rigidly connected with O(A, B, C)] is turned through the same angle. Hence, if before the turning the angle K$zx$ was $\varpi$, it is changed by the turning to $\varpi+\psi$. The angle K$z$A is now $\varpi$, and C is still coincident with $z$. The arc $z$F is a quadrant.

Now let the axes O(A, B, C) be turned through an angle $\theta$ about an axis through O perpendicular to COF; the angle FCK becomes $\varpi$. Finally, when the axes are turned about OC through the angle $\phi$, the angle FCK becomes $\varpi+\phi$. The angle KC$z$ is therefore $\pi-(\varpi+\phi)$.

But the result of all the turnings is to leave OK in its original position, since OK is the fixed axis about which the change of position is produced, and so we have, as at first, the angle K$zx=\varpi$. Hence, angle K$z$C$=\varpi-\psi$, and since the spherical triangle CK$z$ is isosceles, this is also the value of the angle KC$z$. Thus we have

$$\pi-(\varpi+\phi)=\varpi-\psi, \quad \text{or} \quad \varpi=\tfrac{1}{2}(\pi-\phi+\psi).$$

We now tabulate, by means of the schemes in 1, the angular speeds and angular momenta for the different axes, in terms of $\theta$, $\phi$, $\psi$. As before, we denote by $\omega_1$, $\omega_2$, $\omega_3$ quantities taken with reference to the moving axes O(A, B, C), and by $\omega_x$, $\omega_y$, $\omega_z$, the same quantities taken with reference to the fixed axes O$(x, y, z)$. Thus, since the angular speeds about O(E', D, C) are $-\dot\psi\sin\theta$, $\theta$, $\dot\phi+\dot\psi\cos\theta$, and the angular momenta about the same axes are $-\mathrm{A}\psi\sin\theta$, $\mathrm{A}\dot\theta$, $\mathrm{C}(\dot\phi+\dot\psi\cos\theta)(=\mathrm{C}n)$, we get

$$\left.\begin{aligned}\omega_x&=-\dot\psi\sin\theta\cos\theta\cos\psi-\dot\theta\sin\psi+n\sin\theta\cos\psi\\&=-\dot\theta\sin\psi+\dot\phi\sin\theta\cos\psi,\\\omega_y&=-\dot\psi\sin\theta\cos\theta\sin\psi+\dot\theta\cos\psi+n\sin\theta\sin\psi\\&=\dot\theta\cos\psi+\dot\phi\sin\theta\sin\psi,\\\omega_z&=\dot\psi\sin^2\theta+(\dot\phi+\dot\psi\cos\theta)\cos\theta\\&=\dot\psi+\dot\phi\cos\theta.\end{aligned}\right\}\quad\text{.............(1)}$$

From Fig. 12, or otherwise, we get for the angular speeds about the axes O(A, B, C),

$$\left.\begin{aligned}\omega_1&=-\dot\psi\sin\theta\cos\phi+\dot\theta\sin\phi,\\\omega_2&=\dot\psi\sin\theta\sin\phi+\dot\theta\cos\phi,\\\omega_3&=\dot\phi+\dot\psi\cos\theta;\end{aligned}\right\}\quad\text{.....................(2)}$$

and so

$$\left.\begin{aligned}\dot\theta&=\omega_1\sin\phi+\omega_2\cos\phi,\\\dot\psi&=\frac{1}{\sin\theta}(\omega_2\sin\phi-\omega_1\cos\phi),\\\dot\phi&=\omega_3-\frac{1}{\tan\theta}(\omega_2\sin\phi-\omega_1\cos\phi).\end{aligned}\right\}\quad\text{.....................(3)}$$

The angular momenta about the axes O(C, D, E) are $\mathrm{C}(\dot\phi+\dot\psi\cos\theta)$, $\mathrm{A}\dot\theta$, $\mathrm{A}\dot\psi\sin\theta$; and therefore, denoting those about O$(x, y, z)$ by $\mathrm{H}_x$, $\mathrm{H}_y$, $\mathrm{H}_z$, we obtain

$$\left.\begin{aligned}\mathrm{H}_x&=-\mathrm{A}(\dot\psi\sin\theta\cos\theta\cos\psi+\dot\theta\sin\psi)+\mathrm{C}(\dot\phi+\dot\psi\cos\theta)\sin\theta\cos\psi,\\\mathrm{H}_y&=-\mathrm{A}(\dot\psi\sin\theta\cos\theta\sin\psi-\dot\theta\cos\psi)+\mathrm{C}(\dot\phi+\dot\psi\cos\theta)\sin\theta\sin\psi,\\\mathrm{H}_z&=\mathrm{A}\dot\psi\sin^2\theta+\mathrm{C}(\dot\phi+\dot\psi\cos\theta)\cos\theta.\end{aligned}\right\}\quad\text{...(4)}$$

Again, for the axes O(A, B, C), we have

$$\left.\begin{aligned}\mathrm{H}_1&=\mathrm{A}(-\dot\psi\sin\theta\cos\phi+\dot\theta\sin\phi),\\\mathrm{H}_2&=\mathrm{A}(\dot\psi\sin\theta\sin\phi+\dot\theta\cos\phi),\\\mathrm{H}_3&=\mathrm{C}(\dot\phi+\dot\psi\cos\theta).\end{aligned}\right\}\quad\text{.......................(5)}$$

For the three axes O(D, $z$, F) (Fig. 12), which are at right angles to one another (O$z$ vertical and OD, OF horizontal), we have

$$\left.\begin{aligned} &H_{OD}=A\dot{\theta}, \quad H_z=A\dot{\psi}\sin^2\theta+Cn\cos\theta, \\ &H_{OF}=C\dot{\phi}\sin\theta+(C-A)\dot{\psi}\sin\theta\cos\theta=Cn\sin\theta-A\dot{\psi}\sin\theta\cos\theta. \end{aligned}\right\} \quad \ldots(6)$$

3. *Instantaneous axis, axis of figure and axis of resultant* A.M. The reader may verify that any of the sets of equations (2), (3) gives for the resultant angular speed, $\omega$, the equation

$$\omega^2=\dot{\theta}^2+\dot{\psi}^2+\dot{\phi}^2+2\dot{\phi}\dot{\psi}\cos\theta. \quad \ldots\ldots\ldots(1)$$

The instantaneous axis OI makes with O(D, E, C) angles of which the cosines are $(\dot{\theta},\ \dot{\psi}\sin\theta,\ \dot{\phi}+\dot{\psi}\cos\theta)/\omega$.

The component angular speeds about O(D, $z$, F) are $\dot{\theta}$, $\dot{\psi}+\dot{\phi}\cos\theta$, $\dot{\phi}\sin\theta$. Hence OI makes with the axis of figure, OC, the angle $\cos^{-1}\{(\dot{\phi}+\dot{\psi}\cos\theta)/\omega\}$, and with O$z$ the angle $\cos^{-1}\{(\dot{\psi}+\dot{\phi}\cos\theta)/\omega\}$. To the fixed axes, O($x$, $y$, $z$), OI is inclined at the angles of which the respective cosines are

$$(-\dot{\theta}\sin\psi+\dot{\phi}\sin\theta\cos\psi,\ \dot{\theta}\cos\psi+\dot{\phi}\sin\theta\sin\psi,\ \dot{\psi}+\dot{\phi}\cos\theta)/\omega.$$

The resultant A.M., H say, is given by

$$H^2=A^2(\dot{\theta}^2+\dot{\psi}^2\sin^2\theta)+C^2n^2, \quad \ldots\ldots\ldots(2)$$

where $n=\dot{\phi}+\dot{\psi}\cos\theta$, and its axis (the vectorial direction) is inclined to O(D, E, C), at the angles of which the cosines are

$$\{A\dot{\theta},\ A\dot{\psi}\sin\theta,\ C(\dot{\phi}+\dot{\psi}\cos\theta)\}/H,$$

and to O($x$, $y$, $z$), at the angles of which the cosines multiplied by H are

$$\begin{aligned} &Cn\sin\theta\cos\psi-A(\dot{\psi}\sin\theta\cos\theta\cos\psi+\dot{\theta}\sin\psi), \\ &Cn\sin\theta\sin\psi-A(\dot{\psi}\sin\theta\cos\theta\sin\psi-\dot{\theta}\cos\psi), \\ &A\dot{\psi}\sin^2\theta+Cn\cos\theta. \end{aligned}$$

We may compare here the inclinations of OI and OH to the axis of figure. The cosines of the two angles are

$$\frac{n}{(n^2+\dot{\theta}^2+\dot{\psi}^2\sin^2\theta)^{\frac{1}{2}}}, \quad \frac{n}{\{n^2+\frac{A^2}{C^2}(\dot{\theta}^2+\dot{\psi}^2\sin^2\theta)\}^{\frac{1}{2}}}.$$

It follows that, according as C $>$ or $<$ A, the former cosine is less or greater than the latter, that is the former angle is greater or less than the latter [see Figs. 2 and 3]. This is the result already deduced from the consideration of the momental ellipsoid in 5, II. The two axes coincide when A = C, that is when the top is "spherical." When $\dot{\theta}$ is zero, that is, as we shall see when the body is in steady motion, or when the axis is at either of the circles between which it moves on the unit sphere (centre O), the two axes lie in the plane $z$OC.

For the most part in what follows, we shall consider tops which are symmetrical about an axis OC, so that the momental ellipsoid is of revolution about OC. When the ellipsoid is a sphere the top is usually called "spherical"; we shall call it oblate or prolate, according as the momental ellipsoid is oblate or prolate, that is according as C > or < A.

The inclinations of OC and OI to the vertical [see Fig. 4] are of great importance in the theory of the motion of a top under gravity. In other cases of the motion of what may be regarded as a top (for example, that of the earth under the moments, about the principal axes of the terrestrial momental ellipsoid, of the differential forces due to the action of the sun and moon on the matter of the earth) another line takes the place of the vertical, and it is easy to translate the terms and results used and obtained for a gravity top into corresponding expressions for the other. In the case of the earth the moments acting tend to bring the earth's axis of figure, OC, into perpendicularity to the plane of the ecliptic, and OZ, a line at right angles to this plane, drawn towards its northern side from the earth's centre, is the "vertical." But in consequence of the spinning motion of the earth a precessional motion results from the moments of the forces, and the axis OC moves round OZ, at what, apart from periodic disturbances, is approximately a constant angle COZ $(=\theta)$, and a constant angular speed $\dot{\psi}$.

**4.** *Kinematics of precession. Space-cone and body-cone.* As has been already stated, it is from the analogy of the terrestrial top that the angular speed $\dot{\psi}$ is called the "precession" of the gravity top. It has been shown that in the absence of friction the angular speed $n$ of the gravity top about the axis of figure is constant, and that $\dot{\psi}$ is also without variation when $\dot{\theta}=0$.

The inclinations of OC, OI to the vertical are $\theta$, $\cos^{-1}\{(\dot{\psi}+\dot{\phi}\cos\theta)/\omega\}$, while that of OI to OC is $\cos^{-1}\{(\dot{\phi}+\dot{\psi}\cos\theta)/\omega\}$, where

$$\omega=(\dot{\theta}^2+\dot{\phi}^2+\dot{\psi}^2+2\dot{\phi}\dot{\psi}\cos\theta)^{\frac{1}{2}}.$$

In the important case of "steady motion," in which $\dot{\theta}$ is continually zero and $\dot{\psi}$ is constant, we get, putting $\alpha$ for $\angle$COI and $\beta$ for $\angle$IOZ,

$$\alpha=\cos^{-1}\{(\dot{\phi}+\dot{\psi}\cos\theta)/\omega\},\quad \beta=\cos^{-1}\{(\dot{\psi}+\dot{\phi}\cos\theta)/\omega\}, \quad\ldots\ldots\ldots\ldots(1)$$

with $\omega=(\dot{\phi}^2+\dot{\psi}^2+2\dot{\phi}\dot{\psi}\cos\theta)^{\frac{1}{2}}=(n^2+\dot{\psi}^2\sin^2\theta)^{\frac{1}{2}}$.

We easily obtain $\sin\alpha=\dot{\psi}\sin\theta/\omega,\quad \sin\beta=\dot{\phi}\sin\theta/\omega$,

so that $$\dot{\phi}\sin\alpha=\dot{\psi}\sin\beta \quad\ldots\ldots\ldots\ldots\ldots\ldots(2)$$

and $$\tan\alpha=\frac{\sin\theta}{\cos\theta+\frac{\dot{\phi}}{\dot{\psi}}},\quad \tan\beta=\frac{\sin\theta}{\cos\theta+\frac{\dot{\psi}}{\dot{\phi}}}. \quad\ldots\ldots\ldots\ldots(3)$$

We see then that when $\theta$ is constant, and therefore $\dot{\psi}$ is also without variation, the successive positions of the rotation axis OI form the generators of a right circular cone, the axis of which is OZ [Fig. 4] and the semi-vertical angle $\beta$. This cone is fixed in space, and is called the *space-cone*. But the three axes OZ, OC, OI remain in one plane (since $\dot{\theta}=0$), and so, as the body turns about successive positions of OI in space, successive lines in the body, each inclined at the angle $\alpha$ to OC, become the instantaneous axis OI. Thus the locus of OI in the body is a cone of which OC is the axis and $\alpha$ is the semi-vertical angle This is called the *body-cone.*

That the motion of the body-cone is one of pure rolling on the space-cone is verified by the fact that $\dot{\phi}\sin\alpha=\dot{\psi}\sin\beta$; for the angular speed of the plane COI in the body-cone relative to the plane ZOC, which is turning with angular speed $\dot{\psi}$ about OZ, is $\dot{\phi}$, and the radii of the touching circles are, for OC$=$OZ$=1$, respectively $\sin\alpha$, $\sin\beta$.

We shall now indicate the positions of the cones in the different possible cases, taking for all $\dot{\phi}$ as positive. But we have here to remember the result of 18, II, that the precession may be either *direct* or *retrograde.* It is direct when the component $\dot{\psi}\cos\theta$ of $\dot{\psi}$ about the axis of figure is in the same direction as the angular velocity $\dot{\phi}$ about that axis, and retrograde when $\dot{\psi}\cos\theta$ is in the opposite direction to $\dot{\phi}$. In the case of an ordinary top spinning about the lowest point of the axis, so that $\theta$ is an acute angle, both precessions are direct: if the top be *suspended* from the extremity of its axis, so that the angle which the axis makes with the upward vertical is obtuse, the fast precession is retrograde.

If the top be supported at a point O in its axis, in such a way that the precession is produced by the action of gravity on a non-spinning back-weight, carried on the opposite side of the vertical from the part spinning with angular speed $\dot{\phi}$ (which we call the top proper), the slow precession is retrograde when $\theta$ is acute, and both precessions are retrograde when $\theta$ is obtuse.

The precessional motion, the turning of the line of nodes of the earth's orbit in the ecliptic, in the direction of the sun's apparent motion, is retrograde [see X below].

5. *Space-cone and body-cone in different cases.* Equations (2) enable the positions of the space-cone and body-cone to be traced. The path described on a sphere of unit radius with O as centre, by a point on the axis of the top, will be discussed later.

(1) $0<\theta<\frac{1}{2}\pi$; $\dot{\psi}/\dot{\phi}$ *positive. Precession direct.* It is clear that in this case both $\alpha$ and $\beta$ are less than $\theta$. For by (1), 4, the cosines of these angles are both greater than $\cos\theta$. The cones are therefore situated as in Fig. 13: the body-cone is external to the space-cone and rolls round the convex surface of the latter.

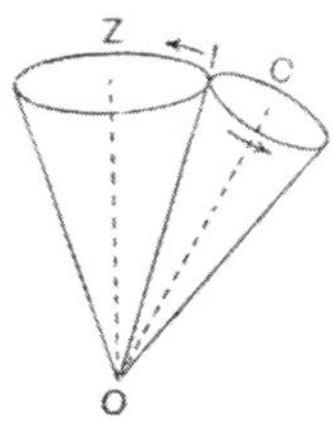

FIG. 13.

When $\dot{\psi}$ is great in comparison with $\dot{\phi}$, $\sin\beta$ is approximately zero while $\sin\alpha$ is approximately equal to $\sin\theta$. The space-cone is then very slender in comparison with the body-cone, and the rotation-axis nearly coincides with the vertical. The reverse is the case when $\dot{\phi}$ is great in comparison with $\dot{\psi}$. Then the body-cone is very slender, and the rotation-axis nearly coincides with OC. For the ordinary top the case of $\dot{\psi}=0$ arises when the couple about OD is zero; but the ratio $\dot{\phi}/\dot{\psi}$ may be very great, so that the top spins rapidly, while the axis turns very slowly round in azimuth.

(2) $0<\theta<\frac{1}{2}\pi$; $\dot{\psi}/\dot{\phi}$ *negative. Precession retrograde.* In the former case, when $\dot{\psi}$ is made smaller and smaller in comparison with $\dot{\phi}$, we have seen that the body-cone becomes more and more slender, so that the rotation-axis approaches without limit of closeness to OC as $\dot{\psi}/\dot{\phi}$ approaches zero. When, however, $\dot{\psi}/\dot{\phi}$ has a small *negative* value the body-cone is still very slender, but is now inside the space-cone [(2), 4], and rolls on the concave surface of the latter [Fig. 14]. The figure in 12, I, in which the space and body-cones are indicated by a ring and a slender cone rolling on it, illustrates this case for terrestrial precession.

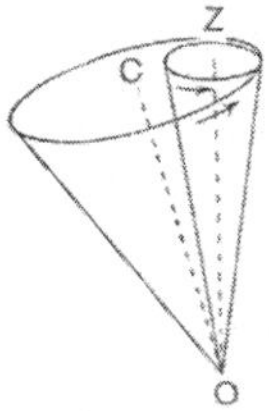

FIG. 14.

As $|\dot{\psi}/\dot{\phi}|$ increases towards the value $\cos\theta$, the body-cone widens out, so that its semi-vertical angle approaches $-(\frac{1}{2}\pi-\theta)$, which is only reached when $\beta$ acquires the value $\frac{1}{2}\pi$. The space-cone has then opened out to a plane, on which the body-cone rolls round O in the direction of retrograde precession.

As $|\dot{\psi}/\dot{\phi}|$ increases beyond the value $\cos\theta$, the space-cone acquires a semi-vertical angle $\beta$ greater than $\frac{1}{2}\pi$, while $\alpha$ takes the value $-(\beta-\theta)$. As $|\dot{\psi}/\dot{\phi}|$ approaches the value $1/\cos\theta$, $\beta$ approaches the value $\frac{1}{2}\pi+\theta$. The value of $\alpha$ then approaches $-\frac{1}{2}\pi$ [see (3), 4]. The interpretation of the sign of $\alpha$ will be clear from Fig. 13, where $\alpha$ and $\beta$ are both positive.

In this case, as indeed in any other, we can replace the cones referred to (which are only composed of single sheets of the complete cones) by the opposite sheets of the complete cones as geometrically defined by their generators. The complete cone fixed in the body may be regarded as rolling on the complete space-cone; and therefore we may take, instead of the space-cone of angle $\frac{1}{2}\pi+\theta$, the sheet of semi-angle $\frac{1}{2}\pi-\theta$, with internal axis directed *upward*, and as the body-cone the sheet of semi-angle $\beta-\theta$, with axis directed oppositely to OC. The latter rolls round the upward directed sheet of the other cone, with convex surface bearing on the latter. [Fig. 15 taken as drawn, and again, when turned right hand for left hand, and at the same time upside down, represents this equivalence.]

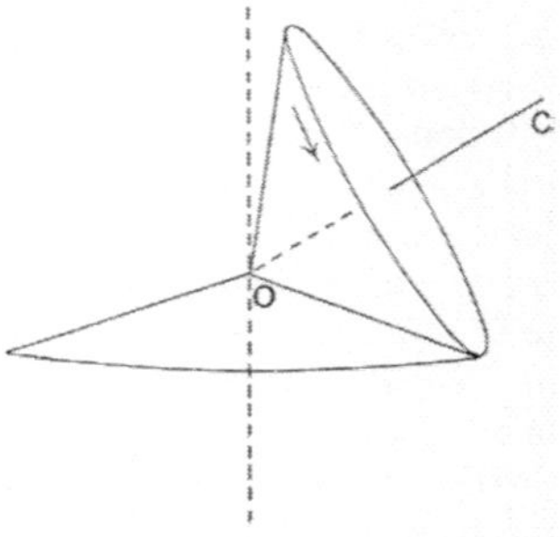

FIG. 15.

As $|\dot{\psi}/\dot{\phi}|$ increases beyond the value $1/\cos\theta$ the value of $\beta$ increases from $\frac{1}{2}\pi+\theta$ towards $\pi$, the value for $\dot{\psi}/\dot{\phi}=-\infty$. At the same time $\alpha$ increases from $\frac{1}{2}\pi$ towards the limiting value $\theta$.

When $\theta>\frac{1}{2}\pi$ the fixed and the moving cone can be determined by proceeding as before, except that the inclination of the axis to the downward vertical is taken as $\theta$. When that direction and left-handed rotation $\dot{\phi}$ round it are taken as positive, and direct precession is as before that for which $\dot{\psi}\cos\theta$ is in the same direction as $\dot{\phi}$, exactly the same results, in the same sequence as above, will be obtained for the different cases.

## CHAPTER V

### THE SIMPLER THEORY OF THE MOTION OF TOPS AND GYROSTATS

1. *Gyrostats. Equations of motion.* We now proceed to the consideration of the motion of a top in more detail and with application to a number of particular cases, such for example as gyrostats and combinations of gyrostats. A gyrostat is a flywheel enclosed in a case which is usually made symmetrical, or nearly so, about an axis of figure coincident with that of rotation of the flywheel. Hence the inertia of the case apart from that of the flywheel must be taken account of. We therefore specify quantities and establish equations of motion as follows. It will be observed that we do not specify the mode of support of the gyrostat, but begin by taking axes G(D, E, C) through the centroid G of the whole system of wheel and case.

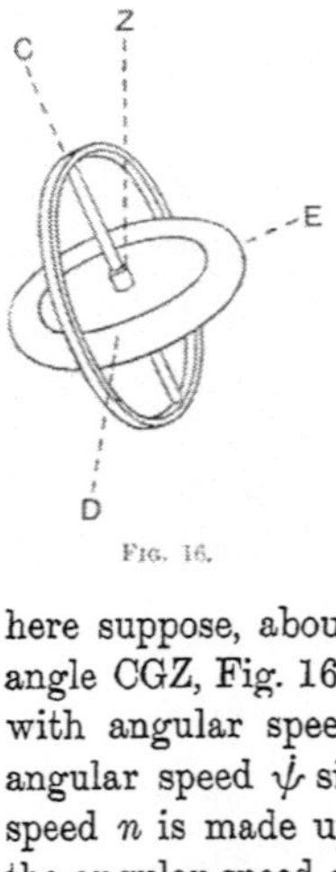

FIG. 16.

Let $n$ be the angular speed of the flywheel, $\omega_1$ that of the case (represented in Fig. 16 by a ring), in each instance about the axis of figure: O is a point, not necessarily fixed, on the axis of figure, and GZ a vertical drawn upward. If the plane ZOC turn, as we here suppose, about the vertical, with angular speed $\dot\psi$, and $\theta$ denote the angle CGZ, Fig. 16, the axes GD, GE are turning about the axis of figure with angular speed $\dot\psi \cos\theta$, while the whole is turning about GE with angular speed $\dot\psi \sin\theta$, and about GD with angular speed $\dot\theta$. The angular speed $n$ is made up of $\dot\psi \cos\theta$, due to the motion of the plane GOZ, and the angular speed $n'$ with which the flywheel turns relatively to that plane. In the same way $\omega_1$ may be put equal to $\dot\psi \cos\theta + \omega_1'$, where $\omega_1'$ is the angular speed of the case relative to the same plane. We put $\omega_2$ for the angular speed of the case about GE. Hence, if the case simply turn with the plane GOZ, $\omega_1'$ is zero, and $\omega_2 = \dot\psi \sin\theta$.

The components of angular momentum of the gyrostat are $Cn + C'\omega_1$ about GC, $A\dot\theta$ about GD, and $A\omega_2$ about GE, where C, C′ are the moments of inertia of the flywheel and case respectively about the axis of figure, and

A is taken for wheel and case together. The axes turn with angular speeds $\dot{\psi}\cos\theta$ about GC, $\dot{\theta}$ about CD, and $\dot{\psi}\sin\theta$ about GE.

Now for the rate of growth of angular momentum about the fixed axis $GD_1$, with which, at time $t$, GD coincides, we have first the term $A\dot{\theta}$. Next, as explained in 5, III, there are two contributions arising from the motions of the axes GC, GE. These are respectively $(Cn+C'\omega_1)\dot{\psi}\sin\theta$, $-A\omega_2\dot{\psi}\cos\theta$. Hence the whole rate of growth is

$$A\ddot{\theta}+(Cn+C'\omega_1)\dot{\psi}\sin\theta-A\omega_2\dot{\psi}\cos\theta.$$

This is equal to the moment of the couple, K say, applied about $GD_1$, and we get

$$A\ddot{\theta}+(Cn+C'\omega_1)\dot{\psi}\sin\theta-A\omega_2\dot{\psi}\cos\theta=K. \quad\text{.................(1)}$$

If, as is sometimes the case, $\omega_2=\dot{\psi}\sin\theta$, this equation becomes

$$A\ddot{\theta}+\{Cn+C'\omega_1-A\dot{\psi}\cos\theta\}\dot{\psi}\sin\theta=K. \quad\text{.................(2)}$$

We have additional equations arising from the motion of G, which we shall write down in each case as it arises. In the important case in which the point O on the axis is fixed, these reduce to

$$\left.\begin{aligned}\frac{d}{dt}(M\dot{\psi}h^2\sin^2\theta)&=Xh\sin\theta,\\ Mh(\dot{\theta}\cos\theta-\dot{\theta}^2\sin\theta-\dot{\psi}^2\sin\theta)&=-Y,\\ Mh(\ddot{\theta}\sin\theta+\dot{\theta}^2\cos\theta)&=Mg-Z,\end{aligned}\right\} \quad\text{.....................(3)}$$

where $h$ is the distance OG, M is the whole mass of the gyrostat, and the axes to which X and Y are referred are parallel respectively to OD and the projection of OE on the horizontal. X is the force along OD, Y the force along the projection of OE, and Z the force in the direction OZ, applied by the constraints at O. The reaction of the table against side slip is $-Y$. [In this case of the gyrostat under gravity or the top we can avoid the introduction of Y and Z by taking moments round the point of support O. Equation (2) requires no change, if it is understood that A now means the moment of inertia about an axis transverse to GC through the point of support, supposed now to lie on the axis, and the moment K is that of the forces about OD (Fig. 4).]

In the same way we obtain the equation of motion for the fixed axis $GE_1$, with which GE coincides at the instant, and about which we suppose the moment of the applied couple to be L. The equation is, as the reader may verify,

$$A\dot{\omega}_2-(Cn+C'\omega_1-A\dot{\psi}\cos\theta)\dot{\theta}=L. \quad\text{.....................(4)}$$

If $\omega_2=\dot{\psi}\sin\theta$, this becomes

$$A\ddot{\psi}\sin\theta-(Cn+C'\omega_1-2A\dot{\psi}\cos\theta)\dot{\theta}=L. \quad\text{.................(5)}$$

For the top $C'=0$; if then the top spin about a fixed point, and A now denote the total moment of inertia about the transverse axis through O, parallel to GE, (5) becomes

$$A\ddot{\psi}\sin\theta-(Cn-2A\dot{\psi}\cos\theta)\dot{\theta}=0, \quad\text{....................(6)}$$

Finally, for the axis of figure we have two equations: first,

$$Cn = \text{constant}, \quad \text{.................................(7)}$$

since we suppose the friction couple on the bearings of the flywheel to be negligible, and second,

$$C'\dot{\omega}_1 = R, \quad \text{.......................................(8)}$$

if R be the moment of the couple acting on the case about the axis of the figure. It is supposed that the forces concerned in this couple do not affect the motion of G, which is given by (3) above.

It may be remarked here that the energy equation is

$$\tfrac{1}{2}(Mv^2 + Cn^2 + C'\omega_1{}^2 + A\dot{\theta}^2 + A\omega_2{}^2) + V = \text{constant}, \quad \text{.............(9)}$$

where M is the total mass, $v$ the resultant speed of G, and A the moment of inertia of flywheel and case about any axis through G transverse to OC.

For the top spinning about a fixed point on the axis of figure there is an equation of constancy of angular momentum, namely

$$A\dot{\psi}\sin^2\theta + Cn\cos\theta = G, \quad \text{..........................(10)}$$

where G is the angular momentum about the vertical, and A is now taken as the moment of inertia about an axis transverse to OC, through O, the fixed point.

For this case C′ is zero and $K = Mgh\sin\theta$, where M is the whole mass, and $h$ is the distance of the centre of gravity from the point of support. Hence (2) becomes

$$A\ddot{\theta} + (Cn - A\dot{\psi}\cos\theta)\dot{\psi}\sin\theta = Mgh\sin\theta. \quad \text{..................(11)}$$

The steady motion obtained when $\dot{\theta}$ is and continues zero [for by (6) $\ddot{\psi}$ is then zero, and therefore $\dot{\psi}$ is constant] has been dealt with already to some extent in (17) and (18), II. We shall return to it again immediately.

Instead of O(D, E, C) axes O(K, L, C) are taken, such that OK, OL are in the plane DOE, and OL is the axis of the resultant, $\omega$, of $\dot{\theta}$ and $\dot{\psi}\sin\theta$ about OD and OE respectively. The reader may prove at once by inspection that if P, Q be the moments of applied forces about OK, OL, the equations of motion are

$$Cn\omega - A(n - \dot{\phi})\omega = P, \quad A\dot{\omega} = Q,$$

where $n$ is constant and has the meaning already assigned to it. These equations have been called the intrinsic equations of the motion of a top.* For general use they are less convenient than those for the axes O(D, E, C).

We now take one or two illustrations of the theory given above.

**2.** *Example* 1. *Gyrostat supported at a point* O *in the axis of spin, and acted on only by a couple of constant moment* K *about the axis* OD. Let the gyrostat spin rapidly so that the A.M. $Cn$ about the axis of the flywheel is very great, and suppose that the moments of inertia and angular speeds are A, B, and $\dot{\theta}$, $\omega$, respectively, for the axes OD, OE, which may be regarded as principal axes and moving with the top. Also we suppose that the A.M. $B\omega$ is small in comparison with $Cn$. The angle $\theta$ is that which an axis $OZ_1$ in the plane COD makes with OC. As we need not here measure $\theta$ from the vertical, since we suppose gravity zero, we may take $OZ_1$ where

* Lamb, *Proc. R.S.E.* 35, Pt. II, 1915.

we please in the plane specified. We shall take it as inclined at an angle $\alpha$ to the vertical through O By (2) and (4), 1, we have, neglecting the motion of the case about OC,

$$A\theta + Cn\omega = K, \quad B\omega - Cn\dot{\theta} = 0. \qquad (1)$$

Eliminating $\theta$, we get

$$AB\ddot{\omega} + C^2n^2\left(\omega - \frac{K}{Cn}\right) = 0. \qquad (2)$$

The complete solution of this equation is, if $\omega = 0$ when $t = 0$,

$$\omega = \frac{K}{Cn} + R\cos\frac{Cn}{(AB)^{\frac{1}{2}}}t,$$

where R is a constant $= -K/Cn$ Hence

$$\omega = \frac{K}{Cn}\left(1 - \cos\frac{Cn}{(AB)^{\frac{1}{2}}}t\right). \qquad (3)$$

The angle $\phi = \int \omega\,dt$ turned through in time $t$ is given by

$$\phi = \frac{K}{Cn}\left\{t - \frac{(AB)^{\frac{1}{2}}}{Cn}\sin\frac{Cn}{(AB)^{\frac{1}{2}}}t\right\} \qquad (4)$$

Initially also we suppose $\dot{\theta} = 0$, and so get

$$\theta = \int\dot{\theta}\,dt = \frac{BK}{C^2n^2}\left(1 - \cos\frac{Cn}{(AB)^{\frac{1}{2}}}t\right), \quad \dot{\theta} = \left(\frac{B}{A}\right)^{\frac{1}{2}}\frac{K}{Cn}\sin\frac{Cn}{(AB)^{\frac{1}{2}}}t. \qquad (5)$$

Thus if the motion start from rest the first effect of the couple is to produce a rate of growth of angular velocity $\dot{\theta}$; but the sidelong motion $\omega$ begins, though by the second equation of motion there is no $\dot{\omega}$ until an angular velocity $\dot{\theta}$ exists. It will be observed that $\phi$ is made up of two parts, a part that grows uniformly at time-rate $K/Cn$, and an oscillatory part of period $2\pi(AB)^{\frac{1}{2}}/Cn$, and amplitude $K(AB)^{\frac{1}{2}}/C^2n^2$. The rate $K/Cn$ is the value of the steady slow precession obtained for the value of the couple in (1) above.

The value of $\theta - \alpha$ returns periodically to the value zero. The expression for $\dot{\theta}$ and the oscillatory part of $\omega$ give to a point P on the axis of spin (say at unit distance from O), motion in an ellipse in the moving plane normal to the plane of the couple K. The axis of spin however swings round, carrying the axes OD, OE with it, and a succession of open loops is described by P on the unit sphere of centre O. This result may be compared with that of 13 below.

The angular range of one of these loops is $2KB/C^2n^2$, and the work done by the couple K in this range is $2K^2B/C^2n^2$. The kinetic energy is then $\frac{1}{2}(Cn^2 + B\omega^2)$, and there is no potential energy. But at the middle of the loop $\omega^2$ is, by (3), $2K/Cn$ and $\dot{\theta} = 0$. Hence at that point the value $\frac{1}{2}B\omega^2$ of the kinetic energy is $2K^2B/C^2n^2$. Thus the work done by the couple accounts for the term depending on $\omega$.

*Example 2. Gyrostatic action of a body of any form:* e.g. ***gyrostatic action of an aeroplane propeller.*** Consider first a thin disk rotating with angular speed $n$ about an axis OC at right angles to its plane, and passing through the centroid O. Refer to two other axes, OA, OB, the other principal axes, in the plane of the disk. Let C, A, B be the moments of inertia for these axes respectively. Further, let the disk be turning with angular speeds $\omega_1$, $\omega_2$ about space axes, coinciding at the instant with axes OD, OE in the plane of the disk, and fixed, we may suppose, with reference to some framework, such as that of an aeroplane, itself in motion, on which the disk is carried. For example, OD may be transverse to the aeroplane so as to be horizontal when the aeroplane is unbanked, and OE will then be vertical if the aeroplane is horizontal fore and aft.

If at the instant considered OA have turned through the angle $\phi$ from coincidence with OD, the angular speeds about OA and OB are respectively $\omega_1\cos\phi + \omega_2\sin\phi$ and

$-\omega_1 \sin\phi + \omega_2 \cos\phi$. Denoting these by $\omega_A$, $\omega_B$ we have, by Euler's equations or directly by first principles, for the equations of motion with respect to O(A, B, C),

$$A\dot{\omega}_A - (B-C)\omega_B n = L, \quad B\dot{\omega}_B - (C-A)n\omega_A = M, \quad C\dot{n} - (A-B)\omega_A\omega_B = N. \quad \ldots\ldots(a)$$

where L, M, N are applied couples.

Multiplying the first by $\cos\phi$, the second by $-\sin\phi$, and adding, we get the equation of motion for OD, and similarly multiplying the first by $\sin\phi$, the second by $\cos\phi$, and adding, we get the equation for OE. These are

$$\left.\begin{aligned} A\dot{\omega}_A\cos\phi - B\dot{\omega}_B\sin\phi - (B-C)\omega_B n\cos\phi + (C-A)n\omega_A\sin\phi &= L\cos\phi - M\sin\phi, \\ A\dot{\omega}_A\sin\phi + B\dot{\omega}_B\cos\phi - (B-C)\omega_B n\sin\phi - (C-A)n\omega_A\cos\phi &= L\sin\phi + M\cos\phi. \end{aligned}\right\} \ldots(b)$$

We may take different cases. An important one for the present illustration is that in which $\dot{\omega}_1 = 0$, $\omega_2 = 0$, $\dot{\omega}_2 = 0$, and the axis OD is fixed at right angles to OC, so that $\dot{\phi} = n$. We have now $\omega_A = \omega_1\cos\phi$, $\omega_B = -\omega_1\sin\phi$, so that the equations of motion just written become

$$-(A-B)\omega_1 n\sin 2\phi = L\cos\phi - M\sin\phi, \quad (A-B)\omega_1 n\cos 2\phi - Cn\omega_1 = L\sin\phi + M\cos\phi. \ldots(c)$$

On the left in (*c*) we have the gyrostatic couples called into play, which, if the motion is to be steady, as supposed, must be balanced by applied couples. In the first the gyrostatic couple is wholly periodic and passes through all its phases twice as $\phi$ varies from 0 to $2\pi$; in the second the gyrostatic couple consists of a similarly periodic part, $(A-B)n\omega_1\cos\phi$, and the constant term $-Cn\omega_1$. The periodic parts vanish if $A = B$.

If there are not applied couples L, M of proper amount, there will be acceleration about OD, OE. If $L = M = 0$, the equations are

$$\left.\begin{aligned} (A\cos^2\phi + B\sin^2\phi)\dot{\omega}_1 - (A-B)n\omega_1\sin 2\phi &= 0, \\ \tfrac{1}{2}(A-B)\dot{\omega}_1\sin 2\phi + (A-B)n\omega_1\cos 2\phi - Cn\omega_1 &= 0. \end{aligned}\right\} \ldots\ldots\ldots\ldots(d)$$

An aeroplane propeller usually consists of two blades, and though these blades are not plane, but have a peculiar helical shape, the propeller may be treated as a disk so as to give, in the manner indicated above, an idea of the varying gyrostatic action set up by its rotation

The body may not be a disk, but be of any form. O will then be the centroid of the body, and O(A, B, C) the principal axes of moment of inertia. The couples applied by gyrostatic action are as found above, and, as in all cases, are unaffected by translational motion of the body as a whole.

Some analogies of gyrostatic action to phenomena of electricity have been discussed by Bogaert in his treatise *L'effet gyrostatique et ses applications*, pp. 55 *et seq.* For example, an angular speed $\omega$ about an axis OE produces an A.M. at rate $Cn\omega$ about OD, and this corresponds to an external couple of moment $Cn\omega$ applied about OD. If no couple about OD is applied from without, the body turns about that axis in the direction to keep the A.M. about OD unchanged. The reaction of the body is thus opposed to the effect for the axis OD produced by the turning about OE. This may be compared with the law of Lenz, that the displacement of a circuit in a magnetic field produces in the circuit an induced current, the electromagnetic action between which and the field opposes the displacement.

The reader should consult the work referred to, for various illustrations set forth in a clear and attractive manner. [See also 18, Chap. XVIII.]

3. *Symmetrical equations of motion.* The following form of the equations of motion is sometimes convenient. [The reader may however pass on to 5.] For a top spinning about a fixed point, equations (2) and (4) of the preceding section become, if we write $\omega_x$ for the angular speed $\dot{\theta}$ about OD, and $\omega_y$ for that, $\dot{\psi}\sin\theta$, [if we take the old axes O(D, E, C)] about OE,

$$\left.\begin{aligned} A\dot{\omega}_x+(Cn-A\omega_y\cot\theta)\omega_y=K, \\ A\dot{\omega}_y-(Cn-A\omega_y\cot\theta)\omega_x=L. \end{aligned}\right\} \quad \text{.........................(1)}$$

Let us suppose that these component couples are produced by the application of a force F at right angles to OC, at a point P on OC distant $h$ from O. Let the components of this force along OD and OE be X and Y. We have $K=-hY$, $L=hX$.

The axes OD and OE are moving axes, turning about OC with angular speed $\dot{\psi}\cos\theta$ ($=\omega_z$, say). If $u$, $v$ be the linear speeds of P (the coordinates of which it is to be noticed are and remain $x=0$, $y=0$) with reference to fixed axes with which OD and OE coincide at the instant under consideration, we have

$$u=h\omega_y, \quad v=-h\omega_x, \quad \dot{u}=h\dot{\omega}_y, \quad \dot{v}=-h\dot{\omega}_x.$$

Substituting in (1), and changing the order of the equations, we get

$$\left.\begin{aligned} \dot{u}-\omega_z v+\frac{Cn}{A}v=\frac{h^2}{A}X, \\ \dot{v}+\omega_z u-\frac{Cn}{A}u=\frac{h^2}{A}Y. \end{aligned}\right\} \quad \text{.............................(2)}$$

The first pair of terms on the left in each of these equations give the $x$ and $y$ accelerations of a particle of unit mass at the point P on the axis of the top; for, since $x$ and $y$ are zero, no terms $\omega_z^2 x$, $\omega_z^2 y$, $\dot{\omega}_z x$, $\dot{\omega}_z y$ exist for the component accelerations of P. If $C=0$, or $n=0$, these equations reduce to

$$\dot{u}-\omega_z v=\frac{h^2}{A}X_0, \quad \dot{v}+\omega_z u=\frac{h^2}{A}Y_0, \quad \text{.......................(3)}$$

and $X_0$, $Y_0$ are the components of force required to produce the accelerations on the left when the A.M. about the axis OC is zero. If the accelerations are specified, equations (3) solve, for the case of zero A.M. about OC, the problem of determining the component forces required to give the specified accelerations to the system, for the given (finite) values of $u$, $v$. As pointed out in 4 below, the quasi-particle considered has different virtual inertias along OD and OE, the axes transverse to the spin axis.

The more general problem of finding the force F,(X, Y), when C is not zero and the values of $\dot{u}$, $\dot{v}$, $u$, $v$ are specified, is solved by (2). If we suppose that $\dot{u}$, $\dot{v}$, $u$, $v$ do not exceed finite limits, and that $Cn$ is comparatively very great, these equations are approximately represented by

$$\frac{Cn}{h^2}v=X-X_0, \quad \frac{Cn}{h^2}u=-(Y-Y_0). \quad \text{.......................(4)}$$

Thus the direction of the resultant force F is approximately at right angles to that of the motion represented by the component velocities $u$, $v$. To specify F exactly, suppose drawn from P the vector V, the components of which are $u$, $v$; then think of the outer extremity J of that vector as being carried round by the angular speed $n$, at right angles to PJ, at speed $n$V in the direction shown in Fig. 17. Draw a vector in this direction, and make it of length $\mathrm{C}n\mathrm{V}/h^2$. The force-vector F is equal and opposite to this latter vector.

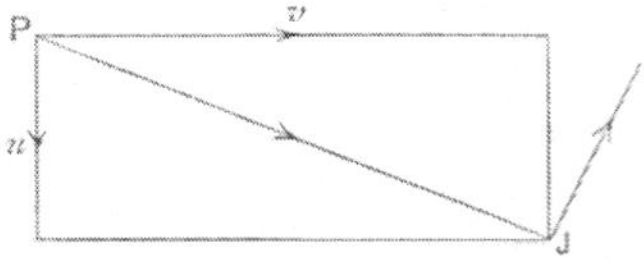

FIG. 17.

From the above investigation it is clear that if the forces X, Y are given, and are of the order of magnitude of C$n$, then the motion consistent with these forces, if comparatively very small, will be directed nearly at right angles to the line of action of F.

**4.** *Motion of a top reduced to motion of a particle.* We have just seen what the equations of motion are if the spinning body may be regarded as a mass concentrated at the point P, while retaining a finite angular momentum C$n$ about the axis of symmetry. For this purpose we have to write a third equation, of which the full form is

$$\dot{w} - \omega_y u + \omega_x v = \frac{h^2}{\mathrm{A}}\mathrm{Z}. \qquad \text{(1)}$$

If, as has been suggested, the body can be regarded, so far as A is concerned, as a particle at P, rigidly connected of course with the fixed point O and turning, with its connection, as a rigid body about that point, we have $h^2/\mathrm{A} = 1/m$, where $m$ is the mass. Thus we get for the three equations of motion, since $w = 0$,

$$\left.\begin{aligned} m(\dot{u} - \omega_z v) + \frac{\mathrm{C}n}{h^2} v &= \mathrm{X}, \\ m(\dot{v} + \omega_z u) - \frac{\mathrm{C}n}{h^2} u &= \mathrm{Y}, \\ m(-\omega_y u + \omega_x v) &= \mathrm{Z}. \end{aligned}\right\} \qquad \text{(2)}$$

The spin about the axis OC gives rates of generation of momentum $\mathrm{C}nv/h^2$, $-\mathrm{C}nu/h^2$ in the directions of $x$ and $y$ respectively, a remarkable result. If we write

$$\alpha = \mathrm{X}/(\mathrm{X} - \mathrm{C}nv/h^2), \quad \beta = \mathrm{Y}/(\mathrm{Y} + \mathrm{C}nu/h^2),$$

we can write the first two equations of (2) in the form

$$\alpha m(\dot{u} - \omega_z v) = \mathrm{X}, \quad \beta m(\dot{v} + \omega_z u) = \mathrm{Y}, \qquad \text{(2')}$$

so that the virtual inertia of the quasi-particle is different in the two directions, and depends on the forces in these directions and on the A.M. C$n$ about the spin axis.

If the motion is one of steady turning about the fixed axis OZ, the vertical, or what corresponds to the vertical, we have $v = 0$ and $\dot{u} = \dot{v} = 0$. Resolving the directed quantities radially inward toward the vertical, that is multiplying the second of (2) by $\cos\theta$, and the third by $\sin\theta$, and subtracting the latter product from the former, we obtain

$$\left(m\omega_z - \frac{\mathrm{C}n}{h^2}\right) u \cos\theta + m\omega_y u \sin\theta = \mathrm{Y}\cos\theta - \mathrm{Z}\sin\theta. \qquad \text{(3)}$$

But $\omega_z = \dot{\psi}\cos\theta$, $u = h\omega_y = h\dot{\psi}\sin\theta$, and therefore the last equation becomes

$$m\dot{\psi}^2 h \sin\theta - \frac{\mathrm{C}n}{h}\dot{\psi}\sin\theta\cos\theta = \mathrm{Y}\cos\theta - \mathrm{Z}\sin\theta. \qquad \text{(4)}$$

Since $\theta$ is constant, we may put R for $\mathrm{Y}\cos\theta - \mathrm{Z}\sin\theta$, and write the equation in the form

$$\dot{\psi}^2 - \frac{\mathrm{C}n}{mh^2}\cos\theta \,.\, \dot{\psi} = \frac{\mathrm{R}}{mh\sin\theta}. \quad \text{.........} \quad (5)$$

Thus there are two values of $\dot{\psi}$ consistent with the steady motion.

5. *Effect of increasing or diminishing the applied couple. Hurrying or retarding precession.* We now inquire as to the effect of increasing or diminishing the applied couple $\mathrm{M}gh\sin\theta$ on the inclination of the axis in steady motion. Going back to (11), 2, let $\mathrm{M}gh$ be increased to $\mathrm{M}gh + \mathrm{N}$ (N positive), so that an additional couple $\mathrm{N}\sin\theta$ about the axis OD is applied, and let, as is possible, steady motion exist under this couple with the same value of $\theta$. Then, if $\mu_1$, $\mu_2$ be the roots of the equation,

$$\mathrm{A}\cos\theta \,.\, \mu^2 - \mathrm{C}n\mu + \mathrm{M}gh = 0,$$

the value $\mu$ of the precession under the enhanced couple is given by

$$\mathrm{A}\cos\theta \,.\, (\mu_1 - \mu)(\mu - \mu_2) = \mathrm{N}\,; \quad \text{.........} \quad (1)$$

so that (since $\cos\theta$ is supposed to be positive) $\mu$ must lie between the two values, the large value $\mu_1$ and the small value $\mu_2$, which correspond to the couple $\mathrm{M}gh\sin\theta$.

The increase of couple thus diminishes the large root and increases the small root. Either of the new values requires for its production the couple $\mathrm{N}\sin\theta$ tending to depress the axis. Thus, if the angular speed of precession be increased from the small value or diminished from the large value, and the corresponding increase of couple be removed, the axis will begin to rise.

Similarly, if N be negative the new values of $\mu$ lie one above $\mu_1$, the other below $\mu_2$, and the removal of N would result in depression of the axis.

If $\theta > \frac{1}{2}\pi$, $\cos\theta$ is negative, but now $(\mu_1 - \mu)(\mu - \mu_2)$ is negative or positive according as N is positive or negative. But $\mu_1$ is negative; and the values of $\mu$ lie, in the first case, one on the negative side of $\mu_1$ and the other on the positive side of $\mu_2$, and, in the second case, one on the positive side of $\mu_1$, the other on the negative side of $\mu_2$. The effect of the removal of N is to be described as before.

Thus we have the rule stated above as to the effect of hurrying and delaying precession, namely, that if an increase of $\dot{\psi}$ from $\mu$ is impressed without additional couple about OD, the axis will rise if $\mu$ is the smaller, and sink if $\mu$ is the larger, of the two roots of the steady motion equation, and the reverse is the case if $\dot{\psi}$ is diminished.

It will be clear from (11), 1, that if an additional couple N is applied to the top in steady motion, the axis will begin to sink if N be positive, and rise if N be negative. For (11), 1, gives $\mathrm{A}\ddot{\theta} = \mathrm{N}$.

We may consider this subject otherwise as follows. Let a couple be applied in the plane of motion of the axis of spin, as it is in almost all cases in which the precession is to be accelerated or retarded. Since we suppose that $\dot{\theta}$ is zero, the couple applied is about the axis OE. If $\mathcal{M}$ be the moment of the couple we have by (5), 1, since $\dot{\theta}=0$,

$$A\ddot{\psi}\sin\theta=\mathcal{M}. \quad\text{.........................................}(2)$$

But after a short interval of time $dt$ has elapsed we have [see the process in 9 below], instead of (6) and (11), 1, the two equations

$$A\,d\dot{\psi}\sin\theta=\mathcal{M}\,dt,\quad A\,d\dot{\theta}+(Cn-2A\mu\cos\theta)d\dot{\psi}\sin\theta=0\,;\quad\text{........}(3)$$

for in the limit $\dot{\theta}$ and $\dot{\psi}$ change by $d\dot{\theta}$ and $d\dot{\psi}$ while $d\theta$ is zero.

First let us suppose that $0<\theta<\frac{1}{2}\pi$. The precession is direct (see 18, II), and so $\mu$ has the same sign as $n$, which we always take as positive. Hence by (3) we see that according as $Cn-2A\mu\cos\theta$ is negative or positive, $d\dot{\theta}$ and $d\dot{\psi}$ have the same or opposite signs. Thus the top begins to rise or to fall according as the couple, $\mathcal{M}$, about OE increases $\dot{\psi}$ or diminishes it.

But if $2A\mu\cos\theta>Cn$ so that $Cn-2A\mu\cos\theta$ is negative the value of $\mu$ is the larger root of the quadratic in $\mu$, which (11) becomes when $\theta$ is constant, and $\mu$ is put for $\dot{\psi}$ (see also 17, II, above). If $2A\mu\cos\theta<Cn$ the value of $\mu$ is the smaller root of the quadratic. Thus, if the steady precession have the larger possible value, the couple here considered causes the axis of the top to rise or fall according as the couple gives a negative or a positive $d\dot{\psi}$.

Next let $\pi>\theta>\frac{1}{2}\pi$, then as we have seen (18, II) the high value of the precession has the retrograde direction, the low value is direct. In the former case $\mu$ is negative, in the latter positive. Hence $2A\mu\cos\theta$ is positive for the fast precession, negative for the slow. Thus, if $Cn-2A\mu\cos\theta$ is negative, a positive value of $d\dot{\psi}$, that is a retardation of the fast precession, means a negative value of $d\dot{\theta}$, or the top begins to rise, while a negative value of $d\dot{\psi}$, that is a hurrying of the fast precession, causes the axis of the top to descend.

When the precession is direct, $Cn-2A\mu\cos\theta$ is always positive (since here $\pi>\theta>\frac{1}{2}\pi$), and so $d\dot{\theta}$ is negative or positive according as $d\dot{\psi}$ is positive or negative. Thus, when $\pi>\theta>\frac{1}{2}\pi$ we have exactly the same result as before.

The effect of resisting or aiding precession by a couple is beautifully illustrated by a piece of apparatus devised long ago by M. Sire of Paris, and given to the author of this book by M. l'Intendant Sire, of Sidi-bel-Abbès, Algiers, the son of the inventor. The axis of a gyrostat wheel coincides with a diameter of a ring, which is grooved round the edge, so that the gyrostat can be hung like a loose pulley by the free end of a cord wound round the groove, as shown in Fig. 18. When the wheel is spinning, and the gyrostat supported by the cord is left to itself, precession in azimuth takes place in the direction to balance the constant couple $Mgh$ applied by gravity. But as precession proceeds, the string, which is held fast at the upper end, is twisted and applies a couple resisting the precession, which becomes slower and slower with turning of the pulley and sinking of the gyrostat as a whole, and turning of the axis of the wheel towards the vertical.

Cessation of precession is accompanied by a sudden fall of the gyrostat, which makes half a turn and so reverses its axis. The couple in the twisted string now aids the precession (in the opposite direction) proper to the new position of the wheel, and the further turning of the axis is checked, fall ceases, and this opposite precession begins and quickly increases. The string, however, untwists and then twists up the other way, opposing the new precession. The result is that the gyrostat again falls, precession in the original direction begins as before, and so on, until the successive falls possible with the length of string chosen have all taken place. Another illustration is found in the solution of the music-hall problem of how to throw down a cheese-shaped body set up on edge. The body contains a rapidly rotating gyroscope of considerable A.M., and resists the attempts of those who attempt the feat, and who of course try to overturn it directly. It can be laid down flat by merely applying a couple about a vertical axis by pressing on one side gently with the little finger.

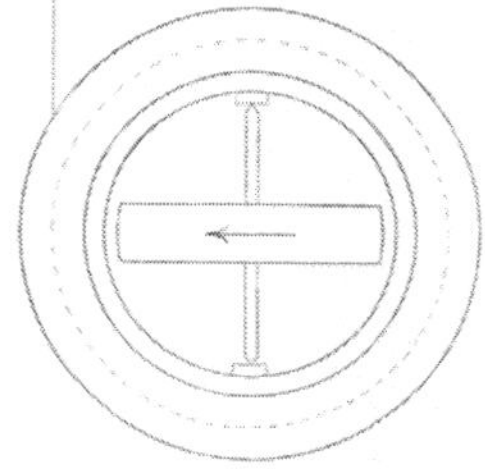

FIG. 18.

**6.** *Greenhill's graphical construction.* The steady motion of a top and the results of hurrying or retarding precession are well illustrated by an elegant graphical construction due to Sir George Greenhill. In Fig. 19 OV is the vertical, OC the axis of the top OV and OC are taken of lengths to represent respectively the A.M.

$$G(=Cn\cos\theta+A\dot{\psi}\sin^2\theta)$$

about the vertical, and the A.M. $Cn$ about the axis of the top. Lines CK and VK are drawn at right angles to OC and OV, and meet in K. Now

$$\begin{aligned}OV&=OC\cos\theta+CK\sin\theta\\&=Cn\cos\theta+CK\sin\theta,\end{aligned}$$

so that

$$CK=A\dot{\psi}\sin\theta.$$

Hence we obtain

$$\begin{aligned}VK&=OC\sin\theta-CK\cos\theta\\&=(Cn-A\dot{\psi}\cos\theta)\sin\theta.\quad\ldots.(1)\end{aligned}$$

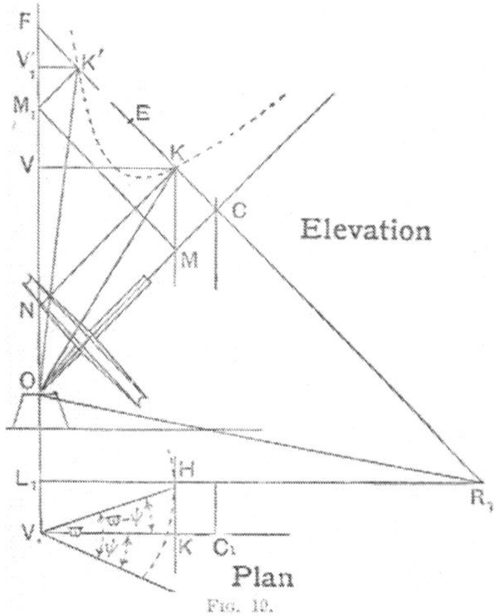


FIG. 19.

Also if KM be drawn parallel to VO to meet OC in M, and KN be drawn parallel to CO,
$$\text{MK} = \text{ON} = \text{CK}/\sin\theta = \text{A}\dot{\psi}, \quad \text{.................(2)}$$
and
$$\text{NV} = \text{C}n\cos\theta + \text{A}\dot{\psi}\sin^2\theta - \text{A}\dot{\psi} = (\text{C}n - \text{A}\dot{\psi}\cos\theta)\cos\theta. \quad \text{.................(3)}$$
Thus, if $\dot{\theta}$ be zero, CK and OC are rectangular components of the A.M., and OK represents the resultant A.M.

In the general case we draw, as shown in the plan (supposed laid in a horizontal plane through V), a horizontal line KH to represent $\text{A}\dot{\theta}$. The resultant A.M. is now completely represented by OH.

If the motion is steady $\dot{\theta}=0$, and H coincides with K. We take $\mu$ as the steady value of $\dot{\psi}$. The point K moves in a horizontal circle about OV as axis, with angular speed $\mu$, and A.M. in the direction of the motion of K grows up at rate VK . $\mu$, that is at rate $(\text{C}n - \text{A}\mu\cos\theta)\mu\sin\theta$. Thus, since $\text{M}gh\sin\theta$ is the moment of the gravity forces about a horizontal axis through O parallel to KH, we have
$$(\text{C}n - \text{A}\mu\cos\theta)\mu = \text{M}gh,$$
which is the quadratic equation for $\mu$ already found.

By the diagram of Fig. 19,
$$\mu = \frac{\text{M}gh}{\text{VK}}\sin\theta = \frac{\text{M}gh}{\text{NK}}, \quad \text{and} \quad \mu = \frac{\text{ON}}{\text{A}} = \frac{\text{MK}}{\text{A}}. \quad \text{.................(4)}$$
Thus we get
$$\text{MK}\,.\,\text{NK} = \text{A}\,.\,\text{M}gh, \quad \text{.................(5)}$$
and so for the given length of OC and inclination of OC to the vertical, K lies on a hyperbola of which OV and OC are the asymptotes. The directions of these lines and the value of M$gh$ determine the hyperbola, so that it is given when $\theta$ and A . M$gh$ are given.

If CK be produced to meet OV in F, and E be the middle point of CF, we get
$$\text{CE}^2 - \text{KE}^2 = \text{CK}\,.\,\text{KF} = \text{KM}\,.\,\text{NK}\sin\theta\tan\theta = \text{A}\,.\,\text{M}gh\sin\theta\tan\theta. \quad \text{.................(6)}$$
Two roots of the quadratic thus exist, one for each of the points KK′, in which CF intersects the hyperbola. The smaller root is CK/A sin $\theta$, the larger is CK′/A sin $\theta$.

When the roots are equal CK touches the hyperbola. Then KM $=\frac{1}{2}$OF, KN $=\frac{1}{2}$OC, and therefore
$$\text{A}\,.\,\text{M}gh = \text{KM}\,.\,\text{NK} = \tfrac{1}{4}\text{OF}\,.\,\text{OC} = \tfrac{1}{4}\text{C}^2n^2\sec\theta. \quad \text{.................(7)}$$
Hence, for equality of roots, we get, remembering that MK $=$ A$\mu$,
$$\text{C}n = 2(\text{A}\,.\,\text{M}gh\cos\theta)^{\frac{1}{2}}, \quad \mu = \frac{\text{M}gh}{\text{NK}} = 2\frac{\text{M}gh}{\text{C}n}. \quad \text{.................(8)}$$

If OC be too short CK will fall below the vertex of the branch of the hyperbola shown dotted in the diagram, and the roots of the quadratic are then imaginary, and steady motion is impossible.

We can trace what happens when the top is started with the given A.M. C$n$ at a given inclination, with $\dot{\theta}$ and $\dot{\psi}$ zero. First, since the term $\frac{1}{2}\text{C}n^2$ of the kinetic energy remains unchanged, and terms $\frac{1}{2}\text{A}\dot{\theta}^2$ and $\frac{1}{2}\text{A}\dot{\psi}^2\sin^2\theta$ are called into existence at the expense of the potential energy, a sinking of the axis below the initial inclination $\theta$ takes place. This sinking continues while $\dot{\theta}$ increases, and $\ddot{\theta}$, at first a maximum, diminishes until, when $\ddot{\theta}$ is zero, $\dot{\theta}$ is a maximum. At that instant $\dot{\psi}$ has the steady motion value $\mu$, corresponding to the point K on the hyperbola. At the starting of the top K lies within the hyperbola, and when $\ddot{\theta}=0$ the value of $\dot{\psi}$ is the smaller root of the steady motion equation. [If, however, an initial value of $\dot{\psi}$ sufficient to make $(\text{C}n - \text{A}\dot{\psi}\cos\theta)\dot{\psi} > \text{M}gh$ is given to the top, A$\ddot{\theta}$ will be negative, and the top when started and left to itself will begin to rise towards the point K′.]

After $\dot{\theta}$ has thus become a maximum, and K has reached the hyperbola, the axis continues to sink, and $\ddot{\theta}$ becomes negative. We have then $(\text{C}n - \text{A}\dot{\psi}\cos\theta)\dot{\psi} > \text{M}gh$, and

$\dot{\psi}$ increases until it attains a maximum value just when the absolute value of $\theta$ is greatest, as we see from (6) and (11) of 1, for then $\ddot{\theta}=0$, and therefore $\ddot{\psi}=0$. Then the absolute value of $\theta$ diminishes, a negative value of $\dot{\theta}$ grows up and the axis rises.

Unless the initial position is such that the line CK intersects the hyperbola, there does not exist a value of $\dot{\psi}$, with which if the top were started it would continue in steady motion.

The diagram shows at once the effect of applying an additional couple N *about OD.* When this is done we obtain a new hyperbola, for which the equation

$$\mathrm{KM}\,.\,\mathrm{NK}=\mathrm{A}\left(\mathrm{M}gh+\frac{\mathrm{N}}{\sin\theta}\right)$$

holds. (Of course on the left the letter N refers to the diagram.) If N be positive the new hyperbola lies within the old, and so the two new intersections of CF with the curve are obviously one above K and the other below K′. If N be negative the new hyperbola lies outside the old. The new position of K is below the new position of K′ above the old.

If then the steady precession be that due to the couple $\mathrm{M}gh\sin\theta+\mathrm{N}$ we see at once that the effect of the withdrawal of N in setting the axis in motion is as stated in 5 above. Initially, we have then $\mathrm{A}\ddot{\theta}=-\mathrm{N}$.

7. *Reaction of ring-guide or space-cone on top.* If, as in the model illustrating the precession of the equinoxes pictured on p. 12, I, and in the well-known toy shown in Fig. 20, the top be furnished with a material cone or axle, fixed round the axis of figure, which rolls on a cone fixed in space, represented by the ring or curved wire of the diagrams, and the point of support be at the centroid, the couple on the top must be applied by the pressure of the fixed against the moving cone. The circle of the points of contact on the moving cone is the polhode on the top, and the fixed ring or curved wire is the herpolhode. [See Chapter XXI below.]

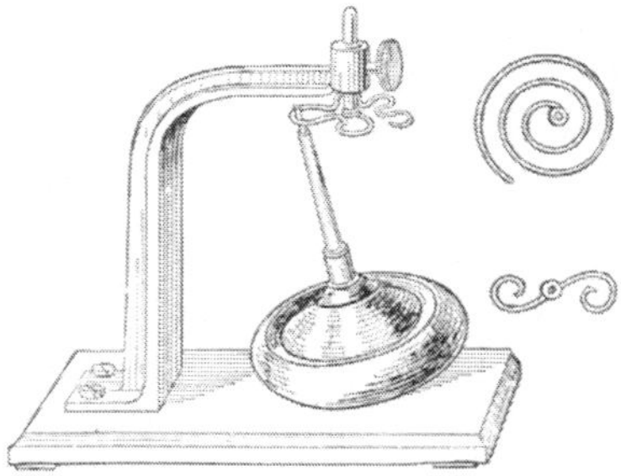

Fig. 20.

The thrust of the axle against the ring-guide, that is the force exerted by one polhode on the other, is to be found from the calculation of the rate of growth of A.M. given in 5, III, above. This is the rate of displacement of the extremity H of the vector OH, representing the A.M.; and clearly, when the motion is steady, H moves about an axis at right angles at once to the axis of figure and to the vertical, an axis therefore which may be represented by OD of Fig. 4. For OH is always in the plane ZOC of Fig. 4, which is perpendicular to the path of the point I of the instantaneous axis along the guide.

But the A.M. grows for the direction OD at rate

$$\mathrm{A}\ddot{\theta}+(\mathrm{C}n-\mathrm{A}\dot{\psi}\cos\theta)\dot{\psi}\sin\theta,$$

and therefore, if N be the couple,

$$A\ddot{\theta}+(Cn-A\dot{\psi}\cos\theta)\dot{\psi}\sin\theta=N; \quad \text{.....................(1)}$$

or, if the motion is steady,

$$(Cn-A\mu\cos\theta)\mu\sin\theta=N. \quad \text{.........................(2)}$$

In this connection this equation is sometimes written as

$$\{C\omega-(A-C)\mu\cos\theta\}\mu\sin\theta=N, \quad \text{......................(3)}$$

where $\omega$ is the rate of turning of the top with respect to the plane ZOC.

If $A=C$ we have the steady precessional motion, under couple N, of a spherical top [3, IV, above]; that is, the equation is

$$C(n-\mu\cos\theta)\mu\sin\theta=N. \quad \text{............................(4)}$$

We shall see below that the term introduced by the inertia of the *case* of a gyrostat enables a similar equation to be obtained for that form of top.

The pressure on the ring is $N/l$, if $l$ denote the distance of the point of contact of the axle with the ring from the point of support.

If a slight push or blow be given to the axis of the top, an impulsive couple is applied which produces an increase of the component $A\dot{\psi}\sin\theta$ of A.M. about the axis OE, that is changes $\dot{\psi}$ to $\dot{\psi}+d\dot{\psi}$, if $\theta$ is kept unchanged by the guide. This increase in $\dot{\psi}$ makes the rate of growth of A.M. about OD more rapid than is accounted for by the couple N, and so the top endeavours to turn about OD in the direction to keep the rate of change of A.M. the same as before, that is so as to press with so much greater force against the guide, that the value of N is enhanced to the extent required for the greater precession.

8. *Explanation of clinging of axle of top to curved guide.* The action of the top shown in Fig. 20 is very curious, but its explanation may be made out easily from the discussion above, which is to be read along with what follows. The axle rolls round the curved guide, following all the convolutions, however sharply curved, and on coming to the end of the guide in one direction turns rapidly round the end of the wire and rolls back on the other side. The axle has been described as clinging to the wire like a piece of iron to a magnet.

For simplicity we have supposed that there is no gravitation couple on the top. The action of the guide may be analysed as follows: Consider a right circular cone, with the vertical through O as axis, and the line OI as a generator: a short element of the guide at the point of contact is at the intersection of the guide and a circular section of the cone. Such a cone may be made to pass through any element of the guide, and $\theta$ is now the semi-vertical angle of that cone. The element will in general give a component of action on the axis of the top in the plane through the axis of the cone.

We have for the couple applied to the axle, in the vertical plane through OI, the equation $A\ddot{\theta}+(Cn-A\dot{\psi}\cos\theta)\dot{\psi}\sin\theta=N.$

Besides this couple N, a couple in the tangent plane to the cone through OI is applied to the top. For clearly a component F of reaction of the guide acts on it at I with or against the direction of motion along the circular section, according to the angle between the edge of the section and the guide, and F and $-F$ inserted at the point of support give a couple of moment N′, the axis of which is at right angles to OI, in the plane COI. This can be resolved into two components $N'\sin\alpha$, $N'\cos\alpha$ ($\alpha=\angle IOC$) about OC and OE (at right angles to OC) in the plane COI. The former couple of comparatively small moment alters the speed of rotation, the latter gives change of $\dot{\psi}$ at numerical rate $A\ddot{\psi}\sin\theta=N'\cos\alpha,$

since $\dot{\theta}$ is negligible. The axle therefore presses more or less on the guide from this cause.

There is also a frictional couple which in general splits into two components, one with or against N, and the other helping or retarding $\dot{\psi}$, according to the direction of the guide.

Now, let the axle come to a discontinuity in the guide, for example one of the ends. The couple N may be regarded as there suddenly annulled, and therefore (since anything like steady motion ceases) $A\ddot{\theta}$ as taking at the same time a value

$$-(Cn-A\dot{\psi}\cos\theta)\dot{\psi}\sin\theta,$$

the value of $A\ddot{\theta}-N$ just before the discontinuity is arrived at. In other words, the motion may be regarded as disturbed by a force $N/l$ applied at I to the axle annulling the moment N. Thus $\dot{\theta}$ grows up very rapidly, and the axle moves across the end of the guide.

But as the axle moves thus owing to $\dot{\theta}$, a rate R of growth of A.M. about OE would be produced were it not for another motion of the top. There is now no couple about OE, and therefore, in order to keep R zero, the top must turn about OE, and in the direction, as will be seen from the figure, to bring the axle against the end of the wire, across which the axle will roll, until the next sharp corner is reached. In this way the axle rolls round the end of the guide, while the space-cone of angle $\theta$ changes position rapidly.

When the end has been rounded the precession becomes again nearly steady, but the axle now presses against the other side of the wire. The precession is now in the opposite direction, and the axle therefore again presses against the wire, but in the opposite direction to that in which it formerly pressed at the same place.

A similar explanation accounts for the hard pressing of the axle against the guide where $\theta$ increases rapidly, as it does in a guide like that of Fig. 20.

**9.** *Stability of motion of a gyrostat or top.* The word "stability" has no quite exactly defined sense in dynamics. For example, it is not sufficient to say that the steady motion of a gyrostat is stable if, when it is deviated from, the motion oscillates about the steady motion. Even with friction present the oscillation may continually increase in amplitude until the range of departure from the steady motion is enormous. In such cases there is not true stability. On the other hand, the existence of friction may cause vibrations about the steady motion to subside continuously, until the deviations from a state of steady motion are imperceptible. Friction cannot be neglected in any really practical discussion of stability.

We now inquire whether the steady motion of a gyrostat or top, which is obviously a possible motion, is stable, and answer this question *for the present* in the following manner: We first suppose that if the steady motion is slightly deviated from, the motion of the gyrostat or top oscillates about the steady motion, and that the vibratory deviation from steady motion approaches the steady motion as a limiting state, when the initial disturbance approaches zero. The logic of this process, and the conditions of what may be described as true stability, will be examined in a later chapter.

We write then, for the disturbed motion, $\theta=\theta_0+\alpha$, where $\theta_0$ is the steady motion value and $\alpha$ is small, and $\dot\psi=\mu+\eta$, where $\mu$ is the steady motion value of the angular speed of precession, and $\mu+\eta$ the disturbed value corresponding to $\theta_0+\alpha$. Since in the steady motion $\ddot\theta_0=0$, equation (11) of 1 becomes

$$\mathrm{A}\ddot\alpha+\mathrm{U}=0, \quad\text{.......................................(1)}$$

where U is what the expression

$$(\mathrm{C}n-\mathrm{A}\dot\psi\cos\theta)\dot\psi\sin\theta-\mathrm{M}gh\sin\theta$$

becomes when $\theta_0+\alpha$ is put for $\theta$, and $\mu+\eta$ for $\dot\psi$. This expression vanishes for $\theta=\theta_0$ and $\dot\psi=\mu$, and therefore, since the changes are small, (1) becomes

$$\mathrm{A}\ddot\alpha+\frac{d}{d\theta}\{(\mathrm{C}n-\mathrm{A}\dot\psi\cos\theta)\dot\psi\sin\theta-\mathrm{M}gh\sin\theta\}\alpha=0, \quad\text{..............(2)}$$

*where, after the differentiation has been carried out, $\theta$ is to be put equal to $\theta_0$ and $\dot\psi$ to $\mu$.* This assumes that the changes are so small that the value of the rate of variation with $\theta$ of the expression in brackets, calculated for $\theta=\theta_0$ and multiplied by $\alpha$, will give the value of U. Performing the differentiation, remembering of course that $\dot\psi$ varies with $\theta$, and that (6) of 1 gives

$$\mathrm{A}\sin\theta_0\frac{d\dot\psi}{d\theta}=\mathrm{C}n-2\mathrm{A}\mu\cos\theta_0, \quad\text{...........................(3)}$$

we obtain easily

$$\mathrm{A}^2\mu^2\ddot\alpha+(\mathrm{A}^2\mu^4-2\mathrm{AM}gh\mu^2\cos\theta_0+\mathrm{M}^2g^2h^2)\alpha=0. \quad\text{................(4)}$$

The quantity within the brackets can be written as the sum of two squares, and is therefore positive. Hence the deviation from steady motion is oscillatory in the real period,

$$T = \frac{2\pi A\mu}{(A^2\mu^4 - 2AMgh\mu^2\cos\theta_0 + M^2g^2h^2)^{\frac{1}{2}}}. \quad \text{..................(5)}$$

The disturbance has been supposed such as to leave the A.M. about the vertical unchanged: for example, a vertical impulse would fulfil the condition. But a similar result would be obtained for any small disturbance, as can easily be verified [see below].

It may be noticed that if the top is spun very fast, so that the smaller value of the angular speed (approximately $Mgh/Cn$) is very small, the period of oscillation about the steady motion, with this value of $\mu$, is approximately $2\pi A/Cn$. This may be verified from (5), or by neglecting all the terms in the brackets in (2) except $Cn$, and using $A \sin\theta_0 d\psi/d\theta = Cn$ instead of (3), when the result follows at once. This approximation to the period is independent of $\theta$.

Again, if the top is spun very fast, the period of oscillation about the steady motion is, for the large value of $\mu$, $2\pi/\mu$, that is the period of revolution of the axis in the precessional cone. This follows from (5), because then the denominator on the right becomes $A\mu^2$ very nearly, so that $T = 2\pi/\mu$. Thus, in one half of the revolution the axis is above the steady motion position, in the other half below it. This period is $2\pi A\cos\theta/Cn$. For $\theta = 0$, the upright position, it agrees with the other period, and varies from that to zero for different values of $\theta$ from 0 to $\frac{1}{2}\pi$.

If we suppose that the small disturbance given to the top is quite general, then *after* the disturbance we have the equations

$$\left.\begin{aligned} A\dot\psi\sin^2\theta + Cn\cos\theta &= G', \\ A\ddot\theta + (Cn - A\dot\psi\cos\theta)\dot\psi\sin\theta &= Mgh\sin\theta. \end{aligned}\right\} \quad \text{..................(6)}$$

The disturbance has changed the value of G to G′, and also changed the value of the energy. Instead of the first of these equations we may write of course, for any instant before or after the disturbance,

$$A\ddot\psi\sin\theta - (Cn - 2A\dot\psi\cos\theta)\dot\theta = 0. \quad \text{....................(7)}$$

If we now suppose, as before, that this motion deviates slightly from a steady motion according to the equations

$$\theta = \theta_0 + \alpha, \quad \dot\psi = \mu + \eta,$$

we may proceed by the process already employed, and arrive at the same equation, (5) above, for the period.

This is not now, however, quite exactly the steady motion from which the top has been disturbed, but a steady motion differing slightly from the former so as to be consistent with the values of the energy and the A.M. about the vertical as modified by the disturbance. For example, if the disturbance be purely one of $\dot\psi$ the steady motion circle will become a limiting circle of the disturbed motion [see below, 14].

10. *Rise and fall of top.* It is very instructive to consider the rise and fall of a top from the point of view of the constancy of A.M. about the vertical, and the constancy of the total energy, combined with the invariability of the angular speed of rotation about the axis of figure. Putting, for brevity,

$$\alpha=(2E-Cn^2)/A,\quad \beta=G/A,\quad z=\cos\theta,\quad a=2Mgh/A,\quad b=C/A,$$

where E is the total energy, we obtain the equations of energy and momentum in the forms,

$$\left.\begin{aligned}\dot\theta^2+\dot\psi^2(1-z^2)&=\alpha-az,\\ \dot\psi(1-z^2)&=\beta-bnz.\end{aligned}\right\}\quad\text{..........(1)}$$

Eliminating $\dot\psi$ between these equations, we get

$$\dot z^2=(\alpha-az)(1-z^2)-(\beta-bnz)^2=f(z).\quad\text{..........(2)}$$

The cubic expression $f(z)$ is negative when $z=-\infty$, $z=\pm1$, and positive when $z=+\infty$. It is also positive when $z$ has its initial value $z_0$, say, which must be between $-1$ and $+1$. Two roots of the equation $f(z)=0$ lie therefore, one, $z_1$, between $-1$ and $z_0$, another, $z_2$, between $z_0$ and $+1$, and the third, $z_3$, between $+1$ and $\infty$. The last is not relevant to the question, since $-1<z<+1$. We have therefore

$$\dot z^2=a(z-z_1)(z_2-z)(z_3-z).\quad\text{..........(3)}$$

The product of the three factors is positive, since $\dot z^2$ is positive; the third factor is obviously positive, and therefore $z$ must lie between $z_1$ and $z_2$. We have taken $z_2$ as the greater of the two roots $z_1$, $z_2$. Hence, as $\dot z$ alters the angle $\theta$ varies also, but is always such that $\cos\theta$ lies between the limits $z_1$, $z_2$.

From (3) $z$ can be found as an elliptic function of the time $t$, and thus the top rises and falls periodically.

11. *Path of point on axis of top. Cusps on path.* By elimination between $\dot z=\{f(z)\}^{\frac12}$ and $\dot\psi=(\beta-bnz)/(1-z^2)$, we get

$$\frac{d\psi}{dz}=\frac{\beta-bnz}{(1-z^2)\{f(z)\}^{\frac12}},\quad\text{..........(1)}$$

from which $\psi$ can be obtained in terms of $z$ by quadrature. In 15, below, the path of a point on the axis of figure under different possible conditions is considered, and illustrated by diagrams showing the path, as seen by an eye situated above the top. We discuss here the occurrence of cusps or loops on the path.

Let a sphere of radius 1 be drawn round O as centre, and denote by $Z_1$ the point in which the vertical through O intersects the sphere [see Fig. 21 (*a*), (*b*), (*c*)]. The radius through the point $Z_1$ is vertical, and if P be a point in which the axis of the top meets the surface, the projection of OP on the radius $OZ_1$ is the value of $z_1$, and the distance of P from the latter radius is $(1-z^2)^{\frac12}$.

The tangent of the inclination $i$ of the line traced by P on the unit sphere to the meridian $Z_1P$ at any position is $-(\beta-bnz)/\{f(z)\}^{\frac{1}{2}}$, in which the value of $\{f(z)\}^{\frac{1}{2}}$ is taken + or − according to the sign of $\dot{z}$. This

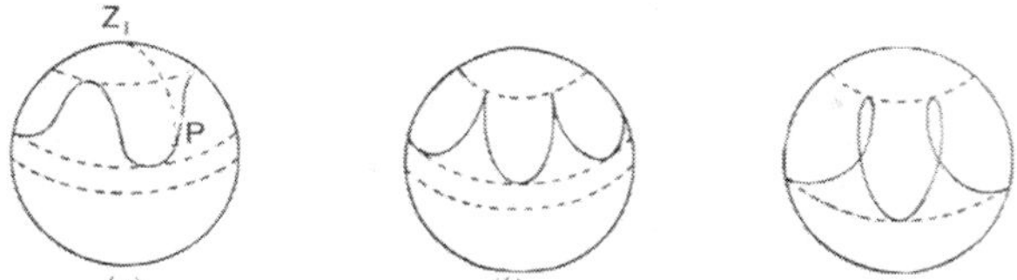


(a) (b) (c)

FIG. 21.

expression shows that the path of P is at right angles to the meridian whenever $z$ has one or other of the values $z_1$, $z_2$, unless it should happen that at either limit $z=\beta/bn$. Thus the path lies between the two parallel circles on the unit sphere corresponding to $z_1$, $z_2$. When $z=\beta/bn$ at one of these circles the path of P has cusps, as shown in Fig. 21 (*b*) [see also 13 below]. As we shall show presently, this can only be the case at the upper circle. The form of the projection on a horizontal plane is shown on a larger scale in Fig. 22 for the second case.

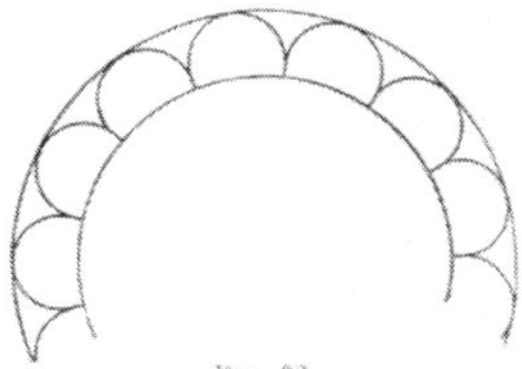

FIG. 22.

If the value of $\beta/bn$ lie outside the limits $z_1$, $z_2$, $\beta-bnz$ must always preserve the same sign as P moves, as in Fig. 21 (*a*). But if $\beta/bn$ lie between $z_1$ and $z_2$ the path will have the form shown in Fig. 21 (*c*), from which the changes of sign and value in $\tan i$ can be traced. To settle whether when $z(=\beta/bn)$ lies between −1 and +1, it also lies between $z_1$ and $z_2$, we have only to consider the value which this gives to $f(z)$, that is $\dot{z}^2$, which of course must be positive. By (2), 10, we have, if $z=\beta/bn$,

$$f(z)=\left(\alpha-a\frac{\beta}{bn}\right)\left(1-\frac{\beta^2}{b^2n^2}\right), \quad\text{..........................(2)}$$

in which the second factor is positive. Hence, if the first factor is positive, $\beta/bn$ lies between $z_1$ and $z_2$. The condition therefore is $\beta/bn < \alpha/a$.

There is however the case in which this factor vanishes: $\beta/bn$ is then equal to one of the limits $z_1$, $z_2$. To find which, we substitute in $f(z)$, $\alpha=a\beta/bn$, and get

$$f(z)=\frac{1}{bn}(\beta-bnz)\{a(1-z^2)-bn(\beta-bnz)\}. \quad\text{.................(3)}$$

If we equate the right-hand side to zero, we see that one root of the equation is given by the first factor. The second factor is zero if

$$\beta-bnz=a(1-z^2)/bn.$$

But $a(1-z^2)/bn$ is positive, and therefore $\beta - bnz$ must also be positive if this equality holds, that is $z < \beta/bn$, with $n$ positive. We see then that if one of the roots $z_1$, $z_2$ of $f(z)=0$ be $\beta/bn$, it must be the greater root. Thus the cusps are at the upper circle, as stated above.

The conclusion thus analytically obtained for the position of the cusps is obvious from considerations of energy. For at either limiting circle the term in the kinetic energy depending on $\dot{\theta}$ is zero. The potential energy, however, has its maximum value at the upper limiting circle, and its minimum value at the lower. At the lower circle therefore it is impossible for $\dot{\psi}$ to be zero, otherwise the kinetic energy would, for the *minimum* of potential energy, be *reduced* to the constant part $\frac{1}{2}Cn^2$, depending on the spin about the axis of figure. But, since the total energy does not vary, the kinetic energy must have its greatest value at the lower limiting circle. Hence the cusps, if they occur, are connected with the reduction of the kinetic energy to the constant part in consequence of the adjustment of the maximum of potential energy to the value $E - \frac{1}{2}Cn^2$.

**12.** *Occurrence of loops on path.* The dynamical reason for the form of the path in Fig. 21 (*c*) is clear from (1) above, or from the second of (1), 10. If the two roots of $f(z)=0$ be such that before reaching the *upper* limit of value, $\cos\theta$ becomes so great that $Cn\cos\theta$ exceeds I or $bnz > \beta$, then $\dot{\psi}(1-z^2)$ must become negative. In other words, the turning about the vertical must be reversed from the direction which it had at the lower limit of $\cos\theta$, where $Cn\cos\theta < \mathrm{I}$.

It can be proved, but not so easily as might be expected, that the positive advance in the value of $\psi$ in each oscillation is, as shown in the diagram, greater than the regression. The fact will be made manifest when an exact discussion of the path is given. But it may be demonstrated by the theory of complex integration as follows. Let a horizontal line (Fig. 23) be the axis of real quantity and O the origin. Let the dotted line through O be the axis of imaginary quantity. Let the positions of the roots $z_1$, $z_2$, $z_3$ of $f(z)=0$, and the points $z=-1$, $z=+1$ be as indicated. Draw a closed contour as shown, consisting of straight lines above and below the axis of real quantity and loops surrounding the critical points of $d\psi/dz$, situated on that axis (the branch points $z_1$, $z_2$ and the poles $-1$, $+1$), and a second closed contour consisting of a loop round the point $z=z_3$, a straight line above the axis to $+\infty$, a circle of infinite radius from O as centre, and finally a straight line below the axis of real quantity from $+\infty$ to the point $z_3$. The integrand in $\int(\beta - bnz)/(1-z^2)\{f(z)\}^{\frac{1}{2}} . dz$ is holomorphic throughout the space between the two contours, and hence the integrals taken round these, in the left-hand direction, say, for the outer, and the right-hand direction for the inner are equal. Now we have at $z=1$ and $z=-1$, $f(z) = -(\beta - bn)^2$,

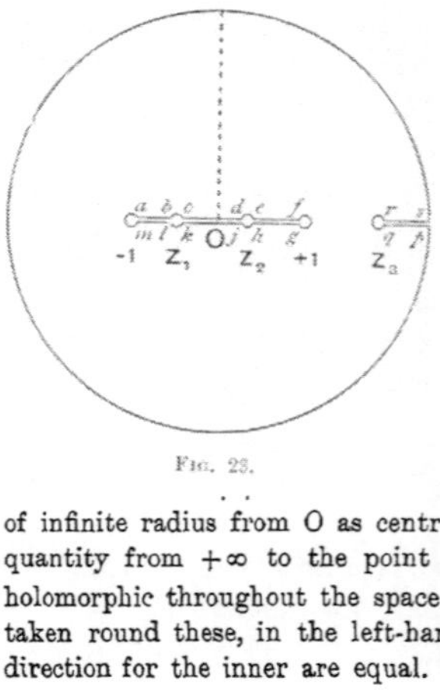

Fig. 23.

and $\beta - bnz$ changes sign between $z_1$ and $z_2$. But the sign of the radical changes also as the integration in the inner contour passes from the point $d$ to the point $j$, and as it passes from the point $k$ to the point $c$, and also in the outer contour, as the integration passes from $r$ to $q$. The residues for the points $z=-1$ and $z=+1$ are equal in value and opposite in sign. The integral from $c$ to $d$ is equal to the integral from $j$ to $k$, and all other parts of the integral along the axis of real quantity ($lm$, $ab$, and $ef$, $gh$) for the inner contour, taken in pairs, above and below the axis, cut one another out. The integral along the large circle is zero, and so the integral from $r$ to $s$ is equal to the integral from $p$ to $q$. Thus we obtain

$$\int_{z_1}^{z_2} \frac{(\beta - bnz)dz}{(1-z^2)\{f(z)\}^{\frac{1}{2}}} = \int_{z_3}^{\infty} \frac{(\beta - bnz)dz}{(1-z^2)\{f(z)\}^{\frac{1}{2}}}.$$

Every element of the integral on the right is positive, and therefore the integral on the left is also positive. The proposition assumed above is therefore established. [This mode of proof is due to M. Hadamard (*Bull d. Sci. Math.* t. XIX, 1895, 1re Partie).]

As the analysis here following will show, the repeated curves on the unit sphere in Fig. 21, may, when the limiting circles are close, and $\theta$ is not zero at either, be regarded as cycloidal.

13. *Motion of the axis of a top between two close horizontal limiting circles. Axis initially at rest.* Let it be supposed that initially $\dot\theta=0$, $\theta=\theta_0$ (and therefore $\cos\theta_0=z_0$), $\dot\psi=0$, and $n$ very great. Then

$$\mathrm{E}=\tfrac{1}{2}\mathrm{C}n^2+\mathrm{M}gh\cos\theta_0. \qquad (1)$$

Thus, initially, $\alpha=az_0$, $\beta=bnz_0$. After the top is started it is left to itself, and the inclination of the axis to the vertical alters. Since E and $\frac{1}{2}\mathrm{C}n^2$ remain constant the axis must sink, so that the terms of the kinetic energy depending on $\dot\theta$ and $\dot\psi$ may come into existence at the expense of the potential energy. Thus, when the inclination of the axis to the vertical is $\theta$, we have

$$\left.\begin{array}{l} \tfrac{1}{2}\mathrm{A}(\dot\theta^2+\dot\psi^2\sin^2\theta)=\mathrm{M}gh(\cos\theta_0-\cos\theta), \\ \text{or} \quad \dot\theta^2+\dot\psi^2(1-z^2)=a(z_0-z). \end{array}\right\} \qquad (2)$$

Also we have
$$\dot\psi(1-z^2)=\beta-bnz=bn(z_0-z). \qquad (3)$$

We have seen that

$$\dot z^2=(\alpha-az)(1-z^2)-(\beta-bnz)^2, \qquad (4)$$

and this equation, since $\alpha=az_0$, $\beta=bnz_0$, can be written

$$\dot z^2=a(z_0-z)\left\{1-z^2-\frac{b^2n^2}{a}(z_0-z)\right\}. \qquad (5)$$

Hence $z=z_0$ is one root of the equation $f(z)=0$; it is the root $z_2$. The second relevant root $z_1$ will cause the factor $1-z^2-b^2n^2(z_0-z)/a$ to vanish. Hence we must have

$$a(1-z_1^2)-b^2n^2(z_0-z_1)=0. \qquad (6)$$

Solving this equation, we get

$$z=\frac{b^2n^2}{2a}\left\{1\pm\left(1-\frac{4a}{b^2n^2}z_0+\frac{4a^2}{b^4n^4}\right)^{\frac{1}{2}}\right\},$$

that is, with $p=b^2n^2/2a$, $z=p\pm(p^2-2pz_0+1)^{\frac{1}{2}}$. ..............................(7)
The smaller of these roots defines the second limiting circle.

The value of $z_1$ lies between $-1$ and $+1$, and it is clear therefore that, if $n$ be great, $z_0-z_1$ must be small. Hence the axis of the top, if the spin be rapid, moves between two close right circular cones with common axis $OZ_1$.

Now consider the value of $\dot\psi$. We have by (3) and (6),

$$\dot\psi=\frac{a}{bn}\frac{z_0-z}{z_0-z_1}\frac{1-z_1^2}{1-z^2}. \qquad (8)$$

The quantities $z_0-z$, $z_0-z_1$ are both small; their ratio depends on the value of $z$. The ratio $(1-z_1^2)/(1-z^2)$ is positive, and very nearly equal to unity, and we know from (2) that $z_0>z$, that is $z_0$ is the value of $z$ for the upper limiting circle on the unit sphere, and $z_1$ is that for the lower.

We notice that when $z=z_0$, that is at the upper circle, $\dot\psi=0$, or the curve of intersection of the axis with the spherical surface is that shown in Fig. 21 (*b*), for it is clear from (2), (3) and (8) that $\dot\psi^2(1-z^2)$ is small in comparison with $\dot\theta^2$ [for $1+\dot\theta^2/\dot\psi^2(1-z^2)=(a/b^2n^2)(1-z^2)/(z_0-z)$] when $z$ approximates to $z_0$, though both $\dot\theta$ and $\dot\psi$ approach zero. The path therefore meets the upper circle in a series of cusps. We have already seen [11] why these cusps may exist at the upper circle but not at the lower.

Again, when $z=z_1$, we have $\dot\psi=a/bn=2Mgh/Cn$. We shall see that the average value of $\dot\psi$ is $Mgh/Cn$ in this case.

**14.** *Top rising and falling through small range. "Strong" and "weak" tops.* We can now complete the solution for a rapidly rotating top. Here $\theta_1-\theta_0$ is small, and $\theta$ lies between $\theta_0$ and $\theta_1$. Let $\theta=\theta_0+\eta$, where $\eta$ is a small quantity. We get $\cos\theta=\cos\theta_0-\eta\sin\theta_0$, that is $z=z_0-\eta\sin\theta_0$. This substituted in the value of $\dot z^2$ gives the approximate expression

$$\dot\eta^2=a\eta\sin\theta_0-b^2n^2\eta^2. \qquad (1)$$

Differentiating this we obtain

$$\ddot\eta+b^2n^2\eta-\tfrac{1}{2}a\sin\theta_0=0. \qquad (2)$$

This gives the complete solution

$$\eta=\frac{a}{2b^2n^2}\sin\theta_0(1-\cos bnt), \qquad (3)$$

in which the constants have been so chosen as to make $\eta=0$ when $t=0$, and $\dot\eta=0$ when $t=\pi/bn$.

Hence $$\theta=\theta_0+\frac{a}{2b^2n^2}\sin\theta_0(1-\cos bnt), \qquad (4)$$

and $$\theta_1=\theta_0+\frac{a}{b^2n^2}\sin\theta_0. \qquad (5)$$

Finally, we have $$\dot\psi=bn(\cos\theta_0-\cos\theta)\frac{1}{\sin^2\theta}, \qquad (6)$$

or approximately, from (4), $\dot\psi=\dfrac{a}{2bn}(1-\cos bnt)$. ..............................(7)

This gives $\dot{\psi}=0$ for $t=0$, and $\dot{\psi}=a/bn$ when $t=\pi/bn(=\pi A/Cn)$. The average value is thus $a/2bn=Mgh/Cn$.

The motion is compounded of a steady motion with constant values $\theta_0+a\sin\theta_0/2b^2n^2$ and $a/2bn$ of $\theta$ and $\dot{\psi}$, and oscillatory variations $-(a\sin\theta_0\cos bnt)/2b^2n^2$, $-(a\cos bnt)/2bn$ of the same two quantities. It is to be noted that the period of oscillation about the steady motion is $2\pi/bn=2\pi A/Cn$, and is therefore shorter the faster the rotation of the top. It is twice the time of passage of the axis from one circle to the other. The circle for which $\theta=\theta_0+a\sin\theta_0/2b^2n^2$ is midway between the two limiting circles.

From the example just considered it follows that, if the top be started spinning at high speed, with the axis exactly vertical and $\dot{\theta}=0$, the axis must remain vertical. For, as we have seen, $\dot{z}^2$ is negative except for values of $z$ which lie between the roots $z_1$, $z_2$. But if the top be started as stated above, equation (5), 13, becomes

$$\dot{z}^2=a(1-z)^2\left(1+z-\frac{b^2n^2}{a}\right), \quad\text{.........................(8)}$$

and two roots of the equation $\dot{z}=0$ are equal to 1. The third root is $b^2n^2/a-1$, and is greater than 1 if $n$ be sufficiently great. Hence the two roots $z_1$, $z_2$ coincide in value, and the two limiting circles have shrunk into the point $Z_1$.

Equation (8) shows that for any value of $z$ less than 1 the value of $\dot{z}^2$ is negative, provided $b^2n^2>2a$. [Here and in what follows $a$ is supposed to be positive.] Hence, if this condition is fulfilled, the top cannot leave the vertical in the course of its own undisturbed motion, as otherwise $\dot{z}^2$ would become negative. The stability of the upright top will be further considered in 15, below, and in 4 ... 19, VI.

If the top fulfils the condition of "sleeping" upright, namely $b^2n^2>2a$, that is $C^2n^2>4AMgh$, it is called a "strong top." If $b^2n^2<2a$, it is called a "weak top" (Klein u. Sommerfeld, *Theorie des Kreisels*, p. 249) It will be observed that the condition $C^2n^2>4AMgh$ can be written

$$\tfrac{1}{2}Cn^2>2\frac{A}{C}Mgh.$$

But $2Mgh$ is the potential energy which would be exhausted if the top were to fall so that the axis turned from the upward to the downward vertical. Thus the kinetic energy of rotation about the axis of figure must exceed A/C times this amount of potential energy, otherwise the top cannot sleep upright. The top sleeps more easily the greater C and the smaller A. Thus a top, tapering to the peg, will spin upright at a lower angular speed on its head than upon its peg; a bolt, with rounded end and massive head, will spin upright head down at a lower speed than when supported on the end.

As we have seen already, 17, II, the roots of the steady motion equation $(Cn - A\mu \cos\theta)\mu = Mgh$ are imaginary if $C^2n^2 - 4AMgh\cos\theta$ is negative. Hence a weak top cannot spin with its axis inclined to the upward vertical at a smaller angle than $\cos^{-1}(C^2n^2/4AMgh)$. [Of course if the couple is not supplied by gravity the equivalent for $Mgh$ must be used on these conditions.]

**15.** *Routh's graphical representation of rise and fall of top.* The rise and fall of a top may be illustrated graphically as follows (Routh, *Adv. Dynamics*, § 202). Recalling that we have $\alpha = (2E - Cn^2)/A$, $\beta = G/A$, $a = 2Mgh/A$, $b = C/A$, we consider the two quantities $A\beta/bnMh$, $A\alpha/aMh$. First we notice that if the

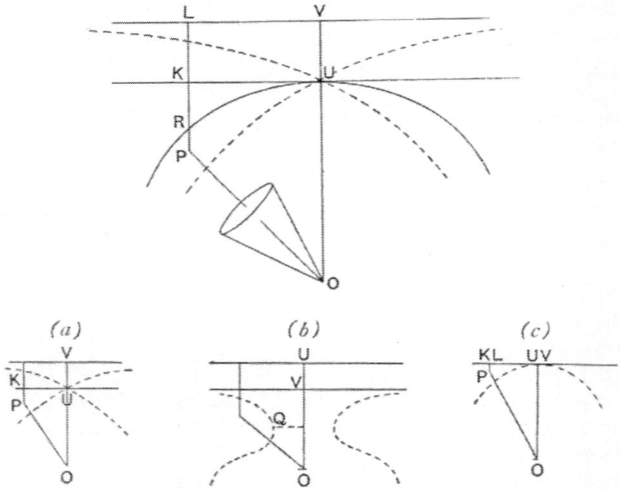


FIG. 24.—It is to be observed that the horizontal speed of P is zero when P is on the level of U, and changes sign when P passes above that level. When P is under that level the azimuthal motion, viewed from above, is in the same direction as that of the spin as viewed from beyond P

V may be called the level of no motion, and it may be either above or below U, or on the same level. These three cases, $c > 0$, $c < 0$, $c = 0$, are shown in the diagrams (*a*), (*b*), (*c*). A horizontal projection of the path of P is shown for three different subcases of each of these in Figs. 25 (1), (2), (3).

top be regarded as a compound pendulum, supported at O, $A/Mh$ is the length, $l$ say, of the equivalent simple pendulum. Hence the quantities just specified may be written $\beta l/bn$ and $\alpha l/a$. Now (Fig. 24), let distances OU, OV, equal to these quantities respectively, be laid off along the upward vertical, and through the points U, V lay two horizontal planes. Along the axis of the top lay off a distance $OP = l$, and through P draw a vertical line, intersecting in K and L the horizontal planes through U and V. We have seen that

$$\dot\psi(1 - z^2) = \beta - bnz. \quad \text{......................................(1)}$$

Hence the speed of P perpendicular to the plane ZOP is

$$l\dot\psi(1 - z^2)^{\frac{1}{2}} = \frac{\beta - bnz}{(1 - z^2)^{\frac{1}{2}}}\,l = \frac{bn}{(1 - z^2)^{\frac{1}{2}}}(OU - lz) = bnl \tan P\hat{U}K. \quad \text{........ .........(2)}$$

Thus we get the horizontal speed of P. The resultant speed is $l\{\dot{\psi}^2(1-z^2)+\dot{\theta}^2\}^{\frac{1}{2}}$, and we have $\dot{\psi}^2(1-z^2)+\dot{\theta}^2=a-az=2g(\text{OV}-lz)/l^2$, by the value assigned to OV. Thus we get

$$l^2(\dot{\psi}^2\sin^2\theta+\dot{\theta}^2)=2g(\text{OV}-l\cos\theta) \quad \ldots\ldots\ldots\ldots\ldots\ldots (3)$$

The resultant speed of P is therefore that which would be acquired by a body falling freely under gravity through a height equal to the distance of P below the plane through V.

The equations here given may be regarded as determining the whole motion. It is obvious that P cannot rise above the horizontal plane through V; for if this were to happen the right-hand side of (3) would become negative.

By (3) we can write $\quad l^2\dot{\theta}^2=2g(\text{OV}-l\cos\theta)-l^2\dot{\psi}^2\sin^2\theta,$

that is $\quad l^2\dot{\theta}^2=2g(\text{PL}-l^2\dot{\psi}^2\sin^2\theta/2g),$

or, by (2),

$$l^2\dot{\theta}^2=2g\left(\text{PL}-\frac{\text{C}^2n^2}{2\text{M}^2gh^2}\,\frac{\text{KP}^2}{\text{UK}^2}\right). \quad \ldots\ldots\ldots\ldots\ldots\ldots (4)$$

Now let a parabola be described through U as vertex, with $\text{C}^2n^2/2\text{M}^2gh^2$ as latus rectum. Then, if R be the point on the curve in which it is met by KP, we have

$$\text{UK}^2=\frac{\text{C}^2n^2}{2\text{M}^2gh^2}\text{KR}. \quad \ldots\ldots\ldots\ldots\ldots\ldots (5)$$

Thus we obtain, if R is above P,

$$l^2\dot{\theta}^2-2g\,.\,\text{KL}=2g\left(\text{KP}-\frac{\text{KP}^2}{\text{KR}}\right)=-2g\frac{\text{KP}\,.\,\text{PR}}{\text{KP}-\text{PR}}, \quad \ldots\ldots\ldots\ldots (6)$$

or

$$\frac{-2g}{l^2\dot{\theta}^2-2g\,.\,\text{KL}}=\frac{1}{\text{PR}}-\frac{1}{\text{KP}}, \quad \ldots\ldots\ldots\ldots\ldots\ldots (7)$$

where all the distances, each indicated here by two capital letters, *e.g.* PR, are taken positive.

Thus, when $\dot{\theta}$ is zero, we have

$$\frac{1}{\text{PR}}=\frac{1}{\text{KP}}+\frac{1}{\text{KL}}, \quad \ldots\ldots\ldots\ldots\ldots\ldots (8)$$

or PR is half the harmonic mean of KP and KL. There are two positions of P consistent with this relation, one above the parabola, the other below it. Between these two positions the axis of the top oscillates.

Let KP$=x$ and UK$=y$, UV$=c$, and $\text{C}^2n^2/2\text{M}^2gh^2=2pl$; then (8) becomes, by (5),

$$y^2(x+c)=2plx^2. \quad \ldots\ldots\ldots\ldots\ldots\ldots (9)$$

The inclinations of the axis to the vertical are therefore given by two positions of P on the cubic curve (9). [Case (*a*) of Fig. 24.]

It is easy to see that if $c$ be positive, that is if V be above U, the curve has, as shown by the dotted lines, a double point at U, and that there are at U two tangents equally inclined to the vertical, which include the vertical angle $2\tan^{-1}(2pl/c)^{\frac{1}{2}}$.

If $c$ is negative the origin is an isolated point on the curve, and there are two branches one on each side of the vertical. [Case (*b*) of Fig. 24.] If we find two points P on either of these, for which PR is half of the harmonic mean of KP, KL, the two lines OP give the limiting inclinations of the axis to the vertical.

If $c=0$, that is if U and V coincide, (9) represents the straight line $x=0$, and the parabola $y^2=2plx$. [Case (*c*) of Fig. 24.] The limiting inclinations of the axis to the vertical are then determined by a point on the straight line and a corresponding point on the parabola, and between the two horizontal circles defined by these inclinations the axis moves.

We give in Chapter VI a full discussion of the upright top, including Klein and Sommerfeld's method, in which a cubic curve is also used. We give here Routh's diagrams for the cases $c > 0$, $c < 0$, $c = 0$. The full theory is contained in the present chapter and in 5 ... 10, VI, below.

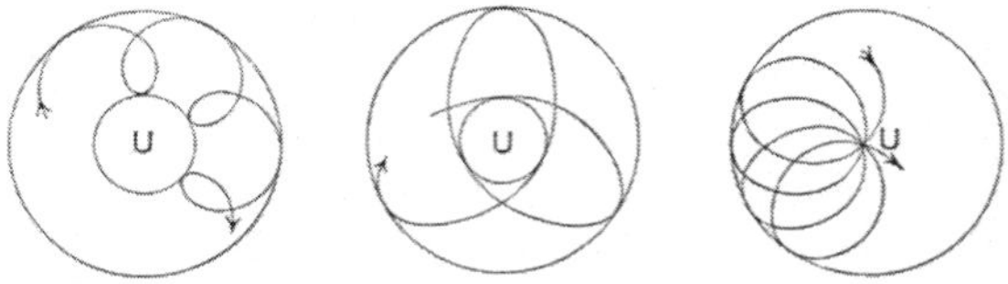


$[a' = OU]$ $l > a'$ $l < a'$ $l = a'$

$c > 0.$

FIG 25(1).—This diagram shows three cases in which $c[=(a/a - \beta/bn)l]$ is positive. The path of P on the sphere described from O with radius $l$ is shown, as seen by an eye placed above the top The small circle is the upper boundary of the motion of P; in the third case, in which $l = a$, it has shrunk to a point. The axis is vertical at U, and so the path passes continually from the lower circle through U to the lower boundary again. If the maximum deviation from the vertical is itself small in this last case, the equation of the spiral path is $\theta = \theta_0 \sin m\psi$, where $m^2 = (p-1)/p$. This is the equation obtained from (5) and (7), 16, VI, by substituting in (5), $t = 2\psi/bn$, from (7)

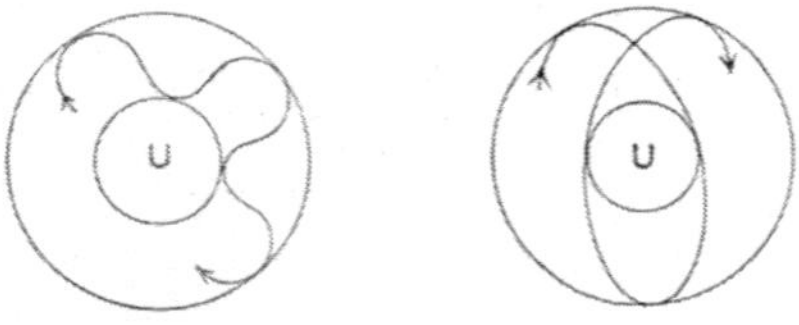


$l > OQ$ $l < OQ$

[See Fig. 24 (b).]

$c < 0.$

FIG. 25 (2).—In Fig. 24 (b) the coordinates of Q at which the tangent to the curve is vertical are given by $x = -2c$, $y^2 = -8plc$. U is an isolated point on the curve. The path of P on the same sphere as before is shown. Both limiting circles are below the level of U, and the axis of the top cannot reach the vertical.

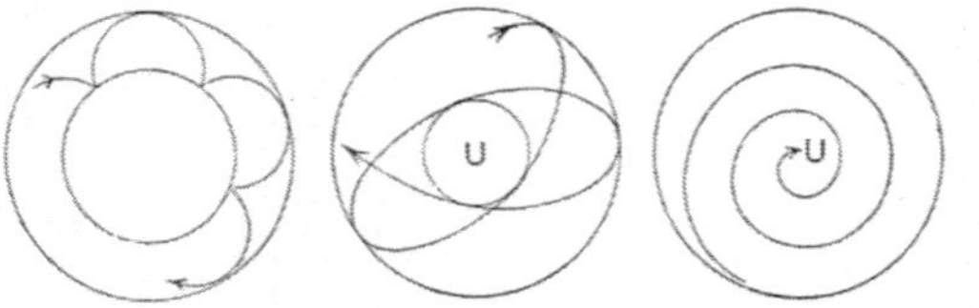


$[a' = OU]$ $l > a'$ $l < a'$ $l = a'$

$c = 0.$

FIG. 25 (3).—In the first of these cases $[l > a']$ one of the limiting circles lies in the plane of UV, and the other on the paraboloid formed by the revolution of the parabola about its axis. The path meets the plane of UV in a series of cusps.

In the second case $[l < a']$ both the limiting circles are on the paraboloid, and the path touches both circles. For either limiting circle we have, as the reader may prove, $\cos^2\theta - 2p\cos\theta = 1 - 2pa'/l$. The greatest and least values of $\theta$ [$\theta_1$ and $\theta_2$] fulfil the relation $\cos\theta_1 + \cos\theta_2 = 2p$, and $p < 1$.

In the third case $[l = a']$ the axis can approach the vertical from a lower position, arriving there after an infinite time by describing a spiral gradually closing on U. The value of $\dot\psi$ approaches the constant value $(gp/l)^{\frac{1}{2}}$. The position of the limiting circle for which $\theta$ is a maximum is given by $\cos\theta_1 = 2p - 1$. At that circle $\dot\psi = (g/lp)^{\frac{1}{2}}$, and $\dot\theta = 0$. But $\theta$ is not zero, as P, if it descends, comes down tangentially to the circle, leaves it tangentially, and begins to ascend.

**16.** *Rise and fall of top started with initial precession.* So far we have considered the top as started with $\dot\psi=0$: we now suppose that it is started at inclination $\theta_0$ of the axis to the vertical, $\dot\theta=0$, and $\dot\psi=\dot\psi_0$.

By (3) of 10 we have

$$(\alpha-az_0)(1-z_0^2)-(\beta-bnz_0)^2=0, \quad\text{.......................(1)}$$

and by (1) of the same section,

$$\dot\psi_0(1-z_0^2)=\beta-bnz_0. \quad\text{............................(2)}$$

But we have likewise

$$\dot z^2=(\alpha-az)(1-z^2)-(\beta-bnz)^2. \quad\text{.......................(3)}$$

From these, eliminating $\alpha$ and $\beta$, we get

$$\dot z^2=(z_0-z)[a(1-z^2)-b^2n^2(z_0-z)+\dot\psi_0(1-z_0^2)\{\dot\psi_0(z_0+z)-2bn\}]. \quad\text{...(4)}$$

Thus one root of $\dot z^2=0$ is $z=z_0$. The second relevant root, $z'$ say, since it makes the quantity in square brackets in (4) vanish, gives

$$z_0-z'=\frac{1}{b^2n^2}[a(1-z'^2)+\dot\psi_0(1-z_0^2)\{\dot\psi_0(z_0+z')-2bn\}]; \quad\text{........(5)}$$

and so $z_0-z'$ can be made as small as may be desired by making $n$ sufficiently great. The root $z'$ will be the smaller or the greater root, that is $z_0$ will define the upper or the lower circle, according to the initial value of $\dot\psi_0$ chosen.

To find the conditions that determine whether $z_0$ defines the upper or the lower circle, we differentiate (4) with respect to $z$, and put $z=z_0$ in the result. We obtain

$$\frac{d\dot z^2}{dz}=(1-z_0^2)\{2\dot\psi_0(bn-\dot\psi_0z_0)-a\}.$$

If $z'$ is the smaller root, that is if $z_0$ defines the upper limiting circle, we obtain, since $d\dot z^2/dz$ must be negative,

$$a+2\dot\psi_0^2z_0>2\dot\psi_0bn,$$

if $\dot\psi_0$ be positive. [Of course $bn$ is taken positive here, as always, unless otherwise stated.]

If $z'$ is the larger root, that is if $z_0$ defines the lower circle, we get, since $d\dot z^2/dz$ must be positive,

$$a+2\dot\psi_0^2z_0<2\dot\psi_0bn,$$

again for $\dot\psi_0$ positive.

We obtain the following results in different cases:

(1) $\dot\psi_0$ positive: $0<\theta_0<\frac{1}{2}\pi$.

$z_0$ defines upper or lower circle according as

$$2bn < \text{ or } > \frac{a+2\dot\psi_0^2z_0}{2\dot\psi_0},$$

that is according as

$$Cn < \text{ or } > \frac{Mgh+A\dot\psi_0^2z_0}{\dot\psi_0}.$$

(2) $\dot\psi_0$ positive: $\pi>\theta_0>\frac{1}{2}\pi$.

$z_0$ defines upper or lower circle according as

$$2bn < \text{ or } > \frac{a-2\dot\psi_0^2|z_0|}{2\dot\psi_0},$$

that is according as

$$Cn < \text{ or } > \frac{Mgh-A\dot\psi_0^2|z_0|}{\dot\psi_0}$$

(3) $\dot\psi_0$ negative: $0 < \theta_0 < \frac{1}{2}\pi$.

$z_0$ defines upper or lower circle according as

$$a + 2\dot\psi_0^2 z_0 > \text{ or } < 2\dot\psi_0 bn.$$

But since $z_0$ is positive and $\dot\psi_0$ negative the second relation is not fulfilled, and the motion must be started at the upper circle.

(4) $\dot\psi_0$ negative: $\pi > \theta_0 > \frac{1}{2}\pi$.

$z_0$ defines upper or lower circle according as

$$a + 2\dot\psi_0^2 |z_0| > \text{ or } < 2bn|\dot\psi_0|;$$

that is according as $\quad \mathrm{M}gh > \text{ or } < -\mathrm{A}\dot\psi_0^2|z_0| + \mathrm{C}n|\dot\psi_0|,$

for $n$ is here positive.

These results are very easily obtainable from (11), 2, V, the equation of motion for the axis OD. At the upper circle $\theta$ is positive, at the lower circle negative. It is instructive to work out the different cases from this equation; but this is left to the reader.

**17.** *Motion under various conditions of starting.* Some further points regarding these different cases are worthy of notice:

($a$) Take Case (1): Let the top be started at the upper circle, and $z_1$, $\dot\psi_1$ be the values of $\cos\theta$ and $\dot\psi$ for the lower circle. We have

$$a + 2\dot\psi_1^2 z_1 < 2\dot\psi_1 bn.$$

Now $|\dot\psi_1| > \dot\psi_0$ since the potential energy has been diminished in the passage to the lower circle, and therefore the kinetic energy must have been augmented by an increase in the numerical value of $\dot\psi_1$ since $\dot\theta = 0$. If $z_1 < \frac{1}{2}\pi$, the quantity on the left is positive, and therefore so also is the quantity on the right. Thus $\dot\psi_1$ is positive.

If however, with the same conditions of starting, the lower circle is in the lower hemisphere, so that $\theta_1 > \frac{1}{2}\pi$, we have $z = -|z|$, and therefore, from the equation of A.M. about the vertical, $\quad \dot\psi_1(1 - z_1^2) = \beta - bnz_1 = \beta + bn|z_1|.$

The right-hand side is positive, since $\beta$ and $bn$ are both positive by the supposition as to starting at the upper circle (in the upper hemisphere of the unit sphere), and therefore also the left-hand side is positive, that is $\dot\psi_1$ is positive. Thus $\dot\psi$ is increased in absolute value by the passage from one circle to the other, and, if positive at the upper circle, is also positive at the lower circle when that is in the lower hemisphere.

It can be seen in the same way that if the top be started at the lower circle with positive $\dot\psi$, the value of $\dot\psi$, ($\dot\psi_2$), that is, is also positive at the upper circle, but is diminished in value.

($b$) Now in Case (2), when the start is made with positive $\dot\psi_0$ at the upper circle, it is immediately clear from similar considerations to those advanced above that $\dot\psi$ is positive and of greater numerical value at the lower.

If the start is made at the lower circle, we have at the upper

$$a + 2\dot\psi^2 z > 2\dot\psi bn,$$

and also $\quad \dot\psi(1 - z^2) = \beta - bnz.$

At the start, since $z_0$ is negative and $\dot\psi_0$ positive, we have

$$\dot\psi_0(1 - z_0^2) - bn|z_0| = \beta,$$

and $\beta$ is constant. At the upper circle $\dot\psi$ is of smaller numerical value, and it is clear that if at the upper circle $z$ is still negative, we have then at that circle

$$\dot\psi(1 - z^2) - bn|z| = \beta.$$

Thus, if $\beta$ be positive, $\dot\psi(1 - z^2)$ must be positive at both circles, and exceed $bn|z|$ by the same amount at both. If $\beta$ be negative it is not impossible, according to these equations, for $\dot\psi$ to be negative at the upper circle. It must however be of smaller numerical value, since the energy term depending on $\dot{\psi}$ must be smaller there than at the lower circle.

If however (still for the same case of start at the lower circle with $\theta > \frac{1}{2}\pi$) $\theta$ at the upper circle be less than $\frac{1}{2}\pi$, then $z$ has become positive, and we have

$$a + 2\dot{\psi}^2 z > 2\dot{\psi} bn,$$

with

$$\dot{\psi}(1 - z^2) = \beta - bnz.$$

It is possible for $\dot{\psi}$ to be negative at the upper circle in this case. Taking this case with the former, we see that at least the lower circle must lie in the lower hemisphere of the unit sphere if the value of $\dot{\psi}$, positive at the lower circle, is negative at the upper.

(*c*) In Case (3) the top is started with $\dot{\psi}_0 = -|\dot{\psi}_0|$. Let the start be made at the upper circle: the condition is

$$a + 2\dot{\psi}_0^2 z_0 > 2\dot{\psi}_0 bn$$

(which of course is fulfilled, since $\dot{\psi}_0$ is negative and $z_0$ is positive). Also we have

$$\dot{\psi}_0(1 - z^2) = \beta - bnz.$$

Now at the lower circle we are to have

$$a + 2\dot{\psi}^2 z < 2\dot{\psi} bn.$$

If $z$ is still positive, that is if the lower circle is in the upper hemisphere, the left-hand side is positive, and therefore so is the right, and so $\dot{\psi}$ is positive.

If however $z$ has become negative, we have

$$\dot{\psi}(1 - z^2) = \beta + bn|z|,$$

and $\dot{\psi}$ is positive if $\beta$ is positive. If $\beta$ be negative we get no decision.

Thus, absolutely in the former case and for positive $\beta$ in the latter, the negative precession of the top is changed in the passage from the upper to the lower circle of positive precession.

(*d*) As regards starting at the lower circle [in Case (3)], the condition is

$$a + 2\dot{\psi}_0^2 z_0 < 2\dot{\psi}_0 bn.$$

Since $z_0$ is positive, the left-hand side of this inequality is positive, and so therefore must also be the right-hand side. Thus $n$ must be negative, that is, if $\dot{\phi}$ be positive, $|\dot{\psi}_0 z| > \dot{\phi}$. The condition at the upper circle is

$$a + 2\dot{\psi}^2 z > 2\dot{\psi} bn.$$

We have seen that $n$ is negative, and it is clear that so far $\dot{\psi}$ may be either positive or negative.

We have also

$$\dot{\psi}(1 - z^2) = \beta - bnz,$$

where $n$ and $\beta$ are both negative. Thus the equation can be written

$$\dot{\psi}(1 - z^2) = -|\beta| + b|n|z,$$

so that again no decision is given as to the sign of $\dot{\psi}$.

(*e*) Finally, we consider Case (4). Here $n = \dot{\phi}_0 + |\dot{\psi}_0||z_0|$, so that $n$ is positive. Hence, for starting at the upper circle, we have

$$a + 2\dot{\psi}_0^2|z_0| > 2bn|\dot{\psi}_0|.$$

At the lower circle

$$a + 2\dot{\psi}^2|z| < 2bn\dot{\psi},$$

and so $\dot{\psi}$ must be positive since $n$ is positive.

If the start is made at the lower circle we have there

$$a + 2\dot{\psi}_0^2|z| < 2bn|\dot{\psi}_0|,$$

and at the upper

$$a + 2\dot{\psi}^2|z| > 2bn|\dot{\psi}|,$$

a condition which may be fulfilled whether $\dot{\psi}$ be positive or negative.

The above are various simple observations suggested by the different cases. The associated modes of motion may be worked out by the reader in a more complete and systematic manner.

# CHAPTER VI

## FURTHER DISCUSSION OF RISE AND FALL OF A TOP WHEN INITIAL PRECESSION IS NOT ZERO

1. *More exact discussion of rise and fall of top through small range.* In this chapter we consider the rise and fall of a top through a limited range of values of $\theta$ in the more general case in which the initial precessional moment $\dot{\psi}_0$ is not zero. Putting, as before, $\theta = \theta_0 + \eta$ so that $z = z_0 - \eta \sin \theta_0$, we obtain from (4), 13, V, neglecting the term in $\eta^3$,

$$\dot{\eta}^2 = \eta \sin \theta_0 (a + 2\dot{\psi}_0^2 z_0 - 2bn\dot{\psi}_0) - \eta^2 b^2 n^2 \left(1 + \frac{\dot{\psi}_0^2(1 - z_0^2) - 2az_0}{b^2 n^2}\right). \quad \text{....(1)}$$

If we neglect $\{\dot{\psi}_0^2(1 - z_0^2) - 2az_0\}/b^2n^2$ in comparison with 1, and write $a'$ for $a + 2\dot{\psi}_0^2 z_0 - 2bn\dot{\psi}_0$, we can write this equation in the form

$$\dot{\eta}^2 = a'\eta \sin \theta_0 - b^2 n^2 \eta^2. \quad \text{..............................(2)}$$

Proceeding precisely as before, we obtain

$$\left.\begin{aligned} \eta &= \frac{a' \sin \theta_0}{2b^2 n^2}(1 - \cos bnt), \\ \text{and} \qquad \theta &= \theta_0 + \frac{a' \sin \theta_0}{2b^2 n^2}(1 - \cos bnt). \end{aligned}\right\} \quad \text{.....................(3)}$$

This value of $\theta$ satisfies the conditions that when $t = 0$, $\theta = \theta_0$ and $\dot{\theta} = 0$. The other limiting value of $\theta$, $\theta_1$ say, is given by

$$\theta_1 = \theta_0 + \frac{a' \sin \theta_0}{b^2 n^2}. \quad \text{..................................(4)}$$

It remains to determine $\dot{\psi}$ at any time $t$. We have

$$bnz_0 + \dot{\psi}_0(1 - z_0^2) = \beta, \quad \text{..............................(5)}$$

and therefore $\qquad bnz + \dot{\psi}(1 - z^2) = \beta = \dot{\psi}_0(1 - z_0^2) + bnz_0,$

so that $\qquad \dot{\psi} = \dfrac{bn(z_0 - z) + \dot{\psi}_0(1 - z_0^2)}{1 - z^2}. \quad \text{..........................(6)}$

But $z = z_0 - \eta \sin \theta_0$, and therefore

$$\dot{\psi} = bn \frac{\eta}{\sin \theta_0} + \dot{\psi}_0(1 - 2\eta \cot \theta_0), \quad \text{.......................(7)}$$

approximately. By the first of (3) this becomes

$$\dot{\psi} = (bn - 2\dot{\psi}_0 \cos\theta_0)\frac{a'}{2b^2n^2}(1-\cos bnt) + \dot{\psi}_0. \quad\text{..............(8)}$$

This gives $\dot{\psi} = \dot{\psi}_0$ when $t=0$, and $\dot{\psi} = \dot{\psi}_1$, where

$$\dot{\psi}_1 = \frac{a'}{bn} + \dot{\psi}_0\left(1 - \frac{2a'\cos\theta_0}{b^2n^2}\right), \quad\text{..........................(9)}$$

when $t = \pi/bn$ and $\theta = \theta_1$.

It will be observed that, when $\theta = \frac{1}{2}(\theta_1 + \theta_0)$, $\dot{\theta}$ has its maximum value $a'\sin\theta_0/2bn$, and that then $\dot{\psi} = \frac{1}{2}(\dot{\psi}_0 + \dot{\psi}_1)$.

2. *Determination of period of rise and fall.* The method used above is instructive. But we can also proceed as follows, and arrive more quickly at some of the results already obtained, and at the same time estimate the error involved in the approximation. As in (3), 10, V, we have

$$\dot{z}^2 = a(z - z_1)(z_2 - z)(z_3 - z);$$

and now $z_1$, $z_2$ are the nearly equal roots of this equation which define the limiting circles of nutation of the top, and $z_1$ is the smaller, that defining the lower circle. The root $z_3$ is greater than 1. Differentiating we obtain

$$\ddot{z} = a(z_m - z)(z_3 - z) - \tfrac{1}{2}a(z - z_1)(z_2 - z), \quad\text{................(1)}$$

where $z_m = \frac{1}{2}(z_1 + z_2)$. From this, putting $\zeta = z - z_m$, we get

$$\ddot{\zeta} + a(z_3 - z_m)\zeta = \tfrac{3}{2}a\zeta^2 - \tfrac{1}{8}a(z_2 - z_1)^2. \quad\text{....................(2)}$$

With $r = z_2 - z_1$, and $m^2 = a(z_3 - z_m)$, (2) becomes

$$\ddot{\zeta} + m^2\left(\zeta - \frac{a}{8m^2}r^2\right) = \tfrac{3}{2}a\zeta^2. \quad\text{.........................(2')}$$

Neglecting the term on the right, which is small, we get the solution

$$\zeta - \frac{a}{8m^2}r^2 = \mathrm{A}\sin mt + \mathrm{B}\cos mt$$

If A, B are determined so that $z = z_2$ when $t=0$, and $\dot{z} = 0$ when $z = z_2$ and when $z = z_1$, this solution becomes

$$\zeta - \frac{a}{8m^2}r^2 = \tfrac{1}{2}r\left(1 - \frac{a}{4m^2}r\right)\cos mt. \quad\text{......................(3)}$$

The period is $2\pi/m = 2\pi/\{a(z_3 - z_m)\}^{\frac{1}{2}}$.

By substituting the value of $\zeta$ from (3) in the term on the right of (2) we might proceed to a second approximation, but for our present purpose this is not worth while. The motion of the axis can be worked out by means of elliptic functions, or by various practical processes of quadrature, as exactly as may be required in any particular case.

To compare this result with those already obtained, we have to find a value for $z_3 - z_m$. Now from the value of $\dot{z}^2$ in (4), 16, V, we obtain, by putting $z_2$ for $z_0$ and $\dot{\psi}_2$ for $\dot{\psi}_0$,

$$a(z - z_1)(z_3 - z) = a(1 - z^2) - b^2n^2(z_2 - z) + \dot{\psi}_2(1 - z_2^2)\{\dot{\psi}_2(z_2 + z) - 2bn\}. \quad\text{...(4)}$$

On the left the coefficient of $z$ is $a(z_3+z_1)$, and on the right it is

$$b^2n^2+\dot{\psi}_2^2(1-z_2^2).$$

Hence we get $$a(z_3+z_1)=b^2n^2+\dot{\psi}_2^2(1-z_2^2),$$

or $$a(z_3-z_m)=b^2n^2+\dot{\psi}_2^2(1-z_2^2)-a(z_m+z_1). \qquad (5)$$

This can be written in the alternative form

$$a(z_3-z_m)=\frac{b^2n^2+\beta^2-2\beta bnz_2}{1-z_2^2}-a(z_m+z_1). \qquad (6)$$

Thus, since $m^2=a(z_3-z_m)$, we get for the period T of oscillation the equation

$$T=\frac{2\pi(1-z_2^2)^{\frac{1}{2}}}{\{b^2n^2+\beta^2-2\beta bnz_2-a(z_1+z_m)(1-z_2^2)\}^{\frac{1}{2}}}. \qquad (7)$$

If we suppose that $z_1=z_2=z_m$, and take the value of $\dot{\psi}$ proper to $z_m$, we can put this equation in the form

$$T=\frac{2\pi A}{(C^2n^2+A^2\dot{\psi}^2\sin^2\theta_m-2aA^2\cos\theta_m)^{\frac{1}{2}}}. \qquad (8)$$

Here we suppose the motion to deviate infinitely little from being steady, that is from fulfilment of the conditions $\dot{\theta}=0$ and $\dot{\psi}=\mu$, a constant. We have by the equation of steady motion $Cn=(Mgh+A\mu^2\cos\theta)/\mu$. Substituting this value of $Cn$ in the expression for the period, we obtain finally

$$T=\frac{2\pi A\mu}{(M^2g^2h^2+A^2\mu^4-2MghA\mu^2\cos\theta)^{\frac{1}{2}}}, \qquad (9)$$

which is the result arrived at in 9, V, above.

Thus, if we lay off two vectors, one of length $Mgh$ along the vertical downwards from the fixed point O, the other of length $A\mu^2$ from O along the axis of the top, the period is the product of $2\pi$ into the ratio of the length of the latter vector to that of the resultant of the two. The length of the equivalent simple pendulum is $g/a(z_3-z_m)$ for the first approximation made, and $A^2\mu^2g/(M^2g^2h^2+A^2\mu^4-2MghA\mu^2\cos\theta)$ for the last.

3. *Error involved in approximations to motion.* By the following process, which was used by Puiseux in his discussion of the similar problem of the motion of a spherical pendulum (see Chap. XV), we can find limits for the error of the result obtained in (3). We have $$\dot{z}=a^{\frac{1}{2}}\{(z-z_1)(z_2-z)(z_3-z)\}^{\frac{1}{2}}, \qquad (1)$$

and as in what follows motion from the lower to the upper circle is alone considered, the radical must be supposed taken with the positive sign. Thus, if $t$ be the time from the lower circle to any position $z$ between the two limiting circles, we have

$$a^{\frac{1}{2}}t=\int_{z_1}^{z}\frac{dz}{\{(z-z_1)(z_2-z)(z_3-z)\}^{\frac{1}{2}}}, \qquad (2)$$

where $z$ lies between the roots $z_1$ and $z_2$. Now, clearly, according as we substitute $z_1$ or $z_2$ for $z$ in the third factor $z_3-z$ of the quantity under the square root sign in the denominator of the integrand, the integral is diminished or increased; and so we get

$$\frac{1}{(z_3-z_1)^{\frac{1}{2}}}\int_{z_1}^{z}\frac{dz}{\{(z-z_1)(z_2-z)\}^{\frac{1}{2}}}<a^{\frac{1}{2}}t<\frac{1}{(z_3-z_2)^{\frac{1}{2}}}\int_{z_1}^{z}\frac{dz}{\{(z-z_1)(z_2-z)\}^{\frac{1}{2}}}. \qquad (3)$$

The integral can be evaluated. Putting $z-z_1=u'+u$, $z_2-z=u'-u$, that is take a new variable $u$ reckoned from the mean circle $z_m$, we obtain

$$\int_{z_1}^{z}\frac{dz}{\{(z-z_1)(z_2-z)\}^{\frac{1}{2}}}=\int_{-u'}^{u}\frac{du}{(u'^2-u^2)^{\frac{1}{2}}}=\tfrac{1}{2}\pi+\sin^{-1}\frac{u}{u'}. \quad\text{.......(4)}$$

We have therefore

$$\frac{1}{\{a(z_3-z_1)\}^{\frac{1}{2}}}\left(\tfrac{1}{2}\pi+\sin^{-1}\frac{u}{u'}\right)<t<\frac{1}{\{a(z_3-z_2)\}^{\frac{1}{2}}}\left(\tfrac{1}{2}\pi+\sin^{-1}\frac{u}{u'}\right). \quad\text{.......(5)}$$

Writing $z_m$ for $z_1$ and $z_2$ we get approximately

$$\{a(z_3-z_m)\}^{\frac{1}{2}}t=\tfrac{1}{2}\pi+\sin^{-1}\frac{u}{u'}. \quad\text{.......(6)}$$

Writing now $t_0$ for $\frac{1}{2}\pi/\{a(z_3-z_m)\}^{\frac{1}{2}}$ and putting $t$ for $t-t_0$, we obtain

$$\left.\begin{aligned} u&=u'\sin[\{a(z_3-z_m)\}^{\frac{1}{2}}t],\\ \text{or}\quad z-z_m&=\tfrac{1}{2}(z_2-z_1)\sin[\{a(z_3-z_m)\}^{\frac{1}{2}}t].\end{aligned}\right\} \quad\text{.......(7)}$$

But $z-z_m$ is what we have called $\zeta$ in 2, VI, above; and thus we have found in an entirely different way an equation practically equivalent to (3) of that section. All the necessary conclusions as to period have already been drawn.

If now $|\tau|$ be the numerical value of the error involved in the approximation, made above, to the time of passage from the lower circle to any value of $z$ arrived at before the upper circle is reached, the ratio of the error to the whole time is $|\tau/t|$, and we obviously have

$$\left|\frac{\tau}{t}\right|<\frac{\dfrac{1}{(z_3-z_2)^{\frac{1}{2}}}-\dfrac{1}{(z_3-z_1)^{\frac{1}{2}}}}{\dfrac{1}{(z_3-z_1)^{\frac{1}{2}}}},$$

that is

$$\left|\frac{\tau}{t}\right|<\left(\frac{z_3-z_1}{z_3-z_2}\right)^{\frac{1}{2}}-1. \quad\text{.......(8)}$$

Thus, if $f$ denote a proper fraction, so far undetermined, we have

$$\tau=\pm f\left\{\left(\frac{z_3-z_1}{z_3-z_2}\right)^{\frac{1}{2}}-1\right\}t, \quad\text{.......(9)}$$

and therefore in the equations above we must replace $t$ by

$$t\left[1\pm f\left\{\left(\frac{z_3-z_1}{z_3-z_2}\right)^{\frac{1}{2}}-1\right\}\right].$$

In the discussion which will be given in a later chapter, of the motion of the axis of a top, the quantity $\{(z_3-z_1)/(z_3-z_2)\}^{\frac{1}{2}}$ will again appear, and we shall then, by means of elliptic functions, give equations for the determination of the motion to any required degree of exactness. Disregard of the error here exhibited enables the inversion of the function $\{a(z_3-z_m)\}^{\frac{1}{2}}t$, in order to give $z$, to be carried out as an affair of circular functions.

We can now find $\dot\psi$. Let, as before, $\dot\psi_m$ be the value of $\dot\psi$ for $z=z_m$. Then

$$bnz_m+\dot\psi_m(1-z_m^2)=bnz+\dot\psi(1-z^2)=\beta, \quad\text{.......(10)}$$

and we get by subtraction

$$\dot\psi=\frac{\dot\psi_m(1-z_m^2)-bnu}{1-z^2}. \quad\text{.......(11)}$$

Now, since $z=z_m+u$, we obtain

$$\begin{aligned}\frac{1}{1-z^2}&=\frac{1}{1-z_m^2}-\frac{1}{1-z_m^2}+\frac{1}{1-z^2}=\frac{1}{1-z_m^2}+\frac{2z_mu+u^2}{1-z_m^2}\,\frac{1}{1-z^2}\\ &=\frac{1}{1-z_m^2}+\frac{2z_mu+u^2}{1-z_m^2}\left(\frac{1}{1-z_m^2}+\frac{2z_mu+u^2}{1-z_m^2}\,\frac{1}{1-z^2}\right)\\ &=\frac{1}{1-z_m^2}+\frac{2z_mu+u^2}{(1-z_m^2)^2}+\frac{(2z_mu+u^2)^2}{(1-z_m^2)^2}\,\frac{1}{1-z^2}. \quad\text{.......(12)}\end{aligned}$$

The process of expansion still further is obvious and easy. For the present we neglect terms in $\dot{\psi}$ which involve the second and higher powers of $u$. Thus we get approximately, by substituting from (7) above for $u$,

$$\dot{\psi}=\dot{\psi}_m+\frac{2\dot{\psi}_m z_m-bn}{1-z_m^2}u'\sin\left[\{a(z_3-z_m)\}^{\frac{1}{2}}t\right]. \quad\text{.......................(13)}$$

This equation may also be written in the form, easily derived from (10),

$$\dot{\psi}=\frac{\beta-bnz_m}{1-z_m^2}+\frac{2\beta z_m-bn(1+z_m^2)}{(1-z_m^2)^2}u'\sin\left[\{a(z_3-z_m)\}^{\frac{1}{2}}t\right]. \quad\text{.................(14)}$$

Both (13) and (14) may be corrected by inserting the corrected value of $u$, both in the terms here included and in any further terms which have to be added from (12) to give a more exact value of $\dot{\psi}$.

Integrating (13) and (14) we get an approximation to the value of the angle turned through by the top about the vertical in time $t$, measured from the instant at which the axis is in the mean circle. Thus

$$\left.\begin{aligned}\psi&=\dot{\psi}_m t-\frac{2\dot{\psi}_m z_m-bn}{p(1-z_m^2)}u'\cos pt,\\ \text{or}\qquad \psi&=\frac{\beta-bnz_m}{1-z_m^2}t-\frac{2\beta z_m-bn(1+z_m^2)}{p(1-z_m^2)^2}u'\cos pt,\end{aligned}\right\}\quad\text{...........................(15)}$$

where $p=\{a(z_3-z_m)\}^{\frac{1}{2}}$.

For the passage from the lower to the upper circle the angle $\psi$ is given by

$$\psi=\frac{1}{p}\left\{\frac{\beta-bnz_m}{1-z_m^2}\pi+\frac{2\beta z_m-bn(1+z_m^2)}{(1-z_m^2)^2}u'\right\}\quad\text{............................(16)}$$

4. *Theory of the upright or "sleeping" top.* When the axis of the top is at rest in coincidence with the upward vertical there is no distinction between the angular speed of rotation $n$ and the angular speed $\dot{\psi}$ about the vertical. The kinetic energy is then $\frac{1}{2}\mathrm{C}n^2$, and the potential energy $\mathrm{M}gh$. The total energy E is thus $\frac{1}{2}\mathrm{C}n^2+\mathrm{M}gh$. This is the case in which the upper and lower limiting circles for a point P on the axis of the top have shrunk to a single point, the zenith of the sphere on which P is always situated, the case in fact of steady motion in the circle $z=1$, of zero radius.

Let now, when the axis is thus at rest in the vertical position, a horizontal impulse be given to the top so that the axis is set suddenly in motion with angular speed $\dot{\theta}_0$: the kinetic energy is increased by the term $\frac{1}{2}\mathrm{A}\dot{\theta}_0^2$. Then the energy equation becomes

$$\mathrm{E}=\tfrac{1}{2}(\mathrm{A}\dot{\theta}_0^2+\mathrm{C}n^2)+\mathrm{M}gh. \quad\text{...........................(1)}$$

Also the A.M. about the vertical is $\mathrm{C}n$, so that when the axis ceases to be coincident with the vertical we have $\beta=\mathrm{C}n/\mathrm{A}=bn$.

When the axis of the top has moved away from the vertical, the term $\frac{1}{2}\mathrm{A}\dot{\psi}^2(1-z^2)$ appears in the kinetic energy, and as the kinetic energy of spin remains unaltered we must have

$$\tfrac{1}{2}\mathrm{A}\dot{\psi}^2(1-z^2)+\tfrac{1}{2}\mathrm{A}\dot{\theta}^2=\tfrac{1}{2}\mathrm{A}\dot{\theta}_0^2+\mathrm{M}gh(1-z) \quad\text{..................(2)}$$

But in the present case $\beta=bn$, and so

$$\dot{\psi}(1-z^2)=\beta-bnz=bn(1-z).$$

Thus (2) becomes $$b^2n^2\frac{1-z}{1+z}+\dot\theta^2=\dot\theta_0{}^2+a(1-z), \qquad \text{.........................(3)}$$

since $a=2Mgh/A$. Hence, if $z$ define the limiting circle to which the axis of the top passes when disturbed from the vertical, we have $\dot\theta^2=0$, and so from (3) $$\{\dot\theta_0{}^2+a(1-z)\}(1+z)-b^2n^2(1-z)=0, \qquad \text{......................(4)}$$

or, writing $y$ for $\dot\theta_0$,

$$y^2(1+z)+\{a(1+z)-b^2n^2\}(1-z)=0. \qquad \text{.....................(4')}$$

5. *Stability of upright top. Graphical representation.* We can lay down a graph of this equation with coordinates $y$, $z$. The graph (Fig. 26)

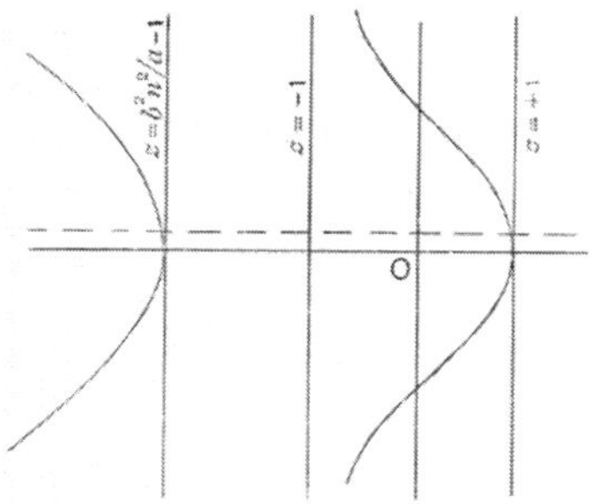


$a < 0$

FIG. 26 (1).

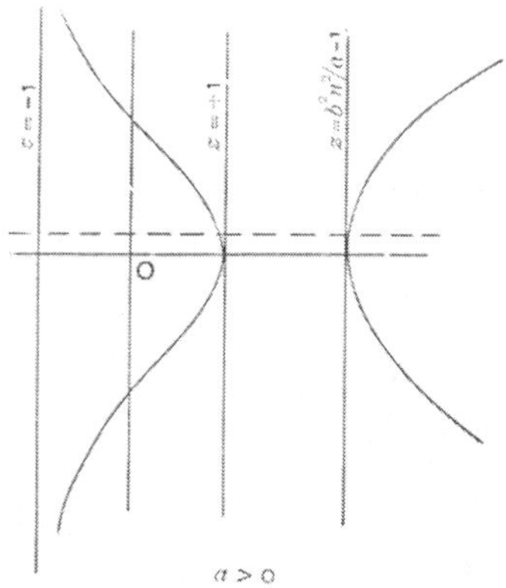


$a > 0$

FIG. 26 (2).

is a cubic curve, of which we can trace the chief properties as follows. Putting $z=1$ we get $$y^2(1+z)=0,$$

so that for this value of $z$ there is a vertical tangent which touches the curve at the point $z=1$, $y=0$. If in (4') we put $y=0$ and reject the factor $1-z$ we get $$a(1+z)-b^2n^2=0,$$

so that there is a vertical tangent which crosses the line $y=0$ at the point $z=-1+b^2n^2/a$.

Besides these there is a vertical asymptote, the equation of which is $z=-1$.

We can now draw the graphs for different cases. First for $a$ negative, that is the case in which the action of the gravity couple is to bring the axis of the top back towards the vertical, when caused to deviate from it, the right-hand or determining branch of the curve extends from the tangent $z=1$ to $z=-1$ [Fig. 26 (1)]. If we draw a line parallel to the axis of $z$, at distance $\dot\theta_0$ from that axis, it meets the curve in a point, the

$z$ of which is very nearly unity, unless $\dot{\theta}_0$ is great. Thus, if $\dot{\theta}_0$ is small, the limiting circle is very near the zenith of the unit sphere, and we see that the axis of figure does not deviate far from the vertical. The dependence of the deviation on the value of $\dot{\theta}_0$ or $y$ is very clearly shown by the graph.

Next we take the case of $a$ positive. Here the result will depend on whether the spin is so great or so small as to make $-1+b^2n^2/a > 1$ or $< 1$. In the first case the vertical tangent corresponding to the value $-1+b^2n^2/a$ of $z$ lies beyond the line $z=1$, and as shown in Fig. 26 (2) the determining branch of the curve extends from the line $z=1$ to the line $z=-1$. The line $y=\dot{\theta}_0$, for $\dot{\theta}_0$ small, therefore cuts the branch at a point the $z$ of which is very nearly unity. This is the case of what has been called in 14, V, above a "strong top."

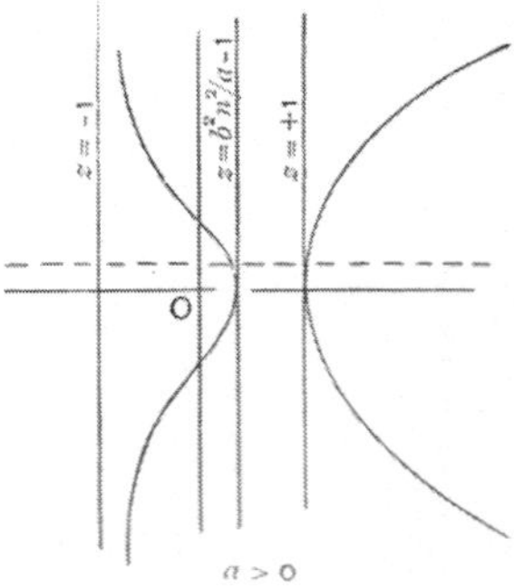

Fig. 26 (3).

In the other case [of $-1+b^2n^2/a < 1$, with $a > 0$] the determining branch of the graph lies between $z=-1$ and $z=b^2n^2/a-1$, and so the line $y=\dot{\theta}_0$ cuts the determining branch at a point the $z$ of which, for a small value of $\dot{\theta}_0$, is only very slightly different from $b^2n^2/a-1$; and there is no possibility of intersection between this value of $z$ and $z=1$. Thus, on the slightest disturbance, the axis of figure passes from the position $z=1$ to a limiting circle at the finite distance $2-b^2n^2/a$ from the point $z=1$. This is the case of the so-called "weak top." [The relative diagram is Fig. 26 (3).]

6. *Analytical discussion of stability of upright top.* The results stated above as to the stability of the upright top may be established very easily in the following way. Let $1-z=x$; then $z=1-x$ and $1+z=2-x$. Hence (4′), 4, becomes

$$ax^2+(b^2n^2-2a+y^2)x-2y^2=0, \qquad \text{.........................(1)}$$

a quadratic in $x$. The roots are real if $(b^2n^2-2a+y^2)^2 > -8ay^2$. This condition is manifestly fulfilled if $a > 0$, and also when $a < 0$ if $b^2n^2-2a$ be not zero, for we can make $y^2$ as small as we please. Solving the equation (1) we get

$$x=-\frac{1}{2}\left(\frac{b^2n^2}{a}-2+\frac{y^2}{a}\right)\pm\frac{1}{2}\left(\frac{b^2n^2}{a}-2+\frac{y^2}{a}\right)\left\{1+\frac{8y^2}{a\left(\frac{b^2n^2}{a}-2+\frac{y^2}{a}\right)^2}\right\}^{\frac{1}{2}}.$$

Now, since $z$ cannot exceed 1, $x$ is essentially positive, and so (when the roots have opposite signs) we have only to consider the positive root. It will be shown later that, in all the cases considered here, the axis comes back periodically to the vertical, in each case in a finite time, even in the case described below as unstable, provided that in that case the disturbance $\dot{\theta}_0$ is not zero. We now take the different cases *seriatim*.

(1) *a is negative.* Here both roots of the quadratic are positive, since the sum and product of the roots are both positive. Denoting $b^2n^2/a-2$ by $-m^2$ we get

$$x=\frac{1}{2}\left(m^2-\frac{y^2}{a}\right)-\frac{1}{2}\left(m^2-\frac{y^2}{a}\right)\left\{1+\frac{8y^2}{a\left(m^2-\frac{y^2}{a}\right)^2}\right\}^{\frac{1}{2}},$$

for we must take the smaller root, inasmuch as the larger far exceeds the diameter of the unit sphere. It is clear from this that $x$ can be made as small as we please by taking $y$ sufficiently small. The motion is therefore stable for the upright position of the top.

(2) *a is positive.* We have here two sub-cases: ($\alpha$) $b^2n^2/a-1>1$, or $b^2n^2/a-2>0$, and ($\beta$) $b^2n^2/a-1<1$, or $b^2n^2/a-2<0$. In the former we put $b^2n^2/a-2=m^2$. Then the equation has a positive and a negative root. The former is

$$x=-\frac{1}{2}\left(m^2+\frac{y^2}{a}\right)+\frac{1}{2}\left(m^2+\frac{y^2}{a}\right)\left\{1+\frac{8y^2}{a\left(m^2+\frac{y^2}{a}\right)^2}\right\}^{\frac{1}{2}};$$

and so, by taking $y$ sufficiently small, we can make $x$ as small as we please. The stability however is not so decided as in the former case, for in that $m^2-y^2/a$, which enters into the denominator of the term $8y^2/a(m^2-y^2/a)^2$, is greater in value, since $|m^2|=|b^2n^2/a|+2$ as against $|m^2|=|b^2n^2/a-2|$, for $a$ positive.

In the sub-case ($\beta$) we make $b^2n^2/a-2=-m^2$. Here the roots have opposite signs (since $a$ is positive). Hence we have

$$x=\frac{1}{2}\left(m^2-\frac{y^2}{a}\right)+\frac{1}{2}\left(m^2-\frac{y^2}{a}\right)\left\{1+\frac{8y^2}{a\left(m^2-\frac{y^2}{a}\right)^2}\right\}^{\frac{1}{2}}.$$

For $y^2$ *however small* this gives $x=m^2$. Thus the smallest disturbance of the top from the vertical gives a limiting circle at the approximate distance $2-b^2n^2/a$ from the upper pole ($z=1$) of the unit sphere.

This last case is of great importance as affording a criterion of distinction, for the motion of a top at least, between stability and instability. The value of $x$ may be small in the unstable case, but its smallness does not depend on the smallness of the disturbance, but on the excess of zero over $b^2n^2-2a$.

In the transition case, of $b^2n^2=2a$, we have

$$x=-\frac{y^2}{2a}\pm\frac{y^2}{2a}(8a+y^2)^{\frac{1}{2}},$$

and, clearly, in this case we must regard the motion as stable, since $x$ can be made as small as we please by making the disturbance sufficiently small.

The transition from stability is gradual, since, as $b^2n^2/a$ is diminished from equality with 2, the distance $2-b^2n^2/a$ of the limiting circle gradually increases.

7. *Time of passage of weak top from limiting circle to any value of z.* In the case of $b^2n^2/a-1<1$ [$a$ positive], the weak top, we can calculate the time of passage of the axis of the top from the limiting circle for the value zero of $\dot\theta_0$. This will be sufficient, as the top being unstable in the upright position the slightest disturbance will be enough to set it in motion towards a lower position of the axis. We get easily from (3) (4) for the value of $\dot z^2$ the equation

$$\dot z^2=\{a(1+z)-b^2n^2\}(1-z)^2. \qquad (1)$$

Now one root of the equation $\{a(1+z)-b^2n^2\}(1-z)^2=0$

is

$$z=\frac{b^2n^2}{a}-1,$$

and the other two roots are both equal to 1. We call these roots $z_1$, $z_2$, $z_3$. The large root of the cubic coalesces with $z_3$, which is 1. Thus we can write the equation for $\dot z^2$ as

$$\dot z^2=a(z-z_1)(1-z)^2;$$

and therefore

$$a^{\frac12}dt=\frac{dz}{(z-z_1)^{\frac12}(1-z)}. \qquad (2)$$

To integrate this we put $(z-z_1)^{\frac12}=u$, and obtain, after reduction,

$$a^{\frac12}dt=\frac{2du}{1-z_1-u^2}. \qquad (3)$$

Thus we find by performance of the integration

$$a^{\frac12}t=\frac{1}{(1-z_1)^{\frac12}}\log\frac{(1-z_1)^{\frac12}+(z-z_1)^{\frac12}}{(1-z_1)^{\frac12}-(z-z_1)^{\frac12}}. \qquad (4)$$

No constant of integration is required if $t$ be reckoned from the instant at which the axis is at the limiting circle $z_1=b^2n^2/a-1$.

Thus we have the result that the time taken by the top to pass from the position given by the limiting circle to the upright position is infinite. Similarly, if the top start from the upright position on account of a vanishingly small disturbance $\dot\theta_0$, the time of passage to the limiting circle is exceedingly great.

8. *Calculation of azimuthal motion for weak top.* We can also calculate the angle $\psi$ turned through in the present case in any time $t$ reckoned from the same zero of time. We have $\dot\psi=bn/(1+z)$. Hence, by the value of $\dot z$ obtained above,

$$d\psi=\frac{bn}{a^{\frac12}}\frac{dz}{(z-z_1)^{\frac12}(1-z^2)}. \qquad (1)$$

By the same transformation as before we get

$$d\psi=\frac{bn}{a^{\frac12}}\frac{2du}{1-(u^2+z_1)^2},$$

or

$$d\psi=\frac{bn}{a^{\frac12}}\left(\frac{du}{1-z_1-u^2}+\frac{du}{1+z_1+u^2}\right). \qquad (2)$$

Integrating, we find

$$\psi=\frac12\frac{bn}{(2a-b^2n^2)^{\frac12}}\log\frac{(1-z_1)^{\frac12}+(z-z_1)^{\frac12}}{(1-z_1)^{\frac12}-(z-z_1)^{\frac12}}+\tan^{-1}\left(\frac{z-z_1}{1+z_1}\right)^{\frac12}. \qquad (3)$$

As time passes and the axis approaches the upright position (without ever reaching it) the value of $\psi$ increases indefinitely, that is the axis describes on the unit sphere an ever narrowing spiral, in consequence of the first term on the right in the value of $\psi$. The term $\tan^{-1}\{(z-z_1)/(1+z_1)\}^{\frac{1}{2}}$ remains finite. If the motion starts from the limiting circle in the opposite direction, an exactly similar spiral (turned however the opposite way) is described. The two branches form a curve symmetrical about the meridian through the starting point.

**9.** *Periodic motion of weak top. If $\dot\theta_0$ is not zero*, there is in this (unstable) case of $a$ positive and $b^2n^2/a-1<1$, periodic return of the axis of the top to the vertical position. In fact the path described by the axis on the unit sphere does not differ very much from that described in the case of stability. We can find the motion to a high degree of approximation by the process already employed. We have now

$$\dot\theta^2(1-z^2)=\dot z^2=[\dot\theta_0^2(1+z)+\{a(1+z)-b^2n^2\}(1-z)](1-z). \quad \text{.................(1)}$$

One root of the equation $\dot z^2=0$ is $z_2=1$, and another is $z_1=b^2n^2/a-1+\epsilon$, and a third is $z_3$. Hence, of course,
$$\dot z^2=a(z-z_1)(1-z)(z_3-z),$$

and
$$dt=\frac{dz}{a^{\frac{1}{2}}\{(z-z_1)(1-z)(z_3-z)\}^{\frac{1}{2}}}. \quad \text{.......(2)}$$

Clearly, if we integrate from $z=z_1$ to any value of $z$ between 1 and $z_1$, we obtain

$$\frac{1}{a^{\frac{1}{2}}(z_3-z_1)^{\frac{1}{2}}}\int_{z_1}^{z}\frac{dz}{\{(z-z_1)(1-z)\}^{\frac{1}{2}}}<t<\frac{1}{a^{\frac{1}{2}}(z_3-1)^{\frac{1}{2}}}\int_{z_1}^{z}\frac{dz}{\{(z-z_1)(1-z)\}^{\frac{1}{2}}}. \quad \text{...............(3)}$$

We take a new variable $u$, such that $z-z_1=u'+u$, $1-z=u'-u$, that is take a new variable $u$ reckoned from the circle midway between the pole $z=1$ and the circle $z=z_1$. Now we have, as in 3,

$$\int_{z_1}^{z}\frac{dz}{\{(z-z_1)(1-z)\}^{\frac{1}{2}}}=\int_{-u'}^{u}\frac{du}{(u'^2-u^2)^{\frac{1}{2}}}=\tfrac{1}{2}\pi+\sin^{-1}\frac{u}{u'}. \quad \text{.....(4)}$$

Thus the inequality written above becomes

$$\frac{1}{\{a(z_3-z_1)\}^{\frac{1}{2}}}\left(\tfrac{1}{2}\pi+\sin^{-1}\frac{u}{u'}\right)<t<\frac{1}{\{a(z_3-1)\}^{\frac{1}{2}}}\left(\tfrac{1}{2}\pi+\sin^{-1}\frac{u}{u'}\right). \quad \text{...............(5)}$$

The approximate value of $t$ reckoned from the instant at which the axis intersects the mean circle [where $z=\frac{1}{2}(1+z_1)=z_m$, say,]

$$t=\frac{1}{\{a(z_3-z_m)\}^{\frac{1}{2}}}\sin^{-1}\frac{u}{u'}. \quad \text{.......................................(6)}$$

The time from $u=-u'$ to $u=u'$ (the half period) is thus $\pi/\{a(z_3-z_m)\}^{\frac{1}{2}}$.

If we write now $m=\{a(z_3-z_m)\}^{\frac{1}{2}}$ we get $u=u'\sin mt$, or

$$z=z_m+u'\sin mt, \quad \text{.....(7)}$$

so that $z$ differs from the mean value $z_m$ by a time-periodic term which passes through all its values in the period $2\pi/\{a(z_3-z_m)\}^{\frac{1}{2}}$, the time of passage from the pole to the limiting circle and back.

**10.** *Determination of azimuthal displacement for periodic motion of weak top.* To find $\dot\psi$ we have, since $\beta=bn$,

$$\dot\psi=bn\frac{1-z}{1-z^2}=\frac{bn}{1+z}. \quad \text{.......................................(1)}$$

But
$$\frac{1}{1+z}=\frac{1}{1+z_m}+\frac{1}{1+z}-\frac{1}{1+z_m}=\frac{1}{1+z_m}-\frac{u}{(1+z_m)^2}+\frac{u^2}{(1+z_m)^2}\frac{1}{1+z}. \quad \text{.................(2)}$$

This expression may be carried as far as may be desired by substitution for $1/(1+z)$ on the right of the expression on the right, and so on. It is sufficient at present to use the first two terms on the right of (2). Thus we get

$$\dot{\psi}=\frac{bn}{1+z_m}\left(1-\frac{u}{1+z_m}\right)=\frac{bn}{1+z_m}\left(1-\frac{u'}{1+z_m}\sin mt\right); \quad \ldots (3)$$

and therefore

$$\psi=\frac{bn}{1+z_m}\left\{t+\frac{1}{m(1+z_m)}u'\cos mt\right\}+\psi_0. \quad \ldots (4)$$

The time $t$ is here reckoned from the instant at which the axis is at the mean circle $z_m$. Hence, at $t=0$, we have

$$\psi=\frac{bnu'}{m(1+z_m)^2}+\psi_0. \quad \ldots (5)$$

After half a period, that is $\pi/m$, the axis is again at the mean circle, and

$$\psi=\frac{bn}{m}\frac{1}{1+z_m}\left(\pi-\frac{u'}{1+z_m}\right)+\psi_0. \quad \ldots (6)$$

In this interval the axis has risen to the vertical and fallen again to the circle $z_m$.

After another half period, during which the axis falls to the limiting circle and rises again to the circle $z_m$, the value of $\psi$ is given by

$$\psi=\frac{bn}{m}\frac{1}{1+z_m}\left(2\pi+\frac{u'}{1+z_m}\right)+\psi_0. \quad \ldots (7)$$

The orthographic projection of the path on a horizontal plane is therefore a "rosette," like that indicated in Fig. 27.

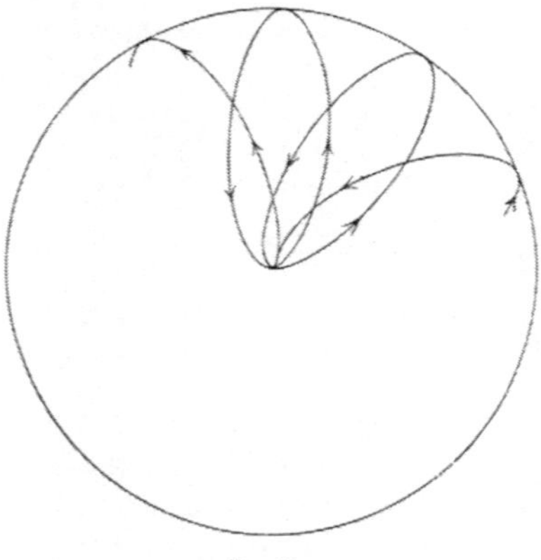

FIG. 27.

**11.** ***Path of weak top with reference to horizontal axes of x and y.*** The equation of the path with reference to horizontal axes of $x$ and $y$ through the pole as origin can be found as follows. Equation (7), 9, can be written

$$\cos\theta-\cos\theta_m=\tfrac{1}{2}(1-\cos\theta_1)\sin mt,$$

or

$$2\sin\tfrac{1}{2}(\theta+\theta_m)\sin\tfrac{1}{2}(\theta_m-\theta)=\sin^2\tfrac{1}{2}\theta_1\sin mt.$$

Since the angles $\theta$, $\theta_m$, $\theta_1$ are all small, we write angles for sines, and obtain

$$\theta^2=\theta_m^2-\tfrac{1}{2}\theta_1^2\sin mt,$$

or, since $\theta_1^2=2\theta_m^2$,

$$\theta^2=\theta_m^2(1-\sin mt). \quad \ldots (1)$$

From this we deduce

$$\theta=\theta_m(\cos\tfrac{1}{2}mt-\sin\tfrac{1}{2}mt), \quad \ldots (2)$$

by choosing the sign of the square root so as to make $\theta$ diminish at the instant $t=0$, as it does in (7), 9 Writing now $x=\theta\cos\psi$, $y=\theta\sin\psi$, we obtain

$$\left.\begin{aligned} x&=\theta_m(\cos\tfrac{1}{2}mt-\sin\tfrac{1}{2}mt)\cos\psi,\\ y&=\theta_m(\cos\tfrac{1}{2}mt-\sin\tfrac{1}{2}mt)\sin\psi. \end{aligned}\right\} \quad\text{.......(3)}$$

At $t=0$, we have, if we suppose that then $\psi=0$,

$$x=\theta_m,\quad \dot{x}=-\tfrac{1}{2}m\theta_m,\quad y=0,\quad \dot{y}=\theta_m\frac{bn}{1+z_m}. \quad\text{.......(4)}$$

**12.** ***Equations of motion of upright top derived from first principles.*** The general equations for the turning of the top about the axes $Ox$, $Oy$ may be established as follows from first principles. The angular momenta about the vertical OZ, and the two horizontal axes, given by the horizontal projection of OE and OD, are respectively,

$$Cn\cos\theta+A\dot{\psi}\sin^2\theta,\quad -Cn\sin\theta+A\dot{\psi}\sin\theta\cos\theta,\quad A\dot{\theta}.$$

These become, with $\sin\theta=\theta$, $\cos\theta=1$,

$$Cn,\quad -Cn\theta+A\dot{\psi}\theta,\quad A\dot{\theta}.$$

Thus, since OD is to be regarded as having turned forward from $Ox$ through the angle $\psi$, the angular momenta about $Ox$, $Oy$ are,

$$(Cn\theta-A\dot{\psi}\theta)\sin\psi+A\dot{\theta}\cos\psi,\quad -(Cn\theta-A\dot{\psi}\theta)\cos\psi+A\dot{\theta}\sin\psi,$$

which can be written $A\dot{x}+Cny$, $A\dot{y}-Cnx$, of which the time rates of change are $A\ddot{x}+Cn\dot{y}$, $A\ddot{y}-Cn\dot{x}$. The moments of the forces are $Mghx$, $Mghy$ about $Ox$ and $Oy$, and so the equations of motion are

$$A\ddot{x}+Cn\dot{y}-Mghx=0,\quad A\ddot{y}-Cn\dot{x}-Mghy=0, \quad\text{.......(1)}$$

as already obtained.

Clearly these equations can be united in one if we put $z=x+iy$. We get

$$A\ddot{z}-iCn\dot{z}-Mghz=0. \quad\text{.......(2)}$$

If we write $z=ke^{i\lambda t}$, this gives

$$A\lambda^2-Cn\lambda+Mgh=0. \quad\text{.......(3)}$$

The roots of this equation are real if $C^2n^2>4MghA$. Putting $\lambda_1$, $\lambda_2$ for the two roots (real if this condition is fulfilled) we have

$$z=k_1e^{i\lambda_1t}+k_2e^{i\lambda_2t}.$$

The constants $k_1$, $k_2$ in their more general form are complex, and may be written $k_1=\alpha_1-i\beta_1$, $k_2=\alpha_2-i\beta_2$. Thus

$$z=(\alpha_1-\beta_1i)e^{i\lambda_1t}+(\alpha_2-\beta_2i)e^{i\lambda_2t}. \quad\text{.......(4)}$$

Supposing $\lambda_1$, $\lambda_2$ to be real, we get by equating real and imaginary parts on the two sides of (4),

$$\left.\begin{aligned} x&=\alpha_1\cos\lambda_1t+\beta_1\sin\lambda_1t+\alpha_2\cos\lambda_2t+\beta_2\sin\lambda_2t,\\ y&=\alpha_1\sin\lambda_1t-\beta_1\cos\lambda_1t+\alpha_2\sin\lambda_2t-\beta_2\cos\lambda_2t. \end{aligned}\right\} \quad\text{.......(5)}$$

For the values of $\lambda$ we have

$$\left.\begin{matrix}\lambda_1\\ \lambda_2\end{matrix}\right\}=\frac{Cn}{2A}\left\{1\pm\left(1-\frac{4AMgh}{C^2n^2}\right)^{\frac{1}{2}}\right\}, \quad\text{.......(6)}$$

from which it will be clear, from the condition of reality of the roots, that both roots have the same sign.

**13.** ***Discussion of the motion.*** It may appear from (4) and (5) that we have here four arbitrary constants for the complete integral of a linear differential equation of the second order, but it is to be observed that the elimination of $x$ or $y$ from (1) would give a linear differential equation of the fourth order, the complete integral of which would involve four constants. Equations (5) are those which we should obtain by finding this integral: we should obtain no new roots, as the four roots of the determinantal equation would be simply $\lambda_1$, $-\lambda_1$, $\lambda_2$, $-\lambda_2$.

As the terms are arranged in (5), those in $\alpha_1$ give a circular motion of period $2\pi/\lambda_1$ and radius $\alpha_1$, those in $\beta_1$ a circular motion of the same period and in the same direction, but of radius $\beta_1$ and of phase $\frac{1}{2}\pi$ behind the former, if $\alpha_1$, $\beta_1$ be positive. Similarly the terms in $\alpha_2$, $\beta_2$ give circular motions of periods $2\pi/\lambda_2$ in the same direction and differing in phase as before, but of radii $\alpha_2$, $\beta_2$ respectively. The values of the radii and the relations of phase depend on the initial conditions.

If the condition for the reality of $\lambda_1$, $\lambda_2$ is not fulfilled, that is if $4\text{AM}gh > \text{C}^2n^2$, we have

$$\left.\begin{matrix}\lambda_1\\ \lambda_2\end{matrix}\right\} = \frac{\text{C}n}{2\text{A}}\left\{1 \pm \iota\left(\frac{4\text{AM}gh}{\text{C}^2n^2} - 1\right)^{\frac{1}{2}}\right\}.$$

Thus

$$\left.\begin{matrix}i\lambda_1\\ i\lambda_2\end{matrix}\right\} = i\frac{\text{C}n}{2\text{A}} \mp \frac{\text{C}n}{2\text{A}}\left(\frac{4\text{AM}gh}{\text{C}^2n^2} - 1\right)^{\frac{1}{2}}.$$

We thus get an oscillation in the period $4\pi\text{A}/\text{C}n$, affected by an exponential factor. Hence in the case of instability the axis of the top does not necessarily deviate far from the vertical. This deviation will be slight for a moderate time if $(4\text{AM}gh - \text{C}^2n^2)/\text{C}^2n^2$ is very small. Obviously however the disturbed motion does not, for an indefinitely small disturbance, coincide with the undisturbed motion, which is the condition fulfilled by the so-called stable motion.

When the axis is nearly coincident with the downward vertical the equations of motion (1) apply with change of the $-$ sign before $\text{M}gh$ in each to $+$. We have then the gyrostatic pendulum, in which both motions are stable without rotation. Thus the roots in (6) are now real in all cases. This case of motion is discussed in 20, 21, VII, below.

It will be observed that in the discussion of the upright top given above the two unstable freedoms which we have without rotation are stabilised by the rotation.

**14.** *Estimate of error in approximation.* As before, we can form an estimate of the amount of error involved in the approximate equations found above. If $\tau$ be the difference between the approximate value of $t$ found in 9 and the true value,

$$\left|\frac{\tau}{t}\right| < \frac{\dfrac{1}{(z_3-1)^{\frac{1}{2}}} - \dfrac{1}{(z_3-z_1)^{\frac{1}{2}}}}{\dfrac{1}{(z_3-z_1)^{\frac{1}{2}}}}$$

or

$$\frac{\tau}{t} < \left(\frac{z_3-z_1}{z_3-1}\right)^{\frac{1}{2}} - 1.$$

Hence, if $f$ denote a proper fraction,

$$\tau = \pm f\left\{\left(\frac{z_3-z_1}{z_3-1}\right)^{\frac{1}{2}} - 1\right\}t$$

or

$$t' = t \pm f\left\{\left(\frac{z_3-z_1}{z_3-1}\right)^{\frac{1}{2}} - 1\right\}t, \quad \text{....(1)}$$

where $t$ is the true value of the time.

It will be observed that this result shows that the equations are entirely inapplicable when $\theta_0 = 0$, for then $z_3 = 1$.

If we go back to (1), 9, we see that the coefficient of $z$, when that of $z^3$ is $a$, is $2b^2n^2 - a$. Now, since $z_2 = 1$, the sum of the products of the roots taken two and two is $z_3(1+z_1) + z_1$. Thus we obtain

$$z_3 = 2\frac{b^2n^2}{a(1+z_1)} - 1. \quad \text{....(2)}$$

Now $z_1 = 1 - 2u'$, and so we have

$$\left.\begin{aligned} z_3 &= \frac{b^2n^2 - a(1-u')}{a(1-u')}, \\ \text{and also} \quad z_3 - z_1 &= \frac{b^2n^2 - 2a(1-u')^2}{a(1-u')}, \quad z_3 - 1 = \frac{b^2n^2 - 2a(1-u')}{a(1-u')}. \end{aligned}\right\} \quad \ldots\ldots(3)$$

Thus

$$\tau = \pm\left\{ f\left(\frac{b^2n^2 - 2a(1-u')^2}{b^2n^2 - 2a(1-u')}\right)^{\frac{1}{2}} - 1\right\}. \quad \ldots\ldots(4)$$

**15.** *Explanation of an apparent anomaly.* In the determination of $\dot{\psi}$ and $\psi$ given in 10 above there is, apart from the fact that the value of $u$ employed is not exact, neglect of the term $u^2/(1+z_m)^2(1+z)$ in $1/(1+z)$. If $\frac{1}{2}bn$ had been taken instead of $bn/(1+z)$ for $\dot{\psi}$, we should have neglected $\frac{1}{2}bn(1-z)/(1+z)$; that is the error would have been greater than zero and less than $\frac{1}{2}bnu'/(1-u')$. As it is, a closer approximation has been employed.

It will have been noticed that the difference between the stable and unstable cases consists in the fact that the difference $1-z_1$, for the former, diminishes to zero with $\dot{\theta}_0$, while for the latter, however small $\dot{\theta}_0$ may be, $1-z_1$ never falls below a lower limit greater than zero. Thus the rosette which pictures the motion of the axis has non-evanescent loops in the case of $b^2n^2/a - 2 < 0$, however slight the disturbance $\dot{\theta}_0$ may be.

The rosette does however depend on the value of $b^2n^2/a - 2$, and if this is numerically very small, the rosette will be very small, and so the rosette (arising in a somewhat different way), which we shall find characterises the stable case, may be actually larger than that for the unstable case.

Now, as has been pointed out by Klein [*Bull. Amer. Math. Society*, 3, 1897], the method of small oscillations, illustrated in 12, does not give this apparent contradiction of instability afforded by the small rosette. He shows however that the complete integral for $t$ includes terms which it is legitimate to neglect when $b^2n^2 - 2a$ is great, but which are comparable with $b^2n^2 - 2a$ when this is small.

**16.** *Stable cases of motion of upright top.*—We now consider the "stable" cases of the upright top under the condition that the disturbance from the vertical is negligibly small. As already remarked, these differ from the unstable case, notably in this, that the deviation of the axis from the vertical is always exceedingly small when the disturbing angular speed is so; while in the unstable case the deviation has the value $2 - b^2n^2/a$ *for the least disturbance*, and a still greater value when $\dot{\theta}_0$ is sensibly greater than zero. The magnitude of $2 - b^2n^2/a$ depends only on the ratio of $b^2n^2$ to $a$.

Taking then $\dot{\theta}_0$ as small, we obtain from (3), 4,

$$\dot{\theta}^2(1+z) + b^2n^2(1-z) - a(1-z^2) = \dot{\theta}_0{}^2(1+z), \quad \ldots\ldots(1)$$

or, as we may write it,

$$\dot{\theta}^2 + b^2n^2\tan^2\tfrac{1}{2}\theta - 2a\sin^2\tfrac{1}{2}\theta = \dot{\theta}_0{}^2. \quad \ldots\ldots(2)$$

Since $\theta$ is very small we can deduce from the last equation the relation

$$\ddot{\theta} + \tfrac{1}{4}(b^2n^2 - 2a)\theta = 0; \quad \ldots\ldots(3)$$

and we have seen in 6 that in the stable cases $b^2n^2 \geqq 2a$. Thus, if $b^2n^2 - 2a > 0$, we have

$$\theta = L_1 \sin\tfrac{1}{2}(b^2n^2 - 2a)^{\frac{1}{2}}t + L_2\cos\tfrac{1}{2}(b^2n^2 - 2a)^{\frac{1}{2}}t. \quad \ldots\ldots(4)$$

If we suppose that $\theta=0$ when $t=0$, this becomes

$$\theta=\mathrm{L}_1\sin\tfrac{1}{2}(b^2n^2-2a)^{\frac{1}{2}}t. \quad\ldots\ldots\ldots\ldots\ldots\ldots (5)$$

The period of variation of $\theta$ is thus real, and of value $4\pi/(b^2n^2+|2a|)^{\frac{1}{2}}$, or $4\pi/(b^2n^2-2a)^{\frac{1}{2}}$, according as $a$ is negative or positive. The value of $\mathrm{L}_1$ is given by $\mathrm{L}_1=2\dot\theta_0/(b^2n^2-2a)^{\frac{1}{2}}$.

In the present case we have also (since the axis passes through the vertical)

$$\dot\psi=bn\frac{1-z}{1-z^2}=bn\frac{1}{1+z}=\tfrac{1}{2}bn\left(1+\frac{1-z}{1+z}\right). \quad\ldots\ldots (6)$$

Thus, subject to an error of amount

$$\tfrac{1}{2}bn\int\frac{1-z}{1+z}dt,$$

the angle $\psi$, turned through in time $t$ from the instant $t=0$, is given by

$$\psi=\tfrac{1}{2}bnt. \quad\ldots\ldots\ldots\ldots\ldots\ldots (7)$$

The orthogonal projection of the path on a horizontal plane has the equations

$$x=\theta\cos\psi,\quad y=\theta\sin\psi, \quad\ldots\ldots\ldots\ldots\ldots\ldots (8)$$

or

$$\left.\begin{aligned}x&=\mathrm{L}_1\sin\tfrac{1}{2}(b^2n^2-2a)^{\frac{1}{2}}t\cos\tfrac{1}{2}bnt,\\ y&=\mathrm{L}_1\sin\tfrac{1}{2}(b^2n^2-2a)^{\frac{1}{2}}t\sin\tfrac{1}{2}bnt,\end{aligned}\right\}\quad\ldots\ldots\ldots\ldots (9)$$

where the origin of coordinates is the projection of the upper pole of the sphere.

17. *Azimuthal turning.* The angle $\psi$ turned through in half a period is

$$\pi\frac{bn}{(b^2n^2\pm|2a|)^{\frac{1}{2}}},$$

where the upper or the lower sign is to be taken according as $a$ is negative or positive. It is thus less than $\pi$ in the former case, and greater than $\pi$ in the latter. But it will be seen [from (5) or (9)] that the half-period is the interval of time between two successive passages of the axis through the vertical position. At an instant of vertical passage the axis is moving along an arc of a meridian, and the angle between the meridian of one passage and that of the next (the difference of "longitude") is

$$\pi bn/(b^2n^2\pm|2a|)^{\frac{1}{2}}\quad(\text{that is }<\pi\text{ or }>\pi)$$

according as $a$ is negative or positive. It is most carefully to be observed that this is the angle at the pole between the line along which the axis leaves the pole and the line along which it returns; in other words, this angle is the supplement of that between the directions in which the axis leaves the pole at the beginning and end of the interval of time specified. It is the angle through which the plane through the vertical and the axis of the top swings round in the interval. This plane is to be regarded also as swinging round through the angle $\pi$ at the instant of the passage of the axis through the vertical [see 19 below].

The intermediate case is that in which $2a$ is zero, that is when the top is acted on by no forces, or has a value of $b^2n^2$ infinitely great in comparison with $2a$. The axis in that case describes and redescribes a cone about a mean position.

An extreme case of stability which may be noticed is that in which $a$ is negative and $bn$ zero, and the axis of the top is in the downward vertical direction We have then $\dot{\psi}=0$, and $-2a=4\mathrm{M}gh/\mathrm{M}(h^2+k^2)=4g/l$, where $l$ is the length of the equivalent simple pendulum. If the axis is slightly disturbed from the vertical the equation of motion is

$$\ddot{\theta}+\frac{g}{l}\theta=0,$$

the equation of the ordinary pendulum vibrating in period $2\pi(l/g)^{\frac{1}{2}}$.

**18.** *Stable motion between two close limiting circles near pole.* Finally we consider the case of motion between two close limiting circles very near the vertical. We shall suppose the top to have been started with A.M. about an axis nearly coincident with the axis of figure, so that the precessional motion is slow. In this case, as we have seen, $b^2n^2$ must be very great in comparison with $2a$. Let $u$ be the distance of the circle in which the axis is at any instant from the circle midway between the two limiting circles. For the mean circle we shall suppose the value of $z$ to be $z_m$, and take the distance between the two limiting circles as $2u'$. The root $z_3$ is here very great, and as the limiting circles are near the pole $z_3-z_m$, $z_3-z_1$, $z_3-z_2$ are all very nearly equal to $z_3-1$. We have, by (2), 10, V,

$$\dot{z}^2=(\alpha-az)(1-z^2)-(\beta-bnz)^2, \quad\ldots\ldots(1)$$

and the limiting circles are determined by the equation

$$(\alpha-az)(1-z^2)-(\beta-bnz)^2=0. \quad\ldots\ldots(2)$$

The coefficient of $z$ in this equation is $2\beta bn-a$. Hence, if $z_3$, $z_2$, $z_1$ be the roots, we have

$$z_3(z_2+z_1)+z_1z_2=\frac{2\beta bn-a}{a},$$

or approximately $$a(z_3-1)=\beta bn-2a=b^2n^2-2a. \quad\ldots\ldots(3)$$

The period of oscillation between the circles is therefore [see 2 above]

$$2\pi/\{a(z_3-1)\}^{\frac{1}{2}}=2\pi/(b^2n^2-2a)^{\frac{1}{2}},$$

nearly, and we have $$z=z_m+u'\sin(b^2n^2-2a)^{\frac{1}{2}}t, \quad\ldots\ldots(4)$$

or, to a less close approximation, $$z=z_m+u'\sin bnt. \quad\ldots\ldots(5)$$

It will be observed that the period here obtained is half of the period for the case of stable motion through the pole. This is as it should be, for in the present case the period is the interval between the two successive intersections of either circle by the axis, while in the case of motion through the pole the half-period is the time of passage from the limiting circle, through the pole, to the circle again. The second circle has coalesced with the pole, and this half-period therefore corresponds to the whole period in the other case.

We have now to consider the precessional motion. We have

$$\dot{\psi}=\frac{\beta-bnz}{1-z^2}.$$

As the axis, though always near the vertical, does not actually pass through it, we cannot put $\beta=bn$ with exactness, and so we write the equation in the form

$$\dot{\psi}=\frac{1}{2}\frac{\beta-bn}{1-z}+\frac{1}{2}\frac{\beta+bn}{1+z}=\dot{\psi}_1+\dot{\psi}_2,\text{ say.} \qquad \text{(6)}$$

Since $$\frac{1}{1+z}=\frac{1}{1+z_m}-\frac{u}{(1+z_m)^2}+\frac{u^2}{(1+z_m)^2}\frac{1}{1+z}$$

we get $$\dot{\psi}_2=\frac{1}{2}\frac{\beta+bn}{1+z_m}-\frac{1}{2}\frac{\beta+bn}{(1+z_m)^2}u, \qquad \text{(7)}$$

or $$\dot{\psi}_2=\frac{1}{2}\frac{\beta+bn}{1+z_m}-\frac{1}{2}\frac{\beta+bn}{(1+z_m)^2}u'\sin mt, \qquad \text{(8)}$$

nearly, where now $m=(b^2n^2-2a)^{\frac{1}{2}}$. The angle $\psi$ turned through in time $t$ by the vertical plane through the axis of the top is made up of two parts $\psi_1$, $\psi_2$, of which the second is given by

$$\psi_2=\frac{1}{2}\frac{\beta+bn}{1+z_m}t+\frac{1}{2m}\frac{\beta+bn}{(1+z_m)^2}u'\cos mt. \qquad \text{(9)}$$

For $\psi_1$ we have $$\psi_1=\frac{1}{2}\int\frac{\beta-bn}{1-z}dt=\frac{1}{2}(\beta-bn)\int\frac{dt}{1-z_m-u'\sin mt}. \qquad \text{(10)}$$

Since the axis does not pass through the pole $1-z_m>u'$, and so we can write

$$\psi_1=\frac{1}{2}\frac{\beta-bn}{1-z_m}\int\frac{dt}{1-\dfrac{u'}{1-z_m}\sin mt}=\frac{1}{2}\frac{\beta-bn}{1-z_m}\int\frac{dt}{1-k\sin mt}, \qquad \text{(11)}$$

where $k<1$. By using the substitution $w=\tan\frac{1}{2}mt$, we get easily

$$\psi_1=\frac{\beta-bn}{m\{(1-z_m)^2-u'^2\}^{\frac{1}{2}}}\tan^{-1}\frac{w-k}{(1-k^2)^{\frac{1}{2}}}. \qquad \text{(12)}$$

Thus we get finally

$$\psi=\frac{\beta-bn}{m\{(1-z_m)^2-u'^2\}^{\frac{1}{2}}}\tan^{-1}\frac{w-k}{(1-k^2)^{\frac{1}{2}}}+\frac{1}{2}\frac{\beta+bn}{1+z_m}t+\frac{1}{2m}\frac{\beta+bn}{(1+z_m)^2}u'\cos mt. \qquad \text{(13)}$$

**19.** *Extension of case of* 18 *to that in which axis passes infinitely near pole.* It is interesting to consider how this investigation can be extended to the case in which the axis passes through (or, as we shall say, infinitely near) the pole. Here $1-z_m=u'+\epsilon$, where $\epsilon$ is small compared with $u'$, which in its turn we shall regard as small compared with 1. The analytical discussion requires very great care; but what happens is easily made out by a consideration of the geometry of the motion. Let the large circle in the diagram (drawn for the case of $u<0$) be the lower limiting circle ($z=z_1$), and let the upper be shrunk down to an exceedingly small circle $c$, corresponding to $z=1-\epsilon$, and therefore of radius $(2\epsilon)^{\frac{1}{2}}$. It will be seen that if $\dot{\theta}_0$ be the rate at which $\theta$ is increasing at a point A at a distance from $c$ large compared with $\epsilon$, but still comparatively far from the lower circle, the value of $\dot{\psi}$ at the small circle, where it is touched by the axis in passing, is approximately $\dot{\theta}_0/(2\epsilon)^{\frac{1}{2}}$. It is obvious that $\dot{\theta}_0$ must, since $\dot{\theta}$ changes very slowly near the pole, be very approximately the value of $\dot{\theta}$ when the axis is passing the pole.

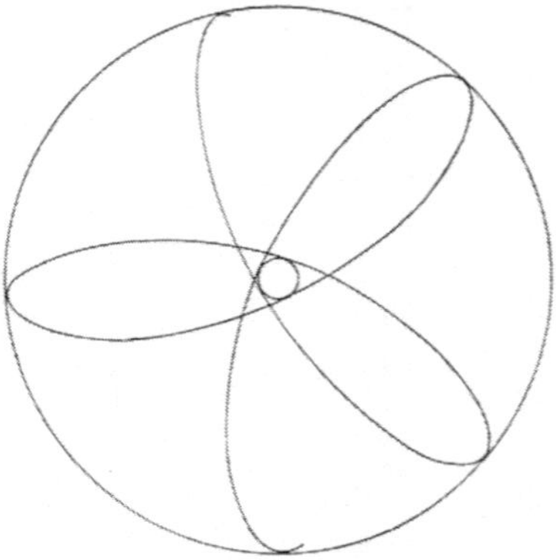

FIG. 28.

Now when the axis actually passes through the pole $\dot{\psi}$ loses its meaning, but we are led by the foregoing to regard it as becoming infinite just at the pole by the vanishing of $2\epsilon$. By this mode of considering the matter we take $\dot{\theta}$ as $\dot{\psi}(2\epsilon)^{\frac{1}{2}}$, and so regard the pole as an infinitely small limiting circle at which there is no $\dot{\theta}$, since that is replaced by the term in $\dot{\psi}$. In the element of time bisected by the instant at which the axis passes the pole, the value of $\psi$ changes abruptly by $\pi$.

It will be observed that the introduction of this value of $\psi$ does not conflict with the condition

$$\dot{\psi}(1-z^2)=\beta-bnz.$$

For $\dot{\psi}=\dot{\theta}_0/(2\epsilon)^{\frac{1}{2}}=\dot{\theta}_0/\theta$, and therefore $\dot{\psi}(1-z^2)=\dot{\theta}_0\theta$, which is vanishingly small when $\dot{\theta}$ is. Thus $\dot{\psi}(1-z^2)$ still vanishes as the upper circle is shrunk infinitely near to zero, and so is consistent with the vanishing of $\beta-bnz$, when $z$ becomes 1.

We have now to inquire whether this discontinuity in the value of $\psi$, as the axis passes the vertical, is given by the expression found above for $\psi_1+\psi_2$. It is clear that no discontinuity is given by $\psi_2$. It remains to examine $\psi_1$. We have

$$\psi_1=\frac{1}{m}\,\frac{\beta-bn}{\{(1-z_m)^2-u'^2\}^{\frac{1}{2}}}\tan^{-1}\frac{w-k}{(1-k^2)^{\frac{1}{2}}},$$

where $k^2=u'^2/(1-z_m)^2$. Now $k=u'/(1-z_m)=u'/(u'+\epsilon)$, and so as the upper limiting circle is shrunk down more and more to zero, radius $1-k^2$ approaches more nearly to zero, and $(w-k)/(1-k^2)$ to $\infty$. But $t$ is reckoned from the circle $z_m$, and when the vertical is reached we have by (4), 18, $t=\frac{1}{2}\pi/m$, and $w=\tan\frac{1}{2}mt=\tan\frac{1}{4}\pi=1$. When $t$ increases beyond this value $w-k$ changes sign from negative to positive. Thus $(w-k)/(1-k^2)^{\frac{1}{2}}$ springs at once from $-\infty$ to $+\infty$. But when $w$ is infinitely near to 1, $(w-k)/(1-k^2)^{\frac{1}{2}}$ is $\{(1-k)/(1+k)\}^{\frac{1}{2}}$, or zero. Hence the sudden change of $(w-k)/(1-k^2)^{\frac{1}{2}}$ from $-\infty$ to $+\infty$ is through the value zero, that is $\tan^{-1}\{(w-k)/(1-k^2)^{\frac{1}{2}}\}$ changes suddenly by the angle $\pi$.

**20.** *Interpretation of discontinuity in $\psi$ at pole.* To complete the identification of this sudden change of $\psi$ with that reckoned above from elementary considerations, we have to show that the coefficient

$$\frac{\beta-bn}{m\{(1-z_m)^2-u'^2\}^{\frac{1}{2}}}$$

has the value 1. First we consider the numerator. We have

$$\begin{aligned}\beta-bn&=\dot{\psi}(1-z^2)-bn(1-z)\\&=\{\dot{\psi}(1+z)-bn\}\epsilon,\end{aligned}$$

where $\epsilon=1-z$, which in the limit is very small. Now we have seen that if $\dot{\theta}_0$ be the rate of variation of $\theta$ near the pole we have $\dot{\psi}=\dot{\theta}_0/\theta$. Thus

$$\beta-bn=\left\{\frac{\dot{\theta}_0}{\theta}(1+z)-bn\right\}\epsilon.$$

But for the lower limiting circle defined by $z_1$ (and corresponding to $\theta_1$) we have by (1), 16,

$$b^2n^2\theta_1^2=\dot{\theta}_0^2(1+z_1)^2,$$

since $bn$ is great in comparison with $a$. Thus

$$bn\theta_1=\dot{\theta}_0(1+z_1),$$

and therefore

$$\dot{\theta}_0=bn\theta_1\frac{1}{1+z_1}.$$

Thus we find

$$\beta-bn=bn\left(\frac{\theta_1}{\theta}\,\frac{1+z}{1+z_1}-1\right)\epsilon=bn\left\{\left(\frac{4u'}{2\epsilon}\right)^{\frac{1}{2}}\frac{2-\epsilon}{2-2u'}-1\right\}\epsilon=bn(2u'\epsilon)^{\frac{1}{2}}$$

very nearly.

Next, consider the denominator, remembering that $m=bn$, approximately. We have

$$bn\{(1-z_m)^2-u'^2\}^{\frac{1}{2}}=bn\{1-z_m-z_m(1-z_m)-u'^2\}^{\frac{1}{2}}=bn\{u'+\epsilon-z_m(u'+\epsilon)-u'^2\}^{\frac{1}{2}}.$$

Since $1-z_m=u'+\epsilon$, the last expression can be written

$$bn\{(u'+\epsilon)^2-u'^2\}^{\frac{1}{2}}=bn(2u'\epsilon)^{\frac{1}{2}},$$

which is the value obtained above for the numerator. In the limit therefore the coefficient is unity.

The angle $\psi$ turned through by the vertical plane through the axis of the top, in a passage from the lower circle past the vertical to the lower circle, and back again past the vertical, is from the point of view just presented (that of just grazing an infinitely small circle surrounding the pole) made up of $2\pi$ given by two discontinuities in $\psi_1$ and the angle

$$\psi_2=\frac{1}{2}\,\frac{\beta+bn}{1+z_m}\,\frac{4\pi}{m},$$

with approximately $1+z_m=2$, $\beta=bn$, $m=(b^2n^2-2a)^{\frac{1}{2}}$. Thus we may write

$$\psi_2=2\pi\left(1+\frac{a}{b^2n^2}\right).$$

The whole angle turned through is therefore $4\pi+2\pi a/b^2n^2$. The change of azimuth of the vertical plane through the centre of the infinitely small circle, corresponding to the passage of the axis from the small circle to the lower circle, and back again to the small circle, is therefore $2\pi+\pi a/b^2n^2$, which, whether $a>$ or $<0$, agrees with the diagram for the stable cases given in Fig. 28. The angle between the directions of two successive lines of leaving the small circle is $\pi a/b^2n^2$.

*Note.*—In addition to the diagrams of path given from time to time in the foregoing chapters, other possible curves will be included in the chapter on *Methods of Calculation and Quadrature.*

# CHAPTER VII

## GYROSTATS AND VARIOUS PHYSICAL APPLICATIONS OF GYROSTATS

1. *Gyrostats.* Before proceeding further with the more abstract theory of a top, it will be useful perhaps to give an account of gyrostats and of some of their applications. The theory of a gyrostat differs from that of an ordinary symmetrical top, inasmuch as the motion of the case or frame which encloses the flywheel of the gyrostat must be taken into account.

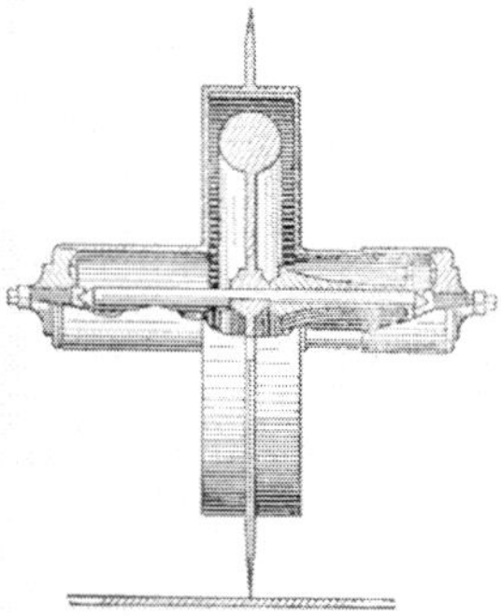

FIG. 29 (*a*).

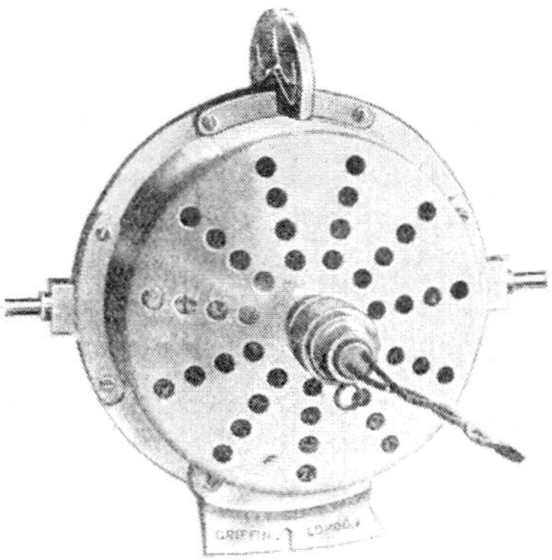

FIG. 29 (*b*).

Some different forms of gyrostat are here pictured [Fig. 29]. The nearly enclosing case of the Kelvin gyrostat [Fig. 29 (*a*)] was designed rather to suggest the idea of a body which, though identical in outward form with other bodies, has yet peculiar properties which are only to be explained by motions of internal parts, the existence of which is inferred from a study of the body's dynamical behaviour.

Fig. 29 (*b*) shows one of Dr. J. G. Gray's motor gyrostats, supported on a skate. The flywheel is the rotor of a continuous current electric motor, which can be driven from an ordinary electric light circuit.

Fig. 30, (*a*) and (*b*), shows an effective form of gyrostat, easily constructed by utilising a bicycle wheel on its ball bearings, as recommended by Sir George Greenhill. In diagram (*a*) the wheel is shown supported at one end of the axle, which rests in a cup. In experiments with such a wheel it is advisable, for safety, always to employ some form of universal joint, *e.g.* the altazimuth suspension shown in (*b*).

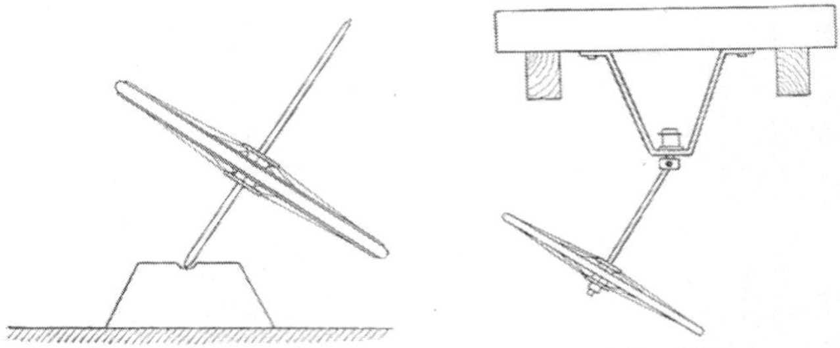

FIG. 30 (*a*). FIG. 30 (*b*).

It has been shown in 2, V, above, how the inertia of the case is taken account of when the axis of the flywheel may be taken as also the axis of symmetry of the whole instrument. Thus equations (1) to (9) of 1, V, contain the elements of the mathematical theory of the subject.

We shall now consider various examples, in the main such as fulfil, or nearly fulfil, the condition that the instrument is supported at a point situated on the axis of symmetry. We repeat here, as the equation mainly to be employed (when a point O of the axis is fixed),

$$A\ddot{\theta}+(Cn+C'\omega_1-A\dot{\psi}\cos\theta)\dot{\psi}\sin\theta=K, \qquad (1)$$

where C′ is the moment of inertia and $\omega_1$ the angular speed of the case about the axis of symmetry, while K is the couple applied about the axis through the point of support, round which the rate of turning is $\dot{\theta}$. This axis is at right angles at once to the axis of symmetry and to the vertical (or what corresponds to the vertical if the couple G is not supplied by gravity). A is the moment of inertia of the whole instrument about this axis, or about any other transverse axis through the point of support.

Along with (1) goes the equation

$$A\ddot{\psi}\sin\theta-(Cn+C'\omega_1-2A\dot{\psi}\cos\theta)\dot{\theta}=L, \qquad (2)$$

for the axis OE at right angles to OC and OD.

If we do not suppose the top or gyrostat supported on a fixed point, A is the moment of inertia about a transverse axis through the centroid, G, while the turnings and couples in (1) and (2) now refer to axes through G parallel to those already specified. But in this case we have to introduce equations for the motion of the centroid. These have the form,

$$M\ddot{x}=X, \quad M\ddot{y}=Y, \quad M\ddot{z}=Z, \qquad (3)$$

where $x$, $y$, $z$ are the Cartesian coordinates of the centroid, and X, Y, Z are the total applied forces (friction included if there is any) along the axes. In the important case in which the point O on the axis is fixed, the values of X, Y, Z are given in (3), 1, V, for axes there specified. The present Z is however equal to the former $Z-Mg$.

2. *Reaction of a top or gyrostat on its support.* We now consider the reaction of a top or gyrostat on its support. First we shall suppose the top to be supported at a fixed point on its axis of figure.

We refer the motion to a vertical axis O$z$, and two horizontal axes, one, O$x$, coincident with OD, the other, O$y$, coincident with the horizontal projection of OE. The components of momentum in the directions of the axes of $x$ and $y$ are respectively $Mh\dot\psi\sin\theta$ and $-Mh\dot\theta\cos\theta$, while that along the upward vertical is $-Mh\dot\theta\sin\theta$.

The rate of production of momentum in the direction of the axis of $x$ is $Mh\ddot\psi\sin\theta+Mh\dot\psi\dot\theta\cos\theta$, from differentiation of $Mh\dot\psi\sin\theta$, together with $Mh\dot\psi\dot\theta\cos\theta$ arising from the motion of the axes. The rates of production of momentum for the directions O$y$, O$z$ are

$$-Mh\ddot\theta\cos\theta+Mh\dot\theta^2\sin\theta+Mh\dot\psi^2\sin\theta,\quad -Mh\ddot\theta\sin\theta-Mh\dot\theta^2\cos\theta$$

respectively. The latter comes entirely from differentiation of $-Mh\dot\theta\sin\theta$, since the axes O$x$, O$y$ remain horizontal; the first two terms of the former are derived from $-Mh\dot\theta\cos\theta$, the third arises from motion of the axis O$x$.

If $-$(X, Y, Z) be the components of the reaction *on the support* for axes O($x$, $y$, $z$), as specified above, we have, as in (3), 1, V,

$$X=Mh(\ddot\psi\sin\theta+2\dot\psi\dot\theta\cos\theta),\quad Y=-Mh(\ddot\theta\cos\theta-\dot\theta^2\sin\theta-\dot\psi^2\sin\theta),$$
$$Z=M(g-h\ddot\theta\sin\theta-h\dot\theta^2\cos\theta).$$

We next take a gyrostat resting on the rim which surrounds its case, as in Fig. 29 ($a$), while the axis of figure is inclined at the angle $\theta$ to the vertical. If $a$ be the distance of the point of support from the centroid, the centre we suppose of the flywheel, the equations just found apply, provided $a$ and $\frac{1}{2}\pi-\theta$ be substituted for $h$ and $\theta$, and the sign of Y be reversed. M includes the mass of the case as well as that of the flywheel.

About G$x$, drawn from the centroid G parallel to OD, the total moment of forces for steady motion ($\dot\psi=\mu$) is $-Ma(g+\mu^2a\sin\theta)g\cos\theta$. If A be the total moment of inertia about a transverse through G to the flywheel axis, the A.M. about G$x$ grows at rate $(Cn-A\mu\cos\theta)\mu\sin\theta$. Hence we get the quadratic for $\mu$,

$$(A-Ma^2)\sin\theta\cos\theta\,.\,\mu^2-Cn\sin\theta\,.\,\mu-Mga\cos\theta=0.$$

3. *The Serson-Fleuriais top for giving an artificial horizon at sea.* The following is a quotation from an article in the *Gentleman's Magazine* for 1754: "Eleven or twelve years ago Mr. Serson, an ingenious mechanick, took a hint from the property of a top set a spinning, that the axis of its rotation affects a vertical position, and got a kind of top made, whose upper surface perpendicular to the axe was a circular plate of polished metal; and

found, as he had expected, that when this top was briskly set in motion, its plain surface would soon become horizontal; that all objects at rest, and reflected by that surface to an eye also at rest, did appear entirely without motion; and that if the whirling plane were disturbed from its horizontal position, it would soon recover it again, and preserve it unless disturbed anew, or that its velocity was too far diminished."

Obviously this device contained a solution of the problem of finding a satisfactory horizon for use in sextant observations at sea, when there was fog round the sea horizon, and with the aid of George Graham, F.R.S., the celebrated instrument maker, it was realised as a kind of top spun in the ordinary way by quickly unwinding a band from an axle surrounding the axis of figure. At first the Admiralty declined to try the instrument, but ultimately it was tried on board an Admiralty yacht at the Nore in 1743, and favourably reported upon. Mr. Serson was sent by the Admiralty to sea in the *Victory* a little later to test a new instrument, but the ship was lost with all on board. Mr. Graham then had one made, and "used frequently to express great indignation at the unaccountable disregard of so promising a discovery, having himself made many tryals of its properties." [*Loc. cit. supra.* See also Short, *Phil. Trans.* 47, 1751-2.]

It may be stated here that a common level, or a pendulum, is of no use for giving a horizontal or vertical direction on a body, like a quickly moving boat or aeroplane, which is constantly undergoing varying acceleration.

Serson's device has been revived (or independently invented) in France within the last twenty or thirty years, and is more or less in use in the French navy in the improved form given to it by Admiral Fleuriais [Fig. 31]. As will be seen from the figure, it is simply a small top with, as shown in the diagram, a certain amount of gravitational stability. M is about 175 grammes, and $Mgh$ is made at least 17·5 grammes. [The value of C for a trial instrument tested in 1887 was 490 gramme-centimetre units.]

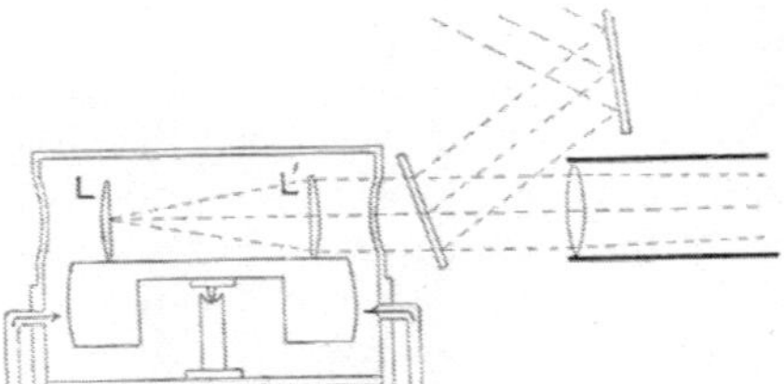


Fig. 31.

The drive is an air-blast acting on "bucket" depressions round the edge of the wheel: the speed is run up to about 80 turns per second before an observation, and the blast is then shut off and the air pump reversed so as to exhaust the chamber in which the top spins. The speed falls to about 60 turns per second during an observation.

The upper circular plane face forms a table which carries at the extremities of a diameter two equal plano-convex lenses, L, L′ as shown, with their plane faces turned towards one another, and at right angles to the diameter. The distance between these faces is the focal length of either lens, and a datum-line is drawn on each parallel to the table and through the axis of the lens arrangement.

The diagram sufficiently explains the use of the instrument for sextant observations. As will be seen from the diagram, it is carried by the sextant behind the small mirror into which the telescope looks. Rays from the mark on one of the lenses are rendered parallel by the other lens, and are then received by the telescope of the sextant, together with rays from the distant object brought into parallelism with the beam from the lenses by two reflections from the mirrors in the usual way. A distinct image is obtained notwithstanding the spin in consequence of the persistence of impressions on the retina. To get good results requires considerable practice.

The spin and moment of inertia of the top about its axis are both so great that any precessional motion caused by disturbances is of very long period. The precession, in fact, is so slow that successive disturbances due to rolling or pitching of the ship have time to annul one another for all practical purposes.

In most descriptions of the action of this instrument which have been given it is said to be subject to an inclinational error, $T \cos l/86160$ (where T is the precessional period of the top and $l$ the latitude), due to the earth's rotation. This holds for a gyrostatic pendulum hung from a universal joint or for (the impossible) case of a top supported by a frictionless peg. In such a case let the ship be in latitude $l$, the angular speed $\omega$ of the earth may be resolved into two components, $\omega \sin l$ about the vertical and $\omega \cos l$ about the northward horizontal at the place. As the axis of spin of the top is very nearly vertical the former component is of little consequence, *provided $Mgh$ is large in comparison with $Cn\omega \sin l$.* [See 6 below.] The latter, however, if the spin $n$ of the top is counter-clockwise, is producing A.M. about an eastward horizontal line at rate $Cn\omega \cos l$. As there is no externally applied couple about this eastward line, the top turns in the direction to neutralise this rate of growth of A.M., that is the upper end of the axis turns towards the north, and, after a little oscillation, relative equilibrium ensues when the gravitational stability gives the couple requisite for the steady production of A.M. at rate $Cn\omega \cos l$. The apparent upward vertical is then north of the true vertical, by an angle $\theta$ given by

$$Cn\omega \cos l = Mgh \sin \theta, \quad \text{..............................(1)}$$

where $h$, the distance of the centroid *below* the point of support, is taken positive. The true latitude is thus south of the apparent latitude as given by observations made with this artificial horizon. Since $T = 2\pi Cn/Mgh$ approximately the inclinational error is as stated above.

But in the case of the Serson-Fleuriais level the vertical is defined by the interaction of the peg and its supporting cup. If the top becomes slightly inclined to the vertical an erecting couple of moment L, depending on the design of the peg and cup, the spin and the weight of the top, comes into play, and the top erects itself in accordance with the equation

$$-\dot{\theta}=\frac{\mathrm{L}}{\mathrm{C}n},$$

where $\theta$ is the tilt from the upward vertical. If

$$\frac{\mathrm{L}}{\mathrm{C}n}>\frac{l\cos\theta}{86160}$$

there is no error due to the earth's rotation.

An exact theory of the inclinational effect of the earth's rotation is given in §6 below for a case in which it has full play, that of Gilbert's bary-gyroscope, which is supported on two knife-edges in a line through the centre of the flywheel. The theory of erection will be given later.

4. *Gyrostatic observation of the rotation of the earth. Foucault's methods.* The famous French experimentalist, Léon Foucault,* suggested two ways of determining the rotation of the earth. One was observation of the apparent turning of the plane of vibration of a long pendulum, suspended so as to be as nearly as possible free from any constraint due to the attachment of the pendulum wire to its fixed support. This classical experiment was carried out with fair success at the Panthéon at Paris, and was repeated under the domes of the cathedrals of Amiens and Rheims.

FIG. 32.

Foucault's other method was based on the fact that a gyrostat, if mounted properly, retains unaltered the direction of the spin-axis when the supports are turned round. Take, for example, the pedestal gyrostat of Fig. 32. The flywheel is within the frame there shown as carried by a vertical rod, which swivels in a vertical socket projecting upward from the supporting stand. Rods with arrow-points are attached, to be set to represent the spin axis, the couple axis, and the direction of rotation. If

* *Comptes rendus*, 35 (1852).

the spinning gyrostat has its spin axis in any chosen direction, and the supporting stand is turned round, a friction couple of some little magnitude is applied to the vertical rod; nevertheless, the direction of the spin-axis is not altered perceptibly. Yet the friction couple is sufficient to carry the gyrostat round with the stand when there is no spin. The spin results in a great increase of virtual inertia for turning displacements, as we shall see quantitatively in one of the arrangements described below in the present chapter.

It is to be understood that when there is no weight hung on the gyrostat case, the axes XX′, YY′, ZZ′ intersect at the common centre of gravity of the flywheel and case, and that there is no friction couple about any of these axes.

When the axis of spin of the pedestal gyrostat is adjusted to be accurately horizontal, and to point north, the horizontal component of the earth's angular velocity will have no gyrostatic effect, and the vertical component will merely turn the base piece round, and enable the earth's rotation to be seen from the changed azimuth of a mark on the base. Such a mark will, since the angular speed about the vertical in latitude $l$, is $\omega \sin l$, make a complete revolution about the vertical in (1 sidereal day)/sin $l$.

Under the conditions stated the axis of spin of the gyrostat, whatever its original direction, would preserve its direction in space as long as the flywheel was kept in rotation. In 5 the effect of non-fulfilment of these conditions in certain respects will be considered.

5. *Gyrostatic balance and gyrostatic dipping needle. Lord Kelvin's methods.* At the British Association meetings at Southport and Montreal, in 1883 and 1884, Lord Kelvin suggested methods of demonstrating the earth's rotation, and of constructing a gyrostatic compass. One of these had reference to the component of rotation about the vertical, the component in fact demonstrated by the Foucault pendulum experiment. If $\omega$ be the resultant angular speed, the component about the vertical at any place in latitude $l$ is $\omega \sin l$, while the companion component about the horizontal there is $\omega \cos l$. Thus at London the component about the vertical is 0·78 of $\omega$, and the period of rotation about the vertical is about 30·77 hours of sidereal time. (One sidereal day = 86,160 seconds, nearly.)

Lord Kelvin's method of measuring $\omega \sin l$ consists in supporting a gyrostat on knife-edges attached to the projecting edge of the case, so that the gyrostat without spin rests with the axis horizontal or nearly so. For this purpose the line of knife-edges is laid through the centre of the flywheel at right angles to the axis, and the plane of the knife-edges is therefore the plane of symmetry of the flywheel perpendicular to the axis. The knife-edges are a little above the centre of gravity of the instrument, which we suppose in or nearly in that plane, so that there is a little gravitational stability. The azimuth of the axis is a matter of indifference, as any couple due to the component of rotation about the horizontal is balanced by an equal couple furnished by the knife-edge bearings. The apparatus described in 6 would serve if the axis of the flywheel were turned through 90° with reference to the knife-edges, and scale-pans were added.

At points in a line at right angles to the line of knife-edges, and passing through it, two scale-pans are attached to the framework, and by weights in these the axis of the gyrostat (without spin) is adjusted, as nearly as may be, in a horizontal position which is marked. The gyrostat is now removed to have its flywheel spun rapidly, and is then replaced. It is found that the weights in the scale-pans have to be altered now to bring the gyrostat back to the marked position. From the alteration in the weights the angular speed about the vertical can be calculated.

To fix the ideas let the gyrostat axis be north and south, and let the spin to an observer, looking at it from beyond the north end, be in the counter-clock, or positive direction The rotation of the earth about the vertical carries the north end of the axis round towards the west, and therefore angular momentum is being produced about a horizontal axis drawn westward, at a rate equal to $Cn\omega \sin l$, where $Cn$ is the angular momentum of the flywheel. If the sum of the increase of weight on one scale-pan and the diminution (if any) in the other be $w$, and $a$ be the horizontal distance between the points of attachment of the scale-pans, we have

$$Cn\omega \sin l = wga.$$

Thus if C and $n$ are known, $\omega \sin l$, or $\omega$, can be calculated.

No figures were given as to the forces to be measured in a practical experiment; but these may be supplied as follows. We may take the mass of a small flywheel as 400 grammes, its radius of gyration as 4 cm , and its speed of revolution if high as 200 revolutions per second. If we take $a$ as 10 cm. we obtain for London the equation

$$400 \times 4^2 \times 400\pi \times \frac{2\pi \times 0{\cdot}78}{86160} = 10 \times 981 \times w.$$

This gives $w = 0{\cdot}047$, and therefore the weight is 47 milligrammes. It would require careful arrangements to carry out the experiment accurately. but the idea is clearly not unpractical. With some of the new gyrostats that we now have, the mass of the wheel is as much as 2,000 grammes, and the radius of gyration is about 7·5 cm. These numbers bring the weight up to 0·82 gramme, at the same speed.

If the gravitational stability of this gyrostatic balance be removed, that is, if the line of knife-edges be made to pass accurately through the centre of gravity of the system of wheel and framework, and the axis of rotation be placed in a truly north and south vertical plane, so that the knife-edges are horizontally east and west, the gyrostat will be in stable equilibrium when the axis is parallel to the earth's axis, and is turned so that the direction of rotation agrees with the rotation of the earth.* For we have then simply the experiment, described in Chapter I and discussed in 7 below, of the gyrostat resting by trunnions on bearings attached to a tray which is carried round by the experimenter. The axis of the gyrostat was at right angles to the tray, and when the tray, held horizontally, was carried round in azimuth the equilibrium of the gyrostat was stable or unstable, according as the two turnings agreed or disagreed in direction. In the present case the tray is the earth, the position of the axis of rotation parallel to the earth's axis replaces the vertical position, and the earth's turning the azimuthal motion. If displaced from the stable position the gyrostat will oscillate about it in the period $2\pi(A/Cn\omega)^{\frac{1}{2}}$, where A is the moment of inertia about the knife-edges, and the other quantities have the meanings already assigned to them [see (1), 7, below].

If the line of knife-edges be north and south, the vertical will be the stable, or unstable, direction of the axis of rotation, and there will be oscillation about the stable position in the period $2\pi(A/Cn\omega \sin l)^{\frac{1}{2}}$.

The gyrostat thus imitates exactly the behaviour of a dipping needle in the earth's magnetic field, and thus we have a gyrostatic model of the dipping needle.

6. *Gilbert's barogyroscope.* It is right to point out that these arrangements were anticipated by Gilbert's barogyroscope, which rests on precisely the same idea, and applies it in a similar manner.

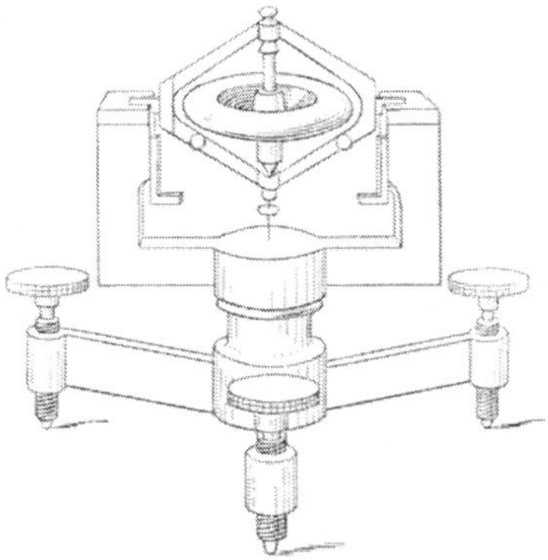

FIG. 33.

In this a gyrostat [Fig. 33] is supported on bearings, fixed horizontally east and west, and has a certain adjustable amount of gravitational stability supplied by placing the centroid below the line of bearings.

* This idea forms the basis of a paper by Sire, published in 1858 (*Arch. des Sci. Phys. et Nat.*, Genève, 1, 1858). Sire proposed to use this principle for the determination of the earth's rotation.

Let $l$ [Fig. 34] be the (north) latitude of the place, and the axis of rotation of the flywheel be inclined at an angle $\theta$ (lower end, say, towards the south) to the vertical at the place P. The angular speed, $\omega$ say, of the earth's rotation can be resolved into two components: one, $\omega \sin(l+\theta)$, about the axis of the flywheel, the other, $\omega \cos(l+\theta)$, about a line at right angles to this axis, and drawn towards the north. If $n$ and $\omega$ be similarly directed, the component $\omega \cos(l+\theta)$ gives a precessional motion, which for a proper value of $\theta$ will equal $\mathrm{M}gh \sin\theta$, the couple about the line of bearings.

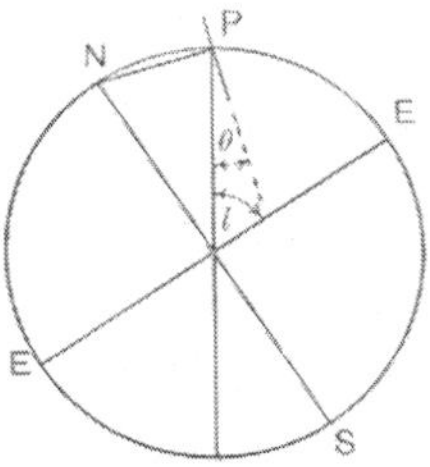

FIG. 34.

If the line of bearings is horizontally east and west, and the flywheel is spun to an angular speed $\phi$, relative to the earth, the speed $n$ when the gyrostat is set up in position will be $\phi \pm \omega \sin(l+\theta)$. At this inclination there will be equilibrium, and then $Cn\omega \cos(l+\theta) = \mathrm{M}gh \sin\theta$. With the understanding stated as to the meaning of $n$ this equation is exact. Hence

$$\tan\theta = \frac{Cn\omega \cos l}{Cn\omega \sin l + \mathrm{M}gh}. \qquad \text{(1)}$$

If the spin be reversed the inclination is to the other side of the vertical, and of amount $\theta'$, given by

$$\tan\theta' = \frac{Cn\omega \cos l}{Cn\omega \sin l - \mathrm{M}gh}. \qquad \text{(2)}$$

This, in point of fact, is an exact solution of the problem of the action of the gyroscopic horizon instrument [3 above], giving the error $\theta$ or $\theta'$ to be taken account of when a gyrostat is used as a clinometer or to give an artificial horizon.

If the line of bearings be horizontal, but inclined at an angle $\phi$ to the east and west horizontal line, and the inclination of the axis of rotation of the gyrostat to the vertical be $\theta$, then, instead of (2), we have

$$\tan\theta = \frac{Cn\omega \cos l \cos\phi}{Cn\omega \sin l + \mathrm{M}gh}. \qquad \text{(3)}$$

The components of $\omega$ about the vertical and the horizontal in the meridian are $\omega \sin l$ and $\omega \cos l$. The latter has a component $\omega \cos l \cos\phi$ about a horizontal axis at azimuth $\phi$ to the east of north, for the angle $\phi$ is here taken towards the north at the west bearing. This, in its turn, gives an angular speed, about an axis perpendicular at once to the line of bearings and to the axis of rotation, of amount $\omega \cos l \cos\phi \cos\theta$. The component $\omega \sin l$, about the vertical, gives a component, $-\omega \sin l \sin\theta$, about the axis last mentioned. The precessional angular speed about that axis is therefore $\omega(\cos l \cos \phi \cos \theta - \sin l \sin \theta)$. Hence, since as before the couple about the line of bearings is $Mgh \sin \theta$, we get

$$Cn\omega(\cos l \cos \phi \cos \theta - \sin l \sin \theta) = Mgh \sin \theta,$$

and therefore
$$\tan \theta = \frac{Cn\omega \cos l \cos \phi}{Cn\omega \sin l + Mgh}. \qquad \text{(4)}$$

Here it is supposed that $n$ and $\omega$ are the same way round. If they are not, the denominator (taking the numerical value of $\theta$) must have the value $Cn\omega \sin l - Mgh$, and the upper end of the axis is turned towards the south, instead of to the north as in the former case.

Taking equation (2), and supposing the flywheel to make N turns per second, we get $n\omega = 2\pi N \times 2\pi/86160 = 4\pi^2 N/86160$, and

$$\tan \theta' = \frac{\cos l}{\sin l - \dfrac{86160 Mgh}{4\pi^2 CN}}. \qquad \text{(5)}$$

If A be the moment of inertia of the instrument about the line of bearings and T its period as a compound pendulum oscillating about that line, we have $T = 2\pi(A/Mgh)^{\frac{1}{2}}$. Hence, if $C = kA$, (5) becomes

$$\tan \theta' = \frac{\cos l}{\sin l - \dfrac{86160}{kNT^2}}. \qquad \text{(6)}$$

Thus the wheel would turn over in latitude $l$ if we could make N as great as $86160/kT^2 \sin l$. As Sir George Greenhill remarks (*R.G.T.* p. 259), if T were 15 (as in the 400-day clock), $l$ were 30°, and $k$ were 2, we should get N = 383. Indeed, apart from the fact that the ring, in which the top is held, adds to A, the value of $k$ is certainly sensibly less than 2 in instruments actually made; nevertheless the speed of inversion is not unattainable. For example, the barogyroscope made by M. Koenigs* may (if there is no question of safety) be run up to 55,000 revolutions per minute, or over 900 per second. Such a speed would however be as much as, if not more than, a brass flywheel, even of small size, would bear. These results are important for various appliances.

7. *Gyrostat with axis vertical, stable or unstable according to direction of azimuthal turning.* We consider here a gyrostat supported (as shown in Fig. 35) by two trunnions screwed to the projecting edge, in the plane of the flywheel, on a wooden frame or tray. The axis of the wheel is very nearly vertical, and the wheel is spinning rapidly in the direction of an arrow (not shown) drawn on the upper side of the case. The centre of gravity of the whole instrument is nearly on the level of the trunnions, so that there is little or no gravity preponderance. The following experiment has already been described in I.

If the tray be carried round horizontally with constant angular speed $\mu$ in the direction of spin, the gyrostat remains quite stable. If however it be carried round in the opposite direction, the gyrostat immediately turns on its trunnions and capsizes, so that the other end of the

* *Revue Générale des Sciences*, Paris, 1891.

axis becomes uppermost, and if the azimuthal motion is continued in the same direction, the gyrostat is now stable. It will be observed that the flywheel is now spinning in the direction of the azimuthal motion. Hence the gyrostat is in stable equilibrium when the azimuthal motion is in the same direction as the rotational motion.

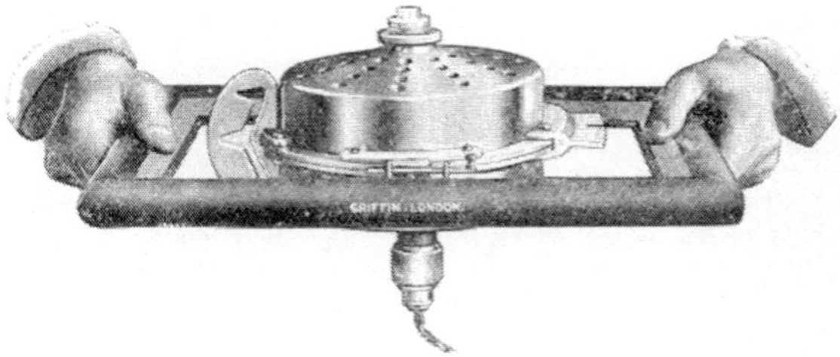


FIG. 35.

This result follows from the $\theta$-equation of motion, which is approximately, for the present case,

$$A\ddot{\theta}+Cn\mu\theta=0. \qquad (1)$$

The solution of this equation, if $n$ and $\mu$ have the same direction so that $n\mu$ is positive, is oscillatory motion of period $2\pi(A/Cn\mu)^{\frac{1}{2}}$ about the vertical position, so that this position is stable according to the definition of stability provisionally adopted in 9, V, above.

On the other hand, if $n$ and $\mu$ have opposite signs the solution of the differential equation is of another form, curiously connected with the former, but representing a different state of things. It shows that if the gyrostat is disturbed from the vertical position of its axis it tends to pass further away from it; the instrument capsizes. The solution in this case, to suit the initial condition of $\theta=0$ when $t=0$, is

$$\theta=B(e^{pt}-e^{-pt}),$$

where $p=(Cn\mu/A)^{\frac{1}{2}}$ and B are constants. This indicates a continuous increase of $\theta$ as $t$ increases.

The two results are indeed indicated by (1). The moment $Cn\mu\theta$, regarded as a couple producing rate of change $A\ddot{\theta}$ of A.M., is in the first case in the direction to check motion away from the vertical position, and to bring the gyrostat back to that position, while, in the other case, $Cn\mu\theta$, having the opposite sign, produces A.M. in the direction away from the vertical [see also below].

It will be seen that in this arrangement of the gyrostat it has one freedom of motion as regards inclination of the axis to the vertical; it can turn about the trunnions, but not about a horizontal axis at right angles to the line of the trunnions. Hence, as we shall show later, the gyrostat cannot have complete dynamical stability.

Even with gravitational stability of the spinning gyrostat, slow azimuthal turning will be consistent with a position of stable equilibrium if in one direction, but not when in the opposite direction.

The inversion of the flywheel brings into play a wrench on the hands of the experimenter. A varying couple, lasting during the time of the inversion, is required to reverse the angular momentum of the wheel in space, and this is applied to the gyrostat by the frame at the trunnions, and to the frame, because that is kept steady, by the hands of the operator. The total change of angular momentum is $2Cn$, and this is the time-integral of the couple.

The couple arises thus. Let the gyrostat axis have been displaced from the vertical through an angle $\theta$ about the trunnion axis. In consequence of the azimuthal motion, at rate $\mu$, the outer extremity of the axis of angular momentum is being moved parallel to the instantaneous position of the line of trunnions, and thus there is rate of production R of angular momentum about that line; but, since there is no applied couple about the trunnions, the gyrostat must begin to turn about the trunnions to neutralise R. This turning tends to erect or to capsize the gyrostat according as the spin and azimuthal motions agree or are opposed in direction. In its turn however this involves production of angular momentum about the vertical for which a couple must be applied by the frame, and of course to the frame by the operator. This couple is greater the greater $Cn$, and therefore if the operator cannot apply so great a couple an azimuthal turning at rate $\mu$ cannot take place. With sufficiently great angular momentum the resistance to azimuthal turning could be made for any stated values of $\theta$ and $\mu$ greater than any specified amount.

The magnitude of this couple, which measures the resistance to turning at a given rate, is greatest when the angle $\theta$ is 90°, that is when the axis of the flywheel is in the plane of the frame.

It is important to notice that if the gyrostat be placed on the trunnions, so that the axis of the wheel is in the plane of the frame, azimuthal turning in one direction causes one end of the axis to rise, and turning in the other direction causes the other end to rise. As the reader will see, this also means a reaction couple in the plane of the frame which must be balanced by a couple applied by the experimenter.

8. *Top supported by a string attached at a point E of the axis.* The string is supposed attached at a fixed point D [Fig. 36]. DO is a downward vertical intersecting EG, the axle, in O. G is the centroid of the top, and is supposed to be coincident with the centre of the wheel or spinning part. We denote the distances DE, EG respectively by $l$, $h$, and the angles $\theta$, $\phi$

are indicated in the diagram. First, we suppose that G is turning about the vertical OD with angular speed $\mu$, and that as the motion is steady the lines DE, EG remain in the vertical plane through OD. Let $A_1$ denote the moment of inertia about a transverse axis through G, and $A = A_1 + Mh^2$, that is let A be the moment of inertia about a transverse axis through E. The equations of motion are, (1) for the centroid,

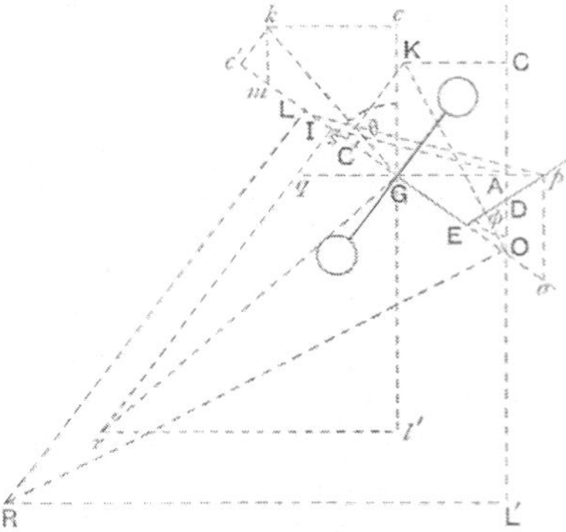

FIG. 36.

$$\left.\begin{aligned} M\mu^2(l\sin\phi + h\sin\theta) &= T\sin\phi, \\ Mg &= T\cos\phi, \end{aligned}\right\} \quad (1)$$

where T is the pull exerted by the cord, and (2) for turning round a vertical through G,

$$(Cn - A_1\mu\cos\theta)\mu\sin\theta = Th(\sin\phi\cos\theta + \cos\phi\sin\theta). \quad \ldots (2)$$

Substituting in (2) the values of $T\sin\phi$, $T\cos\phi$ from (1), we find

$$(Cn - A\mu\cos\theta)\mu\sin\theta = M\mu^2hl\sin\phi\cos\theta + Mgh\sin\theta. \quad \ldots (3)$$

We consider some particular cases. First, if the string were attached at G the moment of forces about any axis through that point would be zero, and there would be no couple changing the direction of the axis of the wheel. The identification of the angular turning of the axis about the vertical OD with that of the centroid about the same line fails, and we see that the centroid goes round the vertical in steady motion with angular speed $\mu$, while the axis of rotation maintains a fixed direction in space.

The value of $\mu$ in this case is given by (1). Putting $h = 0$, we get

$$\mu^2 = \frac{g}{l\cos\phi},$$

that is the motion of the centroid is that of a conical pendulum of height $l\cos\phi$.

If E is above G, equations (1) and (3) become

$$\left.\begin{aligned} M\mu^2(l\sin\phi - h\sin\theta) &= T\sin\phi, \\ Mg &= T\cos\phi, \end{aligned}\right\} \quad \ldots (4)$$

$$(Cn - A\mu\cos\theta)\mu\sin\theta = -M\mu^2hl\sin\phi\cos\theta - Mgh\sin\theta. \quad \ldots (5)$$

The precession (the small root of the quadratic) is now changed in sign.

If E is below O the radial force available for the circular motion of the centroid about the vertical OD acts outwards, and so the motion is not possible without reversal of this horizontal force. The substitution for the string support of a strut acting between a point D′ on the vertical below O and the point E (now to the right of the vertical) will make the motion possible.

A diagram, Fig. 36, similar to that of Fig. 19, can be constructed for a top held up by a string as described above (see also Greenhill, *R.G.T.* p. 12). Along the axle lay off Gc′ to represent $Cn$, and along an upward vertical through G lay off Gc to represent $A_1\mu\sin^2\theta + Cn\cos\theta$. Then, if $c'k$, $ck$ be at right angles to Gc′, Gc respectively, $km$

represents $A_1\mu$ and $c'k$ $A_1\mu\sin\theta$, while $ck$ represents $Cn\sin\theta - A_1\mu\cos\theta\sin\theta$. Moreover the moment of T round G is $Mg\,.\,Gp$ (see Fig. 36) Hence

$$\mu\,.\,ck = Mg\,.\,Gp,\quad \mu\,.\,kn = Mg\,.\,Go,\ \ldots \qquad \ldots\ldots\ldots\ldots\ldots \ (6)$$

if $n$ be the point in which a parallel drawn through $k$ to $Gc'$ intersects $Gc$, and $o$ the intersection of a vertical through $p$ with OG produced backward Of course $kn = mG$.

Let $\lambda$ denote $g/\mu^2$, the length of the equivalent conical pendulum, and draw downward from G a vertical line $Gl' = \lambda$. Through $l'$ draw a horizontal line in the plane of the diagram to meet $Gr$ drawn at right angles to $Gk$ in $r$, and draw $rqs$ perpendicular to $Gc'$, intersecting $pG$ produced in $q$. Also along OC make $ON = (A_1 + Mh_1^2)\mu$, and $OI = h_1 + A_1/Mh_1$, where $h_1$ is the distance OG. Then $ck/kc' = kn/km = Gl'/Gs = g/(\mu^2\,.\,Gs)$, and from these relations we get by (6)

$$Go = \frac{\mu}{Mg}\,kn = \frac{\mu}{Mg}\,\frac{Gl'}{Gs}\,km = A_1\frac{\mu^2}{Mg}\,\frac{Gl'}{Gs}. \ \ldots\ldots\ldots\ldots\ldots\ldots \ (7)$$

From these it follows that $Go\,.\,Gs = Gp\,.\,Gq = OG\,.\,GI = A_1/M$, so that

$$Gs/GI = OG/oG = GA/Gp, \ \ldots\ldots\ldots\ldots \ (8)$$

and $As$ is parallel to $Ip$, and $(p, q)$, $(o, s)$, (O, I) are pairs of convertible centres of suspension and oscillation for the top regarded as a compound pendulum with centroid G, and each centre of suspension and corresponding centre of oscillation on the axis of symmetry.

Thus, if we draw the axis OG at the given angle $\theta$ to the vertical, and make the constructions as described above, then draw $As$ parallel to $pI$ and $sqr$ perpendicular to OG, cutting in $r$ the horizontal through $l'$, we find that $Gk$ is perpendicular to $Gr$.

In the figure OL′ is also made equal to $Gl'(=\lambda)$ and

$$OC' = Cn,\quad OC = (A_1 + Mh_1^2)\mu\sin^2\theta + Cn\cos\theta,$$

the A.M about the vertical through O, and perpendiculars are drawn to OC from C′ and to OC′ from C, meeting in K. OR is drawn perpendicular to OK to meet the horizontal through L′ in R, and RL is drawn at right angles to the axis. Thus we get a diagram for the point O, which is fixed in the steady motion. The reader may prove that

$$ON = \frac{KC'}{\sin\theta},\quad GE = \frac{\mu\,.\,CK}{Mg\sin\theta} = \frac{\mu\,.\,KC'}{Mg\sin\theta}\,\frac{OL'}{OL} = \frac{A_1 + Mh_1^2}{Mg}\,\mu^2\frac{OL'}{OL} = \frac{A_1 + Mh_1^2}{M\,.\,OL}, \ \ldots\ldots \ (9)$$

$$GE\,.\,OL = \frac{A_1 + Mh_1^2}{M} = OG\,.\,OI. \ \ldots\ldots\ldots\ldots\ldots\ldots \ (10)$$

[The reader may refer to Greenhill (*R.G.T.* p. 12) for constructions for particular cases, including that for which the thread must be replaced by a strut, as explained above.]

It may be remarked that though the point O is fixed the motion is not simply that which would exist if O were really the point of support. The moment causing the steady turning would then be $Mgh_1\sin\theta$, which it is not in the present case, unless E is at O, when $h = h_1$. But then we should have in (1) $\sin\phi = 0$, and $\theta = 0$.

**9.** ***Gyrostatic action of the wheels of vehicles and of the rotating parts of machinery. Monorail cars.*** The wheels of a carriage have gyrostatic action when the vehicle turns about an axis perpendicular to the road, this is when changing its direction of motion by turning in azimuth. When the carriage passes over concavities or convexities in an otherwise straight road, no change of direction of the axes of the wheels takes place and there is no gyrostatic action.

Let the carriage have N equal wheels of moment of inertia $mk^2$ each, about parallel axes, and let the speed of the carriage be $v$, in feet per second, so that if the radius of the wheels be $r$ feet, their angular speed is $v/r$. Finally, let the direction of motion be changing at angular speed $\mu$.

The A.M. of the wheels, that is $Cn$, is $Nmk^2v/r$, and to an observer, standing behind the carriage and looking forward, this may be represented by a single vector drawn out from a point O in the carriage to the left, as shown in Fig. 37, for a single wheel. There is stability enough of course to prevent turning of the carriage round either rail, but a capsizing couple which we can estimate is exerted in consequence of the gyrostatic action of the wheels. According as the turning with angular speed $\mu$ is towards the observer's right or his left the outer extremity of the vector $Cn$ turns towards the forward part OA, or the after part OB, of the instantaneous position of a fore and aft line, drawn parallel to the floor and sides of the carriage, through the point O. We suppose here that there are no forces, due to position of the carriage or other cause, balancing the gyrostatic action [3, III]. Everything may be supposed in equilibrium with the carriage moving on the curve with the wheels locked, and then the wheels to be set rotating. Hence, as there is no applied couple about the line OA, we have initially for turning about it the equation $A\ddot{\theta}+Cn\mu=0$; so that the carriage in the first case tilts over to the left on its springs, and equilibrium is finally produced, for steady turning at rate $\mu$, by a reaction of moment $Mgh\theta=Cn\mu$ tending to turn the carriage over to the right, that is towards the inside of the curve in which the carriage is travelling. The carriage is now inclined over through a small angle $\theta$ on its springs, and the equation is $Cn\mu=Mgh\theta$.

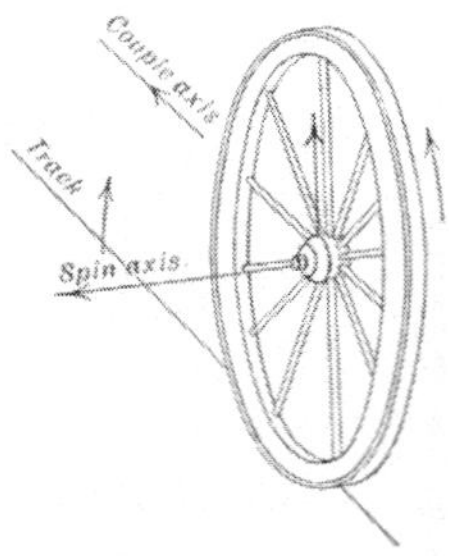

Fig. 37.

In the other case the reacting couple $Cn\mu$ tends to turn the carriage over to the left, that is again towards the inside of the curve.

The moment is $Cn\mu=C\mu v/r$. If R be the mean radius of the curves in which the inside and outside wheels are moving, we have $\mu=v/R$, approximately, and so the capsizing couple due to gyrostatic action is $Cn\mu=Cv^2/Rr$. If P be the force applied to the outside wheels, Q that applied to the inside wheels, at right angles to the road, we get, in gravitation units, and taking $\theta$ small, the wheel-gauge as $2l$, and $h$ as the height of the centre of gravity,

$$P-Q=\frac{Mh}{gRl}v^2+C\frac{v^2}{gRlr}=\frac{Mhv^2}{gRl}\left(1+\frac{C}{Mrh}\right), \qquad (1)$$

which shows that the gyrostatic moment is the fraction $C/Mrh$ of the centrifugal couple. As a rule $C/Mrh$ is very small, and then the gyrostatic couple due to the rotation of the wheels of a vehicle is of little importance.

Super-elevation of the outside rail of a railway curve is not here considered. On high-speed electric railways a carriage on entering or leaving such a curve experiences considerable gyrostatic action. [See a later chapter on *Gyrostatics in Engineering*.]

A monorail car going round a curve of radius R at speed $v$ will, if it be heeled over through an angle $\alpha$ towards the inside of the curve, and contain a gyrostat of A.M. $C'n'$, with axis of rotation at right angles to the sides of the carriage, have the equation of equilibrium,

$$\tan\alpha = \frac{v^2}{gR}\left(1 + \frac{C}{Mrh}\right) + \frac{C'n'v}{gMRh} \cdot \quad \ldots\ldots\ldots (2)$$

In this case the gyrostatic action is all-important for equilibrium. A special arrangement is necessary to enable the carriage to get into this equilibrium position, and this will be described when we come to deal with the more technical applications of gyrostatics. If the gyrostat were fixed in the carriage with its spin axis at right angles to the sides, the carriage when rounding a curve would heel over to the outside of the curve, as in the case of a two-rail carriage, and would be unstable. It would be a system with a single unstable freedom, which could not be stabilised by rotation.

In a motor-car, besides the action of the wheels, there is that of a flywheel placed across the motor. Both actions will come into play when the motor-car is changing the direction of motion, as in turning a corner, while only that of the flywheel has any influence when the car is passing over convexities or concavities of the road. A fast motor-car may by passing over a highly convex part of the road—an old bridge, for example—have the grip of the wheels on the ground dangerously reduced, and the steering action impaired; but this has nothing to do with gyrostatic action.

When the car is rounding a corner, the difference of weights borne by the wheels on the two sides is given as before by equation (1). But now the fore-and-aft axis of the flywheel is turning towards a transverse, and therefore the growth of A.M. about that axis causes the car to tilt the other way, until a couple is developed to balance the gyrostatic couple $C\mu v/r$. Hence more or less weight is thrown on the front wheels than on the back, as compared with the distribution of the weight on these wheels when the car was running straight forward. If the flywheel as seen by an observer, looking from behind, be spinning counter-clockwise, and the car be turning to his left, the gyrostatic couple will bring more weight to bear on the front wheels than on the back, by the amount $C'n'v/gRb$, where $C'n'$ is the A.M. of the flywheel, and $b$ is the length of the wheel-base.

If we take account of the action of the differential gear, we see that ($2l$ being the gauge) the angular speeds of the wheels on the two sides are given by

$$\left.\begin{matrix}\omega_1\\ \omega_2\end{matrix}\right\} = (R \pm l)\frac{\omega}{R}, \quad \ldots\ldots\ldots (3)$$

where R is the mean radius of the curve, and $\omega$ the mean angular speed of the wheels. The gyrostatic couple for the mean speed $v$ is $Cv^2/Rr$, and so, for the weights borne on the two sides, we get

$$\left.\begin{matrix} M_1 \\ M_2 \end{matrix}\right\} = \tfrac{1}{2}M\left\{1 \pm \left(1 + \frac{C}{Mrh}\right)\frac{v^2h}{glR}\right\}. \qquad (4)$$

10. *Gyrostatic action of paddle-wheels and of screw in steamers.* In a paddle-steamer the action of the wheels when the course is being changed is just that of the wheels of a vehicle. But the direction of the axis of rotation of the paddles is also changed by the rolling of the ship, and this gives gyrostatic action about an axis at right angles to the deck. If the angular speed of rolling at any instant is $\omega$, and the A.M. of the wheels be $Cn$, the gyrostatic action is $Cn\omega$, and thus, if at the instant the ship is on an even keel, a rate of growth $Cn\omega$ of A.M. about a vertical axis is produced. If the ship is then rolling to starboard the direction of this $Cn\omega$ is round from starboard to port, that is against the motion of the hands of a watch lying face up on the deck. Hence, if we denote by $\theta$ an angle of turning about the normal to the deck, by A the proper value of the moment of inertia of the ship for the turning, and by L the couple on the ship due to the now oblique motion and to the turning—initially this couple is zero—we have, as in 9,

$$A\ddot{\theta} + Cn\omega = L.$$

Thus, initially at least, $\ddot{\theta}$ is in the direction to neutralise $Cn\omega$, and the ship's head turns to starboard. The rolling from port to starboard and again from starboard to port makes the ship's head turn alternately to starboard and to port—the ship "yaws" in consequence of gyrostatic action. The greater immersion of the paddle-wheel, caused by the rolling, tends to correct this.

In a paddle-steamer pitching has no effect. Changing the course at angular speed $\omega$ gives a rate of growth of A.M. $Cn\omega$ about a fore-and-aft axis, and as a result, since there is no externally applied couple, the ship has an angular acceleration of amount $Cn\omega/A$ in the opposite direction. Thus, if the vessel be turned by the rudder, say to port, the vessel will by gyrostatic action be slightly heeled over to starboard, and the starboard wheel being more deeply immersed will assist the turning action of the rudder. Though the gyrostatic action of the wheels is, owing to their comparatively slow speed of revolution, not very great, calculation shows that it produces an appreciable variation in the immersion of the wheels.

The resistance normal to the course gives the radial acceleration in the curvilinear motion.

In a screw-steamer the action of the engines and propeller is to give a gyrostatic couple about a horizontal transverse axis when the ship is changing its course, and about a vertical transverse axis when the ship is pitching. The magnitude is in each case $Cn\omega$, and the direction is given as in the examples already considered.

11. *Gyrostatic action of turbines.* The use of turbines in screw-steamers has considerably augmented the gyrostatic action of the engines. This is best illustrated by taking a numerical example. Since the axes of the rotors are fore and aft, gyrostatic action only occurs when the ship pitches or when she changes her course. In the Cunard ship *Carmania* the total weight of the rotors of the turbines, two on "wing-shafts" and one in the middle, may be taken as 200 tons ($\frac{4}{5}$ to the wings and $\frac{1}{5}$ to the centre), and the radius of gyration as 4 feet, so that in ton-foot units the moment of inertia of the rotor on each wing-shaft is 1280, and the moment of inertia of the central rotor is 640. The number of revolutions is 200 per minute, and consequently $n$ is $20\pi/3$ in radians per second. The ship's head can be turned about $\frac{3}{4}$ of a degree, or say $\frac{1}{75}$ of a radian, in a second. Hence the moment of the couple which must be applied by the ship to each wing-rotor to give it the precession which the turning of the ship involves, and therefore also the moment of the equal and opposite couple exerted on the ship, is $1280 \times 20\pi \times \frac{1}{3} \times \frac{1}{75} \times \frac{1}{32} = 11{\cdot}2$, in Ton-foot units; that is the moment is that which would be produced by a force of 11·2 Tons, acting at an arm of 1 foot, or a couple of ·28 Ton acting at an arm of 40 feet, the approximate distance between the end-bearings of the turbines. Such a couple cannot have any perceptible effect in producing pitching or in straining the ship. [The word "Ton" is here printed with an initial capital to indicate its use as a unit of force.]

If we take 12° as the range of pitching, and the period as 6 seconds, the maximum angular speed is $2\pi \times 6/(6 \times 57{\cdot}3) = 1/9$, in radians per second, and this is to be substituted for 1/75 in the above calculation. The couple is thus 8·3 times the former couple, or 2·3 Tons at an arm of 40 feet; still quite a small couple from the point of view of breaking the ship. The torpedo-boat destroyer *Cobra* was lost in the North Sea in 1901, and as it was one of the first vessels to be fitted with turbine engines it was thought by several people to have been broken in two by the gyrostatic action of the turbines, a view which the figures given above for the much larger *Carmania* show to be quite untenable. Pitching, the more serious of the two causes of gyrostatic action with turbines, was supposed to have caused the disaster, but pitching would produce with turbines a couple about a vertical axis, and of course it is absurd to suppose that the vessel was destroyed in that way. It was very remarkable that some practical engineers seemed to imagine that pitching might give gyrostatic action about a horizontal axis.

We have referred above to the couple applied to the ship by each of the rotors. If there were only two shafts, one right-handed the other left-handed, the total moment applied to the ship would be zero; but internal stresses of a kind easily understood would be set up in the structure. These would tend to produce alternate extension and compression at the stern, and alternate compression and extension at the bow, athwart-ship in each case, but these strains would be quite negligible.

**12.** *Steering of a bicycle, of a child's hoop, or of a wheel in a wheel-race.* Another example which deserves mention is that of a bicycle (which, if space can be spared, will be more fully treated later in this work). If the rider feels himself beginning to fall over to one side or the other he instinctively turns the bicycle towards that side, and the inertia in the forward movement, assisted by the gyrostatic action of the driving wheel, over which the rider sits, causes the bicycle frame to set itself erect again. The gyrostatic action will be made out very easily—as in Fig. 37, the vector $Cn$ is towards the rider's left, and the frame, if turned to the right or left when inclined over to that side, experiences a gyrostatic couple $Cv^2/r\text{R}$ (where $v$ is the forward speed, $r$ the radius of the wheel, and R that of the curve of turning) tending to turn the frame to the upright position.

A child's hoop, or the wheel in the wheel-race of military sports, affords another example. Each competitor runs forward alongside his wheel and guides it more or less adroitly. By careful experimenting or by reasoning from gyrostatic theory he can compile a set of rules for use in the game.

If he runs with the wheel on his right the vector representing $Cn$ $(=Cv/r)$ will be drawn out towards him from the centre of the hub [Fig. 37]. If he exerts with his hand a downward push on the hub the front of the wheel will turn towards him, if he applies an upward force to the hub the front of the wheel will turn from him. In the first case the couple-axis is drawn backwards, in the second it is drawn forwards, in the plane of the wheel, and in such cases the rule that the A.M. vector turns towards the couple-vector is easily remembered and applied.

Again, as the man pushes from him the front or the back end of a horizontal diameter of the hub, the top of the wheel inclines towards or from him. The same effects are produced by pushing forward or pulling back the hub of the wheel.

Let the wheel be running steadily in a circle of radius R, under the influence of a couple, which we may suppose applied by a weight hung on the inner end of the hub, so that the whole moving mass is M and the centroid is at a distance $h$ nearer the vertical through the centre of the path than is the plane of the edge of the wheel, which we suppose to be vertical. Let X be the radially inward horizontal force applied to the rim of the wheel at its point of contact with the ground, and Z the upward vertical force applied at the same point. Then we have

$$X = M\mu^2(R-h), \quad Z = Mg, \quad\text{..........................(1)}$$

where $\mu$ is the angular speed with which the plane of the wheel is turning about the vertical, that is also the angular speed with which the centroid is turning in its circular path.

Taking moments about the centroid we get

$$Cn\mu = Zh - Xr = Mgh - M\mu^2(R-h)r. \quad \text{.....................(2)}$$

But also we have $\mu = V/R$, $n = V/r$, so that (2) becomes

$$C\frac{\mu}{r}V = Mgh - M\mu rV + M\mu^2 rh.$$

Rearranged this equation is

$$V\left(\frac{C}{Mr}+r\right) = h\left(r\mu + \frac{g}{\mu}\right) = 2h(gr)^{\frac{1}{2}} + h\left\{\left(\frac{g}{\mu}\right)^{\frac{1}{2}} - (r\mu)^{\frac{1}{2}}\right\}^2. \quad \text{..........(3)}$$

Thus V has the smallest possible value for $(g/\mu)^{\frac{1}{2}} = (r\mu)^{\frac{1}{2}}$, that is for $\mu^2 r = g$, or when the angular speed of the wheel round the centre of the curved path is equal to that of a conical pendulum of length (height) equal to the radius of the wheel. Then, since $\mu = (g/r)^{\frac{1}{2}}$,

$$R = \frac{V}{\mu} = \frac{2h}{\dfrac{C}{Mr^2}+1}. \quad \text{.................................(4)}$$

For a graphical representation of this solution the reader may refer to Greenhill, *loc. cit.*

**13.** *One spinning top supported by another. Steady motion.* The axis or stalk of a top which is spinning about a fixed point is prolonged as shown in Fig. 38, and carries at its upper extremity a small cup, in which a second top is supported. It is required to determine the conditions of steady motion, with angular speed $\mu$, and the two axes in the same vertical plane. Denoting the masses by $M$, $M_1$, inclinations of the axes to the vertical by $\theta$, $\theta_1$, moments of inertia about transverse axes by $A$, $A_1$, and A.M. about the axes of figure by $Cn$, $C_1n_1$, all for the lower and upper tops respectively, and putting $l$ for the length of the stalk of the lower top, $h$, $h_1$ for the distances of the centroids from the points of support, and X, Z for the horizontal and vertical components of the forces applied at the cup to the upper top, as shown in Fig. 38, we get for the equation of motion of the upper top

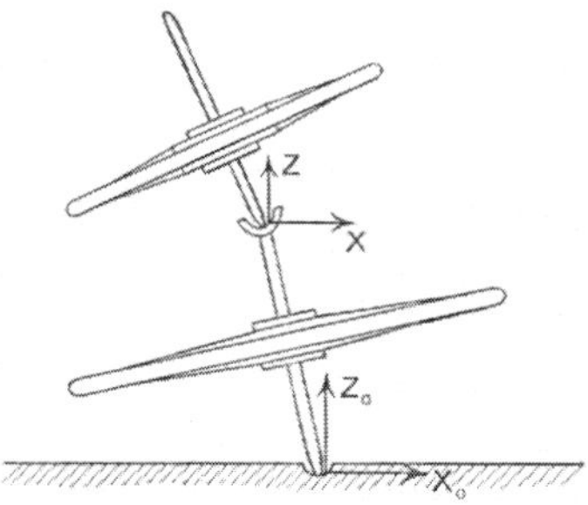

Fig. 38.

$$\mu^2 M_1(l\sin\theta + h_1\sin\theta_1) = X, \quad M_1 g = Z, \quad \text{.......................(1)}$$

$$(C_1n_1 - A_1\mu\cos\theta_1)\mu\sin\theta_1 = Xh_1\cos\theta_1 + Zh_1\sin\theta_1. \quad \text{..............(2)}$$

Equation (2) is by (1),

$$\{C_1n_1 - (A_1 + M_1h_1^2)\mu\cos\theta_1\}\mu\sin\theta_1 = \mu^2 M_1 lh_1\sin\theta\cos\theta_1 + M_1gh_1\sin\theta_1. \quad (3)$$

Similarly we get for the lower top,

$$\{Cn-(A+Mh^2+M_1l^2)\mu\cos\theta\}\mu\sin\theta$$
$$=\mu^2M_1lh_1\sin\theta_1\cos\theta+(M_1l+Mh)g\sin\theta. \quad\text{.................(4)}$$

If $\theta_1$, $\theta_2$ be both small these equations become

$$\left.\begin{array}{l}\{C_1n_1-(A_1+M_1h_1^2)\mu\}\mu\theta_1-M_1gh_1\theta_1-M_1\mu^2lh_1\theta=0,\\ \{Cn-(A+Mh^2+M_1l^2)\mu\}\mu\theta-(M_1l+Mh)g\theta-M_1\mu^2lh_1\theta_1=0.\end{array}\right\} \quad\text{......(5)}$$

Eliminating $\theta$, $\theta_1$ between these we get the equation of condition

$$\{C_1n_1\mu-(A_1+M_1h_1^2)\mu^2-M_1gh_1\}$$
$$\times\{Cn\mu-(A+Mh^2+M_1l^2)\mu^2-(M_1l+Mh)g\}-\mu^4M_1^2l^2h_1^2=0, \quad\text{.......(6)}$$

which is a biquadratic for $\mu$.

A third top might be placed on the upper top of the pair just considered, and supported as before on a small cup carried by a prolongation of the stalk of the top below. This chain of three tops may move in steady motion with the axes in the same vertical plane, under conditions expressed by a sextic equation in $\mu$, which the reader may investigate.

For a double compound pendulum composed of two rigid compound pendulums hinged together, and turning so as to present always the same face to the vertical through the point of support, we have only to reverse $g$, and write (since the $\theta$-angles are both small)

$$Cn=C\mu\cos\theta=C\mu,\quad C_1n_1=C_1\mu\cos\theta_1=C_1\mu.$$

Equation (6) becomes

$$\{(C_1-A_1-M_1h_1^2)\mu^2+M_1gh_1\}$$
$$\times\{(C-A-Mh^2-M_1l^2)\mu^2+(M_1l+Mh)g\}-\mu^4M_1^2l^2h_1^2=0. \quad\text{......(7)}$$

From (4) above we get by putting $M_1=0$, and reversing $g$, the exact equation of motion of a compound conical pendulum.

Supposing this compound pendulum to have plane motion, the reader may prove that if $M_1/M$ be small, and $l$ be not very great, as in the case of a bell and its clapper, and the centres of oscillation of the two pendulums be coincident when the centroids are in line, the two pendulums, started together with $\theta=\theta_1$ and $\dot\theta=\theta_1$, will vibrate together, so that $\theta$ remains equal to $\theta_1$. Thus the bell will not ring.

One way of curing a bell which behaved in this way would be to lengthen the clapper considerably. This is said to have been done for a bell in Cologne Cathedral.

**14.** *Drift of a projectile.* The turbine of a ship moving forward while rotating may be compared to a projectile fired from a rifled gun. As looked at by an observer at the firing point the rotation is right handed, and the shot drifts in its trajectory, which is convex upward, towards the right. But the spin vector is for the rotation specified to be drawn forward, and therefore the bow of the ship, in consequence of a similar turning of the axis due to pitching, would turn towards the left: so that the idea of the projectile as a gyrostat moving forward on a convex track with its axis in the forward direction throws no light on the drift of the projectile.

The cause of this drift is not yet fully understood, but it is connected with the rotation, as reversal of the rotation reverses its direction. It amounts to ·25, 1·1, 4·4, 11·5 metres in ranges of 500, 1000, 2000, 3000 metres respectively. Since the rapidly rotating projectile tends to keep the direction of its axis unchanged, it is presently moving forward on the convex trajectory with its axis in the plane of the trajectory, but pointing a little upward relatively to the path. Thus it has a motion in the direction of the axis together with a lateral component. Hence, as we shall see presently, there is applied by the air a couple, which in the absence of spin would tend to increase this obliquity of the axis of spin to the direction of motion; but as the projectile spins rapidly about its axis, it precesses about the instantaneous position of the axis of the resultant momentum, as explained in 17 below, with of course modification of the resistance in consequence. As a result the projectile moves forward in air, and, relatively to the path, its point is directed slightly upward and to the right, and the shot is continually deflected towards the right by a side thrust applied by the air.

15. *Turning action on a body moving in a fluid.* Before proceeding further with the discussion of the motion of a projectile spinning in air, it will be convenient to discuss some fundamental principles of the dynamics of a body moving in a fluid medium. Consider first a rigid body of mass M moving without rotation parallel to a fixed plane (Fig. 39). Take axes O$x$, O$y$ from any origin in that plane, and let $\dot{x}$, $\dot{y}$ be the speeds of the body parallel to these axes. The momenta of the body in these directions are $M\dot{x}$, $M\dot{y}$, and the body has angular momentum $M(\dot{y}\xi-\dot{x}\eta)$ about an axis of $z$ through the origin, since we may regard the body as replaced by a particle of mass M situated at the centroid (coordinates $\xi$, $\eta$) and moving with the velocity ($\dot{x}$, $\dot{y}$). The time-rate of change of this A.M. is $M(\ddot{y}\xi-\ddot{x}\eta)$, which for the present we shall suppose to be zero through the vanishing of $\ddot{x}$, $\ddot{y}$.

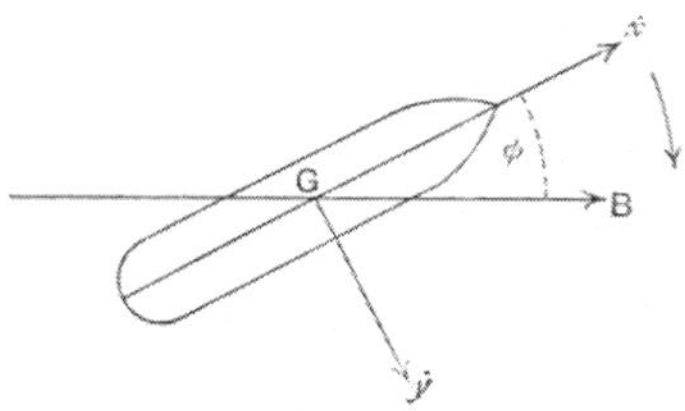


FIG. 39

Now let there be matter set in motion by the body, so that the total momentum in the direction of O$x$ is $M_1\dot{x}$, and that in the direction of O$y$ is $M_2\dot{y}$. Then if we associate these components of momentum with the body, we regard it as having inertia $M_1$ in the direction of O$x$, and inertia $M_2$ in the direction of O$y$. The A.M. about the origin is now $M_2\dot{y}\xi_1-M_1\dot{x}\eta_1$, where $\xi_1$, $\eta_1$ are the coordinates of a point, moving with the body, the position of which it is not necessary here to specify. The rate of change of this A.M. (since $\ddot{x}=\ddot{y}=0$) is $M_2\dot{y}\dot{\xi}_1-M_1\dot{x}\dot{\eta}_1=(M_2-M_1)\dot{x}\dot{y}$, since $\dot{\xi}_1=\dot{x}$, $\dot{\eta}_1=\dot{y}$, and therefore does not depend on $\xi$, $\eta$.

Or, to put the matter in another way, consider a point A of space with which a point B of the body, or moving with the body, coincides at time $t$. By the displacement $\dot{x}\,dt$ of the body in an interval of time $dt$, B is carried this distance parallel to $Ox$ from A, and A.M. $M\dot{y}\dot{x}\,dt$ is produced. Similarly A.M. $-M\dot{x}\dot{y}\,dt$ about A is produced by the displacement $\dot{y}\,dt$ of the body. Thus zero A.M. is produced on the whole. But if the momentum associated with the body be $M_1\dot{x}$ parallel to $Ox$, and $M_2\dot{y}$ parallel to $Oy$, the former gain of A.M. is $M_2\dot{y}\dot{x}\,dt$ and the latter is $-M_1\dot{x}\dot{y}\,dt$, that is A.M. about A is being gained at rate $(M_2-M_1)\dot{x}\dot{y}$. This is independent of the position of A, that is it is the same for all points.

This rate of gain of A.M. about every point is wholly due to the matter set in motion by the body, and is effected by the action of a couple exerted by the body on that matter (the action of a ship, for example, on the water), which therefore exerts an equal and opposite couple on the body.

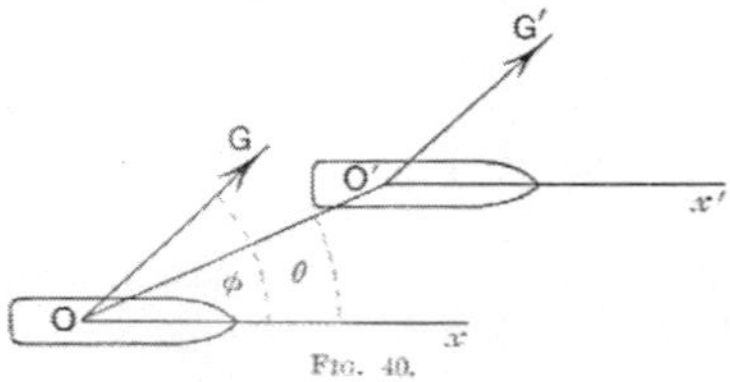

FIG. 40.

The same result may be obtained in another way which is also instructive. Let (Fig. 40) the axis of figure of the body be in the direction $Ox$ at a given instant and remain parallel to this direction as the body moves. Then after an interval of time $t$ the point O has been moved to O′, and the direction of the axis of figure is now $O'x'$. Let the direction of the resultant virtual momentum of the body be OG for the first position, and O′G′ for the second; these resultants (at $t$ and $t'$) are equal and parallel. Now we may suppose the first resultant to have been annulled in time $t$ by a force F acting from the point G to O, and the other resultant to have been brought into existence in the same time by an equal force F acting from O′ towards G′. Thus we have acting a couple of moment $F\,.\,OO'\sin(\phi-\theta)$, and the whole generation of A.M. is $Ft\,.\,OO'\sin(\phi-\theta)$, where $\phi$ is the angle $GOx$, and $\theta$ the angle $O'Ox$. But F is the resultant momentum, which may be regarded as represented by OG. Thus the A.M. generated is $OG\,.\,OO'\sin(\phi-\theta)$.

The identification of this with the result already obtained is easy. Taking $\dot{x}$ and $\dot{y}$ in the directions of $Ox$ and a perpendicular to $Ox$ in the plane of motion, we have

$$OG\cos\phi=M_1\dot{x},\quad OG\sin\phi=M_2\dot{y},\quad OO'\cos\theta=\dot{x}t,\quad OO'\sin\theta=\dot{y}t.$$

Thus we get

$$\begin{aligned} OG\,.\,OO'\sin(\phi-\theta) &= OG\sin\phi\,.\,OO'\cos\theta-OG\cos\phi\,.\,OO'\sin\theta \\ &= M_2\dot{y}\,.\,\dot{x}t-M_1\dot{x}\,.\,\dot{y}t=(M_2-M_1)\dot{x}\dot{y}t. \quad \text{...............(1)} \end{aligned}$$

As usually we put $v$ for the speed of a body in the direction of its axis, we shall in what follows put $v$ for $\dot{x}$ and $w$ for the transverse speed $\dot{y}$. We see then that the couple $(M_2 - M_1)vw$ applied for the time $t$ will generate the change of A.M. which grows up in that time.

**16.** *Turning action of the water on a ship. Why a ship carries a weather helm.* This is the couple that tends to turn a ship at right angles to its course, and that actually sets a ship or plank athwart a stream in which it is allowed to drift. It must be counteracted in the case of the ship by the rudder. A ship set on a course and left with helm lashed would be unstable; the helmsman has continually to prevent the ship from falling off her course, and good steering (steering that is economical of coal) consists in correcting each infinitesimal deviation as it arises. For, considering an elongated symmetrical body immersed in a medium indefinitely extended in each of the directions of motion (so that we are not concerned with reactions from the boundaries), let the speed $\dot{x}$ be that of the body in the direction of its length, and $\dot{y}$ that in the direction at right angles to the length. Let $M_2 - M_1$ be positive. If either $\dot{x}$ or $\dot{y}$ be zero the couple $(M_2 - M_1)\dot{x}\dot{y}$ is zero. Let, for example, $\dot{y}$ be zero. Then if the length be allowed to swerve through the angle $\phi$ from the direction GB (Fig. 39) in which the body is moving, there will now exist a speed $\dot{x}$ in the direction of the length, and a speed $\dot{y}$ in the perpendicular direction, as shown by the arrows, and a couple $(M_2 - M_1)\dot{x}\dot{y}$ in the direction of the curved arrow will be exerted on the matter outside the body but in motion with it. An equal and opposite couple acts on the body and tends to turn it so as to *increase* the angle $\phi$, that is so as to set its length perpendicular to the course. When the length is athwart the course the couple is again zero, but that called into play by a deviation of the body from that position is now such as to send the body back to it. The body's position relative to the direction of motion is therefore one of instability in the first case, and of stability in the second.

A flat dish or plate, if let fall in water, or a card let fall in still air, with its plane horizontal, moves down in stable equilibrium; if it is let fall with its plane vertical, the equilibrium of position in falling is unstable. In this case we must associate $M_1$ with the axial direction, and $M_2$ with a perpendicular direction, and we see that $M_2 - M_1$ is negative, and there is stability therefore in the first case.

The origin of the couple may be seen in a general way as follows. Consider a ship (Fig. 41) advancing with speed $\dot{x}$ in the direction of its length, and making leeway $\dot{y}$, say, to starboard. The bow is continually advancing with speed $\dot{x}$ into undisturbed water, which, on the starboard side of the ship, is given speed $\dot{y}$ to starboard. There is thus a reaction thrust on the bow of the vessel in the direction to port.

We have here the explanation of the fact that a sailing ship generally carries a "weather helm," that is that the rudder must be held turned to leeward to keep the vessel on her course when a wind blows across it. For, as stated above, she makes leeway, that is has a speed $\dot{y}$ to leeward, along with the speed $\dot{x}$ in the direction of her length. Hence, by what has been stated above, the couple $(M_2 - M_1)\dot{x}\dot{y}$, *on the water*, is in the direction of the circular arrow A in Fig. 41, and therefore the reaction-couple, which is of equal moment, tends to turn the ship's head in the direction of the arrow A′, that is to windward, and this tendency (to "gripe," as it is called) must be counteracted by a couple applied to the ship by means of the rudder. The tendency of a ship to "fall off" her course (and thereby convert her forward motion into a component $\dot{x}$ along her length, and another $\dot{y}$ at right angles to her length), which, as explained above, always exists, is therefore augmented by the action of the wind, and the difficulty of steering is increased. This effect of the wind is considerable when the ship is driven by sails, and a steamer using sails as an auxiliary sometimes gripes so badly, especially with canvas on the after masts, as to make it almost impossible to steer. Thus sails on steamers used to be almost entirely confined to the foremast, and are now in large vessels completely discarded.

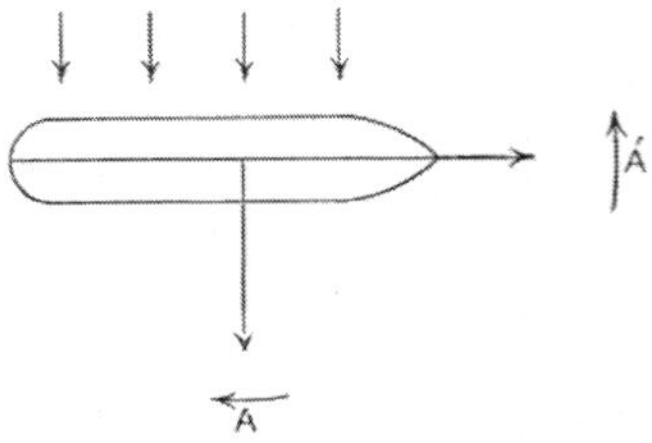

FIG. 41.

A ship sailing slowly shorewards (for example a yacht leaving moorings), when a wind is blowing parallel to the shore, is in danger of running up on the wind, and creeping ashore. The proper remedy, if there is room, is to set more sail so as to give more steerage way, to enable the ship to be steered away from the danger by means of the rudder.

17. *Centre of effort of resistance of a fluid.* We shall now consider the turning action of the fluid and the stability of a projectile a little in detail. The subject of the drift of a projectile will be resumed in Chapter XIII.

It may be noticed here that the existence of this turning action renders it necessary to suppose the resistance to a body moving in a fluid (when regarded as a localised single force) to act at a point E on the axis of figure in front of the centroid O of the external figure of the body. Let F now denote the amount of resistance and $\pi - \phi$ the angle between its direction and the axis of figure. We may apply through O two equal and opposite forces equal and parallel to F. The force-system reduces to a force through O and a couple, and the moment of the couple is $(c_2 - c_1)vw$. Thus we have

$$F \,.\, OE \sin\phi = (c_2 - c_1)vw, \quad \text{.............................(1)}$$

which determines OE when F and $\phi$ are known. E is called sometimes the "centre of effort."

Supposing F to act in the direction opposite to the resultant momentum we have, since now $\tan\phi = c_2 w / c_1 v$,

$$F \,.\, OE \sin\phi = (c_2 - c_1)\frac{c_1}{c_2} v^2 \tan\phi,$$

or

$$F \,.\, OE = (c_2 - c_1)\frac{c_1}{c_2} v^2 \sec\phi. \qquad (2)$$

Sometimes, as in the investigation of the stability of a spinning projectile which follows, it is convenient to write the virtual inertias along and at right angles to the axis of figure as $M + M'\alpha$, $M + M'\beta$, which take the place of $c_1$ and $c_2$. We then have $(c_2 - c_1)c_1/c_2 = M'(\beta - \alpha)(1 + \kappa\alpha)/(1 + \kappa\beta)$, where $\kappa = M'/M$. Thus (2) becomes

$$F \,.\, OE = M'(\beta - \alpha)\frac{1 + \kappa\alpha}{1 + \kappa\beta} v^2 \sec\phi. \qquad (3)$$

A good example is a submarine or an air-ship. If the craft have no buoyancy, then $M' = M$, and so $\kappa = 1$. For a submarine moving at a certain depth there will be a definite horizontal direction of the resultant momentum. Let the axis of the boat be turned in a vertical plane from this direction through an angle $\phi$. If $l$ be the height of the longitudinal metacentre of the boat, we have, taking account of the couple due to the change of direction of the axis,

$$Mgl' \tan\phi = Mgl \tan\phi - M'(\beta - \alpha)\frac{1 + \alpha}{1 + \beta} v^2 \tan\phi.$$

Hence, since $M' = M$, we get

$$l - l' = (\beta - \alpha)\frac{1 + \alpha}{1 + \beta}\frac{v^2}{g}, \qquad (4)$$

which gives the loss of metacentric height.

**18.** *Stability of projectile in fluid.* We now apply the ideas of the foregoing sections to the discussion of the stability of a symmetrical rotating projectile in an unlimited frictionless liquid. Let the projectile rotate about its axis of figure with angular speed $n$, so that its A.M. about that axis is $Cn$. But by the preceding section the projectile will experience a couple depending on its motion with speed $v$ in the axial direction, and in a direction perpendicular to the axis with speed $w$. The moment of the couple is

$$(c_2 - c_1)vw,$$

where we put, for compactness, $c_1$, $c_2$ for the effective inertias in the directions of $v$ and $w$ respectively, that is what are denoted above by $M_1$, $M_2$.

Let now the shot have precessional angular speed about an axis parallel to the resultant momentum, that is the resultant of $c_1 v$ and $c_2 w$. This is the direction of the impulse which would be required to produce these components of momentum. If $\theta$ be the angle (the $\phi$ of 17) which this makes with the axis of figure,

$$\tan\theta = \frac{c_2 w}{c_1 v}. \qquad (1)$$

We suppose the motion to be steady. The shot now "precesses" as if it were an ordinary top (Fig. 4) spinning about a fixed point O with the line of resultant momentum vertical, and endowed with A.M. $Cn$ about the axis of figure, and an *effective* A.M., $A\mu \sin\theta$, about an axis OE at right angles to the axis of figure OC, and in the vertical plane containing OC. The couple N acts about an axis represented in the case of the top by OD.

For steady motion we have the equation

$$(Cn - A\mu\cos\theta)\mu\sin\theta = N. \qquad (2)$$

Now $N = (c_2 - c_1)vw$, and $\tan\theta = c_2 w/c_1 v$, so that we have

$$N = \frac{c_1}{c_2}(c_2 - c_1)v^2 \tan\theta, \qquad (3)$$

and therefore (2) becomes

$$(Cn - A\mu\cos\theta)\mu = \frac{c_1}{c_2}(c_2 - c_1)v^2 \sec\theta; \qquad (4)$$

for we suppose that $\theta$ is not zero. If $\theta$ were zero, it would mean that $w = 0$.

The roots of (4) are real if

$$\frac{n^2}{v^2} > 4\frac{A}{C^2}\frac{c_1}{c_2}(c_2 - c_1),$$

which gives the least value of $n$ compatible with the steady motion, that is

$$n = 2\frac{v}{C}\left\{A\frac{c_1}{c_2}(c_2 - c_1)\right\}^{\frac{1}{2}}. \qquad (5)$$

Now we can put $c_1 = M + M'\alpha$, $c_2 = M + M'\beta$, where $M'$ is the mass of the displaced fluid and $\alpha$, $\beta$ are coefficients depending on the shape of the body. Thus

$$c_2 - c_1 = M'(\beta - \alpha). \qquad (6)$$

If $\gamma$ be the angle of rifling, we have

$$\tan^2\gamma = \frac{1}{4}\frac{n^2}{v^2}d^2 = \frac{A}{C^2}\frac{c_1}{c_2}(c_2 - c_1)d^2, \qquad (7)$$

where $d$ denotes the diameter of the gun at the muzzle, and the minimum value of $n/v$, given by (5), is taken.

If $k_1$, $k_2$ be the radii of gyration of the body about the axis of figure, and the other axis about which the body revolves with angular speed $\mu\sin\theta$, we have, since it is supposed that the rotation about the axis of figure does not set the medium in motion,

$$C = Mk_1^2, \quad A = Mk_2^2 + M'k_2'^2,$$

where $M'k_2'^2$ is the increase of moment of inertia due to the motion of the medium caused by the turning round the axis referred to above as corresponding to the axis OE used for the top. Thus we obtain, putting $M'/M = \kappa$,

$$\tan^2\gamma = (k_2^2 + \kappa k'^2)\frac{d^2}{k_1^4}\frac{1+\kappa\alpha}{1+\kappa\beta}\kappa(\beta - \alpha). \qquad (8)$$

Now we may apply this theory to a shot in air, and in that case $\kappa k'^2$ may be neglected, so that we get approximately

$$\tan^2\gamma = \frac{k_2^{\,2}}{k_1^{\,4}}(\beta - \alpha)\kappa d^2, \qquad (9)$$

which, if $\alpha$ and $\beta$ are known, gives a lower limit to the angle of rifling requited for stability.

So far the quantities $\alpha$, $\beta$ have only been determined in the case of an ellipsoid (see Greenhill, *Encyc. Brit.*, Art. *Hydromechanics*, §§ 44, 51). For a prolate ellipsoid of revolution (an egg-shaped shot) of length of major axis $2a$, and of each transverse axis $2b$, we have the integral

$$\mathrm{I} = \int_0^\infty \frac{ab^2 d\lambda}{2(a^2+\lambda)^{\frac{3}{2}}(b^2+\lambda)}; \qquad (10)$$

from which we obtain $\alpha$, $\beta$ by the equations

$$\alpha = \frac{\mathrm{I}}{1-\mathrm{I}}, \quad \beta = \frac{1-\mathrm{I}}{1+\mathrm{I}} = \frac{1}{1+2\alpha}. \qquad (11)$$

If we write $x$ for the number of calibres contained in the length $l$ of the shot, we have $x = 2a/2b$, and

$$\mathrm{I} = \frac{x}{(x^2-1)^{\frac{3}{2}}}\cosh^{-1}x - \frac{1}{x^2-1}, \qquad (12)$$

and

$$1 - \mathrm{I} = \frac{-x}{(x^2-1)^{\frac{3}{2}}}\cosh^{-1}x + \frac{x^2}{x^2-1}. \qquad (13)$$

Let $\sigma$ be the density of the metal of the shot, and in the case of a shell let the cavity be homothetic with the external ellipsoidal surface, so that each dimension is the fraction $f$ of the corresponding external dimensions. Then

$$\mathrm{M} = \tfrac{1}{6}\pi\sigma x d^3(1-f^3), \quad \mathrm{M}k_1^{\,2} = \tfrac{1}{80}\pi\sigma x d^5(1-f^5), \quad \mathrm{M}k_2^{\,2} = \tfrac{1}{120}\pi\sigma x d^3(l^2+d^2)(1-f^5), \ \ldots (14)$$

and if $\rho$ denotes the density of the air or medium,

$$\mathrm{M}' = \tfrac{1}{6}\pi\rho x d^3, \quad \mathrm{M}' = \frac{1}{1-f^3}\frac{\rho}{\sigma}\mathrm{M}, \quad k_1^{\,2} = \frac{1}{10}\frac{1-f^5}{1-f^3}d^2, \quad k_2^{\,2} = \tfrac{1}{2}(x^2+1)k_1^{\,2}, \qquad (15)$$

$$\tan^2\gamma = 5\frac{\rho}{\sigma}(\beta-\alpha)\frac{1-f^3}{1-f^5}(x^2+1). \qquad (16)$$

The ratio $\sigma/\rho$ may be replaced by 800 times the specific gravity of the metal, since water has about 800 times the density of air.

By means of (16) the following results, which are taken from a larger table given by Greenhill (*loc. cit. supra*), were calculated by A. G. Hadcock.

TABLE OF RIFLING FOR STABILITY OF AN ELONGATED PROJECTILE $x$ CALIBRES LONG, GIVING $\gamma$ THE ANGLE OF RIFLING, AND $p$ THE PITCH IN CALIBRES ($\tan\gamma = \pi/p$).

| $x$ | PALLISER SHELL, $f=\frac{1}{2}$, Sp. Gr. = 8. | | SOLID STEEL SHOT, $f=0$, Sp. Gr. = 8. | |
|---|---|---|---|---|
| $x$ | $\gamma$ | $p$ (cals.) | $\gamma$ | $p$ (cals.) |
| 1 | 0° 0′ | Infinity | 0° 0′ | Infinity |
| 2 | 2 32 | 71·08 | 2 29 | 72·21 |
| 2·5 | 3 23 | 53·32 | 3 19 | 54·17 |
| 3 | 4 13 | 42·79 | 4 09 | 43·47 |
| 3·5 | 5 02 | 35·75 | 4 58 | 36·33 |
| 4 | 5 51 | 30·72 | 5 45 | 31·21 |
| 4·5 | 6 40 | 26·93 | 6 32 | 27·36 |
| 5 | 7 28 | 23·98 | 7 21 | 24·36 |
| 6 | 9 04 | 19·67 | 8 56 | 19·98 |
| 10 | 15 19 | 11·47 | 15 05 | 11·65 |
| ∞ | 90 00 | 0·00 | 90 00 | 0·00 |

Mr. Hadcock also gave the following table:

| $x$ | $\beta - \alpha$ | $\iota$ | $\beta - \alpha$ |
|---|---|---|---|
| 0·0 | $-\infty$ | 4·5 | 0·810 |
| 0·5 | −2·215 | 5 | 835 |
| 1·0 | 0·000 | 6 | 872 |
| 2·0 | 0·494 | 7 | 897 |
| 2·5 | 0·606 | 8 | 915 |
| 3·0 | 0·682 | 9 | 929 |
| 3·5 | 0·737 | 10 | 939 |
| 4·0 | 0·778 | $\infty$ | 1·000 |

[From Greenhill's *R.G.T.*, 1914.]

In the steady motion the centre of the shot moves in a helix with speed parallel to the axis $v\cos\theta + w\sin\theta$, and circumferential speed $v\sin\theta - w\cos\theta$. Thus, as the period of turning is $2\pi/\mu$, the distance travelled by the shot in each turn can be calculated.

**19.** *Motion of a rigid body with altazimuth suspension and containing a gyrostat—gyrostatic pendulum.* (1) *With flywheel clamped.* We take first the case of a symmetrical gyrostat suspended as a compound pendulum by a combined vertical swivel and horizontal axis, O say, as shown in Fig. 30 (*b*). The "altitude" (inclination $\theta$ to the downward vertical) of the axis of figure can thus be changed without changing the azimuth, or the azimuth without changing the altitude, or both may be changed together.

First we suppose the wheel clamped, so that it does not rotate on its axle. Let the total moment of inertia about any transverse axis through O be A, and that for the wheel about its axis be C, while that of the case and axle together about the axis of figure is C′. Let the turning about the downward vertical be at rate $\mu$, in the counter-clock direction as seen from below. This gives angular speed $\mu\cos\theta$ about the axis of figure, OC say, and angular speed $\mu\sin\theta$ about an axis OE at right angles to OC in the same vertical plane. Now consider a horizontal axis OD drawn outward toward the observer. The rate of growth of A.M. about OD due to the motion is $-(C+C')\mu^2\cos\theta\sin\theta$, arising from the rotation $\mu\sin\theta$ about OE, which turns OC with its arm $(C+C')\mu\cos\theta$ away from the instantaneous position of OD, together with $A\mu^2\sin\theta\cos\theta$ produced, in like manner, by the turning of OE towards the same instantaneous position in consequence of the rotation about OC.

To these we must add the rate of growth $-A\ddot{\theta}$ due to acceleration of $\dot{\theta}$. The applied couple is $Mgh\sin\theta$. Thus we obtain the equation of motion

$$A\ddot{\theta} + (C+C'-A)\mu^2\sin\theta\cos\theta + Mgh\sin\theta = 0. \quad \text{.............(1)}$$

The condition for steady motion is therefore

$$(C+C'-A)\mu^2\cos\theta + Mgh = 0, \quad \text{........................(2)}$$

where now $\mu$ and $\theta$ are constants.

To find the equation of small vibrations about steady motion we combine with (1) the equation

$$\mu\{A\sin^2\theta+(C+C')\cos^2\theta\}=G, \quad\text{.......................(3)}$$

which expresses the constancy of A.M. about the vertical through O. Writing U for the terms in (1) which are independent of $\ddot{\theta}$, and putting $\eta$ for the excess of the current value of $\theta$ above the steady motion value, we obtain

$$A\ddot{\eta}+\frac{dU}{d\theta}\eta=0, \quad\text{..................................(4)}$$

where in $dU/d\theta$ the values of $\theta$ and $\mu$ for steady motion are to be used after the differentiation has been performed. It is clearly only necessary to differentiate with respect to $\theta$ the expression on the left of (2), substitute the value of $d\mu/d\theta$ obtained by differentiating (3), and multiply by $\sin\theta$. Thus we obtain, after a little reduction,

$$\ddot{\eta}+\mu^2\frac{C+C'-A}{A}\,\frac{3(C+C'-A)\cos^2\theta-A}{A+(C+C'-A)\cos^2\theta}\,\eta\sin^2\theta=0. \quad\text{.............(5)}$$

The period T of a small oscillation is therefore given by

$$T=\frac{2\pi}{\mu}\left\{\frac{A\{A+(C+C'-A)\cos^2\theta\}}{(C+C'-A)\{3(C+C'-A)\cos^2\theta-A\}\sin^2\theta}\right\}^{\frac{1}{2}}. \quad\text{..........(6)}$$

If $C+C'=0$, $A=Mh^2$, we have the case of a simple conical pendulum oscillating about steady motion, and get the period $2\pi/\mu(3\cos^2\theta+1)^{\frac{1}{2}}$, which may be verified directly. If $\theta$ is very small this period becomes $\pi/\mu$, that is half the period of revolution of the conical pendulum. But then the motion of the bob is that of a particle round a centre towards which it is attracted by a force which varies directly as the distance. The result of a small disturbance is to cause the particle to describe an ellipse about the centre of force, that is to extend one diameter equally at both ends, and to shorten in the same way the diameter at right angles to that, and the period of deviation from the circle is then clearly half the time of describing the latter.

If the body be a straight thin rod, the steady motion equation (2) gives $\mu=(Mgh/A\cos\theta)^{\frac{1}{2}}$, so that the period of revolution is $2\pi(A\cos\theta/Mgh)^{\frac{1}{2}}$. In this case the period of a small vibration about the steady motion is

$$\frac{2\pi}{\mu}\left(\frac{1}{3\cos^2\theta+1}\right)^{\frac{1}{2}}$$

as before.

**20.** *Gyrostatic pendulum with altazimuth suspension.* (2) ***With fly-wheel unclamped.*** We now unclamp the wheel of the gyrostat, and set it turning relatively to the vertical plane of the axis OC with vertical speed $\omega$. The total angular speed of the flywheel is $\omega+\dot{\psi}\cos\theta$, and this can only be changed by a frictional or other couple about its axis. We shall suppose that no such couple exists. The only change to be made in the preceding analysis is the substitution of $C(\omega+\mu\cos\theta)$, or, as we write it, $Cn$, for the A.M. about the axis of the flywheel. The rates of production of A.M. for the axis OD are now $(Cn+C'\mu\cos\theta)\mu\cos\theta$, arising from the rotation about OE, and $-A\mu^2\sin\theta\cos\theta$, arising from the motion of OC. The equation of motion for OD is now

$$A\ddot{\theta}+\{Cn+(C'-A)\mu\cos\theta\}\mu\sin\theta+Mgh\sin\theta=0. \quad \text{.........(1)}$$

Along with this we have for the A.M. about the vertical through O,

$$Cn\cos\theta+(A\sin^2\theta+C'\cos^2\theta)\mu=G. \quad \text{.........(2)}$$

The equation of steady motion is

$$\{Cn-(A-C')\mu\cos\theta\}\mu+Mgh=0. \quad \text{.........(3)}$$

This equation gives two speeds of turning. If $A>C'$, as it will be, the product of the roots is $-Mgh/(A-C')\cos\theta$, and is negative, for we measure $\theta$ from the downward vertical, and it is less than $\pi/2$. The directions of turning about the vertical are therefore opposed. If $n$ be very great the roots differ greatly in numerical value, and the greater is that which agrees in direction with the rotation $n$ when both are looked at from above or from below.

The approximate values of the roots are

$$-\frac{Mgh}{Cn} \quad \text{and} \quad \frac{Cn}{(A-C')\cos\theta}+\frac{Mgh}{Cn}.$$

We can now find the equation of small oscillations about steady motion. We write as before

$$A\ddot{\eta}+\frac{d}{d\theta}[\{Cn-(A-C')\mu\cos\theta\}\mu\sin\theta+Mgh\sin\theta]\eta=0, \quad \text{........(4)}$$

where $\eta$ has the same meaning as before, and $\theta$ and $\mu$ are to have the steady motion values in the result of the differentiation. Calculating $d\mu/d\theta$ from (2) and substituting in (4), we get

$$\ddot{\eta}+\frac{\{Cn-2\mu(A-C')\cos\theta\}^2\sin^2\theta+(A-C')(A\sin^2\theta+C'\cos^2\theta)\mu^2\sin^2\theta}{A\{A-(A-C')\cos^2\theta\}}\eta=0. \quad (5)$$

Thus the period of vibration is

$$2\pi\left[\frac{A\{A-(A-C')\cos^2\theta\}}{\text{Numerator of fraction in (5)}}\right]^{\frac{1}{2}}.$$

**21.** *Gyrostatic pendulum hung by untwistable flexible wire or universal joint.* We have supposed the pendulum hung by an altazimuth suspension. Sometimes however the suspension adopted is a short piece of nearly untwistable steel wire, the upper end of which is rigidly secured in a vertical position to a fixed point, while the lower end is rigidly fixed in line with the axis of the pendulum rod, as in Fig. 42. Such a suspension is kinematically equivalent to a Hooke's universal joint.

The wire is capable of flexure, but resists torsion very much; hence, on the supposition that there is only bending, we can find in the following manner the motion to which it subjects the pendulum rod. On a circle round the case of the gyrostat, coaxial with the pendulum rod, mark two points A, B, and let the arc AB subtend an angle $\phi$ at the centre of the circle. Now let (as shown in Fig. 42) the pendulum rod move clockwise round the vertical in a cone of semi-angle $\theta$, through the point of support, with angular speed $\mu$. If the point A always lay in the vertical plane defined by the axis of the rod at each instant, the rod would turn at angular speed $\mu \cos \theta$ round its axis, and in the counter-clock direction to an eye looking towards O from beyond the gyrostat. But, clearly, in the interval in which the vertical plane through the rod has turned through the angle $\phi$, that plane, if it contained A initially, contains B at the end, and so, to bring A to the position occupied by B, we should have to turn back the pendulum about the axis of figure through an angle $\phi$, equal to that which the vertical plane has turned through.

The angular speed of the pendulum about the axis of figure is thus $-\dot{\phi}+\mu \cos \theta = -\mu(1-\cos\theta)$, in the counter-clockwise direction as viewed from below. We suppose the flywheel to turn in that direction, relatively to the moving vertical plane, at speed $\omega$. Hence the total angular speed $n$ of the flywheel about its axis is $\omega - \mu(1-\cos\theta)$. [Here $\theta$ is the acute angle measured from the vertical drawn downward from the fixed point O: if we measure $\theta$ from the upward vertical point we should have $n = \omega - \mu(1+\cos\theta)$.]

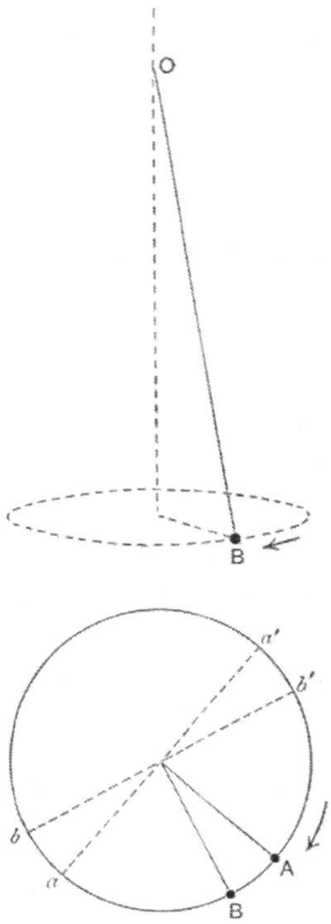

FIG. 42.

Further, the angular speed about OE, drawn to the left at right angles to OB, in the vertical plane containing the latter axis, is $\mu \sin\theta$, counter-clockwise to an eye looking along EO. We can now find the equation of motion for an axis OD drawn out from the paper towards the observer. The turning about OE is carrying OB (here the axis of spin) towards the instantaneous position of OD, and so the rate of production of A.M. is $Cn\mu \sin\theta$ from the flywheel, and $-C'\mu^2(1-\cos\theta)\sin\theta$ from the symmetrical case. Again, as the axis OE moves with the vertical plane of OC, the component turning of that plane about OB gives a rate of production of A.M. $-A\mu^2 \sin\theta\cos\theta$. The applied couple in the counter-clock direction about OD is $-Mgh\sin\theta$. The equation of motion is therefore

$$A\ddot{\theta}+\{Cn-C'\mu-(A-C')\mu\cos\theta\}\mu\sin\theta+Mgh\sin\theta=0. \quad \text{.........(1)}$$

Along with this equation we have for the flywheel

$$n=\omega-\mu(1-\cos\theta)=\text{const.}, \quad \text{..........................(2)}$$

and about the vertical through O the equation of A.M. is

$$(A\sin^2\theta+C'\cos^2\theta)\mu-C'\mu\cos\theta+Cn\cos\theta=G. \quad \text{.............(3)}$$

The condition of steady motion is

$$\{Cn-C'\mu-(A-C')\mu\cos\theta\}\mu+Mgh=0. \quad \text{..................(4)}$$

As before we can obtain the equation of small vibrations about steady motion by differentiating the expression on the left of (4) and inserting the value of $d\mu/d\theta$ from (3); so that we obtain

$$A\ddot{\eta}+\frac{d}{d\theta}[\{Cn-C'\mu-(A-C')\mu\cos\theta\}\mu].\sin\theta.\eta=0,$$

or $A\ddot{\eta}+[\{Cn-2C'\mu-2(A-C')\mu\cos\theta\}\frac{d\mu}{d\theta}+(A-C')\mu^2\sin\theta]\sin\theta.\eta=0.$ ...(5)

But from (3) we get

$$\frac{d\mu}{d\theta}=\frac{-2(A-C')\mu\sin\theta\cos\theta+Cn\sin\theta-C'\mu\sin\theta}{A\sin^2\theta+C'\cos^2\theta-C'\cos\theta}. \quad \text{...........(6)}$$

Substituting in (5) we obtain

$$A\ddot{\eta}+\frac{[K^2-C'\mu K+(A-C')\mu^2\{A-(A-C')\cos^2\theta-C'\cos\theta\}]\sin^2\theta}{A-(A-C')\cos^2\theta-C'\cos\theta}.\eta=0, \quad \text{...(7)}$$

where $K=Cn-C'\mu-2(A-C')\mu\cos\theta$.

We shall refer back to the equations now obtained when we deal in the sequel with the small oscillations of a gyrostatic pendulum.

# CHAPTER VIII

## VIBRATING SYSTEMS OF GYROSTATS. SUGGESTIONS OF GYROSTATIC EXPLANATION OF PROPERTIES OF MATTER

1. *Gyrostatic Spring.* Lord Kelvin suggested in a Royal Institution Lecture, delivered March 4, 1881 (*Popular Lectures and Addresses*, Vol. I. p. 142) that some of the elastic properties of bodies might be capable of explanation by the rotation of their particles. He returned to the subject in his Address as President of Section A of the British Association Montreal Meeting, 1884. As an illustration of the production of elastic quality by motion, he proposed a gyrostatic arrangement which should have the properties of a spiral spring. As shown in Fig. 43, four equal bars freely jointed at their ends form a frame, to the ends of a diagonal of which are attached hooks about which the frame can swivel, so that it can revolve about that diagonal without any turning of the hooks. At the middle of each bar of the frame is placed a flywheel running in its case with its axis in the line of the bar. The case is rigidly connected with the bar. The direction of spin of each wheel is such that if the frame were drawn out into a double line of bars along the diagonal, the directions of turning would coincide for all the wheels. The gyrostats have all the same mass M, moment of inertia of flywheel C, speed of rotation $n$, and moment of inertia A of the whole gyrostat about a transverse axis through its centroid.

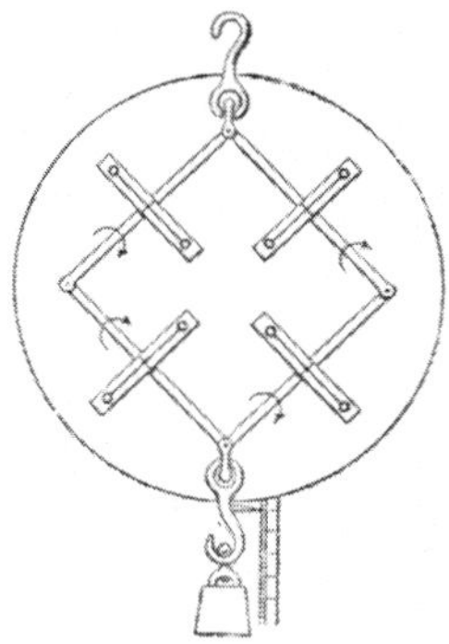

Fig. 43.

When the frame is hung by one hook, so that the line of hooks is vertical and a mass $M_1$ is hung on the lower hook, and the frame is thereafter precessing freely about the swivels, the effect of increasing the weight $M_1$ is to cause elongation of the distance from hook to hook, and vertical vibration of the arrangement about a new equilibrium value of this distance takes place. Similarly if $M_1$ is diminished the diagonal is shortened. We shall see how the elongation or shortening depends on the increase or diminution of $M_1$.

Attempts made to realise this "gyrostatic spring balance," with gyrostats first spun and then placed in position, have failed because of the difficulty of equalising the rotational speeds; but success can be attained by using gyrostats, all precisely alike, in which the flywheels are rotors of electricity driven motors, and driving all with the same current.

We suppose the length of each rod to be $2a$, and that the inclination of each upper rod to the downward vertical is $\theta$. We find first the energy of the arrangement. The distances of the centres of the upper and lower pairs of gyrostats from the level of the upper end of the diagonal are $a\cos\theta$ and $3a\cos\theta$, respectively. Thus the corresponding vertical speeds, taken positive downwards, are $-a\sin\theta\,.\,\dot\theta$ and $-3a\sin\theta\,.\,\dot\theta$. Also the distance of each of these centres from that diagonal is $a\sin\theta$, so that the horizontal speed in each case is numerically $a\cos\theta\,.\,\dot\theta$. Lastly, the vertical speed of the lower hook, and therefore also of the attached mass, is $-4a\sin\theta\,.\,\dot\theta$. Hence we get for the kinetic energy

$$2\{A+Ma^2+4(M+M_1)a^2\sin^2\theta\}\dot\theta^2+2(A+Ma^2)\mu^2\sin^2\theta+R,$$

where $\mu$ is the precessional angular speed and R (a constant) is the kinetic energy due to the spin of the flywheels.

The potential energy, measured from the level of the upper end of the vertical diagonal, is $-4g(2M+M_1)a\cos\theta$. Thus we have the energy equation

$$2\{A+Ma^2+4(M+M_1)a^2\sin^2\theta\}\dot\theta^2$$
$$+2(A+Ma^2)\mu^2\sin^2\theta-4g(2M+M_1)a\cos\theta=K, \quad\text{.............(1)}$$

where K is a constant.

Also there is the equation of constancy of A.M. about the vertical diagonal, which, since we measure $\theta$ from the downward vertical from the fixed point, may be written

$$(A+Ma^2)\mu\sin^2\theta-Cn\cos\theta=G, \quad\text{........................(2)}$$

where G is a constant.

Differentiating (1), putting, for brevity, $\alpha$ for $A+Ma^2$, $\beta$ for $4(M+M_1)a^2$, we get

$$(\alpha+\beta\sin^2\theta)\ddot\theta+\beta\sin\theta\cos\theta\,.\,\dot\theta^2+\alpha\mu\frac{d\mu}{d\theta}\sin^2\theta+\alpha\mu^2\sin\theta\cos\theta$$
$$+g(2M+M_1)a\sin\theta=0. \quad\text{....................................(3)}$$

But, by (2), $\qquad \alpha\mu\dfrac{d\mu}{d\theta}\sin^2\theta=-(Cn\mu+2\alpha\mu^2\cos\theta)\sin\theta,$

so that (3) becomes

$$(\alpha+\beta\sin^2\theta)\ddot\theta+\beta\sin\theta\cos\theta\,.\,\dot\theta^2-(Cn+\alpha\mu\cos\theta)\mu\sin\theta$$
$$+g(2M+M_1)a\sin\theta=0. \quad\text{.................................(4)}$$

The steady motion equation is obtained by deleting in (4) the terms in $\ddot{\theta}$ and in $\dot{\theta}^2$.

For small vibrations about steady motion we take $\dot{\theta}^2$ as negligible, and obtain by the method of 19, VII,

$$(\alpha+\beta\sin^2\theta)\ddot{\eta}-\frac{d}{d\theta}(\mathrm{C}n+a\mu\cos\theta)\,.\,\mu\eta\sin\theta=0, \quad\text{.........................(5)}$$

in which $\mu$ and $\theta$ are to be understood, after the differentiation is performed, as the values for the steady motion, and $\eta$ is the excess of the actual value of $\theta$ above the steady motion value. We have, as before, $d\mu/d\theta=-(\mathrm{C}n+2a\mu\cos\theta)/a\sin\theta$, and (5) becomes

$$\ddot{\eta}+\frac{(\mathrm{C}n+2a\mu\cos\theta)^2+a^2\mu^2\sin^2\theta}{a(\alpha+\beta\sin^2\theta)}\eta=0. \quad\text{..............................(6)}$$

The period of oscillation of the arrangement is therefore

$$2\pi\left[\frac{(\mathrm{A}+\mathrm{M}a^2)\{\mathrm{A}+\mathrm{M}a^2+4(\mathrm{M}+\mathrm{M}_1)a^2\sin^2\theta\}}{\{\mathrm{C}n+2(\mathrm{A}+\mathrm{M}a^2)\mu\cos\theta\}^2+(\mathrm{A}+\mathrm{M}a^2)^2\mu^2\sin^2\theta}\right]^{\frac{1}{2}}.$$

We now investigate the change of equilibrium length of the diagonal produced by altering $\mathrm{M}_1$. For this we take the steady motion form of (4), and differentiate with respect to $\mathrm{M}_1$. We get

$$-(\mathrm{C}n+2a\mu\cos\theta)\frac{d\mu}{d\theta}\frac{d\theta}{d\mathrm{M}_1}+a\mu^2\sin\theta\frac{d\theta}{d\mathrm{M}_1}+ga=0,$$

which, by the value of $d\mu/d\theta$ given above, becomes, after reduction,

$$\frac{d\theta}{d\mathrm{M}_1}=-\frac{ga(\mathrm{A}+\mathrm{M}a^2)\sin\theta}{\{\mathrm{C}n+2(\mathrm{A}+\mathrm{M}a^2)\mu\cos\theta\}^2+(\mathrm{A}+\mathrm{M}a^2)^2\mu^2\sin^2\theta}. \quad\text{...................(7)}$$

Thus the angle $\theta$ diminishes as the load $\mathrm{M}_1$ increases, that is the vertical diagonal increases in length. The length $l$ of the diagonal is $4a\cos\theta$, so that $dl=-4a\sin\theta\,.\,d\theta$. Hence (7) gives

$$\frac{dl}{d\mathrm{M}_1}=\frac{4ga^2(\mathrm{A}+\mathrm{M}a^2)\sin^2\theta}{\{\mathrm{C}n+2(\mathrm{A}+\mathrm{M}a^2)\mu\cos\theta\}^2+(\mathrm{A}+\mathrm{M}a^2)^2\mu^2\sin^2\theta}. \quad\text{...................(8)}$$

This result shows that the action can hardly be described as that of a *spiral spring*. For very fast spin the predominating term in the denominator on the right of (8) is $\mathrm{C}^2n^2$, and the elongation produced by a given increment $d\mathrm{M}_1$ of load is nearly proportional to $\sin^2\theta$. If, however, the frame be shut up so that each side is very nearly horizontal, and the spin be very great, the spiral spring action will, as noticed below, be obtained.

But (6) and (8) show that even when $\mathrm{C}n$ is zero, the arrangement acts as a spring. It is of course very easy to investigate this simple case independently.

A fair idea of the action, and indeed an approximate realisation of the property aimed at, is obtained by means of the arrangement shown in Fig. 44. A gyrostat is hung with its axis horizontal by a cord in the same vertical as the centroid. The flywheel spins, but as there is no couple there is no precession. A weight $mg$ is applied in a vertical line at distance $l$ from the centroid, as indicated by the diagram; a slight, very slight, tilting of the gyrostat is produced, and the gyrostat moves off with not quite steady precession, of average angular speed $\mu$. Neglecting the slight deviation, now set up, of the suspension cord from the vertical, and putting A for the moment of inertia of the gyrostat about a vertical axis through its centre, we get for the kinetic energy of the azimuthal motion the value $\frac{1}{2}A\mu^2 + \frac{1}{2}ml^2\mu^2$. The work done by the weight $mg$ in its descent through the small distance $h$ involved in the tilting is $mgh$. Hence we get

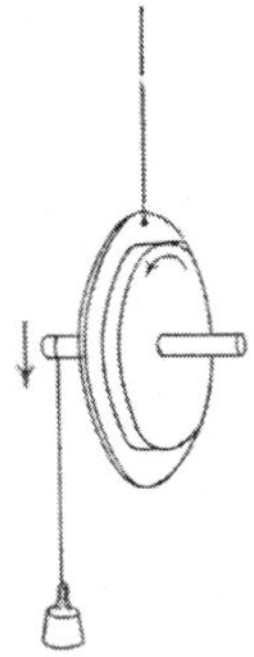

FIG. 44.

$$\tfrac{1}{2}(A + ml^2)\mu^2 = mgh. \qquad (9)$$

As we have already seen, we have in this case $\mu = mgl/Cn$. Substituting in the equation just found this value for $\mu$, and supposing that A is great in comparison with $ml^2$, so that the term $\frac{1}{2}ml^2\mu^2$ may be neglected, we find after a little reduction the equation

$$\frac{h}{m} = \frac{Al^2g}{2C^2n^2}. \qquad (10)$$

It will be seen that this (to a constant factor) agrees with (8) if we suppose $Cn$ very great and $\theta = \frac{1}{2}\pi$. For this value of $\theta$, $\sin^2\theta = 1$, and the rate of variation of $\sin^2\theta$ with $\theta$ is zero, so that proportionality of $h$ to $m$ is for small increments of $m$ fairly accurately obtained. But in strictness the terms in A and M in the denominator of the expression on the right of (8), introducing $\mu\cos\theta$ and $\mu\sin\theta$, affect the result. The rate of variation of $\cos^2\theta$, or of $\sin^2\theta$, with $\theta$ is numerically greatest when $\theta = \frac{1}{4}\pi$.

The action, though it cannot be described as an exact imitation of that of an ordinary spring, is very interesting, and is helpful as furnishing a notion as to how the elastic properties of bodies may possibly be explained by means of a kinetic theory of the constitution of the bodies.

2. *Gyrostat hung by steel wire. Axis horizontal without spin.* The suspension wire in actual experiments made was long, its upper end was fixed and its lower end was attached to the gyrostat rim so that the gyrostat turned with the wire when that turned about its axis, and the gyrostat was free to tilt as shown in the diagram. Let the gyrostat be turning in azimuth so that the wire is twisting or untwisting. Let the wire have torsional rigidity $\tau$, that is the couple required to maintain the lower end in position, when turned round the axis of the wire through an angle $\phi$ from the position of equilibrium, be $\tau\phi$.

As we shall see, the plane of the flywheel will not remain vertical, and we suppose that, at the instant, the inclination of the axis to the horizontal is $\theta$, as shown in the diagram, reckoned positive when the turning is in the counter-clock direction about the horizontal axis OA, as seen from beyond A. If $\phi$ be, as we assume, always small, $\theta$ will always be small also.

Now suppose the angular momentum $Cn$ of the (vertical) wheel to be represented by OB drawn from the centre O of the gyrostat, and that the lower end of the wire is turning in the azimuthal direction indicated by the curved arrow at the top of the diagram. Hence angular momentum is being produced about the horizontal axis OA at rate $-Cn\phi$. The gyrostat must be tilted with the end B of the axis up, through the angle $\theta$ (exaggerated in the diagram), to give a couple for this growth of angular momentum. The moment of the couple is $Mga\theta$ if M be the whole mass of the gyrostat and $a$ the distance of the point of attachment of the wire from the centre of gravity O. The total rate of production of angular momentum about OA is $A'\ddot{\theta}-Cn\dot{\phi}$, where A′ is the moment of inertia of the gyrostat about the point of attachment E of the wire. Putting this rate equal to the moment of the couple in the positive direction, we get the equation of motion

$$A'\ddot{\theta}-Cn\dot{\phi}=-Mga\theta. \quad \text{(1)}$$

But in consequence of the turning at rate $\dot{\theta}$ angular momentum is being produced about the upward vertical at rate $Cn\dot{\theta}$, and the total rate about that axis is $A\ddot{\phi}+Cn\dot{\theta}$. Hence we get the equation

$$A\ddot{\phi}+Cn\dot{\theta}=-\tau\phi. \quad \text{(2)}$$

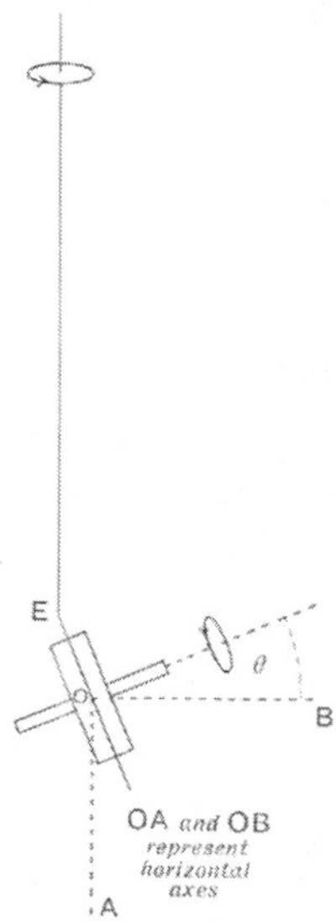


FIG. 45.

It is to be noted that the azimuthal motion of the tilted gyrostat will cause slight deviations of the long suspension wire from the vertical: these are here neglected.

If now we suppose $\theta$ so small, *and the period also so great*, that $\ddot{\theta}$ may be neglected, we have $Cn\dot{\phi}=Mga\theta$, and therefore

$$Cn\ddot{\phi}=Mga\dot{\theta}, \quad \text{or} \quad Cn\dot{\theta}=\ddot{\phi}C^2n^2/Mga.$$

Substituting in the equation (2) we find

$$\left(A+\frac{C^2n^2}{Mga}\right)\ddot{\phi}+\tau\phi=0. \quad \text{(3)}$$

This is the equation of torsional oscillations, and shows that the virtual moment of inertia of the gyrostat as a torsional vibrator is $A+C^2n^2/Mga$. This fact was pointed out by Lord Kelvin [B.A. Meeting, 1884].

A strict solution of the equations (1) and (2) leads to the same result, and to another interesting conclusion. Assuming

$$\theta=\theta_0e^{ikt}, \quad \phi=\phi_0e^{ikt},$$

and substituting in (1) and (2), we obtain the equations

$$\left.\begin{aligned}(A'k^2-Mga)\theta_0+iCnk\phi_0=0,\\ iCnk\theta_0-(Ak^2-\tau)\phi_0=0.\end{aligned}\right\} \quad \text{(4)}$$

Eliminating $\theta_0$, $\phi_0$, we find

$$AA'k^4-(C^2n^2+MgaA+\tau A')k^2+Mga\tau=0, \quad \text{(5)}$$

a quadratic equation for the determination of $k^2$. The roots are certainly real if the angular momentum $Cn$ of the flywheel is made great enough. They are also positive because, since A, A′, M, $g$, $a$, $\tau$ are all positive, the product of the roots $Mga\tau/AA'$ is positive, and the sum of the roots $(C^2n^2+MgaA+\tau A')/AA'$ is also positive.

If now $\tau$ be made very small, that is if the suspension wire be made very long, the product of the roots $Mga\tau/AA'$ becomes very small. But the coefficient of $k^2$ is still numerically great. Thus the equation has then a small root and a comparatively large one. The small root is obtained approximately from

$$-(C^2n^2+MgaA+\tau A')k^2+Mga\tau=0, \quad \text{.....................(6)}$$

and the large root from

$$AA'k^2-(C^2n^2+MgaA+\tau A')=0. \quad \text{........................(7)}$$

Neglecting the term $\tau A'$ in the expression in brackets, we obtain for the small root

$$k^2=\frac{Mga\tau}{C^2n^2+MgaA}. \quad \text{................................(8)}$$

The period of vibration is for this

$$\frac{2\pi}{k}=\frac{2\pi}{\tau^{\frac{1}{2}}}\left(A+\frac{C^2n^2}{Mga}\right)^{\frac{1}{2}}. \quad \text{.................................(9)}$$

Thus, regarding the matter from the point of view of torsional oscillations of a wire of torsional rigidity $\tau$, we see that the moment of inertia of the gyrostat as a torsional vibrator on the wire is virtually $A+C^2n^2/Mga$.

Now take the large root given by the equation

$$k^2=\frac{C^2n^2+MgaA}{AA'}. \quad \text{................................(10)}$$

The period of oscillation is

$$\frac{2\pi}{k}=2\pi\left(\frac{A'}{Mga+\dfrac{C^2n^2}{A}}\right)^{\frac{1}{2}}. \quad \text{...........................(11)}$$

Thus regarded from the point of view of an oscillation in $\theta$, that is of the gyrostat about a horizontal axis through the point of attachment of the wire, and in the plane of the flywheel, the motion takes place in the period which would exist without rotation either under a couple (per unit of $\theta$) $Mga(1+C^2n^2/MAga)$ with moment of inertia A′, or under a couple $Mga$ with moment of inertia $A'/(1+C^2n^2/MAga)$.

The motion of long period and that of short period can exist separately. The general motion, however, when the suspension wire is very long, consists of vibrations of short period, arising from the virtual diminution of moment of inertia just noticed, superimposed on the vibrations of long period due to enhancement of moment of inertia of the torsional vibrator. [These rapid vibrations are naturally more quickly damped out by the action of friction, which is not here considered.] That a virtual enhancement of one moment of inertia A must be accompanied by a virtual diminution, in the same ratio, of the other moment of inertia A′, follows from the fact that the coefficient of $k^4$ in the determinantal equation (5) is AA′. We have thus a general theorem of the effective inertias of systems which have two modes of vibration. Very probably it has been explicitly stated before.

3 *Gyrostat with two freedoms doubly unstable without spin.* We now consider the arrangement of a gyrostat mounted so as to turn about two axes which are at right angles to one another, and may be regarded, in the first place, as both horizontal. We suppose, therefore, that the system has gravitational stability or instability in one or both freedoms. Such an arrangement is shown in Fig. 46, which represents a motor gyrostat mounted on gimbals. In this case the gyrostat, if without spin, is unstable in both freedoms. But it is possible to have one gimbal ring pivoted above the gyrostat, and the other below it, and neither of these may be the ring on which the gyrostat is immediately supported; or the axis carrying the gyrostat may be one about which the frame or case of the gyrostat is free to turn, and the frame or case may be attached to a cross-bar on two vertical legs or stilts. Such an arrangement has one stability and one instability without spin. Both the stability and the instability are gravitational. Thus if M be the mass of the gyrostat and its attachments, and $h$ the distance of the centre of gravity from the axis considered, what has been called the "preponderance," $Mgh$, may be either positive or negative.

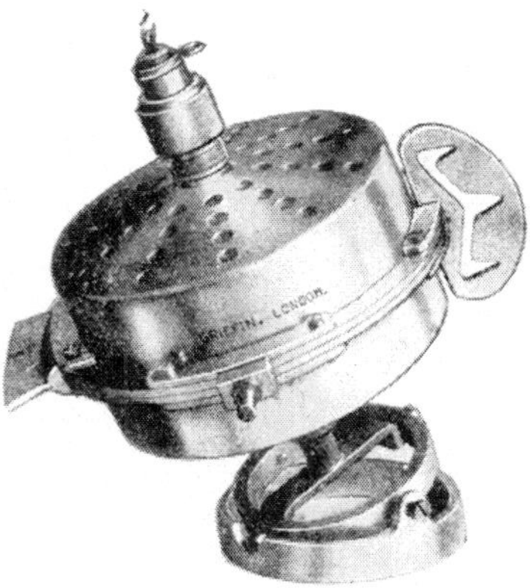


FIG. 46.

Let, then, the masses which turn about the respective knife-edges or axes be M, M′, the heights of the centres of gravity above (or distances from) the axes be $h$, $h'$, the moments of inertia about the axes be A, A′, the respective angular deflections (supposed small) from the vertical be $\phi$, $\psi$, and, as usual, the moment of inertia and angular speed of the flywheel be C, $n$. We get then by the process so often employed above, for the rates of growth of angular momentum about the axes, fixed in the present case,

$$A\ddot{\phi} + Cn\dot{\psi} = Mgh\phi, \quad A'\ddot{\psi} - Cn\dot{\phi} = M'gh'\psi.^* \quad \dots\dots\dots\dots\dots\dots(1)$$

* It may be remarked here that if we multiply the first of these equations by $\dot{\phi}$, the second by $\dot{\psi}$, add and integrate with respect to the time, we get the equation of energy. But the

[If the axes be horizontal and at right angles to one another, we might take them as parallel to axes of $x$ and $y$ drawn from an origin on the vertical through the centre of gravity for the upright position. Each turning is to be taken as positive when it is counter-clockwise to an eye looking at the apparatus towards the origin from a point on the axis of rotation at a positive distance from the origin. The axis of $z$ may be taken downwards.]

If, taking the case of double instability without spin, we write $B = Mgh$, $B' = M'gh'$, Equations (1) become

$$A\ddot{\phi} + Cn\dot{\psi} - B\phi = 0, \quad A'\ddot{\psi} - Cn\dot{\phi} - B'\psi = 0. \quad \text{................(2)}$$

[The reader will observe that the meaning of $\psi$ is here different from that assigned to the same symbol in the theory of a single gyrostat set forth above.]

Now let $$\phi = ae^{ikt}, \quad \psi = be^{ikt}, \quad \text{..................................(3)}$$

where $a$ and $b$ are constants, which are in general complex numbers, that is are of the form $\alpha + i\beta$, $\{i = (-1)^{\frac{1}{2}}\}$ where $\alpha$ and $\beta$ are real quantities. Thus by substitution in (2) we get

$$\left.\begin{aligned} -(k^2A + B)a + ikCnb = 0, \\ -ikCna - (k^2A' + B')b = 0, \end{aligned}\right\} \quad \text{............................(4)}$$

and therefore by elimination of $a$ and $b$

$$AA'k^4 - (C^2n^2 - AB' - A'B)k^2 + BB' = 0. \quad \text{.....................(5)}$$

According to the supposition made above, A, A′, B, B′ are all positive, and the roots of the quadratic in $k^2$ which we have obtained are real and positive if $(C^2n^2 - AB' - A'B)^2 > 4AA'BB'$ and $C^2n^2 > AB' + A'B$.

These are the conditions of dynamical stability, for if they be fulfilled $\phi$ and $\psi$ represent simple harmonic deviations from the equilibrium configuration (unstable in the present case without spin). Each deflection may have either of the two periods given by the two real roots $k_1^2$, $k_2^2$ of (5). The motion is oscillatory and therefore stable, and there are two modes of vibration, which may be taken, either separately or in combination, by the gyrostat. Moreover there are numerically equal positive and negative values of $k$ given by each value of $k^2$.

Now it is clear that the four roots provide for the case in which the sign of $n$ is reversed, that is for both $+n$ and $-n$. To settle what roots go with $+n$ and what with $-n$, we may proceed as follows. Suppose that

gyrostatic terms have disappeared, and they contribute nothing in an explicit form to the energy expression. The same remark is true of all systems of equations containing gyrostatic terms. Hence *from the principle of energy alone* it is impossible to foresee the existence of such terms. It is sometimes asserted that the principle of energy contains all things dynamical. Certain special cases excepted, the principle of energy, by itself, is insufficient for the solution of dynamical problems.

A = A′ and B = B′, then equations (2) can be united in one by writing $\zeta = \phi + i\psi$. Thus multiplying the second of (2) by $i$ and adding, we get

$$\mathrm{A}\ddot{\zeta} - \mathrm{C}ni\dot{\zeta} - \mathrm{B}\zeta = 0. \quad \text{.......(6)}$$

If now we put $\zeta = \mathrm{K}e^{ikt}$,

where K is a constant, we obtain from (6) the condition

$$k^2 - \frac{\mathrm{C}n}{\mathrm{A}}k + \frac{\mathrm{B}}{\mathrm{A}} = 0, \quad \text{.......(7)}$$

which yields

$$k = \frac{1}{2}\frac{\mathrm{C}n}{\mathrm{A}}\left\{1 \pm \left(1 - 4\frac{\mathrm{AB}}{\mathrm{C}^2n^2}\right)^{\frac{1}{2}}\right\}. \quad \text{.......(8)}$$

Thus for $n$ positive $k$ has two positive values, and for $n$ negative has two negative values. The reversal of the direction of rotation reverses the signs of the roots. This will hold also when A and A′, and B and B′, are unequal, as there cannot be any change in the nature of the solution brought about by the equalisation of these quantities.

It will be observed that if the spin be rapid the roots are $\mathrm{C}n/\mathrm{A}$ and $\mathrm{B}/\mathrm{C}n$ nearly. These are the angular speeds of possible circular motions, and agree with the results obtained above for the steady motion of a top.

The roots of the quadratic (5) are given by

$$k^2 = \tfrac{1}{2}f\left\{1 \pm \left(1 - 4\frac{g}{f^2}\right)^{\frac{1}{2}}\right\}, \quad \text{.......(9)}$$

where $f = (\mathrm{C}^2n^2 - \mathrm{AB}' - \mathrm{A}'\mathrm{B})/\mathrm{AA}'$, $g = \mathrm{BB}'/\mathrm{AA}'$. This gives two positive values of $k$ and two negative values, provided $g$ is positive. It will be observed that if B and B′ have not the same sign $g$ is negative, and (9) gives two real roots (equal with opposite signs) and two imaginary roots. We have just seen that the two positive values of $k$ apply to the case of $n$ positive. Now recurring to the case of A = A′, B = B′, we should then have been able to realise the solution very simply by writing

$$\zeta = \phi + i\psi = (\alpha_1 + i\beta_1)e^{ik_1t} + (\alpha_2 + i\beta_2)e^{ik_2t}, \quad \text{.......(10)}$$

where $\alpha_1, \beta_1, \alpha_2, \beta_2$ are supposed all real. We should have had

$$\begin{aligned}\zeta = {} & \alpha_1 \cos k_1t - \beta_1 \sin k_1t + \alpha_2 \cos k_2t - \beta_2 \sin k_2t \\ & + i(\alpha_1 \sin k_1t + \beta_1 \cos k_1t + \alpha_2 \sin k_2t + \beta_2 \cos k_2t),\end{aligned} \quad \text{.......(11)}$$

and therefore should have obtained the real values of $\phi$ and $\psi$ by equating $\phi$ to the first line on the right of this equation, and $i\psi$ to the second line. But if we compare (4) with the equations we should have if A = A′ and B = B′, we see that we must have for positive $n$,

$$\left.\begin{aligned}\phi &= \alpha_1 \cos k_1t - \beta_1 \sin k_1t + \alpha_2 \cos k_2t - \beta_2 \sin k_2t, \\ \psi &= \rho_1(\alpha_1 \sin k_1t + \beta_1 \cos k_1t) + \rho_2(\alpha_2 \sin k_2t + \beta_2 \cos k_2t),\end{aligned}\right\} \quad \text{.......(12)}$$

where $\rho = \{(k^2\mathrm{A} + \mathrm{B})/(k^2\mathrm{A}' + \mathrm{B}')\}^{\frac{1}{2}} = ib/a$, and $\rho_1, \rho_2$ are the positive values of $\rho$ for $k_1$ and $k_2$.

If we put $-i$ in the place of $+i$ in the exponents in (10), (12) becomes

$$\left.\begin{aligned}\phi &= \alpha_1 \cos k_1t + \beta_1 \sin k_1t + \alpha_2 \cos k_2t + \beta_2 \sin k_2t, \\ \psi &= -\rho_1(\alpha_1 \sin k_1t - \beta_1 \cos k_1t) - \rho_2(\alpha_2 \sin k_2t - \beta_2 \cos k_2t).\end{aligned}\right\} \quad \text{.......(13)}$$

Thus we have simply changed the signs of the arguments. In (12) and (13) $k_1$ and $k_2$ are to be taken positive, as the effect of changing from the positive to the negative roots has been taken account of in the signs of the coefficients in (13). These changes correspond to a reversal of the spin $n$, as may be seen from the values of the roots of the biquadratic

(5) as given in (8). It will be seen that each pair of terms of (12), made up of the first term in $\phi$ and the first term of $\psi$, or of the second, third, or fourth term in each expression, would if $\rho$ were unity give a circular motion in the positive direction, and that similarly the corresponding pairs of terms in (13) would represent circular motions in the opposite direction. But opposite circular motions are not superimposed [see 4, below].

It is now obvious that there are two modes of motion for each direction of spin provided the product BB′ is positive. By the spin the two instabilities which existed without spin have been replaced by stabilities. If BB′ be negative there is only one possible mode of motion for each direction of spin, and complete stability has not been attained.

The reader will observe that this analysis of any system with two freedoms, gyrostatically dominated, is applicable, *mutatis mutandis*, to any arrangement, *e.g.* that of the gyrostat on a trapeze, or Lord Kelvin's proposed gyrostatic compass [see 5, VII].

**4.** *Gyrostatic system with two freedoms doubly stable without spin. Gyrostatic pendulum.* Let us now suppose that the two freedoms are both stable without spin. Then in equation (2) we have to change the signs of B and B′, and then suppose both quantities positive. The equations are now

$$A\ddot{\phi}+Cn\dot{\psi}+B\phi=0,\quad A'\ddot{\psi}-Cn\dot{\phi}+B'\psi=0. \quad\ldots\ldots\ldots\ldots\ldots(1)$$

If we had $A=A'$, $B=B'$, we should write the single equation

$$A\ddot{\zeta}-Cni\dot{\zeta}+B\zeta=0, \quad\ldots\ldots\ldots\ldots\ldots\ldots\ldots\ldots\ldots\ldots(2)$$

where $\zeta=\phi+i\psi$. The value $\zeta=(\alpha+i\beta)e^{ikt}$ would give the equation

$$k^2-\frac{Cn}{A}k-\frac{B}{A}=0,$$

so that

$$k=\frac{1}{2}\frac{Cn}{A}\left\{1\pm\left(1+\frac{4AB}{C^2n^2}\right)^{\frac{1}{2}}\right\}. \quad\ldots\ldots\ldots\ldots\ldots\ldots\ldots(3)$$

Reversal of the sign of $n$ would give simply these roots reversed in sign. Thus for a given direction of spin there are two roots $k_1$, $-k_2$ of which the positive numerical values are $k_1$, $k_2$, and for the reversed spin there are the two roots $-k_1$, $k_2$.

We find in the same manner as before

$$\left.\begin{aligned}\phi&=\alpha_1\cos k_1t-\beta_1\sin k_1t+\alpha_2\cos k_2t+\beta_2\sin k_2t,\\ \psi&=\rho_1(\alpha_1\sin k_1t+\beta_1\cos k_1t)-\rho_2(\alpha_2\sin k_2t-\beta_2\cos k_2t).\end{aligned}\right\}\quad\ldots\ldots\ldots\ldots(4)$$

Here $\rho_1$, $\rho_2$ are taken as the positive values of

$$\{(k^2A-B)/(k^2A'-B')\}^{\frac{1}{2}}$$

for the respective values of $k^2$.

If we change the direction of spin we get

$$\left.\begin{aligned}\phi&=\alpha_1\cos k_1t+\beta_1\sin k_1t+\alpha_2\cos k_2t-\beta_2\sin k_2t,\\ \psi&=\rho_1(-\alpha_1\sin k_1t+\beta_1\cos k_1t)+\rho_2(\alpha_2\sin k_2t+\beta_2\cos k_2t).\end{aligned}\right\}\quad\ldots\ldots\ldots\ldots(5)$$

If $\rho$ were unity the first pair of terms, or the second pair of terms, one from $\phi$ and one from $\psi$, in (4), would give a circular motion in the positive direction, and either pair of terms in $k_2$ a circular motion in the negative direction. These circular motions are reversed in (5). There is thus the remarkable difference between this doubly stable case and the former, that now circular motions in opposite directions can be superimposed. The periods of these two motions are $2\pi/k_1$, $2\pi/k_2$. The greater angular speed $k_1$, which for A = A′, B = B′ is Cn/A nearly, is round in the direction of rotation, the smaller angular speed $k_2$, which for A = A′, B = B′ is B/Cn or (Mgh/Cn) nearly, in the case of gravity is in the opposite direction to the rotation $n$. That is as seen from below in both cases.

5. *Illustrations of effect of spin on stability.* The arrangements shown in Fig. 47 illustrate this affair of stabilities very well. It is taken from Thomson and Tait's *Natural Philosophy*, § 345[X]. A gyrostat is hung on a

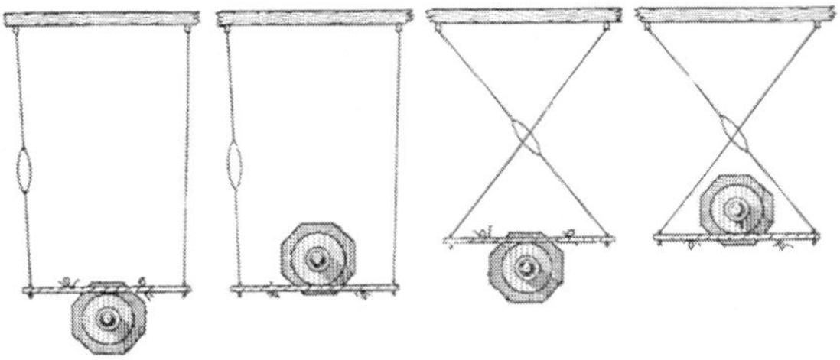

Fig. 47.

bifilar sling, the horizontal bar of which can turn about swivels at its ends. There are four arrangements: (1), (2), (3), (4). In (3) and (4) the cords are crossed by means of a ring placed in one of them. In each case there are three gyrostatic modes of motion, two inclinational and one azimuthal. [The inclinational mode in which the cords move in the vertical plane is supposed annulled by a proper constraint.]

*Without spin of the gyrostat,* these are all three stable in (1); the inclinational mode of shorter period is unstable in (2); the azimuthal mode only is unstable in (3); while the azimuthal mode and one inclinational mode are unstable in (4).

As the following discussion shows, *with sufficiently rapid spin of the gyrostat,* in (1) the motion is wholly stable; (2) has two interlinked modes, the azimuthal and the shorter period inclinational, unstable, and the other inclinational modes stable; for (3) the same result holds; in (4) the motion is again wholly stable.

The theory of the arrangement is shortly as follows. It is supposed that when the cords are in the same vertical plane the arrangement is symmetrical about the line joining the mid-point E of the trapeze, on which the gyrostat is carried, and the mid-point F of the line AB.

Let $\zeta$ denote the azimuthal angle turned through by the trapeze from coplanarity of the cords, $\eta$ the inclination of the axis of the gyrostat to the horizontal, and $\theta$ the angular turning about the line joining the upper ends A, B of the cords. The angle $\theta$ may be taken as the inclination of EF to the vertical in the displaced position.

The turnings $\eta$, $\theta$ we suppose are counter-clockwise when viewed from beyond the right-hand end of the trapeze, and we take the position of stability without spin as the normal position. If M be the total mass suspended and $l$ the length of EF, the total moment about AB will not differ much from $Mgl\theta$, if $\theta$ be small, and $l$ (as in Fig. 47) be great in comparison with the distance $h$ of the line of swivels from the centroid of the gyrostat. Practically the whole mass M is made up of the gyrostat and trapeze. The moment of the forces about the line of swivels is approximately $Mgh\eta$. We denote the moment about the vertical due to the bifilar, when the azimuthal angle is $\zeta$, by $L\zeta$. The equations of motion are approximately

$$\left.\begin{aligned} A\ddot{\zeta} - Cn\dot{\eta} \pm L\zeta = 0, \\ B\ddot{\eta} - Mhl\ddot{\theta} + Cn\dot{\zeta} \pm Mgh\eta = 0, \\ B'\ddot{\theta} - Mhl\ddot{\eta} \pm Mgl\theta = 0, \end{aligned}\right\} \qquad (1)$$

for the suspended mass, apart from that attached to the swivels, is practically zero. Here terms in $\dot{\theta}^2$, $\dot{\eta}^2$ are neglected, all the angles $\zeta$, $\eta$, $\theta$ are taken small, and A, B, B′ denote the moments of inertia respectively about a vertical through the centroid of the gyrostat and trapeze, about the line of swivels, and about the line AB, and the upper or lower sign before L, $Mgh$, or $Mgl$ is to be taken according as the mode of motion is stable or unstable in the absence of spin of the flywheel.

If we suppose that

$$\zeta = ae^{\iota\mu t}, \quad \eta = be^{\iota\mu t}, \quad \theta = ce^{\iota\mu t}, \qquad (2)$$

and substitute in (4), we get, taking for the present the upper signs in the ambiguities, the determinantal equation

$$A(BB' - M^2h^2l^2)\mu^6 - (BB'L + AB'Mgh + ABMgl - LM^2h^2l^2 + B'C^2n^2)\mu^4$$
$$+ (BLl + B'Lh + AMhlg + C^2n^2l)Mg\mu^2 - LM^2hlg^2 = 0. \qquad (3)$$

When L, $Mgh$, or $Mgl$ is to be taken with sign *minus* prefixed, the requisite correction in this equation can be made at once. Since $B > Mh^2$ and $B' > Ml^2$, the coefficient of $\mu^6$ in (3) is positive. We can now consider the different cases specified above.

In case (1) the determinantal equation written as

$$\alpha\mu^6 + \beta\mu^4 + \gamma\mu^2 + \delta = 0$$

has its coefficients $\alpha$, $\beta$, $\gamma$, $\delta$ all real and alternately positive and negative. Hence the roots of the cubic in $\mu^2$ are all real and positive. The motion is therefore oscillatory in a real period. Each value of $\mu^2$ gives two values of $\mu$ numerically equal and opposite in sign. The motion is thus stable. It is easy to show that when there is very rapid rotation of the flywheel, so that the terms $C^2n^2$ dominate the coefficients of $\mu^4$ and $\mu^2$, the positive values of $\mu$ correspond to rotation of the flywheel in one direction, the negative values to rotation in the opposite direction.

In case (2) we suppose the inclinational mode, $\eta$, to be unstable without spin. We shall suppose also that there is complete gyrostatic domination, that is, that $Cn$ is so great that $C^2n^2$ in the coefficients $\beta$, $\gamma$ dwarfs all the other terms into relative insignificance. Since $h$ is now to be taken with the negative sign the determinantal equation (3) is now

$$A(BB' - M^2h^2l^2)\mu^6 - B'C^2n^2\mu^4 + C^2n^2Mgl\mu^2 + LM^2hlg^2 = 0. \qquad (4)$$

Here the sum of the roots of the cubic in $\mu^2$ has a very large positive value

$$B'C^2n^2/A(BB' - M^2h^2l^2),$$

and their product a moderately small negative value $-LM^2g^2hl/A(BB' - M^2h^2l^2)$. The equation has therefore a large positive root and a numerically small negative root. These roots are given approximately by the first two terms and the last two terms of (4). They are respectively

$$\mu^2 = \frac{C^2n^2B'}{A(BB' - M^2h^2l^2)}, \quad \mu^2 = -\frac{LMghl}{C^2n^2}. \qquad (5)$$

The intermediate root is approximately

$$\mu^2 = \frac{Mgl}{B'}. \qquad (6)$$

The small negative value of $\mu^2$ is characteristic of the azimuthal motion in $\zeta$ and the inclinational in $\eta$, now linked by the gyrostatic action. Hence this combination is unstable.

In case (3) the azimuthal mode only is unstable without spin. We get exactly the same roots, and therefore the same result as to stability as before.

In case (4) L and M$gh$ are to be taken with the negative sign prefixed. We see at once that in this case all three values of $\mu^2$ are positive. The motion is therefore wholly stable.

In the above discussion the words *stable* and *unstable* have been used in the restricted sense, that the theorems are only true for a system unaffected by frictional resistance to motion. The subject of gyrostatic domination is resumed again in Chapter XX.

6. *Gyrostatic arrangement doubly stable or doubly unstable without spin.* Fig. 48 shows a device (due to Professor H. A. Wilson) which may be modified in different ways to illustrate the theory given in 3 and 4, and which is easily constructed. The axis of the flywheel of a gyrostat is carried by a ring movable about a diametral axis at right angles to the flywheel axis, and is shown as standing at right angles to the plane of the diagram. The axis of the ring, BB, is carried by a frame which can turn about the line of the bar CC, in one arrangement, which we shall refer to as case ($a$), or about the line C′C′ in another, which we shall call case ($b$). The frame is shown in a vertical position in the diagram. The upper end of the axis BB passes through a bearing in the upper bar of the frame and carries a crank $c$. One end of a spiral spring is attached to this crank, the other end is moored in case ($b$) to a pillar on the upper bar CC as shown in the diagram, and in case ($a$) to a similar pillar on a considerable prolongation of the bar CC to the right.

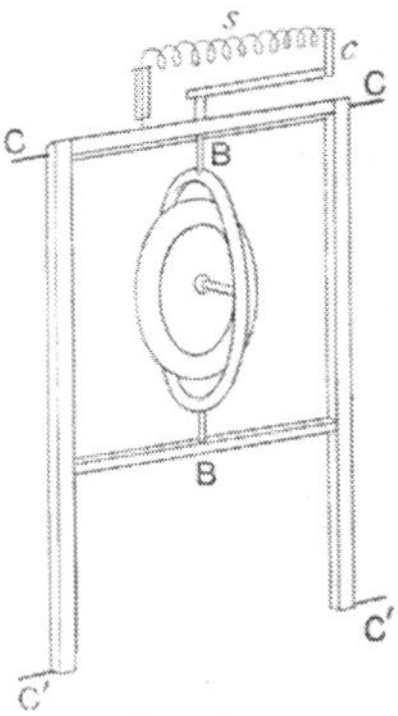

FIG. 48.

It will be seen that there are two freedoms, one of turning of the gyrostat about the axis BB, the other of the whole system, either about the axis in line with CC or about C′C′. With the freedoms about BB and CC we take the spiral spring placed as in case ($a$); its action on the crank, when that is turned out of the plane of the frame by the turning about BB, is then to exert a moment tending to bring the crank, and therefore the plane of the gyrostat flywheel, towards coincidence with the frame. Thus the two freedoms are both stable without spin of the flywheel, the former by the action of the spring and crank, the latter because of gravitation.

In case ($b$) the arrangement is doubly unstable, for the action of the crank is then towards increasing the angle between the planes of the wheel and the frame, and the frame is unstable without spin since the centroid of the system is above the axis C′C′.

Let $\phi$ denote an angle turned through about the axis CC, and let at the instant considered the crank be inclined to the frame at the angle $\theta$, both angles being small. Then the directions of spin and of the angles of turning $\phi$ and $\theta$ when positive, being as indicated, the $\phi$-equation of motion is

$$A\ddot{\phi}+Cn\dot{\theta}+Mgh\phi=0, \qquad (1)$$

where the quantities A, $Cn$, $Mgh$ have the usual significations, that is M is the whole swinging mass, and A is the moment of inertia for the axis CC. The distance $h$ is the height of the axis CC above the centroid of the system, and in Fig. 48 is positive.

If $\theta$ be the small angle turned through by the crank, and the length of the spring be great in comparison with that of the crank, and the spring be considerably stretched when the system is in equilibrium, the force F applied by it will be practically constant, and the moment applied by it will, for length $l$ of the crank, be $Fl\theta$. The $\theta$-equation of motion is therefore

$$B\ddot{\theta}-Cn\dot{\phi}+Fl\theta=0, \qquad (2)$$

where B is the moment of inertia of the gyrostat about the axis BB.

Equations (1) and (2) correspond precisely to equations (1) or (2) of 2 above.

Let now the arrangement be made doubly unstable by pivoting it on the axis C′C′, and carrying the crank to which the spring is attached to the left of B before attachment to the spring, as shown in Fig. 48. The equations of motion are now, if $h$ be the numerical value of the now negative height of C′C′ above the centroid,

$$\left.\begin{aligned} A\ddot{\phi}+Cn\dot{\theta}-Mgh\phi=0,\\ B\ddot{\theta}-Cn\dot{\phi}-Fl\theta=0. \end{aligned}\right\} \qquad (3)$$

All the conclusions derived above hold for these arrangements. We get for equations (1) and (2), the determinantal equation

$$ABk^4-(C^2n^2+MghB+FlA)k^2+MghFl=0, \qquad (4)$$

a quadratic in $k^2$, the roots of which are real and positive. For equations (3), on the other hand, we obtain the quadratic

$$ABk^4-(C^2n^2-MghB-FlA)k^2+MghFl=0, \qquad (5)$$

the roots of which are real and positive if

$$C^2n^2>MghB+FlA+2(MghBFlA)^{\frac{1}{2}},$$

in which F, $h$ are both positive and the positive square root is taken.

In each case, if $Cn$ be very great, there are two roots which are approximately

$$k^2=\frac{C^2n^2+MghB+AFl}{AB}, \quad k^2=\frac{MghFl}{C^2n^2+MghB+AFl} \qquad (6)$$

in one case, and

$$k^2=\frac{C^2n^2-MghB-AFl}{AB}, \quad k^2=\frac{MghFl}{C^2n^2-MghB-AFl} \qquad (7)$$

in the other. The first root given in each case is the larger. The first gives a vibration of shorter, the second of longer period, since of course the period is $2\pi/k$.

If now we suppose that F is zero, that is that there is no spring, the longer period is infinite, and the shorter is $2\pi\{AB/(C^2n^2+MghB)\}^{\frac{1}{2}}$ in one case, and $2\pi\{AB/(C^2n^2-MghB)\}^{\frac{1}{2}}$ in the other. This is also obvious from equations (1) and (2) or (3). For (2) and the second of (3) become

$$B\ddot{\theta}-Cn\phi=0,$$

and it is clear that if $\phi$ is kept zero $\dot{\theta}$ will be zero, so that we have $B\dot{\theta}=Cn\phi$. Equations (1) and the first of (3) therefore become respectively

$$A\ddot{\phi}+\frac{C^2n^2+MghB}{B}\phi=0, \qquad \text{.....................(8)}$$

and

$$A\ddot{\phi}+\frac{C^2n^2-MghB}{B}\phi=0. \qquad \text{.....................(9)}$$

In the former case the length of the equivalent simple pendulum has been diminished from $A/Mh$ to $A/(Mh+C^2n^2/gB)$, that is by the amount $AC^2n^2/Mh(C^2n^2+BMgh)$, which if $C^2n^2$ is very great is approximately $A/Mh$. The effect of increased angular speed of spin is therefore towards reducing the period to zero.

On the other hand there is not in the other case any real period until $C^2n^2$ is greater than $MghB$, and after that continued increase of spin is towards giving a length of equivalent simple pendulum equal to $ABg/C^2n^2$, which again is zero when $Cn$ is infinite.

7. *Gyrostatic control of the rolling of a ship. Controller with gyrostat frame clamped.* The Schlick contrivance for diminishing the rolling of a ship is an example of the first case. If it is left free to precess in the fore and aft direction as the ship rolls from side to side, and there is no couple on the gyrostat in the (fore and aft) plane of precession, the metacentric height is increased by the amount $C^2n^2/BMg$, where B refers to the gyrostat in its precessional motion, about an axis parallel to the deck and athwart ship. As M is the whole mass of the ship this is not a very large increase.

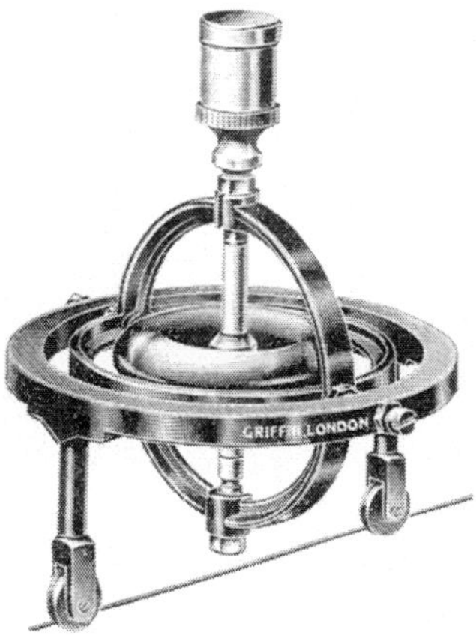

FIG. 49.

We may put $C^2n^2/BMg$ in the form $\frac{1}{2}Cn^2 . 2C/BMg$, and $C/B$ is about 2 for a wheel like a disk.

In the stilt top (Fig. 49) we have an example of the other case. We must have $C^2n^2 > MghB$, that is $\frac{1}{2}Cn^2 > \frac{1}{2}B/C . Mgh$, or the kinetic energy of the flywheel greater than the exhaustion of potential energy involved in the descent of the top from the vertical to the horizontal position.

The theory of the stilt top with frictional resistance to both freedoms is practically identical with that of Schlick's gyrostatic controller of the rolling of a ship. We shall consider this contrivance here, and point out afterwards the bearing of the results of the discussion of its theory on the behaviour of other contrivances. The controller is shown diagrammatically in Fig. 50. A gyrostat G is fixed so that, when the ship is upright and the gyrostat is in equilibrium, the axis of the flywheel is vertical. The wheel is pivoted in a frame F, as shown, which turns on the bearings *bb*, and a weight W gives to the whole arrangement a certain amount of gravitational stability. A brake pulley B surrounds the axis *bb*, about which the frame turns, and friction of a graded amount is applied by a special device. The brake damps out the free oscillations of the system, and also serves to reduce the forced oscillations, or rolling, due to the periodic passage of waves under the ship. But the action of the brake must not be so violent as to prevent the swinging of the gyrostat about the axis *bb*, as that would annul the controlling effect of the arrangement [see 6 above].

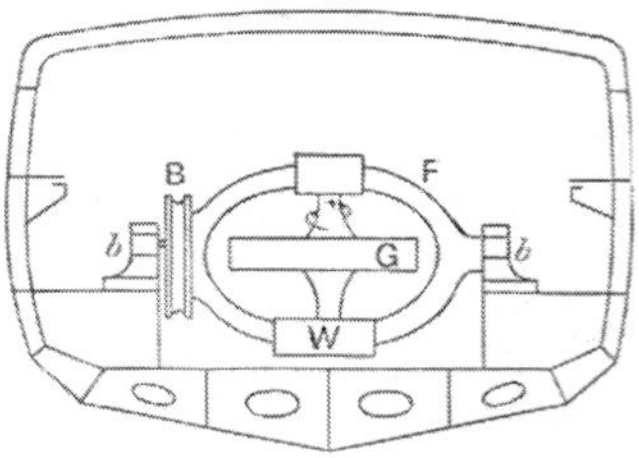

Fig. 50.

We consider first the free oscillations of the ship alone, and then of the ship and gyrostat together when the flywheel is spinning and the gyrostat is free to precess about the axis *bb*. In the former case we shall suppose that there is a resisting force proportional to the angular speed of rolling, and that the gyrostat is on board but clamped so as to be out of action.

The equation of motion of the rolling ship is

$$A\ddot{\phi}+N\dot{\phi}+M\phi=0, \qquad\qquad (1)$$

where A is the moment of inertia of the ship for the longitudinal axis about which she rolls, and M is the righting moment per unit of the angle $\phi$ of heel. N is a coefficient essentially positive which makes $N\dot{\phi}$ the couple resisting the rolling. The solution of this equation is

$$\phi=Ke^{-\frac{1}{2}\frac{N}{A}t}\cos\left\{\frac{(4AM-N^2)^{\frac{1}{2}}}{2A}t-\epsilon\right\}, \qquad\qquad (2)$$

which represents an oscillation the amplitude of which diminishes according to the exponential factor. Of course if disturbances are introduced from time to time the values of K and $\epsilon$ will require modification after each. For oscillation it is necessary that $4AM > N^2$.

If the ship, still with the gyrostat clamped, is rolling in a seaway, the equation of motion becomes

$$A\ddot{\phi}+N\dot{\phi}+M\phi=C'\cos pt, \quad\quad(3)$$

where $p$ is $2\pi$ times the frequency of the periodic disturbance due to the waves passing the ship. We must find a particular solution of this equation, and this, together with the solution (2), will be the complete solution.

Writing D for $d/dt$ in (3), we get

$$(AD^2+ND+M)\phi=C'\cos pt\,; \quad\quad(4)$$

hence $\phi$ is that function which, when the operator $AD^2+ND+M$ is applied to it, generates $C'\cos pt$. Now it is easy to verify that if we apply this operator to $C'\cos(pt+\alpha)$, we obtain

$$C'\{(-Ap^2+M)^2+N^2p^2\}^{\frac{1}{2}}\cos\left(pt+\alpha+\tan^{-1}\frac{Np}{-Ap^2+M}\right),$$

that is, the function is multiplied by $\{(-Ap^2+M)^2+N^2p^2\}^{\frac{1}{2}}$ and advanced in phase by the angle $\tan^{-1}\{Np/(-Ap^2+M)\}$. It is therefore clear that if we write

$$\phi=\frac{C'}{\{(-Ap^2+M)^2+N^2p^2\}^{\frac{1}{2}}}\cos\left(pt-\tan^{-1}\frac{Np}{-Ap^2+M}\right), \quad\quad(5)$$

we have obtained a function which the operator specified will convert into $C'\cos pt$, that is we have obtained a solution of the differential equation (4).

The performance of the inverse operation $(AD^2+ND+M)^{-1}$ on the function $C'\cos(pt+\alpha)$ divides the function operated on by $\{(-Ap^2+M)^2+N^2p^2\}^{\frac{1}{2}}$, and turns the phase back through the angle $\tan^{-1}\{Np/(-Ap^2+M)\}$. This gives an easily remembered rule for applying the symbolical method of treatment, which is the best adapted for the present discussion.

We may note here that since $D^2=-p^2$, the operator has the form $E+FD$, where $E=-Ap^2+M$ and $F=N$. This remark is of practical importance, since operators which are integral functions of D can all be converted into operators of the form $E+FD$ by substitution of $-p^2$ for $D^2$, when we are dealing with functions of the form $\cos(pt+\alpha)$. The performance of the direct operation multiplies by $(E^2+F^2p^2)^{\frac{1}{2}}$, and advances the phase of the function operated on by the angle $\tan^{-1}(Fp/E)$; performance of the inverse operation $(E+FD)^{-1}$ divides by $(E^2+F^2p^2)^{\frac{1}{2}}$, and turns the phase back by the angle $\tan^{-1}(Fp/E)$.

The complete solution of (4) is thus

$$\phi=\frac{C'}{(E^2+F^2p^2)^{\frac{1}{2}}}\cos\left(pt-\tan^{-1}\frac{Fp}{E}\right)+Ke^{-\frac{1}{2}\frac{N}{A}t}\cos\left\{\frac{(4AM-N^2)^{\frac{1}{2}}}{2A}t-\epsilon\right\}. \quad\quad(6)$$

The second part of this solution is continually being extinguished by friction, but starts anew into existence after each disturbance or irregularity of the forced vibration; for example, in consequence of the inequalities of passing waves, or by change of their effective period due to change of the ship's course, or by other disturbances.

Now let the wave-slope be given by $C_1\cos pt$ so that the relative slope $\phi'$ is $\phi+C_1\cos pt$. But $C_1\cos pt$ bears a fixed relation to $C'\cos pt$; the couple changing the relative slope is $M(\phi+C_1\cos pt)$. Hence the couple arising from the slope itself is $MC_1\cos pt$; that is $MC_1\cos pt=-C'\cos pt$, or $C_1=-C'/M$. Thus we have by (5)

$$\phi'=\phi-\frac{C'}{M}\cos pt=C'\left\{\left(\frac{E}{E^2+F^2p^2}-\frac{1}{M}\right)\cos pt+\frac{Fp}{E^2+F^2p^2}\sin pt\right\}=L\cos(pt-\alpha), \quad\quad(7)$$

where

$$L=C'\frac{\{(M-E)^2+F^2p^2\}^{\frac{1}{2}}}{M(E^2+F^2p^2)^{\frac{1}{2}}},\quad \tan\alpha=\frac{FpM}{ME-E^2-F^2p^2}. \quad\quad(8)$$

The values of E and F are stated above.

If $\alpha$ is positive the relative slope is against that of the wave, and if $\alpha$ is negative the relative slope is the other way. But $\tan\alpha$ is positive or negative according as $Ap^2$ is less or greater than $M - F^2/A$.

**8.** *Gyrostatic controller of rolling: gyrostat frame unclamped.* We now consider the free oscillations of the ship with the gyrostat unclamped, in water undisturbed by waves. This is essentially the problem discussed in (6) above. The equations of motion are now $A\ddot{\phi}+Cn\dot{\theta}+M\phi=0, \quad B\ddot{\theta}-Cn\dot{\phi}+Wga\theta=0,$ ......(1)
if we take no account of frictional or other resistances. Here $Cn$ is as usual the A.M. of the flywheel about its axis, B is the moment of inertia of the gyrostat about the bearings $bb$, and $Wga$ is the couple per unit of the precessional deflection $\theta$ of the gyrostat from its equilibrium position, applied by the weight W. [See 10 below for a discussion of the effect of frictional terms.]

If we write $\phi=Ke^{i\kappa t}$, $\theta=ke^{i\kappa t}$, and substitute in (1), we obtain

$$i^2\kappa^2KA+i\kappa kCn+KM=0, \quad i^2\kappa^2kB-i\kappa KCn+Wgak=0, \quad \text{......(2)}$$

so that

$$\frac{k}{K}=\frac{i\kappa Cn}{i^2\kappa^2B+Wga}=-\frac{i^2\kappa^2A+M}{i\kappa Cn}. \quad \text{......(3)}$$

Thus we obtain

$$(A\kappa^2-M)(B\kappa^2-Wga)-C^2n^2\kappa^2=0, \quad \text{......(4)}$$

a quadratic in $\kappa^2$, from which the periods $2\pi/\kappa$ are to be obtained. There are four values of $\kappa$, namely, $\kappa_1$, $\kappa_2$, $-\kappa_1$, $-\kappa_2$, and the complete solution of (1) for the initial conditions $\phi=\phi_0$, $\theta=0$, $\dot{\phi}=\dot{\theta}=0$, is given by

$$\phi=K_1\cos\kappa_1t+K_2\cos\kappa_2t, \quad \theta=k_1\sin\kappa_1t+k_2\sin\kappa_2t, \quad \text{......(5)}$$

where

$$\kappa_1k_1+\kappa_2k_2=0, \quad K_1+K_2=\phi_0. \quad \text{......(6)}$$

Now, by (3), we have

$$\frac{k_1}{K_1}=\frac{i\kappa_1Cn}{i^2\kappa_1^2B+Wga}=-\frac{i^2\kappa_1^2A+M}{i\kappa_1Cn}, \quad \frac{k_2}{K_2}=\frac{i\kappa_2Cn}{i^2\kappa_2^2B+Wga}=-\frac{i^2\kappa_2^2A+M}{i\kappa_2Cn}; \quad \text{......(7)}$$

so that in any case whatever,

$$\frac{K_1}{K_2}=-\frac{(A\kappa_2^2i^2+M)(B\kappa_1^2i^2+Wga)}{i^2\kappa_1\kappa_2C^2n^2}\frac{k_1}{k_2}. \quad \text{......(8)}$$

In the present case $k_1\kappa_1=-k_2\kappa_2$ by (6), and so putting $-1$ for $i^2$, we get

$$\frac{K_1}{K_2}=-\frac{(A\kappa_2^2-M)(B\kappa_1^2-Wga)}{C^2n^2\kappa_1^2}. \quad \text{......(9)}$$

If now for $M/A$, $Wga/B$ we write $4\pi^2F^2$, $4\pi^2f^2$, where F, $f$ are the frequencies of the free oscillations of the ship and gyrostat, the first oscillating with the gyrostat rigidly fixed within it, the second when the ship is at rest, in both cases without rotation of the flywheel, we can write instead of (4),

$$(\kappa^2-4\pi^2F^2)(\kappa^2-4\pi^2f^2)-\frac{C^2n^2}{AB}\kappa^2=0, \quad \text{......(10)}$$

and for (8) and (9),

$$\frac{K_1}{K_2}=\frac{(\kappa_2^2-4\pi^2F^2)(\kappa_1^2-4\pi^2f^2)}{\kappa_1\kappa_2C^2n^2}\frac{k_1}{k_2}=-\frac{(\kappa_2^2-4\pi^2F^2)(\kappa_1^2-4\pi^2f^2)}{\kappa_1^2C^2n^2}. \quad \text{......(11)}$$

**9.** *Gyrostatic controller and ship under forced vibrations.* When the ship rolls in a seaway the main oscillations of the ship are forced oscillations of the period of the waves, and the natural period of the ship is so increased by the gyrostat that any resonance effect, due to near agreement of the period of the waves with that of the ship, which might exist without rotation of the flywheel, is rendered impossible. For it will be noticed that the square of the frequency is now given by (10), being $\kappa^2/4\pi^2$. If we consider only the comparatively slow vibrations, we get the approximate value of the square of the frequency from the small root of (10) as $F^2/(1+F^2/f^2+C^2n^2/4\pi^2 f^2AB)$, where F is the natural frequency of the ship when rolling freely, and $f$ is the (much greater) natural frequency of the gyrostat as a compound pendulum (without spin of the flywheel) under the unital couple $Wga$.

The differential equations of small oscillations are now

$$\left.\begin{array}{r} A\ddot{\phi}+N\dot{\phi}-Cn\dot{\theta}+M\phi=C'\cos pt, \\ B\ddot{\theta}+N'\dot{\theta}+Cn\dot{\phi}+Wga\theta=0, \end{array}\right\} \qquad (1)$$

where $N\dot{\phi}$ is the frictional couple applied by the water to the ship as she rolls, and $N'\dot{\theta}$ is the frictional couple applied to the gyrostat frame by the brake B, or otherwise

To determine the forced oscillations we put

$$\phi=K\cos(pt-\alpha), \qquad (2)$$

and find K and $\alpha$ from (1). The value of $\theta$ is then found from that of $\phi$ by the second of (1) The simplest process is perhaps that which makes use of the result of 7 for the case of forced vibrations there considered. Putting D for $d/dt$, we write (1) in the form

$$\left.\begin{array}{r} (AD^2+M+ND)\phi-CnD\theta=C'\cos pt, \\ (BD^2+Wga+N'D)\theta+CnD\phi=0. \end{array}\right\} \qquad (3)$$

If we operate on the first of these with $BD^2+Wga+N'D$ and on the second with $CnD$, and add, we get

$$\phi=\frac{BD^2+Wga+N'D}{(AD^2+ND+M)(BD^2+N'D+Wga)+C^2n^2D^2}C'\cos pt. \qquad (4)$$

But by (1) and (2) we have $D^2=-p^2$, and therefore $D^4=p^4$. Thus we can write (4) in the form

$$\phi=\frac{E+FD}{E'+F'D}C'\cos pt, \qquad (5)$$

where $E=-Bp^2+Wga$, $F=N'$, $E'=ABp^4-(AWga+MB+NN'+C^2n^2)p^2+MWga$,

$$F'=-\{(AN'+BN)p^2-NWga-MN'\}.$$

We have now

$$(E+FD)\cos pt=(E^2+F^2p^2)^{\frac{1}{2}}\cos\left(pt+\tan^{-1}\frac{pF}{E}\right), \qquad (6)$$

and

$$(E'+F'D)^{-1}\cos pt=\frac{1}{(E'^2+F'^2p^2)^{\frac{1}{2}}}\cos\left(pt-\tan^{-1}\frac{pF'}{E'}\right). \qquad (7)$$

Thus we obtain finally

$$\phi=\frac{(E^2+F^2p^2)^{\frac{1}{2}}}{(E'^2+F'^2p^2)^{\frac{1}{2}}}C'\cos\left(pt+\tan^{-1}\frac{pF}{E}-\tan^{-1}\frac{pF'}{E'}\right),$$

or

$$\phi=\frac{(E^2+F^2p^2)^{\frac{1}{2}}}{(E'^2+F'^2p^2)^{\frac{1}{2}}}C'\cos\left(pt+\tan^{-1}\frac{p(E'F-EF')}{EE'+p^2FF'}\right). \qquad (8)$$

It remains to find $\theta$. By the second of (3) we have

$$\theta=-\frac{CnD}{BD^2+Wga+N'D}\phi=-\frac{CnD}{E+FD}\phi,$$

or

$$\theta=\frac{Cn}{(E'^2+p^2F'^2)^{\frac{1}{2}}}C'\sin\left(pt-\tan^{-1}\frac{pF'}{E'}\right). \qquad (9)$$

In the complete values of $\phi$ and $\theta$ we have to add in (8) and (9) the solutions for free oscillations. These equations however give all that we are concerned with when the ship is rolling under the influence of a regular succession of waves.

For a clear account of the theory of ship control, and numerical examples, the reader should consult a very instructive paper on "The Use of

Gyrostats," by Professor Perry, in *Nature*, March 12, 1908. The biquadratic equation is solved, and numerical results given for the alteration of period and the damping of the rolling of vessels of different sizes, when controlled by gyrostatic gear. As a particular case a vessel of 6000 tons, metacentric height 18 inches, with a natural rolling period of 14 seconds, controlled by a gyrostatic wheel of 10 tons of 6 feet radius, and running at 16 revolutions per second, is considered. We shall give later, in Chapter XXIII, on *Gyrostatics in Engineering*, some of the available practical examples.

**10.** *Theory of two interlinked systems which are separately unstable. Stability in presence of dissipative forces.* For the interlinked motion of ship and gyrostat the complete differential equations, in the case of resistances proportional to the angular speeds, are

$$A\ddot{\phi}+N\dot{\phi}+Cn\dot{\theta}+M\phi=0,\quad B\ddot{\theta}+N'\dot{\theta}-Cn\dot{\phi}+Wga\theta=0. \quad\ldots\ldots(1)$$

If we suppose that $\phi=Ke^{\lambda t}$, $\theta=ke^{\lambda t}$, where, as we suppose the motion to be oscillatory, $\lambda$ is complex, we get, by substitution in (1), the determinantal equation

$$AB\lambda^4+(AN'+BN)\lambda^3+(C^2n^2+AWga+BM+NN')\lambda^2 +(MN'+WgaN)\lambda+MWga=0. \quad\ldots\ldots(2)$$

The stability of the interlinked motions is ensured if the real part of $\lambda$ is negative, and this will be the case if the coefficients of $\lambda^3$, $\lambda^2$, $\lambda$ and the final term $MWga$ are all positive, for AB is positive. This condition is satisfied if the differential equations are as stated in (1), and the coefficients A, B, N, N′, M, $Wga$, $Cn$ are all positive.

But if the motions are individually unstable when not linked together by gyrostatic action equations (1) must be modified by changing $M\phi$, $Wga\theta$ to $-M\phi$, $-Wga\theta$, leaving M and $Wga$ still positive. The determinantal equation becomes now

$$AB\lambda^4+(AN'+BN)\lambda^3+(C^2n^2+NN'-AWga-BM)\lambda^2 -(MN'+WgaN)\lambda+MWga=0. \quad\ldots\ldots(3)$$

The last term is still positive, and that of $\lambda^2$ can be made positive by making $C^2n^2$ sufficiently great; but the coefficient of $\lambda$ has become negative. For stability of the interlinked motion all the coefficients must be positive. Hence the only way in which this condition can be fulfilled is by changing the sign of either N or N′, that is by causing angular *acceleration* either of the controlled body, or of the frame of the controlling gyrostat.

In the present case the controlled body cannot be a ship, as that has gravitational stability, which explains why the controlling gyrostat arrangement is also given gravitational stability, as in Schlick's device—the term $MWga$ must be positive. We may suppose the controlled body to be a monorail carriage, or the frame of a properly mounted stilt top. We shall refer to the body as "the carriage."

Let us then suppose that the angular speed of the gyrostat frame is accelerated. We therefore change the signs of the term AN′, MN′ in (3), leaving N and N′ both still positive. The coefficients of $\lambda^3$ and $\lambda$ are both to be rendered positive. This requires

$$\left.\begin{aligned} &BN-AN'>0, \quad WgaN-MN'<0,\\ \text{or}\quad &BN=AN'+e, \quad WgaN=MN'-f, \end{aligned}\right\}\quad\ldots\ldots(4)$$

where $e$ and $f$ are positive. Thus we are to have

$$\frac{B}{Wga}=\frac{AN'+e}{MN'-f}.$$

The conditions will therefore both be satisfied if

$$\frac{B}{Wga} > \frac{A}{M}. \quad \text{.............................................(5)}$$

In words, this condition asserts that, if the frame of the gyrostat be clamped, the numerical ratio $|\theta/\theta|$ of angular acceleration to angular displacement (friction zero) must be smaller than the corresponding ratio $|\ddot{\phi}/\phi|$ for the carriage.

It is not a practical solution of the problem to accelerate the carriage and retard the gyrostat. If this were done we should have as a condition

$$MN' - WgaN < 0,$$

that is an acceleration coefficient N would be required, sufficient to raise $WgaN$ to a value above MN′, where M is, comparatively, very great.

The acceleration of the gyrostat frame is a means of supplying energy from without to the system, the energy necessary to preserve in operation the functions of the apparatus. The carriage in its oscillatory motion is retarded by friction, the potential energy exhausted is constantly greater than the kinetic energy generated by the displacement of the carriage itself, and so energy must be fed in at the gyrostat frame to maintain the action.

This stabilising action by acceleration of one side of the compound motion is very important; and it is probable that, by analogy, it may be a guide to the explanation of the preservation of the stability of more complicated systems, in the presence of energy dissipating influences, and the breaking down of stability, or death of the system, when energy can no longer be supplied in the manner prescribed for the system by its constitution.

We shall deal with monorail systems in Chapter XXIII, on *Gyrostatics in Engineering.*

# CHAPTER IX

## THE MOTION OF CHAINS OF GYROSTATIC LINKS. MAGNETO-OPTIC ROTATION

1. *Problem of stretched chain of gyrostats.* We now consider some more general vibrational problems, and take first Lord Kelvin's problem of the stretched chain of gyrostats. This problem, as stated and as worked out by Lord Kelvin, was sufficient for the purpose he had in view, the illustration of his theory of magneto-optic rotation as caused by the presence of rapidly rotating particles in the transparent medium. But it is only a particular case of a more general problem which we shall consider, and give some examples of, immediately after the optical problem has been discussed.

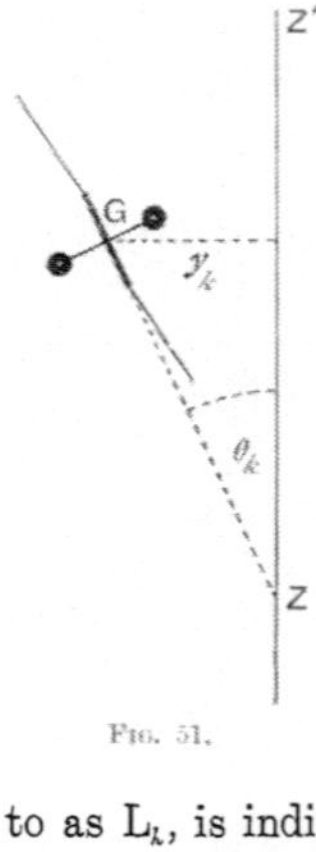

FIG. 51.

The chain consists of equal gyrostatic links alternating with ordinary rigid connecting links all of the same length and supposed to be of negligible mass. In Lord Kelvin's scheme the connection at each junction is made by a universal flexure (Hooke's) joint, and the chain forms an open plane polygon, held at its extremities A, B, in the line ZZ′ (Fig. 51), by joints of the same kind, and the whole turns with uniform angular speed $\mu$, without change of configuration about the line AB. It is required to find the form of the chain, when the inclination of each link to the line AB is very small.

A gyrostatic link, the $k$th in order, which we shall refer to as $L_k$, is indicated in the diagram (Fig. 51) with the connecting links at its ends, which we shall denote by $l_k$, $l_{k+1}$. [It is to be remembered in what follows that the line AB and the line ZZ′ of the diagram are the same.] The length of each L is $a$, and of each $l$ is $b$. We put $\theta_k$ for the inclination of the axis of $L_k$ to AB, $\phi_k$, $\phi_{k+1}$ for the inclinations of $l_k$, $l_{k+1}$ to the same line, $y_k$ for the distance of the centroid (the centre) of $L_k$ from the line AB, and $m$ for the mass of $L_k$. We proceed at once to the case in which the angles here specified are all very small.

We get first the geometrical equation

$$y_{k+1} - y_k = \tfrac{1}{2}a(\theta_{k+1} + \theta_k) + b\phi_{k+1}. \quad \text{.......................(1)}$$

Next, if P be the component parallel to AB of the pull along a link, we easily see that for equilibrium this must be the same for every link. The component of pull at right angles to AB is $P \tan \phi_k = P\phi_k$. We obtain the equation for the acceleration of the centroid of $L_k$ as it moves in its circular path about AB with angular speed $\mu$,

$$m\ddot{y}_k - m\mu^2 y_k = -P(\phi_k - \phi_{k+1}). \quad \text{.........................(2)}$$

Finally we obtain the gyrostatic equation [see (2), 1, V, above]

$$A\ddot{\theta}_k + (Cn + C'\omega_1 - A\mu)\mu\theta_k = \tfrac{1}{2}Pa(\phi_{k+1} + \phi_k - 2\theta_k). \quad \text{.............(3)}$$

The quantity on the right is the approximate value of the moment of the forces on $L_k$ taken about an axis through the centroid of $L_k$ at right angles to the plane of the polygon, an axis corresponding to $GD_1$ of (1), 1, V, above. [The exact value of the moment is

$$\tfrac{1}{2}Pa\{(\tan \phi_{k+1} + \tan \phi_k) \cos \theta_k - 2 \sin \theta_k\}.]$$

We have now to consider the value of the angular speed $\omega_1$ in (3). This is given by the geometry of the Hooke's joint, which is equivalent in its action to a short piece of quite flexible but untwistable wire connecting the adjacent links. Each L when thus joined behaves as if the gyrostatic axis were prolonged to intersect the line AB, and were there held by such a piece of wire in line at one end with BA and at the other with L. The instantaneous axis about which $L_k$ turns bisects the angle supplementary to $\theta_k$. If $\omega$ be the angular speed of $L_k$ about this axis, we have

$$\omega \sin \tfrac{1}{2}(\pi - \theta_k) = \mu \sin \theta_k,$$

or

$$\omega = 2\mu \sin \tfrac{1}{2}\theta_k. \quad \text{..........................(4)}$$

This resolves into a component $2\mu \sin^2 \frac{1}{2}\theta_k$ $[= \mu(1 - \cos \theta_k)]$ about the axis of $L_k$, and a component $\mu \sin \theta_k$ about the axis GE drawn in the plane of the polygon at right angles to the axis of $L_k$. Hence $\omega_1 = 2\mu \sin^2 \frac{1}{2}\theta_k$, which is negligible when $\theta_k$ is very small. Even if this angular speed were not negligible the smallness of C′ would render the term $C'\omega_1$ inappreciable. The precise mode of connection is thus of no consequence. It is, however, to be remembered that the angular speed $\mu$ about AB gives a component $\mu \cos \theta_k$ about the axis of $L_k$, by which the turning of the axis GE is to be reckoned, and a component $\mu \sin \theta_k$ about GE which gives the turning of the axis of the flywheel towards $GD_1$. Equation (3) thus becomes

$$A\ddot{\theta}_k + (Cn - A\mu)\mu\theta_k = \tfrac{1}{2}Pa(\phi_{k+1} + \phi_k - 2\theta_k). \quad \text{...................(5)}$$

Equations (1), (2), and (5) solved by the method of finite differences give the complete solution of the problem proposed, indeed of a more general problem, if it is not supposed that $\mu$ is constant, and the term $A\ddot{\theta}_k$ in (5) is retained. For all three equations are then applicable and changes in $\theta$ will be given according to the imposed conditions. Supposing, however, that $\ddot{\theta}_k=0$, and $\mu$ a constant, the three equations are

$$\left.\begin{aligned} y_{k+1}-y_k-\tfrac{1}{2}a(\theta_{k+1}+\theta_k)-b\phi_{k+1}=0,\\ m\mu^2 y_k+\mathrm{P}(\phi_{k+1}-\phi_k)=0,\\ (\mathrm{C}n-\mathrm{A}\mu)\mu\theta_k-\tfrac{1}{2}\mathrm{P}a(\phi_{k+1}+\phi_k-2\theta_k)=0.\end{aligned}\right\}\quad\text{..........(6)}$$

From these, as we shall see below, we obtain a simple solution if we suppose $y_{k+1}=y_k=0$, that is, that the centres of the gyrostats are all on the line O$z$. We pass over this solution for the present.

In the usual notation of the calculus of finite differences the operator by which a quantity $u_{k+1}$ is derived from $u_k$ is denoted by E, so that $u_{k+1}=\mathrm{E}u_k$. Applying this to the equations in (6) we obtain, if $c$ be written for $(\mathrm{C}n-\mathrm{A}\mu)\mu+\mathrm{P}a$,

$$\left.\begin{aligned} 2(\mathrm{E}-1)y_k-a(\mathrm{E}+1)\theta_k-2b\mathrm{E}\phi_k=0,\\ m\mu^2 y_k+\mathrm{P}(\mathrm{E}-1)\phi_k=0,\\ 2c\theta_k-\mathrm{P}a(\mathrm{E}+1)\phi_k=0.\end{aligned}\right\}\quad\text{..........(7)}$$

Thus we get, eliminating $y_k$, $\theta_k$, and $\phi_k$, the determinantal equation

$$\mathrm{E}^2-2(1-e)\mathrm{E}+1=0, \quad\text{..........(8)}$$

where $e=2m\mu^2(\mathrm{P}a^2+bc)/\mathrm{P}(4c+m\mu^2a^2)$.

If now we suppose that $1-e<1$, a condition fulfilled by making P sufficiently great, and supposing that $\mu$ is the smaller of the two roots of the quadratic in $\mu$ which (3) becomes when $\ddot{\theta}_k=0$, we get from (8), by writing $1-e=\cos\alpha$,

$$\mathrm{E}=\cos\alpha\pm i\sin\alpha, \quad\text{..........(9)}$$

and therefore

$$y_k=f\mathrm{E}^k+g\mathrm{E}^{-k}=f(\cos k\alpha+i\sin k\alpha)+g(\cos k\alpha-i\sin k\alpha). \quad\text{.......(10)}$$

To obtain from this a real solution we put $2f=\mathrm{A}-i\mathrm{B}$, $2g=\mathrm{A}+i\mathrm{B}$, and obtain from (10)

$$y_k=\mathrm{A}\cos k\alpha+\mathrm{B}\sin k\alpha. \quad\text{..........(11)}$$

If $x_k$ denote the coordinate parallel to AB measured from the centre of the link $\mathrm{L}_0$, we have

$$x_k=k(a+b),$$

so that (11) becomes

$$y_k=\mathrm{A}\cos\frac{\alpha x_k}{a+b}+\mathrm{B}\sin\frac{\alpha x_k}{a+b}. \quad\text{..........(12)}$$

Thus the centroids of the gyrostatic links lie on a harmonic curve of wave-length $2\pi(a+b)/\alpha$, which may be regarded as the projection, on a plane through AB, of the helical configuration of the particles in a circularly polarised wave. The period is $2\pi/\mu$, and therefore a circularly polarised wave is propagated along the chain at speed $\mu(a+b)/\alpha$, $=\mathrm{V}$, say. But $1-e=\cos\alpha$, so that $e=2\sin^2\frac{1}{2}\alpha$, or, to terms of the second order inclusive, $\alpha=(2e)^{\frac{1}{2}}$. Thus,

$$\mathrm{V}=\frac{\mu(a+b)}{(2e)^{\frac{1}{2}}}. \quad\text{..........(13)}$$

The exact value of $1/2e$ is [see (8) above]

$$\frac{\mathrm{P}(4c+m\mu^2a^2)}{4m\mu^2(\mathrm{P}a^2+bc)}.$$

But since $a$ is very small we neglect $m\mu^2a^2$ in comparison with $4c$, and obtain, by the value of $c$,

$$\frac{1}{2e}=\frac{P}{m\mu^2(a+b)}\left\{1+\frac{(\beta n-\gamma\mu)\mu a}{aP+b(\beta n-\gamma\mu)\mu}\right\},$$

where $\beta=C/(a+b)$, $\gamma=A/(a+b)$, the moments of inertia, axial and transverse, per unit of length of the chain. Hence, by (13),

$$V^2=\frac{P(a+b)}{m}\left\{1+\frac{(\beta n-\gamma\mu)\mu a}{aP+b(\beta n-\gamma\mu)\mu}\right\}. \qquad (14)$$

But if the links be very small and P be very great, the value of $1/2e$ gives approximately

$$\frac{\mu}{(2e)^{\frac{1}{2}}}=\left\{\frac{P}{m(a+b)}\right\}^{\frac{1}{2}}\left\{1+\frac{1}{2}\frac{(Cn-A\mu)\mu}{P(a+b)}\right\}, \qquad (14')$$

or, if $A\mu$ be neglected,

$$V=\left\{\frac{P(a+b)}{m}\right\}^{\frac{1}{2}}\left\{1+\frac{1}{2}\frac{Cn\mu}{P(a+b)}\right\}. \qquad (15)$$

Thus, but for the gyrostatic influence, the speed of propagation would be $\{P(a+b)/m\}^{\frac{1}{2}}$. Now P is the pull in the chain, and $m/(a+b)$ is the mass per unit of length, so that if the links be very small (making $1/(2e)^{\frac{1}{2}}$ very great, and therefore $a$ very small, and the wave-length $2\pi(a+b)/a$ very great) and there be consequently a large number in the wave-length, the velocity of propagation is that of a wave of transverse displacement along a stretched cord, multiplied by the factor $1+Cn\mu/2P(a+b)$.

If $\lambda$ be the wave-length, then, since $2\pi/\mu$ is the period, we have approximately $\lambda\mu/2\pi=\{P(a+b)/m\}^{\frac{1}{2}}$, or $P(a+b)=m\lambda^2\mu^2/4\pi^2$. Using this in the factor $1+Cn\mu/2P(a+b)$ we get

$$V=\left\{\frac{P(a+b)}{m}\right\}^{\frac{1}{2}}\left(1+2\pi^2\frac{Cn}{\mu m\lambda^2}\right), \qquad (16)$$

or, substituting $2\pi V/\lambda$ for $\mu$,

$$V=\left\{\frac{P(a+b)}{m}\right\}^{\frac{1}{2}}\left(1+\pi\frac{Cn}{mV\lambda}\right)=\left\{\frac{P(a+b)}{m}\right\}^{\frac{1}{2}}+\pi\frac{Cn}{m\lambda} \qquad (17)$$

nearly.

Thus we see that the velocity of propagation of the circularly polarised wave is increased by an amount proportional to the angular momentum $Cn$ of a flywheel, and inversely proportional to the wave-length.

It will be seen from (14) that the gyrostatic term in the value of V changes sign if the direction of the rotation about the line AB is reversed, while that of the rotation of the flywheel remains unaltered, or *vice versa*. For either of these cases we have a new velocity V′ of propagation and a corresponding wave-length given by $2\pi/\mu = \lambda'/V'$. The difference of the velocities is

$$V - V' = 2\pi \frac{Cn}{m\lambda}. \quad \text{.................................}(18)$$

The more exact value of the velocity of propagation in this case ($n$ and $\mu$ in opposite directions) is obtained from (14), which becomes

$$V'^2 = \frac{P(a+b)}{m}\left\{1 - \frac{(\beta n + \gamma\mu)\mu a}{aP - b(\beta n + \gamma\mu)\mu}\right\}. \quad \text{...................}(19)$$

2. *When vibrational motion of gyrostatic chain is possible under thrust or under tension.* Also from (14), 1, we see that, for $n$ and $\mu$ in the same direction, $V^2$ is still positive when for P is substituted $-P$, provided the numerical value of P lies between certain limits, that is the motion is then still possible under thrust. This point is not dealt with in Lord Kelvin's paper. If we examine (14) we see that if we take P as the value of the thrust, reckoned positive, it must lie between the limits $(\beta n - \gamma\mu)\mu(1 + b/a)$ and $(\beta n - \gamma\mu)\mu b/a$. It is assumed of course that $\beta n > \gamma\mu$.

In the case (19) above, in which $n$ and $\mu$ are in opposite directions ($n$ and $\mu$ being the positive numerical values), $V'^2$ is positive when P is greater than $(\beta n + \gamma\mu)\mu(1 + b/a)$, and also when P is less than $(\beta n + \gamma\mu)\mu b/a$. When P has a value between these limits $V'^2$ is negative, and the motion is impossible. The motion in this case is not possible at all under thrust. The limits of tension just given are, it will be seen, in a sense complementary to those of thrust in the other case. These limits of tension and of thrust and their connection do not seem to have been worked out before.

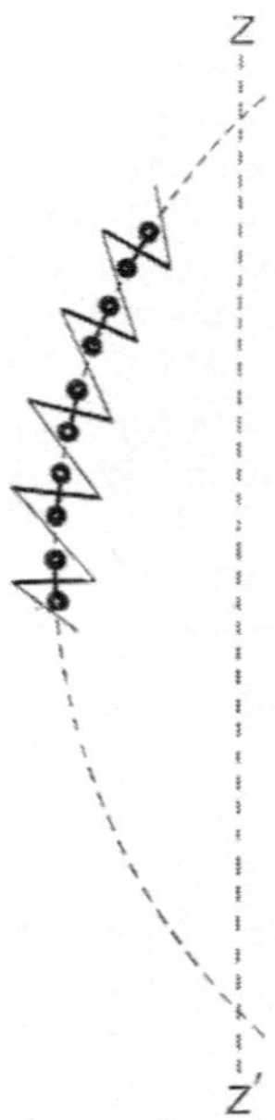

FIG. 52.

It is important to notice that the gyrostatic chain has been supposed to lie at each instant in a plane harmonic curve having the line AB (or ZZ', Fig. 52) as axis [The flywheels appear tangential to the curve, but the arrangement may be different.] It therefore does not directly represent by its motion a circularly polarised wave. For that we should have to suppose the gyrostats to lie at each instant on a helix having AB as axis, and all the gyrostats to turn at the same speed in equal circles round the axis. By a simpler analysis than that given above, the wave velocity may be found for this more general case, instead of that chosen by Lord Kelvin [see 3 below]. We should, however, obtain exactly equations (14) and (19) above; indeed it is not very difficult to see without further analysis that this must be the result. Hence the investigation given above illustrates circularly polarised waves precisely.

Now, returning to equation (18), imagine two chains of gyrostats, in all respects the same as to lengths of links, masses, and angular momenta of flywheels, to exist side by side, each turning in steady motion with angular speed numerically $\mu$, one about a line AB, the other about a parallel line A′B′, but in opposite directions, and having each the configuration shown above to be necessary for steady motion. The wave-lengths are $\lambda$, $\lambda'$ and the wave-velocities V, V′ [see (17) and (19), 1].

Equations (6) give the special solution above referred to, which is obvious otherwise. It is that which we obtain by putting $y_k$, $y_{k+1}$, etc., all zero; in other words it is the solution for which all the gyrostats have their centres on the $z$-axis, and their axles and connecting links form a zigzag wave along the $z$-axis, as shown in Fig. 53.

Fig. 53.

Equations (6) regard the values of $\theta$ and $\phi$ as small. But the exact equations may be written down and used, and the first thing to be observed is that $\phi_k = \phi_{k+1}$, $\theta_k = \theta_{k+1}$, or the angles $\theta$, $\phi$ are the same for each gyrostat.

It is shorter to work out the solution from first principles. The couple on any gyrostat is about an axis at right angles to the paper (see Fig. 53), and has moment $a\mathrm{P}\sin(\theta+\phi)$. The rate of growth of A.M. about the same axis is $(\mathrm{C}n - \mathrm{A}\mu\cos\theta)\mu\sin\theta$. But Fig. 53 shows that if $\theta$, $\phi$ be the (acute) angles of inclination of an axis and a link to ZZ′, and $l = \frac{1}{2}(a\cos\theta + b\cos\phi)$, $l/b = \sin(\theta+\phi)/\sin\theta$. Hence we obtain

$$(\mathrm{C}n - \mathrm{A}\mu\cos\theta)\mu = \mathrm{P}\frac{a}{b}l. \quad \text{.........................} \quad (20)$$

For $\theta = \frac{1}{2}\pi$, that is when the gyrostat axles stand across the $z$-axis, we have

$$\mathrm{C}n\mu = \tfrac{1}{2}a\mathrm{P}\cos\phi \quad \text{.................................} (21)$$

3. *Helical gyrostatic chain.* The investigation for a gyrostatic chain like that just described, but laid with the centres of the gyrostatic links uniformly distributed along a helix, and revolving all with the same uniform angular speed $\mu$, in circles round the axis of a helix, is comparatively simple The following discussion follows Greenhill (*R.G.T.*, p. 262), with some differences of notation and arrangement. It involves only the elementary principles and results as to gyrostatic action set forth in Chapter V.

Let each gyrostat axis make an angle $\theta$, and each connecting link an angle $\phi$, with the axis ZZ′ of the helix. Also we suppose that each gyrostat axis, and each link, is at right angles to the perpendicular let fall from the centre of the gyrostat or link on the axis of the helix.

Now consider the projection of the arrangement at any instant on a plane at right angles to the axis of the helix. The alternate gyrostat axes and links show as alternate sides of a polygon, of lengths $2a\sin\theta$ and $2b\sin\phi$, if we take here $2a$ and $2b$ (instead of as before $a$ and $b$) as the lengths of a gyrostat axis and a link respectively. These subtend, we suppose, angles $2\alpha$ and $2\beta$ at the axis. We can now find the geometrical equations and the equations of motion.

The advance along the axis from the centre of one gyrostat to the centre of the next is $2(a\cos\theta+b\cos\phi)$. The polygonal angle for this is $2(\alpha+\beta)$, and in the time of turning through this angle, that is, in $2(\alpha+\beta)/\mu$, the projection of the helical arrangement on a plane containing the axis seems to make the axial advance just stated. The wave-velocity V is therefore given by

$$V=(a\cos\theta+b\cos\phi)\frac{\mu}{\alpha+\beta}. \quad \text{......(1)}$$

If as before P be the pull along the chain, the force on a gyrostat towards the axis is $2P\sin\phi\sin(\alpha+\beta)$. The distance of the centre from the axis is $a\sin\theta\cot\alpha$, and therefore if $m$ be the mass of a gyrostatic link,

$$m\mu^2 a\sin\theta\cot\alpha=2P\sin\phi\sin(\alpha+\beta). \quad \text{......(2)}$$

But the geometry gives $a\sin\theta\cot\alpha=b\sin\phi\cos\alpha/\sin\beta$, and so the last equation can be written

$$m\mu^2 b\frac{\cos\alpha}{\sin\beta}=2P\sin(\alpha+\beta). \quad \text{......(3)}$$

This may be written $$\frac{m\mu^2 b}{2P}=\frac{\sin\beta\sin(\alpha+\beta)}{\cos\alpha};$$

which is equivalent to $$1-\frac{m\mu^2 b}{2P}=\frac{\cos\beta\cos(\alpha+\beta)}{\cos\alpha}. \quad \text{......(4)}$$

Now the A.M. about a horizontal axis, at right angles to the perpendicular from the centre of the gyrostatic link considered to ZZ (the perpendicular which is the axis of the couple), is $(Cn-A\mu\cos\theta)\sin\theta$, and as this axis and perpendicular revolve at rate $\mu$ round ZZ′, the rate of growth of A.M about the perpendicular is $(Cn-A\mu\cos\theta)\mu\sin\theta$. The moment of the couple producing this is easily seen to be $2aP\{\sin\phi\cos\theta\cos(\alpha+\beta)-\cos\phi\sin\theta\}$, and so we get

$$(Cn-A\mu\cos\theta)\mu\sin\theta=2aP\{\sin\phi\cos\theta\cos(\alpha+\beta)-\cos\phi\sin\theta\}. \quad \text{....(5)}$$

But from (2) we obtain

$$2aP=m\mu^2a^2\frac{\sin\theta\cot\alpha}{\sin\phi\sin(\alpha+\beta)}.$$

Hence (5) becomes

$$(Cn-A\mu\cos\theta)\mu+2aP\cos\phi=m\mu^2a^2\cos\theta\cot\alpha\cot(\alpha+\beta). \quad \text{......(6)}$$

From this it follows, since

$$1+\tan\alpha\tan(\alpha+\beta)=\frac{\cos\beta}{\cos\alpha\cos(\alpha+\beta)},$$

that $$\frac{\cos\beta}{\cos\alpha\cos(\alpha+\beta)}=\frac{(Cn-A\mu\cos\theta)\mu+2aP\cos\phi+m\mu^2a^2\cos\theta}{(Cn-A\mu\cos\theta)\mu+2aP\cos\phi}. \quad \text{......(7)}$$

Also, by (4), the last equation gives

$$\cos^2(\alpha+\beta)=\frac{\{(Cn-A\mu\cos\theta)\mu+2aP\cos\phi\}\left(1-\frac{m\mu^2 b}{2P}\right)}{(Cn-A\mu\cos\theta)\mu+2aP\cos\phi+m\mu^2a^2\cos\theta}; \quad \text{......(8)}$$

and therefore

$$\sin^2(\alpha+\beta)=\frac{Cn\mu - A\mu^2\cos\theta + 2P\frac{a}{b}(a\cos\theta + b\cos\phi)}{Cn\mu - A\mu^2\cos\theta + 2aP\cos\phi + m\mu^2 a^2\cos\theta}\frac{m\mu^2 b}{2P}. \quad \text{......(9)}$$

But by (1),

$$\sin^2(\alpha+\beta)=\sin^2\left\{(a\cos\theta+b\cos\phi)\frac{\mu}{V}\right\},$$

and as the line density $\rho$ is $m/(2a\cos\theta+2b\cos\phi)$, we get from (9) and (1)

$$\frac{(a\cos\theta+b\cos\phi)^2\mu^2}{\sin^2\left\{(a\cos\theta+b\cos\phi)\frac{\mu}{V}\right\}}$$

$$=\frac{Cn\mu - A\mu^2\cos\theta + 2aP\cos\phi + m\mu^2a^2\cos\theta}{Cn\mu - A\mu^2\cos\theta + 2P\frac{a}{b}(a\cos\theta+b\cos\phi)}\left(\frac{a}{b}\cos\theta+\cos\phi\right)\frac{P}{\rho}. \quad \text{....(10)}$$

If now we suppose that $a+b$ is infinitesimal in comparison with the length $V/\mu$, take $\cos\theta$, $\cos\phi$ each unity, and neglect $m\mu^2a^2$, we obtain

$$V^2=\frac{P}{\rho}(a+b)\frac{Cn\mu - A\mu^2 + 2aP}{b(Cn\mu - A\mu^2)+2aP(a+b)}, \quad \text{..................(11)}$$

or

$$V^2=\frac{P}{\rho}\left\{1+\frac{(Cn\mu - A\mu^2)a}{2Pa(a+b)+b(Cn\mu - A\mu^2)}\right\}. \quad \text{..................(12)}$$

Dividing both numerator and denominator of the fraction within the brackets by $2(a+b)$ [since $\beta$, $\gamma$ are now (C, A)/$2(a+b)$], we obtain the equation

$$V^2=\frac{P}{\rho}\left\{1+\frac{(\beta n - \gamma\mu)\mu a}{aP + b(\beta n - \gamma\mu)\mu}\right\}, \quad \text{.....................(13)}$$

which is exactly (14) of 1.

All the other conclusions of that article—the approximations and the conditions under which thrust can be substituted for pull in the chain—hold also in the present case.

4. *More general discussion of gyrostatic chain.* So far only particular cases of the vibrations of a chain of gyrostatic links have been considered. Under the condition that the angles $\theta$, $\phi$ are small, general equations can be developed as follows. Let the direction cosines of the axle of the $k$th gyrostat be $p_k$, $q_k$, 1, and those of the $k$th link be $r$, $s$, 1. Here the cosines taken as unity are $\cos\theta$, $\cos\phi$. The reference is to a set of axes $O(x, y, z)$, drawn from the gyrostat centre O, and such that $Oz$ very nearly coincides with the axes of the gyrostatic links. The values of $p$, $q$, $r$, $s$ are small since the angles they refer to are all nearly $\frac{1}{2}\pi$.

Let $\theta_1$, $\theta_2$, $\theta_3$ be the angular speeds about $O(x, y, z)$. Then, since $p$, $q$, 1 are the coordinates of a point on the gyrostat axis at unit distance from O, $-\dot{q}=\theta_1 z-\theta_3 x=\theta_1$, $\dot{p}=\theta_2 z-\theta_3 y=\theta_2$, nearly. Thus $h_1=A\theta_1=-A\dot{q}$, $h_2=A\dot{p}$.

The geometrical relations fulfilled by the body are

$$x_{k+1}-x_k=a(p_k+p_{k+1})+2br_{k+1},\quad y_{k+1}-y_k=a(q_k+q_{k+1})+2bs_{k+1}.\;\ldots\ldots(1)$$

The equations of motion of the centroid are, if $P_k$ be the stretching force in the $k$th link,

$$m\ddot{x}=P_{k+1}r_{k+1}-P_kr_k,\quad m\ddot{y}=P_{k+1}s_{k+1}-P_ks_k.\;\ldots\ldots\ldots\ldots\ldots\ldots(2)$$

If $h_1$, $h_2$, $h_3$ be the components of A.M about the axes the rate of growth of A.M. about a fixed axis with which O$x$ coincides at the instant is $\dot{h}_1-h_2\theta_3+h_3\theta_2$. We neglect $h_2\theta_3$ and get, by the value of $\theta_2$ obtained above, and since $h_1=-A\dot{q}$, $\dot{h}_1+h_3\theta_2=-A\ddot{q}+Cn\dot{p}$. Similarly for the axis O$y$ the rate of growth of A.M. is $A\ddot{p}+Cn\dot{q}$.

The couple about O$x$ due to the pull of the $k$th link is

$$(P_ky-P_ks_kz)=aP_k(q_k-s_k).$$

The couple about the same axis due to the pull of the $(k+1)$th link is $aP_{k+1}(q_k-s_{k+1})$, so that the total couple is $aP_{k+1}(q_k-s_{k+1})+aP_k(q_k-s_k)$. Similarly we see that the total couple about O$y$ is

$$-aP_{k+1}(p_k-r_{k+1})-aP_k(p_k-r_k).$$

Thus we have the equations of A.M.

$$\left.\begin{aligned}-A\ddot{q}+Cn\dot{p}&=aP_{k+1}(q_k-s_{k+1})+aP_k(q_k-s_k),\\ A\ddot{p}+Cn\dot{q}&=-aP_{k+1}(p_k-r_{k+1})-aP_k(p_k-r_k).\end{aligned}\right\}\;\ldots\ldots\ldots\ldots\ldots(3)$$

We can replace these pairs of equations (1), (2), (3) by three single equations by writing $u=p+iq$, $v=r+is$, $w=x+iy$. Thus we obtain

$$\left.\begin{aligned}w_{k+1}-w_k-a(u_k+u_{k+1})-2bv_{k+1}&=0,\\ m\ddot{w}-P_{k+1}v_{k+1}+P_kv_k&=0,\\ A\ddot{u}_k-iCn\dot{u}_k+a(P_{k+1}+P_k)u_k-a(P_{k+1}v_{k+1}+P_kv_k)&=0.\end{aligned}\right\}\;\ldots\ldots\ldots(4)$$

Thus we have obtained three general equations which we can apply in different cases of nearly straight gyrostatic chains As a first example we take the helical chain, illustrating a circularly polarised wave, which has been dealt with in the preceding article. For this we assume the solution

$$u_k,\ v_k,\ w_k=(M,\ N,\ R)e^{i(\mu t+kc)},\;\ldots\ldots\ldots\ldots\ldots\ldots\ldots\ldots\ldots\ldots(5)$$

where $c$ is the constant step of polygonal angle $2(\alpha+\beta)$, as explained in 3. If we consider the value of the right-hand side of (5) for the $(k+1)$th gyrostat we see that $\mu$ is the angular speed of the steadily moving helical arrangement about the axis. We thus get by substitution in (4), remembering that $P_{k+1}=P_k$,

$$\left.\begin{aligned}R(e^{ci}-1)-aM(e^{ci}+1)-2bNe^{ci}&=0,\\ Rm\mu^2+NP(e^{ci}-1)&=0,\\ M(-A\mu^2+Cn\mu+2aP)-aNP(e^{ci}+1)&=0.\end{aligned}\right\}\;\ldots\ldots\ldots\ldots\ldots\ldots\ldots\ldots(6)$$

Now $\cos c=(e^{ic}+e^{-ic})/2$, and we find by eliminating M, N, R from (6) that

$$\cos c=\frac{(Cn\mu-A\mu^2+2aP)\left(1-\dfrac{m\mu^2b}{P}\right)-m\mu^2a^2}{Cn\mu-A\mu^2+2aP+m\mu^2a^2};$$

and therefore

$$\sin^2(\alpha+\beta)=\frac{Cn\mu-A\mu^2+2\dfrac{a}{b}(a+b)P}{Cn\mu-A\mu^2+2aP+m\mu^2a^2}\,\frac{m\mu^2b}{2P},\;\ldots\ldots\ldots\ldots\ldots\ldots\ldots(7)$$

which agrees with (9), 3. Hence the present process gives the same result as the former one. We might easily have retained $\cos\theta$, $\cos\phi$, and obtained exactly (9), 3.

**5.** *Vertical gyrostatic chain under gravity.* As another example, take a nearly vertical chain of gyrostats stretched by the gravity of the gyrostats themselves. The individual gyrostatic links are not exactly vertical, but they are nearly so. Consider any connecting link. If there are $k$ gyrostats below it the link, according to the reckoning adopted above, is the $(k+1)$th from the bottom. If we measure $z$ from a point $2(a+b)$ below the lowest gyrostat centre the height of the centre of the $k$th gyrostat above this origin is $2k(a+b)$, and this we call $z_k$. Hence $k=z_k/2(a+b)$. The "tension" in the $(k+1)$th connecting link, or $P_{k+1}$, is $kmg$ or $mgz_k/2(a+b)$, and that in the link next below is $mgz_{k-1}/2(a+b)$. The chain is supposed to be revolving about the vertical in steady motion.

Equations (4) of 4 therefore become, if $f=b/a$,

$$\left.\begin{aligned} w_{k+1}-w_k-a(u_{k+1}+u_k)-2bv_{k+1}&=0,\\ \ddot{w}-\frac{a}{g}\,\frac{z_k v_{k+1}-z_{k-1}v_k}{2(1+f)}&=0,\\ \mathrm{A}u_k-i\mathrm{C}n\dot{u}_k+\frac{mg}{2(1+f)}\{(z_k+z_{k-1})u_k-(z_kv_{k+1}+z_kv_k)\}&=0.\end{aligned}\right\} \qquad (1)$$

It is fairly clear that as the tension increases, that is at points higher and higher in the chain, the inclination of the links to the vertical becomes smaller, but this will be made more obvious by supposing the chain to consist of infinitesimal gyrostatic links.

Putting $dz=2a(1+f)$, we get, instead of the preceding equations,

$$\left.\begin{aligned} \frac{dw}{dz}-\frac{u+fv}{1+f}=0, \quad \frac{d^2w}{dt^2}-g\frac{d(zv)}{dz}=0,\\ \gamma\frac{d^2u}{dt^2}-i\beta n\frac{du}{dt}+\frac{g\rho}{1+f}z(u-v)=0,\end{aligned}\right\} \qquad (2)$$

where $f=b/a$, and $\gamma$, $\beta$ are moments of inertia (transverse and axial for a gyrostat) per unit of length of the chain, and $\rho$ is the linear density of the chain, or $m/2a(1+f)$.

Now (see 1 above) the connecting links and the gyrostat centres are helically arranged round the vertical, and adjacent links must be nearly in the same vertical plane. Consequently $w$, $u$, and $fv$, must have the same sign for their real parts, and likewise for their imaginary parts, and the second equation shows that $v$ increases as $z$ diminishes. Thus points lower and lower down on the chain deviate more and more from the vertical, and the second equation shows that this deviation grows with $z$ at a constantly accelerating rate, for by the solution in (4) we have $gd(zv)/dz=-\mathrm{R}\mu^2$. The last of (2) shows that if $n$ is great $d^2u/dt^2$ is also great, that is for the value of $\mu\mathrm{M}$, and also R must be great.

If now we reverse $g$ we get a chain standing on its lower end under thrust in rigid links. As this thrust is due to gravity, it diminishes with increase of distance from the point of support, and so, measuring $z$ upwards as before, we get

$$\frac{dw}{dz}+\frac{u+fv}{1+f}=0. \qquad (3)$$

Thus the instability towards the free end in the former case is replaced by stability at the free end in the latter. A fuller investigation would discuss this instability as an affair of disturbance of the steady motion.

A train of carriages passing round a curve may be regarded as a gyrostatic chain, in which the precession of each carriage is prevented by a couple applied by the rails. If the speed were great enough there would be gyrostatic oscillations of serious amount at the rear end of the train, and it would be necessary to avoid these by pushing the train from behind. This fact is illustrated by a model invented by Dr. J. G. Gray (see Chap. XXIII below).

For steady motion at angular speed $\mu$ about the vertical we have for a continuous chain, from the solution in (5), 4, or from (2) above,

$$\left.\begin{array}{l}\dfrac{dR}{dz}-\dfrac{M+fN}{1+f}=0, \quad R\mu^2+g\dfrac{d(zN)}{dz}=0,\\ M(\beta\mu n-\gamma\mu^2)+(M-N)\dfrac{g\rho}{1+f}z=0.\end{array}\right\}\qquad(4)$$

If we eliminate R and M from these equations we get

$$\frac{1}{N}\frac{d^2(Nz)}{dz^2}+\frac{\mu^2}{g}\frac{f}{1+f}\frac{(\gamma\mu^2-\beta\mu n)f-g\rho z}{\left(\gamma\mu^2-\beta\mu n-\dfrac{g\rho z}{1+f}\right)f}=0.\qquad(5)$$

When $\beta$ and $\gamma$ are zero, or when $\gamma\mu^2=\beta\mu n$, this equation reduces to

$$\frac{1}{N}\frac{d^2(Nz)}{dz^2}=\frac{\mu^2}{g},$$

that is to

$$z\frac{d^2N}{dz^2}+2\frac{dN}{dz}-\frac{\mu^2}{g}N=0,\qquad(6)$$

an equation which gives N as a Bessel function of zero order.

6. *Gyrostat hung by a thread.* We may apply equations (4) of 3 to give the motion of a gyrostat hung by a thread, the problem already discussed in 8, VII above. We make in (4), 3, $P_k=0$, $P_{1+k}=mg$ (since $\theta$ and $\phi$ are supposed to be very small), and, measuring $w$ from the vertical through the point of suspension, taking the length of the thread as $2b$, and the distance of the point of attachment to the gyrostat axis from the gyrostat centre $a$, we obtain

$$w=au+2bv, \quad m\ddot{w}=mgv, \quad A\ddot{u}-iCn\dot{u}+mga(u-v)=0.\qquad(1)$$

If we assume the motion to be steady, and write

$$u,\ v,\ w=(M,\ N,\ R)e^{\iota(\mu t+c)},\qquad(2)$$

we find $\quad R=Ma+N2b, \quad -m\mu^2R=mgN, \quad (-A\mu^2+Cn\mu)M+mga(M-N)=0.\qquad(3)$

From the first two equations we get $N=aM/(g/\mu^2-2b)$, so that the third becomes

$$(Cn\mu-A\mu^2+mga)\left(\frac{g}{\mu^2a}-2f\right)-mga=0.\qquad(4)$$

If we go back to (2), 8, VII, we find that for a single gyrostat supported by a thread

$$(Cn-A\mu\cos\theta)\mu\sin\theta-mga(\sin\phi\cos\theta+\cos\phi\sin\theta)=0.\qquad(5)$$

or, when $\theta$ and $\phi$ are both small,

$$(Cn\mu-A\mu^2)\theta-mga(\phi+\theta)=0.\qquad(6)$$

But (1), 8, VII, gives $\mu^2(2b+a\theta/\phi)=g$, or $\theta/\phi=g/a\mu^2-2f$, so that

$$(Cn\mu-A\mu^2-mga)\left(\frac{g}{\mu^2a}-2f\right)-mga=0.\qquad(7)$$

This result has $-mga$ in the bracket instead of $mga$ as in (5), 8, VII above, but this difference arises from the fact that, owing to difference of arrangement of the links, the term $Ta\cos\phi\sin\theta$ in (2), 8, VII, has the opposite sign to that which it has by the equations now being exemplified.

7. *Continuous elastic medium loaded with small gyrostats.* Consider now a medium endowed with rigidity and containing a uniform distribution of quasi-molecular gyrostats, the axes of which are similarly directed. A plane wave travels along this direction, which we take as that of the axis Oz. The displacements in the wave are transverse to this direction, and at

any point O are supposed to be resolved along two axes $Ox$, $Oy$ at right angles to one another and to $Oz$.

Consider an element of length $dz$, at the centre of which is the point O, and let its cross-section have dimensions $dx$, $dy$. Let the distributed angular momentum (A.M.) be N per unit of volume, so that the A.M. of the element is $N\,dx\,dy\,dz$ (about $Oz$), and the displacements at the centre O be $\xi, \eta$ in the $x$ and $y$ directions respectively. The element is turning with the angular speeds

$$\frac{1}{2}\frac{\partial}{\partial t}\frac{\partial \xi}{\partial z} \text{ about axis of } y, \quad -\frac{1}{2}\frac{\partial}{\partial t}\frac{\partial \eta}{\partial z} \text{ about axis of } x.$$

In consequence of this turning there are rates of growth of A.M.

$$\frac{1}{2}N\frac{\partial}{\partial t}\frac{\partial \xi}{\partial z}\,dx\,dy\,dz \text{ about axis of } x,$$

$$-\frac{1}{2}N\frac{\partial}{\partial t}\frac{\partial \eta}{\partial z}\,dx\,dy\,dz \text{ about axis of } y.$$

Hence there must act on the element couples about these axes given by

$$\left.\begin{aligned} P\,dx\,dy\,dz &= \frac{1}{2}N\frac{\partial}{\partial t}\frac{\partial \xi}{\partial z}\,dx\,dy\,dz, \\ Q\,dx\,dy\,dz &= -\frac{1}{2}N\frac{\partial}{\partial t}\frac{\partial \eta}{\partial z}\,dx\,dy\,dz. \end{aligned}\right\} \quad\text{......................(1)}$$

These are due to tangential stresses, and these stresses it is easy to see must be equally distributed, for the axis of $x$, over the two sets of planes parallel to that axis, that is the tangential stresses must be equally of the types (YZ), (ZY); and similarly for the axis of $y$ the tangential stresses must be equally of the two types (XZ), (ZX).

The tangential stress at the point O is thus $\frac{1}{2}P$ in the direction of $z$, and $-\frac{1}{2}P$ in the direction of $y$ for the planes parallel to the axis of $x$. Similarly we get stresses $\frac{1}{2}Q$ and $-\frac{1}{2}Q$ for the planes parallel to the axis of $y$. We denote the forces in the directions of $Ox$ and $Oy$ by X, Y respectively.

These shearing forces vary from point to point, and clearly the body-force in the direction of $Ox$ is for the element

$$\frac{\partial X}{\partial z}\,dz\,dy\,dx = \frac{1}{4}N\frac{\partial}{\partial t}\frac{\partial^2 \eta}{\partial z^2}\,dx\,dy\,dz. \quad\text{....................(2)}$$

Similarly the body-force in the direction of $Oy$ is

$$\frac{\partial Y}{\partial z}\,dz\,dx\,dy = -\frac{1}{4}N\frac{\partial}{\partial t}\frac{\partial^2 \xi}{\partial z^2}\,dx\,dy\,dz. \quad\text{......................(3)}$$

The resultant forces applied in the directions $Ox$ and $Oy$ by the shearing stresses due to the ordinary rigidity are, if $\mu$ *now denote the rigidity modulus*,

$$\mu\frac{\partial^2 \xi}{\partial^2}\,dx\,dy\,dz, \quad \mu\frac{\partial^2 \eta}{\partial z^2}\,dx\,dy\,dz.$$

Hence we obtain the equations of motion

$$\left.\begin{aligned}\rho\frac{\partial^2\xi}{\partial t^2}&=\frac{1}{4}\mathrm{N}\frac{\partial}{\partial t}\frac{\partial^2\eta}{\partial z^2}+\mu\frac{\partial^2\xi}{\partial z^2},\\ \rho\frac{\partial^2\eta}{\partial t^2}&=-\frac{1}{4}\mathrm{N}\frac{\partial}{\partial t}\frac{\partial^2\xi}{\partial z^2}+\mu\frac{\partial^2\eta}{\partial z^2}.\end{aligned}\right\}\qquad(4)$$

[Equations precisely similar to (4) can be obtained for the propagation of a plane electromagnetic wave in an insulating medium of specific electric and magnetic inductivities K, $\mu$, magnetised by quasi-molecular magnets parallel to O$z$, to magnetic moment C per unit volume. If F, G be $x$ and $y$ components of vector potential, and $\kappa$ a constant, the equations of propagation are *

$$\left.\begin{aligned}\frac{\partial^2\mathrm{F}}{\partial t^2}&=\frac{1}{\mathrm{K}\mu}\frac{\partial^2\mathrm{F}}{\partial z^2}+\frac{\kappa}{4\pi}\mathrm{C}\frac{\partial^3\mathrm{G}}{\partial t\,\partial z^2},\\ \frac{\partial^2\mathrm{G}}{\partial t^2}&=\frac{1}{\mathrm{K}\mu}\frac{\partial^2\mathrm{G}}{\partial z^2}-\frac{\kappa}{4\pi}\mathrm{C}\frac{\partial^3\mathrm{F}}{\partial t\,\partial z^2}.\end{aligned}\right\}\qquad(5)$$

Thus C is to a constant the analogue of N, the A.M. per unit volume in the gyrostatically loaded medium.]

If we write $$\zeta=\xi+i\eta,\qquad(6)$$

we replace the two equations of (4) by the single equation

$$\rho\frac{\partial^2\zeta}{\partial t^2}=-\frac{1}{4}i\mathrm{N}\frac{\partial}{\partial t}\frac{\partial^2\zeta}{\partial z^2}+\mu\frac{\partial^2\zeta}{\partial z^2}.\qquad(7)$$

It will be observed that geometrically $\zeta$ denotes the vector of which the components are $\xi$, $\eta$, and that if $\zeta$ be proportional to $e^{\iota(nt-mz)}$, and, as we assume for the present, $n$ be positive, the argument $nt$ in $\zeta$ is positive. [Here $n$ is used as the so-called "speed" of the simple harmonic motion, and is not to be confused with the angular speed of a flywheel.] Of course $mz$ may have either sign in each case. It is important to observe that if we suppose $$\zeta=(\alpha+i\beta)e^{\iota(nt-mz)}\qquad(8)$$

we obtain a definite quadratic equation for $m^2$ ($n$ is supposed given by an impressed vibration). Thus we get two values, equal with opposite signs, for the ratio $n/m$, which is the speed of propagation. One of these is the speed in the positive direction, the other is the speed in the negative direction.

From (8) we obtain, by substitution in (7),

$$\mu m^2-\rho n^2=-\tfrac{1}{4}\mathrm{N}m^2n,$$

or

$$\frac{n^2}{m^2}=\frac{\mu}{\rho}\left(1+\frac{1}{4}\frac{\mathrm{N}n}{\mu}\right).\qquad(9)$$

* A. Gray, Note on the electromagnetic theory of the rotation of the plane of polarised light.—*British Association Report*, 1891, p. 558.

But $n/m = \mathrm{V}$, the speed of propagation of the wave. Hence, if $\mathrm{N}n$ be small in comparison with $\mu$ we get

$$\mathrm{V} = \pm\left(\frac{\mu}{\rho}\right)^{\frac{1}{2}}\left(1+\frac{1}{8}\frac{\mathrm{N}n}{\mu}\right). \qquad (10)$$

The value of V given by the upper sign on the right corresponds to the value of $\zeta$ given by (6) and also by (8): the other value of V corresponds to $\zeta$ as given by (6) and also by

$$\zeta = (\alpha + i\beta)e^{\iota(nt+mz)},$$

which also satisfies the differential equations.

If the sign of N (or that of $n$) be reversed, we get the quadratic

$$\frac{n^2}{m^2} = \frac{\mu}{\rho}\left(1-\frac{1}{4}\frac{\mathrm{N}n}{\mu}\right). \qquad (9')$$

Either quadratic gives two roots, one positive, the other negative. The negative value of $n$ for either is numerically equal to the positive value for the other. Thus, if we fix the value of $m$, the positive value of $n$ gives the frequency for one direction of rotation, the other positive value of $n$ gives the frequency for the other direction of rotation.

To realise the solution for two waves existing together, one travelling in the positive direction, the other in the negative, but both having $n$ positive, we take

$$\zeta = (\alpha + i\beta)e^{\iota(nt-mz)} + (\alpha' + i\beta')e^{\iota(nt+mz)}. \qquad (11)$$

This gives

$$\begin{aligned}\zeta = {} & \alpha\cos(nt-mz) + \alpha'\cos(nt+mz) - \beta\sin(nt-mz) - \beta'\sin(nt+mz) \\ & + i\{\beta\cos(nt-mz) + \beta'\cos(nt+mz) + \alpha\sin(nt-mz) + \alpha'\sin(nt+mz)\}. \qquad (12)\end{aligned}$$

Thus we obtain

$$\left.\begin{aligned}\xi &= \alpha\cos(nt-mz) + \alpha'\cos(nt+mz) - \beta\sin(nt-mz) - \beta'\sin(nt+mz), \\ \eta &= \beta\cos(nt-mz) + \beta'\cos(nt+mz) + \alpha\sin(nt-mz) + \alpha'\sin(nt+mz).\end{aligned}\right\} \qquad (13)$$

In the compound wave the corresponding displacements in which the time argument is reversed are obtained by substituting $-n$ for $n$ in the foregoing expressions, and $m_2$ for $m$, on the understanding that

$$\left|\frac{n}{m_2}\right| = \left\{\frac{\mu}{\rho}\left(1-\frac{1}{8}\frac{\mathrm{N}n}{\mu}\right)\right\}^{\frac{1}{2}}, \qquad (14)$$

where on the right $n$ is supposed to be positive, and the quantity on the left is the numerical value of the ratio.

These results can be superimposed in different ways. For example, the displacements in two circularly polarised waves of amplitudes $a_1$, $a_2$ may be written ($n$ positive)

$$\left.\begin{aligned}\xi &= a_1\cos(nt-mz) + a_2\cos(nt-m_2z), \\ \eta &= a_1\sin(nt-mz) - a_2\sin(nt-m_2z).\end{aligned}\right\} \qquad (15)$$

The first terms on the right of (15) taken together constitute a circularly polarised wave in which the circular motion is in the same direction as the rotation of the flywheels; the two second terms give a circularly polarised wave in which the circular motion is in the opposite direction. Both waves travel in the same direction, the first with speed

$$\frac{n}{m}=\left(\frac{\mu}{\rho}\right)^{\frac{1}{2}}\left(1+\frac{1}{8}\frac{\mathrm{N}n}{\mu}\right), \quad\text{.............................(16)}$$

the second with speed $$\frac{n}{m_2}=\left(\frac{\mu}{\rho}\right)^{\frac{1}{2}}\left(1-\frac{1}{8}\frac{\mathrm{N}n}{\mu}\right) \quad\text{.............................(17)}$$

For a plane polarised beam we take $a_1=a_2$. The two oppositely circularly polarised waves would travel unit distance in the respective times

$$\left(\frac{\rho}{\mu}\right)^{\frac{1}{2}}\left(1-\frac{1}{8}\frac{\mathrm{N}n}{\mu}\right),\quad \left(\frac{\rho}{\mu}\right)^{\frac{1}{2}}\left(1+\frac{1}{8}\frac{\mathrm{N}n}{\mu}\right).$$

8. *Dynamical explanation of magneto-optic rotation.* Hence we see that in a maintained train of plane polarised waves of definite frequency $n/2\pi$ one circular motion will gain in phase on the other, per unit distance travelled, the angle

$$\frac{1}{4}\frac{\mathrm{N}n^2}{\mu}\left(\frac{\rho}{\mu}\right)^{\frac{1}{2}},$$

and the plane of polarisation will turn through half this angle.

It is interesting to compare this with the turning of the plane of polarisation suggested by the two chains of gyrostats described at the end of 2. As in that discussion $\mu$ and $n$ were used as the angular speed of the chain about the line AB and the angular speed of the flywheels respectively; while here $\mu$ denotes the rigidity of the continuous medium and $n$ the so-called "speed" of the impressed harmonic motion, we express the two rates of turning in terms of wave-velocity and wave-length, and write L for the angular momentum of a flywheel, so that $\mathrm{C}n=\mathrm{L}$.

The relative turning for the gyrostatic chains is, by (15) or (17), 1,

$$2\pi\mathrm{C}n\mu/\mathrm{P}(a+b)\lambda,$$

which, if $\mathrm{V}^2m$ be put for $\mathrm{P}(a+b)$, becomes $2\pi\mathrm{L}\mu/m\mathrm{V}^2\lambda$. But $\mu\lambda/2\pi=\mathrm{V}$, so that $\mu=2\pi\mathrm{V}/\lambda$. Thus the relative turning for the chains is

$$4\pi^2\frac{\mathrm{L}}{m\mathrm{V}\lambda^2}.$$

The relative turning of the two circularly polarised waves in the gyrostatic medium is $\frac{1}{4}\mathrm{N}n^2/\mu\mathrm{V}$, where the letters $n$ and $\mu$ have the different meanings referred to above. But $\mu=\rho\mathrm{V}^2$ and $n\lambda=2\pi\mathrm{V}$, so that $\mathrm{V}^2=n^2\lambda^2/4\pi^2$, and $\mu=\rho n^2\lambda^2/4\pi^2$. Hence the relative turning in this case is

$$\pi^2\frac{\mathrm{N}}{\rho\mathrm{V}\lambda^2}.$$

The two expressions are thus quite analogous, with the correspondence N, $\rho$ to L, $m$. They differ only by a numerical factor which arises from the fact that in one case we have a chain revolving in free space, and in the other gyrostatic elements of a rigid medium moving under the control of the rigidity.

It will be seen that, in either of the gyrostatic illustrations, if the plane polarised system of vibrations be reflected back after passage in one direction, the turning in the second passage will be in the same direction as in the first, so that the total turning will be twice that for a single passage. This is the characteristic of magneto-optic rotation as distinguished from the rotation produced by a plate of quartz or a solution of sugar, where the turning in the forward passage is annulled in the backward passage. This points to the fact, already referred to above, that the rotation of the plane of polarisation in the latter case is an affair of structure of the medium.

9. *Analogy between motion of the bob of a gyrostatic pendulum and that of an electron in a magnetic field.* It was pointed out by the author, in a Royal Institution Lecture, delivered in 1898, that the motion explicitly stated in equations (2), (3), (4), (5) of 3, VIII above is the gyrostatic analogue of the Zeeman effect. An electron moving in a circular orbit, in a plane at right angles to an impressed magnetic field, would have one period of revolution or another according to the direction of its motion, corresponding precisely to $2\pi/k_1$ and $2\pi/k_2$. In fact the motion is exactly analogous to that of a rigid pendulum in the bob of which is a flywheel, the axis of which is in line with the suspension rod of the pendulum. The pendulum was shown in the lecture, and reference may be made to the description given there of the apparatus and to the curves illustrating the motion. For simplicity, and to bring out the electron analogy more clearly, we suppose the mass to be almost entirely in the bob and flywheel, and that, except for the moment of inertia, C, and angular momentum, C$n$, of the flywheel, the bob may be considered as a particle of mass $m$, the whole mass of wheel and framework combined. If $h$ be the distance of the centre of the bob from the point of support (where we suppose the two axes to be situated so that $\mathrm{A}=\mathrm{A}'$, $\mathrm{B}=\mathrm{B}'$) we have $\mathrm{A}=mh^2$ and $\mathrm{B}=mgh$. Thus, taking axes of $x$ and $y$ coincident with the axes about which the turnings $\phi$, $\psi$ take place, we have from the second and first of (1), 3, VIII, since $h\phi=-y$, $h\psi=x$ (and the flywheel is supposed to be rotating counter-clockwise as seen by an eye looking upwards from the origin),

$$m\ddot{x}+\frac{\mathrm{C}n}{h}\dot{y}+mg\frac{x}{h}=0, \quad m\ddot{y}-\frac{\mathrm{C}n}{h}\dot{x}+mg\frac{y}{h}=0,$$

or

$$\ddot{x}+\gamma\dot{y}+\kappa x=0, \qquad \ddot{y}-\gamma\dot{x}+\kappa y=0, \quad \text{..............(1)}$$

where $\gamma=\mathrm{C}n/mh$, $\kappa=g/h$.

These are exactly the equations of motion of an electron E supposed moving in the direction of the arrow in a magnetic field directed upward through the paper (Fig. 54).

The electron is charged negatively with a charge of $\epsilon$ units and has an effective mass $m$. The intensity of the magnetic field is H, and the magnetic inductivity of the medium is $\mu$. We suppose the mass of the electron to be acted on by a force $\kappa r$, where $r$ is the distance of the electron from O. The convection currents are here $\epsilon\dot{x}$, $-\epsilon\dot{y}$ in the direction of $x$ and $y$ respectively. Hence the electromagnetic forces in the direction of $y$ and $x$ are $-\mu\epsilon H\dot{x}$ and $\mu\epsilon H\dot{y}$ respectively. The value of $\gamma$ is thus $\mu\epsilon H/m$. The other outward component forces are then $-\kappa x$, $-\kappa y$, and the equations of motion are (1), which, by putting $z=x+iy$, we can unite in the single equation

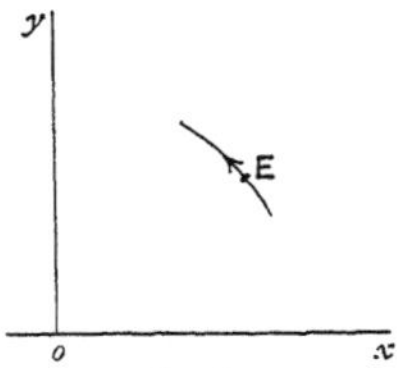

FIG. 54.

$$\ddot{z}-i\gamma\dot{z}+\kappa z=0.$$

Assuming that $z=Ke^{ikt}$, (K = a constant), we obtain

$$k^2-\kappa-\gamma k=0.$$

Again, by reversing the sign of $k$, we obtain

$$k^2-\kappa+\gamma k=0,$$

so that we have

$$k^2=\kappa\pm\gamma k,$$

and if $\gamma$ is small the numerical value of $k$ is given by

$$k=\kappa^{\frac{1}{2}}\pm\tfrac{1}{2}\gamma.$$

The frequency of the original vibration was $\kappa^{\frac{1}{2}}/2\pi$: the imposition of the magnetic field has produced two new frequencies $(\kappa^{\frac{1}{2}}+\frac{1}{2}\gamma)/2\pi$, $(\kappa^{\frac{1}{2}}-\frac{1}{2}\gamma)/2\pi$, one higher, the other lower than the original frequency. The former frequency is that of electrons describing orbits in the direction shown in Fig. 54; the latter frequency is that of electrons moving in the opposite direction. Electrons moving along the line of the force H are not affected; hence the spectrum is modified by the production of two satellitic lines, one above the ordinary line, the other below it, in the spectrum.

A gyrostatic theory of the structure of the electron has been suggested; but we do not discuss that here.

## CHAPTER X

### THE EARTH AS A TOP. PRECESSION AND NUTATION. GYROSTATIC THEORY OF MOTION OF THE NODES OF THE MOON'S ORBIT

1. *Potential of gravitation.* As has been stated above, the earth's axis has a conical motion analogous to that of the axis of a top, and arising from a similar cause. In the case of the terrestrial top the motion of the axis is produced by the action of a couple due to differential attraction exerted by the sun or moon, so that there is a periodically varying rate of production of A.M. about an axis through the centroid of the earth and at right angles to the axis of rotation

The first step in a detailed discussion is the estimation of the couple just referred to. The gravitational potential of any distribution of matter at a point P distant AP from a point A at which an element $dm$ of mass is situated is given by

$$V=\kappa\int\frac{dm}{AP}, \qquad (1)$$

where $\kappa$ is the gravitation constant, and the integral is supposed taken throughout the whole of the attracting matter. We shall take the centroid G of the attracting mass as origin, and refer to axes of which G$x$ is along GP, put $r$ for $(x^2+y^2+z^2)^{\frac{1}{2}}$, R for GP, and suppose that A, the position of $dm$, has coordinates $x$, $y$, $z$. Moreover we shall suppose that GP is great in comparison with any distance GA, of an element of the mass from G. Noticing that $AP^2=R^2+r^2-2Rx$, we obtain

$$V=\kappa\int\frac{dm}{AP}=\kappa\int\frac{dm}{R}\left(1-\frac{2Rx-r^2}{R^2}\right)^{-\frac{1}{2}}, \qquad (2)$$

Expanding the expression on the right and arranging according to descending powers of R, we get

$$V=\kappa\int\frac{dm}{R}\left(1+\frac{x}{R}+\frac{3x^2-r^2}{2R^2}+\frac{5x^3-3xr^2}{2R^3}+\frac{35x^4-30x^2r^2+3r^4}{8R^4}+\cdots\right). \qquad (3)$$

Or, if we write $Z_1=\frac{x}{r}$, $Z_2=\frac{3}{2}\frac{x^2}{r^2}-\frac{1}{2}$, $Z_3=\frac{5}{2}\frac{x^3}{r^3}-\frac{3}{2}\frac{x}{r}, \ldots$,

then $$V=\kappa\int\frac{dm}{R}\left(1+\frac{r}{R}Z_1+\frac{r^2}{R^2}Z_2+\frac{r^3}{R^3}Z_3+\cdots\right). \qquad (4)$$

The quantities $Z_1$, $Z_2$, ... are the successive zonal harmonics taken with respect to the axis OP.

But now it will be noticed that since G, the centroid, is the origin, $\int x\,dm=0$, and so the term containing $Z_1$ vanishes in the integration. If the distribution of matter is symmetrical about any set of rectangular axes, drawn from G as origin, the term $\int Z_3 r^3 dm/R^4$ vanishes also, as the reader may prove.

As an approximation, which is especially close when the attracting system has three axes of symmetry, we obtain the equation

$$V=\kappa\int\frac{dm}{R}\left(1+\frac{r^2}{R^2}Z_2\right). \qquad (5)$$

Here we have

$$2\int dm r^2 Z_2=\int dm(3x^2-x^2-y^2-z^2)=\int dm(2x^2-y^2-z^2).$$

But $2x^2-y^2-z^2=x^2+y^2+y^2+z^2+z^2+x^2-3(y^2+z^2)$. Hence, if A, B, C be the principal moments of inertia with reference to axes passing through G, and I be the moment of inertia about the axis GP, we have

$$2\int dm r^2 Z_2=A+B+C-3I. \qquad (6)$$

Thus, if M denote the whole mass of the distribution,

$$V=\kappa\frac{M}{R}+\kappa\frac{A+B+C-3I}{2R^3}. \qquad (7)$$

If the material system is not symmetrical with respect to a system of axes, the error involved in taking the potential as given by (7) is of the order $(l/R)^3V$, where $l$ is a linear dimension of the distribution; if there is a system of axes of symmetry the error is of the order $(l/R)^4V$.

There is one case however in which this criterion of error is not applicable, in which, in point of fact, the approximation is as close for a distribution whose dimensions are comparable with R as for a distribution of very small dimensions. If the distribution is a homogeneous sphere, or a sphere made up of concentric shells, each of uniform density, its potential at an external point is the same as if the whole mass were collected at the centre. Thus, if P be external to the distribution, the potential given for a distribution of large dimensions, by a chosen process of approximation, is as nearly correct as that obtained by the same process for a point charge of matter at the centroid.

The same remark is true for a distribution which may be regarded as made up of successive confocal ellipsoidal shells of small ellipticity, each of uniform density. The body is thus an ellipsoid and P is an external point. This follows from the theorem of Maclaurin that the attractions of different confocal shells on an external particle are proportional to their masses.

Since the ellipticity is small, we may consider the potential at P due to a small central ellipsoid confocal with the series of shells of which the actual distribution may be regarded as made up. Thus, if $m$ be the mass of this ellipsoid, and A′, B′, C′ be the principal moments of inertia for it, and I′ its moment of inertia about OP, its potential V′ at P is given by

$$V' = \kappa\frac{m}{R} + \kappa\frac{A'+B'+C'-3I'}{2R^3}. \quad\text{.............................(8)}$$

The potential of the actual distribution is V′M/$m$, or

$$V = \kappa\frac{M}{R} + \kappa\frac{M}{m}\frac{A'+B'+C'-3I'}{2R^3}, \quad\text{.........................(9)}$$

and, as the reader may verify, by considering the relations of the semi-axes of the central small ellipsoid and of the whole distribution, and the consequent relations of A, B, C to A′, B′, C′, this reduces to

$$V = \kappa\frac{M}{R} + \kappa\frac{A+B+C-3I}{2R^3}. \quad\text{...................... (10)}$$

2. *Calculation of attractive forces.* The force on a particle at P in the direction GP is $\partial V/\partial R$. Hence the *attraction* is $-\partial V/\partial R$, and we have

$$-\frac{\partial V}{\partial R} = \kappa\frac{M}{R^2} + \frac{3}{2}\kappa\frac{A+B+C-3I}{R^4}. \quad\text{......................(1)}$$

From this it follows that if GP be in such a direction that $A+B+C=3I$, the attraction is more nearly the same as if the whole mass were collected at the centroid than for any other direction of GP. The earth approximately fulfils the condition $A=B$, and in this case we can determine the directions of GP for which $A+B+C=3I$. For we have $2A+C=3I$, and if $\gamma$ be the angle which such a direction of GP makes with the axis of symmetry, we have also

$$I = C\cos^2\gamma + A(1-\cos^2\gamma) = (C-A)\cos^2\gamma + A, \text{ or } 3I = 3(C-A)\cos^2\gamma + 3A.$$

But the left-hand side of this equality is $2A+C$. Thus we obtain

$$C-A = 3(C-A)\cos^2\gamma, \quad\text{or}\quad \cos^2\gamma = 1/3,$$

that is $\gamma = \cos^{-1}(1/3)^{\frac{1}{2}}$. GP thus makes an angle with the axis of figure of 54° 44′. Any radius drawn from the centre to a point in latitude 35° 16′ fulfils this condition.

To find the attraction between one body and another, we have first to calculate the integral $\int V dm'$ for the system, that is calculate V for the first body and a point at which an element of mass $dm'$ of the second is situated, and then integrate the result throughout the second body. This may be interpreted as the potential energy of one body in presence of the other: it is the work which must be done in withdrawing one body to an infinite distance from the first. If $R_{12}$ denote the distance between the centroid of the first distribution and that of the second, and W the mutual energy, we have approximately

$$W=\int V dm'=\kappa\left\{\frac{MM'}{R_{12}}+M\frac{A'+B'+C'-3I'}{2R_{12}^3}+M'\frac{A+B+C-3I}{2R_{12}^3}\right\}, \quad \text{.....(2)}$$

where the accented letters refer to the second distribution and the unaccented letters to the first.

Returning to (7), 1, we may notice that if the body have an axis of symmetry, and GP be inclined at an angle $\theta$ to that axis, we have

$$V=\kappa\frac{M}{R}+\kappa\frac{(C-A)(1-3\cos^2\theta)}{2R^3}, \quad \text{.......................(3)}$$

for now $3I=3\{C\cos^2\theta+A(1-\cos^2\theta)\}$.

Let now the body be a uniform ring of matter of radius $a$. The approximate potential at a distant point on GP, inclined at an angle $\theta$ to the axis of the ring, is

$$V=\kappa\frac{M}{R}+\kappa M\frac{(a^2-\frac{1}{2}a^2)(1-3\cos^2\theta)}{2R^3}. \quad \text{.......................(4)}$$

Thus, if we make the ring of radius $\{2(C-A)/M\}^{\frac{1}{2}}$, it will have the same potential as that given in (3) for the body there considered. The earth may therefore be considered as replaced by a ring of equal mass and of radius $\{2(C-A)/M\}^{\frac{1}{2}}$, without alteration of the potential at a distant point. This radius for the earth is about $\frac{1}{20}$ of the actual radius.

**3.** *Couples applied to the earth by the attractions of the sun and moon.* We can now find the moment of the attraction of the sun or moon about one of the principal axes of moment of inertia of the earth. We shall suppose that the principal axes are taken as coordinate axes, and that of approximate symmetry and moment of inertia C as axis of $z$. We shall calculate the couple about the axis of $y$. Let a line drawn from the sun to the earth's centre make angles $\alpha$, $\beta$, $\gamma$ with the earth's principal axes, and the plane containing this line and the axis of $y$ be inclined at an angle $\theta$ to the plane of $xy$. Consider a point on this line at unit distance from the earth's centre. By projecting this point on the axes of $x$ and $z$ in succession, we obtain

$$\cos\alpha=\sin\beta\cos\theta, \quad \cos\gamma=\sin\beta\sin\theta. \quad \text{....................(1)}$$

But we have $I=A\cos^2\alpha+B\cos^2\beta+C\cos^2\gamma$, and so by the values of $\cos\alpha$ and $\cos\gamma$ just obtained,

$$I=A\sin^2\beta\cos^2\theta+B\cos^2\beta+C\sin^2\beta\sin^2\theta. \quad \text{..................(2)}$$

Now $\beta$ is the sun's longitude measured from ♈ (the first point of Aries), and therefore I varies with $\beta$. Going back to (1), 2, and putting W′ for the third term of the mutual energy in (2), 2, on which alone any turning couple exerted by the second distribution on the first depends, we have

$$W'=\kappa M'\frac{A+B+C-3I}{2R_{12}^3}. \quad \text{.............................(3)}$$

Using the value of I given in (2), multiplying by $d\beta$, and integrating from O to $2\pi$, neglecting variation of $R_{12}$, we get for the $\beta$-average of W′ the equation

$$W'_m = \kappa M' \frac{C-A}{R_{12}^3}(\tfrac{1}{2} - \tfrac{3}{4}\sin^2\theta). \quad \text{.........................(4)}$$

This is not exactly the time-average of W′, but it may be taken as that average, if besides the variation of $R_{12}$ the want of uniformity in the variation of $\beta$ is neglected. Of course in (4) $R_{12}$ is a mean distance.

It is clear that the mutual energy W′, given in (4), may be regarded as produced by a uniform ring of matter placed in the plane of the motion of M′, of total mass M′, and of radius $R_{12}$, with G as centre. The same thing is true when the terms not involving I, which are given in (2), 2, are included.

Now going back to (2), 2, making the accented letters refer to the sun and the unaccented to the earth, we obtain for the moment of the couple on the earth, tending to increase $\theta$,

$$L = \frac{\partial}{\partial\theta}\int V dm' = -\frac{3}{2}\kappa\frac{M'}{R^3}\frac{\partial I}{\partial\theta}, \quad \text{.........................(5)}$$

where we write R instead of $R_{12}$. But by (2) and (1)

$$-\frac{3}{2}\kappa\frac{M'}{R^3}\frac{\partial I}{\partial\theta} = -3\kappa\frac{M'}{R^3}(C-A)\sin^2\beta\sin\theta\cos\theta$$

$$= -3\kappa\frac{M'}{R^3}(C-A)\cos\alpha\cos\gamma. \quad \text{......................(6)}$$

If the line joining the earth's centre with the sun's centre is, at the instant considered, perpendicular to the axis of $y$, $\sin\beta = 1$, and $\theta$ is (as now always) $\omega$, the so-called obliquity of the ecliptic. The axis of $y$ is the intersection of the plane of the ecliptic with the plane of the equator, and the couple is

$$L = -3\kappa\frac{M'}{R^3}(C-A)\sin\omega\cos\omega. \quad \text{.......................(7)}$$

This clearly is the maximum value of the couple. It tends to increase $\omega$.

But $\beta$ is the sun's longitude reckoned from ♈, and the mean value of $\sin^2\beta$ in a complete period of $\beta$ is $\frac{1}{2}$. Hence the average couple for a year is

$$L_m = -\frac{3}{2}\kappa\frac{M'}{R^3}(C-A)\sin\omega\cos\omega, \quad \text{......................(8)}$$

on the inexact supposition that $\beta$ varies uniformly, and taking R as a proper mean distance. The same result is obtained by differentiating (4) above.

Now let the length of the major axis of the earth's orbit be $2R_0$; then $\kappa(M+M')/R_0^3 = n'^2$, where $n'$ is the mean angular speed of the sun round the earth. If we write $\rho = M/M'$, $\kappa M' = n'^2 R_0^3/(1+\rho)$, we obtain for the couple given by (6),

$$L = -3\frac{n'^2}{1+\rho}\left(\frac{R_0}{R}\right)^3(C-A)\sin^2\beta\sin\theta\cos\theta$$

$$= -3\frac{n'^2}{1+\rho}\left(\frac{R_0}{R}\right)^3(C-A)\cos\alpha\cos\gamma. \quad \text{.......................(9)}$$

The maximum couple, which (if we put $R=R_0$) is double the mean couple, is

$$L=-3\frac{n'^2}{1+\rho}\left(\frac{R_0}{R}\right)^3(C-A)\sin\omega\cos\omega. \qquad \text{(10)}$$

For the sun and the earth $\rho$ is exceedingly small and may be neglected. It must be taken account of in dealing in a similar way with the effect of the moon's attraction.

For the moments similarly produced about the axes of $x$ and $z$, we obtain likewise

$$\left.\begin{aligned} L_x&=-3\frac{n'^2}{1+\rho}\left(\frac{R_0}{R}\right)^3(B-C)\cos\beta\cos\gamma,\\ L_z&=-3\frac{n'^2}{1+\rho}\left(\frac{R_0}{R}\right)^3(A-B)\cos\alpha\cos\beta.\end{aligned}\right\} \qquad \text{(11)}$$

4. *Solar couple on earth regarded as due to sun and anti-sun, or to sun's mass distributed round the orbit.* It may be noticed that, since the maximum and mean couples are proportional to $\sin\theta\cos\theta$, we may replace $\theta$ by its complement. Hence either of these couples may be obtained by an arrangement which substitutes the complement of $\omega$ for $\omega$ itself. Thus, if we suppose the sun cut into two halves, and the parts placed, as "sun and anti-sun," on a line through the centre of the earth perpendicular to the ecliptic, at opposite sides of the centre, and at distance equal to the sun's mean distance, a couple equal to the maximum couple will be produced, but in the opposite direction. Thus, with repulsion equal to the attraction, this arrangement would reproduce the couple.

Suppose now that a uniform circular ring of matter of radius R surrounds the body of mass M, and has its centre at the centroid G of the latter. Let the plane of the ring be inclined at an angle $\theta$ to the axis G$z$ of the body, and intersect the plane of $xy$ in a line inclined at an angle $\alpha$ to the principal axis G$x$, and let a plane containing the axis G$z$ and perpendicular to the plane of the ring meet it in a line G$z'$. If $\phi$ be the angle which a radius GF of the ring makes with G$z'$, the projections of GF upon the axes G$x$, G$y$, G$z$ are,

$$R(\sin\phi\cos\alpha+\cos\phi\sin\theta\sin\alpha,\ \sin\phi\sin\alpha-\cos\phi\sin\theta\cos\alpha,\ \cos\phi\cos\theta);$$

and therefore the expressions in the brackets are the three direction cosines of GF, $(\cos\alpha, \cos\beta, \cos\gamma)$. If now I be the moment of inertia of the body about the radius GF, that is $A\cos^2\alpha+B\cos^2\beta+C\cos^2\gamma$, and $m$ be the mass of the ring per unit of its circumference, the whole mutual potential energy, W, of the body and the ring (total mass M′), is

$$W=\int V\,dm'=\kappa\frac{MM'}{R}+\kappa\int_0^{2\pi}\frac{A+B+C-3I}{2R^3}mR\,d\phi. \qquad \text{(1)}$$

But I is a function of $\phi$ given by the values just found for $\cos\alpha, \cos\beta, \cos\gamma$, and so performing the integration we obtain

$$W=\kappa\frac{MM'}{R}-\kappa\frac{A+B+C-3J}{4R^3}M', \qquad \text{(2)}$$

where

$$J=A\cos^2\theta\sin^2\alpha+B\cos^2\theta\cos^2\alpha+C\sin^2\theta. \qquad \text{(3)}$$

But the direction cosines of a normal to the plane of the ring are easily seen to be $\cos\theta\sin\alpha$, $-\cos\theta\cos\alpha$, $-\sin\theta$, for if these are multiplied respectively by $\cos\alpha$, $\cos\beta$, $\cos\gamma$, and the products added, the sum is zero. Hence J is the moment of inertia of the body about a normal to the ring through its centre G.

The couple applied by the ring to the body and tending to increase $\theta$ is $\partial W/\partial\theta$, and we easily find this to be

$$\frac{3}{4}\kappa\frac{M'}{R^3}\frac{\partial J}{\partial\theta}=\frac{3}{2}\kappa\frac{M'}{R^3}(C-A\sin^2\alpha-B\cos^2\alpha)\sin\theta\cos\theta. \quad\text{..........(4)}$$

Thus, if $A=B$, the couple is

$$\frac{\partial W}{\partial\theta}=\frac{3}{2}\kappa\frac{M'}{R^3}(C-A)\sin\theta\cos\theta. \quad\text{........................(5)}$$

If $M'$ be the sun's mass and R the sun's mean distance, the couple tending to increase the angle between the earth's axis and the plane of the ring is given by (5). In 3 above we found the couple tending to increase the angle between the plane of the equator and the plane of the ecliptic, which, as it ought to be, is equal to the couple just found with sign reversed.

If we suppose the mass of the sun distributed in a circular ring in the ecliptic with centre at the earth's centre, and radius equal to the sun's mean distance, we obtain a couple exactly equal to the mean couple calculated above on the supposition that the sun moves uniformly in longitude. It is obvious without calculation that this distribution should give the mean couple. For each equal element of the ring gives a couple proportional to that exerted by the sun when its centre is in that position, and so, if the sun moved round with uniform angular speed, the mean couple would be equal to the couple due to the ring.

If the plane of the ring contain the axis of $y$, then $\alpha=\frac{1}{2}\pi$, and (4) becomes

$$\frac{\partial W}{\partial\theta}=\frac{3}{2}\kappa\frac{M'}{R^3}(C-A)\sin\theta\cos\theta. \quad\text{........................(6)}$$

If $l$, $m$, $n$ denote the direction cosines of the normal to the plane of the ring, we have

$$l=-\cos\theta\sin\alpha,\quad m=\cos\theta\cos\alpha,\quad n=\sin\theta. \quad\text{.................(7)}$$

These give

$$\sin\alpha=-\frac{l}{(l^2+m^2)^{\frac{1}{2}}},\quad \cos\alpha=\frac{m}{(l^2+m^2)^{\frac{1}{2}}},\quad \cos\theta=(l^2+m^2)^{\frac{1}{2}}. \quad\text{.........(8)}$$

If the direction cosines $l$, $m$, $n$ of the normal to the ring are given, and it is required to find the couples turning the body about the principal axes, we may proceed as follows. Considering first the axis of $x$, let a plane be drawn so as to contain that axis and also the normal drawn to the ring from the origin, and find the angle which this plane makes with the plane of $xy$. The $x$-direction cosine of the normal to the plane just specified is zero; let the others be $\mu$, $\nu$. Then we have $\mu m+\nu n=0$, and therefore $\mu/\nu=-n/m$. Thus we get $\mu^2/(\mu^2+\nu^2)=n^2/(m^2+n^2)$, or $\mu=-n/(1-l^2)^{\frac{1}{2}}$. Similarly we get $\nu=m/(1-l^2)^{\frac{1}{2}}$. The angle which the plane containing the axis of $x$ and the normal to the ring makes with the axis of $y$ is $\sin^{-1}\{(-n/(1-l^2)^{\frac{1}{2}}\}$, and that which it makes with the axis of $z$ is $\sin^{-1}\{m/(1-l^2)^{\frac{1}{2}}\}$. If we call the latter angle $\theta_1$, we have

$$J=Al^2+Bm^2+Cn^2=Al^2+(B\sin^2\theta_1+C\cos^2\theta_1)(1-l^2). \quad \text{.........(9)}$$

Thus, for the turning of the body about the axis of $x$, we get

$$\frac{\partial J}{\partial\theta_1}=2(B-C)(1-l^2)\sin\theta_1\cos\theta_1=2(C-B)mn. \quad \text{...........(10)}$$

Similarly, with corresponding angles $\theta_2$, $\theta_3$, for turning about the axes of $y$ and $z$,

$$\left.\begin{aligned}\frac{\partial J}{\partial\theta_2}&=2(C-A)(1-m^2)\sin\theta_2\cos\theta_2=2(A-C)nl,\\ \frac{\partial J}{\partial\theta_3}&=2(A-B)(1-n^2)\sin\theta_3\cos\theta_3=2(B-A)lm.\end{aligned}\right\} \quad \text{..........(11)}$$

The moments about the axes are therefore

$$\left.\begin{aligned}L_x&=\frac{3}{2}\kappa\frac{M'}{R^3}(B-C)(1-l^2)\sin\theta_1\cos\theta_1=\frac{3}{2}\kappa\frac{M'}{R^3}(C-B)mn,\\ L_y&=\frac{3}{2}\kappa\frac{M'}{R^3}(C-A)(1-m^2)\sin\theta_2\cos\theta_2=\frac{3}{2}\kappa\frac{M'}{R^3}(A-C)nl,\\ L_z&=\frac{3}{2}\kappa\frac{M'}{R^3}(A-B)(1-n^2)\sin\theta_3\cos\theta_3=\frac{3}{2}\kappa\frac{M'}{R^3}(B-A)lm.\end{aligned}\right\} \quad \text{.....(12)}$$

If the plane of the ring contain the axis of $y$, $m=0$, and

$$L_y=\frac{3}{2}\kappa\frac{M'}{R^3}(C-A)\sin\theta\cos\theta, \quad \text{.......................(13)}$$

where $\theta$ is now the angle which the plane of the ring makes with the plane of $yz$. This result agrees with that already obtained above.

5. *Mean angular speed of precession.* Returning now to the motion of the earth, and supposing that the axis of $y$ is the axis, in the ecliptic, about which the couple due to the sun's attraction acts, the mean precessional angular speed, $\mu$, of the earth's axis of figure, about a perpendicular to the ecliptic, is given by

$$Cn\mu\sin\omega-A\mu^2\cos\omega\sin\omega=\frac{3}{2}\frac{\kappa S}{R^3}(C-A)\sin\omega\cos\omega, \quad \text{...........(1)}$$

or

$$Cn\mu-A\mu^2\cos\omega=\frac{3}{2}\frac{\kappa S}{R^3}(C-A)\cos\omega. \quad \text{.................(2)}$$

Here S is put for the mass of the sun, as presently we shall denote by M the mass of the moon.

The term $A\mu^2\cos\omega$ is very small, and so we have approximately for the mean precession

$$Cn\mu=\frac{3}{2}\kappa\frac{S}{R^3}(C-A)\cos\omega. \quad \text{.........................(3)}$$

By Kepler's third law we have, if, as before, we put $n'$ for the mean angular speed of the sun round the earth,

$$n'^2R^3=\kappa(S+E)=\kappa S\left(1+\frac{E}{S}\right), \quad\text{.........................(4)}$$

where E is the mass of the earth. Neglecting E/S, which is about 1/324000, we obtain from (3) and (4)

$$\mu=\frac{3}{2}\frac{n'^2}{n}\frac{C-A}{C}\cos\omega \quad\text{..............................(5)}$$

Taking a year as the unit of time we calculate the yearly precession, that is $2\pi\mu/n'$. We have

$$\frac{\mu}{n'}=\frac{3}{2}\frac{n'}{n}\frac{C-A}{C}\cos\omega, \quad\text{..............................(6)}$$

and $n/n'$ is the number of sidereal days in a year, or 366·25. From observations of precession and nutation made by Leverrier and Serret (see Thomson and Tait's *Natural Philosophy*, § 828) it is estimated that (C − A)/C = 306. Also, if $\omega$ be taken as 23°·5, $\cos\omega=\cdot 917$. Thus

$$\frac{\mu}{n'}=\frac{3}{2}\frac{\cdot 917}{366\cdot 25\times 306}. \quad\text{................................(7)}$$

The angle turned through in a year in consequence of the angular speed $n'$ is $2\pi$, or 360 × 3600, seconds. If then $N$ be the number of seconds of angle in the mean annual precession, we have

$$N=\frac{3}{2}\frac{\cdot 917\times 360\times 3600}{366\cdot 25\times 306}=15\cdot 9. \quad\text{.........................(8)}$$

**6.** *Precession due to lunar attraction.* We now consider the mean precession due to the moon, reserving a more complete discussion for a later article of this chapter. Taking a sphere of radius equal to the moon's distance, with centre at the earth's centre G, let it be intersected at Z, Z′ by lines drawn from G at right angles to the ecliptic and the plane of the moon's orbit respectively [Fig. 55]. These are the poles of the ecliptic and the lunar orbit. We consider the lunar orbit as it exists at a given instant; the variation of its position will be taken account of later. Let C be the point in which the earth's axis meets the sphere. Join ZZ′, ZC, Z′C by arcs of great circles, and let ZF be a fixed arc of a great circle on the sphere. The angle Z′ZC is the longitude ☊ of the descending node of the moon's orbit, ZZ′ is $i$ the inclination of that orbit to the ecliptic, and ZC is the obliquity $\omega$ of the ecliptic. The angle FZC may be taken as the angle turned through in time $t$ in consequence of precession. We have, by spherical trigonometry,

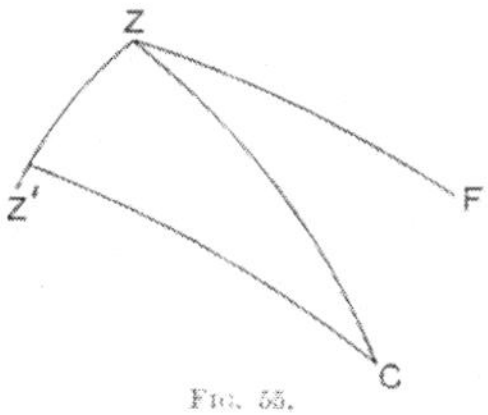

Fig. 55.

$$\cos Z'C=\cos i\cos\omega+\sin i\sin\omega\sin \Omega. \quad\text{.......... ..........(1)}$$

But Z′C is equal to the inclination of the plane of the equator to the plane of the moon's orbit: it corresponds in the present case to the angle denoted in 2 above by $\theta$. The part of the mutual energy of the earth and moon on which the couple producing precession of the earth depends, is therefore [(4), 3]

$$W' = \kappa M \frac{C-A}{R^3}(\tfrac{1}{2} - \tfrac{3}{4}\sin^2\theta), \quad \text{..........................(2)}$$

where M is the moon's mass, and R her proper mean distance from the earth. This gives the mean couple for the different positions of the moon in the orbit as it exists at a given instant, and may be regarded as produced by a uniform circular ring of total mass M, and of radius R, laid in the plane of the orbit about G as a centre. Or it may be regarded as produced according to 4 by two masses, each $\frac{1}{4}$M, one situated at Z′, the other at the diametrically opposite point on the sphere.

We have to calculate the mean value of W′ for a complete revolution of the line of nodes. It is only necessary to find the mean value of $\sin^2\theta$. The mean value of $\cos^2\theta$ is easily found from (1) to be $(\frac{3}{2}\cos^2 i - \frac{1}{2})\cos^2\omega + \frac{1}{2}\sin^2 i$. Hence the mean value of $\sin^2\theta$ is $1 - \{(\frac{3}{2}\cos^2 i - \frac{1}{2})\cos^2\omega + \frac{1}{2}\sin^2 i\}$. Thus we obtain, by (2),

$$W' = \kappa M \frac{C-A}{R^3}\{-\tfrac{1}{4} + \tfrac{3}{4}[(\tfrac{3}{2}\cos^2 i - \tfrac{1}{2})\cos^2\omega + \tfrac{1}{2}\sin^2 i]\}. \quad \text{............(3)}$$

Hence
$$-\frac{\partial W'}{\partial\omega} = \tfrac{3}{2}\kappa M \frac{C-A}{R^3}(\tfrac{3}{2}\cos^2 i - \tfrac{1}{2})\sin\omega\cos\omega. \quad \text{..................(4)}$$

This is the couple tending to diminish $\omega$. It will be observed that if we put $i=0$, that is, suppose that the moon's orbit coincides with the ecliptic, the couple agrees exactly with that given by (8), 3 above, when M is substituted for M′.

Just as the solar precession was found in 5, so we now find the lunar precessional angular speed $\mu'$ to be given by

$$Cn\mu' = \tfrac{3}{2}\kappa \frac{M}{R^3}(C-A)(\tfrac{3}{2}\cos^2 i - \tfrac{1}{2})\cos\omega. \quad \text{.........................(5)}$$

If we introduce the mean angular speed $n''$ of the moon's radius vector we obtain, since $n''^2 R^3 = \kappa(E+M) = \kappa M(1+E/M)$,

$$\mu' = \tfrac{3}{2}\frac{n''^2}{n}\,\frac{1}{1+\dfrac{E}{M}}\,\frac{C-A}{C}(\tfrac{3}{2}\cos^2 i - \tfrac{1}{2})\cos\omega. \quad \text{..................(6)}$$

Thus we obtain for the ratio $\lambda$ of the lunar to the solar precession,

$$\lambda = \frac{\mu'}{\mu} = \frac{n''^2}{n'^2}\,\frac{\tfrac{3}{2}\cos^2 i - \tfrac{1}{2}}{1+\dfrac{E}{M}}. \quad \text{..................................(7)}$$

This gives, if we take $n''/n' = 366\frac{1}{4}/27\frac{1}{3} = 13{\cdot}4$, $i = 5^\circ\ 9'$, and so $\frac{3}{2}\cos^2 i - \frac{1}{2} = 0{\cdot}988$, and $E/M = 81{\cdot}5$,

$$\frac{\mu'}{\mu} = \frac{13{\cdot}4^2 \times 0{\cdot}988}{82{\cdot}5} = 2{\cdot}15. \quad \text{..................................(8)}$$

The solar precession is 16 seconds of angle in a year, and therefore, according to the numerical data here assumed, the lunar precession is 34·4 seconds of angle in a year, giving a total of slightly over 50″. The *Nautical Almanack* gives 50″·4, which, as it happens, is exactly the amount here obtained.

The observed precession, with E/M = 81·5, leads to the value 1/306 for (C − A)/C, given by Thomson and Tait [5, above].

It will be seen that a total annual precession of 50″·4 gives a complete revolution of the equinoxes along the ecliptic in 25,714 years. Now the above discussion gives as the total average couple tending to bring the equator and ecliptic into coincidence,

$$\tfrac{3}{2}(\lambda+1)\frac{\kappa S}{R^3}(C-A)\sin\omega\cos\omega. \qquad (9)$$

If the earth's rotation were to cease, a couple $\frac{3}{2}(\lambda+1)\kappa S(C-A)\sin\theta\cos\theta/R^3$, where $\theta$ is the mutual inclination of the two planes, would cause the earth to swing about the intersection of the two planes, in a period, when the amplitude has been brought down to a small value, given by

$$\ddot{\theta}+\tfrac{3}{2}(\lambda+1)\frac{\kappa S}{R^3}\frac{C-A}{A}\theta=0, \qquad (10)$$

where R is the sun's distance. If, as in 5, we put $n'^2R^3$ for $\kappa S$, and $C\mu n/n'^2\cos\omega$ for $\frac{3}{2}(\lambda+1)(C-A)$, this equation becomes

$$\ddot{\theta}+\frac{Cn\mu}{A\cos\omega}\theta=0, \qquad (11)$$

where $\mu$ is the total precessional angular speed just calculated. Taking as before $\cos\omega = {\cdot}917$, putting C = A, and using the year as unit of time, we obtain as the period of a small oscillation

$$2\pi\left(\frac{A\cos\theta}{Cn\mu}\right)^{\frac{1}{2}}=\left(\frac{{\cdot}917\times 26000}{366{\cdot}25}\right)^{\frac{1}{2}}=8, \text{ nearly.} \qquad (12)$$

The earth would therefore make one small oscillation in a period of about 8 years.

It is to be understood that this would not be the actual period of a small oscillation if the rotation were annulled. The couple assigned is not, for $\theta=\omega$, the couple which really exists, except at particular instants; it is the average couple obtained by considering a much longer interval than the derived period of 8 years.

7. *Precession of the equinoxes, body-cone and space-cone.* We have seen that the couple exerted by the sun or the moon on the earth, and tending to turn it about the axis of $x$, $y$, or $z$, depends on the difference, B − C, C − A, or A − B of the principal moments of inertia, and vanishes when this difference vanishes. And taking the axis of $y$ (which of course, as one of the axes perpendicular to the axis of approximate symmetry, is in the plane of the equator) also in the plane of the ecliptic, we have found the average couple which, as it is usually put, tends to bring the earth's equator

into coincidence with the ecliptic. The translational motion of the earth in its orbit is ignored, as in itself it has no effect; the change in relative position of the earth and sun, which alone matters, is conveniently obtained by supposing the centroid of the earth to be at rest while the sun revolves round it.

The couple found plays the part of the couple about the axis OD (Fig. 4) applied by gravity to the top spinning about a fixed point. The result is the same; just as the top does not fall down, but has an azimuthal motion in virtue of the couple, so that the axis of rotation, if the motion is steady, moves in a right cone, so the earth's axis does not approach perpendicularity to the ecliptic, but relative to the earth's centre regarded as a fixed point, has a conical motion in space about a line drawn from the earth's centre to the pole of the ecliptic, which answers to the vertical in the case of the top (Fig. 4). The mean angular speed of a point on the earth's axis about the axis of the cone is $M/Cn \sin \omega$, where M is a certain mean value of the moment of the couple producing precession, and $\omega$ is the inclination of the earth's axis to a perpendicular to the ecliptic; and this is exactly analogous to the approximate value $Mgh \sin \theta / Cn \sin \theta$, which the theory of the top gives for the precessional motion about the vertical. This conical motion of the earth's axis has, as we have seen, a period of nearly 26,000 years, and causes the astronomical phenomenon of *precession of the equinoxes*, that is, the continual revolution of the line of equinoxes (or *line of nodes*) in the plane of the ecliptic.

How this phenomenon occurs is illustrated by Fig. 56, which shows one of the illustrative models of the Natural Philosophy Department of the University of Glasgow, a terrestrial globe with the lower half cut away, and the upper part loaded so that it can turn about a point of support at the centre, with the pin P in contact with the inside of the horizontal ring RP at the top. The pin P is the upper end of a cone fixed in the body having its vertex at the centre O of the globe; this cone rolls on a cone fixed in space, as indicated in Fig. 4. Here the rolling cone (the body-cone, 4, IV above) rolls round on the inside surface of the fixed- or space-cone (Fig. 14). [In the example of the top spinning about a fixed point the body-cone rolls round the outside of the space-cone as in Fig. 13.] The space-cone in the model is represented by the ring RP, which is enough to guide the moving cone; all the rest is cut away, but it is understood that the vertex in this case is also at the centre.

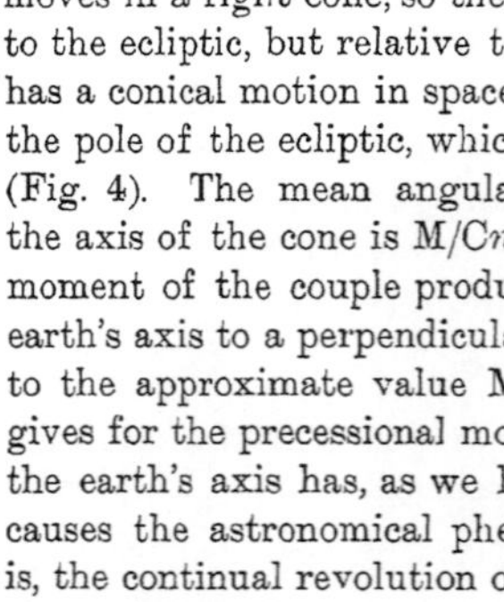

FIG. 56.

As the globe of the model (Fig. 56) turns about the axis of figure, the cone P rolls on the space-cone, and its axis, travelling round the axis of figure, describes a cone in space, in the model a cone of 23° 27′ semi-vertical angle. The equator of the globe is shown by the dark line intersecting a meridian through P in N. The upper surface of the rim, to which the supports of the ring R are attached, represents the plane of the ecliptic, and the point N represents the intersection of the equator with that plane. N therefore represents an equinox. As, imitating the earth, the globe revolves in the counter-clock direction (as seen from beyond P, which may be taken as representing the north pole of the earth), the pin P rolls round the ring in the clock direction, and so the point N moves from right to left along the ecliptic, in the direction to meet the rotation, that is to make the equinoxes occur earlier in time. This is the precession of the equinoxes, the direction and mode of production of which are completely illustrated from a kinematical point of view by the model.

If the model were enlarged to the size of the earth and rotated with the same speed as the earth, the diameter at the north pole of the cone, fixed in the earth with vertex at the centre, which rolling on the internal surface of a cone of semi-vertical angle 23° 27′, with its vertex also at the centre of the earth, would give precessional motion of 25,800 years' period, would be

$$\frac{8000 \times \sin 23^\circ\ 27'}{366\frac{1}{4} \times 25{,}800} \text{ miles} = 21 \text{ inches.}$$

This is the solution of the old Glasgow Natural Philosophy question, sometimes phrased as "Find the diameter of the north pole of the earth" [see 13, I].

In 4, IV, the rolling of the body-cone on the space-cone has been explained. In Fig. 57 the two cones are shown for the case of precession of the earth's axis OZ is the perpendicular to the plane of the ecliptic, OC the earth's axis of figure, OI the instantaneous axis of turning, and OH the axis of resultant A.M. The vertical angle of the body-cone is $2\alpha$, that of the space-cone $2(\theta+\alpha)$. The angular speeds are $\dot{\psi}$ about OZ, $n$ about OC, and $\dot{\psi}\sin\theta$ about OE, which is at right angles to OC. If at the same time the body is turning about an axis OD at right angles to the plane COZ, there is also a component $\dot{\theta}$ to be taken into account. In the case of steady motion the instantaneous axis is OI, and the angular speed about it is $(n^2+\dot{\psi}^2\sin^2\theta)^{\frac{1}{2}}$.

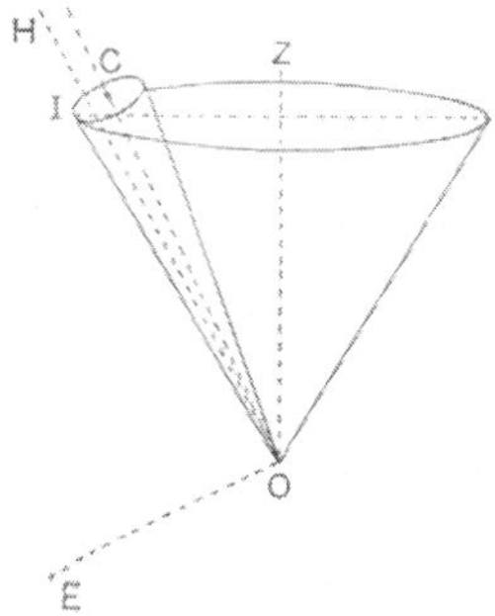

FIG. 57.

By the turning about OI, with angular speed $(n^2+\dot{\psi}^2\sin^2\theta)^{\frac{1}{2}}$, the axial point C, which lies in a plane at right angles to OC and containing the point I, has speed OI $\cos\alpha\sin\alpha\,(n^2+\dot{\psi}^2\sin^2\theta)^{\frac{1}{2}}$, that is OI$n\sin\alpha$, at right angles to the paper. This must be the same as the speed $\mathrm{OI}\dot\psi\cos\alpha\sin\theta$ in the circle of radius $\mathrm{OI}\cos\alpha\sin\theta$. Hence we have

$$n\tan\alpha=\dot\psi\sin\theta.$$

In the case of steady motion the axes OI, OC, OH, and OZ are in one plane, the resultant A.M. is $(\mathrm{C}^2n^2+\mathrm{A}^2\dot\psi^2\sin^2\theta)^{\frac{1}{2}}$, and according as C is $>$ or $<$ A, the axis OH is nearer to or farther from OC than OI. We have $\tan\mathrm{IOC}=\dot\psi\sin\theta/n$, $\tan\mathrm{HOC}=\mathrm{A}\dot\psi\sin\theta/\mathrm{C}n$ If there is turning about an axis OD at right angles to ZOC, the resultant angular speed is $(\dot\theta^2+n^2+\dot\psi^2\sin^2\theta)^{\frac{1}{2}}$, and its direction makes with OD, OC, OE angles whose cosines are $(\dot\theta,\ n,\ \dot\psi\sin\theta)/(\dot\theta^2+n^2+\dot\psi^2\sin^2\theta)^{\frac{1}{2}}$. The resultant A.M. on the other hand is $(\mathrm{A}^2\dot\theta^2+\mathrm{C}^2n^2+\mathrm{A}^2\dot\psi^2\sin^2\theta)^{\frac{1}{2}}$, and, if we call this H, its direction makes angles with OD, OC, OE, the cosines of which are

$$(\mathrm{A}\dot\theta,\ \mathrm{C}n,\ \mathrm{A}\dot\psi\sin\theta)/\mathrm{H}.$$

8 *The free Eulerian precession. Results of theory and of observation.* If the body be under the action of no external forces the direction of OH remains invariable, and the motion of the body *after* any transient disturbance must be consistent with this fact. The equations of motion are (see 8, II)

$$\mathrm{A}\frac{\partial\theta_1}{\partial t}-(\mathrm{B}-\mathrm{C})\theta_2\theta_3=0,\quad \mathrm{B}\frac{\partial\theta_2}{\partial t}-(\mathrm{C}-\mathrm{A})\theta_3\theta_1=0,\quad \mathrm{C}\frac{\partial\theta_3}{\partial t}-(\mathrm{A}-\mathrm{B})\theta_1\theta_2=0,\ \ldots(1)$$

where $\theta_1$, $\theta_2$, $\theta_3$ are the angular speeds about the principal axes for which the moments of inertia are A, B, C. If A = B, which is approximately true for the earth, the third equation gives $\mathrm{C}\theta_3$ = constant. The other equations become, if we write $n$ for the constant $\theta_3$,

$$\mathrm{A}\frac{\partial\theta_1}{\partial t}+(\mathrm{C}-\mathrm{A})n\theta_2=0,\quad \mathrm{A}\frac{\partial\theta_2}{\partial t}-(\mathrm{C}-\mathrm{A})n\theta_1=0.\ \ldots\ldots\ldots\ldots(2)$$

Eliminating $\theta_2$, we obtain

$$\frac{\partial^2\theta_1}{\partial t^2}+n^2\left(\frac{\mathrm{C}-\mathrm{A}}{\mathrm{A}}\right)^2\theta_1=0.\ \ldots\ldots\ldots\ldots\ldots\ldots(3)$$

[Equation (3) may also be obtained from the ordinary equations of the top (*e.g.* the first two of (1), 16, II, or (1) of 9 below) with the right hand sides put equal to zero. The angular speeds $\theta_1$, $\theta_2$ are to be identified with the $\omega_1$, $\omega_2$ of (2), 2, IV, and $\phi$ is to be equated to zero, observing however that, though $\theta_1=-\dot\psi\sin\theta$, $\theta_2=\dot\theta$,

$$\frac{d\theta_1}{dt}=-\ddot\psi\sin\theta-\dot\theta\dot\psi\cos\theta+\dot\theta\dot\phi,\quad \frac{d\theta_2}{dt}=\dot\phi\dot\psi\sin\theta+\ddot\theta.$$

Hence, since $n=\dot\phi+\dot\psi\cos\theta$,

$$-\ddot\psi\sin\theta=\frac{d\theta_1}{dt}-n\dot\theta+2\dot\theta\dot\psi\cos\theta,\quad \ddot\theta=\frac{d\theta_2}{dt}-\dot\phi\dot\psi\sin\theta.$$

These values, substituted in (1), 15, II, give exactly equations (2) obtained above.]

From (3), if $p=n(\mathrm{C}-\mathrm{A})/\mathrm{A}$, we find

$$\theta_1=\mathrm{E}\cos(pt+a),\ \ldots\ldots\ldots\ldots\ldots\ldots\ldots(4)$$

where E and $a$ are constants. This gives also

$$\theta_2=\mathrm{E}\sin(pt+a).\ \ldots\ldots\ldots\ldots\ldots\ldots\ldots(5)$$

If $C-A$ is small the differential equations show that $\theta_1$, $\theta_2$ alter only slowly; their actual values are determined by observation. The values found indicate, since $A=B$ very nearly, a resultant of $A\theta_1$ and $B\theta_2$, of magnitude AE, inclined nearly at 45° to either component in the plane of the equator, and revolving *in the body* in that plane with angular speed $n(C-A)/A$. The invariable line OH lies in the plane of this constant resultant and OC, about which the A.M. $Cn$ is also constant. *After* any disturbance OH remains inclined at a constant angle to OC, and, as the plane HOC revolves in space, describes a cone in the body about OC as axis, and makes one revolution in the period $2\pi A/n(C-A)$. But $2\pi/n$ is one day and $A/(C-A)=306$ nearly. Hence the invariable line, which is fixed in space, describes the cone

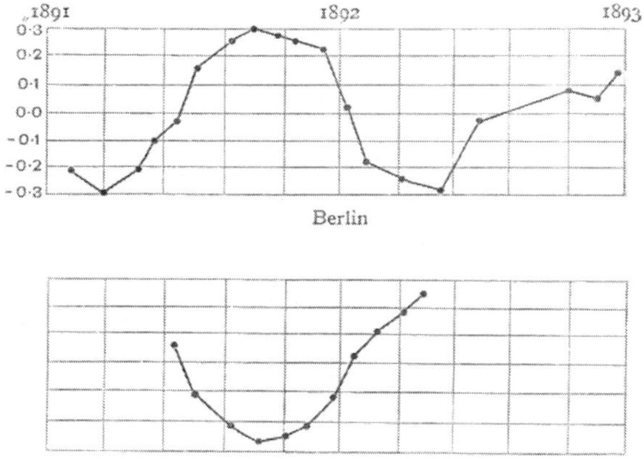


FIG. 58.

just specified in 306 days; in other words, the axis of figure revolves round the invariable line in a cone of semi-vertical angle COH in that period. This angle $\tan^{-1}(AE/Cn)$ is small, and is not to be confounded with the very much smaller mean angle [about 0·00867″], between OC and the instantaneous axis, OI, of rotation, which causes the slow continuous precession.

If the magnitude of E is sensible this revolution of OC should show a variation of latitude of places on the earth's surface, the more considerable the greater the value of E, since $\tan COI = E/n$. But determinations of latitude by observations of the altitude of stars show only a very slight variation, and therefore we must conclude that E is small. Observations made at Berlin, Potsdam, and Prague in 1889-90 to detect this effect showed changes of only about half a second of angle. Though small, this variation was no doubt real, as the latitude of a place can be determined to 1/10 of a second, which corresponds to three yards on the earth's surface. In 1891 a

German expedition was sent to Honolulu in order that observations might be made simultaneously there and at Berlin. Since the two places are 171° apart in longitude, the latitude of Berlin should increase when that of Honolulu diminishes, and *vice versa*. The diagram (Fig. 58) shows how exactly the two variations were found to correspond.

As to the period however, it was found by Mr. S. C. Chandler at Cambridge, Mass., by a discussion of observations of latitude ranging from 1840 to 1891, that the variation ran through its course in 14 instead of 10 months. The course of this variation of latitude, combined with another of yearly period, is shown in the diagram on p. 15. The subject is further dealt with in Chapter XI below.

The free Eulerian precession discussed above is started by some transient external disturbance, and is given by the complementary solution obtained as above when the applied couples are put equal to zero, and the differential equations then solved. We shall see however that the expressions for the couples may contribute terms which require consideration to the oscillatory solutions.

9. *Precession and nutation from instant to instant.* The average precession has been determined above. We now try to trace the progress of events from instant to instant. Taking axes as before, G$y$ along the intersection of the plane of the equator of the earth with the plane of the ecliptic, G$z$ in the direction of the axis of figure, and G$x$ at right angles to the other two, we observe that the axes G$x$, G$y$, G$z$ correspond to the moving axes OE′, OD, OC of the gyrostatic equations, and shown in Fig. 12 (p. 71). In that diagram the sun is supposed to be at S, a point to the left of $x$ on the arc DF produced beyond $x$, and such that the line perpendicular to OS in the plane DOF lies between OD and OE′. OFD is the plane of the ecliptic, DOE that of the equator. If then L be the couple at the instant about the axis of $x$, M that about the axis of $y$, we have by (2) and (5), 1, V, or by applying at once the method of 5, III,

$$\left.\begin{aligned} A\ddot{\psi}\sin\theta-(Cn-2A\dot{\psi}\cos\theta)\dot{\theta}&=-L,\\ A\ddot{\theta}+(Cn-A\dot{\psi}\cos\theta)\dot{\psi}\sin\theta&=\phantom{-}M.\end{aligned}\right\}\quad\ldots\ldots(1)$$

The minus sign is placed before L on the right of the first equation because it will be seen from Fig. 12 that the rate of growth of A.M. for the axis OE′ is the quantity on the left with the sign reversed. The values of the couples are given in 3 above for the sun's attraction. We have by (10) and (11) of 3,

$$\left.\begin{aligned} A\ddot{\psi}\sin\theta-(Cn-2A\dot{\psi}\cos\theta)\dot{\theta}&=\phantom{-}3\frac{n'^2}{1+\rho}\left(\frac{R_0}{R}\right)^3(C-A)\cos\beta\cos\gamma,\\ A\ddot{\theta}+(Cn-A\dot{\psi}\cos\theta)\dot{\psi}\sin\theta&=-3\frac{n'^2}{1+\rho}\left(\frac{R_0}{R}\right)^3(C-A)\cos\alpha\cos\gamma.\end{aligned}\right\}\quad\ldots(2)$$

Now we know from observation that $n$ is very great in comparison with $\dot{\psi}$, and $\dot{\psi}$, itself small, has a long period of variation, and so we neglect all the terms on the left of (2) which do not contain $n$. Thus neglecting also $\rho$ we obtain

$$\left.\begin{aligned}\dot{\theta} &= -3\frac{n'^2}{n}\left(\frac{R_0}{R}\right)^3\frac{C-A}{C}\cos\beta\cos\gamma,\\ \sin\theta\,.\,\dot{\psi} &= -3\frac{n'^2}{n}\left(\frac{R_0}{R}\right)^3\frac{C-A}{C}\cos\alpha\cos\gamma.\end{aligned}\right\}\quad\ldots\ldots(3)$$

But by (1), 3, $\cos\beta\cos\gamma=\sin\beta\cos\beta\sin\theta$, $\cos\alpha\cos\gamma=\sin^2\beta\cos\theta\sin\theta$, so that (3) become

$$\left.\begin{aligned}\dot{\theta} &= -\frac{3}{2}\frac{n'^2}{n}\left(\frac{R_0}{R}\right)^3\frac{C-A}{C}\sin 2\beta\sin\theta,\\ \dot{\psi} &= -\frac{3}{2}\frac{n'^2}{n}\left(\frac{R_0}{R}\right)^3\frac{C-A}{C}(1-\cos 2\beta)\cos\theta.\end{aligned}\right\}\quad\ldots\ldots(4)$$

Let now $l$ be the longitude of the sun measured from a fixed plane perpendicular to the ecliptic, the plane $xOz$ of Fig. 12. Then $xF=\psi$, $xS=l$ (since we measure $l$ in the direction of the apparent motion from the fixed plane $xOz$, where $z$ is the fixed pole of the ecliptic). We put now ☊ for the longitude of the ascending node, D, of the sun's orbit. Then ☊ $=-(\frac{1}{2}\pi+\psi)$. But by definition (3, above), $\beta=-l-$ ☊ $=-l+\frac{1}{2}\pi+\psi$, and therefore $\sin 2\beta=\sin(\pi-2(l-\psi))=\sin 2(l-\psi)$, $\cos 2\beta=-\cos 2(l-\psi)$. Substituting in (4) we obtain

$$\left.\begin{aligned}\dot{\theta} &= -\frac{3}{2}\frac{n'^2}{n}\left(\frac{R_0}{R}\right)^3\frac{C-A}{C}\sin 2(l-\psi)\sin\theta,\\ \dot{\psi} &= -\frac{3}{2}\frac{n'^2}{n}\left(\frac{R_0}{R}\right)^3\frac{C-A}{C}\{1+\cos 2(l-\psi)\}\cos\theta.\end{aligned}\right\}\quad\ldots\ldots(5)$$

To obtain more exact equations we should have to express $1/R^3$ and $l$ in terms of the time by means of the theory of motion in a central orbit. We should obtain $l=n't+\epsilon'+2e'\sin(n't+\epsilon'-f')+\ldots$, and a similar expression for $1/R$, where $e'$ is the eccentricity. For such developments however we must refer to special works on *Celestial Mechanics*, such as Tisserand, *Mécanique Céleste*, t. ii.

We may however substitute for $l$ the approximate value $n't+\epsilon'$, and integrate equations (5) with respect to $t$, assigning to $\sin\theta$, $\cos\theta$, which are nearly constant, values $\sin\alpha$, $\cos\alpha$, which remain unaltered during the period of integration. We get, taking R as constant,

$$\left.\begin{aligned}\theta &= \alpha+\frac{3}{4}\frac{n'}{n}\left(\frac{R_0}{R}\right)^3\frac{C-A}{C}\cos 2(l-\psi)\sin\alpha,\\ \psi &= \psi_0-\frac{3}{2}\frac{n'}{n}\left(\frac{R_0}{R}\right)^3\frac{C-A}{C}\{n't+\tfrac{1}{2}\sin 2(l-\psi)\}\cos\alpha,\end{aligned}\right\}\quad\ldots\ldots(6)$$

where $\psi_0$ is a constant. The term $n't$ in the second equation obviously arises from the nearly uniform precession, the approximate average precession obtained in (5) of 5.

The results we have obtained are only approximate. It is easy to form some idea of the closeness of the approximations. From (1) we obtain, putting $\omega_1$ for $-\dot{\psi}\sin\theta$, and $\omega_2$ for $\dot{\theta}$, the angular speeds about the axes OE′, OD of Fig. 12, p. 71 [to be distinguished from the $\omega_1$, $\omega_2$ referred to in 8 above],

$$\left.\begin{aligned}A\dot{\omega}_1+(Cn-A\dot{\psi}\cos\theta)\omega_2&=L,\\ A\dot{\omega}_2-(Cn-A\dot{\psi}\cos\theta)\omega_1&=M,\end{aligned}\right\} \qquad (7)$$

a symmetrical form which, by the opposite signs of the second terms on the left, well displays the peculiar property of gyrostatic terms. As a first approximation we get, since both $\dot{\psi}$ and $\dot{\theta}$ are small,

$$\omega_1=-\frac{M}{Cn},\quad \omega_2=\frac{L}{Cn}. \qquad (8)$$

Using these results in the first terms on the left of (7) we obtain

$$\omega_1=-\frac{M}{Cn}+\frac{A}{C^2n^2}\frac{dL}{dt},\quad \omega_2=\frac{L}{Cn}+\frac{A}{C^2n^2}\frac{dM}{dt}. \qquad (9)$$

A repetition of this process gives

$$\omega_1=-\frac{M}{Cn}+\frac{A}{C^2n^2}\frac{dL}{dt}+\frac{A^2}{C^3n^3}\frac{d^2M}{dt^2},\quad \omega_2=\frac{L}{Cn}+\frac{A}{C^2n^2}\frac{dM}{dt}-\frac{A^2}{C^3n^3}\frac{d^2L}{dt^2}. \qquad (10)$$

Thus, as elimination of $\omega_2$ from (7), and substitution in L and M of the values of $l$ and $1/R$ indicated above, give $\omega_1$ of the form $\Sigma K\cos(n't+f)$, the adoption of the approximations in (8) means the neglect of terms of the order $Mn'^2/n^2$ or $Ln'^2/n^2$. [See Routh, *Advanced Dynamics*, §§525–550, where the motion of the ecliptic is also considered.]

10. *Graphical representation of effects of variable parts of $\dot{\theta}$ and $\dot{\psi}$.* The results embodied in (5) and (6) are capable of simple graphical representation. Thus the uniform precession, of angular speed

$$\frac{3}{2}\frac{n'^2}{n}\frac{C-A}{C}\cos\theta,$$

may be represented by the motion of a point $P_1$, in a circle of radius equal to the mean value of $\theta$. Taking the parts of $\theta$ and $\psi$, depending on the varying parts of $\dot{\theta}$ and $\dot{\psi}$, we get the equations

$$\dot{\theta}=\ \frac{3}{4}\frac{n'}{n}\frac{C-A}{C}\cos 2(l-\psi)\sin\theta, \qquad (1)$$

$$\sin\theta\,.\,\dot{\psi}=-\frac{3}{4}\frac{n'}{n}\frac{C-A}{C}\sin 2(l-\psi)\sin\theta\cos\theta. \qquad (2)$$

If we denote the quantities on the left by $y$ and $x$ respectively, we obtain

$$\frac{x^2}{K^2\sin^2\theta\cos^2\theta}+\frac{y^2}{K^2\sin^2\theta}=1, \qquad (3)$$

where K denotes the common multiplier of the circular functions on the right of (1) and (2).

Clearly then, if we describe round $P_1$ as centre an ellipse of semi-axes $K\sin\theta\cos\theta$, $K\sin\theta$, the latter in the direction of $CP_1$, the former at right angles to this direction, then a second particle $P_2$ describes this ellipse (which changes in position as $P_1$ moves) in a period, given by the terms $\sin 2(l+\psi)$, $\cos 2(l+\psi)$, equal to half the periodic time of the disturbing body.

Successive further approximations would be given by successive ellipses, the first round $P_2$, the second round the point describing this, and so on. This kind of graphical representation is characteristic of successive approximations by periodic terms.

11. *Mean couple tending to bring equator and moon's orbit into coincidence.* It is proved in (8), 3, that, if R be the mean distance of the sun from the earth, the mean couple exerted on the earth by the sun in a year is given by

$$L = -\frac{3}{2}\frac{S}{R^3}(C-A)\sin\omega\cos\omega. \qquad (1)$$

Here S is the sun's mass. Substituting for $\omega$ the inclination I of the moon's orbit to the plane of the equator, and in place of S and R for the sun putting M′ and R for the moon, retaining M for the earth, we get for the mean couple, tending to produce turning of the plane of the equator into coincidence with the plane of the moon's orbit,

$$L = -\frac{3}{2}\kappa\frac{M'}{R^3}(C-A)\sin I\cos I, \qquad (2)$$

or, since $\kappa(M+M') = \kappa M'(1+\rho) = n''^2R^3$,

$$L = -\frac{3}{2}\frac{n''^2}{1+\rho}(C-A)\sin I\cos I. \qquad (3)$$

This couple acts about the line of intersection of the two planes, and so the earth's axis turns towards the descending node of the moon's orbit on the equator, with angular speed $\psi'\sin I$ given by

$$\sin I\,.\,\psi' = -\frac{3}{2(1+\rho)}\frac{n''^2}{n}\frac{C-A}{C}\sin I\cos I. \qquad (4)$$

This is an average and uniform precessional motion; the inequalities of fortnightly period are neglected. The turning indicated is about an axis at right angles to the earth's axis in the plane of that axis and the perpendicular to the moon's orbit. We resolve it into two components, one about an axis through G (Fig. 55) at right angles to the earth's axis and in the plane of that axis and the perpendicular to the ecliptic, the other about an axis through G at right angles to the plane just specified. The first component is what we have already denoted by $\dot\psi\sin\theta$, except that it is now produced by the moon, and so we have

$$\dot\psi\sin\theta = -\frac{3}{2(1+\rho)}\frac{n''^2}{n}\frac{C-A}{C}\sin I\cos I\cos ZCZ'. \qquad (5)$$

The second component is $\dot{\theta}$, and is given by

$$\dot{\theta} = -\frac{3}{2(1+\rho)}\frac{n''^2}{n}\frac{C-A}{C}\sin I\cos I\sin ZCZ'. \quad \text{.............(6)}$$

Now, from Fig. 55, putting $\theta$ for the inclination of the earth's instantaneous axis of rotation to the perpendicular to the ecliptic, and denoting the inclination of the moon's orbit to the plane of the ecliptic, that is the arc $ZZ'$, by $i$, we obtain by spherical trigonometry

$$\left.\begin{aligned} \cos I &= \cos i\cos\theta + \sin i\sin\theta\cos Z'ZC, \\ \sin I\cos ZCZ' &= \frac{\cos i - \cos I\cos\theta}{\sin\theta}, \\ \sin I\sin ZCZ' &= \sin i\sin Z'ZC. \end{aligned}\right\} \quad \text{.............(7)}$$

Equations (7) give instead of (5) and (6) the results

$$\left.\begin{aligned} \dot{\psi}\sin\theta = &-\frac{3}{2(1+\rho)}\frac{n''^2}{n}\frac{C-A}{C}\{(\cos^2 i - \tfrac{1}{2}\sin^2 i)\sin\theta\cos\theta \\ &-\sin i\cos i\cos 2\theta\cos Z'ZC - \tfrac{1}{2}\sin^2 i\sin\theta\cos\theta\cos 2Z'ZC\}, \\ \dot{\theta} = &-\frac{3}{2(1+\rho)}\frac{n''^2}{n}\frac{C-A}{C}(\sin i\cos i\cos\theta\sin Z'ZC + \tfrac{1}{2}\sin^2 i\sin 2Z'ZC). \end{aligned}\right\} \quad (8)$$

The first line of the first equation of (8) gives exactly the result already obtained in (3), 6, above, for the mean lunar precession, except that the signs of the two results are opposed. This was to be expected however, as the result of 6 was obtained from a consideration of a couple opposed in sign to that considered above, which is equivalent to taking $\dot{\psi}\sin\theta$ about an axis opposed in direction to that used here. The results therefore agree.

The $\angle CZZ'$ diminishes at rate $m$ because of the motion of $Z'$, that is, the revolution of the line of nodes, and increases at rate $p$ because of the motion of C. The value of $p$ is

$$\frac{3}{2}\frac{n''^2}{n(1+\rho)}\frac{C-A}{C}(\cos^2 i - \tfrac{1}{2}\sin^2 i)\cos\theta.$$

The motion of the line of nodes is a regression at rate $m$, which can be calculated from the fact that the line of nodes turns through an angle $2\pi$ in 18 years 7 months, or about ·02817 radian per month. We can write

$$\angle Z'ZC = (-m+p)t + \epsilon - \psi_0, \quad \text{.........................(9)}$$

where $\epsilon$ is a constant.

Integrating (8) we obtain

$$\left.\begin{aligned} \psi = \psi_0 - pt &- \frac{p}{m-p}\frac{\sin i\cos i\cos 2\theta}{(\cos^2 i - \frac{1}{2}\sin^2 i)\sin\theta\cos\theta}\sin Z'ZC \\ &- \frac{p}{4(m-p)}\frac{\sin^2 i}{\cos^2 i - \frac{1}{2}\sin^2 i}\sin 2Z'ZC, \\ \theta = \theta_0 &- \frac{3}{2(1+\rho)}\frac{n''^2}{n}\frac{C-A}{C}\frac{\sin i}{m-p}(\cos\theta\cos i\cos Z'ZC \\ &+ \tfrac{1}{4}\sin i\sin\theta\cos 2Z'ZC), \end{aligned}\right\} \quad \text{.........(10)}$$

where $n$ is the angular speed of rotation of the earth.

The coefficient of cos 2Z′ZC in the second of (10) is only about 1/100 of that of cos Z′ZC, so that the former term is comparatively unimportant. Measured in seconds of angle the amplitudes of sin Z′ZC and cos Z′ZC, in the first and second of (10), are respectively 17·4 and 9·3. The ratio of $m$ $p$ is great, and so instead of $m-p$ we may write simply $m$ in (10).

To the lunar precession and nutation falls to be added the solar precession and nutation found above. We have seen that the lunar precession is about 34″ and the solar about 16″. The ratio of the amplitude of solar nutation to that of lunar nutation is about 2/15.

12. *Gyrostatic theory of the regression of the line of nodes of the moon's orbit.* We now give an application of gyrostatic theory to the calculation of the rate of regression of the line of nodes of the moon's orbit on the ecliptic produced by the attraction of the sun. For this, the attracted body, the moon, which moves comparatively quickly in its orbit, is supposed distributed round its orbit, in a ring revolving about its axis of figure with the moon's mean angular speed about the earth, and to react, as if it were a flywheel, against the couple applied to it by the attraction of the sun. The process here adopted is a modification of the device, first used by Newton, of disposing the attracted body in a ring round its orbit. It is found, as was indeed found by Newton in his calculation of the same quantity, that unless the relative motion of the nodes and the sun is taken account of, the period of regression comes out too short by about 7 months. Thus it is not possible to suppose the matter of the sun disposed in a ring.

Hence we keep in the following discussion the sun undisturbed, and imagine the mass of the moon equally distributed in a circular ring of radius equal to the moon's mean distance from the earth. It is assumed that the eccentricity of the moon's orbit does not seriously affect the revolution of the line of nodes. This eccentricity is considerable, as the ratio of the greatest to the least distance of the moon from the earth is about 40 to 35. We shall afterwards add some remarks in justification of what seems a remarkable process.

In the ring moon we have C = 2A and so C − A = A. Thus we get by (10), 4, for the couple acting,

$$L = -3n'^2 A \sin^2\beta \sin\theta \cos\theta. \quad \text{.........................(1)}$$

Here $n'$ is the angular speed of the sun in the ecliptic round the centre of the ring moon, $\theta$ is the inclination of the moon's orbit to the plane of the ecliptic, and $\beta$ is the inclination to the line of nodes of a line drawn from the sun to the centre of the lunar orbit. It is in fact $l-☊$ if $l$ be the sun's longitude and ☊ the longitude of the ascending node of the moon's orbit in the ecliptic.

Now the regression of the line of nodes which we are dealing with is the precessional motion of the ring gyroscope formed by the ring moon under the influence of the couple given in (1), and $\theta$ is the inclination of the axis of the ring to the perpendicular to the ecliptic. Then if $n_1$ be the angular speed of the ring,

$$\mathrm{C}n_1 \sin\theta \frac{d\Omega}{dt} = -3n'^2 \mathrm{A} \sin\theta \cos\theta \sin^2(l-\Omega), \quad \ldots\ldots(2)$$

or, since C = 2A,

$$\frac{d\Omega}{dt} = -\frac{3}{4}\frac{n'^2}{n_1}\cos\theta\{1-\cos 2(l-\Omega)\}. \quad \ldots\ldots(3)$$

The first term on the right gives the mean rate of variation of $\Omega$. Denoting it by $-p$, we obtain

$$\frac{d\Omega}{dt} = -p + p\cos 2(l-\Omega), \quad \ldots\ldots(4)$$

or, since $dl/dt = n'$,

$$\frac{d}{dt}(l-\Omega) = n' + p - p\cos 2(l-\Omega). \quad \ldots\ldots(5)$$

We can write this in the form

$$\frac{d(l-\Omega)}{n'+p-p\cos 2(l-\Omega)} = dt,$$

which is integrable at once by the substitution $u = \tan(l-\Omega)$. Thus we obtain

$$\left(\frac{n'+2p}{n'}\right)^{\frac{1}{2}} \tan(l-\Omega) = \tan\{(n'^2+2n'p)^{\frac{1}{2}}t\}. \quad \ldots\ldots(6)$$

Therefore, as $l-\Omega$ increases from 0 to $2\pi$, $(n'^2+2n'p)^{\frac{1}{2}}t$ increases from 0 to $2\pi$. Hence the time taken by the line of nodes to make one revolution relative to the sun is

$$t = \frac{2\pi}{(n'^2+2n'p)^{\frac{1}{2}}}. \quad \ldots\ldots(7)$$

But if a year be taken as the unit of time we have $2\pi/n' = 1$, and we can write

$$t = \frac{2\pi}{n'}\frac{n'}{(n'^2+2n'p)^{\frac{1}{2}}} \quad \ldots\ldots(8)$$

This, as we shall see, is about 18/19 of a year.

The real rate of regression of the line of nodes is therefore the angular speed $(n'^2+2n'p)^{\frac{1}{2}} - n'$, and the time of the revolution is

$$\frac{2\pi}{(n'^2+2n'p)^{\frac{1}{2}} - n'} = \frac{2\pi}{n'}\frac{n'}{(n'^2+2n'p)^{\frac{1}{2}} - n'}, \quad \ldots\ldots(9)$$

that is the number of years is $n'/\{(n'^2+2n'p)^{\frac{1}{2}} - n'\}$. But

$$\frac{n'}{(n'^2+2n'p)^{\frac{1}{2}} - n'} = \frac{1}{\left(1+2\frac{p}{n'}\right)^{\frac{1}{2}} - 1} = \frac{n'}{2p}\left\{\left(1+\frac{2p}{n'}\right)^{\frac{1}{2}} + 1\right\} = \frac{n'}{p} + \frac{1}{2} - \frac{p}{4n'}. \quad (10)$$

Now $$p=\frac{3}{4}\frac{n'^2}{n_1}\cos\theta=\frac{3\pi}{2}\frac{27\cdot 3}{365\frac{1}{4}}\cdot 996$$

and $$\frac{n'}{p}=\frac{4\times 365\frac{1}{4}}{3\times 27\cdot 3\times\cdot 996}=18,\text{ nearly.} \quad\text{.............(11)}$$

Thus the period of revolution of the line of nodes is, in years,

$$\frac{n'}{p}+\frac{1}{2}-\frac{p}{4n'}=18\cdot 5,\text{ nearly.} \quad\text{......(12)}$$

According to Neison, the period of regression is 18·5997 years. [Young, *General Astronomy*, § 455 ]

13. *Estimation of periodic term.* The magnitude of the periodic term may be estimated in the following manner (see Greenhill, *R.G.T.* p. 177). Taking as the mean angular speed of the line of nodes relative to the sun $(n'^2+2n'p)^{\frac{1}{2}}$, and putting $n'+p'$ for this, we write

$$l-\Omega=(n'+p')t-\Delta\Omega.$$

Hence $$\tan\Delta\Omega=\tan\{(n'+p')t-(l-\Omega)\}=\frac{p'\tan(l-\Omega)}{n'+(n'+p')\tan^2(l-\Omega)}, \quad\text{.......(1)}$$

by (6), 12. Multiplying numerator and denominator by $\cos^2(l-\Omega)$, we obtain after reduction

$$\Delta\Omega=\frac{p'\sin 2(l-\Omega)}{2n'+p'-p'\cos 2(l-\Omega)}=\frac{\sin 2(l-\Omega)}{\frac{2n'}{p'}+1},\text{ nearly.} \quad\text{.........(2)}$$

But, by (12), 12, $2n'/p'+1=38$, and so the amplitude of $\tan\Delta\Omega$ is 1/38, or in angle about 1·5°.

The couple producing change of the inclination $i$ of the orbit to the ecliptic is [see (5), 9], $$L=-3n'^2A\sin i\sin(l-\Omega)\cos(l-\Omega),$$

or $$L=-\tfrac{3}{4}n'^2C\sin i\sin 2(l-\Omega). \quad\text{.....(3)}$$

But this is approximately $-Cn_1 di/dt$, and so we obtain

$$\frac{di}{dt}=\frac{3}{4}\frac{n'^2}{n_1}\sin i\sin 2(l-\Omega). \quad\text{.....(4)}$$

Integrating with respect to the time, using the approximation

$$l-\Omega=(n'+p')t,$$

we find $$\Delta i=\frac{3}{8}\frac{n'^2}{n_1(n'+p')}\sin i\cos 2(l-\Omega). \quad\text{.....(5)}$$

The amplitude of this variation of $i$ is about 9′. The addition to $i$ is thus 9′ when the sun is in the line of nodes, and −9′ when $l-\Omega=90°$. Thus the inclination is 5° 18′ in the former case and 5° in the latter. When the sun is halfway between these positions the inclination has the mean value.

**14.** *Effect of equatorial belt of earth on motion of moon's nodes.* The equatorial excess of matter in the earth exerts also an influence on the motion of the moon's nodes on the ecliptic. If the earth were a sphere, either homogeneous or made up of concentric spherical shells each of uniform density, it would have no effect on the position of the line of nodes The earth may be regarded as made up of a homogeneous sphere surrounded by an equatorial ring, and we have only to consider the effect of the latter. The inclination I of the moon's orbit to the plane of the equator varies with the revolution of the nodes of the moon's orbit on the ecliptic from $\omega+i$ to $\omega-i$, through a range in fact of $2i$, or about $10^\circ\ 18'$. In order however to obtain an approximate estimate we suppose I to have its mean value $\omega$.

The mass $m$ of the equatorial ring can be expressed by means of the moments of inertia of the earth. Let the radius of the ring be $r$; then we have, if I be the moment of inertia of the spherical part about a diameter,

$$mr^2+\mathrm{I}=\mathrm{C},\quad \tfrac{1}{2}mr^2+\mathrm{I}=\mathrm{A},$$

and therefore

$$m=2\frac{\mathrm{C}-\mathrm{A}}{r^2},\quad \mathrm{I}=2\mathrm{A}-\mathrm{C}. \qquad (1)$$

From (8), 3, we may take as the couple exerted by one ring on the other, tending to bring them into coincidence,

$$\mathrm{L}=-\frac{3}{2}\kappa\frac{\mathrm{M}}{\mathrm{R}^3}(\mathrm{C}-\mathrm{A})\sin\omega\cos\omega, \qquad (2)$$

or by (1),

$$\mathrm{L}=-\frac{3}{4}\kappa\frac{\mathrm{M}mr^2}{\mathrm{R}^3}\sin\omega\cos\omega, \qquad (3)$$

where M is the mass of the moon.

We eliminate $\kappa$ as before by the relation $n''^2\mathrm{R}^3=\kappa(\mathrm{E}+\mathrm{M})$, and obtain

$$\mathrm{L}=-\frac{3}{2}\frac{n''^2}{1+\rho}(\mathrm{C}-\mathrm{A})\sin\omega\cos\omega, \qquad (4)$$

where $\rho=\mathrm{E}/\mathrm{M}$. This acts on the ring moon, and produces the turning round of the line of nodes exactly as that due to the action of the sun was caused.

The couple producing the turning due to the sun was found to be

$$\mathrm{L}_s=-3n'^2\mathrm{A}_1\sin^2\beta\cos\theta\sin\theta, \qquad (5)$$

where $\mathrm{A}_1$ is the moment of inertia of the ring moon about a diameter. The mean couple is therefore

$$\mathrm{L}_s=-\tfrac{3}{2}n'^2\mathrm{A}_1\sin i\cos i, \qquad (6)$$

where $i$ is the inclination of the moon's orbit to the ecliptic, $n'$ the sun's angular speed round the earth. The mean couple $\mathrm{L}_s$ produces precession of the nodes of amount $\mathrm{L}_s/2\mathrm{A}_1 n_1 \sin i$, and the couple L precession of the nodes of amount $L/2A_1 n_1 \sin\omega$, where $n_1$ is as before the angular speed of spin of the ring moon. If then T be the period in years of revolution of the nodes due to the action of the earth alone, we have

$$\frac{T}{18 \cdot 6} = \frac{\frac{3}{4} n'^2 \cos i}{\frac{3}{2} \frac{n''^2}{1+\rho} \frac{C-A}{2A_1} \cos\omega} = \frac{n'^2}{n''^2} \frac{A_1(1+\rho)}{C-A} \frac{\cos i}{\cos\omega}. \qquad (7)$$

For $A_1/(C-A)$, we may write $C/(C-A) . A_1/C = 306 . MR^2/\frac{2}{5}Er^2$ (that is putting $C = \frac{2}{5}Er^2$, a value which is no doubt too great), so that we obtain

$$\frac{T}{18 \cdot 6} = 306 \frac{5}{4} \frac{n'^2}{n''^2} \frac{E+M}{E} \frac{R^2}{r^2} \frac{\cos i}{\cos\omega}. \qquad (8)$$

Hence $$T = 306 \frac{5}{4} \frac{27 \cdot 3^2}{365^2} 60^2 \frac{\cdot 998}{\cdot 917} 18 \cdot 6 = 156000, \text{ nearly}. \qquad (9)$$

This number is rather too small owing to the value assigned to C. Thus the effect of the earth's equatorial belt in producing motion of the moon's nodes is exceedingly slight.

15. *Remarks on gyrostatic method of calculating motion of moon's nodes.* The gyrostatic determination of the regression of the nodes of the moon's orbit on the ecliptic requires perhaps some justification from first principles. It is stated by Klein and Sommerfeld (*Theorie des Kreisels,* Bd III, S. 644) that the gyroscopic process only gives a mode of calculating the rate of regression of the nodes on the presupposition of the existence of the regression, that in fact an "Existenzbeweis" is wanting, and that we neglect the effect of the eccentricity of the lunar orbit. It is no doubt assumed in this process that the eccentricity of the moon's orbit has no great effect on the motion of the nodes; but this assumption will be seen to be justified when we examine, as we shall now do, how the motion of the nodes arises, and consider the result which the process gives.

With regard to the question of proof of existence, it seems sufficient to urge that, if a legitimate process of computation of the action of forces, which undoubtedly exist, gives as a result such a motion as that of the moon's nodes, no such proof is needed; the effect is a consequence of the configuration of matter and the forces between the different parts. That there must be such a motion follows at once from the forces acting on the moon. The general nature of these has been sufficiently discussed above.

We consider then the moon moving in a circular orbit, the plane of which is inclined at an angle $i$ to the plane of the ecliptic. Through the centre O of the orbit draw two normals OZ, OZ′ at right angles to the plane of the ecliptic and the plane of the orbit respectively. Along OZ′ we may lay off a length equal to $M\omega R^2$, the A.M. of the moon, taken of mass M moving in a circle of radius R with angular speed $\omega$ about O. We take this length as OZ′. Whatever the position of the sun may be, there is during the 27·3 days of a lunation a certain average couple exerted on the moon producing A.M. about the line of nodes. Thus, taking the effect produced in a short interval of time, the result is that the line which at the end of that time represents the A.M. in the orbit is now in a slightly different position, having moved from $OZ'_1$, say, to $OZ'_2$ in the plane containing OZ′, and the line of nodes, that is it has moved towards the axis of the couple. The A.M. generated is very small in comparison with $M\omega R^2$, *and is about an axis at right angles to* OZ′ and in the plane just referred to. The length of OZ′ is not changed, its direction only has been altered through the small angle $Z'_2OZ'_1$, in a plane at right angles to that of the angle $ZOZ'_1$. The angle $ZOZ'_2$ is therefore very nearly equal to Z′OZ, that is the inclination of the plane of the orbit to the plane of the ecliptic has not been perceptibly altered. The line of nodes however is now at right angles to the plane of $ZOZ'_2$, that is, it has turned through the angle between the planes $Z'_1OZ$ and $Z'_2OZ$.

In the same way, in the next lunation the line $OZ'_2$ is displaced to the position $OZ'_3$, and it will be observed that this displacement is in a plane at right angles to the plane $Z'_2OZ$. The displacement is not quite the same as in the former case, as the position of the sun with reference to the moon's orbit has altered, but the effect is the same; the inclination of the plane of the orbit to the ecliptic is not altered, the line of nodes has turned through a further angle in the same direction as before.

This process, which we have regarded as consisting of short finite steps, really goes on continuously, and so the A.M. about OZ′ in its continuously changing position is never altered in amount. The step of change vanishes with the interval of time, and when graphically represented is at right angles to the A.M. previously existing.

When the method set forth is examined it appears that each element of the ring is an infinitesimal representative of the real moon, with some limitations imposed by the assumed rigidity. As in the case of the real moon the couple about the line of nodes varies with the position of the element, but the rigidity condition compels the element to remain in the plane of the ring. In the actual case of the moon, or in that of a free element of a ring of particles (*e.g.* one of Saturn's rings), the production of A.M. about the line of nodes causes small periodic deviations from motion in a plane; but for each element these deviations are nearly the same as those for the moon as a whole. Thus, to the degree of exactness with which the moon's actual orbit is a plane curve, we may suppose the ring of free elements to be rigid. The rapidity of the moon's motion in fact justifies this supposition.

The sun in a revolution round the earth puts the couple on the moon through a cycle of changes, and hence the distribution of the mass of the moon in a ring and that of the sun in another about the same centre, but laid in the ecliptic and of radius equal to that of the sun's apparent orbit, gives an average, though not a quite accurate one, for the changes of position of the moon in a lunation and of the sun in a year. The slow motion of the sun however makes it necessary to take account of the relative motion of the sun and the line of nodes in order that a satisfactory approximation may be obtained, as we have seen above.

Now it seems fairly clear that the eccentricity of the moon's orbit could in no way alter the general result at which we have arrived, and the motion of the line of nodes can legitimately be calculated by regarding the moon as a circular ring revolving like a flywheel in its orbit. It is possible to assign limits between which the error due to the neglect of the eccentricity must lie. The solidity or rigidity of the ring however has no sensible influence on the result: a set of discrete particles having the A.M. $M\omega R^2$ is practically as effective as a flywheel. Currents of liquid flowing round closed channels within a solid body can be, and no doubt are, effective in producing gyrostatic action.

# CHAPTER XI

## FREE PRECESSION OF THE EARTH. FURTHER DISCUSSION

1. *Ellipticities of terrestrial spheroid.* In the last chapter we have seen that the period of the free precession is equal to that of rotation multiplied by the reciprocal of the "ellipticity" of the terrestrial spheroid, that is by A/(C − A). This ellipticity, which we denote by $\epsilon$, is related to the eccentricity $e$ of a principal elliptic section of the spheroid, supposed of uniform density, by a plane containing the axis of symmetry, by the equation

$$e^2 = \frac{2\epsilon}{1+\epsilon}, \qquad (1)$$

which it is easy to prove. For we have $A = \frac{1}{5}M(a^2+c^2)$, $C = \frac{2}{5}Ma^2$, and the result follows.

The equation of the spheroid is $$\frac{x^2+y^2}{a^2} + \frac{z^2}{c^2} = 1, \qquad (2)$$

or, as we may write it, if we suppose $e$ to be small,

$$\frac{1}{r^2} = \frac{1}{c^2}(1 - e^2\cos^2\vartheta), \qquad (3)$$

where $\vartheta$ is the vectorial angle measured from the equator in the plane of section. Since $\epsilon$ is also small, we can write (3) in the form

$$r = c(1+\epsilon\cos^2\vartheta). \qquad (4)$$

This is the original spheroid. Now consider a sphere of the same volume. The radius $r_1 = (a^2c)^{\frac{1}{3}}$. But we have, as the reader may verify,

$$c = r_1(1 - \tfrac{2}{3}\epsilon).$$

Hence (4) becomes $$r = r_1\{1+\epsilon(\cos^2\vartheta - \tfrac{2}{3})\}. \qquad (5)$$

Thus a sphere of the same volume, that is of radius $r_1$, will, when converted into the ellipsoid (4), have the equation (5).

Now, if (4) had another small ellipticity $\epsilon'$, the equation corresponding to (5) would be

$$r = r_1\{1+\epsilon'(\cos^2\vartheta - \tfrac{2}{3})\}\,; \qquad (6)$$

and if the ellipsoid (4) had ellipticity $\epsilon+\epsilon'$ the equation would be

$$r = r_1\{1+(\epsilon+\epsilon')(\cos^2\vartheta - \tfrac{2}{3})\}. \qquad (7)$$

But now, after the ellipticity $\epsilon$ has been imposed with an axis of symmetry OC, let ellipticity $\epsilon'$ be imposed about an axis OC′ inclined to the former at an angle $\alpha$. Any chosen radius vector $r$, inclined at an angle $\vartheta$ to the equator for the axis of symmetry OC, will be inclined at the angle $\vartheta+\alpha$ to the equator for the axis OC′. Hence, by superposition, we have for the resulting surface

$$r = r_1\{1+\epsilon\cos^2\vartheta + \epsilon'\cos^2(\vartheta+\alpha) - \tfrac{2}{3}(\epsilon+\epsilon')\}. \qquad (8)$$

This can also be put in the form (5), so that it is also an ellipsoid of revolution. To find its principal axes we have

$$-\frac{dr}{d\vartheta}=r_1\{2\epsilon\cos\vartheta\sin\vartheta+2\epsilon'\cos(\vartheta+\alpha)\sin(\vartheta+\alpha)\}=0.$$

If $\alpha$ be small this reduces to $\tan 2\vartheta=-\frac{2\epsilon'\alpha}{\epsilon+\epsilon'}$. ....(9)

Now $\vartheta$ is the angle which a principal axis makes with the equatorial plane of the original spheroid, and by the result just obtained is either $-\epsilon'\alpha/(\epsilon+\epsilon')$ or $\frac{1}{2}\pi-\epsilon'\alpha/(\epsilon+\epsilon')$. Thus we get two principal axes, one OC′ as shown in Fig. 59, and another at right angles to OC′ in the plane of the diagram. The axis of symmetry, OC′, for the ellipsoid compounded as explained above is, as shown in the diagram, on the right of the axis of symmetry of the first ellipsoid (that of ellipticity $\epsilon$) and inclined to it at the angle $\epsilon'\alpha/(\epsilon+\epsilon')$.

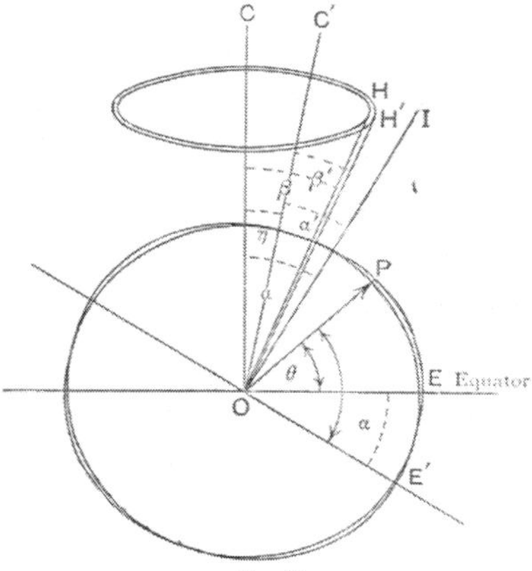

Fig. 59.

**2.** *Period of free precession in terms of ellipticity.* We now suppose that the second ellipticity $\epsilon'$ is produced by rotation about an axis OI, inclined at the angle $\alpha$ to the first axis of symmetry OC. Such ellipticity can only arise through yielding of the body. On the other hand, if there were complete yielding, as in the case of a fluid body, the axis of symmetry of the revolving body would be OI. As it is the angle of deflection of the axis of symmetry is less than $\alpha$ and greater than zero.

Let us now assume for a moment that the body is perfectly unyielding. The period of free precession can be found in two ways:

(1) The axes OH (of resultant A.M.) and OI revolve in the body about the axis of symmetry OC, in the period T. The three axes are and remain in one plane, that of the diagram. The component of angular speed about the axis of symmetry is $\omega\cos\alpha$, if $\omega$ be the resultant angular speed about OI. Hence

$$T=\frac{2\pi}{\omega\cos\alpha}\frac{A}{C-A}=\frac{2\pi}{\omega}\frac{A}{C-A},$$

since $\alpha$ is small.

(2) The radius of the circle described by the extremity H of the vector of A.M. is $H\sin\beta$ (see Fig. 59), and the circumference is $2\pi H\sin\beta$, if H be used also to denote the length of the vector. But the point H moves tangentially to this circle at rate $H\omega\sin(\alpha-\beta)$. Thus

$$T=\frac{2\pi}{\omega}\frac{\sin\beta}{\sin(\alpha-\beta)}=\frac{2\pi}{\omega}\frac{\beta}{\alpha-\beta}, \text{ nearly.}$$

Thus we get

$$\frac{\beta}{\alpha-\beta}=\frac{A}{C-A}=\frac{1}{\epsilon}.$$

**3.** *Positions of axis of figure, instantaneous axis, and axis of resultant* A.M. As the earth turns the position of the axis OI turns in it. Let now the earth yield to the forces called into play by the rotation, and assume that the distribution about

OI is at each instant that which gives the ellipticity $\epsilon'$. The axis of resultant A.M. has not quite the same position in space as it would have if the earth were perfectly unyielding, and we have now angles $\alpha'$, $\beta'$ instead of $\alpha$, $\beta$, and $\epsilon+\epsilon'$ instead of $\epsilon$ [see Fig. 59]. Let $\alpha=\alpha'+\eta$ : writing $\eta=\epsilon'\alpha/(\epsilon+\epsilon')$ we have

$$\therefore \eta = \frac{\epsilon'\alpha}{\epsilon+\epsilon'} = \frac{\epsilon'}{\epsilon+\epsilon'}(\alpha'+\eta) = \frac{\epsilon'}{\epsilon}\alpha'.$$

The axis of symmetry OC′ now also changes its position in the body, always being found in the plane defined by OI and the former axis of symmetry. The period of the free precession is the time of revolution of the new OH in the body about the former axis of symmetry. Thus we get, by Fig. 59 and the same process as before,

$$\mathrm{T}=\frac{\eta+\beta'}{\alpha'-\beta'}\frac{2\pi}{\omega}=\left(\frac{\epsilon'}{\epsilon}\frac{\alpha'}{\alpha'-\beta'}+\frac{\beta'}{\alpha'-\beta'}\right)\frac{2\pi}{\omega}.$$

But, since $\alpha'-\beta'$ is very small in comparison with $\alpha'$ or $\beta'$, we obtain approximately

$$\mathrm{T}=\left(\frac{\epsilon'}{\epsilon}\frac{1}{\epsilon+\epsilon'}+\frac{1}{\epsilon+\epsilon'}\right)\frac{2\pi}{\omega}=\frac{1}{\epsilon}\frac{2\pi}{\omega}.$$

The period of the free precession is thus equal to that of rotation multiplied by the reciprocal of the ellipticity of the non-rotating terrestrial spheroid, that is, the elastic yielding of the body to the rotation has no influence on the period of the free precession.

4. *Influence of internal constitution of the earth.* The ordinary obvious tidal phenomena disprove the old notion that the earth consists of a liquid interior surrounded by a solid crust. If this were the earth's structure, the ebb and flow of the water on the surface would be practically non-existent; the tides would be in the main distortions of the enclosing shell. But as regards precession and nutation, such an earth, if an oblate spheroid covered by a crust possessed of a high degree of elastic rigidity, would behave like a solid spheroid. A perfectly unyielding shell enclosing a fluid interior would, as regards the fortnightly and semi-annual nutations, show a difference of behaviour; but the slow continuous precession, and (less accurately) the nutation depending on the 19 year period of revolution of the nodes of the moon's orbit, would be unaltered. [See Lord Kelvin's paper *loc. cit.* below, also *Popular Lectures*, II, p. 246.]

The fluid theory is however given up by all students of geophysics, and an approximation to the facts of the case is obtained by postulating perfect incompressibility and high elastic rigidity for the whole mass. The earth is not of uniform density, but consists of a dense central part, perhaps to a considerable extent, as has been thought, composed of iron, and an outer shell of relatively smaller density. But it is no doubt at each point practically homogeneous, and therefore possesses the two principal moduli of a homogeneous solid, $k$ the bulk-modulus, and $n$ the shape-modulus or modulus of rigidity. The approximation, stated above, consists then in taking $k$ as infinite and $n$ as comparable with the value of the shape-modulus for steel. This makes Poisson's ratio —the ratio of lateral contraction to longitudinal extension in a laterally free bar under longitudinal pull, which is expressed by $(3k-2n)/(6k+2n)$—have the value $\frac{1}{2}$.

5. *Results of elastic solid theory.* By a theory given by Lord Kelvin* the yielding of the elastic solid earth to the forces applied to it in consequence of its rotation has been calculated, with the result that for a value of the modulus $n$ equal to that for steel the free Eulerian nutation should have a 15 months' period. As the period is 14 months, in spite of the lengthening due to the tide of 14 months' period which the precession must produce, the elasticity of the material of which the earth is composed must, on the whole, be decidedly greater than that of steel.

* *Math. and Phys. Papers*, Vol. III, Art. 45. Love, *Elasticity*, second edition, Chap. X.

We give a brief account, mainly of results, of Lord Kelvin's discussion, referring the reader to the sources mentioned above for further information. He may also refer to a version of this investigation given in Klein and Sommerfeld's *Theorie des Kreisels*, in which several simplifications are made, which considerably shorten the analysis and do not seriously affect the conclusions.

There are two main physical actions which resist the distortion of a rotating body, caused by the forces arising from the inertia of the matter, the so-called centrifugal forces. These are the mutual gravitation of the parts and the elasticity of shape. Compressibility has but little effect. The first can be dealt with by supposing the earth to be fluid. If it is also taken as of uniform density, then we have, by the theory of the equilibrium of an oblate ellipsoid of revolution rotating slowly about its axis of symmetry, for the small ellipticity, $\epsilon$, produced from the spherical form by rotation, the equation

$$\epsilon = \frac{5}{4}\frac{\omega^2 R}{g}, \qquad (1)$$

where $\omega$ is the angular speed of rotation, R the radius of the sphere, and $g$ the acceleration due to gravity at the surface of the sphere. This result was given by Clairaut, in his treatise, *La Figure de la Terre*, 1743. See also Thomson and Tait, *Nat. Phil.* §§ 794, 800.

The proof is not difficult. It can be simplified somewhat by taking the potential at an external point, as that due to a homogeneous sphere encircled by an equatorial ring, of such mass $m$ as to make the ellipticity, $(C-A)/A$, calculated from the moments of inertia, what it is in the actual case. We have $m = 2(C-A)/R^2 = 2A\epsilon/R^2$.

Next the ellipticity due to elastic yielding is to be found. The theoretical discussion leads to an ellipticity

$$\epsilon = \frac{15}{38}\frac{\rho\omega^2 R^2}{E}, \qquad (2)$$

where E denotes the Young's modulus and $\rho$ the density, which is supposed to be the same throughout the sphere, and equal to the mean density, 5·5, of the earth. In the present case E is three times the rigidity modulus.

We denote these two ellipticities by $\epsilon_1$, $\epsilon_2$. It is important to notice that, if experiments are made by placing a sphere on a whirling table, the ellipticity $\epsilon_1$ is, in consequence of the smallness of the sphere, in comparison with the earth for example, the only one which discloses itself. For we have

$$\epsilon_1 = \frac{19}{6}\frac{E}{\rho R g}\epsilon_2, \qquad (3)$$

and as $g$ is now the gravitational acceleration at the surface due to the matter of the experimental sphere, and E in C.G.S. units is about $2{\cdot}2 \times 10^{12}$ for structural steel, it is clear that $\epsilon_1$ is the only ellipticity of sensible amount. It is otherwise in the case of the earth, where $R = 2 \times 10^9/\pi$ (cms.), and $g$ has $R/r$ (that is $2 \times 10^9/\pi r$) times the value it has for a sphere $r$ cms. in radius and of the same mean density. Thus, calculating for the earth, we get

$$\epsilon_1 = \tfrac{1}{231}, \quad \epsilon_2 = \tfrac{1}{463}. \qquad (4)$$

Thus $\epsilon_1 = 2\epsilon_2$, nearly.

**6.** ***Period of free precession for earth as rigid as steel.*** Now however the question arises whether both of these ellipticities are to be attributed to the earth. The earth was no doubt originally at such a high temperature that it could have no rigidity. It would therefore take an oblate figure of revolution, such that the centrifugal forces were balanced by gravitation. In that state it solidified, and the balance was not disturbed. On this supposition there are no elastic forces in play in the rotating earth so long as the angular speed remains that corresponding to the ellipticity.

There is a discrepance between the actual ellipticity, about 1/300, and that, 1/231, calculated above. But the latter was found on the supposition of uniform density, and it is fairly clear that the surface ellipticity of the earth, if the density is much higher toward the centre than at the surface, must be less than the value for uniform density.

This theory of the earth with centrifugal forces balanced by gravitation, and free from elastic strain, is no doubt very imperfect. But assuming it we have now, in order to calculate the free precession, to inquire what ellipticity the earth would assume if the rotation were stopped. [See 5 above.] The rotation acting against elastic forces alone would produce ellipticity $\epsilon_2$. Let $\epsilon_3$ be the ellipticity arrived at. The elastic forces called into play are those which correspond to the difference $\epsilon_1 - \epsilon_3$, and balance the gravitational forces which remain for $\epsilon_3$. The mathematical theory referred to above gives for this difference the equation

$$\epsilon_1 - \epsilon_3 = \epsilon_1\epsilon_2/(\epsilon_1 + \epsilon_2), \quad \text{or} \quad \epsilon_3 = \epsilon_1^2/(\epsilon_1 + \epsilon_2).$$

Therefore the free period of precession, calculated for the elasticity of steel, is $306(1 + 231/465)(=458)$ days, nearly, taking 306 days for an unyielding earth. This is over 15 months. The elasticity of the earth appears therefore to be greater than that of steel.

It will be observed that the supposition that the earth, in its state of rotation, is not under elastic strain, does not mean that the period of free precession is independent of elastic yielding. For that period has the value calculated from the figure taken when the rotation is zero, and here the elastic yielding is of importance. The factor $1 + 231/465$ expresses the effect of such yielding.

7. *Rise or fall of earth's surface for Eulerian precession.* As we have seen in 15, I, the mean displacement of the axis OI in the Eulerian precession is about 4 metres (13 feet) on the surface of the earth. In order to calculate what rise or fall of the earth's surface takes place in consequence, we shall consider the change of position of the principal axis OC through the institution of rotation $\omega$ about an axis OI such that the distance CI, on the earth's surface, is 4 metres. By 1 above we have here $\alpha = \pi 4/(2\times 10^7)$, since the earth's radius in metres is $2\times 10^7/\pi$. Hence [Fig. 59] $\alpha = 2\pi/10^7$. The initial ellipticity is $\epsilon$, say. The new ellipticity is $\epsilon_1$, that is the ellipticity $\epsilon_1 - \epsilon$ has been added. We call this $\epsilon'$. Now, by (9), 1, we see that the principal axis has been shifted through the angular distance $\alpha\epsilon'/(\epsilon+\epsilon') = 2\pi\epsilon'/(\epsilon+\epsilon')\times 10^7$ by the value just obtained for $\alpha$. Thus we can find $\epsilon'$; for by 2 we have, in days, $1/\epsilon$ for the actual period of variation of latitude and $1/\epsilon_1$ for the Eulerian period. Hence $\epsilon/\epsilon_1 = \epsilon/(\epsilon+\epsilon') = 5/7$, and $\epsilon' = \frac{2}{5}\epsilon$.

If now we take a radius vector in latitude $\vartheta$, we get by (5), 1, for the change of radius

$$r_1\epsilon'\{\cos^2\vartheta - \cos^2(\vartheta + \alpha)\} = r_1\epsilon'\alpha \sin 2\vartheta.$$

This is a maximum if $\vartheta = 45°$. Hence, for this latitude the rise or fall is, in millimetres, since $\epsilon = 1/428$,

$$r_1 \frac{2}{5}\frac{1}{428}\frac{2\pi}{10^4} = \frac{4}{1\cdot 07}.$$

8. *Positions of earth's principal axes as affected by annual transfers of matter.* The annual melting and re-formation of polar ice and snow no doubt produce a sensible, if small, displacement of the earth's principal axes of moment of inertia. Other efficient causes of such changes are to be found in ocean currents, and in the periodic changes in the atmosphere surrounding the earth. The atmosphere is carried round by the earth's rotation, and large changes of this outside rotating mass will react on the central body. Atmospheric pressure is higher in winter than in summer, and so there is an excess of mass of air over the northern hemisphere in winter, and a defect over the southern hemisphere. This excess and defect are very considerable, amounting on the average, according to Spitaler, to the equivalent of about 300 cubic kilometres of mercury [Petermann's *Mitteilungen*, 137 (1901)]. See also H. Jeffrey's papers *Month. Not.* June 1915 and April 1916, where the annual term is explained by systematic high barometric pressure in winter over Asia.

To estimate such changes we must notice that as the result of the transfer of a mass $\mu$ from coordinates $X_0$, $Z_0$ to coordinates X, Z, giving rise as it does to changes $a$, $b$, $c$ of moments of inertia, and producing products of inertia $d$, $e$, $f$, the equation of the momental ellipsoid, supposed originally symmetrical about the axis of $z$, becomes

$$(A+a)x^2+(A+b)y^2+(C+c)z^2-2dyz-2ezx-2fxy=1. \quad \ldots\ldots(1)$$

Of this surface the principal axes are given by the equations

$$(A+a-\kappa)x-fy-ez=0, \quad -fx+(A+b-\kappa)y-dz=0, \quad -ex-dy+(C+c-\kappa)z=0. \ldots(2)$$

The elimination of $x$, $y$, $z$ gives the usual cubic for $\kappa$.

Supposing that $\kappa$ has a value which satisfies the cubic equation, we may find values of $x$, $y$, $z$ which, if we choose the scale of magnitude so that $x^2+y^2+z^2=1$, may be taken as the direction cosines of the principal axes. If the axes are only slightly shifted by the transfer of matter, we may, in considering presently the axis OC, put $x=y=0$, $z=1$ in the small terms of the formulae. Thus, in (2), $ex$, $dy$, $fx$, $fy$ are small terms. The third and first of (2) give $\kappa=C+c$, and

$$(A-C)x=e, \quad (A-C)y=d. \quad \ldots\ldots(3)$$

Of course $$a=\mu(Y^2+Z^2-Y_0^2-Z_0^2), \ldots, \quad d=\mu(YZ-Y_0Z_0)\ldots, \quad \ldots\ldots(4)$$

so that $a$, $b$, $c$, $d$, $e$, $f$ can easily be calculated in any given case.

Let now the transference be along the meridian in the plane AOC, then, denoting the latitude by $\lambda$, we have $X=R\cos\lambda$, $Y=R\sin\lambda$, $d=0$,

$$e=\mu(XZ-X_0Z_0)=\tfrac{1}{2}\frac{\mu}{R^2}(\sin 2\lambda-\sin 2\lambda_0). \quad \ldots\ldots(5)$$

Let $\lambda_0=\frac{1}{2}\pi$ and $\lambda=\frac{1}{4}\pi$, so that the matter is transferred from the poles to latitude 45°; we have

$$e=\tfrac{1}{2}\frac{\mu}{R^2}. \quad \ldots\ldots(5')$$

By (3), $x=\frac{1}{2}\mu A/(A-C)\,.\,1/AR^2$. But $A/(A-C)$ is for the earth about $-305$, and A about $\frac{1}{3}M/R^2$, where M is the earth's mass. Thus

$$x=-457\frac{\mu}{M}. \quad \ldots\ldots(6)$$

This is the angle through which the axis OC has been turned by the transfer, and the turning is, as it should be, towards the equator. For a turning of a single second of angle, that is for $x=\pi/180\times 3600$, we have (irrespective of sign)

$$\mu=\frac{\pi M}{180\times 3600\times 457}=10^{-8}M. \quad \ldots\ldots(7)$$

This is about the quantity of matter contained within a sphere of 17 miles in diameter, and of the same average density as the earth, or, if the density is that of water, a sphere of 30 miles diameter. This quantity of matter would suffice for an ice cap of radius 5°, or about 350 miles, about 65 yards in thickness.

9. *Effect of annual transfers of matter in accentuating free precession.* It will now be evident that the positions of the principal axes of the earth may be regarded as affected to a determinate, if not large, extent by transference of matter brought about by annual meteorological changes. We can now establish equations for any case in which a portion of matter, say of mass $\mu$, undergoes progressive change of position, and shall find that of the two parts into which the changes of A.M. fall, the part due to change of position of the principal axes, and the part due to the motion of the shifting mass, the former is the more important. We shall suppose that the axes chosen are the principal axes of an invariable part, together with the mass $\mu$ supposed situated, at rest relative to the earth, at a point of coordinates $X_0$, $Y_0$, $Z_0$, and denote the moments of inertia for that distribution by A, B, C. The products of inertia

for that distribution are zero. We denote by $u$, $v$, $w$ the components of A.M. due to the motion of the mass $\mu$ relative to the earth. If the coordinates of $\mu$ be X, Y, Z we have by (4), 2, II,

$$u=\mu(\text{Y}\dot{\text{Z}}-\text{Z}\dot{\text{Y}}), \quad v=\mu(\text{Z}\dot{\text{X}}-\text{X}\dot{\text{Z}}), \quad w=\mu(\text{X}\dot{\text{Y}}-\text{Y}\dot{\text{X}}). \qquad (1)$$

But in consequence of its motion with the earth the mass $\mu$ has components of A.M.

$$ap-fq-er, \quad -fp+bq-dr, \quad -ep-dq+cr,$$

where $a=\mu(\text{Y}^2+\text{Z}^2-\text{Y}_0^2-\text{Z}_0^2), \ldots, \quad d=\mu(\text{YZ}-\text{Y}_0\text{Z}_0), \ldots.$

The total components, $u'$, $v'$, $w'$, say, of the A.M. of the travelling mass are thus

$$u'=u+ap-fq-er, \quad v'=v-fp+bq-dr, \quad w'=w-ep-dq+cr,$$

or, if we neglect the terms in $p$ and $q$ in comparison with those in $r$,

$$u'=u-er, \quad v'=v-dr, \quad w'=w+cr. \qquad (1')$$

The equations of motion are, if $\text{L}=\text{A}p$, $\text{M}=\text{B}q$, $\text{N}=\text{C}r$,

$$\left.\begin{aligned}\frac{d}{dt}(\text{L}+u')+q(\text{N}+w')-r(\text{M}+v')&=0,\\ \frac{d}{dt}(\text{M}+v')+r(\text{L}+u')-p(\text{N}+w')&=0,\\ \frac{d}{dt}(\text{N}+w')+p(\text{M}+v')-q(\text{L}+u')&=0,\end{aligned}\right\} \qquad (2)$$

since there are no external forces. These become, if A=B and terms in $w'$ are neglected,

$$\left.\begin{aligned}\text{A}\dot{p}-(\text{A}-\text{C})qr&=-\frac{d}{dt}(u-er)+r(v-dr),\\ \text{A}\dot{q}-(\text{C}-\text{A})rp&=-\frac{d}{dt}(v-dr)-r(u-er),\\ \text{C}\dot{r}\qquad\qquad&=-\frac{d}{dt}(w+cr).\end{aligned}\right\} \qquad (3)$$

Here it is to be noticed that by simply omitting the terms in $r$, we obtain the equations for the case in which $u$, $v$, $w$ are the total components of A.M. of $\mu$.

If we put $z=p+iq$, $-\xi=u'+iv'$, we can write the first two equations of (3) in the single equation

$$\text{A}\dot{z}+i(\text{A}-\text{C})rz-\dot{\xi}-ir\xi=0. \qquad (4)$$

where of course $z$ is not to be confounded with the $z$ used in 8 above.

Now if, as we suppose, the changes are unaccompanied by the action of external forces, the direction of the axis OH of resultant A.M. is not changed in space, nor is the value H of this A.M. affected. Also both $p$ and $q$ are small, and so the instantaneous axis is, and remains, nearly coincident with the axis OC and with OH. Thus $r$ is nearly equal to $\omega=(p^2+q^2+r^2)^{\frac{1}{2}}$, and since in the circumstances $\omega$ can only change slightly the variation of $r$ is small. Neglecting this in small quantities of the second order we have, putting $\omega$ for $r$,

$$\text{A}\dot{z}+i(\text{A}-\text{C})\omega z-\dot{\xi}-i\omega\xi=0, \qquad (5)$$

where $\omega$ is to be treated as a constant. A forced periodic change in $z$ will accompany the periodic variation in $\xi$, and $z$ will also be subject to periodic change in the free period of variation of position of the earth's instantaneous axis. The transport of matter will give the necessary disturbance for the production of oscillations, which will then proceed until they are damped out by friction. If $2\pi/c$ be the forced period and $2\pi/n$ the free period, we have the solutions

$$\xi=\alpha\epsilon^{ict}, \quad z=ae^{ict}, \quad z=be^{int}. \qquad (6)$$

By substitution in the differential equation we obtain

$$a=\alpha\frac{c+\omega}{\text{A}c-(\text{C}-\text{A})\omega}=\alpha\frac{c+\omega}{\text{A}(c-n)}, \qquad (7)$$

since $n=(\text{C}-\text{A})\omega/\text{A}$.

Likewise $\xi = a'e^{-ict}, \quad z = a'e^{-ict}$ ........ ........ . . . ...(8)

is a forced solution which gives

$$a = a'\frac{c-\omega}{Ac+(C-A)\omega} = a'\frac{c-\omega}{A(c+n)} \quad \text{..........} \quad (9)$$

The complete solution is the sum of the two forced vibrations and the free vibration.

The coefficients $a$, $a'$ are complex; putting $a=\beta+i\gamma$, $a'=\beta'+i\gamma'$, so that $\beta$, $\beta'$, $\gamma$ $\gamma'$ are real, we find finally for the most general forced vibrational solution

$$Az = (\beta+i\gamma)\frac{c+\omega}{c-n}(\cos ct + i\sin ct) + (\beta'+i\gamma')\frac{c-\omega}{c+n}(\cos ct - i\sin ct). \quad \text{.......(10)}$$

This gives

$$\left.\begin{aligned} Ap &= \left(\frac{c+\omega}{c-n}\beta + \frac{c-\omega}{c+n}\beta'\right)\cos ct - \left(\frac{c+\omega}{c-n}\gamma - \frac{c-\omega}{c+n}\gamma'\right)\sin ct, \\ Aq &= \left(\frac{c+\omega}{c-n}\beta - \frac{c-\omega}{c+n}\beta'\right)\sin ct + \left(\frac{c+\omega}{c-n}\gamma + \frac{c-\omega}{c+n}\gamma'\right)\cos ct. \end{aligned}\right\} \quad \text{...........(11)}$$

To complete the solution we have to take account of the free vibration given by

$$(\beta+i\gamma)(\cos nt + i\sin nt).$$

Thus we have to add to the expression for $Ap$ the terms

$$\beta\cos nt - \gamma\sin nt,$$

and to the expression for $Aq$ the terms

$$\beta\sin nt + \gamma\cos nt.$$

From (10) we can trace the effect of near coincidence of the forced and free periods on the amplitude of $Ap$ and $Aq$. Take the solution $z = ae^{ict}$, so that only the first terms in the brackets on the right of (11) come into the account. Let the period $2\pi/\omega$ be that of the earth's rotation, and $2\pi/c$ be 12 while $2\pi/n$ is 14, when a month is the unit of time. We find that $(c+\omega)/(c-n)$ is about 7 times $(c+\omega)/c$, that is 7 times the value it would have if $n$ were very small. For the solution $a'e^{-ict}$ the amplitude is about 7/13 of what it would be if $n$ were very small. The amplitudes of the former solution are thus greatly magnified by the approximation of $n$ to $c$.

**10.** *Comparison of terms.* The values of $u$, $v$, due to the annual displacement relative to the earth, are small in comparison with the terms $\omega d$, $\omega e$; for the frequency of the earth's rotation, $\omega/2\pi$, is about 366 times that of the annual change. This can be easily proved by (1), 9, assuming that the coordinates X, Y of the mass $\mu$ are

$$X = X_0 + h\sin ct, \quad Y = Y_0 + k\sin ct, \quad Z = Z_0. \quad \text{..........................(1)}$$

Then again $\omega^2 d$, $\omega^2 e$ are great in comparison with $\omega d$, $\omega e$, and so the equations of motion become

$$A\dot{p} - (A-C)qr = -\omega^2 d, \quad A\dot{q} - (C-A)rp = \omega^2 e, \quad \text{.........................(2)}$$

or by (3), 8,

$$\frac{\dot{p}}{\omega} + n\left(\frac{q}{\omega} - y\right) = 0, \qquad \frac{\dot{q}}{\omega} - n\left(\frac{p}{\omega} - x\right) = 0, \quad \text{...........................(2')}$$

where $n=(C-A)\omega/A$. Now by (3), 8, $x$, $y$ are, with reference to the original axes OA, OB, the direction cosines of the principal axis which has superseded OC in consequence of the transference of matter, and $p/\omega$, $q/\omega$, 1, are the direction cosines of the instantaneous axis. Denoting the two last by $\lambda$, $\mu$, we get

$$\dot{\lambda} + n(\mu - y) = 0, \quad \dot{\mu} - n(\lambda - x) = 0. \quad \text{...........................(3)}$$

This equation is due to Sir George Darwin [*Phil. Trans. R.S.* 167 (1877)].

Obviously by the theory of revolving axes the interpretation of these equations is that the axis of rotation is turning at each instant with angular speed $n$, in the clockwise direction, round the instantaneous position of the changing principal axis of moment of inertia.

Instead of the two equations (3) we may write (as we no longer require $z$ in either of its former significations)

$$\dot{\zeta} - in\zeta + inz = 0, \qquad \ldots (4)$$

where $\zeta = \lambda + i\mu$, $z = x + iy$. Putting

$$\left.\begin{aligned} \zeta = \alpha e^{ict}, \quad z = a e^{ict}, \\ \alpha = \frac{n}{n-c} a. \end{aligned}\right\} \qquad \ldots (5)$$

we get

Now let $a = \beta + i\gamma$, and we get

$$z = (\beta + i\gamma)(\cos ct + i \sin ct), \qquad \ldots (6)$$

that is

$$x = \beta \cos ct - \gamma \sin ct, \quad y = \beta \sin ct + \gamma \cos ct. \qquad \ldots (6')$$

Hence

$$\zeta = \frac{n}{n-c}(\beta + i\gamma)(\cos ct + i \sin ct), \qquad \ldots (7)$$

and we have values of $\lambda$, $\mu$ corresponding to $x$, $y$, which are simply $x$, $y$ multiplied by $n/(n-c)$.

Another solution is

$$\zeta = \alpha e^{-ict}, \quad z = a e^{-ict}, \ldots \qquad \ldots (8)$$

with the condition, given by the differential equation,

$$\alpha = \frac{n}{n+c} a. \qquad \ldots (9)$$

From this we get

$$z = (\beta' + i\gamma')(\cos ct - i \sin ct), \quad \zeta = \frac{n}{n+c}(\beta' + i\gamma')(\cos ct - i \sin ct).$$

Thus, combining the solutions, we obtain

$$\left.\begin{aligned} x &= (\beta + \beta')\cos ct - (\gamma - \gamma')\sin ct, \quad y = (\beta - \beta')\sin ct + (\gamma + \gamma')\cos ct, \\ \lambda &= \frac{n}{c^2 - n^2}[\{(c+n)\gamma + (c-n)\gamma'\}\sin ct - \{(c+n)\beta - (c-n)\beta'\}\cos ct], \\ \mu &= \frac{-n}{c^2 - n^2}[\{(c+n)\gamma - (c-n)\gamma'\}\cos ct + \{(c+n)\beta + (c-n)\beta'\}\sin ct]. \end{aligned}\right\} \qquad \ldots (10)$$

**11.** *Numerical illustrations.* As an example, Klein and Sommerfeld take the case of a simple harmonic variation of position of the third axis of movement of inertia. For this we may suppose $\gamma = \gamma' = 0$, $\beta = -\beta'$. We get $x = 0$, $y = 2\beta \sin ct$. Hence

$$\lambda = -\frac{cn}{c^2 - n^2} 2\beta \cos ct, \quad \mu = -\frac{n^2}{c^2 - n^2} 2\beta \sin ct. \qquad \ldots (1)$$

We may take $c/n = 14/12$, roughly. Hence, on the supposition that the transport of matter takes place in the yearly period, we have

$$\lambda = -3{\cdot}2 \,.\, 2\beta \cos ct, \quad \mu = -2{\cdot}8 \,.\, 2\beta \sin ct. \qquad \ldots (2)$$

The period $2\pi/c$ of the driving "pendulum" is here shorter than the free period of the driven, and so we have opposition of phase. If the reverse were the case, that is if $n > c$, we should have agreement of phase.

The result in (2) is shown graphically in Fig. 60 ($a$). The numerals show corresponding points in the displacements of the matter and of the axis. These displacements are in opposite phases.

For a perfectly unyielding earth the period would be about 10 months. With this (1) would give

$$\lambda = 2{\cdot}7 \,.\, 2\beta \cos ct, \quad \mu = 3{\cdot}3 \,.\, 2\beta \sin ct, \qquad \ldots (3)$$

a result which is illustrated in Fig. 60 ($b$). The displacements of matter and of the axis are now in the same phase.

It will be noticed that if $c$ were very small, that is if the period of the transfer of matter were very great, we should have, instead of (1),

$$\lambda = \frac{c}{n} 2\beta \cos ct, \quad \mu = 2\beta \sin ct. \qquad \text{....(4)}$$

On the other hand, if $c$ were very great in comparison with $n$, that is if the forced period were very small in comparison with the free period, we should have

$$\lambda = -\frac{n}{c} 2\beta \cos ct, \quad \mu = -\frac{n^2}{c^2} 2\beta \sin ct. \qquad \text{....(5)}$$

Thus in the former case $\lambda$ would be very small, while $\mu$ would be identical with $y$, that is the curve for the rotation axis would agree with the curve of transfer of the inertia axis.

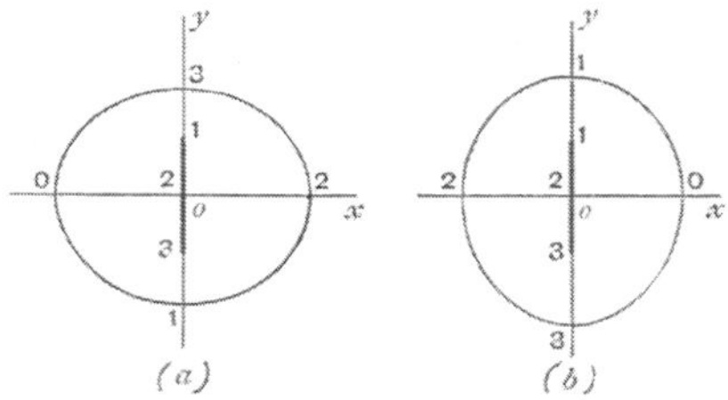


FIG. 69.

In the latter case neither $\lambda$ nor $\mu$ would be appreciable, though $\mu$ would be of a higher degree of smallness than $\lambda$. Hence a very short forced period would have practically no effect in disturbing the axis of rotation. The magnification of amplitude due to the approximation of the periods, 12 to 14 in one case and 12 to 10 in the other, is very obvious by comparison of (2) and (3) with (4) and (5).

In the case of a displacement of the inertia axis represented by an elliptic curve, we can resolve this into two linear components, and find for each an elliptic curve for the corresponding progressive displacement of the rotation axis. Thus we get two elliptic curves which compounded give the path of the rotation axis. This would give, according to (10), a curve for the general case, depending on the different components and their phases.

**12.** *Systematic observations of variation of latitudes at different observatories.* Since the publication of the chart given in 15, I above, for the five years from 1890 to 1895, many observations of latitude have been made, with the result that all the previous conclusions as to period, etc., have been confirmed and made more precise.

In the first place, observations have been made at a number of stations on the parallel 39°·8 of north latitude. These stations and their longitudes are

| | |
|---|---|
| Mizusawa, Japan, - - - - - - - - | $\lambda = -141°\ 8'$ |
| Tschardgui, Turkestan, - - - - - - - | $-\ 61\ \ 29$ |
| Carloforte, Sardinia, - - - - - - - - | $-\ \ 8\ \ 19$ |
| Gaithersburg, U.S.A., - - - - - - - - | $+\ 77\ \ 12$ |
| Cincinnati, „ - - - - - - - - | $+\ 84\ \ 25$ |
| Ukiah, „ - - - - - - - - | $+123\ \ 13$ |

The sign − indicates east longitude, the sign + west longitude.

This arrangement for systematic observations at stations differing considerably in longitude enables simultaneous control observations to be made at a number of different stations. Take, for example, the station Ukiah in the United States and Tschardgui in Turkestan. These differ in longitude by 184° 42′, that is they lie nearly in the same plane through the earth's axis. Hence, if a diminution of latitude is observed at one station, an equal increase of latitude ought to be observed at the other.

Again, the two stations Carloforte in Sardinia and Cincinnati are 92° 44′ distant in longitude. Hence, when there is a maximum or minimum of latitude at one of these stations, there ought to be zero change at the other.

The importance of such simultaneous observations was made clear in 1891, when a German expedition was sent to Honolulu to make simultaneous observations of apparent

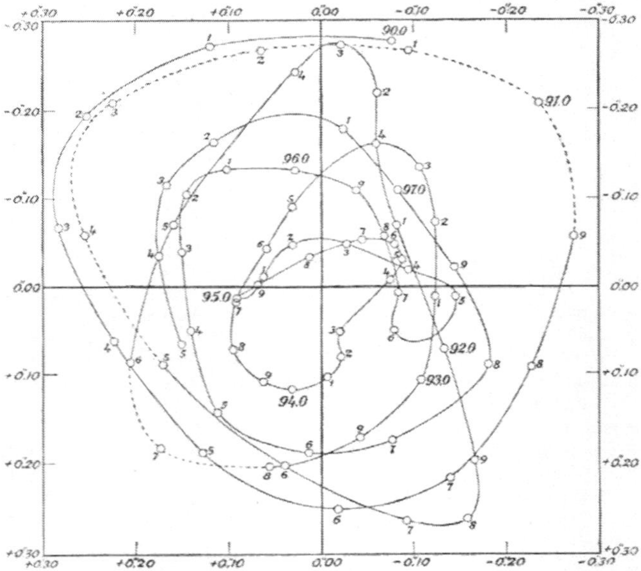


Albrecht, *Astr. Nachr.* No. 3489; *Nature*, May 12, 1898.

Fig. 63.

latitude in conjunction with observers in Berlin. Honolulu lies about 171° west of Berlin, so that the two places are about $4\frac{1}{2}°$ on the two sides of a mean meridian. Thus a maximum at Berlin should coincide with a minimum at Honolulu, and *vice versa*. Fig. 58 (p. 213), which shows some of the results obtained at the two places, proves in a very striking manner that this is the case. The changes of latitude observed, it will be seen, vary in magnitude from zero to ·3 of a second of angle, and there can be no doubt, from the striking agreement of the maxima and minima, that the variation of latitude does take place, and is measured with great accuracy.

For each station in the list given above very systematic series of observations have been carried out for several years. The difference from the mean latitude for each station is equated to the form $x\cos\lambda + y\sin\lambda + z$ for each $\frac{1}{10}$ year, where $x$ denotes the movement of the pole on the Greenwich meridian, and $y$ the movement in the perpendicular direction. It will be clear that for two stations differing by 180° in $\lambda$, $\cos\lambda$ and

$\sin\lambda$ will have opposite signs, while if $\lambda$ differs by $\frac{1}{2}\pi$ at the stations, the value $x\cos\lambda+y\sin\lambda$ at one will correspond to $-x\sin\lambda+y\cos\lambda$ at the other. The term $z$ has caused a good deal of discussion: it is believed to be due partly to meteorological causes.*

The results are given in the *Resultate des Internationalen Breitendienstes* published by the Zentralbureau der Internationalen Erdmessung, Berlin.†

**13.** *Diagrams and tables of later results.* Albrecht's diagram of the motion of the pole for the five years 1890–1895 is given on p. 15. It is repeated here for comparison with his diagram for the six years 1906–1912, which we also give.

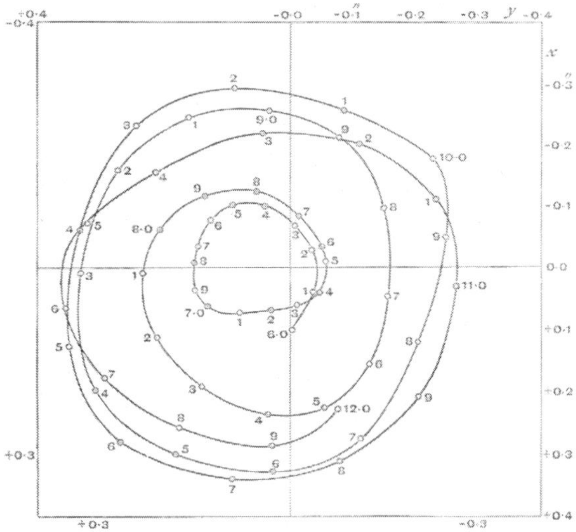


Fig. 62.

The following table kindly supplied by the Astronomer Royal, Sir Frank Dyson, gives the dates of maximum variation of latitude for Greenwich during the last twenty years:

| | | | |
|---|---|---|---|
| 1897·2 | 1·1 | 1907·8 | 1·2 |
| 8·3 | 1·2 | 9·0 | 1·2 |
| 9·5 | 1·1 | 10·2 | 1·1 |
| 1900·6 | 1·5 | 11·3 | 1·1 |
| 2·1 | 1·0 | 12·4 | 1·4 |
| 3·1 | 1·2 | 13·8 | 1·3 |
| 4·3 | 1·1 | 15·1 | 1 05 |
| 5·4 | 1·0 | 16·15 | |
| 6·4 | 1·2 | | |

The differences in the second columns give the intervals between the successive maxima. The mean interval, it will be seen, is 1·172 year, or 428 days.

* The $z$ term is discussed by H. S. Jones in *The Observatory*, Feb. 1916.

† See also *Astronomische Nachrichtung*, 192 (1912).

The curve of Fig. 63 shows the variation of latitude at Greenwich for the interval from 1911·7 to 1916·0. It will be seen that the interval of 4 years and 3½ months includes

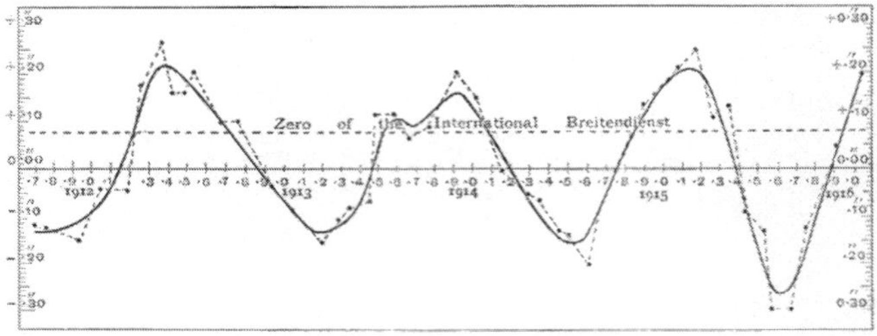


FIG. 63.

rather over 3½ periods. The observations made at Greenwich are systematic and complete, and the co-operation of another observatory 90° distant in longitude would suffice to enable a complete and accurate record of the variation to be obtained.

# CHAPTER XII

## CALCULATION OF THE PATH OF THE AXIS OF A TOP BY ELLIPTIC INTEGRALS

1. *Euler's parameters.* It has been proved in 2, IV that if the system of axes O(A, B, C) be regarded as turned to its position from coincidence with O($x$, $y$, $z$) by rotation about an axis OK, the angle K$zy$ is $\frac{1}{2}(\phi-\psi)$. Now consider the triangle K$zx$ on the unit sphere, centre O. If the direction angles of OK with reference to O($x$, $y$, $z$) be $a$, $b$, $c$, the sides K$x$, $xz$, $z$K of the triangle are $a$, $\frac{1}{2}\pi$, $c$, and we get at once, by the fundamental formula of spherical trigonometry,

$$\cos a=\sin c\sin\tfrac{1}{2}(\phi-\psi),\quad \cos b=\sin c\cos\tfrac{1}{2}(\phi-\psi). \quad\text{.........(1)}$$

But from the spherical triangle K$z$C, which has its two sides K$z$, KC each equal to $c$, the base ZC equal to $\theta$, and the angle, $\vartheta$, between the two equal sides, we get also

$$\sin\tfrac{1}{2}\theta=\sin c\sin\tfrac{1}{2}\vartheta.$$

Eliminating $\sin c$ from each of (1) by this relation, we find

$$\left.\begin{aligned}\cos a\sin\tfrac{1}{2}\vartheta&=\sin\tfrac{1}{2}\theta\sin\tfrac{1}{2}(\phi-\psi),\\ \cos b\sin\tfrac{1}{2}\vartheta&=\sin\tfrac{1}{2}\theta\cos\tfrac{1}{2}(\phi-\psi).\end{aligned}\right\} \quad\text{.........(2)}$$

The isosceles triangle K$z$C gives, since $\angle$ K$z$C $=\frac{1}{2}(\pi-\phi-\psi)$, the two relations,

$$\left.\begin{aligned}\cos c\sin\tfrac{1}{2}\vartheta&=\cos\tfrac{1}{2}\theta\sin\tfrac{1}{2}(\phi+\psi),\\ \cos\tfrac{1}{2}\vartheta&=\cos\tfrac{1}{2}\theta\cos\tfrac{1}{2}(\phi+\psi).\end{aligned}\right\} \quad\text{.........(3)}$$

It is usual to denote the left-hand sides of equations (2) and (3) by $\xi$, $\eta$, $\zeta$, $\chi$. They are in fact parameters used by Euler for the specification of the coordinates $x_1$, $y_1$, $z_1$ of a point with reference to the axes O(A, B, C) fixed in the moving body, in terms of the coordinates $x$, $y$, $z$ of the point with reference to axes fixed in space. But in 1, IV we have used $\xi$, $\eta$, $\zeta$ for the coordinates there specified. As $\varpi$ is no longer required for the temporary signification given to it in 2, IV, we take for these four parameters the symbols $\varpi$, $\rho$, $\sigma$, $\tau$, and write

$$\left.\begin{aligned}\varpi&=\sin\tfrac{1}{2}\theta\sin\tfrac{1}{2}(\phi-\psi),\quad \rho=\sin\tfrac{1}{2}\theta\cos\tfrac{1}{2}(\phi-\psi),\\ \sigma&=\cos\tfrac{1}{2}\theta\sin\tfrac{1}{2}(\phi+\psi),\quad \tau=\cos\tfrac{1}{2}\theta\cos\tfrac{1}{2}(\phi+\psi).\end{aligned}\right\} \quad\text{.........(4)}$$

It will be observed that

$$\varpi^2+\rho^2+\sigma^2+\tau^2=1. \quad\text{.........(5)}$$

Now let us express the coordinates $x, y, z$ with respect to the fixed space axes $O(x, y, z)$ in terms of the coordinates $x_1, y_1, z_1$ with respect to the body axes O(A, B, C). We have, as stated in 1, IV,

$$x = (\cos\theta\cos\phi\cos\psi - \sin\phi\sin\psi)x_1 - (\cos\theta\sin\phi\cos\psi + \cos\phi\sin\psi)y_1 + \sin\theta\cos\psi \, . \, z_1,$$

with similar expressions for $y, z$. Also we have

$$\cos\phi\cos\psi = \tfrac{1}{2}\cos(\phi+\psi) + \tfrac{1}{2}\cos(\phi-\psi),$$
$$\sin\phi\sin\psi = \tfrac{1}{2}\cos(\phi-\psi) - \tfrac{1}{2}\cos(\phi+\psi),$$

and similar equations. From these we obtain for the part of $x$ which depends on $x_1$,

$$x_1\{2(\tau^2 - \rho^2) - \cos\theta\}.$$

But $\cos\theta = \sigma^2 + \tau^2 - \varpi^2 - \rho^2$, by (4); hence the terms just written are

$$x_1(\varpi^2 - \rho^2 - \sigma^2 + \tau^2).$$

Similarly, we get for the terms in $y_1$,

$$2y_1(\varpi\rho - \sigma\tau),$$

and for the terms in $z_1$, $\quad 2z_1(\varpi\sigma + \rho\tau).$

Proceeding in the same way for the values of $y, z$ we get the following scheme of relations, which is equivalent to that given for the same coordinates in 1, IV:

| | $Ox, x$ | $Oy, y$ | $Oz, z$ |
|---|---|---|---|
| OH, $x_1$ | $\varpi^2 - \rho^2 - \sigma^2 + \tau^2$ | $2(\varpi\rho + \sigma\tau)$ | $2(\varpi\sigma - \rho\tau)$ |
| OB, $y_1$ | $2(\varpi\rho - \sigma\tau)$ | $-\varpi^2 + \rho^2 - \sigma^2 + \tau^2$ | $2(\rho\sigma + \varpi\tau)$ |
| OC, $z_1$ | $2(\varpi\sigma + \rho\tau)$ | $2(\rho\sigma - \varpi\tau)$ | $-\varpi^2 - \rho^2 + \sigma^2 + \tau^2$ |

2. *Quaternion property of Euler's parameters.* By equations (2) and (3) we have also the values

$$\varpi = \cos a \sin \tfrac{1}{2}\vartheta, \quad \rho = \cos b \sin\tfrac{1}{2}\vartheta, \quad \sigma = \cos c \sin\tfrac{1}{2}\vartheta, \quad \tau = \cos\tfrac{1}{2}\vartheta, \;\ldots\ldots(1)$$

which are very important. For another turning $\vartheta'$ about the same axis we should have of course the same formulae with the letters $\varpi, \rho, \sigma, \tau$, and $\vartheta$ accented. The resultant $\vartheta + \vartheta'$ of these two turnings has the parameters

$$\left.\begin{aligned} \varpi'' &= \cos a \sin\tfrac{1}{2}(\vartheta+\vartheta'), \quad \rho'' = \cos b \sin\tfrac{1}{2}(\vartheta+\vartheta'), \\ \sigma'' &= \cos c \sin\tfrac{1}{2}(\vartheta+\vartheta'), \quad \tau'' = \cos\tfrac{1}{2}(\vartheta+\vartheta'). \end{aligned}\right\} \;\ldots\ldots\ldots\ldots\ldots(2)$$

We easily find the relations

$$\left.\begin{aligned} \varpi'' &= \varpi\tau' + \rho\sigma' - \sigma\rho' + \tau\varpi', \quad \rho'' = -\varpi\sigma' + \rho\tau' + \sigma\varpi' + \tau\rho', \\ \sigma'' &= \varpi\rho' - \rho\varpi' + \sigma\tau' + \tau\sigma', \quad \tau'' = \tau\tau' - \varpi\varpi' - \rho\rho' - \sigma\sigma'. \end{aligned}\right\} \;\ldots\ldots\ldots(3)$$

Now, if we multiply together the two quaternions,

$$\tau+i\varpi+j\rho+k\sigma, \quad \tau'+i\varpi'+j\rho'+k\sigma',$$

where $i, j, k$ are unit vectors along the axes $O(x, y, z)$, we get for values of $\tau'', \varpi'', \rho'', \sigma''$, in the product

$$\tau''+i\varpi''+j\rho''+k\sigma'',$$

just the expressions on the right of (3), as the reader may verify by multiplying out and taking account of the relations

$$i^2=j^2=k^2=-1, \quad jk=-kj=i, \quad ki=-ik=j, \quad ij=-ji=k.$$

Of course for each set of parameters the condition (5), 1 holds, which amounts to saying that the tensor of the quaternion is unity.

3. *Klein's parameters.* Four other parameters, $\alpha, \beta, \gamma, \delta$, were introduced by Klein, and have been extensively used in elliptic function analysis. As this is not a treatise on elliptic functions, it would be out of place to devote much space to the discussion of these parameters, though they have been used by Klein and Sommerfeld in numerical calculations regarding the motion of the axis of a top. We may however give here a short explanation of their meaning. They are defined by the equations

$$\varpi=\tfrac{1}{2}(\beta-\gamma), \quad \rho=\frac{1}{2i}(\beta+\gamma), \quad \sigma=\frac{1}{2i}(\alpha-\delta), \quad \tau=\tfrac{1}{2}(\alpha+\delta). \quad \text{.......(1)}$$

With these the relation scheme found above for $\varpi, \rho, \sigma, \tau$ becomes

| | $Ox, x$ | $Oy, y$ | $Oz, z$ |
|---|---|---|---|
| $OA, x_1$ | $\frac{1}{2}(\alpha^2+\beta^2+\gamma^2+\delta^2)$ | $\frac{i}{2}(-\alpha^2-\beta^2+\gamma^2+\delta^2)$ | $i(\alpha\gamma+\beta\delta)$ |
| $OB, y_1$ | $\frac{i}{2}(\alpha^2-\beta^2+\gamma^2-\delta^2)$ | $\frac{1}{2}(\alpha^2-\beta^2-\gamma^2+\delta^2)$ | $-\alpha\gamma+\beta\delta$ |
| $OC, z_1$ | $-i(\alpha\beta+\gamma\delta)$ | $-\alpha\beta+\gamma\delta$ | $\alpha\delta+\beta\gamma$ |

To express these parameters in terms of the Eulerian angles, $\theta, \phi, \psi$, we have the equations (4), 1, which by the defining equations, (1) and (3), 1, give

$$\left.\begin{aligned}\alpha&=\tau+i\sigma=\cos\tfrac{1}{2}\theta e^{\frac{1}{2}i(\phi+\psi)}=\cos\tfrac{1}{2}\vartheta+i\sin\tfrac{1}{2}\vartheta\cos c,\\ \beta&=\varpi+i\rho=i\sin\tfrac{1}{2}\theta e^{-\frac{1}{2}i(\phi-\psi)}=\sin\tfrac{1}{2}\vartheta(\cos a+i\cos b),\\ \gamma&=-\varpi+i\rho=i\sin\tfrac{1}{2}\theta e^{\frac{1}{2}i(\phi-\psi)}=\sin\tfrac{1}{2}\vartheta(-\cos a+i\cos b),\\ \delta&=\tau-i\sigma=\cos\tfrac{1}{2}\theta e^{-\frac{1}{2}i(\phi+\psi)}=\cos\tfrac{1}{2}\vartheta-i\sin\tfrac{1}{2}\vartheta\cos c.\end{aligned}\right\} \quad \text{.........(2)}$$

Clearly these give $\alpha\delta-\beta\gamma=\varpi^2+\rho^2+\sigma^2+\tau^2=1.$ ..........................(3)

It will be seen that if $\alpha, \beta, \gamma, \delta$ can be calculated, we can obtain for any instant the position of the axis of a top. Hence Klein and Sommerfeld (*Theorie des Kreisels*, Heft II) have devoted much space to this calculation. As we shall obtain all the requisite numerical results for the path of a point on the axis of figure, by a simple process which does not involve these parameters, we do not pursue the discussion of their properties here. We shall give only one more set of relations.

**4.** *Expression of Klein's parameters by elliptic integrals.* Referring back to 10, V, we recall that in the general case of a symmetrical top

$$\dot{z}^2 = (a - az)(1 - z^2) - (\beta - bnz)^2 = f(z). \quad\text{.....................(1)}$$

If $z_1$, $z_2$, $z_3$ be the roots of $f(z) = 0$, in *ascending* order of magnitude, we know that $z_1$, $z_2$ lie between $z = -1$ and $z = 1$, and that $z_3 > 1$. Thus writing $f(z) = a(z - z_1)(z_2 - z)(z_3 - z)$, we see that $\{f(z)\}^{\frac{1}{2}}$ is real for $z$ between $z_1$ and $z_2$, and for $z$ between $z_3$ and $\infty$. It is imaginary for $z$ between $-\infty$ and $z_1$ and between $z_2$ and $z_3$. By (3), 10 and (1), 11 of V we have

$$t = \int \frac{dz}{\{f(z)\}^{\frac{1}{2}}}, \quad \psi = \int \frac{(\beta - bnz)dz}{(1 - z^2)\{f(z)\}^{\frac{1}{2}}}, \quad\text{.....................(2)}$$

where we take the positive value of the square root for the passage from $z = z_1$ to $z = z_2$, and the negative value for passage from $z_2$ to $z_1$.

Likewise we have $\phi = n - z\dot{\psi}$, so that, by the value of $\dot{\psi}$,

$$\phi = \int \frac{(bn - \beta z)dz}{(1 - z^2)\{f(z)\}^{\frac{1}{2}}} + (1 - b)nt. \quad\text{.......................(3)}$$

Now, by (2), 3, above,

$$\log \alpha = \tfrac{1}{2}\log(1 + z) + \tfrac{1}{2}i(\phi + \psi) - \tfrac{1}{2}\log 2, \quad\text{...... ............(4)}$$

which we can write in the form of an integral by using the values of $\tau$, $\psi$, $\phi$ just written down. Disregarding the constant $-\frac{1}{2}\log 2$, we have

$$\log \alpha = \int \left\{ \frac{\{f(z)\}^{\frac{1}{2}} + i(\beta + bn)}{1 + z} + in(1 - b) \right\} \frac{dz}{2\{f(z)\}^{\frac{1}{2}}},$$

$$\log \beta = \int \left\{ \frac{-\{f(z)\}^{\frac{1}{2}} + i(\beta - bn)}{1 - z} - in(1 - b) \right\} \frac{dz}{2\{f(z)\}^{\frac{1}{2}}},$$

$$\log \gamma = \int \left\{ \frac{-\{f(z)\}^{\frac{1}{2}} - i(\beta - bn)}{1 - z} + in(1 - b) \right\} \frac{dz}{2\{f(z)\}^{\frac{1}{2}}},$$

$$\log \delta = \int \left\{ \frac{\{f(z)\}^{\frac{1}{2}} - i(\beta + bn)}{1 + z} - in(1 - b) \right\} \frac{dz}{2\{f(z)\}^{\frac{1}{2}}},$$

from which also additive constants have been omitted.

**5.** *Relations of elliptic integrals. Expression of time of motion by an elliptic integral.* By these equations the parameters could be calculated numerically, but this involves the evaluation of elliptic integrals of Legendre's third normal type. The numerical determination of the azimuthal angle $\psi$ involves just such integrals, and so far as the numerical tracing of the path is alone concerned, there does not seem to be any advantage in calculating the parameters. A good deal of elliptic function analysis must be employed below, but only so that formulae may be found for the complete determination of the motion. In the present chapter we propose to discuss, as far as is necessary for practical purposes, the numerical determination of the quantities which determine the position of a point on the axis of figure of a top. We begin with a short statement regarding the Jacobian elliptic integrals which will be employed in the calculations.

There are three normal integrals, which are usually written in the forms,

$$\int_0^x \frac{dx}{\{(1-x^2)(1-k^2x^2)\}^{\frac{1}{2}}}, \quad \int_0^x \{(1-x^2)(1-k^2x^2)\}^{\frac{1}{2}}dx,$$

$$\int_0^x \frac{dx}{(1-px^2)\{(1-x^2)(1-k^2x^2)\}^{\frac{1}{2}}},$$

where $0<k^2<1$ and $0<x<1$. These are called elliptic integrals of the first, second and third kinds.

If we put $y=x^2$, so that $dx=dy/2x$, the integrals take the forms

$$\frac{1}{2}\int_0^y \frac{dy}{\{y(1-y)(1-k^2y)\}^{\frac{1}{2}}}, \quad \frac{1}{2}\int_0^y \left\{\frac{1-y}{y}(1-k^2y)\right\}^{\frac{1}{2}}dy,$$

$$\frac{1}{2}\int_0^y \frac{dy}{(1-py)\{y(1-y)(1-k^2y)\}^{\frac{1}{2}}},$$

where $0<y<1$ and $-\infty<p<+1$.

If we write $x^2=\sin^2\phi$ where $\phi$ is an auxiliary angle, not the $\phi$ of 4, we obtain for the three integrals the forms given to them by Legendre,

$$\int_0^\phi \frac{d\phi}{(1-k^2\sin^2\phi)^{\frac{1}{2}}}, \quad \int(1-k^2\sin^2\phi)^{\frac{1}{2}}d\phi, \quad \int_0^\phi \frac{d\phi}{(1-p\sin^2\phi)(1-k^2\sin^2\phi)^{\frac{1}{2}}}.$$

The first is generally denoted by $F(k, \phi)$, the second by $E(k, \phi)$ and the third by $\Pi(k, \phi, p)$. Sometimes the parameter $p$ appears with the opposite sign.

If the upper limit of the integrals is $\frac{1}{2}\pi$, they are said to be *complete*. Then the symbol K is usually employed for the first integral.

Similar integrals are constructed with the complementary modulus $k'$ defined by the relation $k^2+k'^2=1$. These are conveniently denoted by

$$F(k', \phi), \quad E(k', \phi), \quad \Pi(k', \phi, p), \quad \text{and } K'.$$

The intergal $\displaystyle\int_{z_1}^z \frac{dz}{Z^{\frac{1}{2}}}$ $\left[Z=(z-z_1)(z_2-z)(z_3-z)=\frac{f(z)}{a}\right]$

can be reduced to the forms just given for the first elliptic integral by means of the substitution

$$y=\frac{z-z_1}{z_2-z_1},$$

which gives

$$\int_{z_1}^z \frac{dz}{Z^{\frac{1}{2}}}=\frac{k}{(z_2-z_1)^{\frac{1}{2}}}\int_0^y \frac{dy}{\{y(1-y)(1-k^2y)\}^{\frac{1}{2}}}, \quad\text{......(1)}$$

where $k^2=(z_2-z_1)/(z_3-z_1)$, and therefore $k'^2=(z_3-z_2)/(z_3-z_1)$.

Hence the time $t$ of passage of the axis of the top from the circle $z_1$ to the circle $z$ on the unit sphere is given by

$$t=\frac{k}{\{a(z_2-z_1)\}^{\frac{1}{2}}}\int_0^y \frac{dy}{\{y(1-y)(1-k^2y)\}^{\frac{1}{2}}}=\frac{2k}{\{a(z_2-z_1)\}^{\frac{1}{2}}}F(k, \phi), \quad \ldots\ldots(2)$$

where $\phi=\sin^{-1}(y^{\frac{1}{2}})$, $a$ has the value $2Mgh/A$, and the positive value of the square root is taken in each case.

6. *Relations of elliptic integrals. Double periodicity of elliptic functions.* We are also concerned with the integral

$$\int_{z_1}^{z}\frac{dz}{Z^{\frac{1}{2}}},$$

where the upper limit lies between $z_1$ and $-\infty$. This may be expressed as an integral between the limits $+1$ and a value of $y$ between 0 and $+1$. Writing

$$z=z_3-\frac{1}{y}(z_3-z_1)$$

and substituting, we obtain, with the understanding stated as to the upper limit,

$$\int_{z_1}^{z}\frac{dz}{Z^{\frac{1}{2}}}=-\frac{i}{(z_3-z_1)^{\frac{1}{2}}}\int_1^y\frac{dy}{\{y(1-y)(1-k'^2y)\}^{\frac{1}{2}}}.$$

If $z$, the upper limit of the integral on the left, be $-\infty$, we get at once

$$\int_{z_1}^{-\infty}\frac{dz}{Z^{\frac{1}{2}}}=\frac{i}{(z_3-z_1)^{\frac{1}{2}}}\int_0^1\frac{dy}{\{y(1-y)(1-k'^2y)\}^{\frac{1}{2}}}=\frac{i}{(z_3-z_1)^{\frac{1}{2}}}2K'.$$

We shall in what follows usually denote $K/(z_3-z_1)^{\frac{1}{2}}$, $K'i/(z_3-z_1)^{\frac{1}{2}}$ by $\omega$, $\omega'$, so that

$$2\omega=\int_{z_1}^{z_2}\frac{dz}{Z^{\frac{1}{2}}},\quad 2\omega'=\int_{z_1}^{-\infty}\frac{dz}{Z^{\frac{1}{2}}}.$$

Writing $y=(1-k^2u)/(1-k^2)$, we find that

$$\int_0^1\frac{dy}{\{y(1-y)(1-k^2y)\}^{\frac{1}{2}}}=i\int_1^{1/k^2}\frac{du}{\{u(1-u)(1-k^2u)\}^{\frac{1}{2}}}.$$

Hence

$$2iK'=\int_1^{1/k^2}\frac{du}{\{u(1-u)(1-k^2u)\}^{\frac{1}{2}}}.$$

If $w^2$ be substituted for $u$ this becomes

$$iK'=\int_1^{1/k}\frac{dw}{\{(1-w^2)(1-k^2w^2)\}^{\frac{1}{2}}}.$$

By taking the complete integral of the first kind in its normal form,

$$K=\int_0^1\frac{dx}{\{(1-x^2)(1-k^2x^2)\}^{\frac{1}{2}}},$$

we can show that we have also

$$K=\int_{1/k}^{\infty}\frac{dx}{\{(1-x^2)(1-k^2x^2)\}^{\frac{1}{2}}}.$$

It is only necessary to make the substitution $x=1/ky$ and so obtain $dx=-dy/ky^2$, and therefore

$$\frac{dx}{\{(1-x^2)(1-k^2x^2)\}^{\frac{1}{2}}}=-\frac{dy}{\{(1-y^2)(1-k^2y^2)\}^{\frac{1}{2}}}.$$

But when $x=0$, $y=\infty$, and when $x=1$, $y=1/k$, and therefore

$$K=\int_0^1\frac{dx}{\{(1-x^2)(1-k^2x^2)\}^{\frac{1}{2}}}=\int_{1/k}^{\infty}\frac{dx}{\{(1-x^2)(1-k^2x^2)\}^{\frac{1}{2}}}.$$

This last result shows that

$$\int_{z_1}^{z_2}\frac{dz}{Z^{\frac{1}{2}}}=\int_{z_3}^{\infty}\frac{dz}{Z^{\frac{1}{2}}},$$

since the root $z_3$ of the cubic $Z=0$ corresponds to the point $x=1/k$, as $z_1$, $z_2$ correspond to the points $x=0$, $x=1$ respectively. This can be proved in another way by means of the theory of functions of a complex variable, indeed it is only as a part of the subject matter of this theory that elliptic functions and elliptic integrals can be satisfactorily discussed. The reader should consult the modern treatises for full information regarding the purely mathematical matters touched on here.

The numerical calculation of 2K gives the time of passage of the axis of the top from one limiting circle on the unit sphere to the other as

$$\frac{2K}{\{a(z_3-z_1)\}^{\frac{1}{2}}}.$$

The integrals K, $iK'$ are elements of the periods of the Jacobian elliptic functions $\operatorname{sn} u$, $\operatorname{cn} u$, $\operatorname{dn} u$, where $u$, $\operatorname{sn} u$, $\operatorname{cn} u$, $\operatorname{dn} u$ stand for $F(k, \phi)$, $\sin\phi$, $\cos\phi$, $(1-k^2\sin^2\phi)^{\frac{1}{2}}$ respectively. The pairs of periods for $\operatorname{sn} u$, $\operatorname{cn} u$, $\operatorname{dn} u$ are respectively $4K$, $2iK'$; $4K$, $2K+2iK'$; $2K$, $4iK'$. The appearance of the imaginary element, and the double periodicity of the functions here indicated, are easily explained dynamically. [See Greenhill's *Elliptic Functions*, or *Fonctions Elliptiques* by Appell and Lacour, for full explanations.]

7. *Formulae for numerical calculation of elliptic integrals of first and second kinds.* The values of K and E, the complete elliptic integrals of the first and second kinds, can be obtained for any given modulus by expanding, in ascending powers of $k\sin\phi$, $1/(1-k^2\sin^2\phi)^{\frac{1}{2}}$ in the former case, and $(1-k^2\sin^2\phi)^{\frac{1}{2}}$ in the latter, and integrating term by term. This method, though direct and obvious, is far from being so expeditious as some others, when assisted by various subsidiary tables which have been compiled. Also tables of $F(k, \phi)$, $E(k, \phi)$ were constructed by Legendre* for ranges of moduli and amplitudes proceeding by small differences, and from these it is possible by interpolation to find the integrals for other moduli and amplitudes than those given in the tables. But unfortunately these tables in their complete form are only available to those who have access to a large reference library. The actual numerical values of these integrals are essential in many modern practical physical applications, besides

* *Traité des Fonctions Elliptiques*, Tome II.

those to rotational motion, for example to the determination of the constants of coils in electrical work.

We have

$$(1-k^2\sin^2\phi)^{-\frac{1}{2}}=1+\tfrac{1}{2}k^2\sin^2\phi+\frac{1.3}{2.4}k^4\sin^4\phi+\cdots+\frac{1.3.5\ldots(2r-1)}{2.4.6\ldots 2r}k^{2r}\sin^{2r}\phi+\cdots,$$

$$(1-k^2\sin^2\phi)^{\frac{1}{2}}=1-\tfrac{1}{2}k^2\sin^2\phi-\frac{1.3}{2.4}\frac{k^4}{3}\sin^4\phi-\cdots-\frac{1.3.5\ldots(2r-1)}{2.4.6\ldots 2r}\frac{k^{2r}}{2r-1}\sin^{2r}\phi-\cdots.$$

Multiplying these by $d\phi$ and integrating from 0 to $\phi$, we get $F(k, \phi)$, $E(k, \phi)$. Integrating from 0 to $\frac{1}{2}\pi$, we obtain

$$\left.\begin{aligned} K&=\tfrac{1}{2}\pi\left[1+(\tfrac{1}{2})^2k^2+\left(\frac{1.3}{2.4}\right)^2k^4+\cdots+\left\{\frac{1.3.5\ldots(2r-1)}{2.4.6\ldots 2r}\right\}^2k^{2r}+\cdots\right],\\ E&=\tfrac{1}{2}\pi\left[1-(\tfrac{1}{2})^2k^2-\left(\frac{1.3}{2.4}\right)^2\frac{k^4}{3}-\cdots-\left\{\frac{1.3.5\ldots(2r-1)}{2.4.6\ldots 2r}\right\}^2\frac{k^{2r}}{2r-1}-\cdots\right].\end{aligned}\right\}\quad\ldots(1)$$

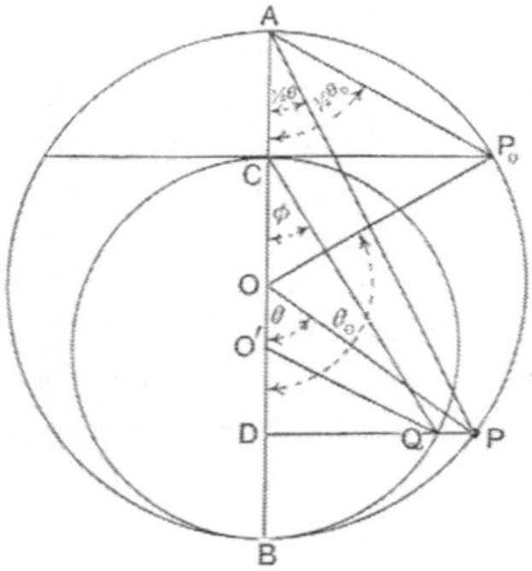


FIG. 64.

Thus for the quarter period of the pendulum vibrating over a finite arc, we have [12, XV, below]

$$\tau=\tfrac{1}{2}\pi\left(\frac{l}{g}\right)^{\frac{1}{2}}\left\{1+(\tfrac{1}{2})^2k^2+\left(\frac{1.3}{2.4}\right)^2k^4+\cdots\right\},\quad\ldots(2)$$

with $k^2=\sin^2\frac{1}{2}\theta_0=CB/2l$, where CB is the diameter of the smaller circle in Fig. 64.

**8.** *Landen's transformation. An elliptic integral expressed as a continued product.* An elliptic integral of the first kind can be transformed into another of a larger modulus and a smaller amplitude, or of a smaller modulus and a larger amplitude. The transformation is that given by Landen [*Phil. Trans.* 1775]. Taking the former case,

$$F(k, \phi)=\int_0^\phi\frac{d\phi}{(1-k^2\sin^2\phi)^{\frac{1}{2}}},\quad\ldots(1)$$

we replace $\phi$ by a new amplitude $\phi_1$, given by

$$\tan\phi=\frac{\sin 2\phi_1}{k+\cos 2\phi_1}.\quad\ldots(2)$$

From this we get

$$\left.\begin{aligned}\frac{d\phi}{(1-k^2\sin^2\phi)^{\frac{1}{2}}}&=\frac{2}{1+k}\frac{d\phi_1}{(1-k_1{}^2\sin^2\phi_1)^{\frac{1}{2}}},\\ \text{where}\qquad k_1{}^2&=\frac{4k}{(1+k)^2}.\end{aligned}\right\}\quad\ldots(3)$$

The modulus $k_1$ is less than 1 and greater than $k$, as the reader may prove. Also $\phi_1<\phi$.

The transformation may be applied any required number of times so as to give from (1),

$$F(k, \phi)=\frac{2}{1+k}\cdot\frac{2}{1+k_1}\cdot\frac{2}{1+k_2}\cdot\ldots\cdot\frac{2}{1+k_{n-1}}F(k_n, \phi_n). \qquad (4)$$

But $2/(1+k)=k_1/k^{\frac{1}{2}}$, $2/(1+k_1)=k_2/k_1^{\frac{1}{2}}$, ..., and therefore the equation just obtained may be written

$$F(k, \phi)=\left(\frac{k_1k_2\ldots k_{n-1}}{k}\right)^{\frac{1}{2}}k_nF(k_n, \phi_n). \qquad (5)$$

A continued product is thus obtained for $F(k, \phi)$ which converges to a limiting value, for which $k_n=1$, and the upper limit of integration is $\Phi$, the value of $\phi_n$ for $n=\infty$. Thus the final integral is $F(1, \Phi)$, that is

$$\int_0^{\Phi}\frac{d\phi}{\cos\phi}=\log\tan(\tfrac{1}{4}\pi+\tfrac{1}{2}\Phi),$$

and we have
$$F(k, \phi)=\left\{\underset{n=\infty}{\mathrm{Lt}}\left(\frac{k_1k_2\ldots k_n}{k}\right)^{\frac{1}{2}}\right\}\log\tan(\tfrac{1}{4}\pi+\tfrac{1}{2}\Phi). \qquad (6)$$

This transformation reversed gives a smaller modulus and an increased amplitude. In this case if $k_1$ be the new modulus and $k$ the old, we have

$$k=\frac{2k_1^{\frac{1}{2}}}{1+k_1}, \quad \text{or} \quad k_1=\frac{1-(1-k^2)^{\frac{1}{2}}}{1+(1-k^2)^{\frac{1}{2}}}=\frac{1-k'}{1+k'}, \qquad (7)$$

and for the amplitude,
$$\tan(\phi_1-\phi)=(1-k^2)^{\frac{1}{2}}\tan\phi. \qquad (8)$$

It is to be noticed that if the amplitude $\phi=\frac{1}{2}\pi$, the new amplitude $\phi_1=\pi$. The amplitude $\phi_2$ will then be $2\pi$, and so on to $\phi_n=2^n(\frac{1}{2}\pi)$

Now we obtain by repetition of the transformation,

$$F(k, \phi)=\frac{1+k_1}{2}F(k_1, \phi_1)=\frac{1+k_1}{2}\cdot\frac{1+k_2}{2}F(k_2, \phi_2). \qquad (9)$$

Hence
$$F(k, \phi)=(1+k_1)(1+k_2)\ldots(1+k_n)\frac{F(k_n, \phi_n)}{2^n},$$

or, by the derivation of the $k$s,

$$\left.\begin{aligned} &F(k_1, \phi)=\frac{1}{k_1k_2\ldots k_n^{\frac{1}{2}}}\frac{k_n}{k}F(k_n, \phi_n),\\ \text{where}\quad &1+k_r=\frac{2}{1+(1-k_{r-1}^2)^{\frac{1}{2}}}, \quad \tan(\phi_r-\phi)=(1-k_{r-1}^2)^{\frac{1}{2}}\tan\phi_{r-1}.\end{aligned}\right\} \qquad (10)$$

Here the limit $k_n$ $(n=\infty)$ is zero, and therefore

$$\underset{n=\infty}{\mathrm{Lt}}F(k_n, \phi_n)=\int_0^{\Phi}d\phi=\Phi. \qquad (11)$$

If the complete elliptic integral is required, we have $\Phi/2^n=\frac{1}{2}\pi$, and therefore

$$K(k)=\underset{n=\infty}{\mathrm{Lt}}(1+k_1)(1+k_2)\ldots(1+k_n)\frac{\pi}{2}. \qquad (12)$$

By means of the last result the complete elliptic integral of the first kind for a given modulus $k$ can be quickly determined. For example, if $k=0{\cdot}5$ $(=\sin 30°)$, we obtain

$$k_1=0{\cdot}0718, \quad k_2=0{\cdot}00129.$$

Using only these we obtain
$$K(\tfrac{1}{2})=1{\cdot}07309\frac{\pi}{2}=1{\cdot}686-\epsilon, \qquad (13)$$

where $0{\cdot}0005>\epsilon>0{\cdot}0001$.

Again let $k=\sqrt{3}/2$ $(=\sin 60°)$. We find at once

$$k_1=\tfrac{1}{3}, \quad k_2=17-12\sqrt{2}=0{\cdot}029438, \quad k_3=0{\cdot}000216,$$

which give
$$K\left(\frac{\sqrt{3}}{2}\right)=2{\cdot}157-\epsilon, \qquad (14)$$

where $0{\cdot}0005>\epsilon>0{\cdot}0001$.

**9.** *Convergent series for elliptic integrals of first and second kinds.* Other formulae of calculation for K and E have been developed. We give here a useful process for K. A similar process is also applicable to E.

Let P, $P_1$, $P_2$, ... be defined by the equations

$$\left.\begin{aligned} P &= 1+(\tfrac{1}{2})^2\kappa^4+\left(\frac{1.3}{2.4}\right)^2\kappa^8+\left(\frac{1.3.5}{2.4.6}\right)^2\kappa^{32}+\cdots,\\ P_1 &= \quad (\tfrac{1}{2})^2\kappa^4+\left(\frac{1.3}{2.4}\right)^2\kappa^8+\left(\frac{1.3.5}{2.4.6}\right)^2\kappa^{32}+\cdots,\\ P_2 &= \qquad\qquad +\left(\frac{1.3}{2.4}\right)^2\kappa^8+\left(\frac{1.3.5}{2.4.6}\right)^2\kappa^{32}+\cdots,\\ P_3 &= \qquad\qquad\qquad\qquad +\left(\frac{1.3.5}{2.4.6}\right)^2\kappa^{32}+\cdots, \end{aligned}\right\} \quad\ldots\ldots\ldots(1)$$

where $\kappa = k'^{\frac{1}{2}}$. Then

$$K = P\log\frac{4}{\kappa^2} - 2\left(\frac{1}{1.2}P_1+\frac{1}{3.4}P_2+\frac{1}{5.6}P_3+\cdots\right). \quad\ldots\ldots(2)$$

The values of P, $P_1$, $P_2$, ... are first calculated, and the value of K is then obtained by the equation just written. It will be observed that when the terms necessary for P have been evaluated, the proper procedure is as follows. Let us suppose that the term

$$\{(1.3.5)/(2.4.6)\}^2\kappa^{32}$$

is the smallest which it has been thought necessary to include: then that is set down as $P_3$. To that is added the next term in order of magnitude to give $P_2$. To $P_2$ is added the next term in magnitude, to form $P_1$; the addition of 1 to $P_1$ gives P. The series in brackets in (2) is then formed, doubled, and subtracted from $P\log(4/\kappa^2)$, and the result is K.

When $k$ is small the value of $k'$ is nearly unity. By using $k$ instead of $k'$ in calculating P, $P_1$, $P_2$, ..., and $\log(4/k)$ instead of $\log 4/k'$, we obtain by (2) the value of K′. That of K is then $\frac{1}{2}P/\pi$.

The formula for E is

$$E = 1+\tfrac{1}{2}\kappa^4\left(\log\frac{4}{\kappa^2}-\frac{1}{1.2}\right)+\frac{1^2.3}{2^2.4}\kappa^8\left(\log\frac{4}{\kappa^2}-\frac{2}{1.2}-\frac{1}{3.4}\right)$$
$$+\frac{1^2.3^2.5}{2^2.4^2.6}\kappa^{12}\left(\log\frac{4}{\kappa^2}-\frac{2}{1.2}-\frac{2}{3.4}-\frac{1}{5.6}\right)+\cdots \quad\ldots\ldots(3)$$

**10.** *Jacobi's Θ-functions. Expression by q-series.* When the value of $k$ is not in the vicinity of zero or unity so that neither $k$ nor $k'$ is small, the integrals may be calculated by other methods which we shall now shortly explain.

If $q$ denote $e^{-\pi K'/K}$ and

$$\left.\begin{matrix}\Theta(u)\\ \Theta_1(u)\end{matrix}\right\} = 1 \mp 2q\cos\frac{\pi u}{K}+2q^4\cos\frac{2\pi u}{K}\mp 2q^9\cos\frac{3\pi u}{K}+\cdots, \quad\ldots\ldots(1)$$

it is proved in treatises on elliptic functions that

$$\mathrm{dn}\,u\,\{=(1-k^2\,\mathrm{sn}^2 u)^{\frac{1}{2}}\} = k'^{\frac{1}{2}}\frac{\Theta_1(u)}{\Theta(u)}. \quad\ldots\ldots(2)$$

If $u=0$, $\mathrm{dn}\,u=1$, and we have

$$k'^{\frac{1}{2}} = \frac{1-2q+2q^4-2q^9+\cdots}{1+2q+2q^4+2q^9+\cdots}. \quad\ldots\ldots(3)$$

We shall now show how K can be calculated when $q$ has been expressed in terms of $k'$.

From (3) we obtain

$$\frac{1-k'^{\frac{1}{2}}}{1+k'^{\frac{1}{2}}} = \frac{2q+2q^9+2q^{25}+\cdots}{1+2q^4+2q^{16}+\cdots}. \quad\ldots\ldots(4)$$

Thus, if powers of $q$ higher than the fourth can be neglected,

$$2q=\frac{1-k'^{\frac{1}{2}}}{1+k'^{\frac{1}{2}}} \quad \text{.....(5)}$$

In the numerical calculations below we shall make the supposition here stated. It is usual to express $k$ as $\sin\alpha$ where $\alpha$ is an auxiliary angle. We have then $k'=\cos\alpha$, and hence by (5) we get the approximate equation

$$2q=\frac{(1+\tan^2\frac{1}{2}\alpha)^{\frac{1}{2}}-(1-\tan^2\frac{1}{2}\alpha)^{\frac{1}{2}}}{(1+\tan^2\frac{1}{2}\alpha)^{\frac{1}{2}}+(1-\tan^2\frac{1}{2}\alpha)^{\frac{1}{2}}}. \quad \text{.....(6)}$$

Equation (3) however gives by expansion the exact equation

$$q=\tfrac{1}{4}\tan^2\tfrac{1}{2}\alpha+\tfrac{1}{16}\tan^6\tfrac{1}{2}\alpha+\tfrac{17}{512}\tan^{10}\tfrac{1}{2}\alpha+\tfrac{45}{2048}\tan^{14}\tfrac{1}{2}\alpha+\cdots, \quad \text{.....(7)}$$

and

$$\log q=2(\log\tan\tfrac{1}{2}\alpha-\log 2)+\log(1+\tfrac{1}{4}\tan^4\tfrac{1}{2}\alpha+\tfrac{17}{128}\tan^8\tfrac{1}{2}\alpha+\cdots). \quad \text{.....(8)}$$

The second part on the right may be regarded as $\log(1+x)$, where

$$x=\tfrac{1}{4}\tan^4\tfrac{1}{2}\alpha+\tfrac{17}{128}\tan^8\tfrac{1}{2}\alpha+\ldots.$$

But

$$\log(1+x)=x-\tfrac{1}{2}x^2+\tfrac{1}{3}x^3-\ldots,$$

and therefore we may write (8) in the form

$$\log q=2(\log\tan\tfrac{1}{2}\alpha-\log 2)+\tfrac{1}{4}\tan^4\tfrac{1}{2}\alpha+\tfrac{13}{128}\tan^8\tfrac{1}{2}\alpha+\tfrac{23}{384}\tan^{12}\tfrac{1}{2}\alpha+\cdots \quad \text{.....(9)}$$

When $\alpha<\frac{1}{4}\pi$, a result correct to five decimal places is obtained by means of (9). The series in the second part is very convergent, and a sufficiently accurate result for most purposes is obtained by including only the terms exhibited.

**11.** *Calculation of complete elliptic integrals by q-series. Numerical example.* It is one of the properties of the Θ-function that $\Theta(u+K)=\Theta_1(u)$, so that $\Theta(K)=\Theta_1(0)$. Thus, by (2), 10,

$$\Theta(0)=k'^{\frac{1}{2}}\Theta(K). \quad \text{.....(1)}$$

But by the theory of these functions we have

$$\Theta(0)=\left(\frac{2k'K}{\pi}\right)^{\frac{1}{2}}. \quad \text{.....(2)}$$

Hence

$$\Theta(K)=\left(\frac{2K}{\pi}\right)^{\frac{1}{2}},$$

that is

$$K=\tfrac{1}{2}\pi(1+2q+2q^4+2q^9+\cdots)^2. \quad \text{.....(3)}$$

Again,

$$\Theta(0)=\left(\frac{2k'K}{\pi}\right)^{\frac{1}{2}}=1-2q+2q^4-2q^9+\cdots \quad \text{.....(4)}$$

By (3) and (4) we get

$$K=\tfrac{1}{2}\pi\left(\frac{2}{1+k'^{\frac{1}{2}}}\right)^2(1+2q^4+2q^{16}+2q^{36}+\ldots)^2. \quad \text{.....(5)}$$

The series on the right is highly convergent.

With observance of the degree of approximation of (5), 10, these become

$$K=\tfrac{1}{2}\pi(1+2q+2q^4)^2, \quad \text{.....(6)}$$

$$\Theta(0)=\left(\frac{2k'K}{\pi}\right)^{\frac{1}{2}}=1-2q+2q^4, \quad \text{.....(7)}$$

$$K=\tfrac{1}{2}\pi\left(\frac{2}{1+k'^{\frac{1}{2}}}\right)^2(1+4q^4). \quad \text{.....(8)}$$

It is interesting to notice that when $q$ is small, as it always is if, say, $\alpha<\frac{1}{3}\pi$, a good approximation to K is given by the equation

$$K=\tfrac{1}{2}\pi\left(\frac{2}{1+k'^{\frac{1}{2}}}\right)^2. \quad \text{.....(6')}$$

The following example of the calculation of $q$ is given in Bertrand's *Calcul Integral*, p. 682, which may be consulted by the reader for a clear statement of the properties of the Jacobian elliptic functions.* We shall find the common logarithm of $q$, or Log $q$. For this we have to multiply the right-hand side of (9), 10, by M($=0\cdot4342944819$), the modulus of the common logarithms. We get, indicating common logarithms by Log instead of log,

$$\text{Log}\, q = 2\,\text{Log}\tan\tfrac{1}{2}\alpha + \bar{1}\cdot3979400 + a\tan^4\tfrac{1}{2}\alpha + b\tan^8\tfrac{1}{2}\alpha + c\tan^{12}\tfrac{1}{2}\alpha + \cdots, \quad\ldots\ldots(9)$$

where $a$, $b$, $c$ are coefficients which have the logarithms,

$$\text{Log}\, a = \bar{1}\cdot0357243, \quad \text{Log}\, b = \bar{2}\cdot64452, \quad \text{Log}\, c = \dot{2}\cdot41518. \quad\ldots\ldots(10)$$

So far this is independent of $\alpha$ and may be used for any determination of $q$ (see for example the calculation in 12 below of the period of a top). Bertrand takes $\alpha = 10°\,23'\,40''$. Then

| | |
|---|---|
| $2\,\text{Log}\tan\tfrac{1}{2}\alpha = \bar{3}\cdot9176842$ | Terms of second part: |
| $\text{Colog}\,4 = \bar{1}\cdot3979400$ | $4\,\text{Log}\tan\tfrac{1}{2}\alpha = \bar{5}\cdot835$ |
| From small terms $\quad 74$ | $\text{Log}\, a = \bar{1}\cdot036$ |
| $\text{Log}\, q = \bar{3}\cdot3156316$ | $\bar{6}\cdot871$ |
| Thus $q = 0\cdot002068$. | $a\tan^4\tfrac{1}{2}\alpha = 0\cdot0000074$ |

The first only of the small terms is taken, as with seven-place logarithms the second small term would contribute nothing to the value of Log $q$.

Now let $\alpha$ be the complement of a small angle. We have $q = e^{-\pi K'/K}$. Let $p = e^{-\pi K/K'}$; then $p$ is the value of $q$ for the small angle $\tfrac{1}{2}\pi - \alpha$. We have

$$\text{Log}\frac{1}{p}\,\text{Log}\frac{1}{q} = M^2\pi^2 = 1\cdot861522835,$$

a result which we reserve also for use in the calculations which follow regarding tops. From it, if $p$ has been found, $q$ can be deduced, and *vice versa*.

Now we have found for the angle $10°\,23'\,40''$ that (if the $q$ for that case be denoted by $p$)

$$\log\frac{1}{p} = 3 - 0\cdot3156316 = 2\cdot6843684.$$

But

$$\text{Log}\,\text{Log}\frac{1}{p} + \text{Log}\,\text{Log}\frac{1}{q} = \text{Log}\,1\cdot861522835.$$

Hence for the large complementary angle

$$\text{Log}\,\text{Log}\frac{1}{q} = \bar{1}\cdot841026,$$

and

$$\text{Log}\, q = \bar{1}\cdot3065321,$$

that is

$$q = 0\cdot2024.$$

We now find the value of K for $\alpha = 10°\,23'\,40''$. We have seen that $q = 0\cdot002068$, and that K is given by

$$K = \tfrac{1}{2}\pi\left(\frac{2}{1+k'^{\frac{1}{2}}}\right)^2(1 + 4q^4 + \cdots).$$

Now $k'^{\frac{1}{2}} = 1 - 0\cdot008255$, so that $\{2/(1+k'^{\frac{1}{2}})\}^2 = \{1/(1 - 0\cdot004127)\}^2 = 1\cdot008238$. If we limit the result to six places of decimals we have to neglect $4q^4$. Thus

$$K = 1\cdot570796 \times 1\cdot008238 = 1\cdot583736.$$

* The value of $\alpha$ is stated by Bertrand as $10°\,23'\,46''$, but the logarithms he uses are for the value in the text. If the degree of accuracy for which equation (6), 10, is valid is sufficient, the series should not include higher powers of $\tan\tfrac{1}{2}\alpha$ than the eighth.

**12.** *Numerical calculation for an actual top.* We now consider a top consisting of a disk with massive rim, mounted rigidly on an axle or peg through its centre at right angles to the mean plane. The following data correspond nearly to an actual case:

$$M=200, \text{ in grammes } \quad h=4, \text{ in centimetres.} \quad k=4, \text{ in centimetres.}$$

Then $C=Mk^2=3200$, $Mgh=200\times 981\times 4=784800$, $A=\frac{1}{2}C+200\times 4^2=4800$.

We shall suppose first that the angular speed $n$ is 100 revolutions per second, that is 628·32 radians per second, and that the top on being spun is left to itself with its axis at rest inclined at 30° to the upward vertical, and spins then about a fixed point of the axis. One root of (5), 13, V, is $z_0=3^{\frac{1}{2}}/2$. The other two roots are those of

$$z^2-\frac{b^2n^2}{a}z+\frac{b^2n^2}{a}z_0-1=0.$$

These are given by
$$z=p\pm(p^2-2pz_0+1)^{\frac{1}{2}},$$
where $p=b^2n^2/2a$.

Thus
$$p=\frac{3200^2\times 628{\cdot}32^2}{4\times 4800\times 784800}=\frac{1600}{3}\frac{628{\cdot}32^2}{784800}=268{\cdot}3.$$

This gives
$$(p^2-2pz_0+1)^{\frac{1}{2}}=268{\cdot}3\left(1-\frac{3^{\frac{1}{2}}}{268{\cdot}3}+\frac{1}{268{\cdot}3^2}\right)^{\frac{1}{2}}$$
$$=268{\cdot}3-\frac{3^{\frac{1}{2}}}{2}+0{\cdot}001864.$$

The two roots of the quadratic are thus

$$535{\cdot}736 \quad\text{and}\quad 0{\cdot}8642.$$

The latter of these is the smallest root $z_1$, in the notation of 10, V, the former is the large root $z_3$. The root $z_1$ gives approximately $\theta_1=30°\ 12'$, so that the range of oscillation in $\theta$ is about 12′. The limiting circles are thus only 12′ apart, and the oscillation must be very nearly simple harmonic, as described in 14, V.

To find the period we have by (1), 7, above, since $k^2$ is very small, $K-\frac{1}{2}\pi$. Also by (2), 5, the time of passage from one limiting circle to the other and back is

$$\frac{2\pi}{\{a(z_3-z_1)\}^{\frac{1}{2}}}=\frac{2\pi}{418\ 4}=0{\cdot}01502.$$

The period is thus about 3/200 of a second. The eye will hardly be able to detect the deviation from steady motion, though as a matter of fact in this period the azimuthal angular speed will change from zero to twice its average value.

The average value $\mu$ of the angular speed in azimuth may in this case be taken as $Mgh/Cn$, since the motion is only a slight vibratory deviation from that of steady slow precession. Hence
$$\mu=\frac{784800}{3200\times 628{\cdot}32}=0{\cdot}3903$$
in radians per second. Thus, in the period 0·01502 second, the angle traversed in azimuth is 0·000586 radian.

If the speed of rotation be 20 revolutions per second. $p$ will be $268{\cdot}3/25=10{\cdot}73$. Then, as the reader may verify,

$$(p^2-2pz_0+1)^{\frac{1}{2}}=(10{\cdot}73^2-2\times 10{\cdot}73\times 0{\cdot}866+1)^{\frac{1}{2}}=9{\cdot}877,$$

so that
$$z_3=20{\cdot}607,\quad z_1=0{\cdot}853,\quad k^2=\frac{0{\cdot}013}{19{\cdot}754}=0{\cdot}000626.$$

The value of $\theta_1$ is about 31° 30′, so that the range from highest to lowest is 1° 30′. The oscillation is still very nearly simple harmonic in a period of $2\pi/80{\cdot}37$ (=0·078) second. The azimuthal angular speed is five times what it was before, and the angle traversed in the period is 0·156 radian.

**13.** *Actual top: different speeds of rotation.* A rotational speed of 10 turns per second, with other data as before, gives $p = 2{\cdot}683$, and $(p^2 - 2pz_0 + 1)^{\frac{1}{2}} = 1{\cdot}8846$, and so $z_3 = 4{\cdot}5675$, $z_1 = 0{\cdot}7985$. Thus $\theta_1 = 37°$, nearly. The range from highest to lowest is now much greater, showing how it increases as the speed of rotation is diminished. Here $k^2 = 0{\cdot}0675/3{\cdot}769 = 0{\cdot}01791$, so that $k'^2 = 0{\cdot}9821$. Thus $k'^{\frac{1}{2}} = 0{\cdot}995$, about. We shall take this as unity. It is clear also that if we do not go beyond four places of decimals we may put $1 + 2q^4 + \ldots = 1$. Hence, in the present case, as in those that precede,

$$K = \tfrac{1}{2}\pi.$$

The period of oscillation of the axis of the top is, in seconds,

$$\frac{2\pi}{\{a(z_3 - z_1)\}^{\frac{1}{2}}}, = \frac{2\pi}{(327 \times 3{\cdot}769)^{\frac{1}{2}}} = 0{\cdot}18.$$

If we suppose that here again we may use the average angular speed $Mgh/Cn$ as the mean rate of turning in azimuth, we shall obtain a result not far from the truth. The azimuthal angle turned through in the period is thus

$$\psi = \frac{784800 \times 0{\cdot}18}{3200 \times 62{\cdot}832} = 0{\cdot}706$$

in radian measure. Thus the axis moves once round the sphere in about 9 periods.

Finally, we suppose the speed of rotation to be only 5 turns per second. This will give a good example of the calculation of K.

We have $p = 0{\cdot}67075$. Hence

$$(p^2 - 2pz_0 + 1)^{\frac{1}{2}} = (0{\cdot}67075^2 - 2 \times 0{\cdot}67075 \times 0{\cdot}8660 + 1)^{\frac{1}{2}} = 0{\cdot}5369.$$

Thus $$z_1 = 0{\cdot}67075 - 0{\cdot}5368 = 0{\cdot}1338,$$

and $$\theta_1 = 82°\ 18'.$$

The range of motion from one limiting circle to the other is now 52° 18′.

Also we have $$z_3 = 0{\cdot}67075 + 0{\cdot}5369 = 1{\cdot}2076.$$

Hence $$k^2 = \frac{z_2 - z_1}{z_3 - z_1} = \frac{0{\cdot}8660 - 0{\cdot}1339}{1{\cdot}2076 - 0{\cdot}1339} = 0{\cdot}6819.$$

Also $$k'^2 = 1 - 0{\cdot}6819 = 0{\cdot}3181, \quad k' = 0{\cdot}5639, \quad k'^{\frac{1}{2}} = 0{\cdot}7509,$$

and $$\alpha = \cos^{-1} 0{\cdot}5641 = 55°\ 42'.$$

We can now calculate $q$ from the value, 27° 51′, of $\frac{1}{2}\alpha$. The work may be set forth as follows:

$$2 \text{ Log} \tan 27°\ 51' = \bar{1}{\cdot}4458$$
$$\text{CoLog } 4 = \bar{1}{\cdot}3979$$
$$\text{From small terms} \quad 85, \quad 3$$
$$\text{Log } q = \bar{2}{\cdot}8525$$

Hence $$q = 0{\cdot}0712,$$
$$4 \text{ Log } q = \bar{5}{\cdot}4100;$$
$$\therefore q^4 = 0{\cdot}0000257,$$
$$2q^4 = 0{\cdot}0000514.$$

Calculation of small terms:

$$4 \text{ Log} \tan \tfrac{1}{2}\alpha = \bar{2}{\cdot}8916$$
$$\text{Log } a = \bar{1}{\cdot}0357$$
$$\bar{3}{\cdot}9273$$
$$\therefore a \tan^4 \tfrac{1}{2}\alpha = 0{\cdot}0085$$

$$8 \log \tan \tfrac{1}{2}\alpha = \bar{3}{\cdot}7832$$
$$\text{Log } b = \bar{2}{\cdot}6445$$
$$\bar{4}{\cdot}4277$$
$$b \tan^8 \tfrac{1}{2}\alpha = 0{\cdot}00027$$

Thus we obtain $$K = \tfrac{1}{2}\pi \left(\frac{2}{1{\cdot}7509}\right)^2 (1 + 0{\cdot}0000514) = 2{\cdot}0497.$$

From the value of K thus found we obtain the period of oscillation between the limiting circles as

$$\frac{4K}{\{a(z_3 - z_1)\}^{\frac{1}{2}}} = \frac{8{\cdot}1988}{\left(\frac{784800}{2400} \times 1{\cdot}0737\right)^{\frac{1}{2}}} = \frac{8{\cdot}1988}{18{\cdot}74} = 0{\cdot}4376, \text{ in seconds.}$$

The azimuthal angle turned through in the period cannot in this case be estimated accurately by means of the angular speed $Mgh/Cn$, but it is of the order of 3·4 radians. Thus the axis swings about half-round, or rather more, in the period.

It will be observed that in all the cases here considered, *except the last*, steady motion is possible, that is the condition $C^2n^2 > 4AMgh \cos\theta$ is fulfilled, in fact in each of these cases the top is a "strong" top.

**14.** *Numerical determination of inclination of axis to vertical for actual top.* We have now to consider how the numerical calculation of the inclination $\theta$ of the axis to the vertical at any time $t$, and of the corresponding azimuthal angle $\psi$, is to be carried out. This will involve the computation of the incomplete elliptic integral of the first kind, $F(k, \phi)$, and of the corresponding value of the sum of two elliptic integrals of the third kind. The second part of the discussion is attended with difficulties, but a scheme of calculation by means of the parameters $\alpha$, $\beta$, $\gamma$, $\delta$ has been given by Klein and Sommerfeld (*Theorie des Kreisels*), which it is claimed renders the whole matter systematic and comparatively easy. We find it however much more convenient to compute a sufficient number of ordinates to enable the area of the curve which represents the integral to be obtained by adding the areas of a succession of narrow strips.

The calculation of the incomplete integrals $F(k, \phi)$, $E(k, \phi)$ is important for the determination of a point on the axis of a top, spinning about a fixed point O. Denoting $F(k, \phi)$ by $u$ and $k'^{\frac{1}{2}}$ by $\kappa$, we have

$$\operatorname{dn} u = (1 - k^2 \operatorname{sn}^2 u)^{\frac{1}{2}} = \kappa \frac{\Theta_1(u)}{\Theta(u)}, \qquad (1)$$

that is if $x = \pi u/2K$,

$$\operatorname{dn} u = \kappa \frac{1 + 2q\cos 2x + 2q^4 \cos 4x + \dots}{1 - 2q\cos 2x + 2q^4\cos 4x - \dots}. \qquad (2)$$

The value of $q$ is found by the process explained in 10. The calculation can now be carried out as follows by a process of successive approximation. Writing $\cot\lambda$ for $(\operatorname{dn} u)/\kappa$, we get

$$\frac{\cot\lambda - 1}{\cot\lambda + 1} = \tan(45° - \lambda) = \frac{\operatorname{dn} u - \kappa}{\operatorname{dn} u + \kappa}, \qquad (3)$$

that is by (2),

$$\tan(45° - \lambda) = 2q \frac{\cos 2x + q^8 \cos 6x + q^{24} \cos 10x + \dots}{1 + 2q^4 \cos 4x + 2q^{16} \cos 8x + \dots}, \qquad (4)$$

or

$$\tan(45° - \lambda) = 2q \frac{\cos 2x + q^8(4\cos^3 2x - 3\cos 2x) + \dots}{1 + q^4(4\cos^2 2x - 2) + \dots}. \qquad (5)$$

Thus, to a first approximation,

$$\cos 2x = \frac{\tan(45° - \lambda)}{2q}. \qquad (6)$$

With the value of $\cos 2x$ thus obtained we can calculate the terms involving $q^4$ and $q^8$, and substituting these in (5), then solve the new equation for $\cos 2x$, neglecting all terms involving higher powers of $q$ than $q^8$ in the numerator and $q^4$ in the denominator of the function on the right of (5).

We shall see later that the first and second elliptic integrals, as well as the third, can be readily evaluated by the simple process of calculation of ordinates.

**15.** *Numerical examples of motion of a top.* We shall carry out the calculation for the case of motion specified in 13 above. There

$$k^2 = 0·6819, \quad \kappa(=k'^{\frac{1}{2}}) = 0·7509, \quad \alpha = \cos^{-1} 0·5641 = 55° \, 41'.$$

It was found that $q = 0·0712$, so that $q^4 = 0·0000257$.

Now $$\Theta(K) = 1 + 2q + 2q^4 + \dots = 1·1425,$$

and so by the value of $q$, $$2 \operatorname{Log} \Theta(K) = 0·1157.$$

We shall suppose that the upper limit of $\phi$ is 50°.

Then
$$\text{Log dn}\, u = \tfrac{1}{2}\,\text{Log}(1 - 0\cdot6819 \times \sin^2 50°) = \bar{1}\cdot8890$$
$$\text{Log}\, \kappa = \text{Log}\, 0\cdot7509 = \bar{1}\cdot8756.$$

Hence
$$\lambda = \cot^{-1}\frac{\text{dn}\, u}{\kappa} = 44°\, 7', \quad \text{and} \quad 45° - \lambda = 53'.$$

Equation (3) therefore becomes
$$\frac{\cot\lambda - 1}{\cot\lambda + 1} = \tan(45° - \lambda) = 0\cdot0154.$$

Thus, by (5), we obtain as a first approximation,
$$\cos 2x = \frac{\tan(45° - \lambda)}{2q} = 0\cdot1083,$$

and
$$2x = 83°\, 47'.$$

It is unnecessary to go to the further approximation, as the correction applied would lie beyond the degree of accuracy here aimed at. In radian measure
$$x = 0\cdot7310,$$

so that $\text{Log}\, x = \bar{1}\cdot8640$. But
$$x = \frac{\pi}{2}\frac{u}{\text{K}} \quad \text{and} \quad \text{K} = \pi\Theta^2(\text{K}),$$

and therefore
$$u = x\Theta^2(\text{K}).$$

We have seen that $2\,\text{Log}\,(\Theta)\text{K} = 0\cdot1157$. Hence
$$\text{Log}\, u = \bar{1}\, 9795$$

and
$$u = 0\cdot954.$$

A reference to Legendre's tables shows that this result is nearly correct.

The value of $\theta$ corresponding to the upper limit of $\phi$ is given by the equation $\sin^2\phi = (z - z_1)/(z_2 - z_1)$, where $z = \cos\theta$. In the present case, $\phi = 50°$, and we get $\theta = 53°\, 42'$, which, as it happens, is the value of $\alpha$ $(= \sin^{-1}k)$.

As another example the reader may verify that if $k = \sin 10°$, and the upper limit of $\phi$ be 25°, we have, carrying the calculation to a higher degree of accuracy,
$$q = 0\cdot0019136, \quad \Theta(\text{K}) = 1 + 2q + 2q^4 + \cdots = \text{Log}^{-1} 0\cdot00165925,$$
$$\text{Log dn}\, u = \text{Log}\,\{(1 - \sin^2 10°.\, \sin^2 25°)^{\frac{1}{2}}\} = \bar{1}\cdot9988274,$$
$$\text{Log}\,\frac{\text{dn}\, u}{\kappa} = \text{Log}\cot\lambda = 0\cdot0021516.$$

Hence, to the first degree of approximation, since $45° - \lambda = 8'\, 30''$,
$$\text{Log}\cos 2x = \text{Log}\tan(45° - \lambda) - \text{Log}\, 2q = \bar{1}\cdot81027$$

and
$$x = 24°\, 52'\, 40''.$$

With the small value of $q$ here found the effect of the second approximation is very slight.

**16.** *Calculation of complete and incomplete elliptic integrals of the second kind.* Now consider the evaluation of $\text{E}(k, \phi)$, which is required in some calculations of the azimuthal angle $\psi$. We have
$$\text{E}(k, \phi) = \int_0^\phi (1 - k^2\sin^2\phi)^{\frac{1}{2}}\, d\phi, \quad \ldots\ldots\ldots (1)$$

which, if $u = \text{F}(k, \phi)$, can be written
$$\text{E}(k, \phi) = u - k^2\int_0^u \text{sn}^2 u\, du. \quad \ldots\ldots\ldots (2)$$

Now in treatises on elliptic functions it is proved that if $\Theta(u)$ have the value stated in (1), and $\Theta'(u)$, $\Theta''(u)$ be its first and second derivatives with respect to $u$,

$$k^2\int \mathrm{sn}^2 u\, du = u\frac{\Theta''(0)}{\Theta(0)} - \frac{\Theta'(u)}{\Theta(u)}. \qquad (3)$$

But

$$\left.\begin{aligned}\Theta'(u) &= \frac{\pi}{2\mathrm{K}}\left(4q\sin\frac{\pi u}{\mathrm{K}} - 8q^4\sin\frac{2\pi u}{\mathrm{K}} + 12q^9\sin\frac{3\pi u}{\mathrm{K}} - \cdots\right),\\ \Theta''(u) &= \frac{\pi^2}{4\mathrm{K}^2}\left(8q\cos\frac{\pi u}{\mathrm{K}} - 32q^4\cos\frac{2\pi u}{\mathrm{K}} + 72q^9\cos\frac{3\pi u}{\mathrm{K}} - \cdots\right).\end{aligned}\right\} \qquad (4)$$

Thus

$$\frac{\Theta''(0)}{\Theta(0)} = \frac{2\pi^2}{\mathrm{K}^2}\,\frac{q - 4q^4 + 9q^9 - 16q^{16} + \cdots}{1 - 2q + 2q^4 - 2q^9 + \cdots}, \qquad (5)$$

and

$$\frac{\Theta'(u)}{\Theta(u)} = \frac{\pi}{2\mathrm{K}}\,\frac{4q\sin\frac{\pi u}{\mathrm{K}} - 8q^4\sin\frac{2\pi u}{\mathrm{K}} + \cdots}{1 - 2q\cos\frac{\pi u}{\mathrm{K}} + 2q^4\cos\frac{2\pi u}{\mathrm{K}} - \cdots}. \qquad (6)$$

Again, if $\mathrm{A} = (1-q^2)(1-q^4)(1-q^6)\ldots,$
it can be proved that

$$\Theta(u) = \mathrm{A}\left(1 - 2q\cos\frac{\pi u}{\mathrm{K}} + q^2\right)\left(1 - 2q^3\cos\frac{\pi u}{\mathrm{K}} + q^6\right)\ldots \qquad (7)$$

Hence

$$\frac{\Theta'(u)}{\Theta(u)} = \frac{\pi}{2\mathrm{K}}\left\{\frac{4q\sin\frac{\pi u}{\mathrm{K}}}{1 - 2q\cos\frac{\pi u}{\mathrm{K}} + q^2} + \frac{4q^3\sin\frac{\pi u}{\mathrm{K}}}{1 - 2q^3\cos\frac{\pi u}{\mathrm{K}} + q^6} + \cdots\right\}. \qquad (8)$$

Now

$$\frac{q^n\sin 2x}{1 - 2q^n\cos 2x + q^{2n}} = q^n\sin 2x + q^{2n}\sin 4x + q^{2n}\sin 6x + \cdots \qquad (9)$$

Hence, putting $n = 1, 3, 5, \ldots$ in succession and adding the results, we obtain, with $2x = \pi u/\mathrm{K}$,

$$\frac{\Theta'(u)}{\Theta(u)} = \frac{\pi}{2\mathrm{K}}\left(\frac{4q\sin 2x}{1-q^2} + \frac{4q^2\sin 4x}{1-q^4} + \cdots\right) = \mathrm{S}, \text{ say}. \qquad (10)$$

Thus

$$\mathrm{E}(k, \phi) = u\left(1 - \frac{\Theta''(0)}{\Theta(0)}\right) + \mathrm{S}. \qquad (11)$$

If the complete integral is taken, that is if $u = \mathrm{K}$ and $\phi = \frac{1}{2}\pi$, S is zero, and

$$\mathrm{E}(k, \tfrac{1}{2}\pi) = \mathrm{K}\left(1 - \frac{\Theta''(0)}{\Theta(0)}\right) \qquad (12)$$

$$= \mathrm{K} - \frac{2\pi^2}{\mathrm{K}}\,\frac{q - 4q^4 + 9q^9 - 16q^{16} + \cdots}{\Theta(0)}. \qquad (13)$$

But $\mathrm{K} = \frac{1}{2}\pi\Theta^2(\mathrm{K})$, and since $\Theta(0) = \kappa\Theta(\mathrm{K})$, we have

$$\mathrm{E}(k, \tfrac{1}{2}\pi) = \mathrm{K} - \frac{4\pi q}{\kappa\Theta^3(\mathrm{K})}(1 - 4q^3 + 9q^8 - 16q^{15} + \cdots). \qquad (14)$$

**17.** ***Numerical examples.*** The numerical calculation can now be carried out. First $q$ is found, as explained in 13 above, for the value of $\alpha$ $(=\sin^{-1}k)$ given. Let it be for example 19°. Then the process gives, as may be verified,

$$3\,\mathrm{Log}\,q + \mathrm{Log}\,4 = \bar{7}{\cdot}5357147 + 0{\cdot}6020600 = \bar{6}{\cdot}1377747.$$

Thus $4q^3 = 0{\cdot}000001373$ and $\log(1 - 4q^3) = -4q^3$, so that

$$\mathrm{Log}(1 - 4q^3) = -4q^3 \times \mathrm{M} = -0{\cdot}000000596.$$

The value of $\log\Theta(\mathrm{K})$ is obtained from the value of $q$ $(=0{\cdot}00700226)$ as $0{\cdot}0060399$ (see below). Also that of K is calculated as before and found to be $1{\cdot}61510$.

The table of logarithms required to complete the calculation of $E(k, \frac{1}{2}\pi)$ is now

$$\begin{aligned} \text{Log } 4\pi &= 1{\cdot}0992099 & 3 \text{ Log } \Theta(K) &= 0{\cdot}0181197 \\ \text{Log } q &= \bar{3}{\cdot}8452383 & \tfrac{1}{2} \text{ Log cos } 19^\circ &= \bar{1}{\cdot}9878350 \\ \text{Log}(1-4q^3) &= -0{\cdot}0000006 & & \underline{0\;0059547} \\ & \bar{2}{\cdot}9444476 & & \\ & 0{\cdot}0059547 & & \\ & \bar{2}\;9384929 = \text{Log } 0\;0867946. & & \end{aligned}$$

Thus $$\frac{4\pi q}{\kappa\Theta^3(K)}(1-4q^3) = 0{\cdot}0867946,$$

and $$E = 1{\cdot}61510 - 0{\cdot}086795 = 1\;52831,$$
$$\text{Log } E = 0{\cdot}1842114.$$

Now let the integral be incomplete, and the terminal value of $\phi$ be 27°. Then, by (11) and (12), 16,

$$E(k, \phi) = \frac{u}{K}E + S = \frac{2x}{\pi}E + S,$$

where, (10), 16, $$S = \frac{2\pi}{K}\left(\frac{q \sin 2x}{1-q^2} + \frac{q^2 \sin 4x}{1-q^4} + \dots\right),$$

with $x = \pi u/2K$.

Now putting, as in 14, $\cot \lambda = (\text{dn } u)/\kappa$, we first determine $x$ by the approximate equation

$$2q \cos 2x = \tan(45^\circ - \lambda),$$

which gives $$2x = 52^\circ\; 42'\; 40''.$$

This, by the process explained in 14, is corrected to

$$x = 26^\circ\; 21'\; 32'' = 94892'',$$

so that the radian measure of $x$ is 94892/206264·8. Hence

$$\begin{aligned} \text{Log } x &= \bar{1}{\cdot}6628042 \\ \text{Log } E &= 0{\cdot}1842104 \\ \text{CoLog } \tfrac{1}{2}\pi &= \bar{1}{\cdot}8038801 \\ \text{Log } \frac{2xE}{\pi} &= \bar{1}{\cdot}6508947 \end{aligned}$$

Thus $$\frac{2xE}{\pi} = 0{\cdot}447605.$$

Now to find S we have first $\text{Log}(1-q^2) = -Mq^2 = -0{\cdot}0000213$. The logarithms of $1-q^4$, $1-q^6$, ... may all be taken as zero. Hence, as the reader may verify, we have for the Logs of the terms in $SK/2\pi$, that is $S\Theta^2(K)/4$, since $K = \frac{1}{2}\pi\Theta^2(K)$,

| | | |
|---|---|---|
| $\text{Log } q = \bar{3}{\cdot}8452383$ | $\text{Log } q^2 = \bar{5}{\cdot}69048$ | $\text{Log } q^3 = 7{\cdot}53571$ |
| $\text{Log sin } 2x = \bar{1}{\cdot}9007282$ | $\text{Log sin } 4x = \bar{1}{\cdot}98405$ | $\text{Log sin } 6x = \bar{1}{\cdot}57069$ |
| $\text{CoLog}(1-q^2) = 0\;0000213$ | $\text{Log } 0{\cdot}0^4 4726 = \bar{5}{\cdot}67453$ | $\text{Log } 0{\cdot}0^6 1278 = \bar{7}\;10640$ |
| $\text{Log } 0{\cdot}0055717 = \bar{3}{\cdot}7459878$ | | |

Thus $$\frac{S\Theta^2(K)}{4} = 0{\cdot}005717 + 0{\cdot}0000473 + 0{\cdot}0000001 = 0{\cdot}005619.$$

But $$\Theta(K) = 1 + 2q + 2q^4 + 2q^9 + \dots,$$

and
$$\begin{aligned} \text{Log } 0{\cdot}0056191 &= \bar{3}{\cdot}7496668 \\ \text{Log } 4 &= 0{\cdot}6020600 \\ \text{CoLog } \Theta^2(K) &= \bar{1}{\cdot}9879202 \\ & \bar{2}{\cdot}3396470 \end{aligned}$$

$$S = 0{\cdot}0218598.$$

---

Hence if $k = \sin 19^\circ$, and $\phi = 27^\circ$,

$$E(k, \phi) = 0{\cdot}447605 + 0{\cdot}0218598 = 0{\cdot}469465.$$

**18.** *Formulae for azimuthal angle $\psi$.* In order to calculate the angle turned through in azimuth in any given time $t$ we have to evaluate an elliptic integral of the third kind, and it is desirable to express it in Jacobian functions. The normal form of this integral may be written

$$\int_0^x \frac{dx}{(1+px^2)\{(1-x^2)(1-k^2x^2)\}^{\frac{1}{2}}}.$$

We denote it by I. If we put $\operatorname{sn} u$ for $x$ and $\operatorname{sn}^2 v$ for $-1/p$, we have $dx = \operatorname{cn} u \operatorname{dn} u \, du$, and

$$\mathrm{I} = -\operatorname{sn}^2 v \int_0^u \frac{du}{\operatorname{sn}^2 u - \operatorname{sn}^2 v}. \quad \text{......(1)}$$

Now it is proved in treatises on Elliptic Functions that if H be the function defined by

$$\mathrm{H}(u) = 2q^{\frac{1}{4}} \sin \frac{\pi u}{2\mathrm{K}} - 2q^{\frac{9}{4}} \sin \frac{3\pi u}{2\mathrm{K}} + 2q^{\frac{25}{4}} \sin \frac{5\pi u}{2\mathrm{K}} - \cdots,$$

and if we define $\mathrm{Z}(u)$ by

$$\mathrm{Z}(u) = \frac{d}{du}\{\log \mathrm{H}(u)\},$$

that is if

$$\mathrm{Z}(u) = \frac{\mathrm{H}'(u)}{\mathrm{H}(u)}, \quad \text{......(2)}$$

any elliptic function can be expanded in a series of elements consisting of Z-functions, of which the coefficients are the *residues* of the functions for the respective poles. [This $\mathrm{Z}(u)$ is not Jacobi's zeta-function, which is $\Theta'(u)/\Theta(u)$.]

The following simple case will illustrate the meaning of this statement. A function $f(u)$ is said to have two isolated simple poles at the points $a$, $b$, if it can be written as $\phi_1(u)/(u-a)$, or $\phi_2(u)/(u-b)$, where $\phi_1(u)$, $\phi_2(u)$ are finite and continuous in the vicinity of $u=a$ and $u=b$, respectively. Then $\phi_1(a)$, $\phi_2(b)$ are the *residues* of the function for the points in question, according to the theory of complex integration given by Cauchy. They are in fact the values for $u=a$, $u=b$, of

$$\frac{1}{2\pi i}\int f(u)\,du,$$

when the integral is taken round a circle of infinitesimal radius surrounding in one case the point $a$, in the other the point $b$. We shall call these residues A, B.

The expression of this function in elements is then given by

$$f(u) = \mathrm{A}\mathrm{Z}(u-a) + \mathrm{B}\mathrm{Z}(u-b) + \mathrm{C}. \quad \text{......(3)}$$

Now for

$$f(u) = \frac{1}{\operatorname{sn}^2 u - \operatorname{sn}^2 v}, \quad \text{......(4)}$$

the poles are the points $u = -v$, $u = +v$, and

$$\mathrm{A} = \operatorname*{Lt}_{u=v} \frac{u-v}{\operatorname{sn}^2 u - \operatorname{sn}^2 v}, \quad \mathrm{B} = \operatorname*{Lt}_{u=-v} \frac{u+v}{\operatorname{sn}^2 u - \operatorname{sn}^2 v}.$$

The limiting values are obtained in the usual way by differentiating the numerator and denominator in each case, so that

$$\mathrm{A} = \frac{1}{2\operatorname{sn} v \operatorname{cn} v \operatorname{dn} v}, \quad \mathrm{B} = -\frac{1}{2\operatorname{sn} v \operatorname{cn} v \operatorname{dn} v}. \quad \text{......(5)}$$

Thus we get at once

$$\frac{2\operatorname{sn} v \operatorname{cn} v \operatorname{dn} v}{\operatorname{sn}^2 u - \operatorname{sn}^2 v} = \mathrm{Z}(u-v) - \mathrm{Z}(u+v) + \mathrm{C}. \quad \text{......(6)}$$

To determine the constant C, we put $u=0$, and thereby obtain, since $\mathrm{Z}(-v) = -\mathrm{Z}(v)$,

$$\mathrm{C} = 2\left\{\frac{\mathrm{H}'(v)}{\mathrm{H}(v)} - \frac{\operatorname{cn} v \operatorname{dn} v}{\operatorname{sn} v}\right\}$$

But

$$\operatorname{sn} v = \frac{1}{k^{\frac{1}{2}}} \frac{\mathrm{H}(v)}{\Theta(v)},$$

and therefore logarithmic differentiation gives

$$\frac{\operatorname{cn} v \operatorname{dn} v}{\operatorname{sn} v} = \frac{\mathrm{H}'(v)}{\mathrm{H}(v)} - \frac{\Theta'(v)}{\Theta(v)}, \quad \text{......(7)}$$

so that

$$\mathrm{C} = 2\frac{\Theta'(v)}{\Theta(v)}. \quad \text{......(8)}$$

We have therefore

$$\frac{2 \sin v \operatorname{cn} v \operatorname{dn} v}{\operatorname{sn}^2 u - \operatorname{sn}^2 v} = \mathrm{Z}(u-v) - \mathrm{Z}(u+v) + 2\frac{\Theta'(v)}{\Theta(v)}, \quad \text{......(9)}$$

and

$$2 \operatorname{sn} v \operatorname{cn} v \operatorname{dn} v \int \frac{du}{\operatorname{sn}^2 u - \operatorname{sn}^2 v} = \log \frac{\mathrm{H}(u-v)}{\mathrm{H}(u+v)} + 2u\frac{\Theta'(v)}{\Theta(v)} + \text{const.} \quad \text{......(10)}$$

If we integrate (9) with respect to $v$ we get

$$-2\int \frac{\operatorname{sn} v \operatorname{cn} v \operatorname{dn} v \, dv}{\operatorname{sn}^2 u - \operatorname{sn}^2 v} = \log(\operatorname{sn}^2 u - \operatorname{sn}^2 v)$$

$$= \log\{\mathrm{H}(u-v)\mathrm{H}(u+v)\} - 2\log\Theta(v) + \log c, \quad \text{......(11)}$$

where $c$ is a function of $u$. But if $v=0$, the last result gives

$$\operatorname{sn}^2 u = c\frac{\mathrm{H}^2(u)}{\Theta^2(0)}, \quad \text{or} \quad c = \operatorname{sn}^2 u \frac{\Theta^2(0)}{\mathrm{H}^2(u)}.$$

But

$$\operatorname{sn} u = \frac{1}{k^{\frac{1}{2}}}\frac{\mathrm{H}(u)}{\Theta(u)},$$

and therefore, finally,

$$c = \frac{1}{k}\frac{\Theta^2(0)}{\Theta^2(u)}. \quad \text{......(12)}$$

We obtain from (10) the result

$$\mathrm{I} = -\frac{\operatorname{sn} v}{2 \operatorname{cn} v \operatorname{dn} v}\left\{\log\frac{\mathrm{H}(u-v)}{\mathrm{H}(u+v)} + 2u\frac{\Theta'(v)}{\Theta(v)} + \mathrm{C}\right\}. \quad \text{......(13)}$$

If we suppose the integral to start from $u=0$, we have for the determination of C,

$$\log\frac{\mathrm{H}(-v)}{\mathrm{H}(v)} + \mathrm{C} = 0,$$

and therefore $\mathrm{C} = -\log\{\mathrm{H}(-v)/\mathrm{H}(v)\}$. Hence (13) becomes, since $\mathrm{H}(-v) = -\mathrm{H}(v)$,

$$\mathrm{I} = -\frac{\operatorname{sn} v}{2 \operatorname{cn} v \operatorname{dn} v}\left\{\log\frac{\mathrm{H}(v-u)}{\mathrm{H}(u+v)} + 2u\frac{\Theta'(v)}{\Theta(v)}\right\}. \quad \text{......(14)}$$

The calculation of the two integrals $\mathrm{I}_1$, $\mathrm{I}_2$ required for $\psi$, by these functions is, as has been stated, troublesome. The reader will realise this if he attempts to develope formulae for the purpose. Numerical values for complete and incomplete integrals of the first and second kinds are given by many writers, but not one, so far as we know, gives calculations of integrals of the third kind. For such integrals there are no tables available, and the process of computation by means of Jacobian $\vartheta$-functions, or by Weierstrassian $\sigma$-functions, is troublesome, inasmuch as on the one hand sn $v$ has in one or other integral an imaginary argument, and on the other the calculation of the $\sigma$-functions by means of $q$-series cannot be carried out with brevity.

The same remark applies to the formulae in terms of Weierstrassian $\sigma$-functions and $\zeta$-functions, which are derived for a top by the process exemplified in 17...20, XV, below.*

**19.** ***Elliptic integral expressions for angles $\psi$ and $\phi$.*** We shall now however show how the integrals can be obtained by the prosaic, but direct and practical, process of calculating ordinates. First we set out the integrals to be found. By 4 above we have

$$\psi = \frac{1}{2\mathrm{A}}\left\{(\mathrm{G}+\mathrm{C}n)\int\frac{dz}{(1+z)\mathrm{Z}^{\frac{1}{2}}} + (\mathrm{G}-\mathrm{C}n)\int\frac{dz}{(1-z)\mathrm{Z}^{\frac{1}{2}}}\right\}. \quad \text{......(1)}$$

* Angles are here measured from the upward vertical, and the order of the roots $z_1 < z_2 < z_3$ is the opposite of the Weierstrassian order, $e_1 > e_2 > e_3$, which is used later.

By the substitution explained above this is transformed into

$$\psi=\left\{\frac{1}{2\mathrm{AM}gh(z_3-z_1)}\right\}^{\frac{1}{2}}\left[\frac{\mathrm{C}n+\mathrm{G}}{1+z_1}\int\frac{d\phi}{\left(1+\frac{z_2-z_1}{1-z_1}\sin^2\phi\right)(1-k^2\sin^2\phi)^{\frac{1}{2}}}\right.$$

$$\left.+\frac{\mathrm{C}n-\mathrm{G}}{1-z_1}\int\frac{d\phi}{\left(1-\frac{z_2-z_1}{1-z_1}\sin^2\phi\right)(1-k^2\sin^2\phi)^{\frac{1}{2}}}\right]. \quad \text{.............(2)}$$

Denoting the integrals by $\mathrm{I}_1$, $\mathrm{I}_2$, we have

$$\psi=\mathrm{C}_1\mathrm{I}_1+\mathrm{C}_2\mathrm{I}_2, \quad \text{.........................................(3)}$$

where $$\mathrm{C}_1=\frac{\mathrm{C}n+\mathrm{G}}{1+z_1}\left\{\frac{1}{2\mathrm{AM}gh(z_3-z_1)}\right\}^{\frac{1}{2}}, \quad \mathrm{C}_2=\frac{\mathrm{C}n-\mathrm{G}}{1-z_1}\left\{\frac{1}{2\mathrm{AM}gh(z_3-z_1)}\right\}^{\frac{1}{2}}. \quad \text{.............(4)}$$

In the general case of an ordinary top we have [see 4 above]

$$\dot\psi=\frac{\mathrm{G}-\mathrm{C}nz}{\mathrm{A}(1-z^2)}\frac{dz}{\mathrm{Z}^{\frac{1}{2}}}, \quad \dot\phi=\frac{\mathrm{C}n-\mathrm{G}z}{\mathrm{A}(1-z^2)}\frac{dz}{\mathrm{Z}^{\frac{1}{2}}}+\mathrm{C}n\left(\frac{1}{\mathrm{C}}-\frac{1}{\mathrm{A}}\right); \quad \text{.....................(5)}$$

so that for a spherical top there is a perfectly symmetrical relation between $\dot\psi$ and $\dot\phi$. It will be clear that in the latter case, if an integral has been calculated for $\psi$, the corresponding value for $\phi$ will be obtained by interchanging $\mathrm{C}n$ and G. In the general case the term $\mathrm{C}n\{(\mathrm{A}-\mathrm{C})/\mathrm{AC}\}t$ must be added to the integral obtained as indicated.

**20.** ***Numerical calculation of time in terms of angle θ for actual top.*** Thus by (2), 19, the calculation of the azimuthal angle turned through in any given time requires the evaluation of two elliptic integrals of the third kind. We compute a sufficient number of ordinates of the curve which represents by its ordinates the successive values of the integrand, and thence deduce the area. This process has the great advantage of giving at once, in each case, all the incomplete integrals. To illustrate it we take the case of a top for which calculations have been made by Klein and Sommerfeld [*Theorie des Kreisels*, Kap. IV, § 9, and VI, § 6], using their own method of determining the parameters $\alpha$, $\beta$, $\gamma$, $\delta$ by $q$-series. We shall thus be able to test the direct method here recommended, by comparing the results with those of the more recondite process employed by these writers.

The top employed had the following constants:

$$\mathrm{C}n=4800\rho\pi^2, \quad \mathrm{G}=4200\rho\pi^2, \quad \mathrm{A}=750\rho\pi, \quad \mathrm{M}gh=10000\rho\pi^3,$$

where $\rho$ denotes the density of the material of which the top was composed. For the lower limiting circle $\theta$ was 60°, so that $z_1=0{\cdot}5$. By (4), 16, V we obtain for the determination of the other two roots of the cubic $\mathrm{Z}=0$ an equation which, after a little reduction, takes the form

$$z^2-1{\cdot}8242z+0{\cdot}776=0, \quad \text{.........................................(1)}$$

of which the roots are 0·6759 and 1·1481. Thus the upper limiting circle is given by $z_2=0{\cdot}6759$, and for that $\theta=47°\,31'$. The large root, 1·1481, is $z_3$.

For the modulus $k$ of the Legendre integral of the first kind, by which we express the time for any value of $\theta$, we have

$$k^2=\frac{z_2-z_1}{z_3-z_1}=0{\cdot}271441=0{\cdot}5210^2. \quad \text{.................................(2)}$$

The time $t$ of passage from the lower limiting circle to the value of $\theta$ which corresponds to $\phi$, the upper limit of the integral, is given by

$$t=\left\{\frac{2\mathrm{A}}{\mathrm{M}gh(z_3-z_1)}\right\}^{\frac{1}{2}}\int_0^{\phi}\frac{d\phi}{(1-0{\cdot}271441\sin^2\phi)^{\frac{1}{2}}}. \quad \text{..........................(3)}$$

The connection between $\theta$ and $\phi$ is given as we have seen by

$$\frac{z-z_1}{z_2-z_1}=\sin^2\phi, \quad \text{......................................................(4)}$$

where $z=\cos\theta$.

We calculate first the integral exhibited above, and the process adopted will give it at once for values of $\phi$ rising by successive steps of 5° from $\phi=0$ to $\phi=90°$. It is obvious that no ordinate can differ very much from 1. The common logarithms of $\sin^2 5°$, $\sin^2 10°$, ..., are got from the tables and written down in column with a space between each and the next in order. The logarithm of 0·27441 is then taken on a slip of paper and added to each of the former logarithms so as to give those of the values of $0{\cdot}27441\sin^2\phi$. The corresponding numbers are then taken out and subtracted from unity, and the reciprocals of the results obtained either by logarithmic tables, or directly from a table of reciprocals.

Thus we get the table of ordinates:

| | | | |
|---|---|---|---|
| 0° | 1 | | |
| 5 | 1·0010 | 50° | 1·0906 |
| 10 | 1·0041 | 55 | 1·1057 |
| 15 | 1·0092 | 60 | 1·1205 |
| 20 | 1·0163 | 65 | 1·1344 |
| 25 | 1·0252 | 70 | 1·1468 |
| 30 | 1·0358 | 75 | 1·1572 |
| 35 | 1·0479 | 80 | 1·1650 |
| 40 | 1·0613 | 85 | 1·1699 |
| 45 | 1·0756 | 90 | 1·1715 |

Let us suppose that we wish to find the complete integral. We add the first and last ordinates, and take half the sum, which gives 1·0858. Then we add to this the other 17 ordinates, and get as the sum 19·4527, which on a certain preliminary scale is approximately the area.

This process is equivalent to supposing that the heads of two successive ordinates may, without sensible error, be taken as joined by a straight line, and then taking as the mean ordinate of each 5° strip of the integral the middle ordinate of the strip. It does not add very seriously to the work to take steps of $2\frac{1}{2}°$ each, so as to obtain greater accuracy. But, as will be seen below, we can obtain with 5° steps a result which is right to three decimal places.

We obtain the proper numerical value of the integral by reducing from degrees to radians. Thus we multiply by the factor 5/57·3, and obtain

$$F\left(0{\cdot}521,\ \frac{\pi}{2}\right)=\frac{19{\cdot}4527\times 5}{57{\cdot}3}=1{\cdot}697,$$

which agrees exactly with the result of the calculation by $q$-series, referred to above.

By exactly the same process we can obtain any incomplete integral. For example, we take that from $\phi=0$ to $\phi=50°$, with the same modulus. The half sum of the ordinate for 0° and that for 50° is 1·0453; the sum of the nine intermediate ordinates is 9·2767. Hence, on the preliminary scale, the incomplete integral required is

$$1{\cdot}0453+9{\cdot}2767=10{\cdot}3219.$$

The proper numerical value is therefore

$$F(0{\cdot}521,\ 50°)=\frac{10{\cdot}3220\times 5}{57{\cdot}3}=0{\cdot}9007$$

Summed in the same way the remaining strips of the complete integral give

$$F\left(0{\cdot}521,\ \frac{\pi}{2}\right)-F(0{\cdot}521,\ 50^\circ)=\frac{9{\cdot}1305\times 5}{57{\cdot}3}=0{\cdot}7967.$$

The two areas make up 1·6974, the value already obtained above.

The factor by which the complete or incomplete integral must be multiplied to give the time of motion is stated above in symbols, and in the present case is

$$\left(\frac{2\times 750\times\rho\pi}{10000\rho\pi^3\times 0{\cdot}6481}\right)^{\frac{1}{2}}=\left(\frac{15}{100\times 0{\cdot}6481}\right)^{\frac{1}{2}}\frac{1}{\pi}=0{\cdot}1531.$$

Thus the time, in seconds, of passage from one circle to the other, is $1{\cdot}697\times 0{\cdot}1531=0{\cdot}26$. Similarly the time for passage from $\phi=0$ to $\phi=50^\circ$ is $0{\cdot}9007\times 0{\cdot}1531=0{\cdot}1379$, in seconds.

The angle $\phi$ corresponds to $\theta=\cos^{-1}0{\cdot}60322$, or $\theta=52^\circ\,54'$, and the range $\phi=0$ to $\phi=50^\circ$ is the range $\theta=60^\circ$ to $\theta=52^\circ\,54'$.

**21.** *Numerical calculation of azimuthal angle* $\psi$. We now consider the determination of the angle $\psi$. Going back to (2), 19, we have first to find the integrals which are there represented in symbols. The values $z_1$ and $z_2$ inserted show that these integrals are

$$\int\frac{d\phi}{(1+0{\cdot}11727\sin^2\phi)(1-0{\cdot}271441\sin^2\phi)^{\frac{1}{2}}},$$

$$\int\frac{d\phi}{(1-0{\cdot}3518\sin^2\phi)(1-0{\cdot}271441\sin^2\phi)^{\frac{1}{2}}}.$$

We first evaluate for values of $\phi$ separated by steps, again of 5°, the factors

$$1+0{\cdot}11727\sin^2\phi,\quad 1-0{\cdot}3518\sin^2\phi,$$

and then multiply each of the values so obtained by the corresponding value of

$$(1-0{\cdot}271441\sin^2\phi)^{\frac{1}{2}},$$

and take the reciprocals of the products. The calculations are best carried out by a table of seven-place logarithms, so as to avoid expenditure of time and possible error in reckoning differences.

Here follow the values of the ordinates for the first integral, each multiplied by 100000:

| | | | |
|---|---|---|---|
| 0° | 100000 | | |
| 5 | 100014 | 50° | 102041 |
| 10 | 100058 | 55 | 102510 |
| 15 | 100135 | 60 | 102993 |
| 20 | 100251 | 65 | 103476 |
| 25 | 100413 | 70 | 103923 |
| 30 | 100626 | 75 | 104309 |
| 35 | 100895 | 80 | 104614 |
| 40 | 101223 | 85 | 104796 |
| 45 | 101611 | 90 | 104850 |

Taking half the sum of the first and last, and adding it to the sum of the other 17 values, we get 18·36314. Reducing the abscissae to radians by multiplying by 5/57·3, we get for the integral

$$I_1=\frac{91{\cdot}815}{90}\frac{\pi}{2}=1{\cdot}02017\frac{\pi}{2}.$$

The factor $C_1$, by which the integral must be multiplied to give the corresponding part, $\psi_1$, of the angle $\psi$, is found from the values of A, $Mgh$, $Cn$, G, $z_3$, $z_1$ above to be 1·9243.

Hence we have

$$\psi_1=1{\cdot}9243\,\frac{91{\cdot}815}{90}\,\frac{\pi}{2}=1{\cdot}9631\,\frac{\pi}{2}.$$

In the same way we deal with the other integral, and apply the necessary factors. The successive elements of the integral, each multiplied by 10000, are

| | | | |
|---|---|---|---|
| 0° | 10000 | | |
| 5 | 10037 | 50° | 13744 |
| 10 | 10149 | 55 | 14474 |
| 15 | 10336 | 60 | 15222 |
| 20 | 10599 | 65 | 15955 |
| 25 | 10939 | 70 | 16637 |
| 30 | 11356 | 75 | 17225 |
| 35 | 11850 | 80 | 17685 |
| 40 | 12418 | 85 | 17975 |
| 45 | 13052 | 90 | 18074 |

Again we add half the sum of the first and last ordinates to the sum of the other 17 values, and get 24385. The factor $C_2$ is easily found to be $-\frac{1}{5}C_1$, and so we get for the part $\psi_2$ of the azimuthal angle the equation

$$-\psi_2 = 0{\cdot}3849\,\frac{121{\cdot}845}{90}\,\frac{\pi}{2} = 0{\cdot}52\,\frac{\pi}{2}.$$

Hence we have $$\psi = \psi_1 + \psi_2 = (1{\cdot}9632 - 0{\cdot}52)\frac{\pi}{2} = 1{\cdot}443\,\frac{\pi}{2}.$$

This is exactly the value obtained by Klein and Sommerfeld for the azimuthal angle traversed in the half period of the motion.

**22.** *Numerical calculation of $\psi$ for any step in time.* The reader may, if he please, calculate in a short time the angles traversed in azimuth for each 5° step of $\phi$, and from the elements of the integral in 20 above the corresponding times and the values of $\theta$.

For example, the range from $\phi = 0$ to $\phi = 50°$ gives for the first integral 10·0630 on the preliminary scale. Hence the value of the azimuthal angle is

$$\psi_1 = 1{\cdot}9243\,\frac{50{\cdot}3150}{90}\,\frac{\pi}{2} = 1{\cdot}0717\,\frac{\pi}{2}.$$

The second integral for the same range is 11·1066 on the preliminary scale, and we have

$$-\psi_2 = 0{\cdot}3849\,\frac{55{\cdot}508}{90}\,\frac{\pi}{2} = 0{\cdot}2373\,\frac{\pi}{2}.$$

Thus the whole angle turned through from the time of contact with the lower circle is

$$\psi_1 + \psi_2 = 0{\cdot}8344\,\frac{\pi}{2}.$$

The time from the lower circle where $\phi = 0$ to $\phi = 50°$, that is from $\theta = 60°$ to $\theta = 52°\,56'$, as may be verified from $$\cos\theta = (0{\cdot}6759 - 0{\cdot}5)\sin^2 50° + 0{\cdot}5,$$
is, by the calculation in 20 above, given, in seconds, by

$$t = 0{\cdot}1531 \times 0{\cdot}9007 = 0{\cdot}1379,$$

about half the interval (0·26) from the lower to the higher circle.

Figure 65, taken from Klein and Sommerfeld's treatise, shows our calculated results very well. The path of an axial point is shown in stereographic projection from the lowest point of the sphere, and the successive points marked 0, 1, 2, 3, ..., 9, which start from the contact with the lower circle, show the positions after successive intervals of time, each equal to $\frac{1}{9}$ of the half period of 0·26 second. The rest of the path for any time whatever is given by proper repetitions of the portion here shown.

The reader will have no difficulty in making out these repetitions. The path, after the part numbered 1, 2, ..., 9 has been described, passes on to touch the outer circle a little below the extreme left, then passes inward again, touches the inner circle on the right of the highest point 0 of the diagram, then passes down to touch the outer circle near the lowest point. Thence it passes upward and inwards to touch the inner circle above the centre on the left, thence to touch the outer circle above the extreme right, and so on.

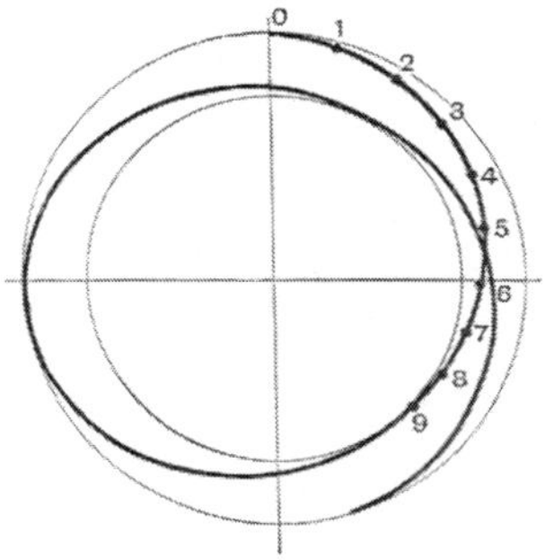


FIG. 65.

The formulae given in 19 for the calculation of the azimuthal motion show the effects of varying the spin and the sidelong motion at the upper limiting circle, in altering the amount of swinging round of the path. The effect, for example, of continual increase of spin from a small value to a large, with the sidelong motion at the upper circle kept zero, will be to give at first extreme cases of Fig. 22, p. 97, with the cusped indentations wide, then smaller and smaller cusped elements, until a regular sequence of microscopic elements is obtained, which simulates but is not really steady motion. It is Klein and Sommerfeld's "pseudo-regular precession." The reader may notice that to the path for infinitely rapid spin no tangent can be drawn except the lower circle, which touches all the undulations.

It is impossible to illustrate the different cases here. A conspectus of diagrams will be given at the end of the book, with descriptive notes on the different cases.

# CHAPTER XIII

## LIQUID GYROSTAT. MISCELLANEOUS INVESTIGATIONS

1. *Rotation of an ellipsoidal case filled with liquid. "Liquid gyrostats."* Lord Kelvin illustrated by what he called a "liquid gyrostat" the fact that an oblate spheroidal shell of rigid material filled with water behaves as regards precession as if its contents were solid.* The gyrostat

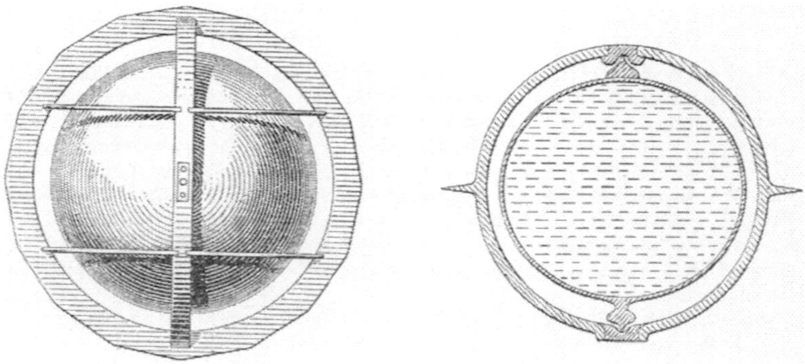

FIG. 66.

with which the experiment was made is shown in Fig. 66. It resembled the ordinary Kelvin gyrostat in being composed of a flywheel mounted in a rigid case surrounded by an equatorial ring with polygonal edge, as shown in the diagram; but the case was an open frame, and the spheroidal globe containing water took the place of the flywheel. When the globe was spun in the ordinary way the liquid gyrostat imitated the behaviour of the solid one in all respects.

The spheroid had an oblateness of about 5 per cent., that is the difference in length of the polar and equatorial diameters was about 5 per cent. of the length of either. But sometime later another liquid gyrostat was made, similar in all respects to the former, except that it was prolate instead of

*Brit. Assoc. 1876; *Nature*, Feb. 1, 1877.

oblate to about the same percentage (Fig. 67). The dynamical behaviour of this was quite different. When an attempt was made to spin it, it was found, as soon as the instrument was removed from the spinning table, that all rotation had disappeared. In consequence of instability of the fluid motion, the energy of rotation had been entirely transformed into heat by turbulent motion of the water, into which in such a case the rotational motion breaks down. Permanent steady rotation of the liquid spheroid is impossible when the axis of figure (the axis of rotation) is prolate.

Oblateness however is not absolutely essential for steady rotational motion of a liquid round the axis of figure in a spheroidal case turning with

FIG. 67.

the liquid. It was shown by Sir George Greenhill * in 1880 that steady motion is possible for a prolate spheroid of a liquid, if it be sufficiently prolate. The axial diameter must, in fact, either be shorter than the equatorial diameter or be more than three times as long. [See 3, below.] A modern elongated projectile, if filled with a liquid, would not rotate steadily about its axis of figure, and therefore would not have a definite trajectory as a rifle bullet has; it would turn broadside on to the direction of motion.

An experiment with a hard-boiled egg and a raw egg, spun together on a table, illustrates very well the stability of the solid prolate spheroid, and the instability of the motion of the liquid prolate spheroid. The egg is placed on its side on a table, and a rapid twist with the fingers sets it spinning about one of the shorter diameters. The solid egg however rises to the position in which its centroid is as high as possible, and then spins stably with the long axis vertical. The experiment does not succeed with the unboiled egg, which remains on its side. By placing the finger on the shell one can bring the egg apparently to rest, but, when the finger is raised immediately after, the shell begins turning again owing to the continued motion of the liquid contents; soon the whole spin has disappeared.

* *Proc. Camb. Phil. Soc.* 1880; *Encycl. Brit.* 10th Edn., Art. "Hydromechanics"; or *Report on Gyroscopic Theory*, 1914.

**2.** *Theoretical discussion of liquid gyrostat.* The following discussion follows Sir George Greenhill's investigation of the stability of the spinning motion of a mass of liquid contained in a spheroidal case. We suppose to begin with that the case is ellipsoidal, fulfilling the equation

$$\frac{x^2}{a^2}+\frac{y^2}{b^2}+\frac{z^2}{c^2}=1. \quad \text{.....(1)}$$

First let the contents of the shell be solid, and in one piece with the shell, and the components of angular velocity about the axes $O(x, y, z)$ be $\xi, \eta, \zeta$. The velocity components $u, v, w$ at any point $x, y, z$ of the containing shell, or its interior, are given by

$$u=z\eta-y\zeta, \quad v=x\zeta-z\xi, \quad w=y\xi-x\eta. \quad \text{.....(2)}$$

Let now the contents be liquefied, and additional components $\Omega_1, \Omega_2, \Omega_3$ impressed on the case. Additional components $u_1, v_1, w_1$ of the velocity of the contents will be found from the velocity potential

$$\phi=-\Omega_1\frac{b^2-c^2}{b^2+c^2}yz-\Omega_2\frac{c^2-a^2}{c^2+a^2}zx-\Omega_3\frac{a^2-b^2}{a^2+b^2}xy, \quad \text{.....(3)}$$

by the usual relations $u_1=-\partial\phi/\partial x$, etc. It will be found on trial that these components give, at any point of contact of the liquid with the shell, the same speed as the shell, together with an additional motion tangential to the shell, of components

$$2a^2\left(\frac{\Omega_3 y}{a^2+b^2}-\frac{\Omega_2 z}{c^2+a^2}\right), \quad 2b^2\left(\frac{\Omega_1 z}{b^2+c^2}-\frac{\Omega_3 x}{a^2+b^2}\right), \quad 2c^2\left(\frac{\Omega_2 x}{c^2+a^2}-\frac{\Omega_1 y}{b^2+c^2}\right).$$

The motion of the case is now derived from the components of angular velocity

$$P=\Omega_1+\xi, \quad Q=\Omega_2+\eta, \quad R=\Omega_3+\zeta, \quad \text{.....(4)}$$

so that if $u, v, w$ be the components of the velocity which the fluid now possesses, the components *relative to the case* $u', v', w'$ are given by

$$\left.\begin{aligned} u'&=u+yR-zQ=2a^2\left(\frac{\Omega_3 y}{a^2+b^2}-\frac{\Omega_2 z}{c^2+a^2}\right),\\ v'&=v+zP-xR=2b^2\left(\frac{\Omega_1 z}{b^2+c^2}-\frac{\Omega_3 x}{a^2+b^2}\right),\\ w'&=w+xQ-yP=2c^2\left(\frac{\Omega_2 x}{c^2+a^2}-\frac{\Omega_1 y}{b^2+c^2}\right).\end{aligned}\right\} \quad \text{.....(5)}$$

It will be noticed that at any internal point

$$u'\frac{x}{a^2}+v'\frac{y}{b^2}+w'\frac{z}{c^2}=0,$$

which expresses the fact that the motion of a particle is always on an ellipsoid similar to the containing shell.

Taking account of the gravitation of the fluid, putting $p$ for the pressure and $\rho$ for the density, we get for the equations of motion with respect to axes moving with the shell, with angular speeds P, Q, R,

$$\frac{1}{\rho}\frac{\partial p}{\partial x}+4\pi\kappa\rho Ax+\frac{\partial u}{dt}-vR+wQ+u'\frac{\partial u}{\partial x}+v'\frac{\partial u}{\partial y}+w'\frac{\partial u}{\partial z}=0, \quad \text{.....(6)}$$

with two similar equations. Here $\kappa$ is the gravitation constant, and the quantities A, B, C for the three equations take account of the attraction of the liquid on a particle of itself at the point $x, y, z$, and are obtained from the equations *

$$A, B, C=\int_0^\infty \frac{abc}{\lambda+a^2,\ \lambda+b^2,\ \lambda+c^2}\,\frac{d\lambda}{2\{(\lambda+a^2)(\lambda+b^2)(\lambda+c^2)\}^{\frac{1}{2}}}. \quad \text{.....(7)}$$

* See a paper "On the Attraction of Ellipsoids," by A. Gray, *Phil. Mag.* May, 1907.

There are also the equations of A.M.

$$\frac{\partial h_1}{\partial t}-h_2\mathrm{R}+h_3\mathrm{Q}=0,\quad \frac{\partial h_2}{\partial t}-h_3\mathrm{P}+h_1\mathrm{R}=0,\quad \frac{\partial h_3}{\partial t}-h_1\mathrm{Q}+h_2\mathrm{P}=0, \quad\ldots\ldots(8)$$

with

$$h_1=\Sigma m(yw-zv)=\Omega_1\frac{b^2-c^2}{b^2+c^2}\Sigma m(y^2-z^2)+\xi\Sigma m(y^2+z^2)=\tfrac{1}{5}\mathrm{M}\left\{\frac{(b^2-c^2)^2}{b^2+c^2}\Omega_1+(b^2+c^2)\xi\right\},\ \ldots(8')$$

and similar values for $h_2$, $h_3$. Here M denotes the mass of the liquid.

It may be verified that the equations of motion (6), reduce to

$$\left.\begin{aligned}\frac{1}{\rho}\frac{\partial p}{\partial x}+4\pi\kappa\rho\mathrm{A}x+\alpha x&=0,\\ \frac{1}{\rho}\frac{\partial p}{\partial y}+4\pi\kappa\rho\mathrm{B}y+\beta y&=0,\\ \frac{1}{\rho}\frac{\partial p}{\partial z}+4\pi\kappa\rho\mathrm{C}z+\gamma z&=0,\end{aligned}\right\}\quad\ldots\ldots(9)$$

where $$\alpha=\frac{4c^2(c^2-a^2)}{(c^2+a^2)^2}\Omega_2{}^2-\left(\frac{c^2-a^2}{c^2+a^2}\Omega_2-\eta\right)^2-\frac{4b^2(a^2-b^2)}{(a^2+b^2)^2}\Omega_3{}^2-\left(\frac{a^2-b^2}{a^2+b^2}\Omega_3+\zeta\right)^2,\quad\ldots\ldots(10)$$

and $\beta$ and $\gamma$ have corresponding values. It might appear that there should be terms in each equation in $x$, $y$ and $z$, but surfaces of equal pressure must be similar quadric surfaces coaxial with the case, and thus by integration we ought to get from (9)

$$\frac{p}{\rho}+2\pi\kappa\rho(\mathrm{A}x^2+\mathrm{B}y^2+\mathrm{C}z^2)+\tfrac{1}{2}(\alpha x^2+\beta y^2+\gamma z^2)=\text{const.}\quad\ldots\ldots(11)$$

The surfaces of equal pressure will be similar to the case if

$$(4\pi\kappa\rho\mathrm{A}+\alpha)a^2=(4\pi\kappa\rho\mathrm{B}+\beta)b^2=(4\pi\kappa\rho\mathrm{C}+\gamma)c^2,\quad\ldots\ldots(12)$$

and the case can then be removed without affecting the motion of the liquid.

The components of angular velocity of the vortex motion of the liquid are $\xi$, $\eta$, $\zeta$, and from a known hydrodynamical theorem we infer from (5) the equations of motion,

$$\left.\begin{aligned}\frac{\partial\xi}{\partial t}-2\left(\frac{a^2}{a^2+b^2}\Omega_3\eta-\frac{a^2}{c^2+a^2}\Omega_2\zeta\right)&=0,\\ \frac{\partial\eta}{\partial t}-2\left(\frac{b^2}{b^2+c^2}\Omega_1\zeta-\frac{b^2}{a^2+b^2}\Omega_3\xi\right)&=0,\\ \frac{\partial\zeta}{\partial t}-2\left(\frac{c^2}{c^2+a^2}\Omega_2\xi-\frac{c^2}{b^2+c^2}\Omega_1\eta\right)&=0.\end{aligned}\right\}\quad\ldots\ldots(13)$$

These give, by integration, $$\frac{\xi^2}{a^2}+\frac{\eta^2}{b^2}+\frac{\zeta^2}{c^2}=\text{const.}\quad\ldots\ldots(14)$$

**3.** ***Limits of instability of prolate ellipsoid. Enclosing shell an ellipsoid of revolution.*** Now let $a=b$, and therefore $\Omega_3=0$, since no value of $\Omega_3$ will affect the motion of the liquid. We get

$$\frac{\partial\xi}{\partial t}=-2\frac{a^2}{a^2+c^2}\Omega_2\zeta,\quad \frac{\partial\eta}{\partial t}=2\frac{a^2}{a^2+c^2}\Omega_1\zeta,\quad \frac{\partial\zeta}{\partial t}=2\frac{c^2}{a^2+c^2}(\Omega_2\xi-\Omega_1\eta),\quad\ldots\ldots(1)$$

and the equations of A.M. (8), 2, become, when the values of $\partial\xi/\partial t$, $\partial\eta/\partial t$, $\partial\zeta/\partial t$ are substituted from the equations just found,

$$\frac{\partial\Omega_1}{\partial t}=\Omega_2\zeta-\frac{a^2+c^2}{a^2-c^2}\eta\zeta,\quad \frac{\partial\Omega_2}{\partial t}=-\Omega_1\zeta+\frac{a^2+c^2}{a^2-c^2}\xi\zeta.\quad\ldots\ldots(2)$$

Multiplying the first of (1) by $\xi$, the second by $\eta$, adding and using the third equation of the set, we obtain by integration

$$\xi^2+\eta^2=\mathrm{L}-\frac{a^2}{c^2}\zeta^2.\quad\ldots\ldots(3)$$

Similarly, multiplying the first of (2) by $\Omega_1$, the second by $\Omega_2$, adding and using again the third equation of (15), we obtain as before

$$\Omega_1^2+\Omega_2^2=\mathrm{M}+\frac{(a^2+c^2)^2}{2c^2(a^2-c^2)}\zeta^2. \qquad (4)$$

By an obvious process we get also

$$\Omega_1\xi+\Omega_2\eta=\mathrm{N}+\frac{a^2+c^2}{4c^2}\zeta^2. \qquad (5)$$

From these results we find

$$\left(\frac{\partial\zeta}{\partial t}\right)^2=\frac{4c^4}{a^2+c^2}(\Omega_2\xi-\Omega_1\eta)^2=\frac{4c^4}{(a^2+c^2)^2}\{(\xi^2+\eta^2)(\Omega_1^2+\Omega_2^2)-(\Omega_1\xi+\Omega_2\eta)^2\}$$
$$=\frac{4c^4}{(a^2+c^2)^2}\left[\mathrm{LM}-\mathrm{N}^2+\left\{\mathrm{L}\frac{(a^2+c^2)^2}{2c^2(a^2-c^2)}-\mathrm{M}\frac{a^2}{c^2}-\mathrm{N}\frac{a^2+c^2}{2c^2}\right\}\zeta^2\right.$$
$$\left.-\frac{(a^2+c^2)^2(9a^2-c^2)}{16c^4(a^2-c^2)}\zeta^4\right]. \qquad (6)$$

L, M, N are constants of integration, and M is not to be confounded with the mass of the liquid.

Thus, if $\mathrm{LM}-\mathrm{N}^2$ is not zero, $\zeta$ is an elliptic function of $t$ unless the term in $\zeta^4$ vanishes or is infinite, that is unless $c=a$, or $c=3a$.

Let us put $\Omega_1=\Omega\cos\phi$, $\Omega_2=-\Omega\sin\phi$, so that $\tan\phi=-\Omega_2/\Omega_1$. We obtain

$$\Omega^2\frac{\partial\phi}{\partial t}=-\Omega_1\frac{\partial\Omega_2}{\partial t}+\Omega_2\frac{\partial\Omega_1}{\partial t}=\Omega^2\zeta-\frac{a^2+c^2}{a^2-c^2}(\Omega_1\xi+\Omega_2\eta)\zeta \qquad (7)$$

and by (4) and (5)

$$\frac{\partial\phi}{\partial t}=\zeta-\frac{a^2+c^2}{a^2-c^2}\,\frac{\mathrm{N}+\frac{a^2+c^2}{4c^2}\zeta^2}{\mathrm{M}+\frac{(a^2+c^2)\zeta^2}{2c^2(a^2-c^2)}}\zeta. \qquad (8)$$

Similarly, if $\xi=\omega\cos\psi$, $\eta=-\omega\sin\psi$, we obtain by (1) and (5)

$$\frac{\partial\psi}{\partial t}=-\frac{2a^2}{a^2+c^2}\,\frac{\Omega_1\xi+\Omega_2\eta}{\omega^2}\zeta=-\frac{2a^2}{a^2+c^2}\,\frac{\mathrm{N}+\frac{a^2+c^2}{4c^2}\zeta^2}{\mathrm{L}-\frac{a^2}{c^2}\zeta^2}\zeta. \qquad (9)$$

Now for a state of steady motion $\partial\zeta/\partial t=0$, and therefore, by (6), $\Omega_2\xi=\Omega_1\eta$. Thus $\phi=\psi$, $\Omega_1\xi+\Omega_2\eta=\Omega\omega$, and so

$$\frac{\omega}{\Omega}=\frac{\mathrm{N}+\frac{a^2+c^2}{4c^2}\zeta^2}{\mathrm{M}+\frac{(a^2+c^2)^2}{2c^2(a^2-c^2)}\zeta^2}. \qquad (10)$$

Thus the values of $\partial\phi/\partial t$, $\partial\psi/\partial t$ (now equal) are

$$\frac{\partial\phi}{\partial t}=\zeta-\frac{a^2+c^2}{a^2-c^2}\frac{\omega}{\Omega}\zeta,\quad \frac{\partial\psi}{\partial t}=-\frac{2a^2}{a^2+c^2}\frac{\Omega}{\omega}\zeta. \qquad (11)$$

Equating these, we obtain by reduction

$$\left(\frac{\omega}{\Omega}-\frac{1}{2}\frac{a^2-c^2}{a^2+c^2}\right)^2=\frac{(a^2-c^2)(9a^2-c^2)}{4(a^2+c^2)^2}. \qquad (12)$$

Steady motion is therefore impossible unless either $c^2<a^2$, or $c^2>9a^2$, that is the shell must be either oblate or so prolate that its axial length is more than three times its equatorial diameter.

It may be noticed that if we represent the quantity given finally on the right of (6) by Z, we have

$$\phi=\int\left\{1-\frac{a^2+c^2}{a^2-c^2}\,\frac{\mathrm{N}+\frac{a^2+c^2}{4c^2}\zeta^2}{\mathrm{M}+\frac{a^2+c^2}{2c^2(a^2-c^2)}\zeta^2}\right\}\frac{\zeta d\zeta}{\mathrm{Z}^{\frac{1}{2}}} \qquad (13)$$

and $$\psi = -\frac{2a^2}{a^2+c^2}\int \frac{N+\frac{a^2+c^2}{4c^2}}{L-\frac{a^2}{c^2}\zeta^2}\frac{\zeta\, d\zeta}{Z^{\frac{1}{2}}}. \qquad (14)$$

These are non-elliptic integrals, since Z is a quadratic function of $\zeta^2$.

**4.** *Cylinder moving in infinite perfect fluid which circulates round it: (1) Case of no forces.* We have not space in which to discuss the gyrostatic aspects of fluid motion, but must content ourselves with one or two comparatively simple cases of rotational motion of solids in a fluid, or in which circulation of a fluid in a cyclical channel in a solid simulates gyrostatic action of a rotating flywheel.

First of all we consider the steady motion of a right circular cylinder of unlimited length, immersed in a combined steady stream and vortex in an unlimited perfect liquid.

If there be no vortex, the motion is irrotational. Let then the cylinder be at rest in a stream which, at all points at a distance from the axis of the cylinder great in comparison with the radius $a$, is in parallel straight lines with speed $-U$ in the direction of the axis of $x$. We suppose the axis of the cylinder to be along the axis of $z$.

Since the motion is irrotational, the velocity components are given by the derivatives of a function $\phi(x, y, z, t)$. Thus

$$u = -\frac{\partial\phi}{\partial x}, \quad v = -\frac{\partial\phi}{\partial y}, \quad w = -\frac{\partial\phi}{\partial z}. \qquad (1)$$

The velocity potential $\phi$ fulfils the equation

$$\frac{\partial^2\phi}{\partial x^2}+\frac{\partial^2\phi}{\partial y^2}+\frac{\partial^2\phi}{\partial z^2}=0 \qquad (2)$$

at every point of the fluid.

The function $$\phi = U\left(\frac{a^2}{r}+r\right)\cos\theta \qquad (3)$$

fulfils the condition of uniplanar motion that there shall be no flow normal to the surface of the immersed cylinder, and satisfies the differential equation (2). It also gives velocity $-U$ for $r=\infty$. The normal and tangential components of velocity at the point $r$, $\theta$, ($x=r\cos\theta$, $y=r\sin t$) are

$$-\frac{\partial\phi}{\partial r}=U\left(\frac{a^2}{r^2}-1\right)\cos\theta, \quad -\frac{\partial\phi}{r\,\partial\theta}=U\left(\frac{a^2}{r^2}+1\right)\sin\theta. \qquad (4)$$

The former vanishes, the latter becomes $2U\sin\theta$, for $r=a$. The tangential velocity is therefore $2U\sin\theta$ in the direction of increasing $\theta$, that is the liquid slips along the surface with this velocity.

Now consider another possible motion of the liquid. Let the velocity at a point on a cylindrical surface of radius $r$ coaxial with the immersed cylinder be tangential to the surface, and of amount $a^2\omega/r$. The liquid thus moves in a sort of whirlpool round the solid cylinder with speed $a\omega$ at the surface. No element of the fluid however has any rotation: it has only *circulation* round the cylinder.

If the immersed cylinder spin about its axis with angular speed $\omega$ in the proper direction, there will be no slipping of the fluid. We suppose the circulation to be counter-clockwise, that of $\theta$, increasing, already considered in the irrotational motion.

Each element of the fluid, besides its motion in a circle round the axis of the cylinder, has obviously spin $\{\omega a^2/(r+dr)-\omega a^2/r\}/dr$ or $-\omega a^2/r^2$, about an axis parallel to that of $z$, which annuls the spin $\omega a^2/r^2$ due to the circular motion. The circulatory motion has thus no elemental rotation, and has a velocity potential $-\omega a^2\theta$.

For the two motions now considered superimposed, the complete velocity potential is

$$\phi=\mathrm{U}\left(\frac{a^2}{r}+r\right)\cos\theta-\omega a^2\theta, \quad\text{......................(5)}$$

which enables us to find the pressure $p$ at any point by the well-known relation

$$\frac{p}{\rho}=\frac{\partial\phi}{\partial t}-\tfrac{1}{2}q^2, \quad\text{..................................(6)}$$

where $q$ is the resultant velocity and $\rho$ the density of the fluid. We shall suppose $\sigma$ to be the density of the solid cylinder, so that the mass between two normal planes at unit distance apart is $\pi a^2\sigma$. It can be shown that the kinetic energy of the fluid, when set in motion by a cylinder moving steadily through it with speed U, is equal to that of the fluid displaced by the cylinder and supposed to be moving with the speed U. We have therefore to add to $\pi a^2\sigma$ the value $\pi a^2\rho$ as the virtual increase of mass of the slice of the cylinder, due to the fluid motion.

Whether the cylinder be at rest and the fluid flowing past it with speed U, or be in motion with speed U through the fluid otherwise at rest, the pressures on the surface of the cylinder will be the same. By (6) and the value of $q^2$ $[=(\partial\phi/r\,\partial\theta)^2$, since $\partial\phi/\partial r=0]$ to be found from (5), we get for the excess of the pressure at a point of coordinates $a$, $-(\pi-\theta)$ over that at the diametrically opposite point $a$, $\theta$ the value $4\mathrm{U}a\rho\omega\sin\theta$. Thus the total thrust from the side of smaller velocity of the fluid towards that of greater velocity is, per unit length of the cylinder,

$$4a^2\mathrm{U}\omega\rho\int_0^\pi\sin^2\theta\,d\theta=2\pi a^2\rho\mathrm{U}\omega. \quad\text{........................(7)}$$

Now suppose the flow at speed U of the fluid annulled, and the solid made to move with that speed in the opposite direction, the forces will not be altered. We shall have, apart from gravity, which for the present we neglect, a transverse force $2\pi a^2\mathrm{U}\omega\rho$ on each unit of length, and a total mass, virtual and real, of $\pi a^2(\sigma+\rho)$, per unit of length. Hence the cylinder receives an acceleration at right angles to the direction of its motion of amount $2\pi a^2\mathrm{U}\omega\rho/\pi a^2(\sigma+\rho)$, that is $\mathrm{U}^2/\{(\sigma+\rho)\mathrm{U}/2\rho\omega\}$, and this is constant. Hence the axis of the cylinder moves in a circle of radius $(\sigma+\rho)\mathrm{U}/2\rho\omega$.

As to the direction of motion: let the circular motion of the fluid be in the counter-clock direction to an observer looking towards the origin, supposed in the plane of the paper, and the cylinder be moving with speed U towards his right. Then the smaller pressure is on the upper half of the surface, and the centre of the circular orbit is above the cylinder. The observer will see the cylinder go round in the counter-clock direction, the direction of the circulation round the cylinder.

5. *Cylinder moving in infinite perfect fluid with circulation:* (2) *Case of extraneous forces.* Let now extraneous forces act on the cylinder, say P in the direction of motion and Q transversely (both taken per unit length of the cylinder), in the same direction as the pressure force $2\pi a^2 \omega\rho U$. The equations of motion are then

$$\left.\begin{aligned} \pi a^2(\rho+\sigma)\frac{dU}{dt} &= P, \\ \pi a^2(\rho+\sigma)U\dot{\vartheta} &= 2\pi a^2\omega\rho U + Q, \end{aligned}\right\} \quad \text{(1)}$$

where $\dot{\vartheta}$ is the angular speed with which a normal to the path is turning round. Clearly $U\dot{\vartheta}$ is the acceleration of the cylinder.

Now take fixed axes in the plane of motion of any chosen point on the axis of the cylinder, and resolve the accelerations and forces along these axes. We get as the equations of motion

$$\pi a^2(\sigma+\rho)\ddot{\xi}+k\dot{\eta}=X, \quad \pi a^2(\sigma+\rho)\ddot{\eta}-k\dot{\xi}=Y, \quad \text{(2)}$$

where $k=2\pi a^2\omega\rho$.

If we put $Y=0$, and $X=\pi a^2(\sigma-\rho)g$, $\kappa=k/(\sigma+\rho)$, and write $g'$ for $(\sigma-\rho)g/(\sigma+\rho)$, the equations of motion become

$$\ddot{\xi}+\kappa\dot{\eta}=g', \quad \ddot{\eta}-\kappa\dot{\xi}=0. \quad \text{(2')}$$

These have the integrals

$$\xi=\frac{g'}{\kappa^2}-\frac{\alpha}{\kappa}+c\cos(\kappa t-f), \quad \eta=\frac{g'}{\kappa}t+\beta+c\sin(\kappa t-f), \quad \text{(3)}$$

(where $\alpha$ and $\beta$ are constants) which show that under the fluid pressure and the force of gravity each point on the axis of the cylinder describes a trochoid. The motion of each point of the axis is therefore periodic in a vertical plane. It is to be observed that no continuously progressive vertical displacement of the cylinder is produced by the action of gravity.

We have assumed the value given above for the total virtual inertia of the cylinder. There is no difficulty in verifying that this value is the correct one to use in the general motion here considered. This can be done by calculating from the value of $\phi$ for the cylinder moving through the fluid (*not* for the case of the fluid flowing past the cylinder reduced to rest) the pressure by the equation

$$\frac{p}{\rho}=\frac{\partial\phi}{\partial t}-\tfrac{1}{2}q^2+F(t). \quad \text{(4)}$$

We take the case of rectilinear motion. It must be borne in mind, that (4) applies to a point of the fluid fixed in space, and that if $r$ and $\theta$ be taken constant in the velocity potential, $\phi$, for the moving cylinder, they and $\phi$ refer to a point which changes position as the cylinder moves. Apart from circulation the value of $\phi$ is now $\mathrm{U}(a^2/r)\cos\theta$; the component of the velocity in the direction of $r$ is $\mathrm{U}\cos\theta$, and in the direction of $\theta$ increasing (counter clockwise) is $-\mathrm{U}\sin\theta$. The angle $\theta$ is supposed to be measured from the direction of motion. As therefore the origin moves the $r$ for a point, $r$, $\theta$, fixed in space is changing at rate $-\mathrm{U}\cos\theta$, and $\theta$ at rate $\mathrm{U}\sin\theta/r$. Thus if, with reference to fixed axes, $d\phi/dt$ denote the time-rate of variation of the potential, so far as it depends on the motion of the cylinder, $\partial\phi/\partial t$ for the point fixed in space is

$$\frac{d\phi}{dt}-\mathrm{U}\cos\theta\frac{\partial\phi}{\partial r}+\frac{\mathrm{U}}{r}\sin\theta\frac{\partial\phi}{\partial\theta}.$$

But
$$\frac{d\phi}{dt}=\dot{\mathrm{U}}\frac{a^2}{r}\cos\theta;$$

and the value of $\partial\phi/\partial t$ to be used in (4) is therefore (with $r=a$)

$$\dot{\mathrm{U}}\frac{a^2}{r}\cos\theta-\mathrm{U}\cos\theta\frac{\partial\phi}{\partial r}+\frac{\mathrm{U}}{r}\sin\theta\frac{\partial\phi}{\partial\theta}.$$

The whole potential $\phi$ is $\mathrm{U}a^2\cos\theta/r-\omega a^2\theta$. From this can be found, by multiplying $p$, from (4), respectively by $a\,d\theta\cos\theta$ and $a\,d\theta\sin\theta$, and integrating from $\theta=0$ to $\theta=2\pi$, the forces against and at right angles to the motion. They are $\pi\rho a^2\dot{\mathrm{U}}$ and $2\pi a^2\rho\mathrm{U}\omega$ [see (7), 4].

This problem was treated as an illustration of the flight of a tennis ball by Lord Rayleigh [*Messenger of Mathematics*, VII, 1877; *Collected Papers*, I, p 344], and in a more general manner by Greenhill [*Mess. Math.* IX, 1880].

6. *Drift of a rotating projectile: Case of a golf ball.* It has been supposed that the deviation of an elongated projectile fired from a rifled gun might be accounted for on a theory like that sketched above. The projectile, dragging air round with it as it spins about its longitudinal axis, would produce differences of velocity round it, especially in that part of the trajectory in which it begins to have serious sidelong motion, and thus differences of pressure would arise. But the observed deviation is in the opposite direction to that given by the theory, and it is necessary to find another explanation.

As Lord Rayleigh remarks (*loc. cit. supra*) it is not clear that in such a case as that of the tennis ball, the pressure is greatest on the side on which the velocity of the fluid is least. Bernoulli's equation is proved on the assumption that the fluid is frictionless, and in the present case the air is dragged round by small roughnesses which may be likened to blades projecting from the surface. "On that side of the ball where the motion of the blades is up stream, their anterior faces are in part exposed to the pressure due to the augmented relative velocity, which pressure necessarily operates also on the contiguous spherical surface of the ball. On the other side the relative motion, and therefore also the lateral pressure, is less, and thus an uncompensated lateral force remains over."

But according to the results of the experiments and observations of the late Professor Tait * on the deflection of a golf ball, the effect of spin is to produce deflection of the ball, as a whole, in the direction in which the front of the ball is moved by the spin. Thus a ball, which spins about a horizontal axis across the path, reaches the ground much sooner if the spin carries the front of the ball downward than when the reverse is the case. If the spin is about a vertical axis, or one in the plane of and nearly normal to the trajectory, the deviation is to the left or the right according as the front of the ball is moving towards the left or the right.

This is so far in accordance with the theory of pressures in a frictionless fluid, supposing that circulation round the ball is produced by the spin. For if we imagine imposed on the ball and fluid, a translatory motion equal and opposite to that of the ball, we see that the relative motions of the ball and fluid are as the theory requires. When the ball, as described above, deviates to the right, the translatory and circulatory motions of the fluid conspire on the right, and are opposed on the left side of the ball, and so in the other cases. For example, if the club strikes the ball on the whole below the centre, so called underspin is given, and the ball soars, so that a much longer drive is obtained than in the case of overspin, which is produced when the ball is "topped." The soaring gives concavity upward in the ascending part of the path, which may even culminate in a cusp.

In a complete theory the effect of friction will have to be taken account of, and the nature of the surface of the ball will no doubt be an important factor. The behaviour of the fluid in the rear of the moving body is very different from that which would occur if the fluid were perfect. There is a wake of eddies of fluid acted on by friction and then left behind by the body. The opinion among golfers appears to be that the "brambling" of the surface increases the "carry" of the ball; probably it is required to make the circulation effective. Careful comparative observations, carried out by an expert player, might give valuable information.

* Tait's conclusions are confirmed by experiments recently made by M. Carrière [*Journ. de Physique*, V, 1916] on light spheres moving much more slowly than golf balls. The balls were cut from the pith of the Jerusalem artichoke (*moelle de topinambour*). They were given rotations varying within a wide range by running them along before projection between two horizontal ribbons moving at different speeds, and were projected horizontally. Observations were made with different rotations and different speeds of projection, and the trajectories were shown by the white balls falling in front of a black background. Appell remarks (*C.r.* Jan. 1918) that these experiments show that the resultant action, of the air on the ball, is inwards along a radius specified by supposing the direction of motion turned through an acute angle, in the direction against the circulation. This is exactly Tait's result, stated in 1893.

7. *Drift of an elongated fast-spinning projectile.* It is however the case that this theory of a frictionless fluid, when applied to elongated projectiles, moving at the speed of rifle bullets or of shells fired from rifled artillery, in no way accounts for the deviations observed. When these projectiles have pointed ends, and are fired from guns rifled with a right-handed screw, they deviate to the right of the vertical plane of fire, as seen from behind. When the rifling is in the opposite direction the drift is reversed. This is the direct opposite of the effect to be expected, according to the theory explained above. But here no doubt the forces applied to the body, in consequence of the combination of sidelong with translatory motion, as explained in 14 and 15, VII, are effective. The translatory motion is of the order of 2500 feet per second in the early part of the flight: the spin about the axis of figure ranges from 100 turns per second, for a large projectile, to over 3000 turns per second for a rifle bullet.

It was proved first by Professor G. Magnus, of Berlin [*Taylor's Scientific Memoirs*, 1853], by carefully arranged experiments, that the action of the air is equivalent to a resultant force which cuts the axis of the shot at a

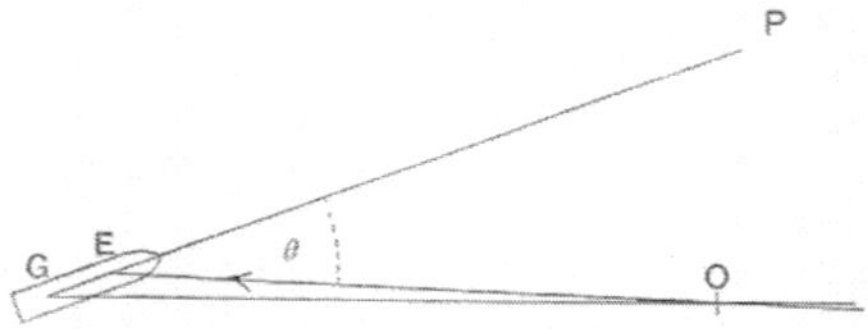


FIG. 68.

point in front of the centre of gravity. Thus when the axis of the projectile coincided nearly with the tangent to the path GO (Fig. 68) of the centre of gravity, deviating a little from that line by a deflection of the point upward and to the right, the action of the air was a force F meeting the axis in E, the so-called centre of effort, and directed as shown.

The theory of the subject requires elucidation in many particulars; but the main cause of drift and the general behaviour of the projectile are tolerably clear [see a paper by Mallock, *Proc. R.S.* June 6, 1907].

The axis of spin follows very closely the direction of motion. As the projectile passes along the path the rate of change of direction of its axis, if that direction adhered to the tangent to the path, would be $\omega = v/R$, where R is the radius of curvature of the path. If $g'$ be the acceleration normal to the path, the value of R is $v^2/g'$, and therefore $\omega = g'/v$. Hence the couple required for this turning of the axis of the shot is $Cn\omega = Cng'/v$, where $Cn$ is the A.M. of the shot about the axis of figure. The couple-axis is along the principal normal to the path.

When the axis of the shot is slightly inclined to the tangent to the trajectory, for example as in Fig. 68, the action of the air is like that of an air jet, in the direction of the arrow, playing obliquely on the face of the shot which, in consequence of the component of sidelong motion, meets the air as the motion proceeds. Thus there is a resultant force F, the line of which OE is inclined at a very small angle $\beta$ to the trajectory OG and intersects the axis of figure at E, which is at a distance $c$ in front of the centre of gravity G. This gives a couple in the plane PEO of moment $Fc\sin\theta$, where $\theta$ is the angle OEP. It also gives a force $F\cos\beta$ retarding the motion of the centroid along the trajectory and a force $F\sin\beta$ at right angles to the trajectory, the effect of which is to contribute a curvature $F\sin\beta/v^2$ in the plane determined by F and the axis of figure EP. [This of course is not the whole curvature.] The horizontal component of the force transverse to the trajectory is effective in producing drift.

As seen by an observer behind the gun, the axis of the couple $Fc\sin\theta$ is towards the right of the plane GOE. The axis of spin is drawn forward through the point of the bullet, if the spin is right-handed, and precesses round towards the axis of the couple and tends to move in a cone round the direction of F as an axis, or, which is practically the same, about the direction of motion. As the axis precesses the force F changes its direction, and if the axis of the shot went round in a right circular cone about the trajectory tangent there would be, on the average, no action at right angles to the tangent, neither in the vertical plane nor horizontally, but only an alternating action giving helical quality to the trajectory. But this, as we shall see, is not the real nature of the *resultant* precessional motion.

8. *Action of friction.* A second couple is exerted on the shot by friction. The shot is rotating, clockwise, as we have supposed, to an observer behind the gun, and so, according to Fig. 68, there is friction between the shot and the air impinging on the forward part of the under side of the body. This gives a couple which we may take as in a plane containing F, and at right angles to GOE. The axis of this couple is nearly at right angles to the trajectory, and points downwards. Hence precession carries the point of the shot downwards, that is the axis if tilted upwards tends to move down towards coincidence with the trajectory.

The projectile is continually engaging air hitherto comparatively undisturbed at the front of its under surface, and losing grip of air underneath it at the near end. But as the air is carried round by friction it leaves the shot in the direction in which the surface, on the side sheltered from the impact of the air by the shot itself, is moving. In this sheltered part eddies are formed, and there is an absence of the closing-in stream lines which would exist if the fluid were "perfect." On the whole, if the forward end is pointed, the tilting action tends to lift the point above the trajectory; for, if not at the outset at some point forward on the path, the point will rise slightly above the tangent, where the path is convex upward, and the axis of spin preserves its direction. The friction couple however, as just explained, prevents this rising from becoming large, and indeed is effective in making the axis of spin follow the trajectory very closely.

As Mr. Mallock suggests [*loc. cit. supra*] this friction couple is no doubt proportional to $n^2$, to the small angle OAP, $\chi$, say [Fig. 69], and to some function $f(v)$ of the speed, and so we have $Cn\omega = Cng'/v = n^2\chi f(v)$, so that

$$\chi = \frac{Cg'}{nvf(v)}.$$

This angle $\chi$ increases as $v$ and $n$ decrease; it is only while $\chi$ is small that the projectile follows the trajectory closely and the motion is steady.

9. *Graphical representation of the motion.* The following representation of the behaviour of a spinning shot is taken, with some slight alteration, from Mr. Mallock's paper. Let XOY be a plane at right angles

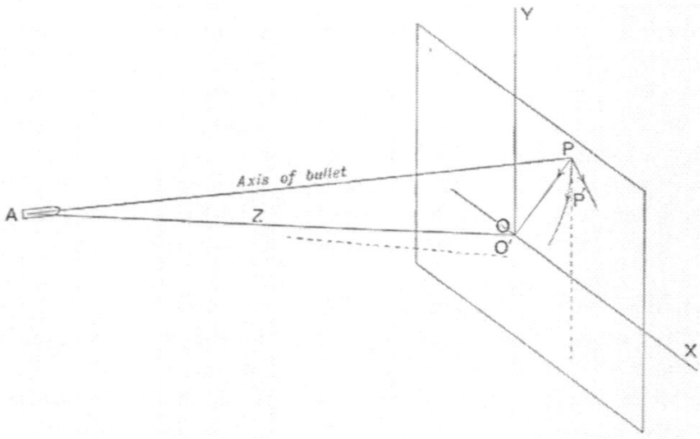


FIG. 69.—AO is direction of motion, O'A is direction of motion, after time $dt$.

to the direction of motion, AZO, set up in front of the firing point, and carried forward so as always to be at the same distance in front of the bullet. Let AP be the direction of the axis and P the point in which that line meets the plane XOY, and let OP be drawn in that plane. Then, in consequence of the first couple discussed above, the point P has a motion at right angles to OP in the plane XOY. In fact OP turns about O in that plane with angular speed $Fc \sin\theta/Cn$, approximately, and the speed of P is therefore $OP \,.\, Fc \sin\theta/Cn$ or $OA \,.\, Fc \sin\theta \sin\chi/Cn$. [Note that $\theta - \beta = \chi$.] But in consequence of the second couple P has a motion towards O, which is also proportional to the length of OP. Hence the path of P is an equiangular spiral, as indicated in the diagram.

If P be so situated that the tangent to the spiral is there parallel to OY $\angle$POY is the constant angle of the spiral, and for a well-shaped projectile giving small drift this angle is small. If the precessional period be T the element of the spiral described in $dt$ is $2\pi OA . \sin\chi\, dt/T \sin\phi$, where $\phi$ is the angle of the spiral.

It is to be observed that in each element of time the direction of motion changes through an angle $\omega\, dt$ from AO to AO′, in the vertical plane containing AO, and that then AP takes the new direction AP′. P′ is now moving in a spiral about a pole in a new position. The sequence of new configurations of AO and AP′ gives the actual motion of the axis with reference to the trajectory. We have

$$\frac{\sin\chi}{\sin\phi}=\frac{\omega T}{2\pi}.$$

If $\omega$ is fairly constant the angle of the spiral must change with $\chi$ so that $\sin\chi/\sin\phi$ is constant. Hence if $\phi$ be small $\chi/\phi$ must remain constant. If $\chi$ remains constant so also will $\phi$.

If the angle $\phi$ is small the couple due to friction is large compared with that due to the tilting action of the air pressure.

We see then that in general an upward vertical force diminishing the effective action of gravity on the curvature, and a sideways force producing drift of the projectile as a whole, are applied. For a right-handed rotation (as seen from behind) the drift is to the right, for left-handed rotation to the left of the observer. But in consequence of the spiral motion of the axis the drift oscillates in amount. If the successive positions of P were recorded, as in Fig. 69, for a great distance along the path, the locus would be a kind of cycloidal curve drawn along a descending line

The action of the sideways force due to friction is described by some writers on ballistics (*e.g.* Cranz *), by saying that a cushion of compressed air is formed by the partly sideways advancing projectile, which rolls sideways on this cushion as it spins.

If the centre of effort is behind the centre of gravity, which is the case for shot of certain shapes, the action described above is reversed. It may happen also that the point of the shot is tilted down either by the recoil acting on the gun so as to throw up the breech, or the streaming of powder gases past the shot, or in some other way, so that left-handed drift is produced in the first part of the trajectory, and right-handed drift later.

Besides Mr. Mallock's paper the reader may consult one by Professor J. B. Henderson [*Proc. R.S.* 1909 (A), 82] which contains a detailed graphical representation of successive positions of the axis of the shot, and shows how, as the axis changes its position in the spiral and the tangent turns down relative to its initial direction, the drift oscillates slightly in value.

* *Komp. d. äussern Ballistik*, or *Encycl. d. Math. Wiss.* IV, 3.

10. *Position of centre of effort for airship.* The diagram given in Fig. 70 shows the position obtained in experiments with a model of an airship in a wind. It is taken from a communication from the Göttingen

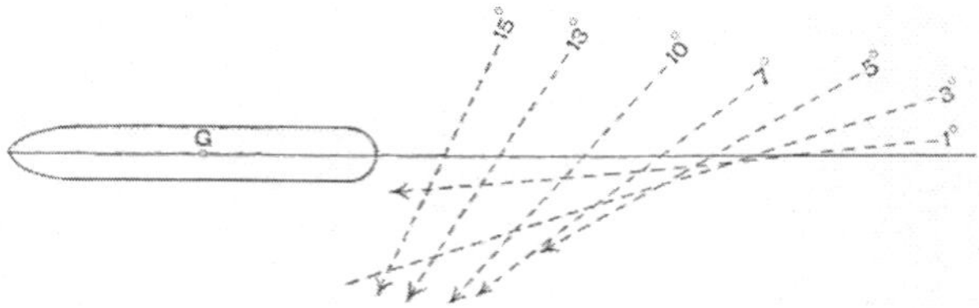


FIG. 70.

Model Testing Institute [*Engineering*, II, 1911]. By means of balances the components of the force due to air pressure, and the moment of the couple on the model were measured, and so the directions of the resultant forces were laid down in the diagram, for different amounts of obliquity of the axis of the ship to the air current. The direction of motion is from left to right. E is at the intersection of the direction of the resultant force and the axis of the ship, and is farther out the smaller the obliquity.

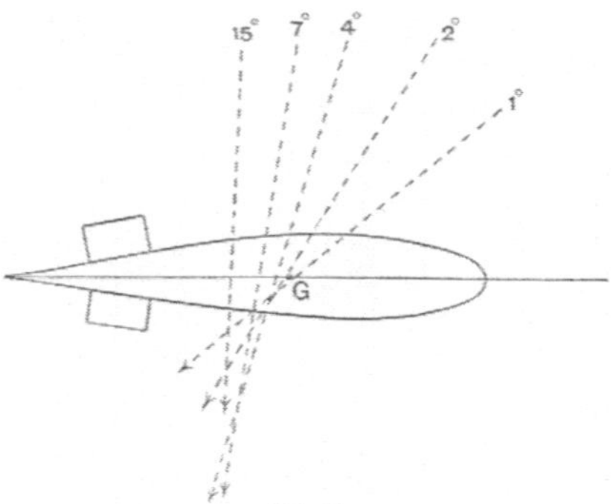


FIG. 71.

By means of vanes in the rear part of the model stabilising forces could be applied. When these were in action the position of E was behind the centre of gravity, and the couple, which without them tended to turn the ship broadside on to the stream, gave, when compounded with the forces on the vanes, a couple bringing the ship back to the direction of the wind. The diagram in such a case is shown in Fig. 71. It will be seen that in this case the greater the inclination of the air current to the axis the farther is E from G.

11. *Motion of a perforated solid in a perfect fluid. Use of Lagrange's equations.* We do not discuss the general motion of a solid in an infinite liquid; for this the reader may refer to Lamb's treatise on *Hydrodynamics*, to Halphen, *Fonctions Elliptiques*, t. II, chap. IV, or to Greenhill, *R.G.T.*, chap. VIII, §27. But the problem of the motion of a perforated solid with circulation through it is of importance from the point of view of gyrostatic action, and therefore we give here an account of its solution, under certain restrictions as to the form of the solid.

We shall suppose that the solid is a circular ring, and refer the motion to axes of coordinates, $O(x, y, z)$, taken along and at right angles to the axis of figure. The choice of the position of the origin on the axis of figure is of importance as regards the form of the terms included in the expression for the kinetic energy, as we shall presently see.

According to a principle first formulated in Thomson and Tait's *Natural Philosophy*, and now generally accepted by mathematicians, the kinetic energy of the motion of an infinite liquid, which contains moving solids, can, in the absence of cyclical motion of the liquid, be expressed in terms of the coordinates and the velocities, which specify the configuration and motion of the solids. When there is cyclosis, and cyclical motion of the liquid, that is when there is circulation of the liquid round re-entrant channels in the solids, the kinetic energy needs only to be modified by addition of a homogeneous quadratic function of the quantities $\kappa_1, \kappa_2, \ldots$, which are the line-integrals

$$\int(u\,dx+v\,dy+w\,dz)$$

of fluid velocity taken round the different channels.

In the absence of cyclosis the motion is that which would be produced throughout the fluid by starting every part of the bounding surface of the fluid, that is every element of the surface of the solids (and if the fluid is of finite extent every element of the bounding surface of the fluid), with the velocity which it has in the actual motion at the instant considered.

When there is cyclosis we have to suppose impulsive pressures properly applied, over diaphragms closing the various channels, by means of mechanism connected with the solids; and corresponding to this of course a system of reactions would be exerted on the solids. These impulsive pressures are required to generate the momenta of the cyclical motions. The reactions would have to be taken account of in the forces applied to the solids.

**12.** *Motion of a ring in a perfect fluid. Equation of energy.* We consider then a solid symmetrical about an axis of figure, along which also it is perforated, moving through a liquid while spinning about its axis of figure and also about axes at right angles to the axis of figure and drawn through the origin. If $u$, $v$, $w$ be the velocities along the axes, which are fixed in the body, and therefore move with it, $p$, $q$, $r$ the angular speeds of the spin, and $\kappa$ the cyclical constant of the circulatory motion, and if the origin is suitably chosen, the total kinetic energy for both fluid and solid is given by

$$2T = Au^2 + B(v^2+w^2) + Pp^2 + Q(q^2+r^2) + K\kappa^2, \qquad (1)$$

where A, B, P, Q and K are constants.

The origin is here chosen so that no products of the form $uq, \ldots$, exist in the expression for T. If the body has three orthogonal planes of symmetry the origin is the centroid. The solid might however have an axis of figure without symmetry otherwise. For example, in the case of $p=0$, $\kappa=0$, we have, on the supposition that the solid moves so that its axis of figure remains always in the same plane, and that it turns about an axis perpendicular to the plane of $u$, $v$, with angular velocity $\omega$, the equation of kinetic energy

$$2T = Au^2 + Bv^2 + Q\omega^2 + 2S\omega v. \quad \text{.........................(2)}$$

The transverse velocity of a point E on the axis at a distance $h$ in front of the origin is $v' = v + h\omega$, so that $v = v' - h\omega$. From this we find, by substituting $v' - h\omega$ for $v$ (dropping the accent) and Q′ for $Q + Bh^2 - 2Sh$, with $h = S/B$ (that is put Q′ for $Q - S^2/B$), that

$$2T = Au^2 + Bv^2 + Q'\omega^2. \quad \text{..............................(3)}$$

This amounts to shifting the origin forward to E through the distance S/B. The point at this distance in front of the centroid has been called by Thomson and Tait the "centre of reaction."

**13.** *Impulse of motion of a solid in a fluid.* No matter how the body and fluid may be moving we may suppose the motion (apart from the axial spin) to be generated by an impulse I on the body in some line EX through the centre of reaction E, and an impulsive couple of amount $Q\omega$ about an axis at right angles to EX. Through E imagine a plane at right angles to the axis of turning, and in that plane take a line $Ox$ parallel to EX, at distance $y$, such that, I being the total momentum of body and fluid, that is $(A^2u^2 + B^2v^2)^{\frac{1}{2}}$ [or (see 14), in the case of circulation through an orifice along the axis of figure of the solid, $\{(Au + \xi)^2 + B^2v^2\}^{\frac{1}{2}}$], $Iy = Q\omega$.

The massless framework required to connect the line of impulse with the body may be imagined, or ignored. $Ox$ is called the line of resultant impulse or resultant momentum.

In the case of a completely symmetrical body the centre of effort E is of course the centroid.

**14.** *Equations of motion of a solid in a perfect fluid: proof (a) by first principles, (b) by the method of Lagrange.* Returning to the more general problem, we consider the case in which the solid moves in a plane containing the axis of symmetry. We put therefore $v^2$ for $v^2 + w^2$, $\omega^2$ for $q^2 + r^2$, and suppose that this turning is about an axis at right angles to the plane of motion, that of $u$, $v$. We have thus, with, where needful, new values of the constants,

$$2T = Au^2 + Bv^2 + Pp^2 + Q\omega^2 + K\kappa^2. \quad \text{.......................(1)}$$

We shall suppose for the present that $p=0$, that is that the solid has no spin about the axis of figure, so that

$$2T = Au^2 + Bv^2 + Q\omega^2 + K\kappa^2. \quad \text{.........................(2)}$$

It will be seen that we have here a case in which, in order to obtain the equation of motion, we must allow for the motion of the axes. There are various ways of proceeding; we take first one which appeals directly to first principles. We suppose, as at 15, VII above, that the virtual inertia of the fluid in the direction of the axis of figure is $\alpha M'$, and in any transverse direction $\beta M'$. The momentum of the fluid in the former direction is $\alpha M'u+\xi$, and in any transverse direction is $\beta M'v$, where $\xi$ is the momentum due to the circulation, which is supposed to have been produced by pressure over a diaphragm closing the orifice.

In consequence of the turning of the transverse axis in the plane of motion with angular speed $\omega$, momentum along the axis of figure is being produced at rate $-\beta M'v\omega$. For in time $dt$ the axis of $y$ has turned through the angle $\omega\, dt$, and a component of momentum in the negative direction of the instantaneous position of the axis of $x$, and of amount $\beta M'v\omega\, dt$, has been produced. The whole rate of production of momentum of the fluid about the axis of $x$ is therefore

$$\alpha M'\dot{u}-\beta M'v\omega.$$

Similarly the total rate of production of momentum about the axis of $y$ is

$$\beta M'\dot{v}+\alpha M'u\omega+\xi\omega.$$

Finally, the rate of production of A.M. about the axis of rotation may be found as follows. Take first the part due to the action of the revolving solid on the fluid. Let $h$, $k$ be the coordinates of a fixed point in the plane of motion. The A.M about an axis through that point at right angles to the plane of motion is $(\alpha M'u+\xi)k-\beta M'vh$. The rate of change of this (terms in $\dot{u}$, $\dot{v}$ omitted, since we are to put presently $h=k=0$) is

$$(\alpha M'u+\xi)\dot{k}-\beta M'v\dot{h}=-\{(\alpha-\beta)M'u+\xi\}v,$$

since $\dot{h}=-u$, $\dot{k}=-v$. Thus we get for the rate of growth of A.M. about the axis of turning, which passes through the point $h=0$, $k=0$, the value $-\{(\alpha-\beta)M'u+\xi\}v$. The total rate of growth of A.M. of the fluid is therefore

$$C'\dot{\omega}-\{(\alpha-\beta)M'u+\xi\}v.$$

If no impressed forces act on the solid, the only forces applied to it are the reactions of the fluid, due to the rates of production of momentum and A.M. just estimated, since these are due to the action of the solid. Hence, if M and C be the inertia and proper moment of inertia of the solid, we obtain

$$\left.\begin{aligned} M(\dot{u}-\omega v)&=-\alpha M'\dot{u}+\beta M'v\omega,\\ M(\dot{v}+\omega u)&=-\beta M'\dot{v}-(\alpha M'u+\xi)\omega,\\ C\dot{\omega}&=-C'\dot{\omega}+\{(\alpha-\beta)M'u+\xi\}v.\end{aligned}\right\}\quad\text{......................(3)}$$

These equations of motion may be written

$$\left.\begin{aligned} A\dot{u}-B\omega v=0,\quad B\dot{v}+(Au+\xi)\omega=0,\\ Q\dot{\omega}-\{(A-B)u+\xi\}v=0,\end{aligned}\right\}\quad\text{......................(4)}$$

where $$A=M+\alpha M',\quad B=M+\beta M',\quad Q=C+C'. \quad\text{..................(5)}$$

The equations thus arrived at from elementary considerations may be deduced by the Lagrangian process from (5), if we make allowance for the motion of the axes. For the expression given for T does not contain the coordinates on which the configuration at time $t$ depends. The generalised components of momentum are

$$\frac{\partial T}{\partial u}+\xi=Au+\xi,\quad \frac{\partial T}{\partial v}=Bv,\quad \frac{\partial T}{\partial \omega}=Q\omega. \quad\text{..................(6)}$$

From these, making the allowances referred to for the motion of the axes, we get, since there are no impressed forces,

$$\left.\begin{aligned}\frac{d}{dt}\frac{\partial T}{\partial u}-\omega\frac{\partial T}{\partial v}=0,\quad \frac{d}{dt}\frac{\partial T}{\partial v}+\omega\left(\frac{\partial T}{\partial u}+\xi\right)=0,\\ \frac{d}{dt}\frac{\partial T}{\partial \omega}-v\left(\frac{\partial T}{\partial u}+\xi\right)+u\frac{\partial T}{\partial v}=0,\end{aligned}\right\} \quad\text{..................(7)}$$

or, by (1),

$$A\dot{u}-B\omega v=0,\quad B\dot{v}+(Au+\xi)\omega=0,\quad Q\dot{\omega}-\{(A-B)u+\xi\}v=0, \quad\text{.....(8)}$$

as before given in (7). We suppose that $A>B$.

**15.** *Vibrations of a ring moving in a fluid. Quadrantal pendulum.* Let us now refer to fixed axes $O'(x', y')$ in the plane of the motion, and take the axis $Ox'$ in the line of the resultant momentum I. This direction is constant, and so if $\theta$ be the angle which the axis of figure makes with $O'x'$, we have

$$Au+\xi=I\cos\theta,\quad -Bv=I\sin\theta.$$

The third equation gives

$$Q\ddot{\theta}+\frac{1}{AB}\{(A-B)I\cos\theta+B\xi\}I\sin\theta=0. \quad\text{..................(1)}$$

This result, (1), was first given by Lord Kelvin in a paper on "Hydrokinetic Solutions and Observations."* He remarked that the ring performs finite oscillations as would a horizontal magnetic needle to which was rigidly attached, parallel to its magnetic axis, a bar of soft iron. Clearly if the needle had a magnetic moment of amount $\xi$, the return couple on the needle suspended in a field of intensity I would be proportional to $\xi I\sin\phi$, and the induced magnetic moment in the soft iron bar would be proportional to $I\cos\theta$, so that, to a constant multiplier, the quantity $(A-B)I^2\sin 2\theta$ would represent the return couple on the soft iron bar.

If $\xi=0$, we get, with $\vartheta=2\theta$,

$$Q\ddot{\vartheta}+\frac{A-B}{2AB}I^2\sin\vartheta=0. \quad\text{..............................(2)}$$

This is an example of a "quadrantal pendulum," in which the law of variation of couple with deflection $\vartheta$ is the same as that of an ordinary pendulum with respect to $\theta$.

By (1) we have for T the period of a small oscillation,

$$T=2\pi\left\{\frac{ABQ}{\{(A-B)I+B\xi\}I}\right\}^{\frac{1}{2}}. \quad\text{..........................(3)}$$

* *Phil. Mag.* XLII, 1871, or *Mathematical and Physical Papers*, IV, p. 69.

It will be seen that the momentum $\xi$ due to the circulation through the ring increases the return couple, that is increases the stability, an effect similar to that of gyrostatic action. This case of small vibrations can only occur if $\theta$ is always small, and therefore $v$ always small and $u$ nearly constant.

The investigation shows that if $B > A$ and $\xi$ be zero or insufficiently large the motion of the solid is unstable. A sufficiently large circulation thus ensures stability.

**16.** ***Elliptic integral discussion of finite oscillations.*** With respect to the new fixed axes of $x$ and $y$ we have, if $x$ and $y$ be the coordinates of the origin before considered, which moves with the body,

$$\left.\begin{aligned}\dot{x}&=u\cos\theta-v\sin\theta=\frac{I}{AB}(A\sin^2\theta+B\cos^2\theta)-\frac{\xi\cos\theta}{A},\\ \dot{y}&=u\sin\theta+v\cos\theta=-\frac{A-B}{AB}I\sin\theta\cos\theta-\frac{\xi\sin\theta}{A}.\end{aligned}\right\}\quad\text{......(1)}$$

Also it is clear from (1), 15, that
$$Q\dot{\theta}=I\dot{y},\quad\text{......(2)}$$
so that, multiplying by $\dot{\theta}$ and integrating, we get
$$\dot{\theta}=\frac{1}{Q^{\frac{1}{2}}}\left(-I^2\frac{A-B}{AB}\sin^2\theta+2\frac{I}{A}\xi\cos\theta+C\right)^{\frac{1}{2}},\quad\text{......(3)}$$
where C is a constant. Of course we have
$$Q\dot{\theta}=Iy,\quad\text{......(4)}$$
without any added constant, since the fixed axis of $x$ is coincident with the line of the resultant momentum I.

If we write $a$ for $I^2(A-B)/ABQ$, $b$ for $2I\xi/AQ$, and $c$ for $C/Q$, we get
$$\dot{\theta}^2=-a\sin^2\theta+b\cos\theta+c.$$
Transforming this by putting $z=\tan\frac{1}{2}\theta$, we obtain
$$4\dot{z}^2=(c-b)z^4+(2c-4a)z^2+b+c=4m^2Z.\quad\text{......(5)}$$
Thus
$$mt=\int\frac{dz}{Z^{\frac{1}{2}}}.\quad\text{......(6)}$$
The reduction to the normal Legendrian form of the elliptic integral is obvious. Thus $\cos\theta$, and therefore $\dot{\theta}$, and, by (5), also $y$, are elliptic functions of the time.

The elliptic function treatment is very simple in any case, but is particularly so if $\xi=0$. We get then, by (3),
$$\dot{\theta}^2=\omega_1^2(1-k^2\sin^2\theta).\quad\text{......(7)}$$
Here $k^2=(A-B)I^2/\omega_1^2QAB$ and $\omega_1^2$, from the constant of integration, is the square of the angular speed when $\theta=0$ and when $\theta=\pi$, if the solid makes complete revolutions. Since the pendulum is quadrantal, this equality of angular speeds was to be expected. We have thus for the time $t$, from $\theta=0$ to any value $\theta$,
$$\omega_1 t=\int_0^\theta\frac{d\theta}{(1-k^2\sin^2\theta)^{\frac{1}{2}}}=F(k,\theta).\quad[\theta=am\,\omega_1 t.]\quad\text{......(8)}$$

If the solid does not make complete revolutions, but only revolves up to amplitude $\alpha$, we have, by 3,
$$\dot{\theta}^2=\frac{A-B}{ABQ}I^2(\sin^2\alpha-\sin^2\theta).\quad\text{......(9)}$$
Writing now $\sin^2\theta=\sin^2\alpha\sin^2\phi$, we obtain
$$\dot{\phi}^2=\frac{A-B}{ABQ}I^2\sin^2\alpha(1-\sin^2\alpha\sin^2\phi),\quad\text{......(10)}$$

so that
$$mt=\int_0^\theta \frac{d\phi}{(1-\sin^2\alpha\sin^2\phi)^{\frac{1}{2}}}=\mathrm{F}(\sin\alpha,\ \phi), \qquad (11)$$
where
$$m=\{(\mathrm{A}-\mathrm{B})\mathrm{I}^2\sin^2\alpha/\mathrm{QAB}\}^{\frac{1}{2}}.$$

The values of the coordinates $x$, $y$, measured along the fixed axes, can now be found for the two cases. When $\xi=0$, we have, from (1) and (2),
$$x=\frac{\mathrm{I}}{\mathrm{B}}\int\sin^2\theta\,dt+\frac{\mathrm{I}}{\mathrm{A}}\int\cos^2\theta\,dt, \qquad (12)$$
$$y=\frac{\mathrm{Q}}{\mathrm{I}}\dot{\theta}.$$

Now, integrating from the instant at which $\theta=0$, we get, when the solid makes complete revolutions,
$$\int\sin^2\theta\,dt=\frac{1}{k^2\omega_1}\{\mathrm{F}(k,\ \theta)-\mathrm{E}(k,\ \theta)\},\quad \int\cos^2\theta\,dt=\frac{k^2-1}{k^2\omega_1}\mathrm{F}(k,\ \theta)+\frac{1}{k^2\omega_1}\mathrm{E}(k,\ \theta).$$
From these we get by the first of (12)
$$x=\left\{\left(\frac{\mathrm{I}}{\mathrm{A}\omega_1}+\frac{\mathrm{Q}\omega_1}{\mathrm{I}}\right)\mathrm{F}(k,\ \theta)-\frac{\mathrm{Q}\omega_1}{\mathrm{I}}\mathrm{E}(k,\ \theta)\right\}, \qquad (13)$$
$$y=\frac{\mathrm{Q}\omega_1}{\mathrm{I}}(1-k^2\sin^2\theta)^{\frac{1}{2}},$$
where it is understood that $x=0$ when $\theta=0$.

When complete revolutions are not performed, we get for the values of $x$ and $y$,
$$x=\frac{\mathrm{I}}{\mathrm{B}m}\mathrm{F}(\sin\alpha,\ \phi)-\frac{\mathrm{Q}m}{\mathrm{I}\sin^2\alpha}\mathrm{E}(\sin\alpha,\ \phi), \qquad (14)$$
$$y=\frac{\mathrm{Q}m}{\mathrm{I}}\cos\phi.$$

**17.** *Graphical representation of motion.* Figure 72, which is reduced from Lamb's *Hydrodynamics* [3rd edition, p. 166], illustrates the motions of a disk for which

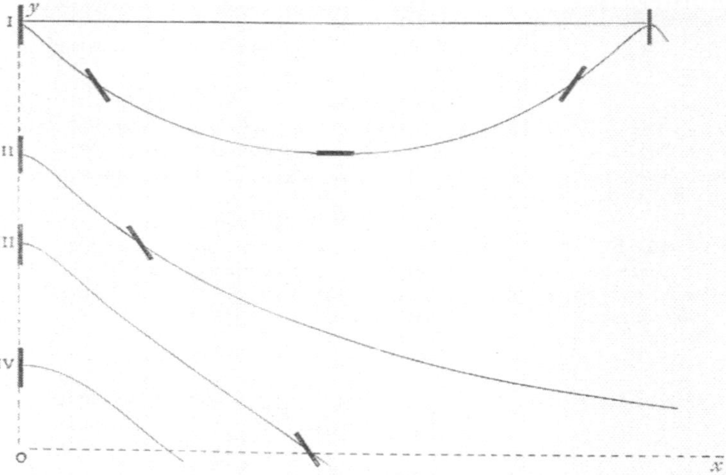


Fig. 72.

$\mathrm{A}=5\mathrm{B}$, projected with different amounts of turning, combined with the same impulse in the direction $Ox$. There are consequently, by (13) and (14), 16, different initial values of $y$, etc. Diagram I shows the path of the centre of the disk and the turning when complete successive revolutions in the same direction are made. In diagram II is shown the motion when complete revolutions are just made; the path is asymptotic to the line of impulse. In diagrams III and IV the path crosses the line $Ox$ at intervals of time equal to half the period of the motion.

The last diagram gives an idea, of course with great exaggeration of amplitude, of the case of small vibrations when the path approximates to a curve of sines.

The following results regarding the separating case II may be verified by the reader either as limiting values of the results already given, or independently. We have $\dot{\theta}=\omega_1\cos\theta$ since $k=1$, and therefore

$$\omega_1 t=\log\tan\left(\tfrac{1}{2}\theta+\tfrac{1}{4}\pi\right).$$

Also
$$\dot{x}=\frac{I}{A}\left(1+\frac{A-B}{B}\sin^2\theta\right),\quad \dot{y}=-I\frac{A-B}{AB}\sin\theta\cos\theta,$$

so that
$$x=\left(\frac{I}{A}+\frac{Q\omega_1^2}{I}\right)t-\frac{Q\omega_1}{I}\tanh\omega_1 t,\quad y=\frac{Q\omega_1}{I}\operatorname{sech}\omega_1 t.$$

It will be seen that there is a remarkable analogy between this periodic motion of a disk or ring in a liquid and the finite oscillations of the axis of a top. It follows from (1), 16, that if the inertia of the disk itself were negligible, and the disk were very thin, so that $A/B=\infty$, the curves in Fig. 72 would meet the line $Oy$ in cusps. Loops are impossible unless [see (1), 16] the value of $\xi$ is sufficiently great to make $\dot{x}$ change sign.

The motion of an infinitely long cylinder, of elliptic cross-section, moving at right angles to its length in a frictionless fluid while turning about its axis of figure, has been worked out by Greenhill [*Mess. of Math.* IX, 1880]. If M be the mass of the cylinder between two cross-sections at unit distance apart, $a$, $b$ the lengths of the semi-axes of the elliptic section, $\sigma$ the density of the cylinder, and $\rho$ that of the fluid, the values of A, B, C are

$$A=M\left(1+\frac{\rho}{\sigma}\frac{a}{b}\right),\quad B=M\left(1+\frac{\rho}{\sigma}\frac{b}{a}\right),$$

$$Q=\tfrac{1}{4}M(a^2+b^2)\left\{1+\frac{\rho}{\sigma}\frac{(a^2-b^2)^2}{2ab(a^2+b^2)}\right\}.$$

**18.** *Stability of a body rotating in a fluid.* Although the subject has been dealt with above [18, VII], we may notice in this connection the gyrostatic stability conferred on the solid by giving it rapid rotation about its axis of figure. Let the angular speed of this rotation be $p$, and the body be in steady motion with this and axial velocity $u$, and all other velocities zero. We now suppose the motion slightly disturbed, and remember in what follows that $u_0$ and $p_0$ are constant steady motion values. The kinetic energy is now given by

$$2T=Au^2+B(v^2+w^2)+Pp^2+Q(q^2+r^2),$$

from which, noticing that $\dot{u}=0$, $\dot{p}=0$, we obtain, by the methods already exemplified, equations of motion

$$B(\dot{v}-p_0w)=-Au_0r,\qquad B(\dot{w}+p_0v)=Au_0q,$$
$$Q\dot{q}+(P-Q)p_0r=-(A-B)u_0w,\quad Q\dot{r}-(P-Q)p_0q=(A-B)u_0v.$$

If now we assume that $\quad v, w, q, r=(\alpha, \beta, \varpi, \rho)e^{i\lambda t},$

we obtain a determinantal equation which is a quadratic in $\lambda^2$, and which splits into the two quadratics
$$BQ\lambda^2\pm B(P-2Q)p_0\lambda-\{B(P-Q)p_0^2+A(A-B)u_0^2\}=0.$$

The condition that the roots should be real is easily found to be that
$$BP^2p_0^2+4A(A-B)Qu_0^2>0.$$

If A > B this condition is always satisfied, but in the absence of spin about the axis it is not satisfied when A < B. Thus, in the latter case, with $p_0=0$, the solid is unstable, and tends to set itself broadside on to the direction of motion. If however $p_0$ is made sufficiently great, the end-on position of the body is stable, even if A < B. The mean motion is now one of precession of the axis of figure round its direction in the steady motion, with very slight deflection.

This contains the theory of the stability of a rifle bullet and of a torpedo, or other air or water craft in which is mounted a sufficiently powerful gyroscope or flywheel, the bearings of which are rigidly attached to the casing or hull of the vessel. It is to be remembered however that in such cases couples may be applied to the body which cause it to be seriously, and permanently, deflected. The body is supposed in the investigation above to be subject to no forces except those called into play by an infinitesimal deflection from the steady straight line motion in the direction of the axis of figure.

# CHAPTER XIV

## EFFECTS OF AIR FRICTION AND PRESSURE. BOOMERANGS

1. *Air friction on rotating body:* (1) *no other applied forces.* It is not possible to give a complete account of the effect of air friction and pressure on the motion of a top or gyrostat, and we have to be content with an approximation based on the assumption, usually made for the motion of a pendulum, of resisting forces simply proportional to the speed of motion. Hence we assume that the action of the air on the revolving body is a couple about the instantaneous axis of rotation, of moment proportional to the angular speed, retarding the rotation and therefore having its axis directed oppositely to the vector representing the angular velocity. Thus, if there is no applied couple except that due to the resistance of the air, the equations of motion may be written

$$A\dot{p}-(A-C)qr=-\lambda p,\quad A\dot{q}-(C-A)rp=-\lambda q,\quad C\dot{r}=-\lambda r,\ \ldots \qquad \text{(1)}$$

where $\lambda$ is a constant.

The third equation gives $$r=r_0 e^{-\frac{\lambda}{C}t}, \qquad \text{(2)}$$

where $r_0$ is the value of $r$ when $t=0$. If we put $p+iq=z$, the first and second equations combine into

$$A\dot{z}-\{i(C-A)r_0 e^{-\frac{\lambda}{C}t}-\lambda\}z=0, \qquad \text{(3)}$$

which is integrable at once. We have

$$A\frac{\dot{z}}{z}=i(C-A)r_0 e^{-\frac{\lambda}{C}t}-\lambda,$$

and therefore $$A\log z=i\frac{C}{\lambda}(C-A)r_0\left(1-e^{-\frac{\lambda}{C}t}\right)-\lambda t+A\log z_0,$$

if $z_0$ be the value, $p_0+iq_0$, of $z$ when $t=0$. Thus we find

$$p+iq=(p_0+iq_0)e^{iC\frac{C-A}{A}\frac{r_0}{\lambda}\left(1-e^{-\frac{\lambda}{C}t}\right)-\frac{\lambda}{A}t}. \qquad \text{(4)}$$

It follows that $$(p^2+q^2)^{\frac{1}{2}}=(p_0^2+q_0^2)e^{-\frac{\lambda}{A}t}. \qquad \text{(5)}$$

Thus the modulus of the vector $p+iq$, the vector of angular velocity, drawn in the equatorial plane of the momental ellipsoid, shrinks in magnitude in geometrical progression as $t$ increases in arithmetical progression.

If now $\phi$ be the amplitude of the vector $(p+iq)/(p_0+iq_0)$, that is the inclination of $p+iq$ to $p_0+iq_0$, we have

$$\phi=C\frac{C-A}{A}\frac{r_0}{\lambda}\left(1-e^{-\frac{\lambda}{C}t}\right) \qquad \text{(6)}$$

Thus, as the axes OA, OB revolve with the varying angular speed $r$ about the axis of figure, the angle $\phi$, starting from zero, continually approaches the value $C(C-A)r_0/\lambda A$ [or an angle of $C(C-A)r_0/2\pi\lambda A$ revolutions] while the modulus $(p^2+q^2)^{\frac{1}{2}}$ continually diminishes exponentially towards zero. These limiting values are arrived at together after an infinite time.

**2.** *Discussion of results of theory.* Let now $\alpha$ denote the inclination of the instantaneous axis OI to the axis of figure OC. We have

$$\tan\alpha = \frac{(p^2+q^2)^{\frac{1}{2}}}{r} = \frac{(p_0^2+q_0^2)^{\frac{1}{2}}}{r_0} e^{-\lambda t\left(\frac{1}{A}-\frac{1}{C}\right)} = \tan\alpha_0 e^{-\frac{C-A}{A}\frac{\lambda}{C}t}. \qquad (1)$$

Thus $\alpha$ increases or diminishes from the value $\alpha_0$ according as C < or > A, that is according as the momental ellipsoid is prolate or oblate. After an infinite time $\tan\alpha$ has become infinite in the former case and zero in the latter, that is the instantaneous axis has, under the influence of air friction, become coincident with the axis of symmetry in the case of a disk top or teetotum, or a spinning oblate ellipsoid of revolution, and perpendicular to the axis of figure in the case of an elongated top, for which the axis of symmetry is the axis of minimum moment of inertia.

It is important to notice that, while the rate of diminution of spin is given by the exponential factor $e^{-\lambda t/C}$, the rate of diminution of $\tan\alpha$ is given by the factor $e^{-\lambda t(1/A-1/C)}$. The latter rate is much smaller than the former. Take the case of the earth, for which $(C-A)/A = 1/304$ nearly. We have

$$\tan\alpha = \tan\alpha_0 e^{-\frac{\lambda t}{C}\frac{1}{304}}. \qquad (2)$$

For example, let $\tan\alpha = \frac{1}{10}\tan\alpha_0$. Then $\lambda t/C = 304\log 10$. Thus

$$r = r_0 e^{-304\log 10}, \qquad (3)$$

that is $r$ has been diminished practically to zero.

The direction of the axis of resultant A.M., OH, being always at right angles to the tangent to the meridional section of the momental ellipsoid, drawn at the point of intersection of the surface by the instantaneous axis, changes direction in the body as the instantaneous axis does so. Finally the axes OI and OH coincide, in one case with the axis of symmetry, in the other with an equatorial axis. The successive positions of OI will trace out on the momental ellipsoid a curve, not generally plane, which the reader may investigate.

Fig. 73.

The action of the air-friction couple necessitates an alteration of position of the axis OH in space as well as in the body. According to the supposition made above the axis of the friction couple is directed oppositely to OI, that is the friction couple axis is OJ, so drawn that the extremity J lies on OI produced backward through O. Thus, as the body turns about OI, which in each of the instantaneous positions of the body-cone (axis OC) is the line of contact with the space-cone (axis OH), the point H moves continually in the direction parallel to IO, that is OJ; and so, as the axes OI and OH approach coincidence, the motion of H approximates more and more to a simple approach of H to O. If $r$ is initially relatively great, it will remain so, and OH will not diverge much from its original position in space. It will however turn round a mean position, and so H will describe a helical curve (see Fig. 73) about this mean line, continually closing in upon it so that ultimately OI and OH there coincide.

**3.** *Air-friction on rotating body: (2) ordinary top under gravity.* So far we have obtained results for the effects of air-friction which hold for any symmetrical body while it is not acted on by other external forces. We now consider a rapidly spinning top under the action of gravity. The method is practically that used by Klein and Sommerfeld [*Theorie des Kreisels*, Bd. III, S. 593] in the discussion of a spherical top, but is here extended to the general case. Referred to the usual axes O(D, E, C), the equations of motion are

$$\left.\begin{aligned} A\ddot{\theta}+(Cn-A\dot{\psi}\cos\theta)\dot{\psi}\sin\theta-Mgh\sin\theta&=-\lambda\dot{\theta},\\ A\ddot{\psi}\sin\theta+(2A\dot{\psi}\cos\theta-Cn)\dot{\theta}&=-\lambda\dot{\psi}\sin\theta,\\ C\dot{n}&=-\lambda n. \end{aligned}\right\} \quad \text{.........(1)}$$

The third equation gives, as before, for $r\ (=n)$,

$$n=n_0e^{-\frac{\lambda}{C}t},\quad Cn=Cn_0e^{-\frac{\lambda}{C}t}\ \text{.........(2)}$$

The angular speed about the vertical through the point of support is $n\cos\dot{\theta}+\dot{\psi}\sin^2\theta$ while the A.M. about the same line is $Cn\cos\theta+A\dot{\psi}\sin^2\theta$. The latter we denote as usual by G. Thus

$$\frac{dG}{dt}=-\lambda(n\cos\theta+\dot{\psi}\sin^2\theta),\ \text{.........(3)}$$

which, of course, could be deduced from the second and third of (1).

For a spherical top C = A, and in that case we get, by (3),

$$G=G_0e^{-\frac{\lambda}{C}t}.\ \text{.........(4)}$$

In the general case we have

$$\begin{aligned}\frac{dG}{dt}&=-\lambda(n\cos\theta+\dot{\psi}\sin^2\theta)\\ &=-\frac{\lambda}{C}(Cn\cos\theta+A\dot{\psi}\sin^2\theta)-\frac{C-A}{C}\lambda\dot{\psi}\sin^2\theta.\ \text{.........(5)}\end{aligned}$$

Thus, integrating, we obtain

$$\left.\begin{aligned} G&=G_1e^{-\frac{\lambda}{C}t},\\ \text{where}\quad G_1&=G_0e^{-\lambda\frac{C-A}{C}\int_0^t\frac{\dot{\psi}\sin^2\theta}{G}dt}. \end{aligned}\right\}\ \text{.........(6)}$$

We now suppose that $\theta$ and $\dot{\theta}$ are both small, so that by the first of (1) we have approximately

$$(Cn-A\dot{\psi}\cos\theta)\dot{\psi}\sin\theta-Mgh\sin\theta=0.\ \text{.........(7)}$$

Now, by the value of G, $Cn\cos\theta+A\dot{\psi}\sin^2\theta$, we find easily that (7), multiplied by $A\sin^3\theta$, can be written in the form

$$(Cn-G\cos\theta)(G-Cn\cos\theta)-MghA\sin^4\theta=0,\ \text{.........(8)}$$

and we have seen that, if $\quad G_1=G_0e^{-\lambda\frac{C-A}{C}\int_0^t\frac{\dot{\psi}\sin^2\theta}{G}dt},$

we have $\quad G=G_1e^{-\frac{\lambda}{C}t}.$

Hence (8) becomes

$$\left(\frac{Cn_0}{G_1}-\cos\theta\right)\left(\frac{G_1}{Cn_0}-\cos\theta\right)=\frac{MghA}{Cn_0G_1}\sin^4\theta\,.\,e^{2\frac{\lambda}{C}t}.\ \text{.........(9)}$$

**4.** ***Discussion of cases:*** **(1)** ***t small,*** **(2)** ***t great.*** We have two cases to consider, (1) $t$ small, (2) $t$ large. If the top is initially set into rapid rotation, $G_1Cn_0$ is great, and the quantity on the right is small if $t$ is not great. Taking case (1), we see that the quantity on the left of (9), 3, is small, and that therefore one of the factors at least is small—the factor $G_1/Cn_0-\cos\theta$. Writing $\cos\theta_1$ for $G_1/Cn_0$, and putting $\cos\theta=\cos\theta_1+\epsilon$, we get, neglecting higher powers of $\epsilon$ than the first,

$$-\epsilon\left(\frac{1}{\cos\theta_1}-\cos\theta_1\right)=\frac{Mgh}{G_1Cn_0}\sin^4\theta\,.\,e^{2\frac{\lambda}{C}t}.$$

By the value of $\cos\theta_1$ we find

$$\epsilon=-\frac{MghA}{C^2n_0^2}\sin^2\theta\,.\,e^{2\frac{\lambda}{C}t},\ \text{.........(1)}$$

and

$$\cos\theta=\cos\theta_1-\frac{MghA}{C^2n_0^2}\sin^2\theta\,.\,e^{2\frac{\lambda}{C}t}.\ \text{.........(2)}$$

Therefore, approximately,

$$\left.\begin{aligned}\theta&=\theta_1+\frac{\mathrm{M}gh\mathrm{A}}{\mathrm{C}^2n_0^{\,2}}\sin\theta_1\,.\,e^{2\frac{\lambda}{\mathrm{C}}t},\\ \text{and}\qquad \sin\theta&=\sin\theta_1+\frac{\mathrm{M}gh\mathrm{A}}{\mathrm{C}^2n_0^{\,2}}\sin\theta_1\cos\theta_1\,.\,e^{2\frac{\lambda}{\mathrm{C}}t}.\end{aligned}\right\}\quad\ldots\ldots\ldots\ldots(3)$$

This result shows that $\theta$ increases with the time, in other words, that the axis begins to descend in consequence of friction, provided $h$ is positive, that is if the centroid is above the point of support. The contrary is the case for the upward vertical if the centroid is below the point of support, as, for example, when a backweight is attached to a gyroscope, or the top is carried by a cap resting on an upturned point, and has gravitational stability from its own weight. This is the arrangement of the small top used in the Fleuriais apparatus described in 3, VII, above, where the question of air-friction is important. The top is driven by a blast of air, which is shut off when the necessary high speed has been attained. The arrangement with centroid above the point of support would be undesirable.

When $t$ is great we have first to consider the effect on the value of G. We have seen that

$$\mathrm{G}=\mathrm{G}_0e^{-\frac{\lambda}{\mathrm{C}}\left\{t+\frac{\mathrm{C}-\mathrm{A}}{\mathrm{A}}\int_0^t\frac{\mathrm{A}\dot\psi\sin^2\theta}{\mathrm{G}}dt\right\}}.\quad\ldots\ldots\ldots\ldots(4)$$

The value of G in the exponent on the right is the current value, that on the left the final value. Suppose G and $\dot\psi$ both positive, A>C, and R$t$ denote the integral; then G will be greater than $\mathrm{G}_0$ if $(\mathrm{A}-\mathrm{C})\mathrm{R}/\mathrm{A}>1$. But in any case $\mathrm{G}_1$ and G must be finite. Considering again (9), with $t$ great, we see that the left-hand side must be finite, and therefore, since $e^{2\lambda t/\mathrm{C}}$ is very great, $\sin^4\theta$ must be very small. Hence $\cos\theta$ is approximately $\pm 1$. If $\cos\theta=1$,

$$-(\mathrm{C}n_0-\mathrm{G}_1)^2=\mathrm{M}gh\mathrm{A}\sin^4\theta\,.\,e^{2\frac{\lambda}{\mathrm{C}}t},\quad\ldots\ldots\ldots\ldots(5)$$

so that $h$ is negative, and, if $\cos\theta=-1$,

$$(\mathrm{C}n_0+\mathrm{G}_1)^2=\mathrm{M}gh\mathrm{A}\sin^4\theta\,.\,e^{2\frac{\lambda}{\mathrm{C}}t},\quad\ldots\ldots\ldots\ldots(5')$$

Thus, in both cases, whatever the initial position, the centroid is finally under the point of support, that is, of course, if the top is such as to admit of this position; if it is simply supported on a peg on a glass plate or a flagstone, it will simply be brought down to the supporting plane with cessation of spin.

**5.** ***Calculation of small term*** $\mathrm{A}\ddot\theta+\lambda\theta$ ***and deduction of more exact values of*** $\sin\theta$, $\cos\theta$. ***Discussion of cases.*** When $\cos\theta=1$, we have by (5), 4,

$$\sin\theta=\theta=\left\{\frac{-(\mathrm{C}n_0-\mathrm{G}_1)^2}{\mathrm{M}gh\mathrm{A}}\right\}^{\frac{1}{4}}e^{-\frac{\lambda}{2\mathrm{C}}t};\quad\ldots\ldots\ldots\ldots(1)$$

and by (5′), 4, when $\cos\theta=-1$,

$$\sin\theta=\pi-\theta=\left\{\frac{(\mathrm{C}n_0+\mathrm{G}_1)^2}{\mathrm{M}gh\mathrm{A}}\right\}^{\frac{1}{4}}e^{-\frac{\lambda}{2\mathrm{C}}t}.\quad\ldots\ldots\ldots\ldots(1')$$

Now the value of $\theta$ in (3), 4, gives by differentiation

$$\mathrm{A}\ddot\theta+\lambda\dot\theta=\frac{2}{\mathrm{C}}\lambda^2\left(1+\frac{2\mathrm{A}}{\mathrm{C}}\right)\frac{\mathrm{M}gh\mathrm{A}}{\mathrm{C}^2n_0^{\,2}}\sin\theta_1\,.\,e^{2\frac{\lambda}{\mathrm{C}}t}.\quad\ldots\ldots\ldots\ldots(2)$$

The condition that $\mathrm{A}\ddot\theta+\lambda\dot\theta$ should be very small compared with $\mathrm{M}gh\sin\theta$ is fulfilled if $\lambda^2$ is small compared with $\mathrm{C}^2n_0^{\,2}$. That this should be so is reasonable, for then $2\pi\lambda$ would be the amount of diminution of C$n$ produced in one revolution, which must be only a small fraction of C$n$ under any ordinary circumstances.

Supposing $t$ to be small, we write, by the result just obtained, the first of (1), 3, and (8), 3,

$$\frac{2}{\mathrm{C}}\lambda^2\left(1+\frac{2\mathrm{A}}{\mathrm{C}}\right)\frac{\mathrm{M}gh\mathrm{A}}{\mathrm{C}^2n_0^{\,2}}\sin\theta_1\,.\,e^{2\frac{\lambda}{\mathrm{C}}t}=-\frac{(\mathrm{C}n_0-\mathrm{G}_1\cos\theta)(\mathrm{G}_1-\mathrm{C}n_0\cos\theta)}{\mathrm{A}\sin^3\theta}e^{-2\frac{\lambda}{\mathrm{C}}t}+\mathrm{M}gh\sin\theta.\quad\ldots(3)$$

Now, by (2), 4,

$$\mathrm{C}n_0-\mathrm{G}_1\cos\theta=\mathrm{C}n_0-\mathrm{G}_1\left(\cos\theta_1-\frac{\mathrm{M}gh\mathrm{A}}{\mathrm{C}^2n_0^{\,2}}\sin^2\theta\,.\,e^{2\frac{\lambda}{\mathrm{C}}t}\right).\quad\ldots\ldots\ldots\ldots(4)$$

Substituting in (3) this value of $Cn_0 - G_1 \cos\theta$, and the value of $\cos\theta$ and $\sin\theta$ from (2) and (3), 4, we obtain a value of $\cos\theta$ in terms of $Cn_0$, $G_1$, $\cos\theta_1$, $\sin\theta_1$, which carries the value of $\cos\theta$ given by (2), 4, to a higher degree of accuracy, and which the reader may work out in detail.

When $t$ is great we calculate $A\theta + \lambda\dot{\theta}$ from (1) or (1′), 5, according as $\cos\theta = \pm 1$, nearly. We obtain

$$A\theta + \lambda\dot{\theta} = \pm \frac{\lambda^2}{4C}\left(\frac{A}{C} - 2\right)\sin\theta. \qquad \text{(5)}$$

From this result and (9), 3, we get

$$\left\{Mgh \pm \frac{\lambda^2}{4C}\left(\frac{A}{C} - 2\right)\right\} A \sin^4\theta = \pm (Cn_0 \pm G_1)^2; \qquad \text{(6)}$$

and therefore

$$\sin\theta = \left\{ \pm \frac{(Cn_0 \pm G_1)^2}{AMgh} \right\}^{\frac{1}{4}} \left\{ 1 \pm \frac{\lambda^2}{4C}\left(\frac{A}{C} - 2\right)\frac{1}{Mgh} \right\}, \qquad \text{(7)}$$

where all the upper signs are to be taken for cos $= -1$, and all the lower signs for $\cos\theta = +1$.

Thus again we have $A\theta + \lambda\dot{\theta}$, a small quantity, as we should, and a nearer approximation to $\sin\theta$. The calculation could be pushed to a higher approximation, but it is needless to do so.

6. *Motion of a flat disk. The boomerang.* The peculiar motions of a light flat disk depend entirely on the action of the air. If a card is held by

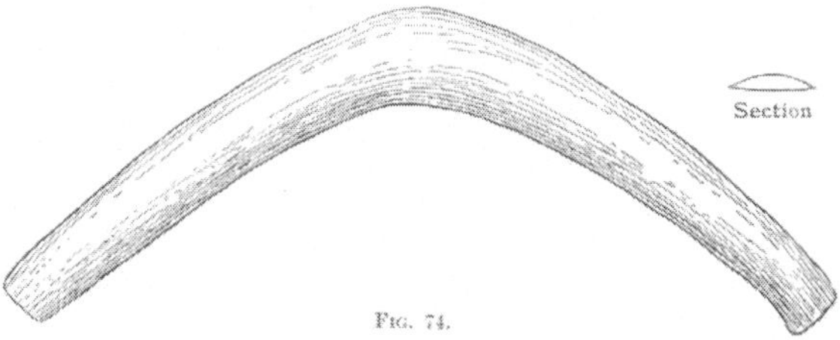


Fig. 74.

one corner in a suitable inclined position to the vertical and is set in translational motion in its own plane, with rotation about a normal to the plane, it will, if thrown adroitly, seem to describe a parabolic path, but on an inclined plane. It is interesting to drop cards under different initial conditions down a wide sheltered shaft, and note motion and paths [see Maxwell, *Sci. Papers*, I, p. 115].

In what follows the reference will be mainly to the Australian aboriginal weapon called the boomerang, but most of the statements made can be illustrated by the flight of cards or of thin disks of wood thrown into the air, with rotation in the plane of the disk in each case. No doubt the shape of the edge of the boomerang, which is very different in different specimens, has some influence on the path, but much would appear to depend on the dexterity of the thrower.

The boomerang is made of hard wood, and is generally of shape similar to that shown in Fig. 74. The angle between the arms varies greatly in different specimens. In section it is slightly convex on one side, and more decidedly convex on the other, as shown also in the figure. It is held by one end, and delivered in the air by the (right-handed) thrower with the nearly flat side vertical and the more convex side turned towards him. The missile is set into rotational motion about a normal to the mean plane, as shown by the short transverse lines in Fig. 75, which gives a plan and elevation of an observed path. The transverse line in each case is the normal, and is drawn to the side of the plane from which the rotation is observed to be in the counter-clock direction.

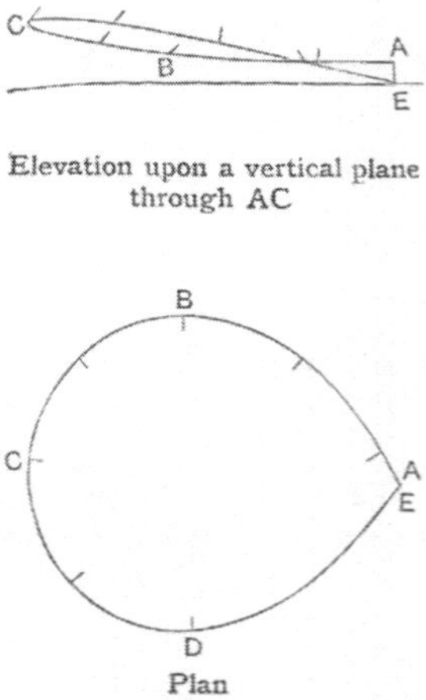


Fig. 75.

One side of the boomerang has been spoken of in what precedes as nearly plane. But in some forms the two arms are not in one plane; they form, in consequence of a slight twist, the two blades of a propeller, which, as it revolves about a normal to what may be called the mean plane, screws itself forward in the direction in which that normal points as a spin axis. According to experiments on the flight of boomerangs, which are confirmed by the mathematical theory referred to below, the return of the path to the starting point depends on the existence either of this propeller action or of the greater convexity of one side of the weapon; or of a combination of the two. Either peculiarity existing by itself may be made to give the effect.

7. *General explanation of motion. Specification of forces.* We shall now try to account for the motion, taking as an example for reference the trajectory shown in Fig. 76. Suppose first, what is not often the case, that the path lies nearly in a vertical plane, so that it is practically represented by an elevation only. We have two cases to consider, the front of the plane of the disk (*a*) tilted down, (*b*) tilted up with regard to the trajectory. The spin keeps the plane of the disk at a constant slope, and in this case at right angles to a vertical plane with which the path nearly coincides.

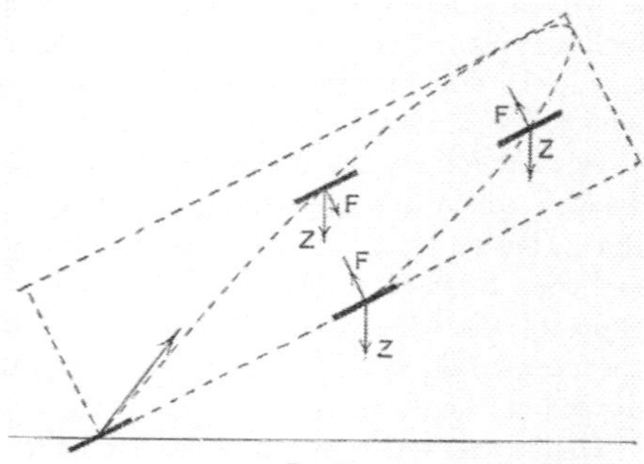


Fig. 76.

In case (*a*) the forward motion and the sideways motion combined give a resultant force due to air pressure, which, in the upward branch is directed towards the under side of the plane. Fig. 76 shows the two branches of the path and the direction of the air pressure force F and of gravity.

A similar diagram for case ($b$) would be like that obtained by interchanging the branches of the curve, and tilting the disk the other way with reference to the path. It will be noticed that in each case the radius of curvature is in the direction of the total force at right angles to the trajectory.

To specify the forces in the general case take as axes of coordinates, at any point O of the path, the downward vertical there, a horizontal line at right angles to the path, and a second horizontal line in a vertical plane containing the tangent to the path at O, choosing these lines so that they form an ordinary system of axes $O(x, y, z)$. Let then $l, m, n$ be the direction cosines of the normal to the disk drawn in the direction of the force F, and $\alpha, \beta, \gamma$ the direction cosines of a forward element of the path. The horizontal force producing drift sideways is $m\mathrm{F}$, the vertically downward force Z is $\mathrm{M}g+l\mathrm{F}$ (in the case illustrated by Fig. 76), the force along the path is $\mathrm{M}g\alpha+\mathrm{F}(l\alpha+m\beta+n\gamma)$. The latter force taken with the reversed sign, and added to the frictional resistance R, is the retarding force along the path, that is

$$\textit{tangential retarding force} = \mathrm{R} - \{\mathrm{M}g\alpha+\mathrm{F}(l\alpha+m\beta+n\gamma)\}.$$

In the case ($b$) the vertically downward force is $\mathrm{M}g-l\mathrm{F}$, and the force normal to the disk and towards the upper side is $\mathrm{F}-l\mathrm{M}g$.

8. *Trajectory of light disk in its own plane.* We shall now try to explain more fully the form of the trajectory in the general case. First let us suppose that the branches give no trace of return to the starting point. If the disk be very light and be projected so as to move very nearly in its own plane, $\mathrm{F}-l\mathrm{M}g$ may be so nearly zero that, though the effective mass of the disk is also small, there is little or no acceleration along the normal. As before, the aspect of the plane is kept constant by the spin.

The force however in the plane of the disk is $\mathrm{M}g(1-l^2)^{\frac{1}{2}}$, that is the disk is a projectile in the gravitational field (a plane inclined to the vertical at the angle $\cos^{-1}l$) of intensity $g(1-l^2)^{\frac{1}{2}}$, and the trajectory is a parabola. This can be illustrated very easily by projecting oblong post-cards or playing-cards so that each moves accurately in its own plane.

9. *Returning boomerang. Propeller action of twist of arms.* The class of cases in which the outward and return branches lie nearly in the same vertical plane is sufficiently dealt with above. We have now to consider how the side-drift may arise and the projectile be made to return to the starting point. Let the boomerang have a certain amount of twist of its arms, and be projected as indicated in Fig. 75, with its mean plane vertical, so that its centroid sets off along the tangent at A to the branch AB. If

the screw-propeller action is such as to give propulsion towards the left, that is along the normal as drawn in the diagram, the trajectory will bend round in the manner shown. The spin is such as to give a force to the left, and the trajectory swings round as in the diagram. The translatory speed and also the spin diminish, and the plane tilts over as the normals indicate. The propeller action also falls off, but, as both speed and propeller action are less, the curvature in the plane is sensibly the same on both sides at corresponding points. The propeller action on the now slightly tilted-over disk contributes to, if it does not wholly produce, the upward concavity of the ascending branch and the straight and gentle downward slope of the last bit of the descending branch, in which the axis of spin has no lateral projection.

10. *Difficulties of complete theory. Approximate constants of motion.* The theory of the motion of a boomerang cannot be worked out with any close approach to accuracy. The action of the air on an element of the surface of a body moving through it is not known exactly, and therefore the forces and couples on a body of any given form can only be roughly estimated; and the figure of a boomerang is so complicated that, even if the action on an element of surface were given, the calculation of the resultant forces and couples would be difficult. Nevertheless the motion has been submitted to calculation by Mr. G. T. Walker, who has arrived at a number of interesting conclusions. We state here the most important of these, and refer the reader for Mr. Walker's analysis to his paper "On Boomerangs" [*Phil. Trans. R.S.* 190, 1897].

If $q$ be the resultant speed of an element of surface, and $a$ the angle between the normal to the element and the direction of $q$, the pressure on the element is taken as $\lambda q^{\mu} \cos a$, and the couple per unit area as having moment $\kappa q^{\nu} \cos a$, about an axis at right angles at once to the normal and to the direction of $q$. Here $\lambda$, $\kappa$, $\mu$, $\nu$ are constants, which are estimated from experiments on bodies of different forms moving in air. Mr. Walker takes $\mu = 3$ and $\nu = 2$

Taking the plane of the arms of the boomerang, or the mean plane if there is twist, as that of XY, the centroid as origin, the projection of the direction of $q$ on this plane as OX, a line drawn towards the more convex side as OZ, and OY to suit these axes, rectangular components of linear and angular velocity of the body are specified. These are U, V, W, and $\Omega_1$, $\Omega_2$, $\Omega_3$, where W is small compared with U, and $\Omega_1$, $\Omega_2$ are small compared with $\Omega_3$. $\Omega_1$ is positive, $\Omega_2$ negative, $\Omega_3$ positive. The time of flight is about 9 seconds, the greatest distance 50 yards; $\Omega_1$ is about $\frac{1}{6}$, $\Omega_2$ about $\frac{1}{3}$, and $\Omega_3$ about 30, in radians per second. U is perhaps 2000 cm./sec. These quantities give the motion of the boomerang with reference to axes travelling with it but not fixed in it. Increase of twist and diminution of round increase the positive value of $\Omega_1$, and the negative value of $\Omega_2$.

With regard to axes fixed in the body the linear and angular speeds are $u, v, w, \omega_1, \omega_2, \omega_3$. These axes, O(A, B, C), are chosen as in Fig. 77, with the centroid as origin, and OA, OB as the mean plane; and Euler's equations of motion are available when the couples and moments of inertia are known.

It will be seen that to an observer in front of the boomerang, looking from X, say, towards O, the turning of the plane about OX is counter-clockwise, that is the normal, which is towards the observer's right at the instant of projection with plane of the boomerang vertical, begins to tilt upwards. The progress of the tilt can be traced from the diagrams. The normal being drawn towards the more convex side the reader can see how that side is situated in the trajectory.

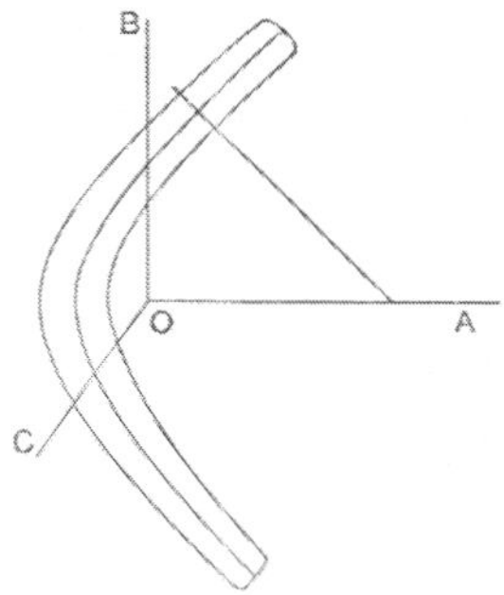

Fig. 77.

11. *Motion with change of direction of plane.* The effects of twist and "rounding" (as Mr. Walker calls the greater convexity) are so important, and the latter is so difficult to focus into a single idea, that it is practically impossible to give anything like a complete explanation of the motions without analysis. But an ordinary unreturning boomerang (such as is commonly used by the Australian aborigines), the arms of which are in one plane, and the faces of which are slightly and equally curved, may be considered. The forward and sideways motions determine the pressure forces, and these of course go through a cycle of values as the missile revolves.

But upward tilt of the front above the trajectory gives lifting force, and tilt round to either side gives side thrust, so that lateral drift from the direction of the initial plane of motion can be accounted for. Alteration of the aspect of the plane of the disk is due to the couples that act on the surfaces. There are two component couples, one due to the position of the centre of effort for the advancing body, a position that changes as the body revolves in its own plane, so that we have an average couple to reckon, and a couple due to the different speeds of different surface elements through the air. These couples will produce precession of the disk by the turning of the axis of spin towards the couple axis in each case.

12. *Twist or "rounding" essential for return. Multiple loops.* Mr. Walker finds that, as stated above, unless either twist or "rounding," or a combination of them exists, the boomerang does not return to the point of projection, but describes an open loop, roughly in a plane inclined to the vertical, or in a vertical plane itself according to the throw. But by properly shaping the boomerang, making, for example, the ends of the arms be nearly at right angles, and properly adjusting twist and "rounding," it may be made to remain

in a nearly circular path for a considerable part of the time it is in the air. Fig. 78 shows the trajectory in such a case. The last figure shows the trajectory of a boomerang thrown by Mr. O. Eckenstein The angle between the arms was greater than a right angle, and there appears to have been "rounding" but no twist. It will be seen that the boomerang rises, passes round a considerable distance with but little change of level, and finally returns, passing over the head of the thrower, and comes to rest after describing a series of diminishing loops. A professional boomerang thrower has been known to throw the missile from the middle of the stage of a theatre, then turn his back till it passed round the front of the upper circle, returned over his head and gently sank in a final small loop to his hand held out to receive it. This apparently could be done as often as required, with perfect safety to the audience and exact repetition of the motion.

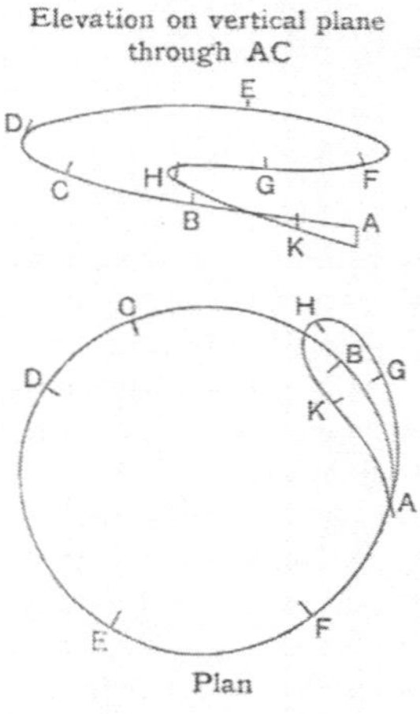


Fig. 78.

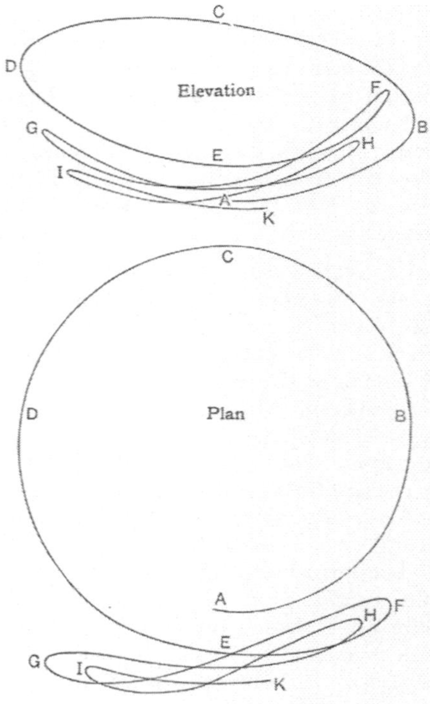


Fig. 79.

The best non-returning boomerangs of the Australian blacks are made with considerable rounding and a little negative twist. They are thrown with the more curved surface uppermost, and the mean plane nearly horizontal. The upward action of the air on the lower surface prolongs the flight, so that the missile may be thrown twice as far as a sphere of the same mass.

# CHAPTER XV

## THE SPHERICAL PENDULUM. MOTION OF A PARTICLE ON A SURFACE OF REVOLUTION

**1.** *Top turning about a fixed point and destitute of* A.M. *about axis of figure.* The spherical pendulum in its simplest form may be regarded as a symmetrical top without A.M. about its axis of figure. This A.M. may be zero, either because the moment of inertia about the axis of figure is negligible or because there is no angular speed about that axis. For the present we shall suppose that the centroid, whether the pendulum is a rigid body or a massive particle, is at a distance $l$ from the point of suspension. The question of rigidity or flexibility of the pendulum will arise when we consider the forces on the supporting stem or thread.

Let A denote the moment of inertia about an axis transverse to the length of the pendulum, and $\dot{\phi}$, $\dot{\psi}$ have the same meanings as for a top. The A.M. about the vertical reduces to $A\dot{\psi}\sin^2\theta$. Thus

$$\dot{\psi}\sin^2\theta = \frac{G}{A} = \frac{h}{l^2}, \quad \text{.................................(1)}$$

where $h$ is a constant. If the pendulum consist of a particle, of mass $m$, hung by a massless thread, we have $A = ml^2$ and $G = mh$. Here and in what follows $\theta$ is measured from a vertical drawn upward from the point of support.

If C be the centroid of the pendulum, so that $OC = l$, the horizontal projection of OC is $l\sin\theta$, which we denote by $\rho$. The rate of turning of this projection round the vertical through O is $\dot{\psi}$. The horizontal speed of C is $\rho\dot{\psi}$, and the moment of this motion about the vertical is $\rho^2\dot{\psi}$. Hence we have

$$\rho^2\dot{\psi} = h, \quad \text{.........................................(2)}$$

which asserts the uniform description of areas by the radius vector $\rho$. If $x$, $y$ be the coordinates of P referred to horizontal axes through O, we have

$$\dot{y}x - \dot{x}y = \rho^2\dot{\psi} = h. \quad \text{.................................(3)}$$

If there be no turning of the top about the axis of figure, $-\dot{\psi}\cos\theta = \dot{\phi}$, and therefore

$$l^2\dot{\phi}\sin^2\theta = -\frac{Gl^2}{A}\cos\theta. \quad \text{..................................(4)}$$

The energy equation is

$$A(\dot\theta^2+\dot\psi^2\sin^2\theta)=2(-mgl\cos\theta+E), \quad\ldots\ldots(5)$$

where E is the total energy. But we have seen that $\dot\psi\sin^2\theta=h/l^2$. Hence

$$A\dot\theta^2=2(-mgl\cos\theta+E)-\frac{Ah^2}{l^4\sin^2\theta}. \quad\ldots\ldots(6)$$

If we write $a$ for $2mgl/A$, $b$ for $2E/A$, $c^2$ for $h^2/l^4$, and $z$ for $\cos\theta$, we get

$$\dot z^2=(-az+b)(1-z^2)-c^2. \quad\ldots\ldots(7)$$

From (7) we obtain $t$ as an elliptic integral in terms of $z$, and $z$ as an elliptic function in terms of $t$ For we have

$$dt=\frac{dz}{\{(-az+b)(1-z^2)-c^2\}^{\frac{1}{2}}}=\frac{dz}{(aZ)^{\frac{1}{2}}}. \quad\ldots\ldots(8)$$

In this equation and those which follow as derived from it, the radical may have either sign according to circumstances. Examples are given below. Equating Z to zero, we get a cubic equation of roots $z_1$, $z_2$, $z_3$, fulfilling the condition $z_1>1>z_2>z_3$.* The roots $z_2$, $z_3$ define limitations of the range of motion of the pendulum. The centroid cannot rise higher than the height $lz_2$ above the fixed point or fall below $lz_3$, for the reason that $\dot z^2$ cannot be negative, and Z must therefore be positive. Of course both of these heights may be negative, that is both limiting levels may be under the centre of the sphere of radius $l$ on which the centroid moves; we shall see that the second is always negative.

Writing Z in the form $(z_1-z)(z_2-z)(z-z_3)$, and supposing $t=0$ when $z=z_3$, we obtain

$$a^{\frac{1}{2}}t=\int_{z_3}^{z}\frac{dz}{\{(z_1-z)(z_2-z)(z-z_3)\}^{\frac{1}{2}}}. \quad\ldots\ldots(9)$$

Thus $t$ is given as an elliptic integral. As we have seen above, the integral can be written

$$mt=\int_0^{\phi}\frac{d\phi}{(1-k^2\sin^2\phi)^{\frac{1}{2}}}=F(k,\phi), \quad\ldots\ldots(10)$$

where $m$ [not the mass of the pendulum]

$$=\{\tfrac{1}{4}a(z_1-z_3)\}^{\frac{1}{2}},\ \tan^2\phi=(z-z_3)/(z_2-z),\ \text{ and }\ k^2=(z_2-z_3)/(z_1-z_3).$$

The period of oscillation T is $2F(k,\tfrac{1}{2}\pi)/m$ [see also (2), 5, XII].

We thus get $z$ in terms of $t$ by the elliptic function equations

$$\left.\begin{aligned} z-z_3&=(z_2-z_3)\,\mathrm{sn}^2 mt, & z_2-z&=(z_2-z_3)\,\mathrm{cn}^2 mt,\\ z_1-z&=(z_1-z_3)\,\mathrm{dn}^2 mt, & z&=z_2\,\mathrm{sn}^2 mt+z_3\,\mathrm{cn}^2 mt.\end{aligned}\right\} \quad\ldots\ldots(11)$$

Returning now to the equation $Z=0$, we observe that the sum of its roots $z_1+z_2+z_3$ is $b/a$, and that the sum of the products of pairs of these roots, $z_3z_2+z_3z_1+z_1z_2$, is $-1$. But we can write this sum of products as

* The order of the roots taken here is the opposite of that used in 4, XII, and elsewhere, that is $z_1$ and $z_3$ are interchanged. This was not observed until the matter was in type, when it did not appear to be worth while to change the formulae.

$z_1(z_2+z_3)+z_2z_3$, so that we have $z_1(z_2+z_3)=-1-z_2z_3$. The quantity $-1-z_2z_3$ is negative whatever the signs of $z_2$, $z_3$ may be, since each is less than 1. Hence, as $z_1$ is positive, $z_2+z_3$ is negative, and the mean of the values of $z_2$ and $z_3$ is negative, that is the horizontal circle on the sphere midway between the circles which limit the motion is below the centre.

2. *Calculation of azimuthal motion.* We can now find the time in the azimuthal motion. We have by (1) and (8), 1,

$$d\psi=\frac{h}{l^2}\frac{dz}{(1-z^2)(aZ)^{\frac{1}{2}}}. \qquad \text{.................................(1)}$$

The numerical value of $\psi$ for any value of $z$, and therefore of the time, is best found by the process of 21 and 22, XII, above.

When $\dot{z}$ is negative, that is when the pendulum is descending, we must give the negative sign to the square root in (8) and in (9), 1. Let us suppose that the pendulum is started at the level $z=z_0$ $[z_2>z_0>z_3]$, with some value of $\dot{z}$; then of course Z is positive. Take $\dot{z}$ negative at starting, $\dot{\psi}$ as the simultaneous angular speed in azimuth. At starting the pendulum has A.M. about a vertical axis, and since this must remain unchanged we have $\rho^2\dot{\psi}=h$. This equation gives $2\rho\dot{\rho}\dot{\psi}+\rho^2\ddot{\psi}=0$, or, since $\rho=l\sin\theta=l(1-z^2)^{\frac{1}{2}}$,

$$\ddot{\psi}=-2\frac{\dot{\rho}}{\rho}\dot{\psi}=-2\frac{z\dot{z}}{1-z^2}\dot{\psi}. \qquad \text{..... .......................(2)}$$

Thus $\ddot{\psi}$ vanishes when $\dot{z}=0$, that is at each limiting circle. It also vanishes when $z=0$, that is when the bob is passing the horizontal plane

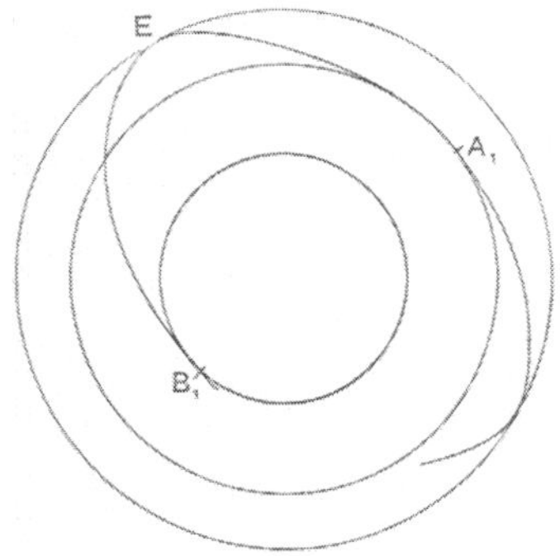


FIG. 80.

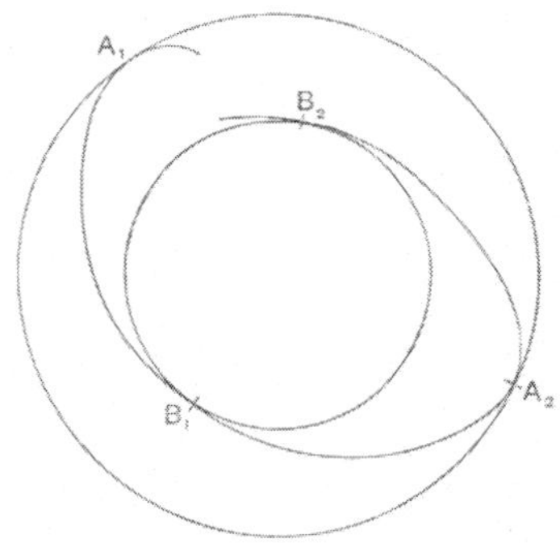


FIG. 81.

through O (if it does pass that plane), and the value of $\rho$ is there stationary as $z$ varies. The path of the bob at a limiting circle, where $\dot{z}=0$, is a tangent to that circle on the sphere, and changes at the point ($A_1$ or $B_1$, Figs. 80 and 81) of touching, from a downward or upward course to an upward or downward, according as the limiting circle is the lower or upper.

There are two cases, *first*, that in which the upper circle is above the centre of the sphere, *second*, that in which both circles are below the centre. The projection of the path on a horizontal plane is shown in Fig. 80. It touches three circles in the first case, the projections respectively of the upper limiting circle at $A_1$, the horizontal great circle of the sphere at E, and the lower limiting circle at $B_1$ It is easy to see that in this case the projection of the path touches the projections of the limiting circles externally, for clearly the path must trend outward from touching either of these to touch the great circle, and thence trend inward again to touch the projection of the lower circle.

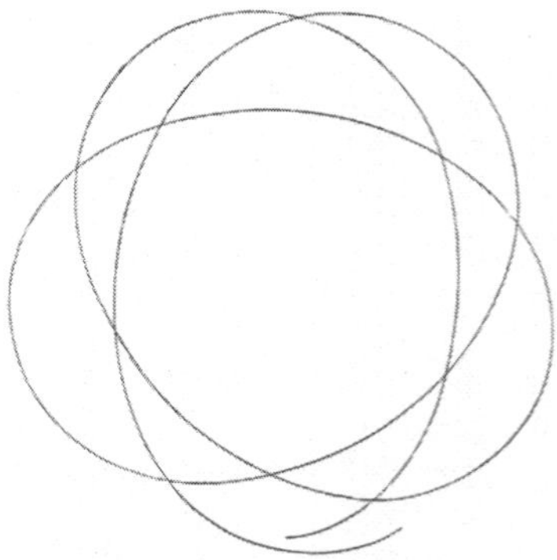
FIG. 82.

The case in which both limiting circles are below the centre is shown in Fig. 81. The projection of the lower limiting circle is shown touched externally by the path, the other is touched internally. It is clear that the path must trend outward from touching the lower circle to touch the upper, and has not begun to turn inward when the latter circle is reached.

Fig. 82 shows a photograph of an actual path. (See the appendix to this chapter for curves obtained by experiment.)

3. *Azimuthal angle described from one limiting circle to the other* $> \frac{1}{2}\pi$. If P, P′ be two successive points of intersection of an intermediate circle by the path, and A the intermediate point of touching a limiting circle, say the upper, the azimuthal angles $\psi_{PA}$, $\psi_{AP'}$ are equal. To prove this we go back to equation (1), 2, and obtain easily

$$\psi_{PA} = \frac{h}{l^2}\int_{z}^{z_2} \frac{dz}{(1-z^2)(aZ)^{\frac{1}{2}}} = -\frac{h}{l^2}\int_{z_2}^{z} \frac{dz}{(1-z^2)(aZ)^{\frac{1}{2}}} = \psi_{AP'}. \quad \text{.........(1)}$$

Therefore the vertical plane through the centre of the sphere and through any turning point of the axis on a limiting circle is a plane of symmetry as regards the portions into which it divides the path on the sphere.

We shall now prove that the angle $\psi_{BC}$, traversed by the projection of the radius vector from a point of contact B with the projection of the circle $z=z_3$ to the next point of contact, C, with the circle $z=z_2$, is greater than $\frac{1}{2}\pi$ and less than $\pi$. The first of these theorems is due to Puiseux (*Journ. des Math.* 7, 1842), the second is due to Halphen (*Fonct. Ellipt.* t. II).

We have
$$\psi_{BC} = c\int_{z_3}^{z_2} \frac{dz}{(1-z^2)(aZ)^{\frac{1}{2}}}, \quad \text{...............................(2)}$$

where
$$aZ = (b - az)(1 - z^2) - c^2 = a(z_1 - z)(z_2 - z)(z - z_3). \quad \text{..............(3)}$$

But we have seen that $z_1=-(1+z_2z_3)/(z_2+z_3)$, which is positive, since both numerator and denominator are negative. Thus

$$z_1-z=-\frac{1+z_2z_3}{z_2+z_3}-z=-\frac{1+z_2z_3+z(z_2+z_3)}{z_2+z_3}, \quad\text{......(4)}$$

which also is positive, even if $z=1$. Also

$$aZ=-\frac{a}{z_2+z_3}\{1+z_2z_3+z(z_2+z_3)\}(z_2-z)(z-z_3). \quad\text{......(5)}$$

But if $z=1$, $aZ$ becomes $-c^2$, so that we have, by the value of $z_1$,

$$-c^2=\frac{a}{z_2+z_3}(1+z_2)(1+z_3)(1-z_2)(1-z_3)=\frac{a}{z_2+z_3}(1-z_2^2)(1-z_3^2), \quad\text{...(6)}$$

which is negative, inasmuch as $z_2+z_3$ is negative. Thus

$$c=\frac{h}{l^2}=\left\{-\frac{a}{z_2+z_3}(1-z_2^2)(1-z_3^2)\right\}^{\frac{1}{2}}, \quad\text{......(7)}$$

and we obtain from (5)

$$\psi_{BC}=\left\{\frac{(1-z_2^2)(1-z_3^2)}{-(z_2+z_3)}\right\}^{\frac{1}{2}}\int_{z_3}^{z_2}\frac{dz}{(1-z^2)\{(z_1-z)(z_2-z)(z-z_3)\}^{\frac{1}{2}}}, \quad\text{......(8)}$$

where $z_1=-(1+z_2z_3)/(z_2+z_3)$. Now, since $\frac{1}{2}(z_2+z_3)$ is negative, $z_3$, which is always negative, is numerically greater than $z_2$. Hence the value of the expression on the right, which is positive, is diminished by giving to $z$, in the factor $z_1-z$ under the square root sign in the integrand, the value of $z_3$. Similarly, the expression is increased by writing $z_1-z_2$ for $z_1-z$. Hence, if I denote the integral,

$$\int_{z_3}^{z_2}\frac{dz}{\{(1-z^2)(z_2-z)(z-z_3)\}^{\frac{1}{2}}},$$

$$\left\{\frac{(1-z_2^2)(1-z_3^2)}{1+z_2z_3+z_2(z_2+z_3)}\right\}^{\frac{1}{2}}I>\psi_{BC}>\left\{\frac{(1-z_2^2)(1-z_3^2)}{1+z_2z_3+z_3(z_2+z_3)}\right\}^{\frac{1}{2}}I.$$

It will be observed that both $1+(z_2+z_3)z_2+z_2z_3$ and $1+(z_2+z_3)z_3+z_2z_3$ are positive, since $1+z_2z_3+z_1(z_2+z_3)=0$, by the value of $z_1$.

Now, as the reader may verify,

$$I=\tfrac{1}{2}\pi\frac{\{(1-z_2)(1-z_3)\}^{\frac{1}{2}}+\{(1+z_2)(1+z_3)\}^{\frac{1}{2}}}{\{(1-z_2^2)(1-z_3^2)\}^{\frac{1}{2}}}=\tfrac{1}{2}\pi\frac{A+B}{\{(1-z_2^2)(1-z_3^2)\}^{\frac{1}{2}}}. \quad\text{......(9)}$$

The preceding inequality therefore becomes

$$\tfrac{1}{2}\pi\frac{A+B}{(1+2z_2z_3+z_2^2)^{\frac{1}{2}}}>\psi_{BC}>\tfrac{1}{2}\pi\frac{A+B}{(1+2z_2z_3+z_3^2)^{\frac{1}{2}}}.$$

It is easy to see from this result that $\psi_{BC}$ lies between two limits, both of which are greater than $\frac{1}{2}\pi$. For, if we square $A+B$, we diminish the result if we substitute $z_3^2$ for $z_2^2$ where the latter occurs. The numerator of the fraction $(A+B)^2/(1+2z_2z_3+z_2^2)$ still exceeds the denominator by

$2(1-z_3^2)+1-z_2^2$, which is positive. Similarly we can show that $(A+B)^2$ exceeds $1+z_2z_3+z_3(z_2+z_3)$. Thus we have the result stated. [See also 7 below.]

**4.** *Azimuthal speed very great. Limiting circles nearly coincident with the horizontal great circle, one above, the other below.* If the azimuthal speed be very great the values of $b$ and $c^2$ in Z are also very great, and the cubic $Z=0$ reduces, approximately, to $z^2=1-c^2/b$, which gives for $z_2$, $z_3$ values equal but of opposite sign. The third root $z_1(>1)$ as $b$ is increased without limit tends towards infinity. The inequality found above shows that then $\psi_{BC}$ reaches the value $\pi$, while the azimuthal motion very closely coincides with the horizontal great circle of the sphere. When the angular speed in azimuth is very great but finite $z_2$ and $z_3$ are opposite in sign, and very nearly equal to 1 in numerical value, and we have, to a first approximation,

$$\psi_{BC}=\tfrac{1}{2}\pi\,.\,2\left(\frac{1-z^2}{1-z^2}\right)^{\frac{1}{2}}=\pi.$$

To a closer approximation we have no need to go at present; it may be worked out by the reader.

**5.** *Pendulum nearly vertical. Theorem of Bravais.* When the pendulum is only slightly deflected from the vertical and is projected with a correspondingly small azimuthal angular speed, the values of $z_2$ and $z_3$ will have the same sign and be nearly equal. The limits between which $\psi_{BC}$ lies become approximately equal, and we find

$$\psi_{BC}=\tfrac{1}{2}\pi\frac{2}{(1+3z_2^2)^{\frac{1}{2}}}=\tfrac{1}{2}\pi\frac{1}{(1-\frac{3}{4}\sin^2\theta_2)^{\frac{1}{2}}}=\tfrac{1}{2}\pi(1+\tfrac{3}{8}\sin^2\theta_2).\quad\ldots\ldots(1)$$

The angle $\theta_3$ might equally well have been used in this approximation instead of $\theta_2$. Hence a somewhat closer approximation would be obtained by writing $\sin\theta_2\sin\theta_3$ instead of $\sin^2\theta_2$ or $\sin^2\theta_3$. Thus we get

$$\psi_{BC}=\tfrac{1}{2}\pi(1+\tfrac{3}{8}\sin\theta_2\sin\theta_3).\quad\ldots\ldots(2)$$

If $\alpha$, $\beta$ be the apsidal distances of the projection on a horizontal plane of the path of a point at distance $l$ from O, we have

$$\psi_{BC}=\tfrac{1}{2}\pi\left(1+\frac{3}{8}\frac{\alpha\beta}{l^2}\right),\quad\ldots\ldots(3)$$

which is Bravais' result [*Journ. de Mathém.*, 19 (1854)]. The fraction, $\frac{3}{8}\alpha\beta/l^2$ of $\frac{1}{2}\pi$, by which $\psi_{BC}$ exceeds $\frac{1}{2}\pi$, is $\frac{3}{4}$ of the ratio of the area $\pi\alpha\beta$ of the ellipse to the area $2\pi l^2$ of the hemisphere below the point O.

The result may also be obtained as follows. By (8) of 3, we have

$$\psi_{BC}=\left\{\frac{(1-z_2^2)(1-z_3^2)}{-(z_2+z_3)}\right\}^{\frac{1}{2}}\int_{z_3}^{z_2}\frac{dz}{(1-z^2)\{(z_1-z)(z_2-z)(z-z_3)\}^{\frac{1}{2}}},\quad\ldots\ldots(4)$$

Now since $z_2$, $z$, $z_3$ are all only slightly greater than $-1$, and $z_1$ is therefore approximately $+1$ (for $z_2z_3+z_3z_1+z_1z_2=-1$), we write

$$z_2=-1+e_2,\quad z_3=-1+e_3,\quad z=-1+e,$$

where $e_2$, $e_3$, $e$ are small positive quantities. In the approximations which follow we neglect products and higher powers of $e_1$, $e_2$, $e_3$ than the first, but retain square roots of products, such as $(e_2e_3)^{\frac{1}{2}}$. We get first

$$\left\{\frac{(1-z_2^2)(1-z_3^2)}{-(z_2+z_3)}\right\}^{\frac{1}{2}}=(2e_2e_3)^{\frac{1}{2}}.\quad\ldots\ldots(5)$$

Now considering the integral, we notice first that $1-z^2=e(2-e)$ and $z_1-z=2-e$. Thus we can write the integral in the form

$$\int_{e_3}^{e_2}\frac{de}{e\{(e_2-e)(e-e_3)\}^{\frac{1}{2}}}\,\frac{1}{(2-e)^{\frac{3}{2}}},$$

and this, to the degree of approximation adopted, can be written

$$\frac{1}{2^{\frac{1}{2}}}\int_{e_3}^{e_2}\frac{de}{e\{(e_2-e)(e-e_3)\}^{\frac{1}{2}}}(1+\tfrac{3}{4}e),$$

so that we have two integrals to evaluate. By the substitution $e=1/u$ we get for the first integral the value $\pi/(e_2e_3)^{\frac{1}{2}}$, while the second is clearly $\frac{3}{4}\pi$. Hence we have

$$\psi_{BC}=\frac{(2e_2e_3)^{\frac{1}{2}}}{2^{\frac{3}{2}}}\pi\left\{\frac{1}{(e_2e_3)^{\frac{1}{2}}}+\frac{3}{4}\right\}. \quad \text{......(6)}$$

But if $\alpha$, $\beta$ be the radii of the limiting circles (on the unit sphere) corresponding to $z_2$, $z_3$, we have approximately

$$\alpha=(2e_2)^{\frac{1}{2}}, \quad \beta=(2e_3)^{\frac{1}{2}}, \quad \text{......(7)}$$

and therefore

$$\psi_{BC}=\tfrac{1}{2}\pi(1+\tfrac{3}{8}\alpha\beta), \quad \text{......(8)}$$

which is the result already obtained above.

Lagrange obtained (*loc. cit. infra*) $\psi_{BC}=\pi\alpha\beta/(\alpha^2+\beta^2)$, and it is very curious that he should not have suspected that his calculation was in error. For, as he remarks, his result gives motion in a curve, consisting of repetitions of a certain number of spirals when $\alpha\beta/(\alpha^2+\beta^2)$ is rational, or a kind of continuous spiral if this is not the case. But it is clear that the motion is simply that of a conical pendulum slightly disturbed from steady motion in the neighbourhood of the downward vertical, which is well known to have as period of disturbance half the period of the steady revolution, *plus* an interval which tends to zero with the inclination of the pendulum to the vertical. This interval causes the motion in an ellipse to be accompanied by revolution of the axes of the ellipse and alteration of the eccentricity as commonly seen with an ordinary pendulum. Bravais remarks, "Tant il est vrai que l'erreur est tellement humaine, qu'elle peut se glisser sous la plume du plus illustre géomètre."

**6.** ***Period between the limiting circles when the pendulum is nearly vertical. Theory of Lagrange.*** As to the period of passage from one limiting circle to the other and back, we see from (10), 1, above that the period of disturbance of the pendulum from a steady motion at an inclination $\cos^{-1}\{\frac{1}{2}(z_2+z_3)\}$ to the upward vertical is

$$\frac{2\pi}{[a\{z_1-\frac{1}{2}(z_2+z_3)\}]^{\frac{1}{2}}}.$$

But as $z_2$ and $z_3$ approach more and more nearly to $-1$ it is clear that $z_1$ approaches more and more closely to 1. For the roots of the cubic, $Z=0$, give $z_2z_3+z_3z_1+z_1z_2+1=0$, and so

$$z_1=-\frac{1+z_2z_3}{z_2+z_3}=\frac{-2+e_2+e_3-e_2e_3}{-2+e_2+e_3}=1+\tfrac{1}{2}e_2e_3, \quad \text{......(1)}$$

if, as before, we write $z_2=-1+e_2$, $z_3=-1+e_3$, where $e_2$, $e_3$ are small in comparison with 1. Thus we obtain for the period

$$\frac{2\pi}{[a\{z_1-\frac{1}{2}(z_2+z_3)\}]^{\frac{1}{2}}}=\frac{2\pi}{[a\{1+\frac{1}{2}(1-\sin^2\theta_2)^{\frac{1}{2}}+\frac{1}{2}(1-\sin^2\theta_3)^{\frac{1}{2}}\}]^{\frac{1}{2}}}$$

$$=\frac{2\pi}{(2a)^{\frac{1}{2}}\{1-\frac{1}{8}(\sin^2\theta_2+\sin^2\theta_3)\}^{\frac{1}{2}}}, \quad \text{......(2)}$$

or, if $\alpha$, $\beta$ have the meanings assigned above for the unit sphere,

$$\text{Period}=2\pi\left(\frac{1}{2a}\right)^{\frac{1}{2}}\{1+\tfrac{1}{16}(\alpha^2+\beta^2)\}. \quad \text{......(3)}$$

If the pendulum be simple, $a=2g/l$, otherwise it is $2mlg/A$. In the former case,

$$\text{Period}=\pi\left[\left(\frac{l}{g}\right)^{\frac{1}{2}}\{1+\tfrac{1}{16}(\alpha^2+\beta^2)\}\right]. \quad \text{......(4)}$$

This result was given by Lagrange [*Méc. Anal.*, Sec. Partie, Section VIII, §§ 21, 22].

7. *Azimuthal angle from one limiting circle to the other always* $> \pi$. When $z_2$ and $z_3$ are nearly $-1$ the value of $\psi_{BC}$ is nearly $\frac{1}{2}\pi$, and the pendulum passes from the upper limiting small circle to the lower or *vice versa* in slightly more than a quarter of the period of revolution. The path has two diameters at right angles to one another at slightly different levels; it is a kind of ellipse traced on the spherical surface.

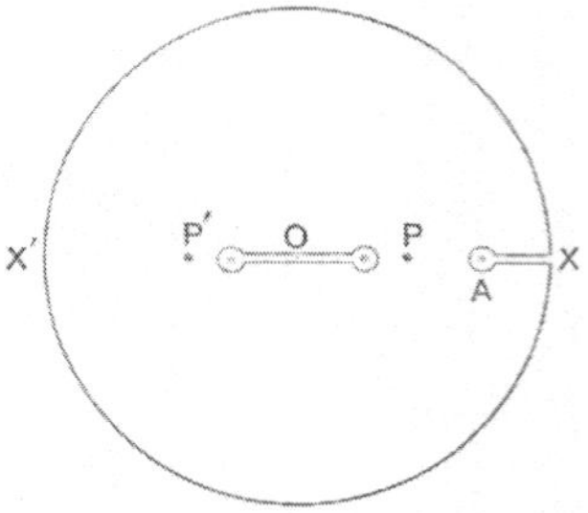

FIG. 83.

The proof that $\psi_{BC} < \pi$ is more difficult. The following process by complex integration is due to M. de Saint Germain (*Bull. des Sci. Math.* 1896, p 114), and is based on the substitution of an integral between limits $z_1$ and $\infty$ for the integral taken from $z_3$ to $z_2$. The same idea is used in the demonstration in 12, V above, due to M. Hadamard, of a property of the top.

From the centre O (Fig. 83) describe a circle of very great radius R in the plane of the paper, and through the centre lay the line X'OX as axis of real quantity. On that line lay down the points A, B, C at distances $z_1$, $z_2$, $z_3$ from O (B, C within the small circles). These represent the roots of the equation $Z=0$. The two points P, P' at $x=1$ and $x=-1$ are the poles of the integral

$$\int \frac{c^{\frac{1}{2}}dz}{(1-z^2)(aZ)^{\frac{1}{2}}}.$$

Considering now $z$ as a complex variable, take the integral

$$\int \frac{dz}{(1-z^2)\{(z_2-z)(z-z_3)F\}^{\frac{1}{2}}}, \quad [F=a(z_1-z)]$$

round the two circuits shown in the figure, that is round the circle from X on the upper side of the loop shown drawn round A to the lower side of the loop, and then round the loop to X. Then take the integral in the opposite direction round the double loop enclosing the points B, C. The sum of these integrals is equal to the sum of the residues for the poles P, P'.

Now, if R be very great, the integral round the circle is zero, and we have only to consider the single loop XA and the double loop BC. The integral round the single loop is

$$2\int_{\infty}^{z_1} \frac{dz}{(1-z^2)\{(z_2-z)(z-z_3)F\}^{\frac{1}{2}}},$$

taken along the axis of real quantity. The integral round the double loop is

$$2\int_{z_3}^{z_2} \frac{dz}{(1-z^2)\{(z_2-z)(z-z_3)F\}^{\frac{1}{2}}},$$

taken also along the axis of real quantity.

By (7), 3, the residue with respect to P (the point $z=1$) multiplied by $2\pi i$ is

$$\frac{2\pi\{-(z_2+z_3)\}^{\frac{1}{2}}}{2\{a(z_2-1)(1-z_3)(z_2+1)(z_3+1)\}^{\frac{1}{2}}} = \frac{\pi\{-(z_2+z_3)\}^{\frac{1}{2}}}{\{a(1-z_2^2)(1-z_3^2)\}^{\frac{1}{2}}} = \frac{\pi}{c},$$

and the residue for P' has the same value. Collecting results, we find by (8), 3,

$$\psi_{BC} = \frac{c}{a^{\frac{1}{2}}}\int_{z_3}^{z_2} \frac{dz}{(1-z^2)\{(z_2-z)(z-z_3)(z_1-z)\}^{\frac{1}{2}}}$$

$$= \pi + \frac{c}{a^{\frac{1}{2}}}\int_{z_1}^{\infty} \frac{dz}{(1-z^2)\{(z_2-z)(z-z_3)(z_1-z)\}^{\frac{1}{2}}}.$$

Since, within the range of integration, $1-z^2$ is negative, the integral on the right is negative, and therefore $\psi_{BC}$ is less than $\pi$.

By substitutions which *increase* the value of

$$\frac{c}{a^{\frac{1}{2}}}\int_{z_1}^{\infty}\frac{dz}{(z^2-1)\{(z_2-z)(z-z_3)(z_1-z)\}^{\frac{1}{2}}},$$

and transform it so that the final result can be evaluated, it can be shown that this integral is less than $\frac{1}{2}\pi$, so that $\psi_{BC} > \frac{1}{2}\pi$, as was shown before by the method of Puiseux.

8. *Possibility of motion of rise and fall on a surface of revolution, with axis vertical.* With respect to the motion of a particle under gravity on any surface of revolution S, the axis of which is vertical, it is pointed out by Routh (*Dynamics of a Particle*, § 550) that with given initial conditions a motion of rise and fall can only take place on certain regions of the surface. Let $v$ be the speed of the particle at any instant when the axial coordinate of the particle is $z$. We can write

$$v^2 = 2g(\mathrm{H}-z).$$

Here $\mathrm{H}-z$ is the "head" which gives the speed $v$, that is the speed is that which would be produced by fall from the plane $z=\mathrm{H}$. Now if $h$ denote $y^2\dot{\psi}$, the A.M. of the particle about the axis, the horizontal speed of the particle is $y\dot{\psi}$, and $y\dot{\psi}$ cannot exceed $v$, but is in general less than $v$. Hence if we have $y^2\dot{\psi}^2 = 2g(\mathrm{H}-z)$, that is if $y^2(\mathrm{H}-z) = h^2/2g$, the motion of the particle is horizontal. In general however $y^2(\mathrm{H}-z) > h^2/2g$. Hence if the cubic surface of revolution, $y^2(\mathrm{H}-z) = h^2/2g$, be described, the particle must in general be at a greater distance from the axis than the points of the cubic surface corresponding to the same $z$. The cubic surface divides the surface S into zones, and thus rise and fall takes place between the two limiting circles of intersection, on a zone of S which lies further from the axis than the corresponding zone of the cubic.

If the particle is started horizontally, the cubic surface for the initial speed is obtained; then the circle on which the particle starts is the boundary of two zones, and the motion lies in that zone of S which is more remote from the axis than the corresponding zone of the cubic surface.

If the cubic surface touch S it will do so in a horizontal circle, which is a circle of possible steady motion. No motion deviating from that circle will be possible on S if the neighbouring parts of the cubic on the two sides of the circle be outside S, but will be possible if the reverse is the case. The motion in the circle is therefore unstable in the former, stable in the latter case.

9. *Motion on a developable surface replaceable by motion in plano.* With regard to a particle moving on a surface, it is important to remark that, if the surface is developable, the motion on the surface may be replaced by motion *in plano*.

By a developable surface is meant one generated by a succession of straight lines, each one of which intersects the line which precedes it in the series. For example take any curve in space, and mark a close succession of points on the curve by the letters $a, b, c, \ldots$ in order, and joining $ab, bc, cd, \ldots$, extend each line both ways beyond the points which define it. The lines thus drawn mark out, or generate, a surface which, by turning about the defining lines in succession, can be brought into a plane which initially contained a consecutive pair of generators. Any curve on the surface becomes a curve in the plane of development, and the curve on the surface, which gave the succession of points $a, b, c, \ldots$, is called the cuspidal edge, or edge of regression, of the surface.

A cone and a cylinder are examples of developable surfaces. In the former the edge of regression is a single point, the vertex of the cone; the cylinder may be regarded as a cone, the vertex of which is infinitely distant.

If a particle move on a developable surface along any path its acceleration will consist of two components, one, $\ddot{s}$ or $v\,dv/ds$, along the path, the other, $v^2/\rho$, along the principal normal to the path towards the centre of curvature. Project this normal on the surface in a plane containing the normal and at right angles to the curve; a component of acceleration along the surface and at right angles to the path is obtained, of amount $v^2 \sin\beta/\rho$, where $\beta$ is the angle between the principal normal and the normal to the surface at the element of the path. The other component is along the normal to the surface. If now the surface is developed into a plane, the latter component becomes unnecessary for any motion of the particle along the path, while, if the motion along successive elements of the path be the same as before, the accelerations $v\,dv/ds$, $v^2 \sin\beta/\rho$, are just those which characterise the motion. The curvature of the plane path is obviously $\sin\beta/\rho$, and $\rho/\sin\beta$ is its radius of curvature at the point considered.

It is only necessary therefore that the forces acting on the particle moving in the plane path should be the same as those in the tangent plane of the surface, corresponding to the element considered, to ensure that the motion at each element may be the same as before.

If the surface be a right circular cone with axis vertical and the motion take place on the inside of the upper sheet, the particle will, in the absence of friction, be under applied force towards the vertex of amount $mg\cos\alpha$, if $\alpha$ be the semi-vertical angle of the cone. The component of gravity normal to the surface is balanced by the reaction of the surface. Thus the motion can be discussed as a case of the motion of a particle in a plane, under a constant force directed to that point in the plane which represents the vertex of the cone. A particular case of motion on the cone is in a horizontal circle with constant speed. This circle, of radius $a$, say, develops into a circle of radius $a/\sin\alpha$, on the plane, and the acceleration in it, $v^2/a$, has component $v^2 \sin\alpha/a$ along the generating line at each point. The force giving the former acceleration was $\mathrm{R}\cos\alpha$, and $\mathrm{R}$ was $mg/\sin\alpha$. Hence the force required in the plane is $\mathrm{R}\cos\alpha\sin\alpha$, that is $mg\cos\alpha$.

**10.** *Motion of a particle between two close circles on a surface of revolution.* The motion of a particle between two near limiting circles on a surface of revolution under force axially directed can be discussed in the following manner. Let the equation of the surface be

$$2az + mz^2 = y^2, \qquad \text{(1)}$$

where $z$ is the axial distance of the particle from the point, the *vertex* of the surface, at which $y=0$. Consider first the particle as in steady motion in the circle at axial distance $z$ from the vertex. If $\dot{\psi}$ be the angular speed of the particle about the axis, and the mass be taken as unity, we have for the components of the reaction of the surface, directed respectively outwards from the axis and axially from the vertex,

$$m\dot{\psi}^2 y = -\mathrm{Y}, \quad mg = -\mathrm{Z},$$

where $g$ is the force per unit mass (gravity) applied to the particle. We shall take the field of force as due to gravity, and the axis as vertical.

If the particle is disturbed very slightly from the steady motion without alteration of the A.M. about the axis, we have

$$y^2\dot{\psi} = h, \qquad \text{(2)}$$

where $h$ is a constant. Let the deviation of the particle from the steady motion circle be $s$, measured along a meridian in the direction from the vertex. The acceleration along the meridian is $\ddot{s}$. If $\theta$ be the inclination, as shown in Fig. 84, to the axis of the normal drawn to the surface at the position P of the particle, the force along the meridian produced by gravity is $g \sin\theta$. The acceleration towards the axis is $y\dot{\psi}^2$, which gives a component $y\dot{\psi}^2 \cos\theta$ along the meridian. But since $y^2\dot{\psi} = h$, we have $y\dot{\psi}^2\cos\theta = h^2\cos\theta/y^3$. Thus the equation of motion is

$$\ddot{s} = \frac{h^2}{y^3}\cos\theta - g\sin\theta. \qquad \text{(3)}$$

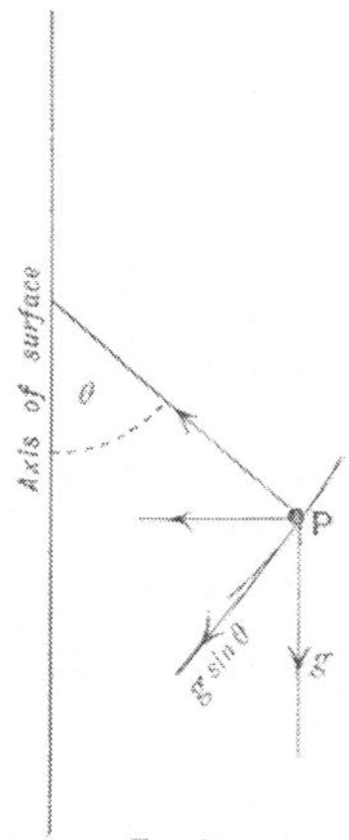


FIG. 84.

Now let $b$ be the radius of the steady motion circle, and $\alpha$ the inclination of the normal to the axis at that circle. Then $y = b + s\cos\alpha$, $\theta = \alpha + s/\rho$, where $\rho$ is the radius of curvature of the meridian at the point. Thus we get

$$\ddot{s} = \frac{h^2\cos\left(\alpha + \frac{s}{\rho}\right)}{b^3\left(1 + 3\frac{s}{b}\cos\alpha\right)} - g\sin\left(\alpha + \frac{s}{\rho}\right). \qquad \text{(3')}$$

Since $s$ is small and $h^2\cos\alpha/b^3 = g\sin\alpha$, this becomes

$$\ddot{s} + \left\{\frac{1}{\rho}\left(\frac{h^2}{b^3}\sin\alpha + g\cos\alpha\right) + 3\frac{h^2}{b^4}\cos^2\alpha\right\}s = 0. \qquad \text{(3'')}$$

The period of a small oscillation about the steady motion is therefore given by

$$\mathrm{T} = 2\pi\left\{\frac{1}{\rho}\left(\frac{h^2}{b^3}\sin\alpha + g\cos\alpha\right) + 3\frac{h^2}{b^4}\cos^2\alpha\right\}^{-\frac{1}{2}}. \qquad \text{(4)}$$

If $r$ be the length of a normal from the steady motion circle, we easily find, since $dy/dz = \cot\alpha$ and $\tan\alpha \,.\, dr = (\rho - r)d\alpha$, that (1), the equation of the surface, is

$$\left.\begin{aligned} r^2(\cos^2\alpha - m\sin^2\alpha) &= a^2, \\ \text{and that} \qquad \rho &= \frac{a}{(\cos^2\alpha - m\sin^2\alpha)^{\frac{3}{2}}}. \end{aligned}\right\} \qquad \text{(5)}$$

Thus the period equation becomes after a little reduction

$$\mathrm{T} = 2\pi\left(\frac{a\cos\alpha}{g}\right)^{\frac{1}{2}}\{(4\cos^2\alpha - m\sin^2\alpha)(\cos^2\alpha - m\sin^2\alpha)^{\frac{1}{2}}\}^{-\frac{1}{2}}. \qquad \text{(6)}$$

If the steady motion circle be very near the vertex, that is if $z$ be very small, we have approximately $\cos^2\alpha = 1 - b^2/a^2$, $\sin^2\alpha = b^2/a^2$, $\cos\alpha = 1 - b^2/2a^2$, where $b^2$ is small in comparison with $a^2$. Thus we get for the period the approximate value

$$T = \pi\left(\frac{a}{g}\right)^{\frac{1}{2}}\left(1 + \frac{4+3m}{8}\frac{b^2}{a^2}\right), \qquad (7)$$

or if $b_1$, $b_2$ be the radii of the upper and lower limiting circles,

$$T = \pi\left(\frac{a}{g}\right)^{\frac{1}{2}}\left(1 + \frac{4+3m}{16}\frac{b_1^2 + b_2^2}{a^2}\right). \qquad (7')$$

The azimuthal angle turned through in the period is the product of the expression on the right of (6) by the angular speed in the steady motion, that is by $(g/r\cos\alpha)^{\frac{1}{2}}$, that is by $\{g(\cos^2\alpha - m\sin^2\alpha)^{\frac{1}{2}}/a\cos\alpha\}^{\frac{1}{2}}$. Hence the angle specified, $\psi_{0,\,T}$ say, is given by

$$\psi_{0,\,T} = 2\pi(4\cos^2\alpha - m\sin^2\alpha)^{-\frac{1}{2}}. \qquad (8)$$

In the special case in which the steady motion circle is very near the vertex we get

$$\psi_{0,\,T} = \pi\left(1 + \frac{4+m}{8}\frac{b^2}{a^2}\right), \qquad (9)$$

or more exactly

$$\psi_{0,\,T} = \pi\left(1 + \frac{4+m}{8}\frac{b_1 b_2}{a^2}\right) \qquad (9')$$

If $m = -1$, the surface is a sphere, and we fall back in (7′) and (9′) on the results already obtained for a particle moving within two close circles near the lowest point.

If $m = 0$, the surface is a paraboloid of revolution. A general discussion of the motion on the sphere and on the paraboloid will be given later by means of elliptic functions.

If the surface is a right circular cone with axis vertical, the semi-vertical angle of the cone is $\frac{1}{2}\pi - \alpha$. Since $\rho = \infty$, and now by (1), $b = y = zm^{\frac{1}{2}} = z\cot\alpha$, (4) above gives

$$T = 2\pi\left(3\frac{h^2}{b^4}\cos^2\alpha\right)^{-\frac{1}{2}} = 2\pi\left(\frac{z}{3g}\right)^{\frac{1}{2}}\frac{1}{\sin\alpha}. \qquad (10)$$

**11.** *Initial motion in the plane of the horizontal great circle.* The particular case of the spherical pendulum in which the initial motion is in the plane of the horizontal great circle (the "equator") is interesting. It will be instructive to deal with it from first principles and then refer it to the general theory. We take the pendulum as "simple" and of length $l$, with bob of mass unity, and put $y$ for the distance of the bob below the plane of the equatorial circle at time $t$, $\psi$ for the azimuthal angle described in the same time, and $v_0$ for the initial speed of the bob. The vertical speed at time $t$ is $\dot{y}$, the radial speed is $d\{(l^2 - y^2)^{\frac{1}{2}}\}/dt = -y\dot{y}/(l^2 - y^2)^{\frac{1}{2}}$. The azimuthal speed is then $(l^2 - y^2)\dot{\psi}$. The energy equation is therefore

$$\dot{y}^2 + \frac{y^2\dot{y}^2}{l^2 - y^2} + (l^2 - y^2)\dot{\psi}^2 - v_0^2 = 2gy. \qquad (1)$$

We have also the equation of A.M.,

$$v_0 l = (l^2 - y^2)\dot{\psi}. \qquad (2)$$

Substituting from (2) in (1) and reducing, we find

$$\dot{y}^2 l^2 = 2gy(l^2 - y^2) - v_0^2 y^2,$$

or

$$dt = \frac{l\,dy}{\{2gy(l^2 - y^2) - v_0^2 y^2\}^{\frac{1}{2}}} \qquad (3)$$

But from (2) we have also

$$dt = \frac{l^2 - y^2}{v_0 l}\,d\psi, \qquad (4)$$

so that

$$d\psi = v_0 l^2 \frac{dy}{\{(l^2 - y^2)2gy(l^2 - y^2) - v_0^2 y^2\}^{\frac{1}{2}}} \qquad (4')$$

If we split the expression on the right into two parts, as shown in

$$d\psi=\tfrac{1}{2}v_0\frac{dy}{\{2gy(l^2-y^2)-v_0{}^2y^2\}^{\frac{1}{2}}}+\tfrac{1}{2}v_0\frac{(l^2+y^2)\,dy}{(l^2-y^2)\{2gy(l^2-y^2)-v_0{}^2y^2\}^{\frac{1}{2}}}\ \ldots\ldots\quad(5)$$

[of which the first is $v_0dt/2l$, by (3)], it will be found that, if $u^2=v_0{}^2y^2/2gy(l^2-y^2)$,

$$d\psi=\frac{1}{2}\frac{v_0}{l}dt+\frac{du}{(1-u^2)^{\frac{1}{2}}}.$$

Hence

$$\psi=\frac{1}{2}\frac{v_0}{l}t+\sin^{-1}\frac{v_0y^{\frac{1}{2}}}{\{2g(l^2-y^2)\}^{\frac{1}{2}}}.\ \ldots\ldots\ldots\ldots\ldots(6)$$

Thus the combination $\psi-\frac{1}{2}v_0t/l$ of elliptic integrals is capable of being expressed in terms of ordinary functions. This result is due to Greenhill.

Referring back to (9) of 1, we find that

$$dt=\frac{dz}{\{-az(z-z_3)(z_1-z)\}^{\frac{1}{2}}},\ \ldots\ldots\ldots\ldots\ldots(7)$$

since here $z_2=0$. But in the present case also $a=2g/l$, and so, remembering that $\theta$ is here measured from the *upward* vertical, so that $-az(z-z_2)(z_1-z)$ is positive, we get

$$dt=\frac{l^{\frac{1}{2}}dz}{\{-2gz(z-z_3)(z_1-z)\}^{\frac{1}{2}}}.\ \ldots\ldots\ldots\ldots\ldots(8)$$

By (3) above we have

$$dt=\frac{l^{\frac{1}{2}}dz}{\left\{-2gz(1-z^2)-\frac{v_0^2}{l}z^2\right\}^{\frac{1}{2}}}\ \ldots\ldots\ldots\ldots\ldots(9)$$

and by (4′)

$$d\psi=\frac{v_0\,dz}{(1-z^2)\{-2gzl(1-z^2)-v_0{}^2z^2\}^{\frac{1}{2}}}.\ \ldots\ldots\ldots\ldots\ldots(10)$$

The quantity $-2gz(1-z^2)-v_0{}^2z^2/l$ is $aZ/l$. Thus to obtain the roots of $Z=0$ we have, after division by $z$, so as to get rid of the zero root,

$$+2glz^2-v_0^2z-2gl=0,$$

and therefore

$$z=\frac{1}{4gl}\{v_0^2\pm(16g^2l^2+v_0^4)^{\frac{1}{2}}\}.\ \ldots\ldots\ldots\ldots\ldots(11)$$

Thus if $v_0$ is very small, the large positive root of $Z=0$ is only slightly greater than 1, while the remaining root is very nearly $-1$. The range of motion is thus from the level of the centre of the sphere very nearly to the lowest point. From the inequality given in 3 it will be found that the range of azimuthal motion between the highest and lowest positions lies between $\frac{1}{2}\pi$ and $\pi/2^{\frac{1}{2}}$.

When $v_0$ is very great the large root given by (11) is very great, the other root is slightly less than zero, so that the limiting circles are very close together. The value of $\psi_{BC}$ is then, as the inequality in 3 shows, very approximately $\pi$.

**12.** *Simple pendulum oscillating through finite range.* We shall now consider the special case of a simple pendulum oscillating through a finite range in a vertical plane. The problem is essentially that of a particle moving without friction on a concave circular ring in a vertical plane, or along the interior of a guide tube bent into a vertical circle. If $l$ denote the radius of the circular path, and $\theta$ the angular deflection of the radius from the lowest position, the equation in both cases is

$$\ddot{\theta}+\frac{g}{l}\sin\theta=0\ \ldots\ldots\ldots\ldots\ldots(1)$$

Multiplying this equation by $\dot{\theta}$ and integrating, we get

$$\tfrac{1}{2}\dot{\theta}^2 = -\int \frac{g}{l}\sin\theta\, d\theta = \frac{g}{l}\cos\theta + \mathrm{C}, \quad\text{.......................(2)}$$

where C is a constant. When $\theta = \theta_0$, $\dot{\theta} = 0$, so that $\mathrm{C} = -g\cos\theta_0/l$, and we have

$$l^2\dot{\theta}^2 = v^2 = 2gl(\cos\theta - \cos\theta_0), \quad\text{........................(3)}$$

where $v$ is the speed of the particle in its path when the deflection is $\theta$.

Equation (3) gives

$$\frac{dt}{d\theta} = \left(\frac{l}{g}\right)^{\frac{1}{2}} \frac{1}{\{2(\cos\theta - \cos\theta_0)\}^{\frac{1}{2}}}. \quad\text{........................(4)}$$

If then $\tau$ denote the time taken by the particle to pass from deflection 0 to deflection $\theta_0$, we have

$$\tau = \left(\frac{l}{g}\right)^{\frac{1}{2}} \int_0^{\theta_0} \frac{d\theta}{\{2(\cos\theta - \cos\theta_0)\}^{\frac{1}{2}}}. \quad\text{........................(5)}$$

Since the pendulum is at rest at the instant $\theta = \theta_0$, $\tau$ is one quarter of the period of the motion.

Writing now

$$\sin\tfrac{1}{2}\theta = \sin\tfrac{1}{2}\theta_0 \sin\phi, \quad\text{........................(6)}$$

where $\phi$ is an auxiliary angle, we obtain

$$d\theta = 2\sin\tfrac{1}{2}\theta_0 \frac{\cos\phi\, d\phi}{(1 - k^2\sin^2\phi)^{\frac{1}{2}}}, \quad\text{........................(7)}$$

where $k = \sin\tfrac{1}{2}\theta_0$. But, by (6),

$$\cos\tfrac{1}{2}\theta \,.\, \dot{\theta} = 2\sin\tfrac{1}{2}\theta_0 \cos\phi \,.\, \dot{\phi}, \quad\text{........................(8)}$$

and the same relation gives

$$\{2(\cos\theta - \cos\theta_0)\}^{\frac{1}{2}} = 2\sin\tfrac{1}{2}\theta_0\cos\phi, \quad\text{........................(9)}$$

so that, if $n^2$ stands for $g/l$, we obtain from (4)

$$\dot{\theta} = 2n\sin\tfrac{1}{2}\theta_0\cos\phi, \quad\text{........................(10)}$$

and by (8)

$$\dot{\phi} = n\cos\tfrac{1}{2}\theta. \quad\text{........................(11)}$$

Thus we have by (6)

$$n\frac{dt}{d\phi} = \frac{1}{\{1 - k^2\sin^2\phi\}^{\frac{1}{2}}},$$

and

$$n\tau = \int_0^{\frac{1}{2}\pi} \frac{d\phi}{(1 - k^2\sin^2\phi)^{\frac{1}{2}}} = \mathrm{K}. \quad\text{........................(12)}$$

Also, of course, we have

$$nt = \int_0^{\phi} \frac{d\phi}{(1 - k^2\sin^2\phi)^{\frac{1}{2}}}. \quad\text{........................(13)}$$

The integral K is the complete elliptic integral of the first kind, of which $k$ is the modulus and $\phi$ is the amplitude.

The time $t_2 - t_1$ for any arc from $\phi = \phi_1$ to $\phi = \phi_2$ described by the pendulum is given by

$$n(t_2 - t_1) = \mathrm{F}(k, \phi_2) - \mathrm{F}(k, \phi_1). \quad\text{........................(14)}$$

It is clear from (12) that, if $k$ (that is $\sin\frac{1}{2}\theta_0$) is very small, we have

$$\tau=\tfrac{1}{2}\pi\left(\frac{l}{g}\right)^{\frac{1}{2}}, \quad \text{and} \quad \mathrm{T}=2\pi\left(\frac{l}{g}\right)^{\frac{1}{2}}, \qquad (15)$$

the result that we obtain at once if we suppose $\theta$ so small that we can write $\theta$ instead of $\sin\theta$ in the fundamental equation (1).

Since $k^2<1$, the factor $(1-k^2\sin^2\phi)^{-\frac{1}{2}}$ can be expanded in a converging series and the integration performed term by term. We obtain

$$\tau=\frac{\pi}{2}\left(\frac{l}{g}\right)^{\frac{1}{2}}\left\{1+(\tfrac{1}{2})^2k^2+\left(\frac{1\,.\,3}{2\,.\,4}\right)^2k^4+\left(\frac{1\,.\,3\,.\,5}{2\,.\,4\,.\,6}\right)^2k^6+\dots\right\}. \qquad (16)$$

For a good many purposes the first two terms of this series form a sufficiently close approximation. By taking a sufficient number of terms of the series we can calculate K to any desired degree of exactness. Tables of the values of K for values of the modulus proceeding by small steps have been constructed by various methods (more expeditious than that here indicated), and the results are available for practical calculation.* [Refer to 20, 21, XII.]

13. *Discussion of force along the supporting cord or rod of pendulum.* If the pendulum start from rest in a position making an angle $\theta_0$ with the downward vertical, the force toward the centre applied by the cord when the deflection $\theta_0$, from the *downward* vertical, has been changed to $\theta$, is $ml\dot{\theta}^2+mg\cos\theta$, or by the value of $\dot{\theta}^2$ given by (3), 12, $mg(3\cos\theta-2\cos\theta_0)$. This gives $mg\cos\theta_0$ for the central force when the pendulum is in a limiting position, and $\cos\theta_0$ is negative when $\theta_0>\frac{1}{2}\pi$. But if, instead of a bob suspended by a string, we have a particle moving on a guiding curve, or in a tube, in the form of a circle of radius $l$ in a vertical plane, the amplitudes may have any value from 0 to $\pi$. Or if the bob be attached to a rod, or a wire capable of bearing thrust along its length, the tube may be dispensed with. In this latter case however the centroid will be at distance $l$ from the point O of support, and if A be the moment of inertia of the whole about the axis through O at right angles to the plane of motion, the length of the equivalent simple pendulum is $\mathrm{A}/ml$. The outward force applied to O is still $mg(3\cos\theta-2\cos\theta_0)$. The pull in the pendulum rod varies from point to point and depends on the distribution of the mass.

Considering however the simple pendulum, we have seen that the outward force applied through the bob to the cord is negative at a turning position when for that $\theta_0>\frac{1}{2}\pi$. It is clear that $3\cos\theta-2\cos\theta_0$ is positive when $\theta_0<\frac{1}{2}\pi$, and also when $\theta_0>\frac{1}{2}\pi$ for all values of $\theta<\frac{1}{2}\pi$. Hence the reversal of the force takes place when $\theta$ has the value $\cos^{-1}(\frac{2}{3}\cos\theta_0)\,(>\frac{1}{2}\pi)$. It is impossible therefore for a particle unless constrained by a tube or

* See the tables of Legendre, *Exerc. de Cal. Int.* t. III, also the *Smithsonian Physical Tables* and the *Funktionentafeln* of Jahnke and Emde [Teubner].

a rigid stem, to vibrate if $\frac{1}{2}\pi < \theta_0 < \pi$. The motion is nevertheless possible if the pendulum makes complete revolutions in a sufficiently short period.

In this latter case let the particle have speed $v_0$ when at distance $l\theta_0$ along the circle from the lowest point; then at deflection $\theta$ the speed is given by

$$v^2 - v_0^2 = 2gl(\cos\theta - \cos\theta_0),$$

and the pull P in the thread is then $mg\cos\theta + mv^2/l$, that is

$$mg(3\cos\theta - 2\cos\theta_0) + mv_0^2/l.$$

Hence, if the particle goes completely round the circle, we have, when $\theta = \pi$, $\mathrm{P} = mg(-3 - 2\cos\theta_0) + mv_0^2/l$, and therefore, if the value of P is not to change sign, we must have $v_0^2 > gl(3 + 2\cos\theta_0)$, and so if $\theta_0 = \pi$, $v_0^2 > gl$. If this condition be fulfilled the particle may be suspended by a string. [See also 16 below.]

**14.** *Graphic representation of finite pendulum motion.* The motion of the pendulum is illustrated by the diagram of Fig. 85. The horizontal line $\mathrm{CP_0}$ represents the level to which the bob rises, the circle $\mathrm{AP_0B}$ is that of the motion, and is

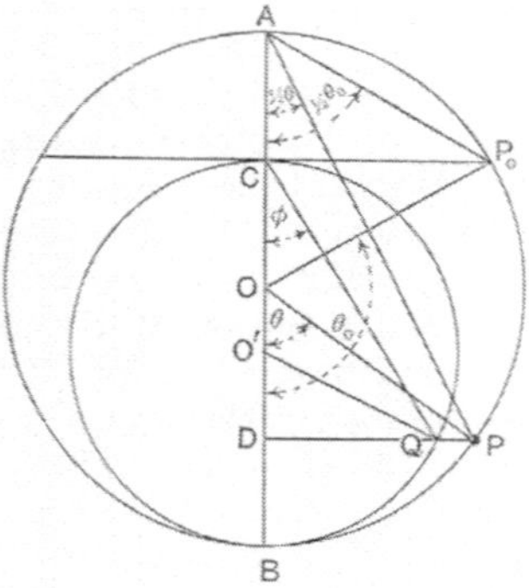


FIG. 85.

therefore of radius $l$, the circle CPB is described on CB as diameter. As we shall show, the angle DCQ is the amplitude $\phi$ for the position P of the particle on the horizontal line DQ. We join AP, OP. Then

$$\angle\mathrm{BOP} = \theta, \quad \angle\mathrm{BAP} = \tfrac{1}{2}\theta, \quad \angle\mathrm{OAP_0} = \tfrac{1}{2}\theta_0.$$

By the diagram $\mathrm{CB} = 2l\sin^2\frac{1}{2}\theta_0$, and therefore $\mathrm{CD} = 2l\sin^2\frac{1}{2}\theta_0\cos^2\mathrm{BCQ}$. Thus, we have $k^2 = \mathrm{CB}/2l$, the ratio to $2l$ of the height CB of the turning points. But also if $u = \mathrm{F}(k, \phi)$,

$$\mathrm{CD} = l - \mathrm{AC} + \mathrm{OD} = l(1 - 2\cos^2\tfrac{1}{2}\theta_0 + \cos\theta) = 2l(\cos^2\tfrac{1}{2}\theta - \cos^2\tfrac{1}{2}\theta_0) = 2lk^2\operatorname{cn}^2 u, \quad \text{......(1)}$$

by (9), 12. Equating these values of CD and reducing we obtain

$$\sin\mathrm{BCQ} = \operatorname{sn} u. \quad \text{............................................(2)}$$

Hence $\mathrm{BCQ} = \phi$.

Also $k = \sin\frac{1}{2}\theta_0 = \mathrm{CP_0}/\mathrm{AP_0}$. If we write $k^2 + k'^2 = 1$, $k'$ is the *co-modulus*, and is therefore represented by $\mathrm{AC}/\mathrm{AP_0}$.

The construction in Fig. 85 replaces the turning of OP, with angular speed $\dot{\theta}$, by the turning of CQ with angular speed $\dot{\phi}$, and the motion of P by that of Q. Q starts from

C when P starts from $P_0$ and coincides with P at B. If the particle just goes completely round in the guiding tube, that is if $P_0$ is infinitely near to A, the larger and smaller circles coincide, and $\phi=\frac{1}{2}\theta$, $\theta_0=\pi$, so that $k=1$.

When $\theta_0=\pi$ and the particle just completes the circle in the guide tube, the time required is infinite, for $k$ being 1, we have

$$\left(\frac{g}{l}\right)^{\frac{1}{2}}\tau=\int_0^{\frac{1}{2}\pi}\frac{d\phi}{(1-k^2\sin^2\phi)^{\frac{1}{2}}}=\int_0^{\frac{1}{2}\pi}\frac{\cos\phi}{1-\sin^2\phi}d\phi=\tfrac{1}{2}\left[\log\frac{1+\sin\phi}{1-\sin\phi}\right]_0^{\frac{1}{2}\pi}=\infty. \quad \text{.........(3)}$$

For a range however from a deflection of 89° on one side of the vertical to a deflection of 89° on the other, the time required is 3·4 times that required for a very small swing from one side of the vertical to the other.

The equation $(g/l)^{\frac{1}{2}}\cos\frac{1}{2}\theta=\dot\phi$, or $(g/l)^{\frac{1}{2}}\,\text{dn}\,u=\dot\phi$, shows that the period always lies between the limits $2\pi/n$ and $2\pi\sec\frac{1}{2}\theta_0/n$. For $\frac{1}{2}\pi>\frac{1}{2}\theta_0>\frac{1}{2}\theta$, and so $\cos\frac{1}{2}\theta>\cos\frac{1}{2}\theta_0$. Thus

$$2n>2\dot\phi>2n\cos\tfrac{1}{2}\theta_0,$$

and the same inequality holds for the mean angular speed of Q. Hence

$$\frac{2\pi}{n}<\text{Period}<\frac{2\pi}{n}\sec\tfrac{1}{2}\theta_0.$$

**15.** *Pendulum making complete revolutions.* Now let the oscillating particle in the circular guiding tube make complete revolutions under gravity. The time of revolution and the time of describing any part of the circle required may be given, or we may be given the speed at top or bottom of the given circle and be required to find the period of revolution, the speed at any point, and the time of its describing any part of the circle. If the speed at top of the circle is known that at the bottom is also known, and *vice versa*. For we have $v\dot v=-g\sin\theta\,.\,l\dot\theta$; ..........................................(1)

and therefore, at any point,

$$v^2=2gl\cos\theta+\text{C}=v_1^2-2gl(1-\cos\theta), \quad \text{...............................(2)}$$

if $v_1$ be the speed at the lowest point. Thus at the highest point where $v=v_2$, we have

$$v_2^2=v_1^2-4gl, \quad \text{................................................(3)}$$

so that for the motion to be possible we must have $v_1^2>4gl$.

In order that a bob suspended by a thread may go completely round it must have at the top of the circle such a speed that the thread is kept taut, that is we must have $v_2^2>gl$, or $v_1^2>5gl$, as we have already shown in 13 above.

Returning to equation (2), we write it in the form

$$v^2=v_1^2-4gl\sin^2\phi, \quad \text{.....................................(4)}$$

if $\phi$ stand for $\frac{1}{2}\theta$. Hence $v=2l\dot\phi$, and so we have

$$dt=\frac{2l\,d\phi}{(v_1^2-4gl\sin^2\phi)^{\frac{1}{2}}}=\frac{2l}{v_1}\int\frac{d\phi}{(1-k^2\sin^2\phi)^{\frac{1}{2}}}, \quad \text{.........................(5)}$$

where $k^2=4gl/v_1^2$, which we know is less than unity. Thus, for the time from the lowest point to the inclination $\theta$, we have

$$t=\frac{2l}{v_1}\int_0^{\phi}\frac{d\phi}{(1-k^2\sin^2\phi)^{\frac{1}{2}}}=\frac{2l}{v_1}u. \quad \left[k=\left(\frac{4gl}{v_1^2}\right)^{\frac{1}{2}}.\right] \quad \text{.........................(6)}$$

The time for any arc from $\phi=\phi_1$ to $\phi=\phi_2$ is thus

$$t_2-t_1=\frac{2l}{v_1}\{u(\phi_2)-u(\phi_1)\}. \quad \text{.........................(7)}$$

This, by (14), 12, above, stands in the constant ratio $2l/v_1\,.\,(g/l)^{\frac{1}{2}}=2(gl)^{\frac{1}{2}}/v_1$ to the time of describing the corresponding arc in the oscillatory motion in a vertical circle of the same radius.

From the lowest point to the highest the time $\tau$ is given by

$$\tau=\frac{2l}{v_1}\mathrm{K}.\quad [k=2(gl)^{\frac{1}{2}}/v_1.] \qquad (8)$$

The value of $\tau$ can be calculated by the series (16), 12 with the value $(2gl)^{\frac{1}{2}}/v_1$ of $k$.

We can now solve various problems, making use of the tables of complete elliptic integrals which have been calculated for a succession of moduli [see 13 above]. Let us suppose for example that the period of revolution $2\tau$ and the length $l$ are given, and it is required to find $v_1$. [By a curious inadvertence this has sometimes been set as a problem in elementary dynamics!] We have from (8), $(g/l)^{\frac{1}{2}}\tau=k\mathrm{K}$. $[k=2(gl)^{\frac{1}{2}}/v_1.]$

Let the circle be 2 feet in radius and be traversed once in half a second by the particle. This gives $\tau=0{\cdot}25$, and therefore $\tau(g/l)^{\frac{1}{2}}=1$. Now if $k=\sin 35°$, $k\mathrm{K}=0{\cdot}993$, about 0 7 per cent. short of 1. But this makes $v_1=2(gl)^{\frac{1}{2}}/k=16/0{\cdot}5736=28$, nearly. Thus at the lowest point the speed is 28 ft./sec., at the top it is 23 ft./sec.

The circumference of the circle is 12·57 feet nearly, and this space is described in half a second. Hence the average value of the speed required is 25·14 ft./sec., so that the result obtained is not very far from the truth.

The reader may verify that a more exact result is obtained by interpolation between the values of K for $k=\sin 35°$ and $k=\sin 36°$. For $k=\sin 35°\ 12'$ we have, to less than $\frac{1}{10}$ per cent of error, $k\mathrm{K}=1$. This makes $k=0{\cdot}57643$, and therefore $v_1=2(gl)^{\frac{1}{2}}/0{\cdot}5764$, that is the speed at the bottom of the circle is 27·76 ft./sec. At the highest point the equation $v_2{}^2=v_1{}^2-2gh$ gives $v_2=22{\cdot}66$, in ft./sec. In this case the particle might be supported by a thread, as the condition $v_1{}^2>5gl$ is obviously fulfilled. In these calculations $g$ has been taken as 32 in ft./sec.$^2$ units.

This may be regarded as the limit of the case of the spherical pendulum in which the terminal circles are very small circles at the zenith and nadir of the sphere, and the bob passes from the left of one to the right of the other, thus giving an azimuthal turning of amount $\pi$.

**16.** *Reaction of surface on a particle moving on it, or force applied at the point of support to a spherical pendulum.* We now consider the stress in the suspension cord, or the force applied at the point of support. When the particle is unsuspended, but moves on a spherical surface, the force found measures the reaction of the surface. In this latter case we suppose that no friction exists. By (5), 1, we have the equation

$$\mathrm{A}(\dot\theta^2+\dot\psi^2\sin^2\theta)=2(-mgl\cos\theta+\mathrm{E}),$$

which we may write also in the form

$$ml(\dot\theta^2+\dot\psi^2\sin^2\theta)=\frac{2lm}{\mathrm{A}}(-mgl\cos\theta+\mathrm{E}). \qquad (1)$$

The expression on the left is the mass acceleration of the pendulum towards the point of support, due to the motion. The equivalent expression on the right can be written in the form $2l^2m^2g(\cos\theta_0-\cos\theta)$ if we put $\mathrm{E}=mgl\cos\theta_0$.

But the gravity force $-mg\cos\theta$ must also be balanced by the action of the point of support on the pendulum, and therefore if the force along the axis applied by the support be R, we have

$$\mathrm{R}=\frac{2l^2m^2g}{\mathrm{A}}(\cos\theta_0-\cos\theta)-mg\cos\theta. \qquad (2)$$

For a simple pendulum $\mathrm{A}=ml^2$, and so we obtain

$$\mathrm{R}=mg(2\cos\theta_0-3\cos\theta). \qquad (3)$$

[It is to be remembered that $\theta$ is here measured from the upward vertical.] For a particle moving on the interior of a surface, this is the normal force which must be exerted on the particle by the surface, and $\theta$ is the angle which the normal at the point of contact makes with the upward vertical.

Now, with the values of $a$, $b$, $c$ stated in 1, we have

$$\dot{z}^2 = (-az+b)(1-z^2) - c^2, \qquad \text{(4)}$$

and the roots of $Z=0$ are $z_1$, $z_2$, $z_3$, as explained above. The two latter roots define the circles which limit the motion, while $z_1 = -(1+z_2z_3)/(z_2+z_3)$, so that since $z_2+z_3$ is negative, $z_1$ is $>0$.

In the general case $\qquad \mathrm{AR} = 2m^2l^2g(z_0 - z) - \mathrm{A}mgz, \qquad \text{(5)}$

so that in order that R should vanish we must have

$$2ml^2(z_0 - z) = \mathrm{A}z. \qquad \text{(6)}$$

But by (1) the quantity on the left is $\mathrm{A}(\dot\theta^2 + \dot\psi^2\sin^2\theta)l/g$, and so $z$ must be positive. Thus the force cannot vanish and change sign unless the pendulum be above the level of the point of support. This holds whether the pendulum be simple or not.

If R is to be zero in the course of the motion we must have by (5) and the value $\frac{1}{2}\mathrm{A}b$ of E

$$z_2 > \frac{\mathrm{A}bl}{(2ml^2+\mathrm{A})g}, \quad z_3 < \frac{\mathrm{A}bl}{(2ml^2+\mathrm{A})g},$$

and these also show that the force cannot change sign if the motion is wholly below the point of support, for then both $z_2$ and $z_3$ are negative.

To carry this discussion further we suppose the pendulum to be simple, and get as the conditions required $z_2 > \frac{1}{3}bl/g$, $z_3 < \frac{1}{3}bl/g$, or

$$z_2 > \frac{2}{3}\frac{b}{a}, \quad z_3 < \frac{2}{3}\frac{b}{a}.$$

But $b/a$ is $z_1+z_2+z_3$, and so we are to have

$$z_2 > \tfrac{2}{3}(z_1+z_2+z_3), \quad z_3 < \tfrac{2}{3}(z_1+z_2+z_3).$$

The value of $z_1$ given above converts these inequalities into

$$z_2 > \frac{2}{3}\,\frac{z_2^2+z_3^2+z_2z_3-1}{z_2+z_3}, \quad z_3 < \frac{2}{3}\,\frac{z_2^2+z_3^2+z_2z_3-1}{z_2+z_3}.$$

The second inequality is satisfied, for we can write it

$$\frac{1}{3}z_3 < \frac{2}{3}\,\frac{z_2^2-1}{z_2+z_3}.$$

The numerator on the right is negative and so also is the denominator, so that

$$\tfrac{2}{3}(z_2^2-1)/(z_2+z_3)$$

is positive. But $z_3$ is negative, and thus the inequality is satisfied.

The first inequality can be written

$$\frac{z_2^2+z_2z_3-2z_3^2+2}{z_2+z_3} > 0.$$

The denominator is negative, and for the fulfilment of the inequality we have so to choose $z_2$ that the numerator is also negative. This will be the case if $z_2$ lie between the roots of the quadratic in $z_2$, $\qquad z_2^2+z_2z_3-2z_3^2+2=0,$

for only for values of $z_2$ in that region is the expression on the left negative. Solving the equation we find the roots $\zeta$, $\zeta'$ given by

$$\left.\begin{matrix}\zeta\\ \zeta'\end{matrix}\right\} = -\tfrac{1}{2}z_3 \pm \tfrac{1}{2}(9z_3^2-8)^{\frac{1}{2}}. \qquad \text{(7)}$$

The roots must be real, and therefore $9z_3^2 \geqq 8$. This gives for the limiting case of equal roots $z_3 = 0{\cdot}942809$, and therefore the pendulum when on the lower limiting circle cannot make a greater angle with the downward vertical than 19° 28′ 16″.

If this condition is fulfilled $z_2$ must lie between $-\frac{1}{2}z_3 \pm \frac{1}{2}(9z_3^2-8)^{\frac{1}{2}}$. At the limits when $z_2$ coincides with the roots we have $b/a = \frac{3}{2}\zeta$ or $b/a = \frac{3}{2}\zeta'$, and between these values $b/a$ has intermediate values. As $z_3$ diminishes from $-(8)^{\frac{1}{2}}/3$ to $-1$, $\zeta$ increases from $2^{\frac{1}{2}}/3$ to $+1$, while $\zeta'$ diminishes from $2^{\frac{1}{2}}/3$ to zero.

When $z_3$ reaches the value $-1$ we have the case of an ordinary pendulum vibrating in a vertical plane, and then the values of $\zeta$ and $\zeta'$ are 1 and zero. Thus the value of $b/a$ must lie between zero and $\frac{3}{2}$.

By (1) the velocity at the height corresponding to $z_2$ is given by

$$v^2 = 2gl\left(\frac{b}{a} - z_2\right), \quad \text{.........................................(8)}$$

so that we may regard the kinetic energy as generated by a fall through the difference of levels $l(b/a - z_2)$, that is through $l(z_1 + z_3)$. Thus we have

$$z_2 - \frac{2}{3}\frac{b}{a} = \frac{z_2 - \frac{v^2}{gl}}{3}. \quad \text{.........................................(9)}$$

The required condition of inequality is thus fulfilled if

$$z_2 > \frac{v^2}{gl}.$$

The foregoing discussion is based on a paper by M. de Saint Germain (*Bull. des Sci. Math.* 1901).

**17. *Motion of bob referred to rectangular axes. Discussion of the motion by elliptic functions.*** If we take now $x, y, \zeta$ as rectangular coordinates of the pendulum bob, $x, y$ being drawn horizontally from O, and $\zeta$ downward from O, we have as the equations of motion

$$m\ddot{x} = -R\frac{x}{l}, \quad m\ddot{y} = -R\frac{y}{l}, \quad m\ddot{\zeta} = -R\frac{\zeta}{l} + mg. \quad \text{.........................(1)}$$

Now, as we have seen,

$$x\dot{y} - y\dot{x} = h \quad \text{.........................................(2)}$$

is the equation of A.M. about the vertical. Also we obtain by (1)

$$m(\ddot{y}x - \ddot{x}y) = -\frac{R}{l}(x\dot{y} - y\dot{x}) = -\frac{R}{l}h. \quad \text{.........................(3)}$$

If then R vanishes at any point P on the trajectory, $\ddot{x} = 0$, $\ddot{y} = 0$, and there is zero curvature of the horizontal projection of the path at P, which is thus a point of inflexion, and the osculating plane is there vertical.

Now $z$ denoting $\cos\theta$ as before, we have seen, (11), 1, that

$$z - z_3 = (z_2 - z_3)\,\text{sn}^2 mt, \quad z_2 - z = (z_2 - z_3)\,\text{cn}^2 mt, \quad z_1 - z = (z_1 - z_3)\,\text{dn}^2 mt.$$

Also we have

$$(l^2 - \zeta^2)\dot{\psi} = h, \quad \text{or} \quad l^2(1 - z^2)\dot{\psi} = h. \quad \text{.........................(4)}$$

The angle turned through by the radius vector of the horizontal projection in any time is thus

$$\psi = \int \frac{h\,dt}{l^2(1 - z^2)}. \quad \text{.........................................(5)}$$

When for $z^2$ is substituted its value in terms of elliptic functions, the integral can be found and $\psi$ determined. Thus determined $\psi$ can be used to give $x$ and $y$ by the relation

$$w = x + iy = \rho e^{i\psi}. \quad \text{.........................................(6)}$$

The integral $\psi$ given by (5) has two critical points $z = 1$, $z = -1$, but $x$ and $y$ are uniform functions of the time.

Now $x^2 + y^2 + \zeta^2 = l^2$, and therefore

$$\dot{x}^2 + \dot{y}^2 + \dot{\zeta}^2 + x\ddot{x} + y\ddot{y} + \zeta\ddot{\zeta} = 0.$$

Hence (1) gives

$$m(\dot{x}^2 + \dot{y}^2 + \dot{\zeta}^2) = -m(x\ddot{x} + y\ddot{y} + \zeta\ddot{\zeta}) = -Rl + mg\zeta. \quad \text{.........................(7)}$$

But we have seen [(8) and (3), 16] that

$$\dot{x}^2 + \dot{y}^2 + \dot{\zeta}^2 = 2gl\left(\frac{b}{a} - z\right), \quad R = mg\left(2\frac{b}{a} - 3z\right). \quad \text{.........................(8)}$$

Now by (1) we get also $$\frac{m\ddot{w}}{w} = -\frac{R}{l} = -\frac{mg}{l}\left(2\frac{b}{a} - 3z\right). \qquad (9)$$

Putting, to avoid confusion with the mass $m$ of the bob, $\mu$ instead of $m$ for $\{\frac{1}{4}a(z_1 - z_3)\}^{\frac{1}{2}}$, and $n = (g/l)^{\frac{1}{2}}$, we find, since
$$z = z_3 + (z_2 - z_3)\,\mathrm{sn}^2\mu t = z_3 + (z_2 - z_3)\,\mathrm{sn}^2 u, \qquad (10)$$
where $u = \mu t$,
$$\frac{d^2w}{du^2} = (6k^2\mathrm{sn}^2 u + h)w, \qquad (11)$$
where $k^2 = (z_2 - z_3)/(z_1 - z_3)$ as before, and $h = 2(3z_3 - 2b/a)/(z_1 - z_3)$, a constant.

Equation (11) of course splits into two equations of the same form, one in which $w$ is replaced by $x$, and the other in which $w$ is replaced by $y$. The equation is a particular case of that known as Lamé's equation, which is of the form
$$\frac{d^2w}{du^2} = \{n(n+1)k^2\mathrm{sn}^2 u + h\}w. \qquad (12)$$
If $n = 2$, this reduces to (11).

Equation (12) may be put in the Weierstrassian form by means of a new variable given by
$$\mathrm{sn}^2 u = \frac{z - z_3}{z_2 - z_3} = \frac{\wp u' - e_3}{e_2 - e_3}; \qquad (13)$$
and with a view to the discussion in 18 and 19 below, we write $z = M\wp u' + N$, where M, N are constants which are determined in 18. We get
$$dz = M\wp' u'\, du',$$
and
$$z_1 - z = M(e_1 - \wp u'), \quad z_2 - z = M(e_2 - \wp u'), \quad z - z_3 = M(\wp u' - e_3),$$
so that, since $Z = (z_1 - z)(z_2 - z)(z - z_3)$,
$$\int_{z_3}^{z_2} \frac{dz}{Z^{\frac{1}{2}}} = 2M^{-\frac{1}{2}} \int_{e_3}^{e_2} \frac{d\wp u'}{\{4(\wp u' - e_1)(\wp u' - e_2)(\wp u' - e_3)\}^{\frac{1}{2}}},$$
that is
$$\int_{z_3}^{z_2} \frac{dz}{Z^{\frac{1}{2}}} = 2M^{-\frac{1}{2}}\omega_1,$$
where $\omega_1$ is the half-period of $u'$, corresponding to the variation of $\wp u'$ from $e_3$ to $e_2$. But
$$\int_{z_3}^{z_2} \frac{dz}{Z^{\frac{1}{2}}} = \frac{2K}{(z_1 - z_3)^{\frac{1}{2}}},$$
and therefore
$$\omega_1 = \frac{M^{\frac{1}{2}}K}{(z_1 - z_3)^{\frac{1}{2}}} = \frac{K}{(e_1 - e_3)^{\frac{1}{2}}}. \qquad (14)$$
Thus, as may be seen otherwise, the variation $du$ has the value $du' . (e_1 - e_3)^{\frac{1}{2}}$. Hence, since
$$k^2\mathrm{sn}^2 u = \frac{e_2 - e_3}{e_1 - e_3}\,\frac{z - z_3}{z_2 - z_3} = \frac{\wp u' - e_3}{e_1 - e_3}, \qquad (15)$$
we have instead of (11)
$$\frac{d^2w}{du'^2} = \{6\wp u' - 6e_3 + h(e_1 - e_3)\}w, \qquad (16)$$
and, instead of (12), the general Lamé equation
$$\frac{d^2w}{du'^2} = \{n(n+1)\wp u' + A\}w, \qquad (17)$$
where A is a constant. In what follows we use the new variable $u'$, writing it however without the accent.

**18. *Calculation of $\theta$ in terms of $t$.*** For the calculation of $\theta$ in terms of $t$, or *vice versa*, we have the relation
$$\dot{z}^2 = (b - az)(1 - z^2) - c^2, \qquad (1)$$
where $a$, $b$, $c^2$ are positive real quantities, and $1 > z > -1$. This equation has, as we have seen, three real roots; one $z_1$ between $+\infty$ and 1, one between 1 and $z_0$ (the initial value

of $z$, for which of course $z^2$ is positive if it is not zero), and a third between $z_0$ and $-1$. Now, transforming as above from (1) to the Weierstrassian form, we get (from the substitution $z=\mathrm{M}s+\mathrm{N}$)

$$\left(\frac{ds}{dt}\right)^2=4s^3-g_2s-g_3, \qquad \text{...(2)}$$

with

$$\mathrm{M}=\frac{4}{a},\quad \mathrm{N}=\frac{b}{3a},\quad g_2=\tfrac{1}{12}(b^2+3a^2),\quad g_3=\frac{a^2b}{24}\left(\frac{b^2}{9a^2}-1\right)+\frac{c^2a^2}{16}. \qquad \text{...(3)}$$

The roots of the Weierstrassian cubic obtained by equating the expression on the right of (2) to zero are given by

$$e_1=\frac{z_1-\mathrm{N}}{\mathrm{M}},\quad e_2=\frac{z_2-\mathrm{N}}{\mathrm{M}},\quad e_3=\frac{z_3-\mathrm{N}}{\mathrm{M}}, \qquad \text{...(4)}$$

and fulfil the condition $e_1+e_2+e_3=0$, since $z_1+z_2+z_3=b/a=3\mathrm{N}$.

If we construct the Weierstrassian function $\wp u$ with $g_2, g_3$ as invariants, we have

$$\wp'^2u=4\wp^3u-g_2\wp u-g_3. \qquad \text{...(5)}$$

With $s=\wp u$, (2) becomes

$$\left(\wp' u\frac{du}{dt}\right)^2=4\wp^3u-g_2\wp u-g_3. \qquad \text{...(6)}$$

Identification of this equation with (5) shows that

$$\left(\frac{du}{dt}\right)^2=1,\quad \text{or}\quad \frac{du}{dt}=\pm 1. \qquad \text{...(7)}$$

We shall suppose that $du$ and $dt$ have the same sign, as it is always possible to change the sign of $u$ without altering the value of $\wp u$. Thus $z$ and $\wp u$ increase and diminish together. As $z$ increases from $z_3$ to $z_2$, $\wp u$ increases from $e_3$ to $e_2$ in the real semi-period $\omega_1$, and for any instant we may write

$$u=t+\omega_3,$$

where $\omega_3$ is the imaginary semi-period and $t$ is real.

In 1 above we have seen how $t$ may be expressed in terms of $\phi\ [=\tan^{-1}\{(z-z_3)^{\frac{1}{2}}/(z_2-z)^{\frac{1}{2}}\}]$ as an elliptic integral of the first Legendrian type to modulus $k=\{(z_2-z_3)/(z_1-z_3)\}^{\frac{1}{2}}$.

Now let $\wp\alpha$, $\wp\beta$ be so chosen that

$$\mathrm{M}\wp\alpha+\mathrm{N}=1,\quad \mathrm{M}\wp\beta+\mathrm{N}=-1. \qquad \text{...(8)}$$

Then we have

$$\mathrm{M}=\frac{2}{\wp\alpha-\wp\beta},\quad \mathrm{N}=-\frac{\wp\alpha+\wp\beta}{\wp\alpha-\wp\beta}, \qquad \text{...(9)}$$

and of course $\alpha$, $\beta$ are imaginary. By (1) and (2) above we have $\mathrm{M}^2\wp'^2\alpha=\mathrm{M}^2\wp'^2\beta=-c^2$, and therefore

$$\wp'\alpha=\pm\wp'\beta=i\frac{c}{\mathrm{M}}. \qquad \text{...(10)}$$

With these values of M and N we calculate the term $h(e_1-e_3)$ in (16) of 17. The value assigned to $h$ in (11), 17, gives

$$h(e_1-e_3)=\frac{6z_3-4\dfrac{b}{a}}{\mathrm{M}}=6e_3+6\frac{\mathrm{N}}{\mathrm{M}}-4\frac{z_1+z_2+z_3}{\mathrm{M}}=6e_3-6\frac{\mathrm{N}}{\mathrm{M}}.$$

Hence, by (9),

$$h(e_1-e_3)=6e_3+3(\wp\alpha+\wp\beta),$$

and (16), 17, becomes

$$\frac{d^2w}{du^2}=\{6\wp u+3(\wp\alpha+\wp\beta)\}w. \qquad \text{...(11)}$$

The values of M and N in (9) give, it may be noted,

$$\left.\begin{aligned}\sin^2\tfrac{1}{2}\theta&=\frac{1-z}{2}=\frac{\wp\alpha-\wp u}{\wp\alpha-\wp\beta},\quad \cos^2\tfrac{1}{2}\theta=\frac{1+z}{2}=\frac{\wp u-\wp\beta}{\wp\alpha-\wp\beta},\\ z&=\frac{2\wp u-\wp\alpha-\wp\beta}{\wp\alpha-\wp\beta}.\end{aligned}\right\} \qquad \text{...(12)}$$

**19.** *Calculation of the azimuthal motion by means of Lamé's equation.* We can now consider the solution of the Lamé equation, that is the determination of $\psi$

in terms of the time. The theory of the subject has been fully discussed by Hermite and by Halphen. The process we give here, though restricted to the dynamical problem, is complete in essentials, and differs in some respects from the usual form. First we suppose that the functions $\wp\alpha$, $\wp\beta$ fulfil the condition [see (10) of 18]

$$\wp'\alpha+\wp'\beta=0. \qquad (1)$$

Next we assume that
$$w=C\frac{\sigma(u+\alpha)\sigma(u+\beta)}{\sigma\alpha\,\sigma\beta\,\sigma^2 u}e^{-u(\zeta\alpha+\zeta\beta)}. \qquad (2)$$

Taking logarithms and differentiating, we get

$$\frac{1}{w}\frac{dw}{du}=\zeta(u+\alpha)+\zeta(u+\beta)-2\zeta u-\zeta\alpha-\zeta\beta=\frac{1}{2}\frac{\wp'u-\wp'\alpha}{\wp u-\wp\alpha}+\frac{1}{2}\frac{\wp'u-\wp'\beta}{\wp u-\wp\beta} \qquad (3)$$

by the addition theorem for the $\zeta$-functions.

Differentiating again, we obtain

$$\frac{1}{w}\frac{d^2w}{du^2}-\left(\frac{1}{w}\frac{dw}{du}\right)^2=-\wp(u+\alpha)-\wp(u+\beta)+2\wp u.$$

By the addition theorem,
$$\wp(u+v)=\frac{1}{4}\left(\frac{\wp'u-\wp'v}{\wp u-\wp v}\right)^2-\wp u-\wp v, \qquad (4)$$

this last result becomes

$$\frac{1}{w}\frac{d^2w}{du^2}=4\wp u+\wp\alpha+\wp\beta-\frac{1}{4}\left\{\left(\frac{\wp'u-\wp'\alpha}{\wp u-\wp\alpha}\right)^2+\left(\frac{\wp'u-\wp'\beta}{\wp u-\wp\beta}\right)^2\right\}+\left(\frac{1}{w}\frac{dw}{du}\right)^2. \qquad (5)$$

But the value of $(dw/du)/w$ given by (3) converts (5) into

$$\frac{1}{w}\frac{d^2w}{du^2}=4\wp u+\wp\alpha+\wp\beta+\frac{1}{2}\frac{\wp'u-\wp'\alpha}{\wp u-\wp\alpha}\frac{\wp'u-\wp'\beta}{\wp u-\wp\beta}. \qquad (5')$$

In the differential equation (11) of 18 the left-hand side of (5′) has the value

$$6\wp u+3\wp\alpha+3\wp\beta.$$

Hence, in order that the right-hand side of (5) may reduce to this value, we must have

$$4(\wp u+\wp\alpha+\wp\beta)=\frac{\wp'u-\wp'\alpha}{\wp u-\wp\alpha}\frac{\wp'u-\wp'\beta}{\wp u-\wp\beta}. \qquad (6)$$

We have therefore to find the condition which, with (1), must be satisfied by $\wp\alpha$, $\wp\beta$ in order that (6) may hold.

By (1) the product on the right may be written

$$\frac{\wp'^2u-\wp'^2\alpha}{(\wp u-\wp\alpha)(\wp u-\wp\beta)},$$

and therefore (6) becomes

$$\begin{aligned}4(\wp u-\wp\alpha)(\wp u-\wp\beta)(\wp u+\wp\alpha+\wp\beta)&=\wp'^2u-\wp'^2\alpha\\&=4\wp^3u-g_2\wp u-g_3-(4\wp^3\alpha-g_2\wp\alpha-g_3),\end{aligned} \qquad (7)$$

by the well-known property of the $\wp$-functions.

The coefficient of $\wp u$ on the left and on the right must be the same, and so also must be the absolute terms. Each of these conditions gives the relation

$$\wp^2\alpha+\wp^2\beta+\wp\alpha\wp\beta-\tfrac{1}{4}g_2=0, \qquad (8)$$

the fulfilment of which renders (7) an identity.

Now consider the quadratic equation

$$\xi^2-k\xi+k^2-\tfrac{1}{4}g_2=0. \qquad (9)$$

Let $\xi_1$, $\xi_2$ be its roots. Then

$$\xi_1+\xi_2=k,\quad \xi_1\xi_2=k^2-\tfrac{1}{4}g_2.$$

Hence
$$\xi_1\xi_2=(\xi_1+\xi_2)^2-\tfrac{1}{4}g_2,$$

that is
$$\xi_1^2+\xi_2^2+\xi_1\xi_2-\tfrac{1}{4}g_2=0, \qquad (10)$$

which is exactly the relation (8) fulfilled by $\wp\alpha$, $\wp\beta$. The conditions fulfilled by $\wp\alpha$, $\wp\beta$ are, therefore, that they should be the roots of an equation of the form (9), and that $\wp'\alpha$, $\wp'\beta$ should have the opposite signs.

**20.** *Calculation of azimuthal motion continued.* Let us now suppose that $w=x+iy$ [$=\rho e^{i\psi}$, see (6), 17], so that by (2), 19,

$$x+iy=C\frac{\sigma(u+a)\sigma(u+\beta)}{\sigma a\,\sigma\beta\,\sigma^2 u}e^{-u(\zeta a+\zeta\beta)}. \quad\ldots\ldots(1)$$

We know that $$\frac{\sigma(u+a)}{\sigma a\,\sigma u}e^{-v\zeta a}=(\wp a-\wp u)\frac{\sigma a\,\sigma u}{\sigma(u-a)e^{v\zeta a}},$$

and $$\frac{\sigma(u+\beta)}{\sigma\beta\,\sigma u}e^{-u\zeta\beta}=(\wp\beta-\wp u)\frac{\sigma\beta\,\sigma u}{\sigma(u-\beta)e^{u\zeta\beta}},$$

so that we obtain

$$x+iy=C(\wp a-\wp u)(\wp\beta-\wp u)\frac{\sigma a\,\sigma\beta\,\sigma^2 u}{\sigma(u-a)\sigma(u-\beta)e^{u(\zeta a+\zeta\beta)}}. \quad\ldots\ldots(2)$$

Hence, if we can take $$x^2+y^2=K(\wp a-\wp u)(\wp\beta-\wp u), \quad\ldots\ldots(3)$$

where K is a constant, we get

$$x-iy=\frac{x^2+y^2}{x+iy}=\frac{K}{C}\frac{\sigma(u-a)\sigma(u-\beta)e^{u(\zeta a+\zeta\beta)}}{\sigma a\,\sigma\beta\,\sigma^2 u}. \quad\ldots\ldots(4)$$

Now, by the equation of the sphere,

$$x^2+y^2=l^2-\zeta^2=l^2(1-z^2)=l^2(1-z)(1+z),$$

and so by (12), 18, we get

$$x^2+y^2=-\frac{4l^2}{(\wp a-\wp\beta)^2}(\wp a-\wp u)(\wp\beta-\wp u), \quad\ldots\ldots(5)$$

or $$x^2+y^2=-M^2l^2(\wp a-\wp u)(\wp\beta-\wp u), \quad\ldots\ldots(5')$$

where $M=2/(\wp a-\wp\beta)$ as found in (9), 18. Thus $K=-M^2l^2=-4l^2/(\wp a-\wp\beta)^2$.

For the sake of symmetry we might choose a new constant E, such that $C=EMl$. We should have then for (2) and (4),

$$\left.\begin{aligned} x+iy&=EMl\frac{\sigma(u+a)\sigma(u+\beta)}{\sigma a\,\sigma\beta\,\sigma^2 u}e^{-u(\zeta a+\zeta\beta)},\\ x-iy&=-\frac{Ml}{E}\frac{\sigma(u-a)\sigma(u-\beta)}{\sigma a\,\sigma\beta\,\sigma^2 u}e^{u(\zeta a+\zeta\beta)}.\end{aligned}\right\} \quad\ldots\ldots(6)$$

Multiplying the two equations together, we verify that they give, as they ought,

$$x^2+y^2=-M^2l^2(\wp u-\wp a)(\wp u-\wp\beta).$$

Also, by the addition theorem,

$$\wp u-\wp v=-\frac{\sigma(u+v)\sigma(u-v)}{\sigma^2 u\,\sigma^2 v},$$

we get $$\wp a-\wp\beta=-\frac{\sigma(a+\beta)\sigma(a-\beta)}{\sigma^2 a\,\sigma^2\beta},$$

and therefore the value of M in (9), 18 gives

$$Ml=-2l\frac{\sigma^2 a\,\sigma^2\beta}{\sigma(a+\beta)\sigma(a-\beta)}. \quad\ldots\ldots(7)$$

Dividing the first of (6) by the second, and supposing that $x=\rho\cos\psi$, $y=\rho\sin\psi$, we get

$$e^{2i\psi}=-E^2\frac{\sigma(u+a)\sigma(u+\beta)}{\sigma(u-a)\sigma(u-\beta)}e^{-2u(\zeta a+\zeta\beta)}, \quad\ldots\ldots(8)$$

by which $\psi$ can be calculated. The value of $E^2$ falls to be determined from the initial circumstances of the case.

**21.** *Direct determination of azimuthal motion.* We have thus found the azimuthal motion of the spherical pendulum as a solution of Lamé's equation. The following direct process leads to the same result, and we give it here for the sake of the conclusions to which it leads as to the angles turned through in successive half-periods of the motion. In 2 above we obtained

$$\dot{\psi}=\frac{h}{l^2}\frac{1}{1-z^2}=\tfrac{1}{2}c\left(\frac{1}{1+z}+\frac{1}{1-z}\right). \quad [c=h/l^2.] \qquad (1)$$

Writing $M\wp u+N=z$, $M\wp\alpha+N=1$, $M\wp\beta+N=-1$, and remembering that $du=dt$, we get

$$\frac{d\psi}{du}=\frac{c}{2M}\left(\frac{1}{\wp u-\wp\beta}-\frac{1}{\wp u-\wp\alpha}\right). \qquad (2)$$

Now $$\dot{z}^2=\left(\frac{dz}{du}\right)^2=M^2\wp'^2u=(-az+b)(1-z^2)-c^2, \qquad (3)$$

and $u=\alpha$ or $\beta$ according as $z=1$ or $-1$. Hence

$$\wp'^2\alpha=\wp'^2\beta=-\frac{c^2}{M^2}, \qquad (4)$$

and $\wp'\alpha$, $\wp'\beta$ may have the same or contrary signs. In accordance with the choice already made above we take

$$\wp'\alpha=-\wp'\beta=i\frac{c}{M} \qquad (5)$$

Equation (2) becomes $$-2i\frac{d\psi}{du}=\frac{\wp'\beta}{\wp u-\wp\beta}+\frac{\wp'\alpha}{\wp u-\wp\alpha}. \qquad (6)$$

By the addition theorem of the $\zeta$-functions, we have from this,

$$-2i\frac{d\psi}{du}=-\zeta(u+\beta)+\zeta(u-\beta)+2\zeta\beta$$
$$-\zeta(u+\alpha)+\zeta(u-\alpha)+2\zeta\alpha. \qquad (7)$$

Integrating we obtain $$e^{2i\psi}=-E^2\frac{\sigma(u+\alpha)\sigma(u+\beta)}{\sigma(u-\alpha)\sigma(u-\beta)}e^{-2u(\zeta\alpha+\zeta\beta)}. \qquad (8)$$

From this and the equation of the sphere we can deduce $x+iy$, $x-iy$; but the values of these quantities are given above.

**22.** *Proof of Bravais' theorem by elliptic functions.* In this connection the determination of what are called the *complete* integrals,

$$\int_0^{2\omega}\zeta(u+v)\,du, \quad \int_0^{2\omega/i}\zeta(iu+v)\,du,$$

is important. For the present we take only the first. We have

$$\int_0^{2\omega}\zeta(u+v)\,du=\log\frac{\sigma(2\omega+v)}{\sigma v}. \qquad (1)$$

But $$\sigma(2\omega+v)=-e^{2\eta(v+\omega)}\sigma v,$$

and thus $$\log\{\sigma(2\omega+v)\}=2(\omega+v)\eta+(2n+1)\pi i+\log(\sigma v). \qquad (2)$$

Now let the step of azimuthal angle described by the spherical pendulum when $u$ is altered by $2\omega_1$ be $2\psi_0$. Applying (2) to (8), 21, we get

$$4\psi_0 i=2\eta_1(\alpha+\omega_1)-2\eta_1(-\alpha+\omega_1)+2\eta_1(\beta+\omega_1)-2\eta_1(-\beta+\omega_1)-4\omega_1(\zeta\alpha+\zeta\beta),$$

that is $$i\psi_0=-\omega_1(\zeta\alpha+\zeta\beta)+\eta_1(\alpha+\beta). \qquad (3)$$

Following a process due to Greenhill [*Elliptic Functions*, App.], we find an approximation to the value of the expression on the right of (3). We divide the expression into two parts $\psi_1$ and $\psi_2$, so that we have

$$i\psi_1=\alpha\eta_1-\omega_1\zeta\alpha, \quad i\psi_2=\beta\eta_1-\omega_1\zeta\beta. \qquad (4)$$

Now $\omega_1$ corresponds to the largest root $z_1$ of the cubic equation in $z$, and $\omega_3$ to the smallest root $z_3$. Let the two roots $z_3$, $z_2$ (corresponding to $\omega_3$ and $\omega_1+\omega_3$, between which the variation of $u$ takes place in the half-period) be both approximately equal to $-1$. The variation of $u$ in the integration considered is through the whole period $2\omega_1$, so we now suppose $\omega_1$ written for $\omega$ in (4), and $\eta_1$ written for $\eta$ to correspond.

The value of $u$ for $z=+1$ is $\alpha$, and for $z=-1$ is $\beta$ [see (8), 18]. If then $q$ and $s$ be small quantities, we get by a period-rectangle, reckoning arguments from the upper circle,

$$\alpha=\omega_3(1-s),\quad \beta=\omega_1+q\omega_3. \qquad (5)$$

Developing $\zeta\alpha$, $\zeta\beta$ by Taylor's theorem and using the values

$$\frac{d}{du}(\zeta u)=-\wp u,\quad \wp\omega_3=e_3,\quad \wp\omega_1=e_1,$$

we find $$\zeta\beta=\eta_1-q\omega_3\wp\omega_1=\eta_1-e_1q\omega_3,\quad \zeta\alpha=\eta_3+s\omega_3\wp\omega_3=\eta_3+e_3s\omega_3. \qquad (6)$$

and therefore, by (4),

$$i\psi_2=q\omega_3(\eta_1+e_1\omega_1),\quad i\psi_1=\omega_3\eta_1-\omega_1\eta_3+s\omega_3(\eta_1+e_3\omega_1)=\tfrac{1}{2}i\pi-s\omega_3(\eta_1+e_3\omega_1). \qquad (7)$$

It can be proved that $$\mathrm{dn}^2\{u(e_1-e_3)^{\frac{1}{2}}\}=\frac{e_1-\wp(u+\omega_3)}{e_1-e_3}.$$

Multiplying by $du$ and integrating from 0 to $\omega_1$, we get

$$\int_0^{\omega_1}\mathrm{dn}^2\{u(e_1-e_3)^{\frac{1}{2}}\}du=\omega_1+\frac{\eta_1+\omega_1e_3}{e_1-e_3}, \qquad (8)$$

so that $$(e_1-e_3)^{\frac{1}{2}}\mathrm{E}=e_1\omega_1+\eta_1; \qquad (9)$$

where E is the complete Legendrian integral of the second kind to modulus

$$\{(e_2-e_3)/(e_1-e_3)\}^{\frac{1}{2}}.$$

But we know also that $$\mathrm{K}(e_1-e_3)^{\frac{1}{2}}=(e_1-e_3)\omega_1, \qquad (10)$$

where K is the complete elliptic integral of the first kind corresponding to E. From (9) and (10), by subtraction, we find

$$(e_1-e_3)^{\frac{1}{2}}(\mathrm{E}-\mathrm{K})=\eta_1+e_3\omega_1 \qquad (11)$$

Equation (9) used in the first of (7), and (11) used in the second, give

$$\left.\begin{array}{l} i\psi_2=q\omega_3(e_1-e_3)^{\frac{1}{2}}\mathrm{E},\\ i\psi_1=\tfrac{1}{2}i\pi+s\omega_3(e_1-e_3)^{\frac{1}{2}}(\mathrm{K}-\mathrm{E}).\end{array}\right\} \qquad (12)$$

Now, in 17 above, we have seen that $\mathrm{M}\wp\alpha+\mathrm{N}=1$, $\mathrm{M}\wp\beta+\mathrm{N}=-1$, $\mathrm{M}\wp u+\mathrm{N}=z$. Hence if $\gamma_2$, $\gamma_3$ be the angles which the axis of the pendulum makes with the downward vertical when in the positions $z_2$, $z_3$ respectively, we have by (12), 18,

$$\cot^2\tfrac{1}{2}\gamma_2=\frac{e_3-\wp\alpha}{\wp\beta-e_3}=\frac{e_3-e_3-\frac{1}{2}s^2\omega_3^2\wp''\omega_3+\cdots}{e_1-e_3+\frac{1}{2}q^2\omega_3^2\wp''\omega_1+\cdots}. \qquad (13)$$

But $\wp''\omega_3=2(e_1-e_3)(e_2-e_3)$, $k^2=(e_2-e_3)/(e_1-e_3)$, $k'^2=(e_1-e_2)/(e_1-e_3)$. Hence approximately, by (13),

$$\cot^2\tfrac{1}{2}\gamma_2=-(e_2-e_3)s^2\omega_3^2=-k^2s^2\omega_3^2(e_1-e_3). \qquad (14)$$

Similarly, $$\cot^2\tfrac{1}{2}\gamma_3=\frac{e_2-\wp\alpha}{\wp\beta-e_2}=\frac{e_2-e_3}{e_1-e_2}=\frac{k^2}{k'^2}. \qquad (15)$$

Thus we find by multiplication,

$$(e_1-e_3)s^2\omega_3^2=-\frac{k'^2}{k^4}\cot^2\tfrac{1}{2}\gamma_2\cot^2\tfrac{1}{2}\gamma_3, \qquad (16)$$

and by substitution in (12),

$$\psi_1=\tfrac{1}{2}\pi+\frac{k'}{k^2}(\mathrm{K}-\mathrm{E})\cot\tfrac{1}{2}\gamma_2\cot\tfrac{1}{2}\gamma_3. \qquad (17)$$

Again, by (1) of 19 above, which relates $\alpha$ to $\beta$,

$$1=-\frac{\wp'\beta}{\wp'\alpha}=-\frac{\wp'(\omega_1+q\omega_3)}{\wp'(\omega_3-s\omega_3)}=\frac{q\wp''\omega_1}{s\wp''\omega_3}=\frac{q}{s}\frac{e_1-e_2}{e_2-e_3}=\frac{q}{s}\frac{k'^2}{k^2}.$$

Hence $$q=s\frac{k^2}{k'^2}, \qquad (18)$$

and therefore by (12) $$i\psi_2=s\omega_3(e_1-e_3)^{\frac{1}{2}}\frac{k^2}{k'^2}\mathrm{E}. \qquad (19)$$

But [(16), above] we have seen that

$$(e_1 - e_3)^{\frac{1}{2}} s\omega_3 = i\frac{k'}{k^2}\cot\tfrac{1}{2}\gamma_2 \cot\tfrac{1}{2}\gamma_3,$$

and thus we get $$\psi = \psi_1 + \psi_2 = \tfrac{1}{2}\pi + \frac{1}{k^2 k'}\{k'^2(\mathrm{K} - \mathrm{E}) + k^2\mathrm{E}\}\cot\tfrac{1}{2}\gamma_2\cot\tfrac{1}{2}\gamma_3. \quad \text{(20)}$$

Substituting now from the expansions of K and E in ascending powers of $k^2$, putting $k'^2 = 1 - k^2$, $1/k' = 1 + \frac{1}{2}k^2$, we get easily

$$\psi = \psi_1 + \psi_2 = \tfrac{1}{2}\pi\{1 + (\tfrac{3}{2} + \tfrac{15}{16}k^2 + \cdots)\cot\tfrac{1}{2}\gamma_2\cot\tfrac{1}{2}\gamma_3\}. \quad \text{(21)}$$

A rougher approximation is

$$\psi = \tfrac{1}{2}\pi(1 + \tfrac{3}{2}\cot\tfrac{1}{2}\gamma_2\cot\tfrac{1}{2}\gamma_3), \quad \text{(22)}$$

which may also be obtained by noticing that as $k$ approaches zero, $k'$ approaches 1, without limit of closeness, and that the limiting value, for $k=0$, of $(\mathrm{K}-\mathrm{E})/k^2$ is $\frac{1}{4}\pi$, while that of E is $\frac{1}{2}\pi$. Thus $$\psi = \tfrac{1}{2}\pi(1 + \tfrac{3}{2}\cot\tfrac{1}{2}\gamma_2\cot\tfrac{1}{2}\gamma_3) = \tfrac{1}{2}\pi(1 + \tfrac{3}{8}\sin\gamma_2\sin\gamma_3). \quad \text{(23)}$$

This is Bravais' result, which was found above by another process, not involving the use of elliptic functions.

If we replace the factor 1 in the value of $\psi_2$, found from (19) and (16), by

$$-(\mathrm{G} - \mathrm{C}n)/(\mathrm{G} + \mathrm{C}n),$$

we obtain $$\psi_2 = \left(\frac{\mathrm{G} - \mathrm{C}n}{\mathrm{G} + \mathrm{C}n}\right)^2 \frac{\pi}{2}\cot\tfrac{1}{2}\gamma_2\cot\tfrac{1}{2}\gamma_3;$$

which with (17) gives the approximate azimuthal turning for the corresponding case of the motion of a top. G is the A.M. about the upward vertical.

**23.** *Consideration of special cases of the spherical pendulum.* Going back to the exact equation, (3), 22, we may now reckon arguments from the *lower* limiting circle so that $\alpha = \omega_1 + q\omega_3$, $\beta = \omega_3(1 - s)$, $d\alpha = i d\alpha'$. Thus differentiating with respect to $\alpha$, we get $$\frac{d\psi_0}{d\alpha'} = \omega_1\wp\alpha + \eta_1 + (\omega_1\wp\beta + \eta_1)\frac{d\beta}{d\alpha}. \quad \text{(1)}$$

But we have here to take $\wp'\alpha + \wp'\beta = 0$, and therefore have

$$\wp''\alpha\, d\alpha + \wp''\beta\, d\beta = 0, \quad \text{or} \quad \frac{d\alpha}{d\beta} = -\frac{\wp''\beta}{\wp''\alpha}. \quad \text{(2)}$$

Thus we find $$\wp''\beta\frac{d\psi_0}{d\alpha'} = (\omega_1\wp\alpha + \eta_1)\wp''\beta - (\omega_1\wp\beta + \eta_1)\wp''\alpha. \quad \text{(3)}$$

But direct differentiation of the fundamental equation of the $\wp$-functions, that is of

$$\wp'^2 = 4\wp^3 u - g_2\wp u - g_3,$$

gives $$\wp'' u = 6\wp^2 u - \tfrac{1}{2}g_2,$$

and so we have $$\wp''\alpha = 6\wp^2\alpha - \tfrac{1}{2}g_2, \quad \wp''\beta = 6\wp^2\beta - \tfrac{1}{2}g_2.$$

In virtue of these values of $\wp''\alpha$, $\wp''\beta$, equation (3) takes the form

$$-\frac{\wp''\beta}{\wp\alpha - \wp\beta}\frac{d\psi_0}{d\alpha'} = \omega_1(6\wp\alpha\,\wp\beta + \tfrac{1}{2}g_2) + 6\eta_1(\wp\alpha + \wp\beta). \quad \text{(4)}$$

Now we have seen in 19 that $\wp\alpha$, $\wp\beta$ are the roots of a certain quadratic equation which, if $-\kappa$ be the sum of the roots, may be written

$$\xi^2 + \kappa\xi + \kappa^2 - \tfrac{1}{4}g_2 = 0.$$

Substituting, therefore, in (4), $-\kappa$ for $\wp\alpha + \wp\beta$, and $\kappa^2 - \frac{1}{4}g_2$ for $\wp\alpha\,\wp\beta$, we get

$$-\frac{\wp''\beta}{\wp\alpha - \wp\beta}\frac{d\psi_0}{d\alpha'} = 6\{\omega_1(\kappa^2 - \tfrac{1}{6}g_2) - \eta_1\kappa\}. \quad \text{(5)}$$

Also we have $$\wp\alpha + \wp\beta + \kappa = 0.$$

By the addition theorem of the $\wp$-functions we have, since $\wp'\alpha = -\wp'\beta$,

$$\left.\begin{array}{l}\wp\alpha+\wp\beta+\wp(\alpha-\beta)=0,\\ \wp(\alpha-\beta)=\kappa.\end{array}\right\} \qquad (6)$$

so that

It is important to notice that, by (6), $\alpha+\beta+\alpha-\beta$, that is $2\alpha$, is a period of the argument $u$.

We know that $\wp''\beta$, $\wp\alpha-\wp\beta$ are positive, and it can be proved [see Halphen, *Fonctions Elliptiques*, t. i. p. 315] that the expression $(\kappa^2-\frac{1}{6}g_2)\omega_1-\eta_1\kappa$, which appears in (5), is negative. Thus $d\psi_0/d\alpha'$ is positive, that is $\psi_0$ increases with increase of $\alpha'$. The imaginary part of $\alpha$ is $i\alpha'$, and $\alpha'$ may have any value from 0 to the full half-period value $\omega_3/i$, so that $\alpha$ lies between $\omega_1$ and $\omega_1+\omega_3$; also $\beta$ lies between 0 and $\omega_3$ The values of $\alpha$ and $\beta$ are situated as indicated in the inequalities

$$e_1>\wp\alpha>e_2, \quad e_3>\wp\beta>-\infty,$$

and $\wp\beta$ reaches the value $-\infty$ when $\beta=0$. Thus we can write

$$\alpha=\omega_1+q\omega_3, \quad \beta=(1-s)\omega_3,$$

where $q$ and $s$ are proper fractions. In each of the extreme cases specified in the table below $s$ is zero, and it will be seen that $\wp'\alpha+\wp'\beta=0$, and that $2\alpha$ is a period.

Again, if $\tau$ be the time value of the half-period, and $t$ ($<\tau$) the current value of the time reckoned from the lower limiting circle, the equation for $u$ is

$$u=\omega_1\frac{t}{\tau}+\omega_3.$$

As an example, let the pendulum very nearly reach the lowest point of the sphere. Then we have, also very nearly, $q=0$, that is $\wp\alpha=e_1$ and $\alpha=\omega_1$. But now to the same degree of approximation $z_3=-1$, and so also $z_1=1$. Hence $\wp\beta=e_3$ and $\beta=\omega_3$, that is the purely imaginary part of $\beta$ is zero.

The following table shows the values of $\alpha'$, $\alpha$, $\beta$, when the azimuthal angle turned through in half a period has the extreme values $\frac{1}{2}\pi$ and $\pi$. This angle has been calculated, from (3) of 22, that is from

$$i\psi_0=-\omega_1(\zeta\beta+\zeta\alpha)+\eta_1(\alpha+\beta)$$

| | $\alpha'$ | $\alpha$ | $\beta$ | $\psi_0$ |
|---|---|---|---|---|
| I. | 0 | $\omega_1$ | $\omega_3$ | $\frac{1}{2}\pi$ |
| II. | $\frac{\omega_3}{i}$ | $\omega_1+\omega_3$ | $\omega_3$ | $\pi$ |

In the first case we get, by the values of $\beta$ and $\alpha$,

$$i\psi_0=\eta_1\omega_3-\eta_3\omega_1,$$

and in the second case

$$i\psi_0=2(\eta_1\omega_3-\eta_3\omega_1).$$

The quantity $\eta_1\omega_3-\eta_3\omega_1$ is $\pm\frac{1}{2}i\pi$, according as $\omega_3/i\omega_1$ is positive or negative. Thus, in the present case, we have the values stated in the table

The meaning of these results may be shortly stated. When $\alpha=\omega_1$ and $\beta=\omega_3$, that is when one limiting position of the pendulum bob is at the lowest point of the sphere and the other infinitesimally higher, the azimuthal angle turned through in the passage from one limiting level to the next is $\frac{1}{2}\pi$.

In the other extreme case we have, by (6),

$$\wp(\alpha-\beta)=-(\wp\alpha+\wp\beta)=\frac{2N}{M},$$

by (8) of 18. But $\alpha-\beta=\omega_1$, and so we have

$$\wp\omega_1=\frac{2N}{M}.$$

This is the case in which [see 15, above] the limiting circles are shrunk into the extremities of the vertical diameter, when $\psi_0=2\pi$.

In the case of motion described in 4 above, the value of $z_1$ is infinite, and so we have $\omega_1=0$, as the defining integral for $\omega_1$ shows at once. Here by (10), 18, $\wp'\alpha=\infty$.

**24.** *Motion of a particle on a concave surface of revolution.* Now let us suppose that a particle moves on a concave surface of revolution, the axis of which is vertical. We suppose that the origin of coordinates $(\zeta, \rho)$ is taken on the axis, and that $\zeta$ is measured upwards, $\rho$ horizontally, and that $\zeta=f(\rho)$ is the equation of the surface. The energy equation is

$$\tfrac{1}{2}m\{\dot\rho^2(1+f'^2)+\rho^2\dot\psi^2\}=-mgf(\rho)+h. \quad\text{......................(1)}$$

For clearly the expression on the left is the value of the kinetic energy, $+mgf(\rho)$ is the potential energy, and $h$ is the constant sum of these two.

Since the action of the surface on the particle is always directed towards the axis of figure and the gravity force upon it is parallel to the axis the A.M. $m\rho^2\dot\psi$ is constant. We have

$$m\rho^2\dot\psi^2=2\{-mgf(\rho)+h\}-m\dot\rho^2(1+f'^2)=m\frac{c^2}{\rho^2}. \quad\text{................(2)}$$

Thus we get
$$t-t_0=\pm\int_{\rho_0}^{\rho}\rho\,d\rho\left\{\frac{1+f'^2}{2\rho^2\left\{-gf(\rho)+\frac{h}{m}\right\}-c^2}\right\}^{\frac{1}{2}}, \quad\text{.........................(3)}$$

and
$$\psi-\psi_0=\pm\int_{\rho_0}^{\rho}c\frac{d\rho}{\rho}\left\{\frac{1+f'^2}{2\rho^2\left\{-gf(\rho)+\frac{h}{m}\right\}-c^2}\right\}^{\frac{1}{2}}, \quad\text{....................(4)}$$

where $t-t_0$ is the time of passage from the distance $\rho_0$ from the axis to the distance $\rho$, and $\psi-\psi_0$ is the angle turned through by the horizontal projection of the radius vector in the same time.

It will be noticed that $c^2$ has not here the meaning assigned to it in 1. There it denoted $\dot\psi^2\sin^4\theta$, here it denotes $\dot\psi^2\rho^4$.

It is clear that the value of $c$ is $v\rho\sin\alpha$, where $v$ is the resultant speed of the particle at any instant and $\alpha$ the inclination of the direction of motion to the meridian (a section of the surface by a plane containing the axis) on which the particle is situated at the instant. Thus we have

$$c=v\rho\sin\alpha=v_0\rho_0\sin\alpha_0. \quad\text{..............................(5)}$$

If the particle describe a parallel (a section of the surface made by a horizontal plane), the vertex of the cone the generators of which are the normals to the surface drawn from points of the parallel must be above the level of the parallel. This is necessary in order that the reaction may balance $mg$. If the semi-vertical angle of the cone be $\beta$, the condition necessary for equilibrium is clearly

$$\tan\beta = \frac{g\rho}{v^2}.$$

But $\cot\beta = d\zeta/d\rho = f'(\rho)$. Thus

$$v = \{g\rho f'(\rho)\}^{\frac{1}{2}}. \qquad \text{(6)}$$

Thus $v$ is a possible equilibrium speed only if $f'(\rho)$ is positive.

If the particle be under no force, except the reaction of the surface, we have, putting $c'^2$ for $c^2m/2h$,

$$\left.\begin{aligned} t - t_0 &= \pm\frac{c'}{c}\int_{\rho_0}^{\rho}\rho\,d\rho\left(\frac{1+f'^2}{\rho^2-c'^2}\right)^{\frac{1}{2}}, \\ \psi-\psi_0 &= \pm c'\int_{\rho_0}^{\rho}\frac{d\rho}{\rho}\left(\frac{1+f'^2}{\rho^2-c'^2}\right)^{\frac{1}{2}}. \end{aligned}\right\} \qquad \text{(7)}$$

In this case, since the force on the particle is always at right angles to the direction of motion, the value of the resultant speed remains constant, and therefore we have by (5), as a condition fulfilled by the path on the surface, the equation

$$\rho\sin\alpha = \text{const.}$$

This condition is characteristic of a geodesic on a surface of revolution, that is of a line so drawn on the surface that its osculating plane contains the normal at each point. We have thus obtained a dynamical proof of this characteristic property of geodesics.

As a particular case, consider the motion of a particle on the interior concave surface of a hollow right circular cylinder. Here $\rho$ is constant, and the inclination of the path to the successive generating lines of the cylinder is everywhere the same, so that the path is a helix on the surface. This is also obvious from the facts that whatever axial speed the particle may be given initially will be preserved unaltered, and that the A.M. of the particle about the axis also remains constant. Of course if the axial speed is zero, or if the angular speed about the axis is zero, we have the limiting cases of the helix, a circle about the axis, or a generating line of the cylinder.

**25.** *Motion on a paraboloid with axis vertical.* Now let the surface of revolution be a paraboloid with axis vertical. The equation of the surface, if the vertex—the lowest point—be taken as origin, and $\zeta$ be measured upwards, is

$$\rho^2 = 4a\zeta. \qquad \text{(1)}$$

In this case, if $m$ be taken as unity, the equation for $t$ becomes

$$t - t_0 = \int_{\zeta_0}^{\zeta}\frac{(a+\zeta)^{\frac{1}{2}}d\zeta}{(2h\zeta-2g\zeta^2-c^2/4a)^{\frac{1}{2}}}. \qquad \text{(2)}$$

This may be written in the form

$$t - t_0 = \int_{\zeta_0}^{\zeta}\frac{(\zeta+a)\,d\zeta}{\{(\zeta+a)(2h\zeta-2g\zeta^2-c^2/4a)\}^{\frac{1}{2}}}. \qquad \text{(2')}$$

Let $\zeta_2$, $\zeta_3$ be the roots, in *ascending* order of magnitude, of the quadratic

$$2g\zeta^2 - 2h\zeta + \frac{c^2}{4a} = 0, \qquad \text{(3)}$$

that is if
$$\left.\begin{matrix}\zeta_2\\ \zeta_3\end{matrix}\right\}=\frac{1}{2g}\left\{h\mp\left(h^2-\frac{c^2g}{2a}\right)^{\frac{1}{2}}\right\},\qquad\qquad(4)$$

we have
$$t-t_0=\left(-\frac{2}{g}\right)^{\frac{1}{2}}\int_{\zeta_0}^{\zeta}\frac{(\zeta+a)\,d\zeta}{\{4(\zeta+a)(\zeta-\zeta_2)(\zeta-\zeta_3)\}^{\frac{1}{2}}}.\qquad\qquad(5)$$

Now let
$$\zeta=\mathrm{M}s+\mathrm{N}.\qquad\qquad(6)$$

We get for the cubic expression in the denominator of (5) the form
$$4\mathrm{M}^3(s-e_1)(s-e_2)(s-e_3),$$

where
$$e_1=-\frac{\mathrm{N}+a}{\mathrm{M}},\quad e_2=\frac{-\mathrm{N}+\zeta_2}{\mathrm{M}},\quad e_3=\frac{-\mathrm{N}+\zeta_3}{\mathrm{M}}.$$

Hence, if $e_1+e_2+e_3=0$,
$$\mathrm{N}=\frac{-a+\zeta_2+\zeta_3}{3},\qquad\qquad(7)$$

and
$$e_1=-\frac{2a+\zeta_2+\zeta_3}{3\mathrm{M}},\quad e_2=\frac{a+2\zeta_2-\zeta_3}{3\mathrm{M}},\quad e_3=\frac{a-\zeta_2+2\zeta_3}{3\mathrm{M}}.\qquad\qquad(8)$$

We suppose that $\zeta$ lies between $\zeta_2$ and $\zeta_3$, so that $s$ lies between $e_2$ and $e_3$. We have now
$$t-t_0=\left(-\frac{2}{g}\mathrm{M}\right)^{\frac{1}{2}}\int_{e_3}^{s}\frac{(s-e_1)\,ds}{\{4(s-e_1)(s-e_2)(s-e_3)\}^{\frac{1}{2}}}.\qquad\qquad(9)$$

Let $s=\wp u$. We get
$$t-t_0=\left(-\frac{2}{g}\mathrm{M}\right)^{\frac{1}{2}}\int(\wp u-e_1)\,du.\qquad\qquad(10)$$

To determine the limits of integration we observe that as $\zeta$ lies between $\zeta_3$ and $\zeta_2$ so also $\wp u$ must lie between $e_2$ and $e_3$, and that when $\wp u=e_3$ we may take $u=\omega_3$. Thus
$$t-t_0=\left(-\frac{2}{g}\mathrm{M}\right)^{\frac{1}{2}}\int_{\omega_3}^{u}(\wp u-e_1)\,du=\left(-\frac{2}{g}\mathrm{M}\right)^{\frac{1}{2}}\{-\zeta u+\zeta\omega_3-e_1(u-\omega_3)\}.\qquad\qquad(11)$$

The value of M may be chosen at convenience. If, for example, we take $\mathrm{M}=-1$, we have $e_1>e_2>e_3$, and get
$$t-t_0=\left(\frac{2}{g}\right)^{\frac{1}{2}}\{-\zeta u+\zeta\omega_3-e_1(u-\omega_3)\}.\qquad\qquad(12)$$

If we choose $\mathrm{M}=-(a+\zeta_3)$, we have $e_1-e_2=(a+\zeta_2)/(a+\zeta_3)$, $e_1-e_3=1$.

In (11) and (12) $\zeta$ is the Weierstrassian function usually denoted by that letter. Elsewhere, unless the contrary is stated, it denotes axial distance as defined above.

For the azimuthal motion we have $\rho^2 d\psi=c\,dt$, that is
$$d\psi=\frac{c}{4a\zeta}dt.\qquad\qquad(13)$$

But we have just seen that
$$dt=\left(-\frac{2\mathrm{M}}{g}\right)^{\frac{1}{2}}(\wp u-e_1)\,du\,;$$

and, by (6) and (7),
$$\zeta=\mathrm{M}\wp u-\frac{a-\zeta_2-\zeta_3}{3}=\mathrm{M}\left(\wp u-\frac{a-\zeta_2-\zeta_3}{3\mathrm{M}}\right).$$

Thus (13) becomes
$$d\psi=\frac{c}{4a}\left(-\frac{2}{g\mathrm{M}}\right)^{\frac{1}{2}}\frac{(\wp u-e_1)\,du}{\wp u-\dfrac{a-\zeta_2-\zeta_3}{3\mathrm{M}}},\qquad\qquad(14)$$

which may be written as
$$d\psi=\frac{c}{4a}\left(-\frac{2}{g\mathrm{M}}\right)^{\frac{1}{2}}\left\{du+\left(\frac{a-\zeta_2-\zeta_3}{3\mathrm{M}}-e_1\right)\frac{du}{\wp u-\dfrac{a-\zeta_2-\zeta_3}{3\mathrm{M}}}\right\}.\qquad\qquad(15)$$

Integrating, we find

$$\psi-\psi_0=\frac{c}{4a}\left(-\frac{2}{gM}\right)^{\frac{1}{2}}\left\{u-\omega_3+\frac{a}{M}\int_{\omega_3}^{u}\frac{du}{\wp u-\frac{a-\zeta_2-\zeta_3}{3M}}\right\} \qquad (16)$$

We now define an argument $\alpha$ by the equation

$$\wp\alpha=\frac{a-\zeta_2-\zeta_3}{3M}=\frac{a-\frac{h}{g}}{3M}, \qquad (17)$$

and calculate $\wp'\alpha$ by the relation

$$\wp'^2\alpha=4(\wp\alpha-e_1)(\wp\alpha-e_2)(\wp\alpha-e_3).$$

We get

$$\wp'\alpha=\frac{c}{(2gM^3)^{\frac{1}{2}}}, \qquad (18)$$

and therefore by (16),

$$2i(\psi-\psi_0)=-2(u-\omega_3)\frac{c}{a(8gM)^{\frac{1}{2}}}-\int\frac{\wp'\alpha\,du}{\wp u-\wp\alpha}. \qquad (19)$$

Now

$$\int\frac{\wp'\alpha\,du}{\wp u-\wp\alpha}=2u\zeta\alpha+\log\sigma(u-\alpha)-\log\sigma(u+\alpha),$$

where $\zeta$ now denotes the Weierstrassian $\zeta$-function.

Hence, taking as limits of the argument $u$ and $\omega_3$, we find

$$e^{2i(\psi-\psi_0)}=e^{2(u-\omega_3)\left\{-\frac{c}{a(8gM)^{\frac{1}{2}}}-\zeta\alpha\right\}}\frac{\sigma(u+\alpha)\sigma(\omega_3-\alpha)}{\sigma(u-\alpha)\sigma(\omega_3+\alpha)}.$$

**26.** *Cases integrable by elliptic functions.* It is stated in a paper by Gustaf Kobb [*Acta Math.*, 10, 1887] that the integration of the motion of a particle on a surface of revolution, under the action of a constant external force parallel to the axis (*e.g.* gravity), can be effected in terms of elliptic functions if the equation has one of the five forms, in which $\zeta$ denotes axial distance and $\rho$ radial distance from the axis,

$$\rho=m\zeta,\quad \rho^2+\zeta^2=a^2,\quad \rho^2=4a\zeta,$$
$$9a\rho^2=\zeta(\zeta-3a)^2,\quad 2\rho^4+3a^2\rho^2=2a^3\zeta.$$

To these a sixth,

$$(\rho^2-a\zeta-\tfrac{1}{2}a^2)^2=a^3\zeta,$$

was added by Stäckel [*Math Annalen*, 41, 1893], and this was supposed to complete the list of algebraic surfaces for which the integration by elliptic functions is possible. Yet another integrable case has been reported by Salkowski [*Diss. Jena*, 1904]. The equation is

$$\rho^6-8a^3\zeta\rho^2+2a^6=0.$$

**27.** *Ball rolling on a concave spherical surface.* We now suppose that the particle moving on the spherical surface is replaced by a ball which rolls on the surface without slipping. Let $r$ be the radius of the surface (denoted by $l$ above) and $a$ the radius of the ball. Take axes drawn from the point of contact O, (1) OD drawn towards the observer at right angles to the vertical plane containing the centre of the sphere and the point O, (2) OE tangential to the sphere and in the vertical plane just specified, and (3) the line OG joining O with the centre G of the sphere. These axes are drawn so that, as shown in Fig. 86, they constitute an ordinary right-handed system of axes. The inclination of GO to the downward vertical is $\theta$, and so the positive (counter-clock) turning of OG about OD is $\dot\theta$. We suppose that the vertical plane containing the centres of the sphere and ball is turning in azimuth with angular speed $\dot\psi$.

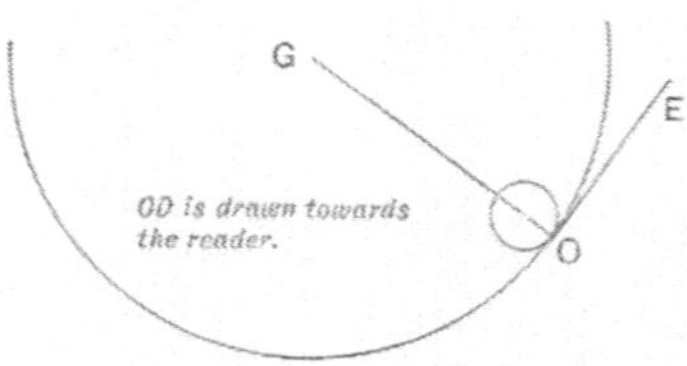

Fig. 86.

Let the azimuthal motion be zero, and the centre of the ball have turned through an angle $\theta$ about an axis parallel to OD, drawn through the centre of the sphere, and $\phi$ be the angle (measured of course the opposite way round) between the radii, the extremities of which were in contact at the beginning and end of this turning The ball has turned through the angle $\phi-\theta$, and its counter-clock angular speed is $-(\dot\phi-\dot\theta)$. But clearly $a\phi=r\theta$, and therefore

$$\dot\phi-\dot\theta=\frac{r-a}{a}\dot\theta.$$

The rate of production of angular momentum about OD due to this turning is

$$-m(k^2+a^2)(r-a)\ddot\theta/a.$$

This is the part due to acceleration.

Again, the components of angular velocity about OG and OE are $\dot\psi\cos\theta$, $-\dot\psi\sin\theta(r-a)/a$, due to the azimuthal motion, while if the angular speed of the spin about OG be $n$, the corresponding components of A.M. are $mk^2n$, $-m(k^2+a^2)\{(r-a)/a\}\dot\psi\sin\theta$. The rates of growth of A.M. about OD due to the motion are thus

$$mk^2n\dot\psi\sin\theta,\quad m(k^2+a^2)\{(r-a)/a\}\dot\psi^2\sin\theta\cos\theta.$$

The moment of forces about OD is $mga\sin\theta$. Hence we get for the equation of motion

$$(k^2+a^2)(r-a)\ddot\theta-\dot\psi\sin\theta\{(k^2+a^2)(r-a)\dot\psi\cos\theta+k^2na\}+ga^2\sin\theta=0. \quad\ldots\ldots(1)$$

The same process yields for the axis OE the equation

$$-(k^2+a^2)(r-a)\frac{d}{dt}(\dot\psi\sin^2\theta)-ak^2n\sin\theta\,.\,\dot\theta=0 \quad\ldots\ldots(2)$$

The constant A.M., G, say, about the downward vertical through O, is given by

$$(k^2+a^2)(r-a)\sin^2\theta\dot\psi-ak^2n\cos\theta=a\frac{G}{m}, \quad\ldots\ldots(3)$$

or, writing $G_1=G/m$,

$$\dot\psi=\frac{aG_1+ak^2n\cos\theta}{(k^2+a^2)(r-a)\sin^2\theta}. \quad\ldots\ldots(4)$$

Substituting for $\dot\psi$ in (1), we find

$$(k^2+a^2)(r-a)\ddot\theta-\frac{(aG_1+ak^2n\cos\theta)(aG_1\cos\theta+ak^2n)}{(k^2+a^2)(r-a)\sin^3\theta}+ga^2\sin\theta=0, \quad\ldots\ldots(5)$$

which we call the $\theta$-equation of motion. It will be seen that it is of precisely the same form as that used frequently above for the motion of a top.*

Multiplying (1) by $\dot\theta$, (2) by $\dot\psi$, adding and integrating, we obtain

$$(k^2+a^2)(r-a)(\dot\theta^2+\dot\psi^2\sin^2\theta)-2ga^2\cos\theta=\left(\frac{2E}{m}-k^2n^2\right)a, \quad\ldots\ldots(6)$$

where E denotes the total energy when the potential energy is taken as zero for $\theta=\frac{1}{2}\pi$, and therefore as $-mga$ when $\theta=0$.

* The $\theta$-equation of motion is (if $\theta$ is measured, as here, from the downward vertical)

$$-A\ddot\theta+(Cn+A\dot\psi\cos\theta)\dot\psi\sin\theta=Mgh\sin\theta.$$

But if G be the A.M. about the downward vertical, we have

$$G=-Cn\cos\theta+A\dot\psi\sin^2\theta.$$

This gives

$$\dot\psi\sin\theta=\frac{G+Cn\cos\theta}{A\sin\theta},\quad A\dot\psi\cos\theta+Cn=\frac{G\cos\theta+Cn}{\sin^2\theta}.$$

Hence the equation of motion becomes

$$A\ddot\theta-\frac{(G+Cn\cos\theta)(G\cos\theta+Cn)}{A\sin^3\theta}+Mgh\sin\theta=0,$$

which is exactly of the same form as (5).

If we write $z=\cos\theta$ and

$$\alpha=\frac{a(2E-mk^2n^2)}{m(k^2+a^2)(r-a)},\quad \beta=\frac{aG_1}{(k^2+a^2)(r-a)},\quad b=\frac{ak^2}{(k^2+a^2)(r-a)},\quad\text{(7)}$$

and give $a$ the new meaning $2ga^2/(k^2+a^2)(r-a)$, the equations of energy and momentum may be written

$$\left.\begin{aligned} \dot{z}^2+\dot{\psi}^2(1-z^2)^2&=(\alpha+az)(1-z^2),\\ \dot{\psi}(1-z^2)&=\beta+bnz.\end{aligned}\right\}\quad\text{(8)}$$

Eliminating $\dot{\psi}$, we get from these equations

$$\dot{z}^2=(\alpha+az)(1-z^2)-(\beta+bnz)^2=f(z).\quad\text{(9)}$$

In accordance with the notation of Darboux, followed by Greenhill, we can write (4) in the form

$$\dot{\psi}=\frac{a}{r-a}\frac{G_1+G'_1\cos\theta}{(k^2+a^2)\sin^2\theta}=\frac{h-h'}{1+\cos\theta}+\frac{h+h'}{1-\cos\theta},\quad\text{(10)}$$

where $\quad G'_1=k^2n,\quad h=\frac{1}{2}\frac{aG_1}{(r-a)(k^2+a^2)},\quad h'=\frac{1}{2}\frac{aG'_1}{(r-a)(k^2+a^2)}.\quad$ (11)

Thus we have (since $\theta$ is measured from the downward vertical)

$$\dot{\phi}=\omega=n+\dot{\psi}\cos\theta=\left\{1-\frac{ak^2}{(r-a)(k^2+a^2)}\right\}n+\frac{a(G'_1+G_1\cos\theta)}{(r-a)(k^2+a^2)\sin^2\theta},\quad\text{(12)}$$

that is

$$\dot{\phi}=2\left\{\frac{(r-a)(k^2+a^2)}{ak^2}-1\right\}h'+\frac{h'-h}{1+\cos\theta}+\frac{h+h'}{1-\cos\theta}.\quad\text{(13)}$$

From (10) and (12) we obtain

$$\left.\begin{aligned}\tfrac{1}{2}(\dot{\phi}+\dot{\psi})&=\left\{\frac{(r-a)(k^2+a^2)}{ak^2}-1\right\}h'+\frac{h+h'}{1-\cos\theta},\\ \tfrac{1}{2}(\dot{\phi}-\dot{\psi})&=\left\{\frac{(r-a)(k^2+a^2)}{ak^2}-1\right\}h'+\frac{h'-h}{1+\cos\theta}.\end{aligned}\right\}\quad\text{(14)}$$

The motion of the point of contact is thus exactly similar to that of the motion of a point on the axis of a top. In comparing the equations here found with those given in 4, XII, above for the top, it is to be remembered that here we have supposed $\theta$ measured from the downward vertical, which accounts for the appearance of $\alpha+az$ and $\beta+bnz$ in (8), instead of $\alpha-az$, $\beta-bnz$ [*loc. cit.*]. The calculation of $t$ and of $\psi$ in terms of $\theta$ (or in terms of $z$) can be carried through by the same elliptic function analysis as for the top [see 18 and 19, XII].

**28.** ***Reaction of the surface on the rolling ball.*** The spherical surface reacts on the ball with a thrust R which balances the normal component $mg\cos\theta$ of gravity, and supplies the force $m(u^2+v^2)/(r-a)$ required to give the acceleration $(u^2+v^2)/(r-a)$ towards the centre of the surface. But $u=(r-a)\dot{\theta}$, $v=\dot{\psi}(r-a)\sin\theta$, and thus we get

$$R=mg\cos\theta+m(r-a)(\dot{\theta}^2+\dot{\psi}^2\sin^2\theta).\quad\text{(1)}$$

But, by (6) of 27, this becomes

$$R=m\left[g\cos\theta+\left\{\left(\frac{2E}{m}-k^2n^2\right)a+2ga^2\cos\theta\right\}\frac{1}{k^2+a^2}\right],$$

or

$$\frac{R}{m}=\frac{g\cos\theta(k^2+3a^2)+2Ha}{k^2+a^2},\quad\text{(2)}$$

where H is the total energy per unit mass of the ball apart from rotation about the radius OG.

We infer that the reaction along the axis of a top, or gyroscopic flywheel, which is constrained to move about a fixed point in its axis of figure, is given by (2) adapted to suit this case, by the supposition that, while $k$ is the radius of gyration about any axis through the centroid of the top, at right angles to the axis, $a$ is now the distance of the fixed point from the centroid. If we write $l$ for the length of the simple pendulum equivalent to the compound pendulum formed by the top or gyroscope turning about a horizontal axle at the fixed point, we have $l=(k^2+a^2)/a$, and therefore

$$R=g\cos\theta+2\frac{a^2}{k^2+a^2}g\cos\theta+2H\frac{a}{k^2+a^2}$$

$$=g\cos\theta\frac{l+2a}{l}+2\frac{H}{l}. \qquad\qquad (3)$$

The problem of the spherical pendulum and of the motion of a particle on a surface of revolution is of much interest, and more space has been given to it here than is perhaps consistent with the plan of this work. The reader will find the elliptic function aspect of the subject studied very fully, with a wealth of results, in Greenhill's *Report on Gyroscopic Theory*.

## APPENDIX TO CHAPTER XV

In Figures 87, 89 are shown reproductions of photographs of curves actually described by the bob of a spherical pendulum, in experiments made by Professor A. G. Webster, of Worcester, Mass. A small glow lamp was attached to a brass ball which formed the bob

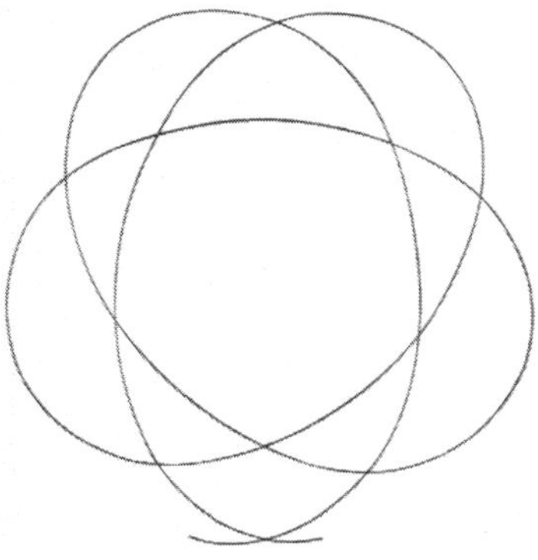

FIG. 87.

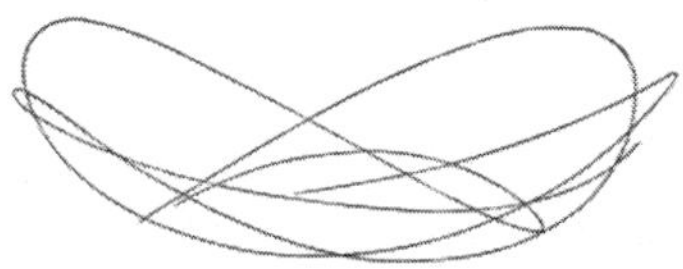

FIG. 88.

of the pendulum, and cameras were placed directly under and to one side of the swinging lamp. The experiments were made in a dark room, so that the cameras were without shutters, and the light was kept on only for a sufficient number of swings to give a complete curve.

In this way the plan (already shown in Fig. 82) and the elevation (Fig. 88) were obtained. Figure 87 shows for comparison a plan calculated from Fig. 82 by measuring the maximum and minimum radii, finding the roots of the cubic equation, and calculating values of $\psi$. It will be seen how closely the curves agree.

A comparison elevation was also calculated which agreed very closely with Fig. 88.

# CHAPTER XVI

## DYNAMICS OF A MOVING FRAME CONTAINING A FLYWHEEL

1. *General equations for moving origin and axes.* We have dealt with a considerable variety of problems, using in each case the system of axes which seemed most convenient for the particular purpose, and establishing the equations of motion by a direct appeal to elementary first principles. It is however desirable to set up formal equations of motion applicable to most of the gyrostatic combinations that occur in practice.

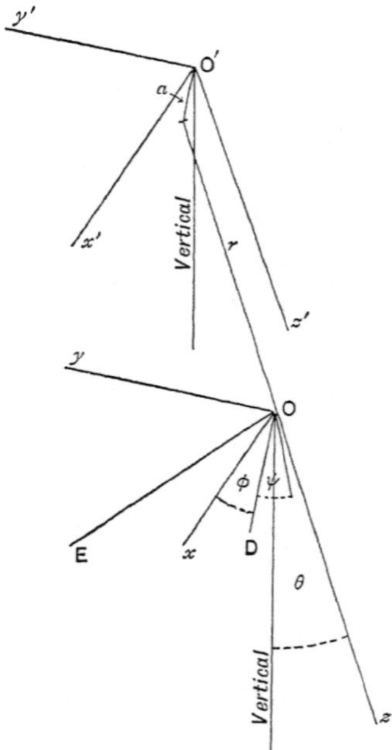


FIG. 89.

Let a frame of rectangular axes, O $(x, y, z)$ [see Fig. 89], be drawn from an origin O which is in motion with speeds $v_1$, $v_2$, $v_3$ along the instantaneous positions of these axes, and let the frame have angular speeds $\omega_1$, $\omega_2$, $\omega_3$ about these positions. A straight line OP is at the instant inclined to the axes at angles the cosines of which are $l$, $m$, $n$. These cosines are clearly the coordinates of a point A on OP at unit distance from O. Then, *apart from the motion of* O, the components of the velocity of A in space, with respect to the fixed axes with which O$(x, y, z)$ coincide, are

$$\dot{l} - m\omega_3 + n\omega_2, \quad \dot{m} - n\omega_1 + l\omega_3,$$
$$\dot{n} - l\omega_2 + m\omega_1.$$

If these are zero the point A has only the motion of O, and so OA remains unchanged in direction. For this the conditions are

$$\dot{l} = m\omega_3 - n\omega_2, \quad \dot{m} = n\omega_1 - l\omega_3, \quad \dot{n} = l\omega_2 - m\omega_1. \quad \text{..............(1)}$$

Now take a fixed origin O′, and let axes O′$(x', y', z')$ be drawn from it which move so as always to be parallel to $O(x, y, z)$. Let $\mathfrak{m}_1$, $\mathfrak{m}_2$, $\mathfrak{m}_3$ denote components of momentum of a body taken parallel to the axes $O'(x', y', z')$ in their instantaneous position. The momentum resolved in the direction OP is

$$\mathfrak{M} = l\mathfrak{m}_1 + m\mathfrak{m}_2 + n\mathfrak{m}_3. \quad \ldots\ldots\ldots\ldots\ldots\ldots\ldots\ldots\ldots(2)$$

If OP is fixed in direction, we have

$$\frac{d\mathfrak{M}}{dt} = l(\dot{\mathfrak{m}}_1 - \omega_3\mathfrak{m}_2 + \omega_2\mathfrak{m}_3) + m(\dot{\mathfrak{m}}_2 - \omega_1\mathfrak{m}_3 + \omega_3\mathfrak{m}_1) + n(\dot{\mathfrak{m}}_3 - \omega_2\mathfrak{m}_1 + \omega_1\mathfrak{m}_2), \ \ldots(3)$$

since the total time rates of variation of $l$, $m$, $n$ are zero, according to (1).

Thus if X, Y, Z be the components of applied force along the axes, the equations of linear momentum are

$$\dot{\mathfrak{m}}_1 - \omega_3\mathfrak{m}_2 + \omega_2\mathfrak{m}_3 = X, \quad \dot{\mathfrak{m}}_2 - \omega_1\mathfrak{m}_3 + \omega_3\mathfrak{m}_1 = Y, \quad \dot{\mathfrak{m}}_3 - \omega_2\mathfrak{m}_1 + \omega_1\mathfrak{m}_2 = Z. \ \ldots(4)$$

It remains to form the equations of A.M. Then we shall be able to calculate in any tractable case the motion of the system, and the reactions on the supporting framework, aeroplane or airship, or whatever it may be. Let $h_1$, $h_2$, $h_3$ be the components of A.M due to the turning about the moving axes $O(x, y, z)$. The components of rate of change of these are

$$\dot{h}_1 - \omega_3 h_2 + \omega_2 h_3, \quad \dot{h}_2 - \omega_1 h_3 + \omega_3 h_1, \quad \dot{h}_3 - \omega_2 h_1 + \omega_1 h_2.$$

To these we have to add the rates of change of A.M. arising from the motion of O. Denoting by $\xi, \eta, \zeta$ the coordinates of O with respect to the axes $O'(\xi, \eta, \zeta)$, and by $v_1$, $v_2$, $v_3$ the components of velocity of O with reference to fixed axes with which $O'(x', y', z')$ at the instant coincide, that is

$$v_1 = \dot{\xi} - \omega_3\eta + \omega_2\zeta, \quad v_2 = \dot{\eta} - \omega_1\zeta + \omega_3\xi, \quad v_3 = \dot{\zeta} - \omega_2\xi + \omega_1\eta, \ \ldots\ldots\ldots(5)$$

and the whole mass of the body by M, the components of momentum due to the motion of O are

$$Mv_1 = M(\dot{\xi} - \omega_3\eta + \omega_2\zeta), \ Mv_2 = M(\dot{\eta} - \omega_1\zeta + \omega_3\xi), \ Mv_3 = M(\dot{\zeta} - \omega_2\xi + \omega_1\eta). \ \ldots(5')$$

If the origins coincide *at the instant*, $Mv_1 = M\dot{\xi}$, $Mv_2 = M\dot{\eta}$, $Mv_3 = M\dot{\zeta}$.

Now taking components of A.M. about parallel axes through the centroid G, and then those of the A.M. arising from the motion of G, putting $\bar{x}+\xi$, $\bar{y}+\eta$, $\bar{z}+\xi$ for the component distances of G from O′, we find that the A.M. about $O'x'$ is

$$H_1 = h_1 + M\{\gamma\eta - \beta\zeta + v_3(\bar{y}+\eta) - v_2(\bar{z}+\zeta)\},$$

where $\alpha, \beta, \gamma = \bar{z}\omega_2 - \bar{y}\omega_3, \ \bar{x}\omega_3 - \bar{z}\omega_1, \ \bar{y}\omega_1 - \bar{x}\omega_2$. $H_2$, $H_3$ can be written down by symmetry.

We now calculate $\dot{H}_1 - \omega_3 H_2 + \omega_2 H_3$, which is the rate of growth of A.M. about $O'x'$. To find that for $Ox$, we have only to put, in the result, $\zeta, \eta, \xi$ equal to zero after the differentiation has been carried out, and equate to the applied couple P; and similarly for $Oy$, $Oz$. The equation for $Ox$ is,

$$\dot{h}_1 - \omega_3 h_2 + \omega_2 h_3 - \left\{\frac{d}{dt}(Mv_2\bar{z}) - Mv_3\bar{x}\omega_3 + Mv_1\bar{y}\omega_2\right\}$$
$$+ \left\{\frac{d}{dt}(Mv_3\bar{y}) - Mv_1\bar{z}\omega_3 + Mv_2\bar{x}\omega_2\right\} + M(-\beta v_3 + \gamma v_2) = P, \ \ldots\ldots(6)$$

If the body be rigid, $\bar{x}$, $\bar{y}$, $\bar{z}$ do not change as the body moves, and therefore the momentum terms affected by differentiation simplify to $-\mathrm{M}\dot{v}_2\bar{z}+\mathrm{M}\dot{v}_3\bar{y}$, with similar terms for the other equations.

If A, B, C be the moments, and D, E, F the products of inertia for the moving axes at the instant, we have

$$h_1=\mathrm{A}\omega_1-\mathrm{F}\omega_2-\mathrm{E}\omega_3,\quad h_2=\mathrm{B}\omega_2-\mathrm{D}\omega_3-\mathrm{F}\omega_1,\quad h_3=\mathrm{C}\omega_3-\mathrm{E}\omega_1-\mathrm{D}\omega_2.\;\ldots(7)$$

We shall find below some applications of these formal equations to practical cases.

2. *Expanding or contracting body. Mean axes.* When the body is invariable the rate of change of A.M. arises only from angular acceleration combined with variation of the motion of the centroid. There are however cases in which the mass of the body or system undergoes change, or in which the configuration is altered in consequence, it may be, of heating or cooling. We have then, in forming the equations of motion, to equate the total time-rate of variation of A.M., from all causes, to the sum of the moments of external forces, together with any rate of change of A.M. directly due to such action between the system and external matter as the interchange of matter bringing or carrying with it A.M.

In the case of a body changing by expansion or contraction, or in any other way, the equations of the form (6), 1, hold, provided the terms involving differentiation with respect to the time are properly estimated. As an example, consider a body rotating about a fixed axis, and receiving matter from external space in such a way that the impacts of the particles have, either individually or in the aggregate, no moment about the axis at any instant. Then, if $\dot{\theta}$ be the angular speed about the fixed axis, and $\Sigma(mr^2)$ the corresponding moment of inertia, we have

$$\frac{d}{dt}\{\dot{\theta}\Sigma(mr^2)\}=\ddot{\theta}\Sigma(mr^2)+\dot{\theta}\frac{d}{dt}\Sigma(mr^2);\;\ldots\ldots\ldots\ldots(1)$$

and, if no external forces which have moments about the axis act upon the body, this is zero. Of course if we consider the body as it exists at any instant, and the infinitesimal layer of matter deposited upon it in time $t$, we see that, to the first order of small quantities, the additional layer of matter has been given the angular speed about the axis that the body has at time $t$, while the small impulse moment required to do that, and applied by the body, involves a reaction moment on the body which produces diminution of the A.M. about the axis. The loss of A.M. by the body is thus equal to the gain of A.M. by the mass added. We have then to reckon $\dot{\theta}d(\Sigma mr^2)/dt$ as a retarding moment acting on the original body.

In the case of a body changing in configuration, in a known manner, we can choose axes of reference, $O(\xi,\eta,\zeta)$, in such a way that, if the body were to become rigid at any instant, the axes would during the subsequent element of time be fixed in the body. Thus if we imagine a rigid body coinciding with the changing body at the time $t$, and having the same moments, A, B, C, and products, D, E, F, of inertia as the body has at the instant for any system of axes, the A.M. about each axis is then the same for the imaginary rigid body as for the changing body.

Let $\omega_1, \omega_2, \omega_3$ be the angular speeds of the axes, and $\xi, \eta, \zeta$ the coordinates of a point; the velocity components of the point are

$$\dot{\xi}-\omega_3\eta+\omega_2\zeta, \quad \dot{\eta}-\omega_1\zeta+\omega_3\xi, \quad \dot{\zeta}-\omega_2\xi+\omega_1\eta.$$

The components of A.M. are given by

$$h_1=\Sigma[m\{\eta(\dot{\zeta}-\omega_2\xi+\omega_1\eta)-\zeta(\dot{\eta}-\omega_1\zeta+\omega_3\xi)\}]$$
$$=\Sigma\{m(\eta\dot{\zeta}-\zeta\dot{\eta})\}+\mathrm{A}\omega_1-\mathrm{E}\omega_3-\mathrm{F}\omega_2\} \quad \ldots\ldots(2)$$

and two similar equations. For axes fulfilling the condition stated above the quantities $\dot{\xi}, \dot{\eta}, \dot{\zeta}$ are zero, and so for these we have

$$h_1=\mathrm{A}\omega_1-\mathrm{E}\omega_3-\mathrm{F}\omega_2, \quad h_2=\mathrm{B}\omega_2-\mathrm{F}\omega_1-\mathrm{D}\omega_3, \quad h_3=\mathrm{C}\omega_3-\mathrm{D}\omega_2-\mathrm{E}\omega_1. \ldots(3)$$

Such axes have been called *mean axes* [see Tisserand, *Mécanique Céleste*, t. 2, chap. 30]. They are not uniquely determinate, as may be seen from an instance suggested by Routh [*Advanced Rigid Dynamics*, 6th edition, § 22]. Let a body be initially at rest, and its parts be set in relative motion by internal changes. The A.M. about any axis fixed in space is zero, and so any set of rectangular axes given in position are mean axes. If then the body and axes be given both the same motion, the axes will remain mean axes for the resultant of the superimposed movements.

In connection with the subject of internal changes certain special cases are of importance, and we give here the equations adapted to the more important of these. Denoting by $\mathfrak{h}_1, \mathfrak{h}_2, \mathfrak{h}_3$ the components, about any other axes $\mathrm{O}(x, y, z)$, due to the terms

$$\Sigma m(\eta\dot{\zeta}-\zeta\dot{\eta}), \quad \Sigma m(\zeta\dot{\xi}-\xi\dot{\zeta}), \quad \Sigma m(\xi\dot{\eta}-\eta\dot{\xi}),$$

we have with reference to that set of axes,

$$\left.\begin{array}{l} h_1=\mathfrak{h}_1+\mathrm{A}\omega_1-\mathrm{E}\omega_3-\mathrm{F}\omega_2, \quad h_2=\mathfrak{h}_2+\mathrm{B}\omega_2-\mathrm{F}\omega_1-\mathrm{D}\omega_3, \\ h_3=\mathfrak{h}_3+\mathrm{C}\omega_3-\mathrm{D}\omega_2-\mathrm{E}\omega_1. \end{array}\right\} \ldots\ldots \ldots.(4)$$

Here $\omega_1, \omega_2, \omega_3$ are now the angular speeds *of the body*, as it exists at the instant, about the axes $\mathrm{O}(x, y, z)$: in the general case the angular speeds of the axes are distinct from these, and may be denoted by $\theta_1, \theta_2, \theta_3$. A, B, C, D, E, F also refer to the axes $\mathrm{O}(x, y, z)$. The equations of motion are

$$\frac{d}{dt}(\mathfrak{h}_1+\mathrm{A}\omega_1-\mathrm{E}\omega_3-\mathrm{F}\omega_2)-(\mathfrak{h}_2+\mathrm{B}\omega_2-\mathrm{F}\omega_1-\mathrm{D}\omega_3)\theta_3$$
$$+(\mathfrak{h}_3+\mathrm{C}\omega_3-\mathrm{D}\omega_2-\mathrm{E}\omega_1)\theta_2=\mathrm{P}, \quad \ldots\ldots\ldots(5)$$

with two similar equations. If then we suppose that $(\theta_1, \theta_2, \theta_3)=(\omega_1, \omega_2, \omega_3)$ and that the axes are mean axes, we get

$$\frac{d}{dt}(\mathfrak{h}_1+\mathrm{A}\omega_1-\mathrm{E}\omega_3-\mathrm{F}\omega_2)-(\mathrm{B}-\mathrm{C})\omega_2\omega_3-\mathrm{D}(\omega_2^2-\omega_3^2)-\mathrm{E}\omega_1\omega_2+\mathrm{F}\omega_1\omega_3=\mathrm{P}, \quad (6)$$

with two similar equations. If the axes are not mean axes the term $-\mathfrak{h}_2\omega_3+\mathfrak{h}_3\omega_2$ must be added on the left of (6), and corresponding terms added in the two equations not here written down.

If at the instant considered the axes of reference and the principal axes of the body coincide, $D=E=F=0$, and the equations of motion (5) become

$$\frac{d}{dt}(\mathfrak{h}_1+A\omega_1)-\dot{E}\omega_3-\dot{F}\omega_2-(\mathfrak{h}_2+B\omega_2)\theta_3+(\mathfrak{h}_3+C\omega_3)\theta_2=P, \quad\text{.........(7)}$$

with two similar equations. The terms $-\dot{E}\omega_3-\dot{F}\omega_2$ are sometimes omitted, but it seems necessary to include them, for though D, E, F are zero at the instant, the axes of reference are separating from the principal axes, and so $\dot{D}$, $\dot{E}$, $\dot{F}$ are not necessarily zero.

The component angular speeds $\alpha_1$, $\alpha_2$, $\alpha_3$ of separation of the axes $O(x, y, z)$ from coincidence with the principal axes, give $\theta_1=\omega_1+\alpha_1$, $\theta_2=\omega_2+\alpha_2$, $\theta_3=\omega_3+\alpha_3$. Then (7) becomes

$$\frac{d}{dt}(\mathfrak{h}_1+A\omega_1)-\dot{E}\omega_3-\dot{F}\omega_2-(B-C)\omega_2\omega_3-B\omega_2\alpha_3+C\omega_3\alpha_2$$
$$-\mathfrak{h}_2(\omega_3+\alpha_3)+\mathfrak{h}_3(\omega_2+\alpha_2)=P. \quad\text{...........................(8)}$$

An important case is that in which the axis of rotation is nearly coincident with a principal axis, say that of $\zeta$, for which the equation of motion is

$$\frac{d}{dt}(\mathfrak{h}_3+C\omega_3)-\dot{D}\omega_2-\dot{E}\omega_1-(A-B)\omega_1\omega_2-A\omega_1\alpha_2+B\omega_2\alpha_1$$
$$-\mathfrak{h}_1(\omega_2+\alpha_2)+\mathfrak{h}_2(\omega_1+\alpha_1)=R. \quad\text{...........................(9)}$$

Then $\omega_1$, $\omega_2$ are both small. If the internal changes are small and take place slowly, $\omega_1$, $\omega_2$ remain small, and $\dot{D}$, $\dot{E}$, $\dot{F}$ are also small. For a set of axes nearly coincident with the principal axes, which we suppose are only slightly displaced in the body by internal changes, $\alpha_1$, $\alpha_2$, $\alpha_3$ are also small. If then $\mathfrak{h}_1$, $\mathfrak{h}_2$, $\mathfrak{h}_3$ are also small and $R=0$, (9) becomes

$$\frac{d(C\omega_3)}{dt}=0, \quad\text{.............................(10)}$$

to quantities of the second order of smallness, and $C\omega_3$ is approximately a constant. Calling this constant $h$, we get for the other two equations of motion,

$$\left.\begin{aligned}\frac{d}{dt}(A\omega_1)-\left(\frac{B-C}{BC}h+\alpha_3\right)B\omega_2+h\alpha_2=P,\\ \frac{d}{dt}(B\omega_2)-\left(\frac{C-A}{CA}h-\alpha_3\right)A\omega_1-h\alpha_1=Q\end{aligned}\right\} \quad\text{..................(11)}$$

Equations similar to (8) were used by Sir George Darwin in his paper "On the Influence of Geological Changes on the Earth's Axis of Rotation" [*Phil. Trans.*, 1876]; equations similar to (11) were given by Lord Kelvin in an appendix to that paper.

Let the motion be referred to the axes $O(x, y, z)$ drawn from the fixed origin O, and the coordinates of a point at unit distance from O on the axis $Oz$ be $\xi$, $\eta$, 1 (which we may call the angular coordinates of $Oz$), so that $Oz$ nearly coincides with $O\zeta$, and $\xi$, $\eta$ are small. The axes $O(x, y, z)$ are moving with respect to the axes $O(\xi, \eta, \zeta)$, but slowly, so that $\alpha_1$, $\alpha_2$, $\alpha_3$ are small. Clearly we have

$$\alpha_1=-\dot{\eta}, \quad \alpha_2=\dot{\xi}. \quad\text{..............................(12)}$$

Let now the body be symmetrical about $Oz$, and A, B, C change so slowly that terms in $\dot{A}$, $\dot{B}$ may be neglected. Putting $\mu = h(C-A)/CA$, and $\nu = -h/A$, we get (11) in the form

$$\dot{\omega}_1 + \mu\omega_2 + \nu\dot{\xi} = \frac{P}{A}, \quad \dot{\omega}_2 - \mu\omega_1 + \nu\dot{\eta} = \frac{Q}{A}, \quad \text{......................(13)}$$

which, it will be observed, are equations with gyrostatic terms $\mu\omega_2$, $-\mu\omega_1$.

We come now to an important gyrostatic application of these results—the investigation of the motion of the resultant axis of rotation of a body like the earth, the axis of figure (practically coincident with the axis of rotation) of which is changing its position in the body in consequence of slow internal changes, the effects we may suppose of the yearly cycle of meteorological phenomena. Let us suppose that the angular coordinates of OC are given by $\xi = p\cos mt$, $\eta = q\sin mt$, so that the point in question moves in a small ellipse about the axis $O\zeta$, the mean position of the axis of figure. Putting $P = Q = 0$ in (13), we get

$$\dot{\omega}_1 + \mu\omega_2 - \nu pm\sin mt = 0, \quad \dot{\omega}_2 - \mu\omega_1 + \nu qm\cos mt = 0. \quad \text{..........(14)}$$

If we differentiate the first of these with respect to $t$, and substitute for $\dot{\omega}_2$ from the second, we get

$$\ddot{\omega}_1 + \mu^2\omega_1 - \nu m(mp + \mu q)\cos mt = 0, \quad \text{......................(15)}$$

of which the complete solution is, with K and $\epsilon$ as arbitrary constants,

$$\omega_1 = \frac{\nu m(mp + \mu q)}{\mu^2 - m^2}\cos mt + K\cos(\mu t + \epsilon). \quad \text{..................(16)}$$

Thus we have

$$\dot{\omega}_1 = -\frac{\nu m^2(mp + \mu q)}{\mu^2 - m^2}\sin mt - \mu K\sin(\mu t + \epsilon), \quad \text{............(17)}$$

which, by the first of (14), gives

$$\omega_2 = \frac{\nu m(\mu p + mq)}{\mu^2 - m^2}\sin mt + K\sin(\mu t + \epsilon). \quad \text{..................(18)}$$

Now the angular coordinates of the instantaneous axis are $\xi + \omega_1/n$, $\eta + \omega_2/n$, where $n\ (= \omega_3)$ is the angular speed of the earth about OC. We get, since $n = \nu - \mu$,

$$\xi + \frac{\omega_1}{n} = \frac{1}{\mu^2 - m^2}\left\{\mu^2 p + m\mu q + \frac{m}{n}(\mu^2 q + m\mu p)\right\}\cos mt + \frac{K}{n}\cos(\mu t + \epsilon), \quad \text{...(19)}$$

where K and $\epsilon$ are the constants of integration.

Similarly we obtain

$$\eta + \frac{\omega_2}{n} = \frac{1}{\mu^2 - m^2}\left\{\mu^2 q + m\mu p + \frac{m}{n}(\mu^2 p + m\mu q)\right\}\sin mt + \frac{K}{n}\sin(\mu t + \epsilon). \quad \text{...(20)}$$

These results are due to Helmert [*Astron. Nachr.*, Bd. 126]. It is to be observed that $\mu$ is small compared with $\nu$, and that $m/(\nu - \mu)$ is small. Hence the second term of the multipliers of $\cos mt$ and $\sin mt$ in (19) and (20) may be omitted. For the earth (supposed unyielding) $2\pi/\mu$ is about 306 days, or 10 months, and $2\pi/\mu$ is a year, while $2\pi/\nu$ is about a day.

The effect of the terms $(\mathrm{K}/n)\cos(\mu t+\epsilon)$, $(\mathrm{K}/n)\sin(\mu t+\epsilon)$ has been discussed in Chapters X and XI. They form what is called the free precession, which, if there were no yielding of the earth, would be accomplished in a period of 306 days.

3. *Expanding or contracting body unacted on by force.* If an expanding or contracting body is not acted on by any force the Eulerian equations of the motion are of the form

$$\frac{d}{dt}(\mathrm{A}\omega_1)-(\mathrm{B}-\mathrm{C})\omega_2\omega_3=0,\quad \frac{d}{dt}(\mathrm{B}\omega_2)-(\mathrm{C}-\mathrm{A})\omega_3\omega_1=0,$$

$$\frac{d}{dt}(\mathrm{C}\omega_3)-(\mathrm{A}-\mathrm{B})\omega_1\omega_2=0,\ \ldots\ldots\ldots\ldots(1)$$

since A, B, C vary with the time. Let us suppose that A, B, C fulfil the condition

$$\mathrm{A},\ \mathrm{B},\ \mathrm{C}=(\mathrm{A}_0,\ \mathrm{B}_0,\ \mathrm{C}_0)f(t).\ \ldots\ldots\ldots\ldots(2)$$

The equations of motion become

$$\mathrm{A}_0\dot{p}f(t)-(\mathrm{B}_0-\mathrm{C}_0)qr=0,\ \text{etc.},\ \ldots\ldots\ldots\ldots(3)$$

where $p, q, r=(\omega_1, \omega_2, \omega_3)f(t)$. If then we take a new independent variable $\tau$, such that $d\tau=dt/f(t)$, we obtain for the equations of motion

$$\mathrm{A}_0\frac{dp}{d\tau}-(\mathrm{B}_0-\mathrm{C}_0)qr=0,\ \text{etc.},\ \ldots\ldots\ldots\ldots(4)$$

which are of the usual form.

We observe that the A.M. of the body remains unaltered while the body changes, but that the kinetic energy increases or diminishes according as $f(t)$ diminishes or increases with $t$. The first result follows from the fact that

$$\mathrm{A}^2\omega_1^{\ 2}+\mathrm{B}^2\omega_2^{\ 2}+\mathrm{C}^2\omega_3^{\ 2}=\mathrm{A}_0^{\ 2}p^2+\mathrm{B}_0^{\ 2}r^2+\mathrm{C}_0^{\ 2}r^2,\ \ldots\ldots\ldots\ldots(5)$$

the second from the relation

$$\tfrac{1}{2}(\mathrm{A}\omega_1^{\ 2}+\mathrm{B}\omega_2^{\ 2}+\mathrm{C}\omega_3^{\ 2})=\tfrac{1}{2}\frac{\mathrm{A}_0p^2+\mathrm{B}_0q^2+\mathrm{C}_0r^2}{f(t)},\ \ldots\ldots\ldots\ldots(6)$$

since from (4) it follows that

$$\mathrm{A}_0p\dot{p}+\mathrm{B}_0q\dot{q}+\mathrm{C}_0q\dot{q}=0,$$

so that the numerator on the right of (6) is constant.

The remark made above as to regarding terms arising from changes of configuration as moments of forces is illustrated here. We can write the equations of motion in the form

$$\mathrm{A}\dot{\omega}_1-(\mathrm{B}-\mathrm{C})\omega_2\omega_3=-\dot{\mathrm{A}}\omega_1=-\mathrm{A}\omega_1\frac{f'(t)}{f(t)},\ \text{etc.}\ \ldots\ldots\ldots\ldots(7)$$

They are therefore the equations of motion of a body acted on by couples (about the principal axes) which are equal and opposite in sign to the corresponding components of A.M., each multiplied by $f'(t)/f(t)$. The couples are actually equal to the components of A.M. if $f(t)=e^{-t}$, and to these components with the sign changed if $f(t)=e^t$.

If we suppose that $f(t)=e^{-\lambda t}$ and that $\tau$ and $t$ start from zero together, we get

$$\tau=\frac{1}{\lambda}(e^{\lambda t}-1). \quad\text{.......................}(8)$$

The equations of motion are then

$$A\dot{\omega}_1-(B-C)\omega_2\omega_3=\lambda A\omega_1,\ \text{etc.} \quad\text{.......................}(9)$$

If $f(t)=1-at$, so that $f(t)=1$ when $t=0$, we have, supposing $\tau=0$ when $t=0$,

$$a\tau=\log\frac{1}{1-at}, \quad \text{and} \quad at=1-e^{-a\tau}. \quad\text{.................}(10)$$

Finally we may take $f(t)=1-a^2t^2$, so that, if $\tau=0$ when $t=0$,

$$\tau=\frac{1}{2a}\log\frac{1+at}{1-at}. \quad\text{.......................}(11)$$

In this latter case $$at=\tanh(a\tau). \quad\text{.......................}(12)$$

4. *Rigid body containing a flywheel and turning about an axle.* We now consider some applications of the system of equations set forth in 1 above. The most important practical case is that of a rigid body turning like a pendulum about an axle in any position while the axle is turning about a fixed vertical. Let $\theta$ be the inclination of the axle to the downward vertical, and $\phi$ the angle which a plane containing the axle and fixed in the body makes with the vertical plane containing the axle. Denote the length of the common perpendicular to the fixed vertical and the axle of the pendulum by $a$, and the distance of O along the axle from that perpendicular by $r$. Figure 89 (or Fig. 12 and (2), 3, III, with $\omega$ increased by $\frac{1}{2}\pi$) gives

$$\omega_1=\dot{\theta}\cos\phi+\dot{\psi}\sin\theta\sin\phi, \quad \omega_2=-\dot{\theta}\sin\phi+\dot{\psi}\sin\theta\cos\phi,$$
$$\omega_3=n=\dot{\psi}\cos\theta+\dot{\phi}. \quad\text{.................}(1)$$

Also, by the figure, the coordinates of O with respect to $O'(x', y', z')$ are given by

$$\xi=a\cos\phi, \quad \eta=-a\sin\phi, \quad \zeta=r. \quad\text{.................}(2)$$

Thus we obtain for the component velocities of the moving origin O

$$\left.\begin{aligned} v_1&=\dot{\xi}-\omega_3\eta+\omega_2\zeta=-r\dot{\theta}\sin\phi+\dot{\psi}(r\sin\theta\cos\phi+a\cos\theta\sin\phi),\\ v_2&=\dot{\eta}-\omega_1\zeta+\omega_3\xi=-r\dot{\theta}\cos\phi+\dot{\psi}(-r\sin\theta\sin\phi+a\cos\theta\cos\phi),\\ v_3&=\dot{\zeta}-\omega_2\xi+\omega_1\eta=-a\mu\sin\theta. \end{aligned}\right\}\text{.....}(3)$$

Since, besides the motion due to the turning about the axes $O(x, y, z)$, the whole body has the component velocities $v_1$, $v_2$, $v_3$, we get, writing $\bar{x}$, $\bar{y}$, $\bar{z}$ as before for the coordinates of the centroid with reference to these axes, the equations of motion [see (6), 1]

$$\dot{h}_1-h_2\omega_3+h_3\omega_2+M\{\dot{v}_3\bar{y}-\dot{v}_2\bar{z}-(v_1\bar{z}-v_3\bar{x})\omega_3+(v_2\bar{x}-v_1\bar{y})\omega_2-\beta v_3+\gamma v_2\}=P, \quad (4)$$

with two similar equations. If A, B, C be the moments and D, E, F the products of inertia with respect to $O(x, y, z)$,

$$h_1=A\omega_1-F\omega_2-E\omega_3, \quad h_2=B\omega_2-D\omega_3-F\omega_1, \quad h_3=C\omega_3-E\omega_1-D\omega_2. \ \ldots(5)$$

We consider the third of the equations typified by (4) in its application to the case in which the system is turning with uniform angular speed $\dot{\psi}=\mu$

about the fixed vertical, while $a$ and $r$ remain unaltered in length. Thus we get

$$C\ddot{\phi}-\tfrac{1}{2}(A-B)\mu^2\sin^2\theta\sin 2\phi+F\mu^2\sin^2\theta\cos 2\phi$$
$$-\mu^2\sin\theta\cos\theta\{(D+M\bar{y}r)\sin\phi-(E+M\bar{x}r)\cos\phi\}+M\mu^2a(\bar{x}\sin\phi+\bar{y}\cos\phi)$$
$$=-Mg\sin\theta(\bar{x}\cos\phi-\bar{y}\sin\phi). \quad\text{......(6)}$$

If the body contains flywheels in rotation, unretarded by frictional couples at the bearings or elsewhere, terms must be added on the left in (4) arising from the components $K_1, K_2, K_3$ of A.M., about the axes $O(x, y, z)$, contributed by the flywheels. The groups of terms to be inserted in the respective equations are

$$-K_2\omega_3+K_3\omega_2,\quad -K_3\omega_1+K_1\omega_3,\quad -K_1\omega_2+K_2\omega_1,$$

for the terms $\dot{K}_1, \dot{K}_2, \dot{K}_3$ are zero. Thus on the left of (6) we must add the expression

$$-K_1\mu\sin\theta\cos\phi+K_2\mu\sin\theta\sin\phi.$$

Of course the fulfilment of the condition that $\mu$ and $\theta$ should be constant requires the application of constraint to the pendulum, and this constraint will give a reaction on the supporting system, to be calculated with the other reactions due to the motion.

When the two axles intersect, and the origin O is at the intersection, the values of $a$ and $r$ are zero, and (6) reduces to

$$C\ddot{\phi}-\tfrac{1}{2}(A-B)\mu^2\sin^2\theta\sin 2\phi-\mu\sin\theta(K_1\cos\phi-K_2\sin\phi)$$
$$+F\mu^2\sin^2\theta\cos 2\phi-\mu^2\sin\theta\cos\theta(D\sin\phi-E\cos\phi)$$
$$=-Mg\sin\theta(\bar{x}\cos\phi-\bar{y}\sin\phi). \quad\text{......(7)}$$

If the axis $Oz$ is a principal axis, $F=0$, and by turning the other two axes round $Oz$ we can cause D and E also to vanish, so that (7) becomes

$$C\ddot{\phi}-\tfrac{1}{2}(A-B)\mu^2\sin^2\theta\sin 2\phi-\mu\sin\theta(K_1\cos\phi-K_2\sin\phi)$$
$$=-Mg\sin\theta(\bar{x}\cos\phi-\bar{y}\sin\phi). \quad\text{......(8)}$$

If $\mu^2$ be very small, this becomes

$$C\ddot{\phi}-\mu\sin\theta(K_1\cos\phi-K_2\sin\phi)=-Mgs\sin\theta. \quad\text{......(8')}$$

where $s=\bar{x}\cos\phi-\bar{y}\sin\phi$.

As an example we take a flywheel pivoted within a ring or case (which is symmetrical about the axis $Oz$), with its rotation axis inclined to $Oz$ (not in general the vertical) at an angle $\alpha$. The centroid is a point G, below O, on the axis of rotation, and the distance of G from O the common centre of the wheel and case taken as origin is $h$. We measure $\phi$ from the position of the system when $Oz$ and OZ (the vertical) are in plane with the axis of the flywheel. We have then, by Fig. 90, $\bar{x}=0$, $\bar{y}=-h\sin\alpha$.

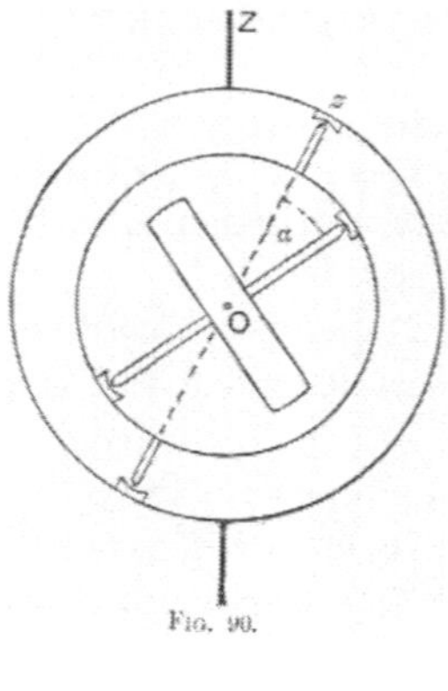

FIG. 90.

If the moment of inertia about an axis through the centre of the wheel, at right angles to the rotation axis, be A, that about the rotation axis be C, while C′ and A′ are moments of inertia of th case about $Oz$, and about an axis through O at right angles to $Oz$, the moment of inertia about $Oz$ of the whole system, with the flywheel clamped so as to be in one rigid piece with the ring or case, is $C\cos^2\alpha + A\sin^2\alpha + C'$. The moment of inertia about $Ox$ in the same circumstances is $C\sin^2\alpha + A\cos^2\alpha + A'$, and that about OY is $A + A'$. [Thus the meanings of A and C are altered.]

But if the flywheel be unclamped and free to rotate about its axis, the moment of inertia about $Oz$ is $A\sin^2\alpha + C'$, while that about $Ox$ is

$$A\cos^2\alpha + A'.$$

As to the products of inertia, these are zero for the case and axes $O(x, y, z)$. Hence only products of inertia arising from the wheel have to be considered. From what has been stated above as to the reckoning of $\phi$, we see that E and F are zero. The product D arises from the flywheel alone, and is given at once by the theorem, that for any given origin the expression

$$AB + BC + CA - D^2 - E^2 - F^2$$

does not depend on the axes chosen, taken along with the fact that $E^2 = F^2 = 0$. Thus we have for the flywheel free to rotate

$$D^2 = A^2\sin^2\alpha\cos^2\alpha. \quad \text{......(9)}$$

[If the flywheel is clamped, $D^2 = (A - C)^2\sin^2\alpha\cos^2\alpha$.]

If K denote the A.M. of the flywheel about its axis, the equation of motion, (7) above, becomes

$$(A\sin^2\alpha + C')\ddot{\phi} + \tfrac{1}{2}A\mu^2\sin^2\alpha\sin^2\theta\sin 2\phi - A\mu^2\sin\alpha\cos\alpha\sin\theta\cos\theta\sin\phi$$
$$+ (Mgh - K\mu)\sin\alpha\sin\theta\sin\phi = 0, \quad \text{......(10)}$$

or, if we neglect $C'$,

$$A\sin\alpha\,.\,\ddot{\phi} + \tfrac{1}{2}A\mu^2\sin\alpha\sin^2\theta\sin 2\phi - A\mu^2\cos\alpha\sin\theta\cos\theta\sin\phi$$
$$+ (Mgh - K\mu)\sin\theta\sin\phi = 0. \quad \text{....(11)}$$

For small values of $\mu$, for example that of the earth's rotation about a given vertical, (10) gives small oscillations of the period

$$2\pi\left\{\frac{A\sin^2\alpha + C}{(Mgh - K\mu)\sin\alpha\cos\theta}\right\}^{\frac{1}{2}},$$

or, if $C'$ be neglected, in the period

$$2\pi\left\{\frac{A\sin\alpha}{(Mgh - K\mu)\sin\theta}\right\}^{\frac{1}{2}}.$$

If $K\mu > Mgh$, this period is imaginary. But then if the flywheel were turned through 180°, oscillations in the shorter period given by these expressions with the sign of $K\mu$ reversed would be performed. With the flywheel thus directed it would then be possible to invert the pendulum, and the period would be

$$2\pi\left\{\frac{A\sin^2\alpha + C'}{(K\mu - Mgh)\sin\alpha\cos\alpha}\right\}^{\frac{1}{2}},$$

where $K\mu$ is taken with the positive sign.

For the relative equilibrium of steady turning about the vertical we have, by (11), either $\sin\phi=0$, so that $\phi=0$, or $\phi=\pi$, or

$$\cos\phi=\frac{A\mu^2\cos\alpha\cos\theta+Mgh-K\mu}{A\mu^2\sin\alpha\sin\theta}. \quad \text{.......................(12)}$$

5. *Gilbert's barogyroscope.* As an example of steady motion and relative equilibrium we apply (11) or (12) to give the theory of Gilbert's barogyroscope, which we have already discussed from first principles in 6, VII. At a place P in latitude $\lambda$ the apparatus (see Fig. 33 above) is supported on trunnions, or, better, on knife-edges, the line of which is horizontal and inclined at an angle $\beta$ to the east and west horizontal line through P. The axis of the gyroscope is in plane with, and at right angles to the line of knife-edges, and contains the centroid. It is inclined at an angle $\phi$ to the vertical at P. We have by (11), since $\mu^2$ is small,

$$Mgh\sin\phi-K\mu\sin\phi=0. \quad \text{.....................................(1)}$$

Here, by the theory given above, $\mu\sin\phi$ is the rate of turning about a line AB at right angles at once to the axis of the flywheel and to the line of knife-edges. Also $Mgh\sin\phi$ is the couple about the line of knife-edges and is the rate of production of A.M. about that line, while $K\mu\sin\phi$ is the same rate of generation of A.M. due to the rotation of the earth about the polar vertical with angular speed $\omega$ and the consequent angular speed about AB. We have thus

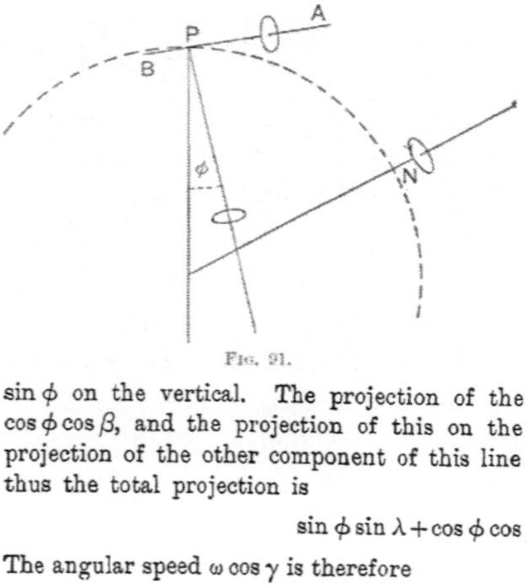


FIG. 91.

$$\omega\cos\gamma=\mu\sin\phi, \quad \text{.............(2)}$$

where $\gamma$ is the inclination of the line AB to the downward polar vertical.

To find $\cos\gamma$ we proceed as follows. A unit distance along AB (see Fig. 91), taken in the direction from A to B, has projection $\cos\phi$ on the horizontal and $\sin\phi$ on the vertical. The projection of the first component along the meridian is $\cos\phi\cos\beta$, and the projection of this on the polar vertical is $\cos\phi\cos\beta\cos\lambda$. The projection of the other component of this line on the polar vertical is $\sin\phi\sin\lambda$; and thus the total projection is

$$\sin\phi\sin\lambda+\cos\phi\cos\beta\cos\lambda=\cos\gamma.$$

The angular speed $\omega\cos\gamma$ is therefore

$$\omega(\sin\phi\sin\lambda+\cos\phi\cos\beta\cos\lambda).$$

If now $K=-Cn$, the rate of production of A.M. about the common perpendicular at P to AB and to the axis of the flywheel is

$$-Cn\,\omega(\sin\phi\sin\lambda+\cos\phi\cos\beta\cos\lambda)=-K\mu\sin\phi,$$

that is

$$\tan\phi=\frac{Cn\omega\cos\beta\cos\lambda}{Mgh-Cn\,\omega\sin\lambda}. \quad \text{.....................................(3)}$$

The directions of turning shown in Fig. 91 fit this equation. If the gyroscope rotate the opposite way the equation will be

$$\tan\phi=\frac{Cn\omega\cos\beta\cos\lambda}{Mgh+Cn\,\omega\sin\lambda}, \quad \text{.....................................(3')}$$

with the gyroscopic axis on the opposite side of the vertical at P.

**6.** *Relative equilibrium of rigid body; oscillations about steady motion.* Returning to 4 above, we see that if the relative equilibrium is stable in the steady motion we can find the period of the oscillatory deviations. We have, writing $q$ for $\dot{\phi}$,

$$\frac{d\ddot{\phi}}{d\phi}=\frac{1}{q}\ddot{q}.$$

Hence, from (10), 4, by differentiating with respect to $\phi$, treating $\theta$ and $\mu$ as maintained constant, we get

$$(\mathrm{A}\sin^2\alpha+\mathrm{C}')\ddot{q}$$
$$+\{(\mathrm{M}gh-\mathrm{K}\mu)\cos\phi+\mathrm{A}\mu^2(\sin\alpha\sin\theta\cos 2\phi-\cos\alpha\cos\theta\cos\phi)\}\sin\alpha\sin\theta\,.\,q=0.$$

Thus, if we write

$$p^2=\frac{\{(\mathrm{M}gh-\mathrm{K}\mu)\cos\phi+\mathrm{A}\mu^2(\sin\alpha\sin\theta\cos 2\phi-\cos\alpha\cos\theta\cos\phi)\}\sin\alpha\sin\theta}{\mathrm{A}\sin^2\alpha+\mathrm{C}'},$$

we have for the period of a small oscillation $2\pi/p$.

For a position of relative equilibrium in which $\phi=0$ or $\phi=2\pi$, that is when the axis of the pendulum and the vertical are in one plane, and $\theta$ is positive, as shown in Fig. 89, we have, if $\mathrm{M}gh-\mathrm{K}\mu$ be positive,

$$p^2=\frac{\mathrm{M}gh-\mathrm{K}\mu}{\mathrm{A}\sin^2\alpha+\mathrm{C}'},$$

and the equilibrium is to be regarded as stable. When however $\phi=\pi$ and $\mathrm{M}gh-\mathrm{K}\mu$ is positive,

$$p^2=-\frac{\mathrm{M}gh-\mathrm{K}u}{\mathrm{A}\sin^2\alpha+\mathrm{C}'},$$

and the relative equilibrium is unstable. This configuration is stable and the other unstable when $\mathrm{M}gh-\mathrm{K}\mu$ is negative. The stability therefore depends on the magnitude and sign of $\mathrm{K}\mu$.

When $\phi=\frac{1}{2}\pi$ we have

$$p^2=-\frac{\mathrm{A}\mu^2\sin^2\alpha\sin^2\theta}{\mathrm{A}\sin^2\alpha+\mathrm{C}'},$$

and when $\phi=\frac{3}{2}\pi$, $p^2$ has the same value. Hence the relative equilibrium is unstable in these cases.

**7.** *Example: Watt's steam-engine governor.* In his *Report on Gyroscopic Theory*, p. 204, Greenhill considers a Watt's steam-engine governor as a case of this arrangement of a rotating pendulum. It is shown in Fig. 92, and consists of two equal arms, $a$, $a$, carrying two equal massive balls, and connected near their lower ends by two rods, $b$, $b$, attached to a short sleeve free to move up or down on the vertical spindle, which supports the joint O. The arms are equally inclined to the spindle on opposite sides of it, and, when the spindle is driven round, the arms with their balls are made to revolve about the vertical through the joint, and the rise and fall of the latter is made to regulate the admission of steam to the engine.

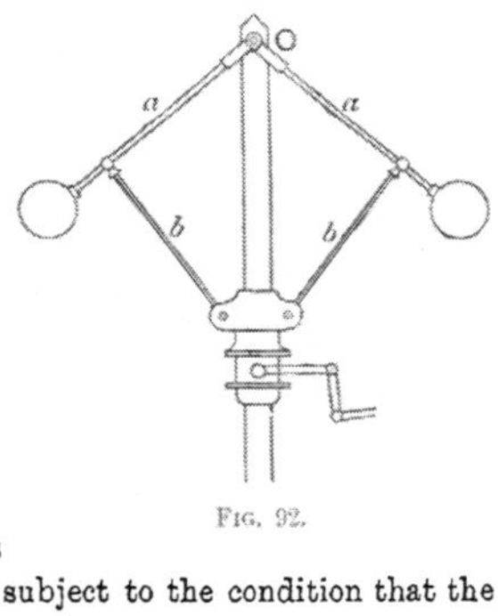

FIG. 92.

Putting aside this use of the arrangement, which depends on variations of speed about the vertical, we may suppose the relative equilibrium when the apparatus is running in steady motion to be slightly disturbed, subject to the condition that the angular speed $\mu$ about the vertical is kept unaltered.

Take three axes, one O$z$ towards the observer through the joint O (Fig. 92), the second O$x$ along the arm on the left in the diagram, and a third O$y$ in the plane of the paper, and (neglecting the moments of inertia of the rods $b$, $b$) let A, B be the moments of inertia about O$x$, O$y$, $mk^2$ that about O$z$, where $m$ is the mass of an arm and attached ball. Let $\phi$ be the inclination of the arm on the left, in Fig. 92, to the downward vertical. Calculating the rate of production of A.M. about O$z$ for the left-hand arm, we have first the term $-mk^2\ddot{\phi}$, which is affected with the negative sign, since the positive direction of rotation about O$z$ diminishes $\phi$. The angular speed about O$y$ is $\mu\sin\phi$, and about O$x$ is $\mu\cos\phi$. The A.M. about O$y$ is $\mathrm{B}\mu\sin\phi$ and about O$x$ is $\mathrm{A}\mu\cos\phi$. By the turning about O$y$ and the consequent motion of O$x$, A.M. about O$z$ is produced at rate $-\mathrm{A}\mu^2\cos\phi\sin\phi$, and by the turning about O$x$ and the consequent motion O$y$, it is produced at rate $\mathrm{B}\mu^2\sin\phi\cos\phi$. The total rate of production of A.M. is thus $-mk^2\ddot{\phi}-(\mathrm{A}-\mathrm{B})\mu^2\sin\phi\cos\phi$, and the couple acting is $mgh\sin\phi$. Thus we get the equation of motion

$$mk^2\ddot{\phi}+(\mathrm{A}-\mathrm{B})\mu^2\sin\phi\cos\phi+mgh\sin\phi=0, \quad \text{.......(1)}$$

or, since in the case of symmetry about O$x$, $\mathrm{B}=mk^2$,

$$\ddot{\phi}+\frac{\mathrm{A}-\mathrm{B}}{\mathrm{B}}\mu^2\sin\phi\cos\phi+\frac{gh}{k^2}\sin\phi=0. \quad \text{.......(2)}$$

If we differentiate (2) with respect to $\phi$, on the supposition that $\mu$ does not vary with $\phi$, we get, putting $q$ for $\dot{\phi}$,

$$\frac{1}{q}\ddot{q}+\frac{\mathrm{A}-\mathrm{B}}{\mathrm{B}}\mu^2(\cos^2\phi-\sin^2\phi)+\frac{gh}{k^2}\cos\phi=0. \quad \text{.......(3)}$$

Now let this value of $\phi$ be that for which $\ddot{\phi}=0$, the value, in fact, for relative equilibrium; then from (2) we get

$$\frac{\mathrm{B}-\mathrm{A}}{\mathrm{B}}\mu^2\cos\phi=\frac{gh}{k^2}=\frac{mgh}{\mathrm{B}}, \quad \text{.......(4)}$$

and so (3) reduces to

$$\ddot{q}+\frac{\mathrm{B}-\mathrm{A}}{\mathrm{B}}\mu^2\sin^2\phi\,.\,q=0. \quad \text{.......(5)}$$

Thus, if $\mathrm{B}>\mathrm{A}$, the motion is stable, and small oscillations are performed about steady motion in the period

$$\left(\frac{\mathrm{B}}{\mathrm{B}-\mathrm{A}}\right)^{\frac{1}{2}}\frac{2\pi}{\mu\sin\theta}, \quad \text{or} \quad \frac{2\pi}{\mu\sin\theta}, \text{ if A be negligible.}$$

If the ball revolve about the arm, or contain within it a flywheel with its axis along the arm, the equations can easily be modified to take account of the additional A.M., K, say, involved. We have only to alter the expression $\mathrm{A}\mu^2\sin\phi\cos\phi$ in (1) and in (2) to $(\mathrm{A}\mu\cos\phi-\mathrm{K})\mu\sin\phi$, so that (2) becomes

$$\ddot{\phi}+\frac{\mathrm{A}-\mathrm{B}}{\mathrm{B}}\mu^2\sin\phi\cos\phi+\frac{mgh-\mathrm{K}\mu}{\mathrm{B}}\sin\phi=0. \quad \text{.......(6)}$$

The alteration here specified determines the sign attributed to the A.M., K. Thus the period of oscillation about the configuration of steady motion is not altered by the rotation of the ball or flywheel. The moments of inertia are of course those for the whole mass moving about the axes.

A flywheel placed with its axis at right angles to the arm, and in the plane of the two arms, produces A.M. about a horizontal axis at right angles to the plane of the arms, at rate $\mathrm{K}\mu\cos\phi$, but has no other effect except that of contributing to the inertia of the moving system.

Returning to the question of stability, it is not difficult to show that, if the condition that $\mu$ should be constant is removed, the system is really unstable when in steady motion. We shall suppose that a flywheel, of moment of inertia I, is fixed on the vertical spindle, and turns with it. Associating $\frac{1}{2}\mathrm{I}$ with one of the arms, we have for the A.M. about the vertical for one half of the arrangement $(\frac{1}{2}\mathrm{I}+\mathrm{B}\sin^2\phi+\mathrm{A}\cos^2\phi)\mu$. We assume that when $\phi$ is changed from the value for steady motion [that is the motion for which $\mu$ and $(\frac{1}{2}\mathrm{I}+\mathrm{B}\sin^2\phi+\mathrm{A}\cos^2\phi)\mu$ are constant, and therefore $\ddot{\phi}$ is zero] by a small amount

$a$, a force $-fa$ is called into play, altering the A.M., and that $f$ is positive. Putting $\gamma$ for the steady motion value of $\phi$, we get the equations

$$\left.\begin{aligned} \frac{d}{dt}\{(\tfrac{1}{2}\mathrm{I}+\mathrm{B}\sin^2\phi+\mathrm{A}\cos^2\phi)\mu\} &= -fa, \\ \mathrm{B}\ddot{\phi}-(\mathrm{B}-\mathrm{A})\mu^2\sin\phi\cos\phi+(mgh-\mathrm{K}\mu)\sin\phi &= 0, \end{aligned}\right\} \quad\ldots\ldots(7)$$

with the steady motion condition $$(\mathrm{B}-\mathrm{A})\mu_0{}^2\cos\gamma=mgh-\mathrm{K}\mu. \quad\ldots\ldots(8)$$

If this condition be slightly deviated from, so that $\mu=\mu_0+\nu$, $\phi=\gamma+a$, where $\nu$ and $a$ are small, equations (7) become, with $p$ written for $d/dt$,

$$\left.\begin{aligned} (\tfrac{1}{2}\mathrm{I}+\mathrm{B}\sin^2\gamma+\mathrm{A}\cos^2\gamma)p\nu+\{(\mathrm{B}-\mathrm{A})\mu_0\sin 2\gamma\,.\,p+f\}a=0, \\ \{(\mathrm{B}-\mathrm{A})\mu_0\sin 2\gamma+\mathrm{K}\sin\gamma\}\nu-\{(\mathrm{B}-\mathrm{A})\mu_0{}^2\sin^2\gamma+\mathrm{B}p^2\}a=0. \end{aligned}\right\} \quad\ldots\ldots(9)$$

Eliminating $a$ and $\nu$ from (9), we get

$$(\tfrac{1}{2}\mathrm{I}+\mathrm{B}\sin^2\gamma+\mathrm{A}\cos^2\gamma)\mathrm{B}p^3+(\mathrm{B}-\mathrm{A})\mu_0\sin\gamma[2\{(\mathrm{B}-\mathrm{A})\mu_0\sin 2\gamma+\mathrm{K}\sin\gamma\}\cos\gamma$$
$$+(\tfrac{1}{2}\mathrm{I}+\mathrm{B}\sin^2\gamma+\mathrm{A}\cos^2\gamma)\mu_0\sin\gamma]p+f\{(\mathrm{B}-\mathrm{A})\mu_0\sin 2\gamma+\mathrm{K}\sin\gamma\}=0. \ldots(10)$$

Unless K is negative and $|\mathrm{K}|$ so great that $|\mathrm{K}|\sin\gamma$ is greater than $(\mathrm{B}-\mathrm{A})\mu_0\sin 2\gamma$, the last term on the left of (10) is positive, and the product of the roots of the cubic in $p$ is negative. Hence the real root is negative. The coefficient of $p$, by a similar condition as to K, is also positive, and so the other two roots are imaginary; and because the sum of the roots is zero the real parts of the roots are positive. The motion is therefore essentially unstable: it is oscillatory, but with increasing amplitude.

As Sir George Airy seems to have first pointed out (Routh, *Advanced Rigid Dynamics*, 6th edition, p. 73), stability may be obtained by connecting the opening or closing arms of the governor with a dash-pot arrangement, which brings into play a retarding force proportional to the angular speed $\dot{\phi}$. If this force be $k\dot{\phi}$ we get an additional term in (10), $$(\tfrac{1}{2}\mathrm{I}+\mathrm{B}\sin^2\gamma+\mathrm{A}\cos^2\gamma)kp^2,$$

which is positive. Thus, if the roots are all real they are all negative, since all the coefficients in (10) are positive. If there are two imaginary roots the real root is negative, and it is easy to see that if $k$ be sufficiently great the real parts of the imaginary roots will be negative. The motion is then stable.

Regulation for a given load may be obtained in this way, but if the load is considerably altered a new setting of the governor, if of the simple form, is required. For clearly, if the load is diminished, the supply of steam will be too great, and the arms of the governor will diverge to a new angle for steady motion. The steady speed of revolution must then be greater for a smaller rate of supply of steam. For most kinds of work a certain speed is required and must be adhered to within limits. Thus, from an important point of view the unmodified form of Watt's governor is seriously defective [see Routh, *loc. cit.*, for further particulars].

**8.** *Watt's governor. Elliptic function discussion.* Referring now to (6) of 7, multiplying by $\dot{\phi}$ and integrating we get, with $\mu$ constant as before,

$$\dot{\phi}^2=-p^2\cos^2\phi+2n^2\cos\phi+h, \quad\ldots\ldots(1)$$

where now $p^2=(\mathrm{B}-\mathrm{A})\mu^2/\mathrm{B}$, $n^2=(mgh-\mathrm{K}\mu)/\mathrm{B}$, and $h$ is constant.

If we can put $\cos a\cos\beta=-h/p^2$, and $\cos a+\cos\beta=2n^2/p^2=2\cos\gamma$, so that [(6), 7] $\gamma$ is the value of $\phi$ for relative equilibrium, we can determine two angles $a$ and $\beta$ $(\beta<\phi<a)$ between which the inclination of the arms to the vertical lies. For we have

$$\dot{\phi}^2=p^2(\cos\phi-\cos a)(\cos\beta-\cos\phi), \quad\ldots\ldots(2)$$

and $\dot{\phi}$ is zero for $\phi=a$, and for $\phi=\beta$. Thus a point at unit distance from O on the axle of the pendulum vibrates from a height $\cos\gamma-\cos a$ above, to an equal vertical distance below the position of relative equilibrium.

From equation (2) we can determine the time $t$ in terms of the corresponding angle $\phi$. For assuming an angle $\chi$ given by the equation,

$$\tan^2\tfrac{1}{2}\alpha\cos^2\chi+\tan^2\tfrac{1}{2}\beta\sin^2\chi=\tan^2\tfrac{1}{2}\phi, \qquad (3)$$

we are able to reduce (2) to the form

$$\dot\chi^2=p^2\sin^2\tfrac{1}{2}\alpha\cos^2\tfrac{1}{2}\beta\left(1-\frac{\tan^2\frac{1}{2}\alpha-\tan^2\frac{1}{2}\beta}{\tan^2\frac{1}{2}\alpha}\sin^2\chi\right). \qquad (4)$$

Thus, if we write $k^2=\dfrac{\tan^2\frac{1}{2}\alpha-\tan^2\frac{1}{2}\beta}{\tan^2\frac{1}{2}\alpha}$, $m=p\sin\tfrac{1}{2}\alpha\cos\tfrac{1}{2}\beta$ ......(5)

(where of course $m$ no longer denotes the mass of an arm of the governor), we obtain

$$m\,dt=\frac{d\chi}{(1-k^2\sin^2\chi)^{\frac{1}{2}}}, \qquad (6)$$

and so, starting the integration from $\phi=\alpha$, find

$$mt=\int_0^\chi\frac{d\chi}{(1-k^2\sin^2\chi)^{\frac{1}{2}}}. \qquad (7)$$

If T be the period of the motion, the time of passage of $\phi$ from the value $\alpha$ to the value $\beta$ is $\frac{1}{2}$T, and we get

$$\tfrac{1}{2}m\mathrm{T}=\int_0^{\frac{1}{2}\pi}\frac{d\chi}{(1-k^2\sin^2\chi)^{\frac{1}{2}}}=\mathrm{K}, \qquad (8)$$

where K is the complete elliptic integral to modulus $k$.

The frequency N of this finite oscillation is given by

$$\mathrm{N}=\frac{1}{\mathrm{T}}=\frac{m}{2\mathrm{K}}. \qquad (9)$$

When the range $\alpha-\beta$ is very small, this reduces to

$$\mathrm{N}=\frac{1}{\mathrm{T}}=\frac{p\sin\gamma}{2\pi}=\frac{n}{2\pi}\frac{\sin\gamma}{(\cos\gamma)^{\frac{1}{2}}}, \qquad (10)$$

since, with $\alpha$ and $\beta$ very nearly equal, $m=\frac{1}{2}p\sin\frac{1}{2}(\alpha+\beta)=\frac{1}{2}p\sin\gamma$ by (5), and $p^2=n^2/\cos\gamma$.

9. *Watt's governor. Case in which the arms reach the vertical.* If $\beta=0$ the arms reach the vertical. But then $k^2=1$ and $\mathrm{K}=\infty$. Thus the time of passing from inclination $\alpha$ to the lowest position is infinite, and the position is asymptotically approached. We have in fact

$$m\,dt=\frac{d\chi}{\cos\chi}, \qquad (1)$$

which, if we substitute $x$ for $\tan\chi$, reduces to

$$m\,dt=\frac{dx}{(1+x^2)^{\frac{1}{2}}}. \qquad (2)$$

Thus, by integration, we obtain $\cosh mt=\dfrac{\cot\frac{1}{2}\phi}{\cot\frac{1}{2}\alpha}$. ......(3)

Here of course also $m=p\sin\frac{1}{2}\alpha$.

In the case in which (1), 8, thrown into the form (2) of that article, gives $|\cos\beta|>1$, we may write $\cosh\beta$, instead of $\cos\beta$. We get then

$$\cosh\beta=2\frac{n^2}{p^2}-\cos\alpha,\quad \tanh^2\tfrac{1}{2}\beta=\frac{n^2-p^2\cos^2\frac{1}{2}\alpha}{n^2+p^2\sin^2\frac{1}{2}\alpha}. \qquad (4)$$

Using now the transformation $\tan^2\frac{1}{2}\alpha\cos^2\chi=\tan^2\frac{1}{2}\phi$, ......(5)

we find

$$m\,dt=\frac{d\chi}{(1-k^2\sin^2\chi)^{\frac{1}{2}}}, \qquad (6)$$

where

$$k^2=\frac{\tan^2\frac{1}{2}\alpha}{\tan^2\frac{1}{2}\alpha+\tanh^2\frac{1}{2}\beta},\quad m^2=p^2\frac{\tan^2\frac{1}{2}\alpha+\tanh^2\frac{1}{2}\beta}{(1-\tanh^2\frac{1}{2}\beta)(1+\tan^2\frac{1}{2}\alpha)}. \qquad (7)$$

The equation for $m^2$ reduces to $m^2 = n^2 - p^2 \cos\alpha$, ..........(7')
by (4), as the reader may verify.

When $\cosh^2\beta = \infty$ and $p^2 = 0$, so that $p^2\cosh^2\beta$ is finite, the equation for $\dot\phi$ reduces to

$$\dot\phi^2 = 2\frac{g}{l}(\cos\phi - \cos\alpha),$$

where $2g/l$ has been written for $p^2\cosh^2\beta$. Thus we have

$$m\,dt = \frac{d\chi}{(1 - k^2\sin^2\chi)^{\frac{1}{2}}},$$

where $k^2 = \sin^2\frac{1}{2}\alpha$, and $m = p\cosh\beta/2^{\frac{1}{2}} = (g/l)^{\frac{1}{2}}$. The motion thus reduces to the plane vibration of a pendulum of length $l$, through a finite angle $\alpha$ on each side of the vertical. So also in the general case in which $p$ is not zero, we see, putting

$$m^2 = n^2 - p^2\cos\alpha = \frac{g}{l}, \quad \text{and} \quad k^2 = \frac{\tan^2\frac{1}{2}\alpha}{\tan^2\frac{1}{2}\alpha + \tan^2\frac{1}{2}\beta} = \sin^2\tfrac{1}{2}\theta_0,$$

that the motion corresponds exactly to that of a pendulum of length $l = g/(n^2 - p^2\cos\alpha)$, vibrating through a finite angle $\theta_0$ on each side of the vertical.

There remains the case in which both the angles $\alpha$ and $\beta$ are unreal. This corresponds to the motion of a pendulum performing complete revolutions in a vertical circle. Here we write $-\cosh\alpha'$ for the quantity which takes the place of $\cos\alpha$, and note that the motion just discussed and that about to be considered will agree if $\cos\alpha = -\cosh\alpha' = -1$. We have

$$\dot\phi^2 = p^2(\cosh\beta - \cos\phi)(\cos\phi + \cosh\alpha'). \quad \text{..........(8)}$$

If we assume that

$$\tan\tfrac{1}{2}\phi = \tanh\tfrac{1}{2}\beta\tan\chi, \quad \text{..........(9)}$$

we find

$$m\,dt = \frac{d\chi}{(1 - k^2\sin^2\chi)^{\frac{1}{2}}}, \quad \text{..........(10)}$$

where

$$k^2 = 1 - \tanh^2\tfrac{1}{2}\alpha'\tanh^2\tfrac{1}{2}\beta, \quad m^2 = \frac{p^2}{(1 - \tanh^2\frac{1}{2}\alpha')(1 - \tanh^2\frac{1}{2}\beta)}. \quad \text{..........(11)}$$

When $\alpha = \pi$, and therefore $\alpha' = 0$, we get from (4),

$$k^2 = 1, \quad m^2 = \frac{p^2}{1 - \tanh^2\frac{1}{2}\beta} = n^2 + p^2, \quad \text{..........(12)}$$

in agreement with (7').

Equations (7), above, give the same values of $k^2$ and $m^2$ for $\alpha = \pm\pi$. Hence the two results agree, as they ought, when $\alpha' = 0$, $\alpha = \pm\pi$. In this case of the motion complete revolutions are just achieved. The angular speed at the lowest point is then given by

$$\dot\phi^2 = 4m^2\tanh^2\tfrac{1}{2}\beta.$$

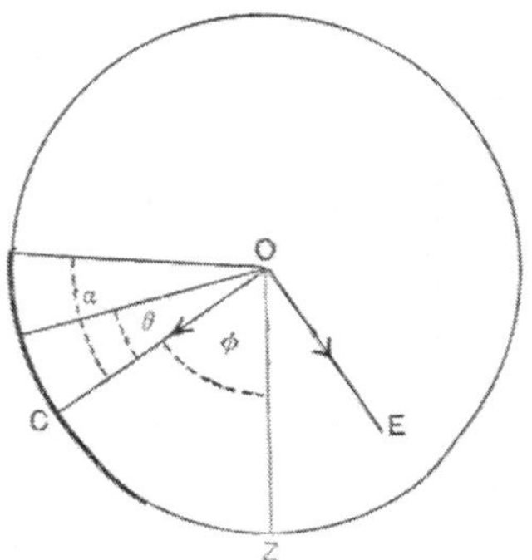

FIG. 93.

**10.** ***Example: Liquid filament in a revolving vertical circular tube.*** We now consider an example which forms an interesting variant of the problem of the Watt governor. A filament of mercury is enclosed in a uniform circular glass tube, the plane of which is vertical and revolves with uniform angular speed about the vertical, OZ, through the centre of the circle. It is required to find the motion of the mercury in the tube. Let (Fig. 93) a line drawn from the centre of the circle to the centre C of the filament make an angle $\phi$ with the vertical, and the line drawn to an element of the filament make an angle $\phi + \theta$ with the vertical. Let $m$ be the mass of the filament per unit of length, and $l$ its radius. Take three axes of coordinates OC, OD, drawn from the paper upwards, and OE in the plane of the diagram perpendicular to OC.

If $\mu$ be the angular speed about the vertical, we have $\mu\cos\phi$, $\mu\sin\phi$ for the components about OC and OE. Calling the angle subtended by the filament at the centre $2\alpha$, we have the following

$$\text{Components of A.M.}\left\{\begin{array}{l} 1.\ \text{About OD} = -2ml^3\alpha\dot{\phi}. \\ 2.\ \text{About OC} = \int_{-\alpha}^{\alpha} ml^3\mu\cos\phi\sin^2\theta d\theta = ml^3\mu\cos\phi(\alpha - \tfrac{1}{2}\sin 2\alpha). \\ 3.\ \text{About OE} = \int_{-\alpha}^{\alpha} ml^3\mu\sin\phi\cos^2\theta\, d\theta = ml^3\mu\sin\phi(\alpha + \tfrac{1}{2}\sin 2\alpha). \end{array}\right\} \quad \ldots\ldots (1)$$

Hence the rate of growth of A.M. about OD is $-2ml^3\alpha\ddot{\phi}$, due to the time-rate of increase of the first component, $-ml^3\mu^2\sin\phi\cos\phi(\alpha - \frac{1}{2}\sin 2\alpha)$, due to the turning of the vector OC about OE, and finally $ml^3\mu^2\sin\phi\cos\phi(\alpha + \frac{1}{2}\sin 2\alpha)$, due to the turning of OE about OC. The total rate of growth of A.M. about OD is therefore

$$-ml^3(2\alpha\ddot{\phi} - \mu^2\sin\phi\cos\dot{\phi}\sin 2\alpha).$$

The moment of forces about OD is easily found by integration to be $2\alpha ml^2 g\sin\alpha\sin\phi/\alpha$. Thus the equation of motion is

$$l(2\alpha\ddot{\phi} - \mu^2\sin\phi\cos\phi\sin 2\alpha) + 2g\sin\alpha\sin\phi = 0. \quad \ldots\ldots (2)$$

The filament will be in relative equilibrium if $\ddot{\phi} = 0$ and $\dot{\phi} = 0$. When this is the case we have

$$\mu^2 = \frac{g}{l\cos\alpha\cos\phi}. \quad \ldots\ldots (3)$$

As a particular case of this result we see that if a filament of angular length $4\alpha$ be symmetrically situated with respect to the vertical, it will (if there be no capillary forces to be taken account of) just break in two when $\mu^2$ is slightly increased beyond the value $g/l\cos^2\alpha$. Or, since the fluidity of the filament does not influence the dynamical result, except through capillary action which is here neglected, if two curved rods of the same material, fitting the tube and each subtending an angle $2\alpha$ at the centre, have their adjacent ends in contact at the vertical through the centre of the circle, they will, unless prevented by friction, separate when $\mu^2$ exceeds the value given in (2).

Moreover, if $\mu^2 = 0$, the equation of motion becomes

$$\ddot{\phi} + \frac{g}{l}\frac{\sin\alpha}{\alpha}\sin\phi = 0. \quad \ldots\ldots (4)$$

The motion is therefore one of oscillation in the period of a pendulum of length $l\alpha/\sin\alpha$.

Supposing $\mu^2$ kept constant while the motion is slightly disturbed from one of relative equilibrium, we get from (2), since

$$\frac{d\dot{\phi}}{d\phi} = \frac{1}{q}\frac{d^2q}{dt^2},$$

$$\frac{2\alpha l}{q}\frac{d^2q}{dt^2} + \mu^2 l\sin 2\alpha(\sin^2\phi - \cos^2\phi) + 2g\sin\alpha\cos\phi = 0. \quad \ldots\ldots (5)$$

But in this we must use $\mu^2 = g/l\cos\alpha\cos\phi$, the value for steady motion. Thus we find

$$\frac{d^2q}{dt^2} + \frac{g}{l}\frac{\sin\alpha\sin^2\phi}{\alpha\cos\phi}q = 0. \quad \ldots\ldots (6)$$

Small oscillations about steady motion are therefore performed along the tube by the filament (or circular rod) in the period of a pendulum of length $l\alpha\cos\phi/\sin\alpha\sin^2\phi$. Hence no steady motion can exist if the centre of the filament is above the horizontal line through the centre of the circle.

We may take a vertical circular tube erected on the surface of the earth, and turning with it, as an example. If $\omega$ be the angular speed of the earth and $\lambda$ the latitude of the place, the component of angular speed about the vertical is $\omega \sin \lambda$. Then in order that a particle may not be in stable equilibrium at the lowest point of the tube, we must have (since now $a=0$) $\omega^2 \sin^2 \lambda > g/l$, or $l > g/\omega^2 \sin^2 \lambda$. Thus the lower limit of the radius of the tube, set up at either pole, would be, in feet, about $32 \cdot 2 \times 86160^2/4\pi^2$, or 1,120,000 in miles. In latitude $\lambda$ the radius would be this divided by $\sin^2 \lambda$.

11. *Example: Ball containing a gyrostat and rolling without slipping on a horizontal table.* An example which may be taken here of the motion of a body containing a revolving piece, is the problem, proposed and treated by Bobylev [*Moscow Math. Rec.*, 1892], of a hollow sphere containing a flywheel mounted on an axis along a diameter. The centre of the flywheel is at the centre of the spherical case, which is also the centroid of the whole. The sphere rolls without slipping on a horizontal plane.

Let $a$ be the radius of the sphere, A the moment of inertia of the sphere and flywheel together, about any diameter of the sphere at right angles to the axis of the wheel, C the moment of inertia of the sphere, not including the flywheel, about the flywheel axis OC (the axis of symmetry), M the total mass, and K the A.M. of the wheel, supposed constant. We take two sets of axes, (1) a set $O(x, y, z)$, of which $Oz$ is vertical, and $Ox$, $Oy$ horizontal, the former in the vertical plane containing the axis of symmetry, and the latter at right angles to that plane; (2) a set OD, OE, OC, of which OD is at right angles to, and OE is in the plane $COz$. The angle $COz$ we denote as usual by $\theta$.

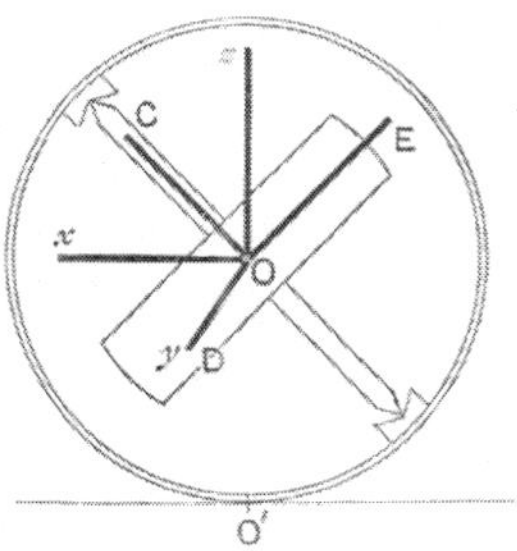

FIG. 94.

We suppose the angular speeds about $Ox$, $Oy$, $Oz$ to be $\omega_x$, $\omega_y$, $\omega_z$, and those about OD, OE, OC to be $\dot{\theta}$, $\dot{\psi} \sin \theta$, $\dot{\phi} + \dot{\psi} \cos \theta$ $(=n)$. Thus $\dot{\psi}$ is the angular speed about $Oz$, and $\dot{\phi}$ the angular speed with which a plane fixed in the body is turning with reference to the plane $COz$.

The arrangement of axes are as shown in the diagram [Fig. 94], and from that we obtain

$$\omega_x = n \sin \theta - \dot{\psi} \sin \theta \cos \theta, \quad \omega_y = \dot{\theta}, \quad \omega_z = \dot{\psi}. \quad \text{.................(1)}$$

The horizontal components $u$, $v$ and the vertical component $w$, of the velocity of the centroid with reference to the axes $O(xyz)$, are given by

$$u = a\omega_y, \quad v = -a\omega_x = -a\dot{\phi} \sin \theta, \quad w = 0. \quad \text{.................(2)}$$

The angular momenta about O(D, E, C) are

$$A\dot{\theta}, \quad A\dot{\psi} \sin \theta, \quad Cn + K,$$

so that the components about $O(xyz)$ are

$$h_x = (Cn + K - A\dot{\psi} \cos \theta) \sin \theta, \quad h_y = A\dot{\theta}, \quad h_z = A\dot{\psi} \sin^2 \theta + (Cn + K) \cos \theta. \quad (3)$$

Clearly $h_z$ is a constant: in what follows we shall denote it by G.

The equations of motion are

$$\left.\begin{array}{ll} M(\dot{u}-\dot{\psi}v)=X, & M(\dot{v}+\dot{\psi}u)=Y, \\ \dot{h}_x-\dot{\psi}h_y=aY, & \dot{h}_y+\dot{\psi}h_x=-aX. \end{array}\right\} \qquad (4)$$

Eliminating X, Y we obtain

$$\dot{h}_x-\dot{\psi}h_y=aM(\dot{v}+\dot{\psi}u), \quad \dot{h}_y+\dot{\psi}h_x=-aM(\dot{u}-\dot{\psi}v), \qquad (5)$$

which may be written

$$\left.\begin{array}{l} \frac{d}{dt}(h_x+a^2M\dot{\phi}\sin\theta)-\dot{\psi}(h_y+a^2M\dot{\theta})=0, \\ \frac{d}{dt}(h_y+a^2M\dot{\theta})+\dot{\psi}(h_x+a^2M\dot{\phi}\sin\theta)=0. \end{array}\right\} \qquad (6)$$

Equations (6) are obviously true. The angular momenta about horizontal axes parallel to $Ox$, $Oy$ and drawn through the point of contact with the table are $h_x+Ma^2\dot{\phi}\sin\theta$, $h_y+Ma^2\dot{\theta}$ respectively, and the total rates of growth of A.M. about them are the expressions on the left of (6), which are zero, since no couples act about these axes.

The second equation of (6) is important, and can be established at once as follows. Written in full it is

$$(A+Ma^2)\ddot{\theta}+(Cn+K+Ma^2n)\dot{\psi}\sin\theta-(A+Ma^2)\dot{\psi}^2\sin\theta\cos\theta=0. \quad (7)$$

If we write the second and third terms as

$$(Cn+K+Ma^2\dot{\phi}\sin^2\theta)\dot{\psi}\sin\theta-(A\dot{\psi}\sin\theta-Ma^2\dot{\phi}\sin\theta\cos\theta)\dot{\psi}\cos\theta,$$

we see at once, by taking axes O′(D′, E′, C′) through the point of contact O′ parallel to O(D, E, C), that they constitute the rate at which A.M. is being produced about the axis O′D′ by the motion; for the speed of the centroid at right angles to the plane EOC is $-a\dot{\phi}\sin\theta$, and the angular momenta about O′C′ and O′E′, arising from this motion, are

$$Ma^2\dot{\phi}\sin^2\theta \quad \text{and} \quad -Ma^2\dot{\phi}\sin\theta\cos\theta$$

respectively. Besides this there is for O′D′ rate of growth of A.M. $(A+Ma^2)\ddot{\theta}$; and the sum of these two rates must be zero since there is no moment of forces about O′D′.

Equations (6) give by integration

$$(h_x+Ma^2\dot{\phi}\sin\theta)^2+(h_y+Ma^2\dot{\theta})^2=H^2, \qquad (8)$$

where H is a constant. Expanded, this equation is

$$(A+Ma^2)\dot{\theta}^2+\{Cn+K+Ma^2n-(A+Ma^2)\dot{\psi}\cos\theta\}^2\sin^2\theta=H^2. \qquad (8')$$

By (6) and (7) we can write, making $\psi=0$, that is $Oy$ coincident with OD, when $\dot{\theta}=0$,

$$h_y+Ma^2\dot{\theta}=-H\sin\psi, \quad h_x+Ma^2\dot{\phi}\sin\theta=H\cos\psi, \qquad (9)$$

and so obtain

$$\frac{d}{dt}[\{Cn+K+Ma^2n-(A+Ma^2)\dot{\psi}\cos\theta\}\sin\theta]=(A+Ma^2)\dot{\theta}\dot{\psi}, \qquad (10)$$

or

$$A\sin\theta\frac{d}{dt}\{(Cn+K+Ma^2n)\sin\theta\}-(A+Ma^2)\cos\theta\frac{d}{dt}(A\dot{\psi}\sin^2\theta)=0. \qquad (11)$$

But, since $A\dot{\psi}\sin^2\theta=G-(Cn+K)\cos\theta$, performance of the differentiations gives

$$\{AC+Ma^2(A\sin^2\theta+C\cos^2\theta)\}\dot{n}-(Cn+K-An)Ma^2\dot{\theta}\sin\theta\cos\theta=0. \qquad (12)$$

Multiplying by $Cn+K-An$ and integrating, we find, putting $A(C+Ma^2)E^2$ for the constant of integration, $z$ for $\cos\theta$, and $k$ for $(A-C)Ma^2/A(C+Ma^2)$,

$$Cn+K-An=\frac{E}{(1-kz^2)^{\frac{1}{2}}},\quad n=\frac{K}{A-C}-\frac{E}{A-C}\frac{1}{(1-kz^2)^{\frac{1}{2}}}. \quad\text{.....................(13)}$$

Also by (9)

$$H\sin\theta\cos\psi=\left\{(C+Ma^2)(1-z^2)+\frac{A+Ma^2}{A}Cz^2\right\}n$$
$$+K(1-z^2)+\frac{A+Ma^2}{A}Kz^2-\frac{A+Ma^2}{A}Gz \quad\text{.............(14)}$$

By the value of $n$ given in (13) this becomes

$$H\sin\theta\cos\psi=\frac{A+Ma^2}{A-C}K-\frac{A+Ma^2}{A}Gz-\frac{C+Ma^2}{A-C}E(1-kz^2)^{\frac{1}{2}}. \quad\text{............(15)}$$

Again, by (8) and (9), we have

$$\dot{z}^2=\sin^2\theta\,.\,\dot{\theta}^2=\left(\frac{H\sin\theta\sin\psi}{A+Ma^2}\right)^2=\left(\frac{H}{A+Ma^2}\right)^2(1-\cos^2\psi)(1-z^2)\,; \quad\text{.............(16)}$$

and so by (15)

$$\dot{z}^2=\left(\frac{H}{A+Ma^2}\right)^2(1-z^2)-\left\{\frac{K}{A-C}-\frac{G}{A}z-\frac{C+Ma^2}{A+Ma^2}\frac{E}{A-C}(1-kz^2)^{\frac{1}{2}}\right\}^2, \quad\text{........(17)}$$

or [in the notation defined by a comparison of the two forms]

$$\dot{z}^2=H'^2(1-z^2)-\{K'-G'z-E'(1-kz^2)^{\frac{1}{2}}\}^2=Z. \quad\text{...............................(17')}$$

Thus

$$t=\int\frac{dz}{Z^{\frac{1}{2}}}. \quad\text{.......................................(18)}$$

The centroid moves with the component velocities

$$\left.\begin{aligned} u&=a\dot{\theta}=-aH'\sin\psi=-a\left(\frac{Z}{1-z^2}\right)^{\frac{1}{2}},\\ v&=-a\omega_x=-(-a\dot{\psi}\cos\theta+an)\sin\theta.\end{aligned}\right\} \quad\text{..............................(19)}$$

By (8), (9) and (13) we get

$$\left.\begin{aligned} v&=-aH'\cos\psi+a\frac{A-C}{C+Ma^2}\frac{E'\sin\theta}{(1-kz^2)^{\frac{1}{2}}},\\ \text{where}\quad E'&=\frac{C+Ma^2}{(A-C)(A+Ma^2)}(Cn+K-An)(1-kz^2)^{\frac{1}{2}}.\end{aligned}\right\} \quad\text{.......................(20)}$$

**12.** ***Rolling ball containing a gyrostat. Track on table referred to fixed axes.*** If we refer the motion to fixed axes, the velocities $\dot{x}$, $\dot{y}$ with respect to these are

$$\dot{x}=u\cos\psi-v\sin\psi,\quad \dot{y}=u\sin\psi+v\cos\psi, \quad\text{.........................(1)}$$

where $\psi$ is the angle which the plane EOC makes with a fixed vertical plane in which is laid the *fixed* horizontal axis $Ox$, here referred to, which is not the moving axis $Ox$ of Fig. 94 [see Fig. 12. The angle is that between the planes OCE′ and $Ozx$ as stated in 4, IV]. We have

$$\dot{x}=a\frac{A-C}{C+Ma^2}\frac{E'\sin\theta\sin\psi}{(1-kz^2)^{\frac{1}{2}}}=a\frac{A-C}{C+Ma^2}\frac{E'}{H'}\frac{\dot{z}}{(1-kz^2)^{\frac{1}{2}}} \quad\text{.....................(2)}$$

by the first of (9), 11. Again, by (19), 11,

$$\dot{y}=-aH'+a\frac{A-C}{C+Ma^2}\frac{E'\sin\theta\cos\psi}{(1-kz^2)^{\frac{1}{2}}}; \quad\text{.......................................(3)}$$

or if we solve (15) for $\sin\theta\cos\psi$, and substitute in (3),

$$\dot{y}=a\left\{-H'-\frac{A-C}{C+Ma^2}\frac{E'^2}{H'}+\frac{A-C}{C+Ma^2}\frac{E'}{H'}\frac{K'-G'z}{(1-kz^2)^{\frac{1}{2}}}\right\}. \quad\text{.....................(4)}$$

The integration of (2) and (4) is direct. It is carried out by Greenhill (*R.G.T.* p. 216) as follows: First, let $A-C$ and $k$ be positive, then $1-k=C(A+Ma^2)/A(C+Ma^2)$, is positive. Put then $k=\sin^2\alpha$. We get, writing $1-kz^2=(1-q^2)^2/(1+q^2)^2$,

$$z\sin\alpha=\frac{2q}{1+q^2},\quad dz\sin\alpha=2\frac{1-q^2}{(1+q^2)^2}dq. \qquad \text{(5)}$$

The equation for $\dot{x}$ is at once integrable; we get

$$x=2a\frac{A-C}{C+Ma^2}\frac{E'}{H'\sin\alpha}\tan^{-1}q. \qquad \text{(6)}$$

Now by (17′), 11, we have

$$(1+q^2)^2Z=Q=H'^2\left\{(1+q^2)^2-4\frac{q^2}{\sin^2\alpha}\right\}-\left\{K'(1+q^2)-2\frac{G'}{\sin\alpha}q-E'(1-q^2)\right\}^2, \qquad \text{(7)}$$

and therefore, since $dz\sin\alpha=2(1-q^2)dq/(1+q^2)^2$,

$$\sin\alpha\,dt=\sin\alpha\frac{dz}{Z^{\frac{1}{2}}}=\frac{1-q^2}{1+q^2}2\frac{dq}{Q^{\frac{1}{2}}}. \qquad \text{(8)}$$

Finally, from (4) above, since $\sin\alpha\,dt=\sin\alpha\,dz/Z^{\frac{1}{2}}$, we obtain

$$dy=\frac{a}{\sin\alpha}\frac{dz\sin\alpha}{Z^{\frac{1}{2}}}\left[-H'-\frac{A-C}{C+Ma^2}\frac{E'}{H'}\left\{E'-\frac{K'-G'z}{(1-kz^2)^{\frac{1}{2}}}\right\}\right]$$

$$=\frac{a}{\sin\alpha}2\frac{dq}{(1+q^2)Q^{\frac{1}{2}}}\left[-H'(1-q^2)-\frac{A-C}{C+Ma^2}\frac{E'}{H'}\left\{E'(1-q^2)-K'(1+q^2)+2G'\frac{q}{\sin\alpha}\right\}\right]. \qquad \text{(9)}$$

Thus we see that $x$ is given directly by (6) and that $t$ and $y$ are given by elliptic integrals of the third kind.

This conclusion is true also in the following case. Let $C-A$ be positive and $k$ negative. Then we write

$$-k=\sinh^2\alpha,\quad 1-k=\cosh^2\alpha,$$

and obtain, with $c=-a(A-C)E'/(C+Ma^2)H'\sinh\alpha$,

$$\sinh\frac{x}{c}=\sinh\alpha\cos\theta, \qquad \text{(10)}$$

$$(1-q^2)^2Z=Q=H'^2\left\{(1-q^2)^2-4\frac{q^2}{\sinh^4\alpha}\right\}-\left\{K'(1-q^2)-2\frac{G'q}{\sinh\alpha}-E'(1+q^2)\right\}^2. \qquad \text{(11)}$$

$$\sinh\alpha\,dt=\frac{\sinh\alpha\,dz}{Z^{\frac{1}{2}}}=2\frac{1+q^2}{1-q^2}\frac{dq}{Q^{\frac{1}{2}}}. \qquad \text{(12)}$$

$$dy=\frac{a}{\sinh\alpha}2\frac{dq}{(1-q^2)Q^{\frac{1}{2}}}\left[-H'(1+q^2)-\frac{A-C}{C+Ma^2}\frac{E'}{H'}\left\{E'(1+q^2)-K'(1-q^2)+2G'\frac{q}{\sinh\alpha}\right\}\right]. \qquad \text{(13)}$$

We take some special cases. If the spherical shell have little mass, we may suppose that $C=0$, and that A depends almost entirely on the enclosed gyrostat. Then $k=1$, and (7) becomes

$$Q=H'^2(1-q^2)^2-\{K'(1+q^2)-2G'q-E'(1-q^2)\}^2. \qquad \text{(14)}$$

Also by (9)

$$dy=2a\frac{dq}{(1+q^2)Q^{\frac{1}{2}}}\left[-H'(1-q^2)-\frac{A}{Ma^2}\frac{E'}{H'}\{E'(1-q^2)-K'(1+q^2)+2G'q\}\right]. \qquad \text{(15)}$$

If C is not zero but K is, we have a spherical case with a rod along a diameter, round which therefore there is symmetry. If both K′ and C are zero, we have merely a diametrical rod supported on a massless spherical shell. The equations can be easily written down from those given above.

When K and G vanish but not necessarily also C, $\dot{y}$ is constant and $\dot{x}$ is zero if $A-C=0$, and the ball rolls along a straight line.

**13.** *Rolling ball containing a gyrostat. Case of* A − C = 0. If A = C, $k$ is zero and (12), 11 gives
$$A\,(A+Ma^2)\dot{n}=KMa^2\dot{\theta}\sin\theta\cos\theta. \qquad (1)$$
Integrating, we obtain, since $z=\cos\theta$,
$$A\,(A+Ma^2)n+\tfrac{1}{2}Ma^2Kz^2=A\,F, \qquad (2)$$
a constant. In this case we have also
$$u=-aH'\sin\psi,\quad v=-aH'\cos\psi+a\frac{K\sin\theta}{A+Ma^2}, \qquad (3)$$
which give, by the values of $\dot{x}$, $\dot{y}$ in (1), 12,
$$x+a\frac{K}{H}z=\text{const.} \qquad (4)$$
$$\dot{y}=-aH'+\frac{a}{A+Ma^2}K\sin\theta\cos\psi. \qquad (5)$$
Now by 11, (14), since C = A,
$$\sin\theta\cos\psi=\frac{1}{H'}\left\{\frac{(A+Ma^2)n+K}{A+Ma^2}(1-z^2)-G'z+\frac{An+K}{A}z^2\right\},$$
or, by (2),
$$\sin\theta\cos\psi=\frac{1}{H'}\left\{\frac{F+K}{A+Ma^2}+\frac{1}{2}\frac{Ma^2Kz^2}{A(A+Ma^2)}-G'z\right\}. \qquad (6)$$
Substituting in (5), we obtain
$$\dot{y}=-aH'+\frac{aK}{H'(A+Ma^2)}\left\{\frac{(A+Ma^2)n+K}{A+Ma^2}(1-z^2)-G'z+\frac{An+K}{A}z^2\right\}. \qquad (7)$$
We have also
$$H'\sin\theta\sin\psi=\dot{z}. \qquad (8)$$
With (6) this gives
$$\dot{z}^2=Z=H'^2(1-z^2)-\left\{\frac{F+K}{A+Ma^2}+\frac{1}{2}\frac{Ma^2Kz^2}{A(A+Ma^2)}-G'z\right\}^2. \qquad (9)$$
E, F, and K are constants, and we can write the equation in the form
$$\dot{z}^2=Z=E+Fz-(\alpha+\beta z+\gamma z^2)^2, \qquad (10)$$
where $\alpha$, $\beta$, $\gamma$ are constants.

By a proper choice of origin (the "centre of the elastic forces" in the analogue of a bent spring) we can express the rate of variation of the square $r^2$ of the radius vector of the elastic curve* from point to point along the curve by an equation of the form
$$\left(\frac{dr^2}{ds}\right)^2=4\{r^2-(Ar^4+2Br^2+C)^2\},$$
and so find
$$\frac{d\theta}{ds}=\frac{Ar^4+2Br^2+C}{r^2}.$$
The integration of the differential equations of the curve is based on these relations [see Halphen, *Fonct. Ellipt.* t. II, chap. V, and Greenhill, *Math. Ann.* 52], and the analysis is available for the present problem. We shall give it with some other elliptic function calculations in a later chapter if there is space.

It will be found that the path of the point of contact lies between two parallel straight lines on the table, and consists of periodic repetitions of a certain figure, the nature of which is discussed in 15 below. By covering the table with a sheet of coloured transfer paper over a sheet of white paper, and rolling the ball on the former, the curve can be traced out. The weight of the ball is enough to produce an imprint of the path.

* The form of a uniform elastic wire or spring under terminal couples and uniform normal pressure over its whole length.

**14.** *Rolling ball containing a gyrostat. Small oscillations about steady motion.* From the equation (7), 11,

$$(A+Ma^2)\theta+(Cn+K+Ma^2n)\dot\psi\sin\theta-(A+Ma^2)\dot\psi^2\sin\theta\cos\theta=0, \quad\text{.........(1)}$$

we can find the period of a small oscillation about steady motion, that is the motion for which $\dot\theta$ and $\ddot\theta$ are zero, and therefore H and

$$Cn+K+Ma^2n-(A+Ma^2)\dot\psi\cos\theta\ (=\Phi) \quad\text{.........(2)}$$

are zero. We denote the steady motion value of $\dot\psi$ by $\mu$. For a slight deviation $\alpha$ of $\theta$ from the steady motion value we have

$$(A+Ma^2)\ddot\alpha+\mu\sin\theta\frac{d\Phi}{d\theta}_{\alpha=0}.\alpha=0. \quad\text{.........(3)}$$

Carrying out the calculation of the second term on the left, using (12), 11, and the constant value $(Cn+K)\cos\theta+A\dot\psi\sin^2\theta$ of G, to determine $dn/d\theta$ and $d\dot\psi/d\theta$, we see at once that the group of quantities obtained vanishes if we put $A=-Ma^2$, and that in fact the terms amount to $(A+Ma^2)\mu^2\alpha$. Thus we obtain the vibrational equation

$$\ddot\alpha+\mu^2\alpha=0. \quad\text{.........(4)}$$

The period of oscillation is thus $2\pi/\mu$.

In steady motion the sphere rolls so that its centre moves in a circle with angular speed $\mu$. If $c$ be the radius, we have, by Fig. 94, $v$ positive for $\mu$ negative, and so, since $u=a\dot\theta=0$, $-c\mu=v$. By (2), 11,

$$-v=a(n\sin\theta-\mu\sin\theta\cos\theta), \quad\text{.........(5)}$$

so that

$$c=a\left(\frac{n}{\mu}\sin\theta-\sin\theta\cos\theta\right)=\frac{a}{\mu}\phi\sin\theta. \quad\text{.........(6)}$$

The first term in (5), $an\sin\theta$, is the speed with which the centre L of the small circle of contact on the sphere is moving at right angles to the vertical plane, and $an\sin\theta/\mu$, the first term in (6), is the radius of the horizontal circle described by that centre.

Now $n/\mu=(A+Ma^2)n\cos\theta/(Cn+K+Ma^2n)$, and therefore

$$c=a\frac{An-(Cn+K)}{Cn+K+Ma^2n}\sin\theta\cos\theta. \quad\text{.........(7)}$$

Thus the ball will roll steadily in a straight line if $K=-(Cn+Ma^2n)$. The radius is zero if $K=(A-C)n$.

**15.** *Rolling ball containing a gyrostat. Small oscillations about straight line motion. Stability of straight line motion.* Now let the ball roll along a straight line with the axis of the gyrostat horizontal, and let the motion be slightly disturbed without change of the (zero) value of the A.M. about the vertical through the point of contact. The result just obtained is inapplicable, for $\mu$ now vanishes, and there evidently must be a finite period of oscillation about the steady straight line motion of the centre.

Putting then the A.M. about the vertical equal to zero, we get

$$(Cn+K)\cos\theta+A\mu\sin^2\theta=0, \quad\text{.........(1)}$$

where $\theta$ is very slightly different from $\frac{1}{2}\pi$. We have approximately $\mu=-(Cn+K)\cos\theta/A$. Hence substituting in (1), 14, which now is

$$(A+Ma^2)\ddot\theta+(Cn+K+Ma^2n)\mu=0, \quad\text{.........(2)}$$

we find

$$A(A+Ma^2)\ddot\theta-(Cn+K+Ma^2n)(Cn+K)\cos\theta=0. \quad\text{.........(3)}$$

Thus, according to the notation adopted above, we may put $\Phi=\cos\theta$, since the terms in $dn/d\theta$ which occur in $d\Phi/d\theta$ are all affected by the factor $\cos\theta$. Hence

$$d\Phi/d\theta=-\sin\theta=-1,$$

since $\theta=\frac{1}{2}\pi+\alpha$, where $\alpha$ is very small. Equation (3) becomes

$$\ddot{\alpha}+\frac{(Cn+K+Ma^2n)(Cn+K)}{A(A+Ma^2)}\alpha=0. \qquad \text{.................(4)}$$

Thus the period of disturbance of the motion is

$$T=2\pi\left\{\frac{A(A+Ma^2)}{(Cn+K+Ma^2n)(Cn+K)}\right\}^{\frac{1}{2}}. \qquad \text{.................(5)}$$

The ball is rolling forward with the linear speed $na$. Hence the wave length $\lambda$ of the disturbance as traced out on the table is

$$\lambda=2\pi na\left\{\frac{A(A+Ma^2)}{(Cn+K+Ma^2n)(Cn+K)}\right\}^{\frac{1}{2}}. \qquad \text{.................(6)}$$

These expressions for the period and wave length will not be real if the angular speed $n$ and the A.M. K are oppositely directed, and also

$$|(C+Ma^2)n|>|K|>|Cn| \quad \text{or} \quad |K/(C+Ma^2)|<|n|<|K/C|.$$

The ball cannot in this case roll stably in a straight line.

If $\theta$ is neither zero nor $\frac{1}{2}\pi$, we obtain, from (1), 14, neglecting the term in $\dot{\psi}^2$, for a slight deviation from steady rolling along a straight line (at right angles to the plane of the diagram in Fig. 94), the equation

$$(A+Ma^2)\ddot{\theta}+(Cn+K+Ma^2n)\mu\sin\theta=0. \qquad \text{.................(7)}$$

Hence, since the steady value of $\mu$ is zero, we put $\Phi=\mu$, and get

$$(A+Ma^2)\ddot{\alpha}+(Cn+K+Ma^2n)\sin\theta\,.\frac{d\Phi}{d\theta}\alpha=0. \qquad \text{.................(8)}$$

Here
$$\frac{d\Phi}{d\theta}=\frac{d\dot{\psi}}{d\theta}=\frac{(Cn+K)\sin\theta-C\cos\theta\dfrac{\partial n}{\partial\theta}}{A\sin^2\theta},$$
$$\dot{\psi}=\mu$$

and by (12), 11,
$$\frac{dn}{d\theta}=\frac{(Cn+K-An)Ma^2\sin\theta\cos\theta}{AC+Ma^2(A\sin^2\theta+C\cos^2\theta)}. \qquad \text{.................(9)}$$

The reader may verify that from these values we obtain

$$\ddot{\alpha}+\frac{(Cn+K+Ma^2n)\{(C+Ma^2)Cn+(C+Ma^2\sin^2\theta)K\}}{(A+Ma^2)\{AC+Ma^2(A\sin^2\theta+C\cos^2\theta)\}}\alpha=0. \qquad \text{..........(10)}$$

Denoting the second term on the left by $p^2\alpha$, we have for the period of oscillation $2\pi/p$. This agrees with the former result, if $\theta=\frac{1}{2}\pi$.

In the present case, since in the steady motion $\mu$ is zero, we have for the speed of advance of the centroid $v=-an\sin\theta$, so that we may write $n=-v/a\sin\theta$. Thus we find

$$p^2=\frac{\left\{K-(C+Ma^2)\dfrac{v}{a\sin\theta}\right\}\left\{(C+Ma^2\sin^2\theta)K-(C+Ma^2)C\dfrac{v}{a\sin\theta}\right\}}{(A+Ma^2)\{AC+Ma^2(A\sin^2\theta+C\cos^2\theta)\}}.$$

Hence the ball cannot roll along a straight line unless both factors of the numerator have the same sign, or, to put the result in another way, it will not roll stably along a straight line if

$$\frac{Ka\sin\theta}{C+Ma^2}<v<\frac{C+Ma^2\sin^2\theta}{C}\,\frac{Ka\sin\theta}{C+Ma^2}.$$

In using this inequality it must be remembered that if K and $n$ have opposite signs, we may take K positive, and then *for n negative we have v positive.* When K and $n$ have both the same sign $p^2$ is essentially positive.

The value of a wave length of the small deviation from the straight course is $2\pi an\sin\theta/p$.

The reaction against side slip is easily found, in any of these cases of motion, from the couple producing rate of change of A.M. about the $\theta$ axis (OD) through the centroid.

**16.** *Rolling ball containing a gyrostat. Method of solution by direct reference to first principles.* So far the foregoing discussion follows the mode of solution adopted by the proposer of the problem, and by Greenhill (*R.G.T.*) with some differences in the analysis. But the reference to first principles, suggested in 11, may be extended so as to give all the essential equations of motion, in fact practically to solve the whole problem.

We note, in the first place, that if the motion of the sphere be referred to axes O′(D′, E′, C′) parallel to O(D, E, C), the angular momenta about these are as follows:

$$\text{about O}'\text{D}',\quad (\text{A}+\text{M}a^2)\dot{\theta},$$
$$\text{about O}'\text{E}',\quad \text{A}\dot{\psi}\sin\theta-\text{M}a^2\dot{\phi}\sin\theta\cos\theta,$$
$$\text{about O}'\text{C}',\quad \text{C}n+\text{K}+\text{M}a^2\dot{\phi}\sin^2\theta,$$

since a little consideration shows that the speed at right angles to the plane EOC, of the point L (see 14) is $na\sin\theta$, and of O is $(n-\dot{\psi}\cos\theta)a\sin\theta$, that is $\dot{\phi}a\sin\theta$. There is no couple about any axis through O′, and the rates of turning of the axes are respectively $\dot{\theta}$, $\dot{\psi}\sin\theta$, $\dot{\psi}\cos\theta$ Hence the rates of growth of A.M. about the axes as enumerated above are

$$(\text{C}n+\text{K}+\text{M}a^2\dot{\phi}\sin^2\theta)\dot{\psi}\sin\theta-\{\text{A}\dot{\psi}\sin\theta-\text{M}a^2\dot{\phi}\sin\theta\cos\theta\}\dot{\psi}\cos\theta,$$

[or $$(\text{C}n+\text{K}+\text{M}a^2n)\dot{\psi}\sin\theta-(\text{A}+\text{M}a^2)\dot{\psi}^2\sin\theta\cos\theta,]$$

$$\text{A}\dot{\theta}\dot{\psi}\cos\theta-(\text{C}n+\text{K}+\text{M}a^2\dot{\phi}\sin^2\theta)\dot{\theta},$$
$$(\text{A}+\text{M}a^2)\dot{\theta}\dot{\psi}\sin\theta-(\text{A}+\text{M}a^2)\dot{\theta}\dot{\psi}\sin\theta, \text{ or zero.}$$

Now the first of these equated to zero gives the O′D′ equation for steady motion, and thus we have

$$(\text{C}n+\text{K}+\text{M}a^2n)\dot{\psi}\sin\theta-(\text{A}+\text{M}a^2)\dot{\psi}^2\sin\theta\cos\theta=0, \quad\ldots\ldots\ldots\ldots(1)$$

as that equation.

If we consider the A.M. about O′D′ when the motion is not steady, we see at once that the complete equation of motion is

$$(\text{A}+\text{M}a^2)\ddot{\theta}+(\text{C}n+\text{K}+\text{M}a^2n)\dot{\psi}\sin\theta-(\text{A}+\text{M}a^2)\dot{\psi}^2\sin\theta\cos\theta=0. \quad\ldots(2)$$

This is (7) of 11 above.

The magnitude of the couple about OD is clearly

$$\text{A}\ddot{\theta}+(\text{C}n+\text{K})\dot{\psi}\sin\theta-\text{A}\dot{\psi}^2\sin\theta\cos\theta.$$

Hence the horizontal force applied by the table in the plane EOC is

$$\frac{1}{a}\{\text{A}\ddot{\theta}+(\text{C}n+\text{K})\dot{\psi}\sin\theta-\text{A}\dot{\psi}^2\sin\theta\cos\theta\}.$$

This gives the reaction against side slip.

Again, the A.M. about a horizontal line drawn to the left through O′ in the plane EOC, [Fig. 94] the line of the force just calculated, is, by the angular momenta given above,

$$(\text{C}n+\text{K}+\text{M}a^2n)\sin\theta-(\text{A}+\text{M}a^2)\dot{\psi}\sin\theta\cos\theta.$$

Also the rate of growth of A.M. about the positive direction of this horizontal line [the direction to the right from O′ in Fig. 94] due to the turning of the axis O′D′ [with which is associated A.M. $(A+Ma^2)\dot{\theta}$] at rate $\dot{\psi}$ about the vertical, is $(A+Ma^2)\dot{\theta}\dot{\psi}$. Since there is no other rate of production of A.M. for this line and there is no couple, the total rate of growth of A.M. must be zero. Hence we get

$$\frac{d}{dt}\{(Cn+K+Ma^2n)\sin\theta-(A+Ma^2)\dot{\psi}\sin\theta\cos\theta\}-(A+Ma^2)\dot{\theta}\dot{\psi}=0, \quad \ldots(3)$$

which is precisely (10) of 11. From this we find

$$\dot{n}=\frac{(Cn+K-An)Ma^2\sin\theta\cos\theta}{AC+Ma^2(A\sin^2\theta+C\cos^2\theta)}\dot{\theta}.$$

With this and the value of $\dot{\psi}$ as found from

$$G=(Cn+K)\cos\theta+A\dot{\psi}\sin^2\theta$$

we have all the equations required for the discussion of the motion. The discussion of the different cases of oscillation about steady motion proceeds as in 14 above.

**17.** *Example: A cylinder containing a gyrostat and rolling on a horizontal plane.*

*Example.* [Problem restated from Math. Tripos paper, 1908.] The ship in Fig. 50 is replaced by a hollow right circular cylinder which rests on a horizontal table, and the gyrostat axis $b$, $b$ is at right angles to and intersects the axis of the cylinder. There is no gravitational stability of the gyrostat ring or wheel, and the cylinder, supposed in position with $b$, $b$ horizontal and the spin axis of the wheel vertical, has an attached mass $m$ at the centre of its highest generating line. A is the moment of inertia of the flywheel and ring about the axle $b$, $b$, B that of the flywheel and ring about the axis through the common centre parallel to the planes of rotation of the wheel, I the moment of inertia of the cylinder and attached mass about the line of contact with the table, and $Cn$ the A.M. of the flywheel, counter clockwise as seen from above. [As we suppose all angular deflections produced in the motion to be very small, B remains nearly enough constant.] Prove that the time of a small oscillation of the cylinder is

$$2\pi\left\{\frac{A(B+I)}{C^2n^2-mgaA}\right\}^{\frac{1}{2}}.$$

Let the cylinder be rolling at rate $\dot{\phi}$ towards the left [Fig. 50]. This produces A.M. at rate $Cn\dot{\phi}$ about an axis across the wheel to the left. Hence the top of the spin axis turns towards the reader with acceleration $\ddot{\theta}$, and we have $A\ddot{\theta}=Cn\dot{\phi}$, and therefore $A\dot{\theta}=Cn\phi$. This motion of the spin axis produces A.M. about the cylinder's line of contact at rate $Cn\dot{\theta}$, and the total rate of generation of such A.M. is $(I+B)\ddot{\phi}+Cn\dot{\theta}$, that is, since $A\dot{\theta}=Cn\phi$, $(I+B)\ddot{\phi}+C^2n^2\phi/A$. The applied couple is $mga\phi$. Hence we have

$$A(I+B)\ddot{\phi}+(C^2n^2-mgaA)\phi=0.$$

Thus if $C^2n^2>mgaA$ there is stability and the system oscillates in the period stated above.

# CHAPTER XVII

## MOTION OF AN UNSYMMETRICAL TOP

1. *General equations of motion.* A straightforward process of analysis enables the equations for the discussion of the stability of a body of any form resting on a horizontal plane to be investigated. We start from the equations of motion for moving axes established in 1, XVI. Referring to principal axes GA, GB, GC drawn from the centroid G we suppose that the coordinates of the point of contact O are $x, y, z$ and that the equation of the surface of the body is $f(x, y, z)=0$. The components of linear velocity of G will be denoted by $u, v, w$, those of angular velocity about the axes by $\omega_1, \omega_2, \omega_3$, the direction-cosines of the outward normal at O from the surface of the body by $p, q, r$, and the force components (arising from normal reaction and friction), applied to the body at O by the supporting surface, by X, Y, Z. The equations of rotational motion are

$$\left.\begin{aligned} A\dot{\omega}_1-(B-C)\omega_2\omega_3&=yZ-zY=R(\gamma y-\beta z),\\ B\dot{\omega}_2-(C-A)\omega_3\omega_1&=zX-xZ=R(\alpha z-\gamma x),\\ C\dot{\omega}_3-(A-B)\omega_1\omega_2&=xY-yX=R(\beta x-\alpha y),\end{aligned}\right\}\quad\ldots\ldots\ldots\ldots\ldots\ldots(1)$$

where R is the reaction at O, and $\alpha$, $\beta$, $\gamma$ the direction-cosines of its line of action taken outward from the surface.

For the motion of G we have, taking the mass of the body as unity,

$$\left.\begin{aligned} \dot{u}-v\omega_3+w\omega_2&=gp+X,\\ \dot{v}-w\omega_1+u\omega_3&=gq+Y,\\ \dot{w}-u\omega_2+v\omega_1&=gr+Z.\end{aligned}\right\}\quad\ldots\ldots\ldots\ldots\ldots\ldots\ldots\ldots\ldots(2)$$

The direction $p, q, r$ is fixed, and so we have

$$\dot{p}-q\omega_3+r\omega_2=0,\quad \dot{q}-r\omega_1+p\omega_3=0,\quad \dot{r}-p\omega_2+q\omega_1=0.\ldots\ldots\ldots\ldots(3)$$

Equations might be established for O as origin. The moments of forces would then be those of the forces applied at the centroid. Further the motion of O would have to be brought into the account.

There are three cases to be considered: (1) no sliding at the point of contact, (2) no friction, (3) sliding with friction. The conditions for case (1) are clearly

$$u-y\omega_3+z\omega_2=0,\quad v-z\omega_1+x\omega_3=0,\quad w-x\omega_2+y\omega_1=0.\ \ldots\ldots\ldots(4)$$

In case (2) the action of the plane on the body is a normal force R, so that, since here $\alpha, \beta, \gamma = p, q, r$,

$$X = -pR, \quad Y = -qR, \quad Z = -rR. \quad \text{.....................(5)}$$

The *minus* sign is required since R acts towards the body and $p, q, r$ are defined for the outward normal. In this case the point G moves parallel to the plane at uniform speed. If $PG = \zeta$, we can write the single equation

$$\ddot{\zeta} = +R - g \quad \text{or} \quad R = g + \ddot{\zeta}. \quad \text{..........................(6)}$$

Equation (3) holds also in case (2), since the direction of the vertical does not change.

In case (3) the resultant of the forces X, Y, Z makes an angle with the normal equal to $\tan^{-1}\mu$ where $\mu$ is the coefficient of friction. Thus

$$\frac{(pX + qY + rZ)^2}{X^2 + Y^2 + Z^2} = \cos^2(\tan^{-1}\mu) = \frac{1}{1 + \mu^2}. \quad \text{.....................(7)}$$

Also the slipping (of components $u', v', w'$) must take place in the opposite direction to the horizontal component of X, Y, Z. The direction-cosines of a line, perpendicular to the normal and to the resultant of X, Y, Z, are proportional to $qZ - rY$, $rX - pZ$, $pY - qX$. Hence we have

$$u'(rY - qZ) + v'(pZ - rX) + w'(qX - pY) = 0. \quad \text{...............(8)}$$

2. *Stability of a body spinning about a nearly vertical principal axis.* Now taking case 1, that of no slipping, let the body, supposed of any form, be set up with the principal axis GC vertical and its extremity C in contact with the horizontal plane, and be made to spin with angular speed $n$ about GC. The body is then slightly disturbed: it is required to find the motion.

Here initially $p, q$ are small and $r = 1$, nearly. Also $\omega_3 = n$, and $\omega_1, \omega_2, u, v, w$ are all small. A will be taken as less than B. We shall suppose that the body does not deviate far from its original position, so that these small quantities remain small. The following discussion will indicate under what conditions the supposition is justified.

In the first place we see from equations (2) that X, Y are small, and that $Z = -g$, nearly. Hence to the degree of approximation involved in neglecting the squares and products of small quantities we see that (1), (2), (3), (4) become

$$\left.\begin{array}{lll} A\dot{\omega}_1 - (B - C)n\omega_2 = -gy - hY, & B\dot{\omega}_2 - (C - A)n\omega_1 = hX + gx, & C\dot{n} = 0, \\ \dot{u} - nv = gp + X, & \dot{v} + nu = gq + Y, & \\ \dot{p} + \omega_2 - nq = 0, & \dot{q} - \omega_1 + np = 0, & \\ u - ny + h\omega_2 = 0, & v + nx - h\omega_1 = 0. & \end{array}\right\} \quad \text{...(1)}$$

The equation of the surface may be written in the form

$$z = h - \frac{1}{2}\left(\frac{x^2}{a} + \frac{2xy}{b} + \frac{y^2}{c}\right), \quad \text{..........................(2)}$$

where $a, b, c$ depend on the curvatures of the principal sections of the surface through the point P. We have from (2), neglecting squares and products of $x$, $y$, which are small,

$$p=-\frac{dz}{dx}=\frac{x}{a}+\frac{y}{b}, \quad q=-\frac{dz}{dy}=\frac{x}{b}+\frac{y}{c}, \quad r=1. \quad \text{.................(3)}$$

Eliminating $\omega_1$, $\omega_2$, $u$, $v$, X, Y from these equations, we get

$$\left.\begin{array}{l}(\mathrm{A}+h^2)\ddot{q}+(\mathrm{A}+\mathrm{B}+2h^2-\mathrm{C})n\dot{p}-\{(\mathrm{B}-\mathrm{C})n^2+hg+h^2n^2\}q \\ \quad +(g+hn^2)y-hn\dot{x}=0, \\ (\mathrm{B}+h^2)\ddot{p}-(\mathrm{A}+\mathrm{B}+2h^2-\mathrm{C})n\dot{q}-\{(\mathrm{A}-\mathrm{C})n^2+hg+h^2n^2\}p \\ \quad +(g+hn^2)x+hn\dot{y}=0.\end{array}\right\} \quad \text{......(4)}$$

[To take account of the mass M it is only necessary to multiply the last pair of terms in each equation, and every $h$ or $h^2$ in the others, by M. It is a good exercise, of no very great difficulty, to account for all the terms in (4) by first principles in the manner often exemplified above, and in that way establish the equations. The evaluation of the couples requires a little care.]

If we substitute for $x$ and $y$ in (4) from (3), the terms affected are

$$\left.\begin{array}{l}(g+hn^2)y-hn\dot{x}=-\dfrac{cb(g+hn^2)}{ac-b^2}(bq-ap)+\dfrac{abhn}{ac-b^2}(b\dot{p}-c\dot{q}), \\ (g+hn^2)x+hn\dot{y}=-\dfrac{ab(g+hn^2)}{ac-b^2}(bp-cq)-\dfrac{bchn}{ac-b^2}(b\dot{q}-a\dot{p}).\end{array}\right\} \quad \text{...(5)}$$

If the equation of the surface be

$$z=h-\frac{1}{2}\left(\frac{x^2}{a}+\frac{y^2}{c}\right), \quad \text{...............................(2')}$$

$a$ and $c$ are the radii at P of the lines of curvature, and the tangents to these at P are parallel to the principal axes at G. The terms written down become in this case

$$c(g+hn^2)q-nha\dot{p}, \quad a(g+hn^2)p+hnc\dot{q}.$$

In the general case the equations are intractable; but when the terms $x$, $y$, $\dot{x}$, $\dot{y}$ in (4) have the simple form just found the equation of motion (4) can be made to take the usual form for a gyrostatic vibrator possessing two freedoms. For we have then

$$\left.\begin{array}{l}(\mathrm{A}+h^2)\ddot{q}+(\mathrm{A}+\mathrm{B}-\mathrm{C}+2h^2-ah)n\dot{p} \\ \quad -\{(\mathrm{B}-\mathrm{C})n^2+gh+h^2n^2-c(g+hn^2)\}q=0, \\ (\mathrm{B}+h^2)\ddot{p}-(\mathrm{A}+\mathrm{B}-\mathrm{C}+2h^2-ch)n\dot{q} \\ \quad -\{(\mathrm{A}-\mathrm{C})n^2+gh+h^2n^2-a(g+hn^2)\}p=0.\end{array}\right\} \quad \text{........(6)}$$

If we assume the solutions

$$p=\alpha e^{i\lambda t}, \quad q=\beta e^{i\lambda t},$$

and substitute in (6), we get from each of these equations a relation between $\alpha/\beta$ and $\lambda$. Eliminating the ratio $\alpha/\beta$, we find the quadratic in $\lambda^2$

$$(\mathrm{L}\lambda^2+\mathrm{N})(\mathrm{L}'\lambda^2+\mathrm{N}')=\lambda^2n^2(\mathrm{M}-ha)(\mathrm{M}-hc), \quad \text{....................(7)}$$

where $L=A+h^2$, $L'=B+h^2$, $N=(B-C)n^2+(h-c)(hn^2+g)$,
$N'=(A-C)n^2+(h-a)(hn^2+g)$, $M=A+B+2h^2-C$.

[If it is desired to introduce the mass of the body (here supposed unity) into these formulae, it is only necessary to consider A, B, C as the ratios of the corresponding moments of inertia to the mass.]

The roots of the quadratic (7) in $\lambda^2$ are real if the inequality

$$\{LN'+L'N-n^2(M-ah)(M-ch)\}^2>4LL'NN'$$

is satisfied.

Let the solid be of revolution, and the radius of gyration about the axis of figure be $k$, and that about an axis through the point of support at right angles to the axis of figure be $k'$. Then since the mass is taken as unity,

$$k^2=C,\quad k'^2=A+h^2=B+h^2,\quad L=L'=k'^2,\quad M=2k'^2-k^2,$$
$$N=N'=(k'^2-k^2-ch)n^2+(h-c)g.$$

Thus we obtain, after a little reduction, for the expression in this case of the inequality stated above, $(k^2+ch)^2n^2>4k'^2(h-c)g$, or

$$n>2\frac{\{k'(h-c)g\}^{\frac{1}{2}}}{k^2+hc}. \quad\text{.........(8)}$$

The values of $\lambda^2$ which satisfy (7) are real when (8) is satisfied, and the condition (8) also shows that they are positive. If $\lambda_1^2$, $\lambda_2^2$ be the values of $\lambda^2$, each gives two values of $\lambda$ numerically equal and of opposite sign.

3. *General case: lines of curvature of surface at contact not parallel to the principal axes through the centroid.* In the more general case in which

$$z=h-\frac{1}{2}\left(\frac{x^2}{a}+2\frac{xy}{b}+\frac{y^2}{c}\right) \quad\text{.........(1)}$$

we have, using $n$ to denote the angular speed about the principal axis which joins the centroid with the point of support, and is vertical or nearly so,

$$\left.\begin{aligned}A'\ddot{q}-achkn\dot{q}+(D+abhk)n\dot{p}+ackg'p-\{(B'-C')n^2+(h+bck)g'\}q=0,\\ B'\ddot{p}+achkn\dot{p}-(D+bchk)n\dot{q}+ackg'q-\{(A'-C')n^2+(h+abk)g'\}p=0,\end{aligned}\right\}\text{...(2)}$$

where (using $k$ now in a different sense)

$$A'=A+h^2,\quad B'=B+h^2,\quad C'=C+h^2,\quad D=A'+B'-C'+h^2,$$
$$g'=g+hn^2,\quad k=b/(ac-b^2).$$

Let us suppose that

$$p=Ke^{i\lambda t},\quad q=Le^{i\lambda t}, \quad\text{.........(3)}$$

we obtain, by substitution in (2),

$$\begin{aligned}&\{A'\lambda^2+(B'-C')n^2+(h+bck)g'+iachnk\lambda\}\\ &\quad\times\{B'\lambda^2+(A'-C')n^2+(h+abk)g'-iachnk\lambda\}\\ &\quad+\{i(D+abkh)n\lambda+ackg'\}\{i(D+bckh)n\lambda-ackg'\}=0. \quad\text{.........(4)}\end{aligned}$$

If $b=\infty$, we fall back on the simpler case in which the terms affected by the multiplier $acknh$ are zero, that is (4) becomes, when $b$ is large, approximately

$$\{A'\lambda^2+(B'-C')n^2+(h+bck)g'\}\{B'\lambda^2+(A'-C')n^2+(h+abk)g'\} \\ -(D+abkh)(D+bckh)n^2\lambda^2=0, \quad \text{.......(5)}$$

where $bck$ and $abk$ are approximately $-c$ and $-a$, respectively. This is of the form

$$A'B'\lambda^4-P\lambda^2+Q=0, \quad \text{.......(5')}$$

where P and Q are independent of $\lambda$.

Equation (5) corresponds to the case in which a principal plane of moment of inertia for the centroid coincides with a principal plane of normal section at the point of contact O. For let $\theta$ denote the (acute) angle between the plane GOA, and the plane of normal section for which the radius of curvature at O is $\rho_1$, taken positive from the former plane to the latter (that is clockwise in Fig. 95), we have, assuming that $\rho_1 > \rho_2$,

$$\frac{1}{b}=\left(\frac{1}{\rho_2}-\frac{1}{\rho_1}\right)\sin\theta\cos\theta. \quad \text{.......(6)}$$

The multiplier $ack$ is $abc/(ac-b^2)$, and, as can easily be proved, is negative if $b$ be positive, as $b$ is on the supposition that $\rho_1 > \rho_2$. For, since we suppose the two curvatures turned the same way, the indicatrix at the point of contact of a tangent plane with the surface is an ellipse, and $ac-b^2$ is negative. The multiplier $ack$ is zero when $b$ is infinite, that is when $\theta=0$.

If however $\theta$ be not zero, we have, omitting terms in $k^2$,

$$A'B'\lambda^4-P\lambda^2+Q+iachnk(B'-A')(\lambda^2-n^2)\lambda=0. \quad \text{.......(7)}$$

The roots of (5′) are real if $P^2 > 4A'B'Q$, and are positive if P and Q are both positive. We shall suppose that these conditions are satisfied. The roots are given by

$$\lambda'^2=\frac{1}{2A'B'}\{P\pm(P^2-4A'B'Q)^{\frac{1}{2}}\}. \quad \text{.......(8)}$$

Let a root of (7) be $\lambda'^2+\kappa$, where $|\kappa|$ is small; then we have from (7)

$$(2A'B'\lambda'^2-P)\kappa+iachnk(B'-A')(\lambda^2-n^2)\lambda=0. \quad \text{.......(9)}$$

Hence $\kappa$ is imaginary; let it be $i\kappa'\lambda$. Then

$$(2A'B'\lambda'^2-P)\kappa'=achnk(B'-A')(n^2-\lambda^2), \quad \text{.......(10)}$$

and by (8), if $\lambda_1'^2$, $\lambda_2'^2$ be the two values of $\lambda'^2$, and $\kappa_1'$, $\kappa_2'$ the corresponding values of $\kappa'$, we obtain

$$\left.\begin{aligned}\kappa_1'(P^2-4A'B'Q)^{\frac{1}{2}}&=achnk(B'-A')(n^2-\lambda_1^2),\\ \kappa_2'(P^2-4A'B'Q)^{\frac{1}{2}}&=achnk(B'-A')(\lambda_2^2-n^2).\end{aligned}\right\} \quad \text{.......(11)}$$

But since $|\kappa|$ is small

$$\lambda'=(\lambda^2-i\kappa'\lambda)^{\frac{1}{2}}=\lambda-\tfrac{1}{2}i\kappa',$$

and hence

$$p=Ke^{i\lambda' t}=Ke^{(\frac{1}{2}\kappa'+i\lambda)t}; \quad \text{.......(12)}$$

so that the motion would increase exponentially if $\kappa'$ were positive. Thus $\kappa'$ must be negative for stability. Hence in (11), $\kappa_1'$, $\kappa_2'$ are both negative.

Now we suppose that $B' > A'$, and, in (6), $\rho_1 > \rho_2$. This involves taking $b$ positive and therefore $k$ negative for positive $\theta$, since $b$ is large and $ac - b^2 < 0$. Let then $\lambda_1^2 > n^2 > \lambda_2^2$, if that be possible without affecting the reality of the roots. We see that $nk$ must be positive, that is that $n$ must be negative in order that (11) may be satisfied. This means that the turning must be in the direction shown in the diagram by the arrow. [See experimental results in 5 below.]

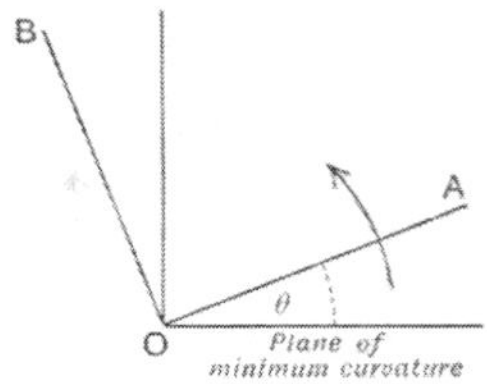


FIG. 95.

4. *A body spinning steadily in one direction on a horizontal plane, unsteadily in the opposite direction.* If $n^2 < \lambda_2^2$, turning in either direction will be consistent with one of equations (11) and inconsistent with the other. The turning indicated in the diagram will be accompanied by diminution of the amplitude of the vibrations of period $2\pi/\lambda_1$, and increase of the amplitude of those of period $2\pi/\lambda_2$. The reverse will happen if the turning is in the opposite direction. But since $\lambda_1^2 - n^2$, $\lambda_2^2 - n^2$ are of very different magnitudes, it is manifest that there will be much less increase of vibration and consequent unsteadiness when the turning is in the direction indicated in the diagram, than in the other case. The effect thus of a small value of $k$ might be hardly perceptible for spin in the direction taken as positive in the diagram, that is from OA to OB, and so great for the opposite spin as to render continuance of the motion impossible.

5. *Experiments with a top supported on an adjustable curved surface.* The results found in 3 were first given by Mr. G. T. Walker in a paper published in the *Quarterly Journal of Mathematics*, 28 (1896). They were confirmed experimentally by him by means of a simple form of top, which admitted of the axes of least moment and least curvature being placed in any desired relative positions. This top is in general shape a flat ellipsoid. In it a circular disk is embedded, so that the plane upper face of the disk rests against the face of a cavity cut in the body and the lower face continues (but not exactly except in one position) the curved contour of the lower surface of the top. The disk turns about a central axle which is normally vertical, and can be clamped in any position. In one clamped and marked position the disk was shaped with the lower surface, so that at any time the line of least curvature can be made to coincide with the longer axis of symmetry, or be placed at any required angle to that axis.

A massive nut which can be screwed up or down on the axle enables the position of the centroid to be altered. When the nut is removed or is screwed low down, the frequency of a transverse oscillation is greater than that of a longitudinal one; but when the nut is placed high enough, the transverse period is made greater, and the dynamical behaviour of the top otherwise altered.

The following are some of the results obtained with this apparatus. When the periods were adjusted to be considerably different, the disk was turned so as to make the angle $\theta$ about 2°, and the top was then gently spun about a vertical axis. It was found that when the direction of spin was opposite to the angular displacement given to the disk, the rotation lasted about four times as long as it did when the spin was in the other direction. The rotation was stopped in the former case by the growing up of oscillations of the slower type, in the latter case by the production of the quicker vibrations, as explained above. With only a very small angle the spin either way was little interfered with; when it stopped however, the top was found to be oscillating in the period appropriate to the direction of spin.

With an angle $\theta$ of 4°, a spin in the direction of $\theta$ set up violent oscillations which stopped and then reversed the spin. It will be observed how this took place. The spin energy disappeared, being replaced by energy of the vibrational motion. But in the oscillations, forces of reaction were brought into play which had moments round the line GO, which joined the centroid with the point of contact, and these reversed the rotation.

When the spin was thus reversed the motion was stopped, though more slowly, by the growing up of the vibrations of longer period; then these vibrations reversed the spin, and another cycle similar to the first was started, and so on, with proper arrangements, to a considerable number of reversals.

When purely longitudinal or transverse vibrations were started by tapping the edge of the top on an axis of symmetry, the top began to rotate in the direction for which such disturbing oscillations would die out. The difference between the times of oscillation with and without spin could be marked as the top settled down into oscillation about static equilibrium. Except when the angle $\theta$ was very small, one reversal or more occurred before this equilibrium was finally reached.

When the periods—transverse and longitudinal—were nearly equal, a tap given at a point near the edge, from 25° to 30° in azimuthal distance from the longer axis of symmetry, produced a spin the direction of which depended on the azimuth of the point struck, but not on the value of $\theta$ or the difference of periods. The direction was always that from the diameter struck to the longer axis.

Mr. Walker found that similar results were obtained by spinning the "celts," or stone axes, which are to be found in museums and private collections. When these bodies are laid in equilibrium on a table it is found in many cases that the principal axes of moment of inertia, which are parallel or nearly so to the table, do not coincide with the lines of curvature of the surface of the body at the point of support. In such cases the celt can be made to spin smoothly one way, but if it is made to turn the other way it begins to "chatter" as it turns and the spin ceases.

6. *Effect of oscillations of the body on the curved surface in producing azimuthal turning.* We shall now investigate by Mr. Walker's method the oscillations of the solid on the supposition that these are small, and endeavour to trace their effect in producing rotation of the solid in azimuth. Taking $\omega_1$, $\omega_2$, $\omega_3$, $u$, $v$, $w$, $p$, $q$, X, Y, all so small that their squares and products may be neglected, we have by 1,

$$A\ddot{q}=-gy-hY,\quad -B\ddot{p}=gx+hX,\quad C\dot{\omega}_3=0 \text{ or } C\omega_3=Cn. \qquad (1)$$

The motion of the centroid is given by

$$\dot{u}=gp+X,\quad \dot{v}=gq+Y,\quad w=g+Z=0. \qquad (2)$$

But $\dot{u}=-h\dot{\omega}_2=h\ddot{p},\quad \dot{v}=h\dot{\omega}_1=h\ddot{q},$

and so we obtain $X=h\ddot{p}-gp,\quad Y=h\ddot{q}-gq. \qquad (3)$

The equation of the surface, $z=h-\frac{1}{2}\left(\frac{x^2}{a}+\frac{2xy}{b}+\frac{x^2}{c}\right),$

gives $x=\frac{ab(bp-cq)}{b^2-ac},\quad y=\frac{bc(bq-ap)}{b^2-ac}.$

Thus the first two of the equations of motion above become

$$A'\ddot{q}+g(\alpha q-\beta p)=0,\quad B'\ddot{p}+g(\gamma p-\beta q)=0, \qquad (4)$$

where $\alpha=\frac{b^2c}{b^2-ac}-h,\quad \beta=\frac{abc}{b^2-ac},\quad \gamma=\frac{ab^2}{b^2-ac}-h. \qquad (5)$

Let $p=Ke^{i\lambda t},\quad q=Le^{i\lambda t}. \qquad (6)$

Substituting in (4) and eliminating K, L we get

$$(A'\lambda^2-\alpha g)(B'\lambda^2-\gamma g)-\beta^2g^2=0. \qquad (7)$$

Consider the roots of the equation in $\lambda^2$

$$(A'\lambda^2-\alpha g)(B'\lambda^2-\gamma g)=0. \qquad (8)$$

These are $\lambda_1'^2=\frac{\gamma}{B'}g,\quad \lambda_2'^2=\frac{\alpha}{A'}g. \qquad (9)$

But if $\lambda'^2$ be a root of (8) and $\lambda'^2+\kappa$, where $\kappa$ is small, be the corresponding root of (7), we have

$$\{2A'B'\lambda'^2-g(A'\gamma+B'\alpha)\}\kappa-\beta^2g^2=0. \qquad (10)$$

There are two values of $\kappa$, one corresponding to the root $\lambda_1'^2$, the other to the root $\lambda_2'^2$, given in (9). Call these $\kappa_1$, $\kappa_2$. We get by (9)

$$\left.\begin{aligned}\kappa_1&=\frac{\beta^2g^2}{2A'B'\lambda_1'^2-g(A'\gamma+B'\alpha)}=\frac{\beta^2g}{A'\gamma-B'\alpha},\\ \kappa_2&=\frac{\beta^2g^2}{2A'B'\lambda_2'^2-g(A'\gamma+B'\alpha)}=\frac{\beta^2g}{B'\alpha-A'\gamma}.\end{aligned}\right\} \qquad (11)$$

Thus the roots of (7) are

$$\lambda_1^2=\lambda_1'^2+\frac{\beta^2g}{A'\gamma-B'\alpha},\quad \lambda_2^2=\lambda_2'^2-\frac{\beta^2g}{A'\gamma-B'\alpha}. \qquad (12)$$

Considering now the third equation of (1) and including quantities of the second order of smallness, we have to find the reduced form of

$$C\dot{n}-(A-B)\omega_1\omega_2=xY-yX. \qquad (13)$$

We have first $\omega_1=\dot{q}+np,\quad \omega_2=nq-\dot{p}, \qquad (14)$

and
$$\left.\begin{aligned}xY-yX&=\frac{ab}{b^2-ac}(bq-cp)Y-\frac{bc}{b^2-ac}(bq-ap)X,\\ \text{with}\quad Y&=\dot{v}-gq,\quad X=\dot{u}-gp.\end{aligned}\right\} \qquad (15)$$

Substituting in (13) from (14) and (15), having regard to (5) and (4), we find to the degree of approximation adopted

$$C\ddot{n}=(A-B)(-\dot{p}\dot{q}+h\ddot{p}\ddot{q}/g)+Bpq-A\ddot{q}p. \quad \text{.....(16)}$$

If we assume $$p=a_1\cos\lambda_1 t+a_2\cos\lambda_2 t+b_1\sin\lambda_1 t+b_2\sin\lambda_2 t, \quad \text{.....(17)}$$

we obtain by (4),

$$g(\beta q-\gamma p)=-B'\{\lambda_1^2(a_1\cos\lambda_1 t+b_1\sin\lambda_1 t)+\lambda_2^2(a_2\cos\lambda_2 t+b_2\sin\lambda_2 t)\}, \quad \text{.....(18)}$$

which gives $q$.

Let the initial values of $p, q, \dot{q}$ be zero, so that the body is oscillating about an axis parallel to $Oy$. We get then by (17) and (18), and the values of $\lambda_1^2$, $\lambda_2^2$ in (9),

$$a_1=-a_2=0, \quad \lambda_2 b_2=\frac{A'B'\beta^2}{\Delta^2}\lambda_1 b_1, \quad \text{.....(19)}$$

where $\Delta\,(=-B'\alpha+A'\gamma)$ is supposed great in comparison with $\beta^2 g$. The exact denominator in the second of (19) is $\Delta^2+\beta^2 g$. Hence, with the initial conditions stated above, we get from (18) and (19)

$$p=b_1\left(\sin\lambda_1 t+\frac{A'B'\beta^2}{\Delta^2}\frac{\lambda_1}{\lambda_2}\sin\lambda_2 t\right), \quad q=\frac{b_1\beta B'}{\lambda_2\Delta}(-\lambda_2\sin\lambda_1 t+\lambda_1\sin\lambda_2 t). \quad \text{.....(20)}$$

7. *Summary of results.* When these values of $p$ and $q$ are substituted in (16), it is found that the mean value of $\dot{\omega}_3$ is proportional to $1/(\lambda_1'^2-\lambda_2'^2)$, so that a longitudinal oscillation will give rise to rotation in the direction of steady turning or in the contrary direction according as $\lambda_1'^2-\lambda_2'^2$ is positive or negative, that is according as the period of longitudinal oscillations is shorter or longer than the period of transverse oscillations.

Exactly in the same way a transverse oscillation will set up azimuthal turning proportional in mean value to $1/(\lambda_2'^2-\lambda_1'^2)$. The turning is thus opposite to that produced in the former case. Here of course $p, q, \dot{p}$ are supposed to be initially zero, and the initial angular speed is $\dot{q}$.

In all cases the rule stated above holds, that the azimuthal turning is in the direction which is stable for oscillations of corresponding frequency, provided corresponding oscillations about rest and about steady motion are each of greater or each of less frequency.

For the general case, in which either $\beta$ or $\Delta$ may be numerically small and initial angular speeds exist about the axes $Ox$ and $Oy$, the reader is referred to Mr. Walker's paper [*loc. cit.*, 5, *supra*]. It is found that if $\beta$ be small compared with $\Delta$, the direction of the spin set up changes sign with $\beta$ and with $\Delta$, but not with either of the initial speeds $\Omega_1$, $\Omega_2$ about the axes $Ox$, $Oy$. If $\Omega_2^2/\Omega_1^2-\alpha/\gamma$ be small, the direction of the rotation changes sign with $\Omega_1$ or $\Omega_2$ but not with $\beta$ or $\Delta$.

8. *Hess's particular solution of the problem of an unsymmetrical top.* An interesting particular solution of the equations of motion for the solid of three unequal principal moments of inertia may be given here. If $\alpha_1$, $\beta_1$, $\gamma_1$ be the direction-cosines of the line joining the centroid G and the point of contact O, these cosines are proportional to $x$, $y$, $z$, the coordinates of the point of contact O, and we have by (1), 1,

$$\alpha_1 A\dot{\omega}_1+\beta_1 B\dot{\omega}_2+\gamma_1 C\dot{\omega}_3-\{\alpha_1(B-C)\omega_2\omega_3+\beta_1(C-A)\omega_3\omega_1+\gamma_1(A-B)\omega_1\omega_2\}=0. \quad \text{.....(1)}$$

This equation expresses the fact that the rate of growth of A.M. about the line fixed in space, with which GO coincides at the instant, is zero.

It is clear without analysis that this quantity must be zero, for the resultant of gravity at G′ (the projection of G on the horizontal plane) and the reaction at O both intersect the line OG.

If further the part of (1) in brackets { } is zero, we have

$$\alpha_1 A\dot{\omega}_1 + \beta_1 B\dot{\omega}_2 + \gamma_1 C\dot{\omega}_3 = 0. \quad \text{.........(2)}$$

This, taken by itself, means that the rate of growth of A.M. about any line through G and fixed in the body, and directed by cosines $\alpha_1$, $\beta_1$, $\gamma_1$ with respect to the principal axes, is zero. [The rate of growth of A.M for an instantaneously coincident line, the direction-cosines of which vary, and which therefore is moving in the body, is

$$A\alpha_1\dot{\omega}_1 + B\beta_1\dot{\omega}_2 + C\gamma_1\dot{\omega}_3 + A\dot{\alpha}_1\omega_1 + B\dot{\beta}_1\omega_2 + C\dot{\gamma}_1\omega_3.$$

If this is zero, then for that line

$$\alpha_1 A\omega_1 + \beta_1 B\omega_2 + \gamma_1 C\omega_3 = c, \quad \text{.........(3)}$$

where $c$ is a constant.]

Equation (3) is true for OG, since $\alpha_1$, $\beta_1$, $\gamma_1$ are invariable. Let $c$ be zero; then

$$\alpha_1 A\omega_1 + \beta_1 B\omega_2 + \gamma_1 C\omega_3 = 0 \quad \text{.........(4)}$$

states that the A.M. about GO, as that line moves, remains zero. If then the second part of (1), with the *minus* sign prefixed, is equivalent to

$$A\dot{\alpha}_1\omega_1 + B\dot{\beta}_1\omega_2 + C\dot{\gamma}_1\omega_3,$$

equation (3) will hold, and also the particular case (4).

The expressions $(B-C)\omega_2\omega_3$, $(C-A)\omega_3\omega_1$, $(A-B)\omega_1\omega_2$ are clearly proportional to the direction-cosines of a normal to the plane in which lie both the vectors $(A\omega_1, B\omega_2, C\omega_3)$, $(\omega_1, \omega_2, \omega_3)$, that is the plane of the axis of resultant A.M. and the instantaneous axis of rotation. The equation

$$\alpha_1(B-C)\omega_2\omega_3 + \beta_1(C-A)\omega_3\omega_1 + \gamma_1(A-B)\omega_1\omega_2 = 0 \quad \text{.........(5)}$$

thus states that the line OG lies in this plane, and, if the equation always held, OG would always lie in the plane specified, and be at the same time perpendicular to the resultant of A.M.

Equation (5) is thus the condition that (3) or (4) may be a particular solution of the equations of motion.

Let us now suppose that $\beta_1$ is continually zero, that is that the line OG lies permanently in the plane of the principal axes GA, GC, then the particular solution (4) becomes

$$\alpha_1 A\omega_1 + \gamma_1 C\omega_3 = 0. \quad \text{.........(6)}$$

The condition (5) is in this case

$$\alpha_1(B-C)\omega_3 + \gamma_1(A-B)\omega_1 = 0, \quad \text{.........(7)}$$

which, if (6) is to hold, may be written

$$A\alpha_1^2(B-C) = C\gamma_1^2(A-B). \quad \text{.........(8)}$$

In order that (8) may be possible we must have either $A > B > C$ or $A < B < C$. But since $\alpha_1^2 + \gamma_1^2 = 1$, we get from (8)

$$\alpha_1^2 = \frac{C}{B}\frac{A-B}{A-C}, \quad \gamma_1^2 = \frac{A}{B}\frac{B-C}{A-C}, \quad \text{.........(9)}$$

so that the line OG is fixed with reference to the axes GA, GC, and moves with the body. Then $\dot{\alpha}_1 = \dot{\gamma}_1 = 0$, and (3) holds with $\beta$ and $c$ both zero.

If $x$, $z$ be, as indicated above, the coordinates of O, (8) becomes

$$x^2 A(B-C) = z^2 C(A-B),$$

or

$$x^2\left(\frac{1}{C} - \frac{1}{B}\right) = z^2\left(\frac{1}{B} - \frac{1}{A}\right). \quad \text{.........(10)}$$

Now consider the ellipsoid of which the equation is

$$\frac{\xi^2}{A} + \frac{\eta^2}{B} + \frac{\zeta^2}{C} = 1, \quad \text{.........(11)}$$

which is the reciprocal of the momental ellipsoid,

$$A\xi^2 + B\eta^2 + C\zeta^2 = 1.$$

The equation of the surface may be written in the form

$$\frac{1}{B}(\xi^2+\eta^2+\zeta^2)+\left(\frac{1}{A}-\frac{1}{B}\right)\xi^2+\left(\frac{1}{C}-\frac{1}{B}\right)\zeta^2=1.$$

Hence if $A>B>C$, $$\xi^2\left(\frac{1}{A}-\frac{1}{B}\right)-\zeta^2\left(\frac{1}{B}-\frac{1}{C}\right)=0 \qquad\text{.......(12)}$$

represents a pair of planes which give the circular sections of the surface. Comparing this equation with (10), we see at once that the line OG is perpendicular to one of the planes of circular section of the reciprocal ellipsoid.

It follows that in this case of motion the axis of resultant A.M., which is always perpendicular to the line OG, lies, as the body changes position and $A\omega_1$, $B\omega_2$, $C\omega_3$ change in amount, always in a plane of circular section.

The particular solution (7) of the equations (1) of motion which has just been discussed appears to have been first given by Hess in *Math. Ann.*, Bd. 27, 1890; it was also given by G. Kolosov of Petrograd in 1898 [*Gott. Nachr.* 1898]. A geometrical account of the solution by Sommerfeld follows Kolosov's paper [*loc. cit.*]. The reader interested in the theory of the unsymmetrical top may also refer to papers by Hess [*Acta Math.*, 37].

9. *Other particular solutions for an unsymmetrical top.* Another particular solution of the equations [(1), 1] can be obtained when the top is symmetrical in a certain special way about an axis and spins about a fixed point O, which lies in the plane through the centroid G perpendicular to the axis of symmetry. The motion is referred to axes $Oxyz$ drawn from O and fixed in the body; of these $Oz$ is the axis of symmetry, and the symmetry is such that $A=B=2C$.

The equations are applicable with the meanings of A, B, C proper for the new axes [which are drawn from O instead of from G], provided $x$, $y$, $z$ are now the coordinates of the centroid, $p$, $q$, $r$ the direction-cosines of the downward vertical. From the position assigned to the fixed point O, we see that $z=0$. We can also so choose the axes that $y=0$. The equations of motion are then, if we write $Rx/C=c$,

$$2\dot\omega_1-\omega_2\omega_3=0,\quad 2\dot\omega_2+\omega_3\omega_1=-cr,\quad \dot\omega_3=cq. \qquad\text{.......(1)}$$

Multiplying the first of these by $\omega_1$, the second by $\omega_2$, and the third by $\omega_3$, and adding, we get

$$2(\omega_1\dot\omega_1+\omega_2\dot\omega_2)+\omega_3\dot\omega_3=-c(r\omega_2-q\omega_3)=c\dot p.$$

This gives by integration $$2(\omega_1^2+\omega_2^2)+\omega_3^2=2cp+f, \qquad\text{.......(2)}$$

where $f$ is a constant.

Again multiplying the first equation of (1) by $p$, the second by $q$, and the third by $r$ adding and taking account of the values of $\dot p$, $\dot q$, $\dot r$ as given by (3), 1, we easily find

$$2\frac{d}{dt}(p\omega_1+q\omega_2)+\frac{d}{dt}(\omega_3 r)=0,$$

so that $$2(p\omega_1+q\omega_2)+\omega_3 r=2l, \qquad\text{.......(3)}$$

where $l$ is another constant.

Finally we obtain by multiplying the second of (1) by $i$ and adding to the first

$$2(\dot\omega_1+i\dot\omega_2)+i\omega_3(\omega_1+i\omega_2)=-icr; \qquad\text{.......(4)}$$

also we obtain similarly from the first two equations of (3), 1,

$$\dot p+i\dot q+i\omega_3(p+iq)-ir(\omega_1+i\omega_2)=0. \qquad\text{.......(5)}$$

Eliminating $r$ between (4) and (5) we get

$$\frac{d}{dt}[\log\{(\omega_1+i\omega_2)^2-c(p+iq)\}]=-i\omega_3. \qquad\text{.......(6)}$$

In the same way we find

$$\frac{d}{dt}[\log\{(\omega_1 - i\omega_2)^2 - c(p - iq)\}] = i\omega_3. \quad \text{.......(7)}$$

By addition and integration (6) and (7) give

$$\{(\omega_1 + i\omega_2)^2 - c(p+iq)\}\{(\omega_1 - i\omega_2)^2 - c(p-iq)\} = \text{const.} \quad \text{.......(8)}$$

In (2), (3), (8) we have three integrals of the equations of motion for this special case of symmetry. The third integral was discovered by Madame Kowalevski [*Acta Math.*, 14]. Its complete discussion is not of sufficient interest for the theory of a top to find a place here, but the reader may refer to the paper just cited, and to papers by Kötter [*Acta Math.*, 17] and Kolosov [*Math. Ann.*, 56].

**10.** *Tshapliguine's integral.* For a symmetrical top turning about a fixed point O under gravity, and having moments of inertia A = B = 4C, about principal axes through O, and its centroid G continually at the point of coordinates $h$, 0, 0 with reference to these axes, with the line OG directed by the cosines $p$, $q$, $r$, a particular integral, subject to the condition that the A.M., N say, about the vertical is zero, is

$$\omega_3(\omega_1^2 + \omega_2^2) - \frac{gh}{C}\omega_1 r = \text{const.} \quad \text{.......(1)}$$

This can be verified as follows. The equations of motion are

$$4C\dot\omega_1 - 3C\omega_2\omega_3 = 0, \quad 4C\dot\omega_2 + 3C\omega_1\omega_3 = ghr, \quad C\dot\omega_3 = -ghq, \quad \text{.......(2)}$$

if the mass is taken as unity, and $p$, $q$, $r$ be the direction-cosines of OG. From these, if

$$N = 4Cp\omega_1 + 4Cq\omega_2 + Cr\omega_3 = 0, \quad \text{.......(3)}$$

we have

$$\frac{gh}{C}(\dot\omega_1 r + \omega_1\dot r) = \frac{gh}{C}(\tfrac{3}{4}\omega_2\omega_3 r + \omega_1\omega_3 p - \omega_1^2 q)$$

$$= \frac{gh}{4C^2}\omega_2 N + \frac{gh}{C}\{\tfrac{1}{2}\omega_2\omega_3 r - (\omega_1^2 + \omega_2^2)q\}$$

$$= 2\omega_3(\dot\omega_2\omega_2 + \dot\omega_1\omega_1) + \dot\omega_3(\omega_1^2 + \omega_2^2). \quad \text{.......(4)}$$

Thus we have found that

$$2\omega_3(\dot\omega_1\omega_1 + \dot\omega_2\omega_2) + \dot\omega_3(\omega_1^2 + \omega_2^2) - \frac{gh}{C}(\dot\omega_1 r + \omega_1\dot r) = 0, \quad \text{.......(5)}$$

and hence by integration obtain equation (1).

This particular solution is due to M. Tshapliguine [*Moscow Collections*, 1901. See also a paper by Kolosov (*Mess. Math.*, 1902)].

**11.** *The principal invariants of an unsymmetrical top.* The equations of motion of an unsymmetrical top spinning about a fixed point O, under the action of the gravity M$g$ of the top, are [see 16, II, where the notation is explained and the equations established]

$$\left.\begin{aligned} A\dot p - (B - C)qr &= \beta c - \gamma b, \\ B\dot q - (C - A)rp &= \gamma a - \alpha c, \\ C\dot r - (A - B)pq &= \alpha b - \beta a. \end{aligned}\right\} \quad \text{.......(1)}$$

For brevity $p$, $q$, $r$ are used instead of the $\omega_1$, $\omega_2$, $\omega_3$ of 16, II, and $a$, $b$, $c$ denote $Mg(\xi, \eta, \zeta)$. Besides (1) we have also

$$\dot\alpha - \beta r + \gamma q = 0, \quad \dot\beta - \gamma p + \alpha r = 0, \quad \dot\gamma - \alpha q + \beta p = 0. \quad \text{.......(2)}$$

By the table of relations at the top of p. 73 we have

$$\alpha = -\sin\theta\cos\phi, \quad \beta = \sin\theta\sin\phi, \quad \gamma = \cos\theta, \quad \text{.......(3)}$$

and by (3), 2, IV

$$\dot\psi = \frac{q\sin\phi - p\cos\phi}{\sin\theta} \quad \text{.......(4)}$$

The potential energy V of the top may be taken as given by

$$V=a\alpha+b\beta+c\gamma. \quad\text{.....(5)}$$

The equation

$$V=E-T, \quad\text{.....(6)}$$

where T is the kinetic energy, that is

$$T=\tfrac{1}{2}(Ap^2+Bq^2+Cr^2), \quad\text{.....(7)}$$

expresses the conservation of energy if we take E as a constant.

There is constancy of A.M. about the vertical through O, that is

$$G=\alpha Ap+\beta Bq+\gamma Cr, \quad\text{.....(8)}$$

where G is a constant.

Two other quantities, S and U, may be here introduced. They are defined by

$$S=aAp+bBq+cCr, \quad U=\tfrac{1}{2}(A^2p^2+B^2q^2+C^2r^2). \quad\text{.....(9)}$$

Clearly S is the product of the A.M. about the line joining the fixed point O with the centroid G by $Mgh$, where $h$ is the length of that line, and U is half the square of the resultant A.M. S, T, U have been called the principal invariants of the top.

From (7) and (9) and the fundamental equation (1) we can find $\alpha$, $\beta$, $\gamma$. Thus we obtain, writing $l^2$ for $a^2+b^2+c^2$,

$$(2l^2U-S^2)\alpha=V(2aU-ApS)+G(l^2Ap-aS)+\dot{U}(bCr-cBq), \quad\text{.....(10)}$$

with two similar equations which can be written down by symmetry. Further, we find

$$\left.\begin{aligned}\dot{S}&=(B-C)aqr+(C-A)brp+(A-B)cpq,\\(2l^2U-S^2)\dot{T}&=(SV-Gl^2)\dot{S}+\{2l^2T-(ap+bq+cr)S\}\dot{U},\\\dot{U}^2&=l^2(2U-G^2)-S(S-2GV)-2UV^2.\end{aligned}\right\}\quad\text{.....(11)}$$

Equations (1) give also

$$\dot{U}=\alpha(bCr-cBq)+\beta(cAp-aCr)+\gamma(aBq-bAp). \quad\text{.....(11')}$$

**12.** *The Hess-Schiff equations of motion of a top.* Equations (10) and (11) are equivalent to the relations derived by Hess in 1882* and by Schiff in 1903,† and may be called the Hess-Schiff equations.‡ By the first of (9) and the first and second of (11) we can determine $p$, $q$, $r$, and from the three equations of which (10) is the type find $\alpha$, $\beta$, $\gamma$. The question then arises whether the values of $p$, $q$, $r$, $\alpha$, $\beta$, $\gamma$ thus obtained satisfy (1) and (2) in all cases, that is whether (10) and (11) may be regarded as always an exact substitute for (1) and (2). This question was discussed by Stäckel in an important paper in the *Mathematische Annalen*, 67 (1909). We state here his chief results, with indications of the manner in which they are obtained. It will be seen that the equivalence of the two sets of equations is by no means unconditional.

* "Über das Problem der Rotation," *Math. Ann.*, 20 (1882).

† "Über die Differentialgleichungen der Bewegung eines schweren starren Körpers," *Moscow Math. Coll.*, 24 (1903).

‡ Also equations (1) and (2) may be called the Euler-Poisson equations. They seem to have been first published by Poisson in his *Traité de Mécanique*, t. II (1811). They were given by Lagrange in his *Mécanique Analytique*, 1815 edition, but they do not appear in the 1811 edition.

In the first place, it is clear from (10) that $2l^2U - S^2$ must not vanish. Now we easily find from (9)

$$2l^2U - S^2 = (bCr - cBq)^2 + (cAp - aCr)^2 + (aBq - bAp)^2, \quad \text{.................(1)}$$

so that the right-hand side will vanish when $p = q = r = 0$, or when $p : q : r = a/A : b/B : c/C$. The former case need not be considered, the latter is that of motion about an instantaneous axis fixed in the body, a case in fact of the motion called by von Staude a *permanent turning* [see 16 below].

Let us now write

$$\left.\begin{aligned} L &= A\dot{p} - (B - C)qr - \beta c + \gamma b, \\ M &= B\dot{q} - (C - A)rp - \gamma a + \alpha c, \\ N &= C\dot{r} - (A - B)pq - \alpha b + \beta a, \end{aligned}\right\} \quad \text{..............................(2)}$$

so that L, M, N vanish when equations (1) are satisfied. The first of (11), 12, becomes

$$aL + bM + cN = 0. \quad \text{..........................................(3)}$$

If we multiply the three equations of the form (10) of 11 by

$$bCr - cBq, \quad cAp - aCr, \quad aBq - bAp$$

respectively and take account of (1), we get

$$ApL + BqM + CrN = 0. \quad \text{...................................(4)}$$

But if (10), 11, be multiplied by $br - cq$, $cp - ar$, $aq - bp$ respectively, we get

$$pL + qM + rN = 0. \quad \text{........................................(5)}$$

Equations (3), (4), (5) can only hold if

$$L = M = N = 0, \quad \text{or} \quad \begin{vmatrix} a, & b, & c \\ Ap, & Bq, & Cr \\ p, & q, & r \end{vmatrix} = 0. \quad \text{..........................(6)}$$

The first alternative is that equations (1) are fulfilled, the second is $\dot{S} = 0$. *Hence, if* S *is not constant equations* (10) *and* (11) *of* 11 *are equivalent to* (1).

Now to find when equations (2), 11, are satisfied we write $\lambda$, $\mu$, $\nu$ for the expressions $\dot{\alpha} - \beta r + \gamma q$, etc. We easily find, since $\alpha\dot{\alpha} + \beta\dot{\beta} + \gamma\dot{\gamma} = 0$,

$$\alpha\lambda + \beta\mu + \gamma\nu = 0. \quad \text{.......................................(7)}$$

Multiplying the three equations of (10), 11, by $a$, $b$, $c$ respectively, we obtain $a\alpha + \beta b + \gamma c = V$, and therefore $a\dot{\alpha} + b\dot{\beta} + c\dot{\gamma} + \dot{T} = 0$, that is, by (1), 11,

$$a\lambda + b\mu + c\nu = 0. \quad \text{.........................................(8)}$$

Again we have

$$Ap\lambda + Bq\mu + Cr\nu = Ap\dot{\alpha} + Bq\dot{\beta} + Cr\dot{\gamma} + \alpha(B - C)qr + \beta(C - A)rp + \gamma(A - B)pq.$$

Also by (2)

$$\alpha L + \beta M + \gamma N = A\dot{p}\alpha + B\dot{q}\beta + C\dot{r}\gamma - \alpha(B - C)qr - \beta(C - A)rp - \gamma(A - B)pq.$$

Hence by addition we obtain

$$Ap\lambda + Bq\mu + Cr\nu + \alpha L + \beta M + \gamma N = \frac{d}{dt}(Ap\alpha + Bq\beta + Cr\gamma) = \dot{G}.$$

But $\dot{G} = 0$, and so $\quad Ap\lambda + Bq\mu + Cr\nu = -(\alpha L + \beta M + \gamma N).$

Now we have seen that if $\dot{S}$ is not zero, $L = M = N = 0$. Hence, on this supposition

$$Ap\lambda + Bq\mu + Cr\nu = 0. \quad \text{...................................(9)}$$

From (7), (8), (9) we see that either $\lambda = \mu = \nu = 0$ or the variables satisfy the determinantal equation

$$\begin{vmatrix} \alpha, & \beta, & \gamma \\ a, & b, & c \\ Ap, & Bq, & Cr \end{vmatrix} = 0. \quad \text{................................(10)}$$

Now $$\dot{U}=A^2p\dot{p}+B^2q\dot{q}+C^2r\dot{r}=Ap\,.\,A\dot{p}+Bq\,.\,B\dot{q}+Cr\,.\,C\dot{r}$$
$$=Ap\{L+(B-C)qr+\beta c-\gamma b\}+\dots,\qquad(11)$$
that is if (10) holds $$\dot{U}=ApL+BqM+CrN.\qquad(12)$$

If, as we suppose, $L=M=N=0$, the satisfaction of the determinantal equation (10) leads to $\dot{U}=0$. Hence we have the result: *If neither* S *nor* U *is constant equations* (1) *and* (2) *follow from* (10) *and* (11).

**13.** *The Hess-Schiff equations are not applicable to a symmetrical top under gravity.* Under the conditions found above, which are convenient for the investigation of special cases of motion, the quantities S, T, U are the parameters employed in the Hess-Schiff equations. If no forces act on the top we have $S=0$, not because there is zero A.M. about the line OG, but because $Mg=0$. If the top is symmetrical and under the action of gravity, with O on the axis of figure, so that $a=b=0$, we have $S=Ccr$, and $r$ is constant. Thus $\dot{S}=0$, and the top is excluded from the cases for which the equations in terms of S, T, U are equivalent to the Eulerian equations.

**14.** *Instantaneous axis fixed in the body and resultant angular speed constant. Staude's cone.* Let now S, T, U be all constant, and the top be unsymmetrical and under gravity, then $p$, $q$, $r$ are also constant. The instantaneous axis is fixed in the body, and the resultant angular speed is constant. The motion is then what O. Staude has called a permanent turning. From the constancy of S we have
$$(B-C)aqr+(C-A)brp+(A-B)cpq=0.\qquad(1)$$
Thus the instantaneous axis may be any generator of the cone of which the equation is
$$(B-C)ayz+(C-A)bzx+(A-B)cxy=0.\qquad(2)$$
Since for constant values of S, T, U, $p$, $q$, $r$ are also constants, we may write
$$p=e\omega,\quad q=f\omega,\quad r=g\omega,\qquad(3)$$
where $\omega$ is the resultant angular speed (and $g$ is no longer the gravity acceleration, but a coefficient). Thus $e^2+f^2+g^2=1$. Going back to the Euler-Poisson equations we find, since $\dot{p}=\dot{q}=\dot{r}=0$,
$$(C-B)fg\omega^2=\beta c-\gamma b,\quad(A-C)ge\omega^2=\gamma a-ac,\quad(B-A)ef\omega^2=ab-\beta a,\qquad(4)$$
where $\alpha$, $\beta$, $\gamma$ are now constants. But since now by (10), 11, $\dot{\alpha}=\dot{\beta}=\dot{\gamma}=0$, we have
$$\alpha:\beta:\gamma=e:f:g,$$
or, since $\alpha^2+\beta^2+\gamma^2=e^2+f^2+g^2=1$,
$$\alpha,\ \beta,\ \gamma=e,\ f,\ g.\qquad(5)$$
Thus the axis of rotation is vertical, and there are for each instantaneous axis in the body two speeds, $\omega$ and $-\omega$, of rotation.

As regards the values of $\omega$ we have the third of (11), 11,
$$l^2(2U-G^2)-S^2+2GS(E-T)-2U(E-T)^2=0.\qquad(6)$$
Substituting in this the constants
$$\left.\begin{aligned}S&=(aAe+bBf+cCg)\omega,\\ T&=\tfrac{1}{2}(Ae^2+Bf^2+Cg^2)\omega^2,\\ U&=\tfrac{1}{2}(A^2e^2+B^2f^2+C^2g^2)\omega^2,\end{aligned}\right\}\qquad(7)$$
we obtain an equation of the sixth degree for $\omega$. But since E and G are arbitrary constants the values of $\omega$ are only given by the roots of the sextic when E and G are known.

The Eulerian angles $\theta$ and $\phi$ are now also constant, and we have by (4), 11,
$$\psi=\psi_0+\frac{f\sin\phi-e\cos\phi}{\sin\theta}\omega t\qquad(8)$$
if $t$ be reckoned from the instant when $\psi=\psi_0$.

The hitherto excluded cases are $p=q=r=0$, and $p:q:r=a/\mathrm{A}:b/\mathrm{B}:c/\mathrm{C}$. No doubt the case of rest may be considered as a permanent turning; but in the other case $\omega$ is not constrained to be constant, though if it varies its components must all vary in the same ratio, that is the relations (3), with $e, f, g$ constants, must hold. We shall see that this varying motion is excluded by the Euler-Poisson equations.

The only motions possible with S, T, U all constant are the so-called permanent turnings.

**15.** *Case of resultant A.M. of constant amount.* For the case in which U only is constant we must refer to Stäckel's paper [*Math. Ann.*, 67 (1909)]. The conclusions arrived at are however as follows: *The Eulerian equations* (1) *are a consequence of* (10) *and* (11) *of* 11 *above. Poisson's equations,* (2), 11, *however are not satisfied unless, besides* (10) *and* (11) *of* 11, *the supplementary condition*

$$\{2\mathrm{U}(\mathrm{E}-\mathrm{T})-\mathrm{GS}\}\{(ap+bq+cr)\mathrm{S}-2l^2\mathrm{T}\}-\{\mathrm{S}(\mathrm{E}-\mathrm{T})-\mathrm{G}l^2\}^2=0 \quad\text{.............(1)}$$

*is fulfilled.* This equation of course exists along with that obtained by putting $\dot{\mathrm{U}}=0$ in the third of (11), 11. It forms a substitute for the second of (11), 11, which becomes an identity when $\dot{\mathrm{U}}=0$.

But it is to be observed that the equation obtained from the third of (11), 11, by putting $\dot{\mathrm{U}}=0$, that is

$$l^2(2\mathrm{U}-\mathrm{G}^2)-\mathrm{S}^2+2\mathrm{GS}(\mathrm{E}-\mathrm{T})-2\mathrm{U}(\mathrm{E}-\mathrm{T})^2=0 \quad\text{.......................(2)}$$

and (1) are independent equations. If for all values of A, B, C, $a$, $b$, $c$ one equation were a consequence of the other, then this should be true for the special values $a=1$, $b=c=0$. We should then have $\mathrm{S}=\mathrm{A}p$, and (1) and (2) would become

$$\left.\begin{aligned}\{2\mathrm{U}(\mathrm{E}-\mathrm{T})-\mathrm{GS}\}(\mathrm{S}^2-2\mathrm{A}l^2\mathrm{T})-\mathrm{A}\{\mathrm{S}(\mathrm{E}-\mathrm{T})-\mathrm{G}l^2\}^2=0,\\ l^2(2\mathrm{U}-\mathrm{G}^2)-\mathrm{S}^2+2\mathrm{GS}(\mathrm{E}-\mathrm{T})-2\mathrm{U}(\mathrm{E}-\mathrm{T})^2=0.\end{aligned}\right\} \quad\text{................(3)}$$

These are two quadratics in $\mathrm{E}-\mathrm{T}$, and if we form their resultant we obtain an integral rational function of S of the eighth degree, of which the first term is $\mathrm{A}^2\mathrm{S}^8$. This is inconsistent with the value $\mathrm{S}=\mathrm{A}p$. Thus the equations are independent, and we see that if the values of A, B, C, $a$, $b$, $c$ are left quite general, the condition $\dot{\mathrm{U}}=0$ involves also $\dot{\mathrm{S}}=0$, $\dot{\mathrm{T}}=0$. Thus if $\mathrm{U}=$ const., the only motions are the permanent turnings.

Next considering the effect of supposing $\dot{\mathrm{S}}=0$, Stäckel comes to the conclusion that there are two possibilities, according as the plane $\mathrm{S}=$ const. cuts Staude's cone in a curve, or is one of two planes represented by the equation of the cone. In the first case the second of (11), 11, is rendered an identity by the first: in the other case the second equation remains independent, and the motion is that discovered by Hess and described in 9, above.

In point of fact, in the general case the Eulerian equations are, when $\dot{\mathrm{S}}=0$, not a consequence of the equations (10) and (11) of 11, given by Hess and Schiff, and a supplementary equation is required. The conclusion reached is that here $\dot{p}$, $\dot{q}$, $\dot{r}$ are to be found from the equations

$$\mathrm{A}p\dot{p}+\mathrm{B}q\dot{q}+\mathrm{C}r\dot{r}=\dot{\mathrm{T}},\quad a\mathrm{A}\dot{p}+b\mathrm{B}\dot{q}+c\mathrm{C}\dot{r}=0,\quad \mathrm{P}\dot{p}+\mathrm{Q}\dot{q}+\mathrm{R}\dot{r}=0, \quad\text{.............(4)}$$

where

$$\mathrm{P}=(\mathrm{C}-\mathrm{A})br+(\mathrm{A}-\mathrm{B})cp,\quad \mathrm{Q}=(\mathrm{A}-\mathrm{B})cp+(\mathrm{B}-\mathrm{C})ar,\quad \mathrm{R}=(\mathrm{B}-\mathrm{C})aq+(\mathrm{C}-\mathrm{A})bp.$$

**16.** *Case in which* $S=0$. As an example take the case in which Staude's cone breaks down into two planes of which the plane $\mathrm{S}=0$ is one. Then two of the principal moments of inertia are equal, and the axes may be so chosen that a principal plane passes through G, the centroid. The equation $\mathrm{S}=0$ may now be written as

$$a\mathrm{A}x+b\mathrm{B}y=0. \quad\text{..........................................(1)}$$

The equation of the cone is $(B-C)ay+(C-A)bx=0,$ ..........(2)

so that we have $(B-C)a/Bb=(C-A)b/Aa$, that is

$$A(B-C)a^2=B(C-A)b^2. \qquad (3)$$

Equation (1) shows that the vector of A.M. is at right angles in this case to the line OG, joining the fixed point to the centroid.

It will be seen that this is the case of motion discovered by Hess and discussed in 8, above.

It is interesting to notice that the second of (11), 11, is in the present case no consequence of the first, and takes the form $U\dot{T}=T\dot{U}$, which gives

$$T=\lambda U, \qquad (4)$$

where $\lambda$ is some constant. Using this relation in the third equation, we get

$$\dot{U}^2=l^2(2U-G^2)-2U(E-\lambda U)^2. \qquad (5)$$

Thus

$$t=\int\frac{dU}{\{l^2(2U-G^2)-2U(E-\lambda U)^2\}^{\frac{1}{2}}}, \qquad (6)$$

that is $t$ is an elliptic integral of the first kind in U, and U is an elliptic function of the time.

If in the equations $aAp+bBq=0,\quad T-\lambda U=0$ ..........(7)

we put $\lambda=1/C$, we obtain the distribution of matter required in the motion discovered by Hess, and the first equation shows that the axis of rotation has the direction which that motion requires. The equation $CT=U$ ..........(8)

is the supplementary equation required in the present case, to render the Hess-Schiff equations equivalent to the Euler-Poisson equations.

**17.** *Motions when $S=0$.* When $S=0$ the only possible motions are the motion discovered by Hess, permanent turnings, and permanent or continued pendulum motions. We proceed to consider the pendulum motions.

Let us find first the conditions under which the resultant vector of A.M. remains in a plane fixed in the body. Going back to 14, and supposing the instantaneous axis OI fixed in the body, we have $p, q, r=(e, f, g)\omega$, with $e^2+f^2+g^2=1$. These values of $p, q, r$ used in (1), 11, multiplied respectively by $a, b, c$, give

$$(aAe+bBf+cCg)\dot{\omega}=\{(B-C)afg+(C-A)bge+(A-B)cef\}\omega^2. \qquad (1)$$

In the same way, when the equations (1), 11, are multiplied by $\alpha, \beta, \gamma$, they give

$$(\alpha Ae+\beta Bf+\gamma Cg)\dot{\omega}=\{(B-C)\alpha fg+(C-A)\beta ge+(A-B)\gamma ef\}\omega^2. \qquad (2)$$

In order that (1) and (2) may hold simultaneously we must have either $\dot{\omega}=0$ and $\omega=0$, or

$$\begin{vmatrix}(B-C)afg+(C-A)bge+(A-B)cef, & aAe+bBf+cCg, \\ (B-C)\alpha fg+(C-A)\beta ge+(A-B)\gamma ef, & \alpha Ae+\beta Bf+\gamma Cg,\end{vmatrix}=0. \qquad (3)$$

The second alternative must of course be taken.

But also we have $e\dot{\alpha}+f\dot{\beta}+g\dot{\gamma}=0$, and so

$$e\alpha+f\beta+g\gamma=\kappa, \qquad (4)$$

where $\kappa$ is a constant. The direction cosines $\alpha, \beta, \gamma$ satisfy the three equations (3), (4), and $\alpha^2+\beta^2+\gamma^2=1$, and are therefore constants. Hence by (2) of 11 we have

$$\beta g-\gamma f=0,\quad \gamma e-\alpha g=0,\quad \alpha f-\beta e=0,$$

and therefore $\alpha=e, \beta=f, \gamma=g$, that is OI is vertical. We have thus a permanent turning about the vertical of kinetic energy given by

$$T=\tfrac{1}{2}(Ae^2+Bf^2+Cg^2)\omega^2.$$

But if we insert the values of $e, f, g$ in (2) we get

$$(Ae^2+Bf^2+Cg^2)\dot{\omega}=0. \quad \ldots\ldots(5)$$

This result is obviously true, for since the body turns about a vertical axis through O, fixed in the body, the level of the centroid G does not change, and so the potential energy is constant.

Again from (3) and the vanishing of $(B-C)\alpha fg+(C-A)\beta ge+(A-B)\gamma ef$, as an identity, by the values of $\alpha, \beta, \gamma$, we get, since $fg=qr/\omega^2, \ldots, \ldots,$

$$(B-C)aqr+(C-A)brp+(A-B)cpq=0,$$

that the axis is a generator of Staude's cone.

These results are independent of any assumption regarding the distribution of mass.

**18.** *Motion when $S=0$. Pendulum motions.* Now writing $\rho$ for $(B-C)afg+(C-A)bge+(A-B)cef$ and $\sigma$ for $aAe+bBf+cCg$ we have for (3), 17, the form

$$\{\sigma(B-C)fg-\rho Ae\}\alpha+\{\sigma(C-A)ge-\rho Bf\}\beta+\{\sigma(A-B)ef-\rho Cg\}\gamma=0. \quad \ldots\ldots(1)$$

If we put $\kappa$ in (4), 17, equal to zero, we have also

$$e\alpha+f\beta+g\gamma=0. \quad \ldots\ldots(2)$$

The expression on the left of the last equation does not vanish term by term.

Equation (1) must either be a consequence of (2), or the left-hand side must vanish term by term. Let us assume the former. Then, if M be a multiplier, we get the equations

$$\left.\begin{aligned} eM-(B-C)fg\sigma+Ae\rho=0,\\ fM-(C-A)ge\sigma+Bf\rho=0,\\ gM-(A-B)ef\sigma+Cg\rho=0. \end{aligned}\right\} \quad \ldots\ldots(3)$$

Thus we have the determinantal equation

$$\left.\begin{aligned} (B-C)^2f^2g^2+(C-A)^2g^2e^2+(A-B)^2e^2f^2=0,\\ \text{which gives} \quad (B-C)fg=0, \quad (C-A)ge=0, \quad (A-B)ef=0. \end{aligned}\right\} \quad \ldots\ldots(4)$$

Hence either $B=C$ and $e=0$, or $e=f=0$ and $g=1$. The former gives a symmetrical top, which we need not consider; the latter by (1), 17, gives $c=0$, and therefore $\sigma=0$. It follows by (3) that $M=0$. Thus (1) is not a consequence of (2); it must therefore, by (3), vanish term by term. Hence we get

$$\sigma(B-C)fg=\rho Ae, \quad \sigma(C-A)ge=\rho Bf, \quad \sigma(A-B)ef=\rho Cg.$$

Multiplying these in order by $e, f, g$ and adding we get

$$\rho(Ae^2+Bf^2+Cg^2)=0,$$

and therefore $\rho=0$. It follows that $\sigma$ is also zero. Hence we get the equations

$$\left.\begin{aligned} aAe+bBf+cCg=0,\\ (B-C)afg+(C-A)bge+(A-B)cef=0. \end{aligned}\right\} \quad \ldots\ldots(5)$$

The first of these shows that the vector of A.M. is perpendicular to the line OG, the second that the instantaneous axis OI lies in the plane containing the vector of A.M. and the line OG.

Now take the case of $e=f=0$ and $c=0$. Then $r=\omega$ and the A.M. is $C\omega$. The axis OI and that of A.M. coincide. The third equation of (1), 11, becomes [since by (2) $\gamma Cg=0$, so that $\gamma=\cos\theta=0$]

$$C\dot{\omega}=-(a\sin\phi+b\cos\phi). \quad \ldots\ldots(6)$$

The top swings as a pendulum about the horizontal line through O, which is at right angles to the principal plane containing the centroid. By this motion equations (1) and (2) are satisfied.

We have thus obtained the theorem: *If the centroid of a top under gravity lies in a principal plane through O, the equations of motion* (1) *and* (2) *of* 11 *are satisfied by ordinary pendulum motion about the horizontal normal to this plane.* This theorem expresses the results arrived at by Mlodzjejowskij, who investigated these motions [*Moscow, Phys. Sect. of the Imp. Russian Assoc. of Friends of Nat. Sci.*, 7 (1894)]. The method of proof here followed is due to Stäckel [*Math. Ann.*, 67, 1909]. Stäckel completes the discussion by proving that the pendulum motion specified in the theorem just stated is the only permanent motion of this sort. Moreover for this motion it is shown that the Hess-Schiff equations are not equivalent to the Euler-Poisson equations. The reader must consult the original paper for these and other interesting results.

**19.** *Motion when S=0. Distinctions between cases.* As has been stated, the three motions possible when S is and remains zero are the permanent turnings, the continued pendulum motions, and the case of motion discovered by Hess. To distinguish between the different cases, Stäckel calls a top, the centroid of which lies in a principal plane, a *planar top*, if Hess's distribution of mass is not fulfilled, and a *Hessian top* if the condition is satisfied. In all other cases of unsymmetry he called the top a *general top*.

For a general top Staude's cone is a veritable surface of the second degree, and the plane S=0 either passes through the vertex, or touches it along a generator, or cuts it in two generators. In the first of these cases the top is at rest, in the two others the motions are permanent turnings.

In the case of a planar top the cone splits into two planes, one of which contains the centroid, while the second is perpendicular to the first. We may call these the Staude planes. The plane S=0 is distinct from these and intersects each in a straight line. The intersection with the first plane is an axis of permanent turning, the other line of intersection is an axis of pendulum motion. About this a permanent turning of infinite angular speed is possible.

In the case of the Hessian top, the intersection of the plane S=0 with the first of the two planes into which the cone has degenerated, is again an axis of permanent turning. The second of these is however identical with the plane S=0, and gives therefore an infinite number of axes of permanent turning all passing through O. Further the plane S=0 is perpendicular to the principal plane which contains the centroid, and therefore contains the third principal axis. For this axis the angular speed of the permanent turning is infinite, hence there exists an infinite number of continued pendulum motions about it. Moreover there is a family of Hessian motions, for which the instantaneous axis turns in the second Staude plane about O.

Finally the equation of the Staude cone is identically satisfied by the line

$$p:q:r=a/\mathrm{A}:b/\mathrm{B}:c/\mathrm{C}.$$

Hence every unsymmetrical top may have a motion of permanent turning about this line as axis. For a planar top $c=0$, and therefore also $r=0$; hence the line just referred to is not an axis of pendulum motion.

**20.** *Steady motion of an unsymmetrical top. Vibrations about steady motion.* The conditions of steady motion of an unsymmetrical top have also been considered by Routh (*Adv. Dynamics*, p. 164, sixth edition). He finds that if no two of the principal moments of inertia are equal, steady motion is not possible unless the axis of rotation be vertical and the centroid (coordinates $h$, $k$, $l$) lie in the vertical straight line of which the equations are

$$g\frac{x}{\alpha}-\mathrm{A}\omega^2=g\frac{y}{\beta}-\mathrm{B}\omega^2=g\frac{z}{\gamma}-\mathrm{C}\omega^2,$$

where $\omega$ is the angular speed about the vertical, and the mass is taken as unity. This is the result given in (4), 14. To construct the straight line, measure along the vertical a length $OV = g/\omega^2$ [$g$ = acceleration due to gravity], and draw a horizontal plane through V to touch an ellipsoid confocal with the ellipsoid of gyration. The centroid must lie on the normal at the point of contact.

The investigation of the oscillations about steady motion involves an equation of the form $\alpha = \cos\alpha' + P_0 \sin\mu t + P_1 \cos\mu t$ for $\alpha$, and similar equations for $\beta$, $\gamma$. Substitution in the equations (1), (2), 11, gives twelve linear equations for eleven ratios, elimination of which gives an equation for $\mu$. For stability this must have real roots.

The reader may refer to Routh (*loc. cit.*) for a discussion of a variety of other problems connected with the motion of unsymmetrical tops.

**21.** *Homogeneous ellipsoid spinning on a horizontal plane.* A homogeneous ellipsoid spins without friction on a horizontal plane, with a principal axis inclined at a small angle $\theta$ to the vertical. It is required to discuss the stability of the motion.

Let the coordinates of the point of contact with the horizontal plane be $x$, $y$, $z$, and the axis of $z$ be the downward nearly vertical axis. We refer to (1) 1, as the equations of motion. The equation of the ellipsoid

$$\frac{x^2}{a^2}+\frac{y^2}{b^2}+\frac{z^2}{c^2}=1$$

gives for the direction cosines of the normal at the point $x$, $y$, $z$

$$\alpha=\frac{\varpi x}{a^2},\quad \beta=\frac{\varpi y}{b^2},\quad \gamma=1=\frac{\varpi z}{c^2}, \quad\text{.........................(1)}$$

where $\varpi$ is the length of the perpendicular from the centre on the tangent plane at $x$, $y$, $z$. But here $\varpi = c$, so that

$$x=\frac{\alpha a^2}{c},\quad y=\frac{\beta b^2}{c},\quad z=c. \quad\text{..............................(2)}$$

Thus $$\gamma y-\beta z=\frac{b^2-c^2}{c}\beta,\quad \alpha z-\gamma x=\frac{c^2-a^2}{c}\alpha,\quad \beta x-\alpha y=\frac{a^2-b^2}{c}\alpha\beta. \quad\text{......(3)}$$

Now when $\theta$ is small we see from (1), 1, that $\omega_2\omega_3$ is small. Hence we get, approximately, from the third of the equations of motion, (1), 1,

$$\dot\omega_2=0. \quad\text{.........................................(4)}$$

But, adopting here the arrangement of axes and angular velocities $\omega_1$, $\omega_2$, $\omega_3$ explained in 4, XVI, we obtain, also approximately,

$$\omega_3=\dot\phi+\dot\psi, \quad\text{....................................(5)}$$

and so $$\left.\begin{aligned}\omega_1&=\theta(\omega_3-\dot\phi)\sin\phi+\dot\theta\cos\phi=\dot\beta+\alpha\omega_3,\\ \omega_2&=\theta(\omega_3-\dot\phi)\cos\phi-\dot\theta\sin\phi=-\dot\alpha+\beta\omega_3.\end{aligned}\right\}\quad\text{...................(6)}$$

Moreover, since $z$ is approximately constant, the vertical acceleration of the centroid may be taken as zero, and so

$$R=-Mg. \quad\text{....................................(7)}$$

By using (6) and (3) in the equations of motion we find, after reduction,

$$\left.\begin{aligned} &A\ddot{\beta}+(A+B-C)\omega_3\dot{\alpha}+\left\{(C-B)\omega_3{}^2+Mg\frac{b^2-c^2}{c}\right\}\beta=0,\\ &B\ddot{\alpha}-(A+B-C)\omega_3\dot{\beta}+\left\{(C-A)\omega_3{}^2+Mg\frac{a^2-c^2}{c}\right\}\alpha=0;\end{aligned}\right\} \quad \text{..........(8)}$$

or $$A\ddot{\beta}+D\dot{\alpha}+E\beta=0, \quad B\ddot{\alpha}-D\dot{\beta}+F\alpha=0, \quad \text{....................(9)}$$

where D, E, F have the values indicated in (8).

If we assume, for an oscillatory solution of period $2\pi/n$,

$$\alpha=Ke^{int}, \quad \beta=Le^{int}, \quad \text{..............................(10)}$$

and substitute in (9), we get, by eliminating K and L from the resulting equations, the quadratic for $n^2$,

$$ABn^4-(AF+BE+D^2)n^2+EF=0. \quad \text{....................(11)}$$

The roots of this equation are real if

$$(AF+BE+D^2)^2>4ABEF,$$

and are positive if $EF>0$ and $AF+BE+D^2>0$.

They must, for stability, be both real and positive.

We consider two cases, (1) $\omega_3$ very small, and (2) $\omega_3$ as great as may be required for stability.

It is easy to verify that in case (1) the first inequality is satisfied, and that the two others become

$$(a^2-c^2)(b^2-c^2)>0, \quad A(a^2-c^2)+B(b^2-c^2)>0.$$

These will both be satisfied by taking $a>c$ and $b>c$. Thus, if the shortest axis of the ellipsoid be vertical, the body will be in stable equilibrium, with no spin at all or only a slight spin about the vertical axis.

In case (2) a little analysis shows that the inequalities are

$$\left.\begin{aligned} &(a^2+b^2)^2\left(\frac{c\omega_3{}^2}{5g}\right)^2+2(a^4+b^4-2c^4)\frac{c\omega_3{}^2}{5g}+(a^2-b^2)^2>0,\\ &(a^2-c^2)(b^2-c^2)>0,\\ &\frac{c\omega_3{}^2}{5g}>\frac{c^4-a^2b^2}{c^4+a^2b^2}.\end{aligned}\right\} \quad \text{.........(12)}$$

These are clearly all satisfied if the shortest axis of the ellipsoid is vertical. If the intermediate axis is vertical the second inequality is not satisfied, and the ellipsoid cannot spin in equilibrium with that axis vertical. Finally, if the longest axis be vertical the spin must be great enough to satisfy the last inequality, and must also be such as to satisfy the first inequality. A value of $c\omega_3{}^2/5g$ which is greater than the larger root, or smaller than the smaller root, of the equation

$$(a^2+b^2)^2u^2+2(a^4+b^4-2c^4)u+(a^2-b^2)^2=0 \quad \text{...............(13)}$$

will satisfy this inequality. Hence a value of $c\omega_3{}^2/5g$ which satisfies the third inequality and does not lie between the roots of the quadratic equation just written will enable the ellipsoid to be stable with the longest axis vertical.

The roots of the quadratic (13) are

$$\left\{\frac{(c^4-a^4)^{\frac{1}{2}}\pm(c^4-b^4)^{\frac{1}{2}}}{a^2+b^2}\right\}^2.$$

If we substitute $(c^4-a^2b^2)/(c^4+a^2b^2)$ for $u$ in the quadratic we obtain an expression which vanishes when $c^4=0$, when $c^4=a^4$ and when $c^4=b^4$, and as we easily see can be written

$$-4\frac{c^4(c^4-a^4)(c^4-b^4)}{(c^4+a^2b^2)^2}.$$

This expression is essentially negative, and hence we conclude that the limiting value of $c\omega_3^2/5g$ lies between the roots of the quadratic. To satisfy all the conditions necessary for real and positive roots we must therefore have for stability with the longest axis vertical

$$\frac{c\omega_3^2}{5g}>\left\{\frac{(c^4-a^4)^{\frac{1}{2}}+(c^4-b^4)^{\frac{1}{2}}}{a^2+b^2}\right\}^2$$

or

$$\omega_3>\frac{(5g)^{\frac{1}{2}}}{c^{\frac{1}{2}}}\frac{(c^4-a^4)^{\frac{1}{2}}+(c^4-b^4)^{\frac{1}{2}}}{a^2+b^2}. \quad \text{.....................(14)}$$

It is interesting to observe the dependence of the minimum value thus obtained for the angular speed upon the dimensions of the body. If all the dimensions be increased or diminished in the same ratio, no change will be produced except in the factor $1/(c)^{\frac{1}{2}}$ of the expression on the right of (14). Thus, if $c$ be changed to $kc$ in this way, $\omega_3$ will become $\omega_3/k^{\frac{1}{2}}$. The limiting speed of rotation thus depends on the actual, not the relative, lengths of the axes of the body.

**22.** *Stability of any solid with a principal axis normal to the horizontal surface.* The foregoing discussion of the stability of a spinning ellipsoid follows, with some modifications of notation and analysis, an important memoir by Puiseux in Liouville's *Journal de Mathématiques*, 17 (1852). In the same memoir Puiseux considered also the problem of the stability of a solid of any form and distribution of matter, consistent with having one at least of the principal axes of moment of inertia normal to the bounding surface. The body is placed with that axis vertical.

The discussion of this case proceeds on the same lines. We have as before

$$\left.\begin{aligned}\omega_1&=\dot\theta\cos\phi+\dot\psi\sin\theta\sin\phi, \quad \omega_2=-\dot\theta\sin\phi+\dot\psi\sin\theta\cos\phi,\\ \alpha&=\sin\theta\sin\phi, \quad \beta=\sin\theta\cos\phi, \quad \gamma=\cos\theta.\end{aligned}\right\} \quad \text{..............(1)}$$

Instead of the equation of the ellipsoid we have, taking the origin at the centroid,

$$z=c-\tfrac{1}{2}(lx^2+2mxy+ny^2), \quad \text{..................................(2)}$$

and the direction cosines of the normal with reference to the principal axes are given by

$$\frac{lx+my}{\alpha}=\frac{mx+ny}{\beta}=\frac{1}{\gamma}. \quad \text{........................................(3)}$$

If we give to $\gamma$ its approximate value 1 we get

$$x=-\frac{m\beta-n\alpha}{ln-m^2},\quad y=-\frac{m\alpha-l\beta}{ln-m^2}. \quad\text{.....(4)}$$

For the point of contact we have approximately $z=c$, and since the deflection of the axis of $z$ from verticality is supposed to be very small and the curvature is continuous, with the axis of $z$ normal to the surface, we have $\dot{z}=0$, and so, since $z$ is downward,

$$R=-Mg. \quad\text{.....(5)}$$

Going back to the Eulerian equations of motion, we note that now

$$\gamma y-\beta z=c\beta+\frac{m\alpha-l\beta}{ln-m^2},\quad \alpha z-\gamma x=-c\alpha-\frac{m\beta-n\alpha}{ln-m^2}, \quad\text{.....(6)}$$

and since $\omega_1$, $\omega_2$ are both of the first order of small quantites $\omega_1\omega_2$ is negligible, and we have

$$\dot{\omega}_3=0. \quad\text{.....(7)}$$

Thus the equations of motion become, by the same process as that used in 21,

$$\left.\begin{aligned}A\ddot{\beta}+D\dot{\alpha}+E\beta-G\alpha=0,\\ B\ddot{\alpha}-D\dot{\beta}+F\alpha-G\beta=0,\end{aligned}\right\} \quad\text{.....(8)}$$

where

$$\left.\begin{aligned}&D=(A+B-C)\omega_3,\quad G=Mg\frac{m}{ln-m^2},\\ &E=(C-B)\omega_3{}^2+Mg\left(\frac{l}{ln-m^2}-c\right),\\ &F=(C-A)\omega_3{}^2+Mg\left(\frac{n}{ln-m^2}-c\right).\end{aligned}\right\} \quad\text{.....(9)}$$

If $\rho_1$, $\rho_2$ be the principal radii of curvature at the point of support, and $\vartheta$ be the angle which the plane $Gzx$ makes with the principal section of curvature $1/\rho_1$, we get

$$\left.\begin{aligned}l=\frac{1}{\rho_1}\cos^2\vartheta+\frac{1}{\rho_2}\sin^2\vartheta,\quad n=\frac{1}{\rho_1}\sin^2\vartheta+\frac{1}{\rho_2}\cos^2\vartheta,\\ m=\left(\frac{1}{\rho_1}-\frac{1}{\rho_2}\right)\sin\vartheta\cos\vartheta.\end{aligned}\right\} \quad\text{.....(10)}$$

Thus (9) become

$$\left.\begin{aligned}&E=(C-B)\omega_3{}^2+Mg\{(\rho_1-c)\sin^2\vartheta+(\rho_2-c)\cos^2\vartheta\},\\ &F=(C-A)\omega_3{}^2+Mg\{(\rho_1-c)\cos^2\vartheta+(\rho_2-c)\sin^2\vartheta\},\\ &G=Mg(\rho_2-\rho_1)\sin\vartheta\cos\vartheta.\end{aligned}\right\} \quad\text{.....(11)}$$

For (8) we assume the oscillatory solution $\alpha=h\cos nt+h'\sin nt$, $\beta=k\cos nt+k'\sin nt$, and substituting we get by equating to zero the coefficients of $\cos nt$ and $\sin nt$ in the two resulting equations

$$\left.\begin{aligned}(An^2-E)k-Dnh'+Gh=0\\ (An^2-E)k'+Dnh+Gh'=0,\\ (Bn^2-F)h+Dnk'+Gk=0,\\ (Bn^2-F)h'-Dnk+Gk'=0.\end{aligned}\right\} \quad\text{.....(12)}$$

Solving for $h'$ and $k'$ from the first and third of (12), and substituting in either of the other equations, we obtain the relation

$$(An^2-E)(Bn^2-F)-D^2n^2-G^2=0. \quad\text{.....(13)}$$

The roots of this quadratic in $n^2$ must be real and positive if the motion is as assumed. For this three conditions must be fulfilled,

$$\left.\begin{aligned}(AF+BE+D^2)^2-4AB(EF-G^2)>0,\\ AF+BE+D^2>0,\\ EF-G^2>0.\end{aligned}\right\} \quad\text{.....(14)}$$

These inequalities can be written in the form

$$\left.\begin{array}{l} C^2(A+B-C)^2\omega_3^4+H\omega_3^2+I>0, \\ (C-A)(C-B)\omega_3^4+H'\omega_3^2+I'>0, \\ \{(C-A)(C-B)+AB\}\omega_3^2+K>0, \end{array}\right\} \qquad (14')$$

where the coefficients H, I, H′, I′, K do not depend on $\omega_3$. From these it is clear that if the principal axis which is made vertical be that of greatest or of least moment of inertia, the motion can be made stable by sufficiently increasing the value of $\omega_3$.

If the spin be made very small the conditions of stability become

$$I>0, \quad I'>0, \quad K>0.$$

It will be found that these last inequalities can be written

$$[(A\cos^2\vartheta+B\sin^2\vartheta)(\rho_1-c)-(A\sin^2\vartheta+B\cos^2\vartheta)(\rho_2-c)]^2 \\ +4(A-B)^2\sin^2\vartheta\cos^2\vartheta(\rho_1-c)(\rho_2-c)>0,$$

$$(\rho_1-c)(\rho_2-c)>0,$$

$$(A\cos^2\vartheta+B\sin^2\vartheta)(\rho_1-c)+(A\sin^2\vartheta+B\cos^2\vartheta)(\rho_2-c)>0.$$

The second and third conditions are satisfied if $\rho_1$ and $\rho_2$ are both greater than $c$. Thus we have the result, which of course could have been established by a more direct process, and is indeed obvious without any analysis at all, that a body placed on a horizontal plane with the centroid on the normal at the point of support will be in stable equilibrium if both the principal radii of curvature at the point of support are greater than the distance $c$ of the centroid above the plane.

# CHAPTER XVIII

## THE RISING OF A SYMMETRICAL TOP SUPPORTED ON A HORIZONTAL SURFACE.

1. *A top supported on a rounded peg.* We now consider a top supported, not necessarily at a point on the axis, but at a point on a surface of revolution about the axis of figure as shown in Fig. 96. This is the case of a solid of revolution spinning and rolling on a horizontal plane under the influence of gravity and frictional forces, in short the case of a top spinning on a rounded peg. At first we introduce no condition as to slipping at the point of contact. Let G be the centroid, GC the axis of figure. Draw a vertical through G meeting the horizontal plane in M, and a perpendicular from the point of contact O to the axis of figure, meeting that line in N. Denote GN by $z$ and ON by $x$.

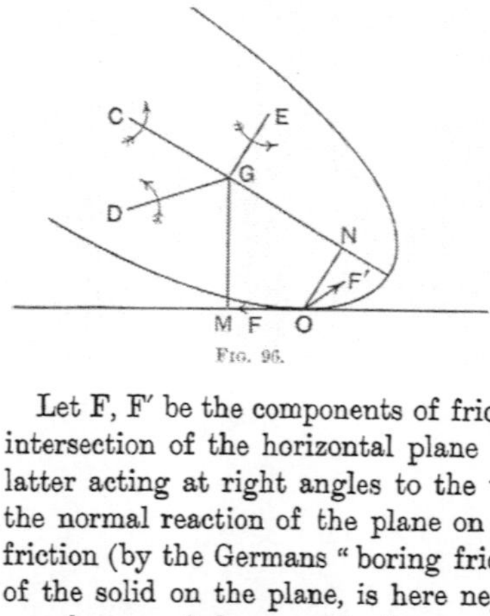

Fig. 96.

Let F, F′ be the components of friction at O, the former acting along the intersection of the horizontal plane and the vertical plane GOM, and the latter acting at right angles to the vertical plane as shown, and let R be the normal reaction of the plane on the solid at O. What is called pivot friction (by the Germans "boring friction"), the resistance due to spinning of the solid on the plane, is here neglected. For a body resting on what may be regarded as practically a point, its moment is very small. Let $\theta$ be the inclination of GC to the upward vertical.

Now take axes at G, one along GC, and the others, GD, GC, at right angles respectively to the plane GOC, and to the line GE in the plane GOC, all as shown in Fig. 96. If the solid turn about the vertical at speed $\dot{\psi}$, counter-clockwise to an observer looking downward, the A.M. about the axes just specified are $Cn$, $A\dot{\theta}$, $A\dot{\psi}\sin\theta$. Hence for the rate of growth of A.M. about GD we get, by the process used above, the expression

$$A\ddot{\theta}+(Cn-A\dot{\psi}\cos\theta)\dot{\psi}\sin\theta.$$

But the moment of force about GD is clearly

$$R(z\sin\theta - x\cos\theta) - F(z\cos\theta + x\sin\theta).$$

Hence we have

$$A\ddot{\theta} + (Cn - A\dot{\psi}\cos\theta)\dot{\psi}\sin\theta = R(z\sin\theta - x\cos\theta) - F(z\cos\theta + x\sin\theta). \quad (1)$$

Again, as in (2), 1, VII, we see that the rate of growth of A.M. about GE is $A\ddot{\psi}\sin\theta + (2A\dot{\psi}\cos\theta - Cn)\dot{\theta}$, and the only force with moment about GE is F′. Hence

$$A\ddot{\psi}\sin\theta + (2A\dot{\psi}\cos\theta - Cn)\dot{\theta} = F'z. \quad \text{.....................}(2)$$

The motion of the axes produces no change of A.M. about GC, and therefore

$$C\dot{n} + F'x = 0. \quad \text{.................................}(3)$$

The force F′ if positive thus gives a moment diminishing $Cn$; to this must be added in actual practice the couple due to air friction. F′ will be positive or negative according to the motion of the point O of the top [see 8, below].

The total couple given by F′ is F′. OG; thus besides the component specified in (3), there is a component couple of moment F′. GN. The axis of this is in the plane OGN and points upwards at right angles to GN. The extremity of the axis of resultant A.M. will move in the direction in which points the resultant of the three couples which have now been specified. The motion due to the first and third has a vertically upward component and a horizontal component, and hence, unless there is rapid loss of spin, the axis will rise. The motion of the axes may now be discussed in detail.

Let $u$, $v$ be the speeds of the centroid parallel to OM and perpendicular to the plane GOC respectively when the rolling is pure. Then since O, taken as a point on the top, is supposed to be at rest,

$$u = MG\,.\,\dot{\theta} = (z\cos\theta + x\sin\theta)\dot{\theta}, \quad v = xn - \dot{\psi}z\sin\theta, \quad \text{..............}(4)$$

where $v$ is taken in the direction DG. Equations (1), (2), (3) hold also when there is slipping of the point of contact.

The rates of growth of momentum in these directions and along MG are given according to the equations of moving axes by

$$M(\dot{u} + v\dot{\psi}) = F, \quad M(\dot{v} - u\dot{\psi}) - F', \quad M\ddot{\zeta} = R - Mg, \quad \text{..............}(5)$$

where $\zeta$ is the vertical height of the centroid above the horizontal plane.

Equations (1), ... (5) give the whole motion. A relation between $x$ and $z$ is of course given by the form of the surface. For steady motion $\ddot{\theta} = 0$; $\dot{\theta} = 0$, $\dot{u} = 0$, $\ddot{\zeta} = 0$, $\ddot{\psi} = 0$. Thus if $\dot{\psi} = \mu$, a constant,

$$F' = 0, \quad -F = M\mu(\mu z\sin\theta - nx), \quad R = Mg.$$

Hence (1) becomes

$$\{(C + Mx^2)n - (A + Mz^2)\mu\cos\theta\}\mu\sin\theta + Mxz(n\mu\cos\theta - \mu^2\sin^2\theta)$$
$$= Mg\,(z\sin\theta - x\cos\theta). \quad \text{...}(6)$$

Of course F is constant when the motion is steady, and (4) gives

$$v = xn - \dot{\psi}z\sin\theta = F/M\mu.$$

Thus $v$ is constant. The direction of $v$ turns round with uniform angular speed $\mu$, and therefore G moves in a circle and M in a parallel circle in the horizontal plane. But $\mu v$ is the acceleration of M towards the centre of the circle. Hence if $r$ be the radius of the circle, $\mu v = v^2/r$ and $r = v/\mu$. Thus

$$r = \frac{nx}{\mu} - z\sin\theta. \quad \text{..............................(7)}$$

This result, (7), is obvious without calculation. For in the steady motion the position of the point N will move parallel to the horizontal plane with speed $nx$, in consequence of the spin $n$ about the axis of figure, and the fact that the rolling is pure. The angular speed about the centre of the circular path of N is $\mu$, and thus the radius of that path is $nx/\mu$, and we obtain (7).

But from (6), if $\mu$ and $x$ be small we have (including the A.M. of the flywheel)

$$(\mathrm{K} + \mathrm{C}n)\mu = \mathrm{M}gz. \quad \text{..............................(8)}$$

From (7), by the value found above for F, that is $\mathrm{M}\mu(nx - \mu z\sin\theta)$, we get

$$r = \frac{\mathrm{M}gk}{\mathrm{M}\mu^2} = \frac{gk}{\mu^2} = \frac{k}{g}\frac{(\mathrm{K}+\mathrm{C}n)^2}{\mathrm{M}^2 z^2}. \quad \text{.........................(9)}$$

The azimuthal motion has the counter-clock direction to an observer looking from above on the solid, and therefore the circle in which O moves has the position shown in Fig. 98. This circle and the projection of that in which G moves are both indicated in Fig. 98.

2. *Varying motion of a top on a rounded peg.* Now returning to the unsteady motion, we notice first that

$$\dot{x} = \rho\cos\theta\,.\,\dot{\theta}, \quad \dot{z} = -\rho\sin\theta\,.\,\dot{\theta}, \quad \text{........................(1)}$$

where $\rho$ is the radius of curvature at O of the section of the body which contains O and the axis of figure. By means of these values of $\dot{x}$, $\dot{z}$, and those of $u$, $v$, $\zeta$, and noticing that $\zeta = z\cos\theta + x\sin\theta$, and that by (1) $\dot{z}\sin\theta + \dot{x}\cos\theta = 0$, we find easily

$$\left.\begin{aligned}
\mathrm{M}(\dot{u} + v\dot{\psi}) &= \mathrm{M}\{(z\cos\theta + x\sin\theta)\ddot{\theta} - (z\sin\theta - x\cos\theta)\dot{\theta}^2 \\
&\qquad + (xn - z\sin\theta\,.\,\dot{\psi})\dot{\psi}\} = \mathrm{F}, \\
\mathrm{M}(\dot{v} - u\dot{\psi}) &= \mathrm{M}\{\rho(n\cos\theta + \dot{\psi}\sin^2\theta)\dot{\theta} + x\dot{n} - z\sin\theta\,.\,\ddot{\psi} \\
&\qquad - (x\sin\theta + 2z\cos\theta)\dot{\theta}\dot{\psi}\} = \mathrm{F}', \\
&\mathrm{M}\{g + \rho\dot{\theta}^2 - (z\sin\theta - x\cos\theta)\ddot{\theta}\} = \mathrm{R}.
\end{aligned}\right\} \quad \text{.........(2)}$$

Substituting the values of F and R from these equations in (1), 1, we get

$$\begin{aligned}
&\{\mathrm{A} + \mathrm{M}(x^2+z^2)\}\ddot{\theta} - \mathrm{M}\rho(z\sin\theta - x\cos\theta)\dot{\theta}^2 - \mathrm{M}\{(z^2 - x^2)\sin\theta\cos\theta \\
&\qquad - xz(\cos^2\theta - \sin^2\theta\}\dot{\theta}^2 - \{(\mathrm{A} + \mathrm{M}z^2)\cos\theta + \mathrm{M}xz\sin\theta\}\dot{\psi}^2\sin\theta \\
&\quad + \mathrm{M}x(z\cos\theta + x\sin\theta)n\dot{\psi} + \mathrm{C}n\dot{\psi}\sin\theta - \mathrm{M}g(z\sin\theta - x\cos\theta) = 0. \quad \text{.......(3)}
\end{aligned}$$

Again, by (2) and (3), 1, we get

$$\mathrm{C}z\dot{n} + \mathrm{A}x\sin\theta\,.\,\ddot{\psi} + (2\mathrm{A}\cos\theta\,.\,\dot{\psi} - \mathrm{C}n)x\dot{\theta} = 0, \quad \text{................(4)}$$

or

$$\mathrm{C}z\frac{dn}{d\theta} + \mathrm{A}x\sin\theta\frac{d\dot{\psi}}{d\theta} + (2\mathrm{A}\cos\theta\,.\,\dot{\psi} - \mathrm{C}n)x = 0. \quad \text{..............(5)}$$

By eliminating $\ddot{\psi}$ and F′, by means of the second of equations (2) above and (2) and (3) of 1, we find

$$\left(\frac{\mathrm{AC}}{\mathrm{M}}+\mathrm{A}x^2+\mathrm{C}z^2\right)\dot{n}-\mathrm{A}(x-\rho\sin\theta)x\sin\theta\,.\,\dot{\theta}\dot{\psi}+\mathrm{A}\rho x\cos\theta\,.\,n\dot{\theta}-\mathrm{C}nxz\dot{\theta}=0,\quad(6)$$

or
$$\left(\frac{\mathrm{AC}}{\mathrm{M}}+\mathrm{A}x^2+\mathrm{C}z^2\right)\frac{dn}{d\theta}-\mathrm{A}(x-\rho\sin\theta)x\sin\theta\,.\,\dot{\psi}+\mathrm{A}\rho xn\cos\theta-\mathrm{C}nxz=0.\quad(7)$$

Finally from (2), 1, and the second of (2) we get a value of $\dot{n}$ free from F′, and (6) gives another expression for the same quantity. Equating these we obtain

$$\left(\frac{\mathrm{AC}}{\mathrm{M}}+\mathrm{A}x^2+\mathrm{C}z^2\right)(\sin\theta\,.\,\ddot{\psi}+2\cos\theta\,.\,\dot{\theta}\dot{\psi})$$
$$+\mathrm{C}(x-\rho\sin\theta)z\sin\theta\,.\,\dot{\theta}\dot{\psi}-\mathrm{C}z\rho\cos\theta\,.\,n\dot{\theta}+\frac{\mathrm{C}+\mathrm{M}x^2}{\mathrm{M}}\mathrm{C}n\dot{\theta}=0.\quad\ldots\ldots(8)$$

We may of course divide this by $\dot{\theta}$, and write $d\dot{\psi}/d\theta$ for $\ddot{\psi}/\dot{\theta}$.

It will be observed that if $x$ and $\rho$ be zero, we are thrown back on the equations of a top supported on a point and spinning without friction.

**3.** *A top supported on a circular edge round the axis of figure.* From (3), (4), and (6) we get the equation of energy, but further integration seems only possible in particular cases. For example, let the body be supported by a disk, on a sharp edge of which it rolls. A smooth-edged coin, a wine-glass rolling in an inclined position on the

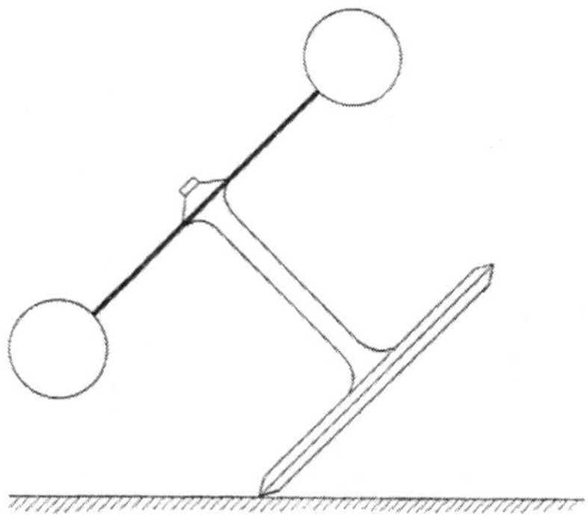

FIG. 97.

edge of its disk-shaped support, or a top, as shown in section in Fig. 97, and a child's hoop made of thin but rigid wire, are cases in point. In such a case as this we have approximately $\rho=0$, and $x$, $z$ constant. Then (7), 2, becomes [Greenhill, *R.G.T.*, p. 241]

$$\left(\frac{\mathrm{AC}}{\mathrm{M}}+\mathrm{A}x^2+\mathrm{C}z^2\right)\frac{dn}{d\theta}-\mathrm{A}x^2\sin\theta\,.\,\dot{\psi}-\mathrm{C}nxz=0.\quad\ldots\ldots\ldots\ldots(1)$$

Eliminating $\dot{\psi}$ between this and (5), 2, we obtain

$$\frac{d^2n}{d\theta^2}+\cot\theta\frac{dn}{d\theta}-\mathrm{N}\left(1+\frac{z}{x}\cot\theta\right)n=0,\quad\ldots\ldots\ldots\ldots(2)$$

where
$$\mathrm{N}=\frac{\mathrm{C}x^2}{\frac{\mathrm{AC}}{\mathrm{M}}+\mathrm{A}x^2+\mathrm{C}z^2}.$$

The differential equation (2) is that of a hypergeometric series, and can be solved by the theory of such series. If we make the substitution $w=n(\sin\theta)^{\frac{1}{2}}$, (2) is transformed to

$$\frac{1}{w}\frac{d^2w}{d\theta^2}=\mathrm{N}\left(1+\frac{z}{x}\cot\theta\right)-\tfrac{1}{4}\cot^2\theta-\tfrac{1}{2}. \quad\ldots\ldots(3)$$

But the most useful substitution is that which holds when the circular base on which the body rolls contains the centroid G. Then $z=0$, and $x$ is the radius of the circular base. We take a new variable $\xi$ defined by $2\xi=1-\cos\theta$, or $\xi=\sin^2\frac{1}{2}\theta$. Equation (3) becomes then

$$\xi(1-\xi)\frac{d^2n}{d\xi^2}+(1-2\xi)\frac{dn}{d\xi}-\mathrm{N}n=0. \quad\ldots\ldots(4)$$

This is the particular case of the differential equation of the hypergeometric series

$$\frac{d^2y}{d\xi^2}+\frac{\gamma-(\alpha+\beta+1)\xi}{\xi(1-\xi)}\frac{dy}{d\xi}-\frac{\alpha\beta}{\xi(1-\xi)}y=0, \quad\ldots\ldots(5)$$

for which $\alpha+\beta=1$, $\gamma=1$, $\alpha\beta=\mathrm{N}$.

It is well known [see Forsyth's *Differential Equations*, Chap. VII] that if $\beta$ be positive and $\gamma>\beta$, a solution of (5) is

$$y=\mathrm{B}\int_0^1 v^{\beta-1}(1-v)^{\gamma-\beta-1}(1-v\xi)^{-\alpha}dv, \quad\ldots\ldots(6)$$

where B is a constant. The definite integral in (6) is a generalised form of Euler's first integral

$$\int_0^1 v^{l-1}(1-v)^{m-1}\,dv,$$

where $l$, $m$ are positive constants. By expanding $(1-v\xi)^{-\alpha}$, and calculating the values of the coefficients of different powers of $\xi$, we obtain

$$\mathrm{SB}\int_0^1 v^{\beta-1}(1-v)^{\gamma-\beta-1}\,dv,$$

where $$\mathrm{S}=1+\frac{\alpha\beta}{1.\gamma}\xi+\frac{\alpha(\alpha+1)\beta(\beta+1)}{1.2.\gamma(\gamma+1)}\xi^2+\frac{\alpha(\alpha+1)(\alpha+2)\beta(\beta+1)(\beta+2)}{1.2.3.\gamma(\gamma+1)(\gamma+2)}\xi^2+\cdots,$$

the hypergeometric series. By inserting the special conditions we get a particular solution of the top problem stated for these conditions. [See Greenhill, *loc. cit.*]

4. *A top on a rounded peg and containing a flywheel.* If the body contain a flywheel, the centroid and axis of which coincide respectively with the centroid and axis of figure of the remainder of the body, the equations of motion given above require correction by the substitution of $\mathrm{K}+\mathrm{C}n$ for $\mathrm{C}n$, where K is the A.M. of the flywheel about its axis. M then denotes the whole mass of the body, including the flywheel, and A the moment of inertia of the whole about an axis through the centroid transverse to the axis of figure.

With this change the equation (6), 1, for steady motion becomes

$$\{\mathrm{K}+(\mathrm{C}+\mathrm{M}x^2)n-(\mathrm{A}+\mathrm{M}z^2)\mu\cos\theta\}\mu\sin\theta+\mathrm{M}xz(n\mu\cos\theta-\mu^2\sin^2\theta)$$
$$-\mathrm{M}g(z\sin\theta-x\cos\theta)=0. \quad\ldots(1)$$

This may be written in the form

$$[2\{(\mathrm{A}+\mathrm{M}z^2)\cos\theta+\mathrm{M}xz\sin\theta\}\mu\sin\theta-\{\mathrm{K}+(\mathrm{C}+\mathrm{M}x^2)n\}\sin\theta+\mathrm{M}xzn\cos\theta]^2$$
$$=[\{\mathrm{K}+(\mathrm{C}+\mathrm{M}x^2)n\}\sin\theta+\mathrm{M}xzn\cos\theta]^2$$
$$-4\mathrm{M}g\{(\mathrm{A}+\mathrm{M}z^2)\cos\theta+\mathrm{M}xz\sin\theta\}(z\sin\theta-x\cos\theta)\sin\theta. \quad(2)$$

The quantity on the right in (2) must be positive for steady motion, since the first line is a perfect square. The roots of the quadratic, (1), in $\mu$ are imaginary when the quantity on the right of (2) is negative, and steady motion is then impossible.

If $p$ denote the length of the perpendicular let fall from the centroid on the horizontal plane, and $q$ the distance of the foot of this perpendicular from the point of contact O, then

$$p = z\cos\theta + x\sin\theta, \quad q = z\sin\theta - x\cos\theta.$$

The A.M. G about the vertical through G is given by

$$G = (Cn + K)\cos\theta + A\mu\sin^2\theta, \quad\text{.........................(3)}$$

while that about a horizontal line drawn through G in the meridian plane is

$$H = (Cn + K)\sin\theta - A\mu\sin\theta\cos\theta. \quad\text{.....................(4)}$$

Hence the equation of steady motion can be written

$$H\mu + M(\mu^2 pr - gq) = 0, \quad\text{...............................(5)}$$

where $r$ denotes the radius of the circle described by G in the steady motion (see above, (7), 1).

In this form the equation is obvious. The inward radius from O to the centre of the path of O on the horizontal plane, about which the A.M. is $H + M\mu rp$, is turning with angular speed $\mu$. The rate at which the extremity of the vector is moving in the horizontal plane is $(H + M\mu rp)\mu$, which is the rate of growth of A.M. about a horizontal line through O perpendicular to the radius, the line towards which the vector is turning. The moment of the gravity forces about this line is $Mgq$. Hence equation (5).

From the value of H it is clear that when $\theta$ is small, that is when the axis of the body is nearly vertical, we have

$$\mu^2 pr = gq. \quad\text{....................................(6)}$$

If the body is a flat disk, or to a great extent consists of a flat disk, the edge of which rolls on the table, $p$ is small when $\theta$ is small. But then $q$ is large and therefore $\mu$ is now large. This can be well seen when a coin is spinning with its axis nearly vertical.

If the body is spinning upright (like a prolate ellipsoid or egg-shaped body, with its longest diameter vertical) supported on a base well rounded, we have both $\theta$ and $x$ very small, and $x = \rho\sin\theta$. The equation of steady motion becomes

$$\{K + Cn - (A + Mz^2)\mu\}\mu + M\rho zn\mu - Mg(z - \rho) = 0, \quad\text{..............(7)}$$

where $\rho$ is as before the radius of curvature of an axial section at the point of contact. Steady motion is therefore possible if

$$\{K + (C + M\rho z)n\}^2 > 4Mg(z - \rho)(A + Mz^2). \quad\text{................(8)}$$

This condition is always fulfilled if $z < \rho$; hence the body is stable in the upright position with or without a flywheel, and even if in the latter case the spin is vanishingly small.

From equations (3), (5), 7 we might find that of small oscillations about steady motion. For this we shall suppose $\theta$ to represent the inclination of the axis to the vertical in steady motion, and $\theta+\alpha$ the same angle for a small deviation from the motion. Thus, if $\Phi$ denote the quantity on the left of (1), 4, we may write the general equation, (3) of 2 for the latter case, in the form

$$\{A+M(x^2+z^2)\}\ddot{\alpha}+\frac{d\Phi}{d\theta}\alpha=0, \qquad (9)$$

where after the differentiation the values of the quantities for steady motion are to be used in the second term on the left in the equation just written. This term involves $d\mu/d\theta$ and $dn/d\theta$, and the values of these quantities are to be inserted from (5) and (7) of 4.

**5**. *Problem of a disk or hoop on a horizontal plane.* The general equation thus obtained is very cumbrous, and on the whole it is more convenient and instructive to deal with various special problems separately.

The vibration equation for the case of $\dot{n}=0$ will be given later [(7) below], but we may, for the sake of illustration of what precedes, consider the problem of a disk or hoop rolling on a horizontal plane and performing oscillations about steady motion. We have here $z=0$, and $x=a$, the radius of the circular edge, and the equations of motion for the disk, with a coaxial flywheel included as specified above, are

$$\left.\begin{aligned} A\ddot{\theta}+(Cn+K-A\dot{\psi}\cos\theta)\dot{\psi}\sin\theta &= -a(R\cos\theta+F\sin\theta), \\ A\ddot{\psi}\sin\theta+(2A\dot{\psi}\cos\theta-Cn-K)\dot{\theta} &= 0, \\ C\dot{n} &= -F'a. \end{aligned}\right\} \qquad (1)$$

These equations can easily be established from first principles.

With equations (1) we have also

$$M(\dot{u}+v\dot{\psi})=F, \quad M(\dot{v}-u\dot{\psi})=F', \quad M\ddot{\zeta}=R-Mg. \qquad (2)$$

If the rolling is pure $\quad u=a\dot{\theta}\sin\theta, \quad v=an,$

and, since $\zeta=a\sin\theta$, we get

$$R=Mg+Ma(\ddot{\theta}\cos\theta-\dot{\theta}^2\sin\theta),$$

$$Ma(\ddot{\theta}\sin\theta+\dot{\theta}^2\cos\theta+n\dot{\psi})=F, \quad Ma(\dot{n}-\dot{\theta}\dot{\psi}\sin\theta)=F'. \qquad (3)$$

Hence we obtain for the first of (1), rejecting of course terms in $\dot{\theta}^2$,

$$(A+Ma^2)\ddot{\theta}+\{(C+Ma^2)n+K-A\dot{\psi}\cos\theta\}\dot{\psi}\sin\theta+Mga\cos\theta=0. \quad (4)$$

The other two equations of (1) are

$$\frac{d}{dt}(A\dot{\psi}\sin^2\theta)-(K+Cn)\dot{\theta}\sin\theta=0, \quad (C+Ma^2)\dot{n}-Ma^2\dot{\theta}\dot{\psi}\sin\theta=0. \quad (5)$$

Now let the motion be steady, that is $\ddot{\psi}=0$, $\dot{\theta}=0$, $\ddot{\theta}=0$, $u=0$, $\dot{u}=0$, $\dot{v}=0$, $\dot{n}=0$, and we get

$$\{K+(C+Ma^2)n-A\mu\cos\theta\}\mu\sin\theta+Mga\cos\theta=0, \qquad (6)$$

with $v=na$.

[For a uniform circular disk without flywheel $C=\frac{1}{2}Ma^2$, $A=\frac{1}{4}Ma^2$, and for a hoop $C=Ma^2$, $A=\frac{1}{2}Ma^2$.]

Let $n'$, $\theta'$, $\mu$ be the steady values of $n$, $\theta$, $\dot\psi$, and $n'+\nu$, $\theta'+\alpha$, $\mu+\dot\beta$ be the values at any instant for a slight deviation from steady motion. Then equations (4) and (5) become by the process suggested in (9), 4,

$$\left.\begin{aligned}(A+Ma^2)\ddot\alpha+\{K+(C+Ma^2)\nu-A\dot\beta\cos\theta'+A\alpha\mu\sin\theta'\}\mu\sin\theta'\\ +\{K+(C+Ma^2)n-A\mu\cos\theta'\}(\dot\beta\sin\theta'+\mu\alpha\cos\theta')-Mga\alpha\sin\theta=0,\\ A\ddot\beta\sin\theta'+(2A\mu\cos\theta'-K-Cn)\dot\alpha=0,\\ (C+Ma^2)\dot\nu-Ma^2\dot\alpha\mu\sin\theta'=0.\end{aligned}\right\}\quad\text{...(7)}$$

From the last equation we get

$$(C+Ma^2)\nu=Ma^2\alpha\mu\sin\theta', \quad\text{.........(8)}$$

since the constant of integration is zero. Substituting in the first equation we obtain

$$(A+Ma^2)\ddot\alpha+(Ma^2\alpha\mu\sin\theta'-A\dot\beta\cos\theta'+A\alpha\mu\sin\theta')\mu\sin\theta'$$
$$+\{K+(C+Ma^2)n-A\mu\cos\theta'\}(\dot\beta\sin\theta'+\mu\alpha\cos\theta')-Mga\alpha\sin\theta=0. \quad\text{...(9)}$$

Putting now $\alpha=r\sin(pt-f)$, $\beta=s\cos(pt-f)$, .........(10)

and substituting in the last equation, we obtain

$$\frac{s}{r}=\frac{\{K+(C+Ma^2)n-A\mu\cos\theta'\}\mu\cos\theta'+(A+Ma^2)(\mu^2\sin^2\theta'-p^2)-Mga\sin\theta'}{\{K+(C+Ma^2)n-2A\mu\cos\theta'\}p\sin\theta'} \quad\text{.........(11)}$$

Again substituting in the second of (7), we find

$$\frac{s}{r}=\frac{2A\mu\cos\theta'-K-Cn}{Ap\sin\theta'}. \quad\text{.........(12)}$$

Equating these two expressions for $s/r$ and writing L for

$$K+Cn-2A\mu\cos\theta',$$

we obtain

$$p^2=\frac{L(L+Ma^2n)+A(L+Ma^2n+A\mu\cos\theta')\mu\cos\theta'+A(A+Ma^2)\mu^2\sin^2\theta'-MAga\sin\theta'}{A(A+Ma^2)}. \quad\text{.........(13)}$$

Here $\theta'<\frac{1}{2}\pi$, and therefore, even if there be no flywheel, $p^2$ is positive if the rotation $n$ be sufficiently rapid. The flywheel adds to the value of M and A, but nevertheless, if K be sufficiently great, controls the stability of the arrangement.

6. *Coin spinning on a table.* As one example let $n=0$ and $\theta=\frac{1}{2}\pi$, so that the body turns in azimuth with its axis horizontal, as a coin spins with its plane vertical on a table. We have

$$p^2=\frac{K^2+A(A+Ma^2)\mu^2-AMga}{A(A+Ma^2)}, \quad\text{.........(1)}$$

which is evidently positive whatever $\mu$ may be if K be sufficiently great. If there be no flywheel $K=0$, and we have $p^2$ positive only if $\mu$ be sufficiently great. If the disk or coin be spinning very rapidly $p^2=\mu^2$, so that the period of an oscillation about the vertical position is approximately the period of the azimuthal rotation.

As another example take $n=0$ and $\mathrm{K}=0$. Then for any value of $\theta'$ we have by (6), 5, $\mathrm{A}\mu^2\sin\theta'=\mathrm{M}ga$, and (13), 5, becomes, with $\mathrm{A}=\mathrm{M}k^2$,

$$p^2=ga\frac{3k^2\cos^2\theta'+a^2\sin^2\theta'}{k^2(k^2+a^2)\sin\theta'}, \quad\text{.........................(1')}$$

and the time of a small oscillation about the steady motion is $2\pi/p$. The reader may establish this result directly.

For a hoop, equation (13), 5, for $p^2$ reduces to

$$p^2=\frac{8na(n-\mu\cos\theta')+3a\mu^2-2g\sin\theta'}{3a}, \quad\text{..................(2)}$$

and this, if $n=0$ and $\theta'=\frac{1}{2}\pi$, that is if the hoop is spinning about the vertical, is

$$p^2=\mu^2-\frac{2}{3}\frac{g}{a}. \quad\text{......................................(3)}$$

In the steady motion of a hoop the radius of the circle in which the centroid moves is $na/\mu$, which agrees with (7), 1, since here $z$ is zero. The point of contact with the horizontal plane therefore moves in a circle of radius

$$\frac{na}{\mu}+a\cos\theta=a\frac{n+\mu\cos\theta}{\mu}. \quad\text{..........................(4)}$$

This also holds for a circular disk in motion in the same way. The result here obtained agrees with that of (7), 1, given for O in Fig. 96.

We can now find the condition that a disk or hoop may roll upright in a straight line. It is only necessary to put in (13), 5, $\mu=0$, $\theta'=\frac{1}{2}\pi$. We obtain for the case in which the disk carries a coaxial flywheel

$$p^2=\frac{(\mathrm{K}+\mathrm{C}n)(\mathrm{K}+\mathrm{C}n+\mathrm{M}a^2n)-\mathrm{AM}ga}{\mathrm{A}(\mathrm{A}+\mathrm{M}a^2)}. \quad\text{.....................(5)}$$

The condition is therefore the inequality

$$(\mathrm{K}+\mathrm{C}n)(\mathrm{K}+\mathrm{C}n+\mathrm{M}a^2n)>\mathrm{AM}ga. \quad\text{........................(6)}$$

Let us take K positive, and consider the values of $n$ which satisfy the quadratic

$$(\mathrm{K}+\mathrm{C}n)(\mathrm{K}+\mathrm{C}n+\mathrm{M}a^2n)=\mathrm{AM}ga,$$

that is

$$\mathrm{C}(\mathrm{C}+\mathrm{M}a^2)n^2+\mathrm{K}(2\mathrm{C}+\mathrm{M}a^2)n+\mathrm{K}^2-\mathrm{AM}ga=0. \quad\text{...............(7)}$$

If $\mathrm{AM}ga>\mathrm{K}^2$ there will be a positive value of $n$ and a numerically greater negative value which both satisfy the quadratic. For the satisfaction of the inequality the value of $n$ must either be greater than this positive value or less than this negative value.

On the other hand, if $\mathrm{K}^2>\mathrm{AM}ga$ both roots of the quadratic are negative, and any value of $n$ which does not lie between these roots will enable the disk to roll upright in a straight line.

If K be zero the condition becomes $n^2 > \mathrm{AM}ga/\mathrm{C}(\mathrm{C}+\mathrm{M}a^2)$, which for a hoop is $n > (g/4a)^{\frac{1}{2}}$, and for a uniform disk is $n > (g/3a)^{\frac{1}{2}}$.

Thus the hoop is more stable than the disk, requiring for the same radius less speed of rotation in the ratio of $3^{\frac{1}{2}}$ to 2, in order to remain upright.

7. *Rolling of a disk on a table: calculation from first principles.* We may verify (5) from first principles. For a small angular deviation $\alpha$ from the vertical $\dot{\alpha}$ and $\dot{\psi}$ are both small. Hence the total rate of production of A.M. about a tangent to the trace of the point of contact on the horizontal plane is approximately

$$(\mathrm{A}+\mathrm{M}a^2)\ddot{\theta}+\{\mathrm{K}+(\mathrm{C}+\mathrm{M}a^2)n\}\dot{\psi}=\mathrm{M}ga\alpha. \quad \text{.................(1)}$$

Also when the disk is rolling upright we have $\dot{n}=0$, and when it is tilted over through the angle $\alpha$, $(\mathrm{C}+\mathrm{M}a^2)n$ is changing at rate $\mathrm{M}a^2\dot{\alpha}\dot{\psi}\cos\alpha$, which may be neglected.

We have also, from the constancy of A.M. about the vertical through the point of support and the fact that $\dot{\psi}$ is now denoted by $\dot{\beta}$,

$$\left.\begin{aligned} \mathrm{A}\ddot{\beta}\cos\alpha+(2\mathrm{A}\mu\alpha-\mathrm{C}n-\mathrm{K})\dot{\alpha}&=0,\\ \text{or} \qquad \mathrm{A}\ddot{\beta}-(\mathrm{C}n+\mathrm{K})\dot{\alpha}&=0.\end{aligned}\right\} \quad \text{.......................(2)}$$

Now writing as before $\alpha=r\sin(pt-f)$, $\beta=s\cos(pt-f)$, we obtain

$$\left.\begin{aligned} -(\mathrm{A}+\mathrm{M}a^2)p^2r-\{\mathrm{K}+(\mathrm{C}+\mathrm{M}a^2)n\}sp&=\mathrm{M}gar,\\ -\mathrm{A}p^2s-(\mathrm{K}+\mathrm{C}n)rp&=0.\end{aligned}\right\} \quad \text{..............(3)}$$

Thus we get
$$\frac{r}{s}=-\frac{\{\mathrm{K}+(\mathrm{C}+\mathrm{M}a^2)n\}p}{(\mathrm{A}+\mathrm{M}a^2)p^2+\mathrm{M}ga}=-\frac{\mathrm{A}p^2}{(\mathrm{K}+\mathrm{C}n)p}. \quad \text{.................(4)}$$

Therefore
$$p^2=\frac{(\mathrm{K}+\mathrm{C}n)(\mathrm{K}+\mathrm{C}n+\mathrm{M}a^2n)-\mathrm{AM}ga}{\mathrm{A}(\mathrm{A}+\mathrm{M}a^2)}, \quad \text{......................(5)}$$

which agrees with (13), 5.

8. ***Rising of a top when spinning on a rounded peg. Elementary discussion.*** We see by experiment that a top supported on a rounded peg rises under certain circumstances, and a prolate ellipsoid of revolution, or a hard-boiled egg, if spun as described above [16, I], rises from a lowest position of the centre of gravity and spins stably on one end. Initially the top is spun in some one of various ways. Generally it is thrown from the hand so that it alights on its peg. As a rule the speeds $u$, $v$, $\dot{\psi}$ are small, and the speed $n$ of rotation is large. The result is that the friction F′ is, for the counter-clockwise turning indicated in Fig. 98, in the direction there shown, and is as great as the reaction R can make it; for the point of contact of the solid owing to the rapid rotation slips back on the plane and the friction is not limited to that required for pure rolling.

The centroid G of the top and the point of contact O in the supporting horizontal plane move in concentric circles which slowly change with the inclination of the axis to the vertical. Figure 99 shows the latter circle and the projection of the former on the supporting plane. The line OG′ is the projection of the axis of figure on the horizontal. The spin, which is counter-clockwise when regarded from a point outside the top on the axis of

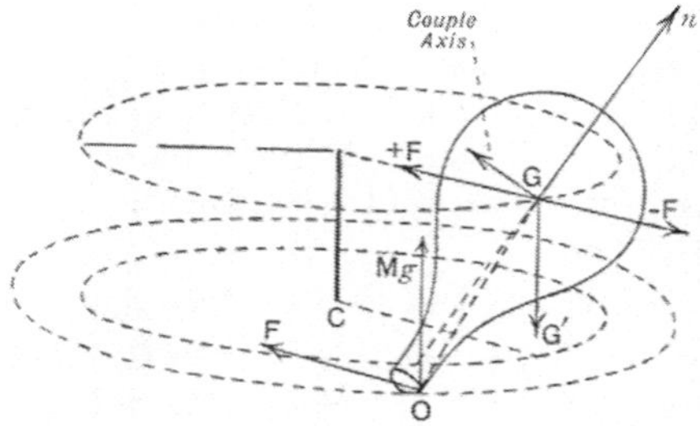


FIG. 98.

figure produced, gives a velocity $v$ of slipping of the point of the top in contact with the plane, represented in magnitude and direction by the line marked $v$. Besides this motion of the contact point of the top due to slipping, there is a component of amount $v'$, due to precession, indicated in the diagram by the line marked $v'$. The vector $v$ is at right angles to OG′, and the resultant $v_r$ of $v$ and $v'$ is inclined to OG′ at an angle exceeding (for

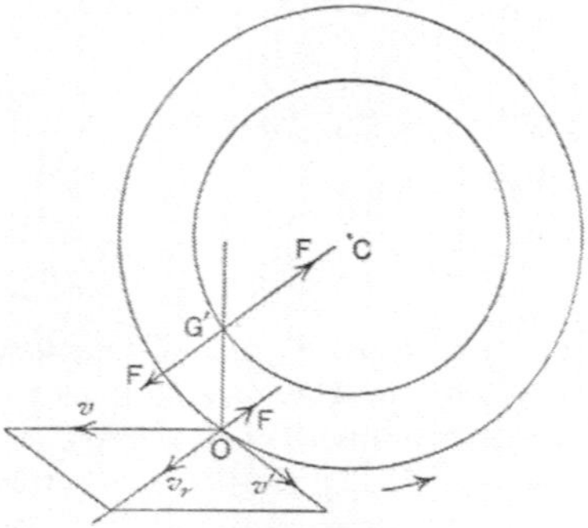


FIG. 99.

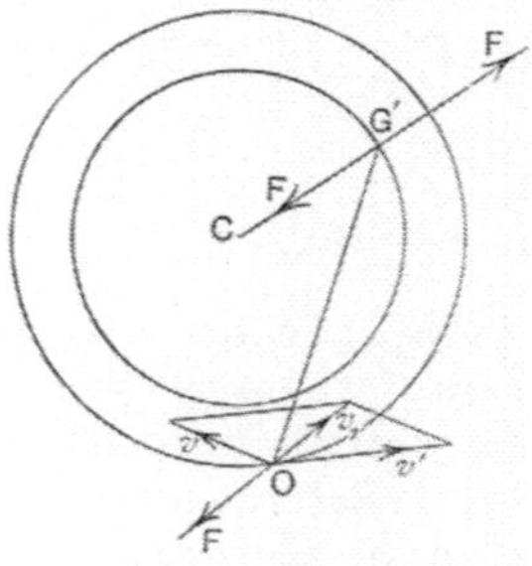


FIG. 100.

the particular case dealt with in Fig. 99) a right angle. The directions $v_r$ and CG′ are parallel. The direction of the resultant friction F (not the F of 1 above), due to the action between the top and the plane, is opposed to this resultant, and, if transferred to G without change of direction, must give a force towards the centre of the circle in which G moves. This is only strictly true in the case of steady precessional motion.

As G moves in a circle the axis OG moves on a hyperboloid of one sheet coinciding with successive rectilineal generators, and if the top rises or falls this surface varies in the manner described in Chapter XXI.

The couple $Fl \sin \phi$, where $\phi$ is the inclination of the *whole friction* F to OG, and $l$ is the distance OG, has its axis perpendicular to the plane of OG and F, and [Fig. 99], when $n$ is great enough, acts so as (in the usual phrase) to hurry the precession, and the axis of A.M. moving towards it raises the axis of the top towards the vertical. The top rises and the points O, C, G approximate more and more to coincidence until when the top reaches the upright position (which it will do if the spin is fast enough) it turns only about the axis of figure and "sleeps."

The couple $F'x$, specified in 1, is part of this couple, and tends to slow down the rotation of the top, and this action is aided, and generally exceeded, by the couple due to air friction. The rotation diminishes, hence the upright

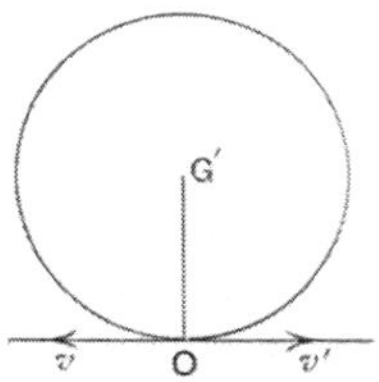


FIG. 101.

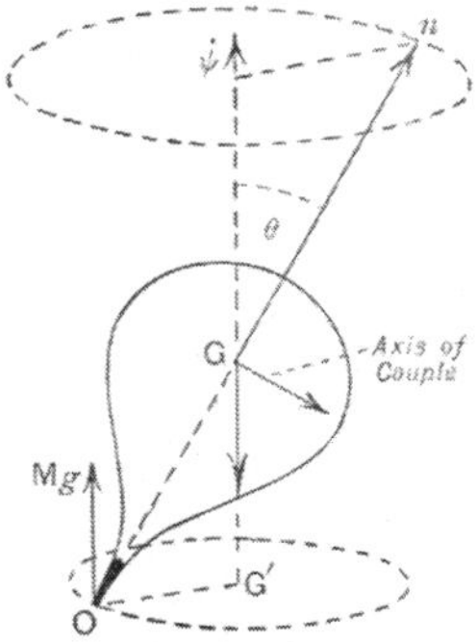


FIG. 102.

position becomes unstable and is deviated from. The speed $v$ of slipping is now so small that Fig. 99 has changed to Fig. 100. The couple which formerly hurried precession now delays it, and when the axis of figure becomes more inclined to the vertical, the top falls.

If the speed of rotation is not great enough the axis will not rise to the vertical. The diagram will be intermediate between the rising diagram of Fig. 99 and the falling diagram of Fig. 100, that shown in plan and elevation by Figs. 101, 102, in which the centroid G is above the centre of the circles, and the top rolls round on a small circle of points of contact symmetrical about the axis of figure.

The theory of the rising of a top, as caused by slipping of the rapidly rotating peg on a rough supporting plane, seems to be due to Archibald Smith (*Camb. Math. Journ.*, I, 1847). The following mathematical theory is in the main that given by Jellett in his *Theory of Friction*. Jellett assumed that the top rotated so rapidly that the force of friction might be taken as perpendicular to the vertical plane determined by the centroid and the point of support. The result given in (3) below, however, does not involve this condition, a fact pointed out by Routh [*Adv. Rigid Dyn.*, 5th edition, p. 167].

9. *Centre of gravity of a top raised by friction.* If we suppose the rounded end of the axis, the "peg," to be spherical and of small radius $\rho$, with centre O′ on the axis, we can trace the process of rising of the axis. From (2) and (3), 1, we have

$$A\ddot{\psi}\sin^2\theta + 2A\dot{\psi}\sin\theta\cos\theta\,.\,\dot{\theta} - Cn\sin\theta\,.\,\dot{\theta} = -C\dot{n}\frac{z}{\rho}, \qquad (1)$$

which is an exact equation. Denoting O′G by $h$, we have $z = h + \rho\cos\theta$, and so the equation just written becomes

$$A\ddot{\psi}\sin^2\theta + 2A\dot{\psi}\sin\theta\cos\theta\,.\,\dot{\theta} - Cn\sin\theta\,.\,\dot{\theta} + C\dot{n}\cos\theta = -C\dot{n}\frac{h}{\rho}. \qquad (2)$$

Thus we obtain by integration

$$A\dot{\psi}\sin^2\theta + Cn\left(\cos\theta + \frac{h}{\rho}\right) = N, \qquad (3)$$

where N is a constant. This equation is also exact. It can be obtained directly by observing that the time-rate of increase of the A.M. about the vertical through G (that is of the quantity $A\dot{\psi}\sin^2\theta + Cn\cos\theta$) is $F'h\sin\theta$, and then substituting $-C\dot{n}/x$, or $-C\dot{n}/\rho\sin\theta$, for F′.

If, initially, $n = n_0$ and $\dot{\psi} = 0$, (3) becomes

$$A\dot{\psi}\sin^2\theta + C(n\cos\theta - n_0\cos\theta_0) = -\frac{h}{\rho}C(n - n_0). \qquad (3')$$

But if $\dot{\psi}$ be small we get approximately

$$C(n\cos\theta - n_0\cos\theta_0) = -\frac{h}{\rho}C(n - n_0). \qquad (4)$$

The quantity on the right is positive since $n < n_0$. Thus $\cos\theta > \cos\theta_0$ and $\theta < \theta_0$. The axis has therefore risen through the angle $\theta_0 - \theta$.

If $z/\rho = 20$, $\theta_0 = 60^c$, which are possible numbers, since we may have $z = 2{\cdot}5$ and $\rho = \frac{1}{8}$, in inches, we have, very nearly,

$$2\cos\theta - 1 = 40(n_0 - n)/n_0. \qquad (5)$$

Thus $\theta$ will be zero when $(n_0 - n)/n_0$ is 1/40, that is when $\frac{1}{40}$ of the original spin has been destroyed.

The percentage loss of angular speed of spin required to bring the axis of the top to the vertical is $\rho(1 - \cos\theta_0)/100(\rho + h)$, and is thus proportional directly to the radius of the peg and inversely to the distance of G from the point.

On the supposition of negligible $\dot{\psi}$ the time-rate of variation of $\theta$ can also be found approximately. From (2) we have, writing $\kappa$ for $h/\rho$,

$$\dot{\theta} = \frac{\kappa + \cos\theta}{n\sin\theta}\dot{n}, \qquad (6)$$

or, since $\dot{n} = -F'\rho\sin\theta/C$,

$$\dot{\theta} = -\frac{\kappa + \cos\theta}{Cn}F'\rho. \qquad (7)$$

Thus for a given value of F′ the rate of diminution of $\theta$ is for any given value of the angle proportional to $\rho$. Hence the sharper the peg the slower is the rise, and if the peg be very sharp indeed the top will not rise at all. Also for various applications it is important to notice that if $Cn$ be great the rate of rising is correspondingly slow. A fast spinning top, brought to the vertical by friction alone, is thus a slowly acting contrivance.

10. *Condition of minimum kinetic energy.* Equation (3), 9, means that the A.M. about the line OG remains constant as the top moves. For we can write the equation in the form

$$A\dot{\psi}\sin\theta\frac{\rho\sin\theta}{l}+Cn\frac{h+\rho\cos\theta}{l}=N\frac{\rho}{l},\quad\text{.....................(1)}$$

where $l$ represents the distance OG. The components of A.M. about the axes GC and GE (Fig. 96) are $Cn$ and $A\dot{\psi}\sin\theta$, and $\rho\sin\theta/l$, $(h+\rho\cos\theta)/l$ are the cosines of the angles GON, NGO.

Our object is to find the condition that the kinetic energy may be a minimum. One condition obviously is that G should be at rest, and that $\dot{\theta}=0$. For the former the kinematic condition is

$$n\rho\sin\theta=\dot{\psi}\sin\theta(h+\rho\cos\theta),$$

or

$$\frac{n}{h+\rho\cos\theta}=\frac{\dot{\psi}\sin\theta}{\rho\sin\theta}.\quad\text{..............................(2)}$$

That this gives minimum kinetic energy subject to constancy of N may be proved as follows We have the expressions for the kinetic energy and N,

$$\left.\begin{aligned}T&=\tfrac{1}{2}(Cn^2+A\dot{\psi}^2\sin^2\theta),\\ \rho N&=A\rho\dot{\psi}\sin^2\theta+Cn(h+\rho\cos\theta).\end{aligned}\right\}\quad\text{......................(3)}$$

These, with the condition $\dot{\theta}=0$, give for T a minimum, and N a constant,

$$Cn\,dn+A\dot{\psi}\sin\theta\,d(\dot{\psi}\sin\theta)=0,\quad C(h+\rho\cos\theta)dn+A\rho\sin\theta\,d(\dot{\psi}\sin\theta)=0,$$

and therefore the relation (2), which, it will be noticed, is the condition for pure rolling about OG. When (2) is satisfied the points on OG are instantaneously at rest.

Equation (2) can be written

$$\frac{Cn(h+\rho\cos\theta)}{C(h+\rho\cos\theta)^2}=\frac{A\dot{\psi}\sin^2\theta}{\rho A\sin^2\theta},$$

which gives for each ratio in (2) the value

$$\frac{Cn(h+\rho\cos\theta)+\rho A\dot{\psi}\sin^2\theta}{C(h+\rho\cos\theta)^2+\rho^2A\sin^2\theta}=\frac{N\rho}{Il^2},\quad\text{..........................(4)}$$

where I is the moment of inertia of the top about OG. Thus by (1), the common value of the ratio in (2) is the product of the ratio of the A.M. about OG to the moment of inertia about that line by the multiplier $1/l$.

Now consider the energy of the top. Equation (2) gives by (4)

$$\frac{n^2}{n(h+\rho\cos\theta)}=\frac{\dot{\psi}^2\sin^2\theta}{\rho\dot{\psi}\sin^2\theta}=\frac{Cn^2+A\dot{\psi}^2\sin^2\theta}{Cn(h+\rho\cos\theta)+A\rho\dot{\psi}\sin^2\theta}=\frac{N\rho}{Il^2}.\quad\text{......(5)}$$

If T be the part of the kinetic energy due to the rotation about GE and the spin about the axis of figure, we have by (1)

$$\frac{2T}{\rho N}=\frac{Cn^2+A\dot{\psi}^2\sin^2\theta}{Cn(h+\rho\cos\theta)+A\rho\dot{\psi}\sin^2\theta}.$$

Combining this with (5) we get

$$T=\frac{1}{2}\frac{\rho^2N^2}{Il^2}, \qquad (6)$$

so that T varies inversely as $Il^2$.

Initially $T=\frac{1}{2}Cn_0^2$, for $\dot{\psi}=0$ at the beginning of the motion. Then also if $\cos\theta_0=z_0$, we get by (3)

$$\rho N=Cn_0(h+\rho z_0). \qquad (7)$$

The potential energy may be taken as $Mgh(z-z_0)$ where $z=\cos\theta$, and so the least energy, $E_m$ say, which the top can have for any value of $\theta$, is given (with this value of N) by

$$E_m=Mgh(z-z_0)+\frac{1}{2}\frac{\rho^2N^2}{Il^2}. \qquad (8)$$

The kinetic energy here taken account of is the whole kinetic energy with the exception of $\frac{1}{2}A\dot{\theta}^2$, that due to the rate of change of $\theta$.

When the axis of the top has become vertical the minimum energy for that position is the whole energy. We have then $z=1$, and $Il^2=C(\rho+h)^2$. Thus (8) becomes

$$E=Mgh(1-z_0)+\frac{1}{2}Cn_0^2\frac{(h+\rho z_0)^2}{(\rho+h)^2}. \qquad (9)$$

This value of E must be less than the initial value $\frac{1}{2}Cn_0^2$, and thus we have the inequality

$$\frac{1}{2}Cn_0^2>Mgh(1-z_0)+\frac{1}{2}Cn_0^2\frac{(h+\rho z_0)^2}{(\rho+h)^2}. \qquad (10)$$

If $h/\rho=k$ we can write this inequality in the form

$$n_0^2>2\frac{Mgh}{C}\frac{(k+1)^2}{2k+1+z_0}. \qquad (11)$$

When $\rho$ is small compared with $h$, as it usually is, $k$ is great, and so $n_0^2$ fulfils the inequality through the preponderance of $(k+1)^2$ over $2k+1+z_0$. If $\rho$ be extremely small the initial spin $n_0$ is very great, and tends to infinity as $\rho$ is diminished towards zero. Thus we have another proof that a sharp-pointed top will not rise.

**11.** *Minimum kinetic energy is necessary but is not sufficient for the erection of a top.* We have thus obtained an inferior limit for $n_0^2$, when the top is started with its axis at inclination $\cos^{-1}z_0$ to the vertical. Satisfaction of this inequality is necessary, but is not sufficient to ensure that the top will rise to the vertical.

This question has been discussed in a very interesting manner by Mr. E. G. Gallop [*Camb. Phil. Trans.*, XIX, 1904], from the point of view of the variation of energy which takes place as the top erects itself. As we have seen [(8), 10], the minimum value of E for inclination $\theta = \cos^{-1} z$ is given by the equation

$$E_m = Mgh(z - z_0) + \tfrac{1}{2}\frac{\rho^2 N^2}{C(h+\rho z)^2 + A\rho^2(1-z^2)}. \quad \text{.............(1)}$$

Suppose a curve drawn with values of $z$ as abscissae and the corresponding values of $E_m$, the minimum energy, as ordinates. The energy of the top, while the axis is rising, will be greater than E [(9), 10], for, in order that a couple "hurrying" the precession may exist, it is necessary that the top should have an angular speed, as explained in 8, over and above the angular speed required for pure rolling about OG. Also the value of the energy will continually diminish, for, since there is dissipation of energy, the gain

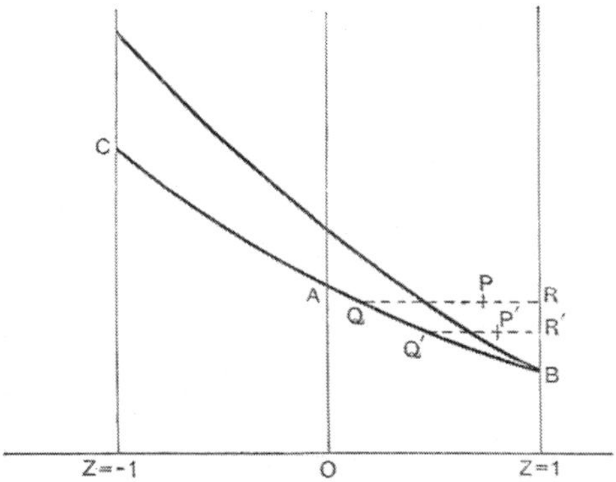


FIG. 103.

of potential energy must be less than the loss of kinetic. The curve of variation of the real energy must start from some point above the curve of minimum energy, and only meet the latter curve at the upright position as shown in Fig. 103. For, if the curves met earlier, there would be cessation of loss of energy at the point of meeting, as the rolling would have become pure, and the top could not erect itself.

For consider the energy E for any point on the line QR in the diagram. The inclination of the axis to the vertical may have any value between that for the point Q and zero, and the top will oscillate so that the succession of values of $\cos^{-1} z$ is given by the abscissae of points in QR. As the process goes on more and more energy will be dissipated, and the point P will take position in the line Q'R'. Finally the point will have reached B, and the top will be at constant inclination $\theta = 0$.

Thus the top cannot reach the vertical unless the energy $E - E_a$ is dissipated, where $E_a$ is the minimum energy for $\theta = 0$, and it is clear that as energy is being continually dissipated, owing to the fact that pure rolling has not supervened, the erect position must ultimately be reached with reduction of the energy precisely to $E_a$.

The energy curve must be above the curve of minimum energy as shown in the diagram. To ensure that this can always happen the minimum energy must also continually diminish as $z$ increases. Now from (1) we have

$$\frac{dE}{dz} = Mgh - \tfrac{1}{2}\rho^2 N^2 \frac{f'(z)}{\{f(z)\}^2}, \quad \text{.........(2)}$$

where

$$f(z) = C(h+\rho z)^2 + A\rho^2(1-z^2). \quad \text{.........(3)}$$

Hence we are to have

$$\tfrac{1}{2}\rho^2 N^2 \frac{f'(z)}{\{f(z)\}^2} > Mgh, \quad \text{.........(4)}$$

and the conditions will be fulfilled if, when $z=1$,

$$\tfrac{1}{2}\rho^2 N^2 \frac{f'(1)}{\{f(1)\}^2} > Mgh; \quad \text{.........(5)}$$

and also

$$\frac{d}{dz}\frac{f'(z)}{\{f(z)\}^2} < 0. \quad \text{.........(6)}$$

Hence we are to have

$$\{f'(z)\}^2 - \tfrac{1}{2}f(z)f''(z) > 0. \quad \text{.........(7)}$$

But

$$\left.\begin{aligned} f(z) &= (C-A)\rho^2 z^2 + Ch^2 + A\rho^2 + 2Ch\rho z, \\ f'(z) &= 2(C-A)\rho^2 z + 2C\rho h, \quad f''(z) = 2(C-A)\rho^2, \end{aligned}\right\} \quad \text{.........(8)}$$

and we find easily

$$\{f'(z)\}^2 - \tfrac{1}{2}f(z)f''(z)$$
$$= 3(C-A)^2\rho^4 z^2 + 6C(C-A)\rho^3 hz + C(3C+A)\rho^2 h^2 - A(C-A)\rho^4. \quad \text{...(9)}$$

On the right the coefficient of $z^2$ is positive. The whole expression will therefore be positive if the roots of the quadratic, obtained by equating the expression to zero, are imaginary. The condition for this is easily found after a little reduction to be

$$C\rho^2 h^2 - (C-A)\rho^4 > 0,$$

or with $k^2 = h^2/\rho^2$,

$$Ck^2 - (C-A) > 0. \quad \text{.........(10)}$$

If $C < A$ this inequality is fulfilled, if $C > A$ it is fulfilled if

$$k^2 > 1 - A/C.$$

For an ordinary top $k^2 > 1$, so that the inequality is always fulfilled.

**12.** *Summary of conditions for the rise of a top.* Summing up we have the condition (10), 11, and also

$$Mgh - \tfrac{1}{2}\rho^2 N^2 \frac{f'(1)}{\{f(1)\}^2} < 0, \quad \text{.........(1)}$$

for which it is necessary that $f'(1) > 0$, that is, by (8), 11,

$$C(k+1) > A, \quad \text{.........(2)}$$

and also, since $f'(1) = 2\rho^2\{C(k+1) - A\}$, $f(1) = C(\rho+h)^2 = C\rho^2(1+k)^2$,

$$N^2 > Mgh \frac{C^2(k+1)^4}{C(k+1)-A}. \quad \text{.........(3)}$$

If, initially, the motion consists of a spin $n_0$ about the axis of figure, we have

$$N = Cn_0(k+z_0),$$

and the last condition becomes

$$n_0^2 > \frac{(k+1)^4 Mgh}{\{C(k+1)-A\}(k+z_0)^2}. \quad \text{.........................(4)}$$

The conditions (10), 11, and (2) refer to the construction of the top, condition (3) must be fulfilled by the motion. When all are fulfilled the top can and will reach the vertical, provided the energy has been reduced by sliding friction from the initial value to the absolute minimum $E_a$. The conditions are all fulfilled by a hard-boiled egg, or by a nearly egg-shaped ellipsoid of revolution, as in the experiment described in 16, I, above. Without a sufficient amount of sliding the top will not rise. This explains the failure, which the reader may verify, of an egg-shaped solid, made with a very rough surface, to rise when spun.

The limiting value of $n_0$ given by (4) is greater than that given by (11), 10. For the ratio of the former to the latter is

$$\tfrac{1}{2}(k+1)^2 \frac{C(2k+1+z_0)}{\{C(k+1)-A\}(k+z_0)^2}.$$

Now, by (2), $C(k+1)-A$ is positive, and the ratio is diminished by substituting $C(k+1)$. Also it is diminished by substituting $z_0$ for 1 in the numerator of the fraction. These changes make the ratio $(k+1)/(k+z_0)$, which is greater than 1. Thus the ratio though diminished is still greater than 1. Hence (4) covers also the condition (11), 10.

It is interesting to compare (4) with

$$C^2n^2 > 4AMgh,$$

the condition found in 17, II, above, as that of stability of a top in the upright position. The inequality (4) is equivalent to

$$C^2n^2 > C^2Mgh\frac{(k+1)^2}{C(k+1)-A}.$$

The limit here indicated is greater than the former. For, neglecting the common factor $Mgh$, we have

$$\frac{C^2(k+1)^2}{C(k+1)-A} - 4A = \frac{\{C(k+1)-2A\}^2}{C(k+1)-A},$$

which is positive since $C(k+1)-A$ is positive. Thus (4) covers the previously found condition.

**13.** *Numerical examples.* Mr. Gallop (*loc. cit.*) takes as a numerical example a top for which $A=C=3M$, $h=4$, $z_0=0{\cdot}866$, and $\rho=0{\cdot}2$ in cms. Thus $k=20$. Hence the limiting value of $n_0^2$ is $21^4 \times 981 \times 4/60 \times 20{\cdot}866^2$, or $n_0=171$, in radians, very nearly. The condition (11), 10, gives $n_0=166$, in radians.

If we take the initial value of $n$ as 200 the initial energy, when the potential energy is reckoned from the starting position, is 60000M, on the supposition of zero precessional motion at starting. At the vertical the rotational energy is, by (9) of 10,

$$\tfrac{1}{2}Cn_0^2\frac{(h+\rho z_0)^2}{(\rho+h)^2}.$$

Thus the rotational angular speed has been reduced from 200 to 200 × 4·1732/4·2, that is to 198·7, or about 0·65 p.c.

The initial energy 60000M has been diminished to

$$M\left(981\times 4\times\cdot 134+60000\,\frac{4\cdot 1732^2}{4\cdot 2^2}\right),$$

that is to 59763M. The diminution is thus rather less than 0·4 p.c. of the original energy. When this small amount of energy has been dissipated the axis has become vertical.

Mr. Gallop gives a table by which the values of $\theta$ on the limiting curve of minimum energy can be traced for increasing amounts of energy dissipated. Of this we give an abridgment for the sake of one or two points of interest.

| Loss of Energy. | Value of $z$. | Value of $\theta$. |
|---|---|---|
| 0 | 0·8444 | 32° 23′ [Initial value 30°] |
| $\frac{1}{10}(E_0-E_1)$ | 0·8593 | 30° 46′ |
| $\frac{2}{10}(E_0-E_1)$ | 0·8743 | 29° 2′ |
| $\frac{4}{10}(E_0-E_1)$ | 0·9048 | 25° 12′ |
| $\frac{6}{10}(E_0-E_1)$ | 0·9359 | 20° 38′ |
| $\frac{8}{10}(E_0-E_1)$ | 0·9676 | 14° 37′ |
| $E_0-E_1$ | 1·000 | 0° 0′ |

It will be seen from the table that at starting the top first falls a little. This must happen, since, if there is no azimuthal motion at the starting, the large couple applied to the top by gravity comes into play, and it is only after the top has fallen a little that it has a precessional motion, compensating the production of A.M. about the vertical which arises in consequence of the alteration of direction of the axis of spin. Also the frictional couple comes into play and helps to check the fall.

Thus if the top rise very slowly, with near approach to pure rolling, it will follow the curve of minimum energy very closely; but the curve of variation of energy will begin with a horizontal part, in which *z diminishes* without perceptible loss of energy, until the curve of diminishing energy is nearly reached.

Finally it is to be noticed that, though it is true that the axis will have reached the vertical when the energy has been reduced from the initial value $E_0$ to the absolute minimum, it does not follow that this energy $E_0-E_a$ will be dissipated and the top rise. For if the plane be very rough the top may be brought to a state of pure rolling, or nearly so, very quickly, and the axis will then move in a cone round the vertical.

In all that has been stated above, no account has been taken of the action of the air on the top. This may reduce the spin below the necessary limit before the friction couple acting at the peg has brought the top to the vertical.

**14.** *A top in form of a sphere loaded symmetrically about a diameter.* As a particular case let the top be a sphere loaded symmetrically about a diameter; and let it be placed with the line of symmetry horizontal, and spun counter-clockwise with speed $n_0$ about the upward vertical diameter. The A.M. is $An_0$, and, if $h$ be the distance of the centroid from the centre, the linear speed of the centroid is $nh$, and the initial energy $E_0$ is given by

$$E_0=\tfrac{1}{2}(A+h^2)n_0^2. \qquad (1)$$

The final energy, by (9), 10, is, since the erection causes no change of A.M. about the vertical,

$$E_1 = Mgh + \tfrac{1}{2}\frac{A^2 n_0^2}{C(k+1)^2}. \quad \text{.......................................(2)}$$

The condition $E_0 > E_1$ leads to

$$\tfrac{1}{2}\left\{A + h^2 - \frac{A^2}{C(k+1)^2}\right\} n_0^2 > Mgh. \quad \text{.................................(3)}$$

It is to be remembered that in this case $k$ is not very great. The coefficient of $n_0^2$ is positive if the condition, (2), 12, $C(k+1) > A$ is fulfilled, and we see that the top must be started so that

$$n_0^2 > \frac{2Mgh}{A + h^2 - \dfrac{A^2}{C(k+1)^2}}. \quad \text{.......................................(4)}$$

The condition that $dE/dz$ should be negative gives [(3), 12]

$$n_0^2 > \frac{Mgh}{A^2}\,\frac{C^2(k+1)^4}{C(k+1) - A}, \quad \text{.......................................(5)}$$

and this includes condition (4). The condition (10), 11 must also be fulfilled.

We conclude that the method just discussed is applicable to this outwardly spherical top, if $C(k+1) - A$ and $Ck^2 - (C - A)$ are both positive and $n_0^2$ satisfies (5). By Fig. 103 it is clear that if the energy is reduced from $E_0$ to $E$, where $E$ is such that the equation

$$E = Mghz + \tfrac{1}{2}\frac{A^2 n_0^2}{A(1-z^2) + C(k+z)^2}$$

has a root $\zeta$ between $-1$ and $+1$, the inclination of the axis to the vertical is less than $\cos^{-1}\zeta$.

**15.** *Uniform sphere loaded by an additional spherical distribution.* In an example, also given by Mr. Gallop, a uniform sphere, radius R and mass M, is loaded by replacing a spherical portion with denser material giving an additional mass $m$. The distance between the centres of the spherical surfaces is $c$. It is found that $Ck^2 - (C - A) > 0$ is satisfied at once, and $C(k+1) - A > 0$ if $c/R > \frac{2}{5}(1 + mr^2/MR^2)$.

With $R = 10$, $r = 2$, $c = 2$, density of load 21 times that of the sphere, and initial angular speed $n_0 > 19{\cdot}14$ (radians per second), permanent rotation with axis vertical is attained with 1·3 per cent. loss of spin. For proper slope of the guiding curve $n_0 > 24{\cdot}8$ is required. With $n_0 = 30$, the initial energy is 15865M, and the final 15471M, in C.G.S. units.

Also it is found that according as the initial spin, 30, is about the vertical through the centre of the spherical surface, or about the vertical through G, the maximum value of $\theta = \cos^{-1} z$ is given by

$$0{\cdot}82\, z^3 - 51{\cdot}38\, z^2 + 131{\cdot}7\, z + 10{\cdot}26 = 0,$$

or

$$0{\cdot}82\, z^3 - 51{\cdot}27\, z^2 + 131{\cdot}3\, z + 2{\cdot}641 = 0.$$

Roots, relevant to the problem, are

$$-0{\cdot}0756 = \cos 94°\,20', \quad -0{\cdot}02 = \cos 91°\,9'.$$

The reader may verify these results.

**16.** *Example 1. Heterogeneous sphere with centre of mass at the centre of figure.* The principal moments of inertia at G are A, B, C, and the angular speeds about the principal axes $\omega_1, \omega_2, \omega_3$. Show that the A.M. about the vertical is

$$(-A\omega_1\cos\phi + B\omega_2\sin\phi)\sin\theta + C\omega_3\cos\theta = N,$$

a constant, where $\theta$ is the inclination of the C axis to the vertical, and $\phi$ the inclination of the plane through the A and C axes to the vertical plane containing the C axis.

Show also that the minimum value of the energy E is

$$\frac{1}{2}\frac{N^2}{(A\cos^2\phi+B\sin^2\phi)\sin^2\theta+C\cos^2\theta},$$

corresponding to a motion of pure rotation about the axis of figure, and that when $\phi$ is allowed to vary

$$E=\frac{1}{2}\frac{N^2}{A\sin^2\theta+C\cos^2\theta}.$$

Writing $x$ for $\cos^2\theta$ and denoting the quantity on the right of this equation for E by $y$, $\frac{1}{2}N^2/\{A+(C-A)x\}$ by $F(x)$, show that the curve $y=F(x)$ is a hyperbola, and, if $C>A$, slopes towards the axis of $x$, like the limiting curve in Fig. 103.

Hence show that the C axis will be brought into the vertical when the energy is reduced to $N^2/2C$, so that sliding friction causes the axis of greatest moment to approach the vertical.

When the energy in the motion considered in the preceding examples has a given value E, show that

$$\theta<\cos^{-1}\left\{\frac{N^2-2AE}{2E(C-A)}\right\}^{\frac{1}{2}},$$

and verify the conclusion in the last example.

**17.** *Further examples.* (1) Writing $\xi=h+\rho\cos\theta$, $\eta=\rho\sin\theta$, $\omega_1=\dot\psi\sin\theta$, show that the equation of energy is, if there is no slipping,

$$\{A+M(\xi^2+\eta^2)\}\dot\theta^2+A\omega_1^2+Cn^2+M(\xi\omega_1+\eta n)^2=2(E_0-V),$$

where E is the initial energy.

(2) Also using the equation of A.M.

$$-A\eta\omega_1+C\xi n=\rho N,$$

show that if $v$ be the resultant speed of G,

$$v^2=2\frac{C\xi^2+A\eta^2}{AC+C\xi^2+A\eta^2}\left(E_0-V-\frac{1}{2}\frac{\rho^2N^2}{C\xi^2+A\eta^2}\right)+\frac{A(C-A)}{AC+C\xi^2+A\eta^2}\eta^2\dot\theta^2.$$

Show that if $\dot\theta=0$ and $z=\cos\theta$

$$v^2=\frac{\rho^2\{C(k+z)^2+A(1-z^2)\}}{AC+\rho^2\{C(k+z^2)+A(1-z^2)\}}\{E_0-F(z)\},$$

where $$F(x)=Mgh(z-z_0)+\frac{1}{2}\frac{N^2}{C(k+z)^2+A(1-z^2)}.$$

(3) Also show that if (4), 12, is fulfilled $E_0-F(x)$ increases as the top rises. Prove also that, if $C(k+1)-A>0$, $C(k+z)^2+A(1-z^2)$ increases as $x$ increases. Finally show that therefore $v^2$ is greater when G is in the highest position than when it is in the lowest.

(4) Deduce conclusions as to the possibility of the top's attaining a state of steady motion with the axis vertical.

[See Gallop, *loc. cit. supra*, from whose paper the results stated in the preceding examples are taken.]

**18.** *Energy relations for a top spinning about a fixed point, and taking up steady motion.* As we have seen, a top spinning rapidly about a fixed point will, if left to itself with its axis of figure at rest inclined at some angle $\theta_0$ to the vertical, fall away from the vertical, acquiring at the same time azimuthal speed $\dot\psi$. This azimuthal turning grows up to a maximum and the speed $\dot\theta$ diminishes to zero, and the axis then begins to return to its former value, which it reaches only to begin the same variation of motion again at, of course, a new azimuth. We may consider here the energy relations involved.

If $Cn$ is great this falling away is small, the maximum azimuthal speed is twice the average speed with which the axis turns round the vertical [see 1, VI, above]. This oscillatory motion will be damped out by the resistance of the air, and the top will settle down to an approximately steady motion about the vertical. We shall suppose that the average motion about the vertical is not changed (which amounts to supposing that the A.M. of spin remains constant) but that the oscillation is damped out. Let $2\delta\theta$ be the maximum angle of dip of the axis from the initial inclination $\theta_0$. By this sinking of the axis potential energy $2Mgh \sin \theta_0 \delta\theta$ is lost. If $2\dot{\psi}$ be the azimuthal angular speed taken, the kinetic energy gained is $2A\dot{\psi}^2 \sin^2\theta$. The energy of spin is not altered, and so we have

$$2A\dot{\psi}^2 \sin^2\theta_0 = 2Mgh \sin \theta_0 \, . \, \delta\theta$$

or

$$A\dot{\psi}^2 \sin^2\theta_0 = Mgh \sin \theta_0 \, . \, \delta\theta.$$

After the oscillation has been damped out and the average azimuthal speed $\dot{\psi}$ has been taken up the energy E of the top will be given by

$$E = E_0 + \tfrac{1}{2}A\dot{\psi}^2 \sin^2\theta - Mgh \sin \theta_0 \, . \, \delta\theta,$$

where $E_0$ is the initial energy. By the preceding result we get

$$E = E_0 - \tfrac{1}{2}Mgh \sin \theta_0 \, . \, \delta\theta.$$

Thus energy equal to half the work done by gravity in the change $\delta\theta$ of inclination, that is equal to the kinetic energy added in the azimuthal motion, has been dissipated.

This recalls a theorem in electrodynamics. Let, for example, two mutually influencing circuits on which are impressed electromotive forces, say those due to batteries in the circuits, be left to themselves. They will be relatively displaced by their mutual action, and the currents in the circuits will in general be changed. The electrokinetic energy will be increased by an amount of $dT$, and work $dW$ will be done in overcoming the ordinary inertia of the circuits, and therefore in producing molar kinetic energy, or in otherwise doing external work, or in both ways. The batteries thus are called upon to supply energy $dT + dW$, in addition to that consumed in heat in the circuits during the displacement.

If the circuits move from rest to rest again, so that the changes in the currents are zero, and the molar kinetic energy is zero, and if the change in configuration of the system be one which excludes change of form of the individual circuits, the whole energy furnished by the batteries is $2\gamma_1\gamma_2 dM$, where $\gamma_1$, $\gamma_2$ denote the currents in the circuits, and $dM$ the change in their mutual inductance. But the change $dT$ in the electrokinetic energy is now $\gamma_1\gamma_2 dM$, so that

$$dT = dW = \gamma_1\gamma_2 dM.$$

The energy furnished by the batteries is thus equally divided between the increment of electrokinetic energy and the external work done in the displacement.

If the work $dW$ is done against external frictional forces the equivalent energy $\gamma_1\gamma_2 dM$ will be expended in producing heat, while an equal amount goes to augment the electrokinetic energy, and we shall have an exact analogue of the gyrostatic example discussed above. The circuits pass from a configuration in which a certain coordinate, $x$ say, has a value $x_0$ to another in which it has a value $x_1$, and initially and finally $\dot{x}$ is zero. In the gyrostatic example $\dot{\theta}$ is initially and finally zero.

The descent of this top through falling off of spin is discussed in Chapter XIV.

## CHAPTER XIX

### GENERAL DYNAMICS OF GYROSTATIC AND CYCLIC SYSTEMS

1. *General equations of dynamics. Holonomous and not holonomous systems.* If we suppose the equations of motion for each of a system of particles to be written down, we easily obtain, by combining them, the so-called variational equation, which forms the foundation of Lagrange's treatment of the dynamics of a connected system of particles. Three typical equations of motion are

$$m\ddot{x} = \mathrm{X}, \quad m\ddot{y} = \mathrm{Y}, \quad m\ddot{z} = \mathrm{Z}, \quad \text{.............................(1)}$$

where X, Y, Z are the total forces from all causes acting in the three coordinate directions on the representative particle. Multiplying the first of these by $\delta x$, the second by $\delta y$, the third by $\delta z$, and treating similarly the triad of equations for each of the other particles of the system, we get

$$\Sigma\{m(\ddot{x}\,\delta x + \ddot{y}\,\delta y + \ddot{z}\,\delta z)\} = \Sigma(\mathrm{X}\,\delta x + \mathrm{Y}\,\delta y + \mathrm{Z}\,\delta z). \quad \text{...............(2)}$$

This is the variational equation. One or two remarks regarding it may be made.

(1) Only active or working forces appear in the equation, The X, Y, Z on the right are *not* the total forces acting on the particles. Groups of forces which do no work on the whole have disappeared.

(2) The system is in general subject to kinematical conditions or "constraints," and $\delta x$, $\delta y$, $\delta z$ are typical of displacements which are possible without violation of these conditions, as they exist at the instant $t$ considered. These conditions are expressed, we shall suppose, by $m$ equations connecting the coordinates, and it is clear that $m < 3n$, otherwise there would be no freedom of motion for the system.

The kinematic equations enable the coordinates to be determined in terms of the $3n - m$ others, so that there are $3n - m$ independent coordinates.

(3) The kinematic equations may or may not involve the time $t$ explicitly, that is they may be either relations between explicit functions of the coordinates alone, or of the coordinates and $t$. Thus in the general case they are of the form

$$f(x_1, y_1, z_1, x_2, y_2, z_2, \ldots, x_n, y_n, z_n, t) = 0, \quad \text{..................(2')}$$

and give the integrable differential relations

$$\Sigma\left(\frac{\partial f}{\partial x}\delta x+\frac{\partial f}{\partial y}\delta y+\frac{\partial f}{\partial z}\delta z\right)+\frac{\partial f}{\partial t}\delta t=0. \qquad \text{.......................(3)}$$

If $t$ thus appears the kinematical condition is said to be variable: if the coordinates alone appear, so that $\partial f/\partial t=0$, the condition is said to be invariable.

Such a system of equations of condition, with or without the explicit appearance of $t$, is characteristic of what is called a *holonomous system*. If the conditions are, however, in whole or in part non-integrable relations of the coordinates, the system is not holonomous. As we shall see the Lagrangian equations of motion do not hold without modification for a system which is not holonomous.

**2.** *Generalised coordinates.* We shall now suppose that the $3n-m\ (=k)$ independent coordinates have been chosen by means of the equations of condition. Let these coordinates, or $k$ distinct functions of them, be taken to express the motion of the system. We denote them by $q_1, q_2, \ldots, q_k$. Now let the variations $\delta x, \delta y, \delta z$, at time $t$, of the Cartesian coordinates of a specimen particle be given by the equations

$$\left.\begin{aligned}\delta x&=\frac{\partial\phi}{\partial q_1}\delta q_1+\frac{\partial\phi}{\partial q_2}\delta q_2+\cdots+\frac{\partial\phi}{\partial q_k}\delta q_k,\\ \delta y&=\frac{\partial\chi}{\partial q_1}\delta q_1+\frac{\partial\chi}{\partial q_2}\delta q_2+\cdots+\frac{\partial\chi}{\partial q_k}\delta q_k,\\ \delta z&=\frac{\partial\psi}{\partial q_1}\delta q_1+\frac{\partial\psi}{\partial q_2}\delta q_2+\cdots+\frac{\partial\psi}{\partial q_k}\delta q_k,\end{aligned}\right\} \qquad \text{.......................(4)}$$

where $\phi, \chi, \psi$ are three functions of $q_1, q_2, \ldots, q_k$, by which $x, y, z$ are given for the representative particle. We write these equations more concisely in the form

$$\left.\begin{aligned}\delta x&=a_1\delta q_1+a_2\delta q_2+\ldots+a_k\delta q_k,\\ \delta y&=b_1\delta q_1+b_2\delta q_2+\ldots+b_k\delta q_k,\\ \delta z&=c_1\delta q_1+c_2\delta q_2+\ldots+c_k\delta q_k.\end{aligned}\right\} \qquad \text{.......................(4')}$$

There are $n$ such triads of functions, and $3n$ such equations for the expression of all the coordinates of the system. If $\delta x, \delta y, \delta z, \ldots$ are all zero, the quantities $\delta q_1, \delta q_2, \ldots, \delta q_k$ must all be zero, for if it were otherwise the functional determinant of any $k$ of the equations would be zero.

We now denote actual displacements of the system in the element of time $dt$ by $dx, dy, dz$, so that we have

$$\left.\begin{aligned}dx&=a_1dq_1+a_2dq_2+\ldots+a_kdq_k+\alpha dt,\\ dy&=b_1dq_1+b_2dq_2+\ldots+b_kdq_k+\beta dt,\\ dz&=c_1dq_1+c_2dq_2+\ldots+c_kdq_k+\gamma dt,\end{aligned}\right\} \qquad \text{.......................(5)}$$

or, as we may also write the equations,

$$\left.\begin{aligned}\dot x&=a_1\dot q_1+a_2\dot q_2+\ldots+a_k\dot q_k+\alpha,\\ \dot y&=b_1\dot q_1+b_2\dot q_2+\ldots+b_k\dot q_k+\beta,\\ \dot z&=c_1\dot q_1+c_2\dot q_2+\ldots+c_k\dot q_k+\gamma.\end{aligned}\right\} \qquad \text{.......................(5')}$$

If $\alpha$, $\beta$, $\gamma$ are zero for each triad of equations, the actual displacement is one consistent with the conditions fulfilled by the system at time $t$. If however $\alpha$, $\beta$, $\gamma$ be not zero, then, if we suppose $\delta x$, $\delta y$, $\delta z$, ... in (4′) to coincide in value with $dx$, $dy$, $dz$, ..., we get

$$\left.\begin{aligned} a_1(dq_1-\delta q_1)+a_2(dq_2-\delta q_2)+\cdots+a_k(dq_k-\delta q_k)+\alpha\,dt=0,\\ b_1(dq_1-\delta q_1)+b_2(dq_2-\delta q_2)+\cdots+b_k(dq_k-\delta q_k)+\beta\,dt=0,\\ c_1(dq_1-\delta q_1)+c_2(dq_2-\delta q_2)+\cdots+c_k(dq_k-\delta q_k)+\gamma\,dt=0.\end{aligned}\right\}\quad\text{(6)}$$

These equations cannot be satisfied by $dq_1=\delta q_1$, $dq_2=\delta q_2$, ..., unless either $dt=0$, or the functional determinant, formed from any $k$ variables vanishes.

Whether or not the kinematical conditions are invariable, we have

$$\Sigma\{m(\ddot{x}\,\delta x+\ddot{y}\,\delta y+\ddot{z}\,\delta z)\}=\Sigma(\mathrm{X}\,\delta x+\mathrm{Y}\,\delta y+\mathrm{Z}\,\delta z)=\delta q_1\Sigma(a_1\mathrm{X}+b_1\mathrm{Y}+c_1\mathrm{Z})$$
$$+\delta q_2\Sigma(a_2\mathrm{X}+b_2\mathrm{Y}+c_2\mathrm{Z})+\ldots+\delta q_k\Sigma(a_k\mathrm{X}+b_k\mathrm{Y}+c_k\mathrm{Z})=\Sigma(\mathrm{Q}\,\delta q),\quad\text{(7)}$$

which for an actual motion becomes

$$\Sigma\{m(\ddot{x}\dot{x}+\ddot{y}\dot{y}+\ddot{z}\dot{z})\}=\Sigma(\mathrm{X}\dot{x}+\mathrm{Y}\dot{y}+\mathrm{Z}\dot{z})$$
$$=\dot{q}_1\Sigma(a_1\mathrm{X}+b_1\mathrm{Y}+c_1\mathrm{Z})+\dot{q}_2\Sigma(a_2\mathrm{X}+b_2\mathrm{Y}+c_2\mathrm{Z})$$
$$+\ldots+\dot{q}_k\Sigma(a_k\mathrm{X}+b_k\mathrm{Y}+c_k\mathrm{Z})=\Sigma(\mathrm{Q}\dot{q}).\quad\text{(8)}$$

3. *Lagrange's equations of motion.* We can now prove Lagrange's equations. We have for any parameter $q$

$$\Sigma(a_1\mathrm{X}+b_1\mathrm{Y}+c_1\mathrm{Z})=\Sigma\{m(a_1\ddot{x}+b_1\ddot{y}+c_1\ddot{z})\}=\mathrm{Q}_1.\quad\text{(1)}$$

We have as many equations of this form as there are parameters $q$.

It will be observed that, since any Q is the coefficient of $\delta q$ in the expression for the work done in a possible arbitrary variation of the parameter $q$, Q does not include any of the forces such as those due to guides and constraints which are invariable. We have now to consider the equation

$$\Sigma\{m(a\ddot{x}+b\ddot{y}+c\ddot{z})\}=\mathrm{Q}.\quad\text{(2)}$$

Here it is to be understood for the present that $a$, $b$, $c$ are the partial differential coefficients $\partial x/\partial q$, $\partial y/\partial q$, $\partial z/\partial q$, that is, we suppose that the system is holonomous. We can prove that

$$a\ddot{x}=\ddot{x}\frac{\partial x}{\partial q}=\frac{d}{dt}\left(\dot{x}\frac{\partial \dot{x}}{\partial \dot{q}}\right)-\dot{x}\frac{\partial \dot{x}}{\partial q}.\quad\text{(3)}$$

For we have
$$\ddot{x}\frac{\partial x}{\partial q}=\frac{d}{dt}\left(\dot{x}\frac{\partial x}{\partial q}\right)-\dot{x}\frac{d}{dt}\frac{\partial x}{\partial q},$$

and by (5′), 2,
$$\frac{\partial \dot{x}}{\partial \dot{q}}=\frac{\partial x}{\partial q},$$

which gives the first part of (3). Again we have

$$\frac{d}{dt}\frac{\partial x}{\partial q}=\frac{\partial}{\partial q_1}\frac{\partial x}{\partial q}\dot{q}_1+\frac{\partial}{\partial q_2}\frac{\partial x}{\partial q}\dot{q}_2+\cdots+\frac{\partial}{\partial q_k}\frac{\partial x}{\partial q}\dot{q}_k+\frac{\partial}{\partial t}\frac{\partial x}{\partial q}$$
$$=\frac{\partial}{\partial q}\left(\frac{\partial x}{\partial q_1}\dot{q}_1+\frac{\partial x}{\partial q_2}\dot{q}_2+\cdots+\frac{\partial x}{\partial q}\dot{q}_k+\frac{\partial x}{\partial t}\right)=\frac{\partial \dot{x}}{\partial q},\quad\text{(4)}$$

which proves the second part of the transformation in (3). Similar results can be obtained for $b\ddot{y}$, $c\ddot{z}$.

Considering now the kinetic energy equation,

$$T = \tfrac{1}{2}\Sigma\{m(\dot{x}^2+\dot{y}^2+\dot{z}^2)\},$$

we see that if the values of $\dot{x}$, $\dot{y}$, $\dot{z}$ be inserted from (5′), 1, we get

$$\frac{d}{dt}\frac{\partial T}{\partial \dot{q}} - \frac{\partial T}{\partial q} = Q, \qquad (5)$$

which is a typical Lagrangian equation of motion.

There is nothing in the proof here given affected by the presence or absence of $t$ as a variable in the kinematical equations. The equations therefore hold in either case. When the time does not appear explicitly in the equations of constraint, the kinetic energy is a homogeneous quadratic function of the velocities $\dot{q}_1, \dot{q}_2, \ldots, \dot{q}_k$, with coefficients which are functions of the coordinates. In the other case the kinetic energy expression consists of three parts, $T_2+T_1+T_0$, a homogeneous quadratic function of the velocities, a linear function, and a function of the coordinates alone. The reader may write down these three parts at once from the values of $\dot{x}$, $\dot{y}$, $\dot{z}$.

The distinction between the two cases may be made to vanish formally by regarding $t$ as a coordinate like $q_1, q_2, \ldots$, and taking a new independent variable, $\chi$ say. If a new variable is thus introduced there is of course a corresponding momentum component $\partial T/\partial \dot{\chi}$. [See 6, below.]

**4**. *Condition to be fulfilled by generalised coordinates.* It is generally stated that the coordinates chosen must be such as to define the configuration of the body or system with respect to some fixed axes of reference. This is true but not quite in the sense in which the statement is usually understood. It is enough to assume a set of axes, the geometrical relations of which to the axes which have been adopted can be expressed. If this condition is fulfilled, any system of axes will serve, even a system which coincides at the instant considered with moving axes, say the principal axes in the case of a rigid body, to which the motion is referred. What must be done is to correct the result of specialising the axes, by which some coordinates are rendered zero and others appear as constants in quantities, which, in consequence of the motion of the axes of reference with respect to coincident fixed axes, are subject to differentiation.

An example will make this clear. A rigid body turns, with centroid O fixed, about its principal axes O(A, B, C) with the angular speeds $p$, $q$, $r$. The axes of reference are the principal axes, which move the body, and there is nothing to fix the position of the body in space at any instant. If A, B, C be the principal moments of inertia, the kinetic energy is given by

$$T = \tfrac{1}{2}(Ap^2+Bq^2+Cr^2). \qquad (1)$$

The application of the Lagrangian rule for the formation of the equations of motion, without precautions to take account of the motion of the axes, would give the erroneous equations

$$A\dot{p}=L, \quad B\dot{q}=M, \quad C\dot{r}=N, \qquad (2)$$

where L, M, N are the applied couples.

Consider now a second system of axes O(D, E, C), of which the first two are in the plane of OA and OB, with OA between OD and OE, and OD on the side of OA remote from OB as shown in Fig. 104, while OC coincides with the third principal axis.

Let $\omega_1$, $\omega_2$ be the angular speeds about OD and OE: the angular speed about OC is $r$ in both systems. If $\phi=\angle EOB$ we have

$$p=\omega_1\cos\phi+\omega_2\sin\phi,\quad q=-\omega_1\sin\phi+\omega_2\cos\phi. \qquad\text{(3)}$$

Also if we regard OD and OE as fixed, for the moment, we see that the rate at which OB is increasing its angular distance from OE is $r$, and so obtain the equation $r=\dot\phi$.

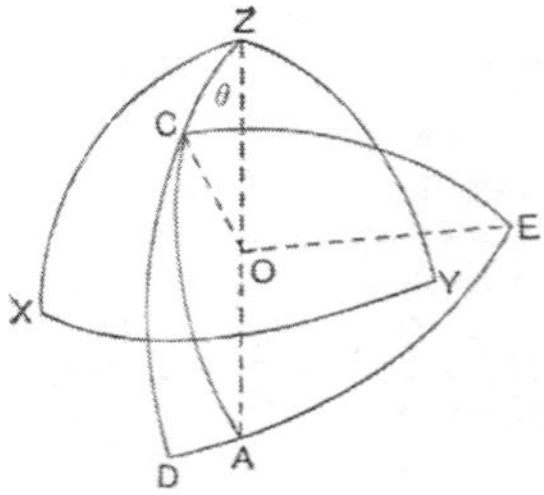

FIG. 104.

The kinetic energy is now given by putting (1) in the form

$$T=\tfrac{1}{2}\{A(\omega_1\cos\phi+\omega_2\sin\phi)^2 + B(-\omega_1\sin\phi+\omega_2\cos\phi)^2+C\dot\phi^2\}. \qquad\text{(1')}$$

Hence we obtain

$$\frac{\partial T}{\partial\dot\phi}=C\dot\phi,\quad \frac{\partial T}{\partial\phi}=(A-B)pq. \qquad\text{(4)}$$

Thus since $\dot\phi=r$, the equation of motion for the axis OC is given by the Lagrangian process as

$$C\dot r-(A-B)pq=L, \qquad\text{(5)}$$

and by symmetry we have similar equations for the axes OA and OB.

It is now easy to correct the procedure which led to the erroneous equations of motion (2). The axes O(A, B, C) coincide with the fixed axes O(D, E, C) at the instant. The former axes are in motion and the coincidence does not continue: OB separates from OE at rate $r=\dot\phi$. Consider

$$\frac{\partial}{\partial\phi}\{\tfrac{1}{2}(Ap^2+Bq^2)\}=Ap\frac{\partial p}{\partial\phi}+Bq\frac{\partial q}{\partial\phi}. \qquad\text{(6)}$$

The ordinary process assumes that there is no variation of $\frac{1}{2}Ap^2+\frac{1}{2}Bq^2$ with variation of $\phi$, since $\phi$ does not appear explicitly. The correct view is that $\phi$ is only apparently absent, owing to its having been given the special value 0. The values of $\partial p/\partial\phi$, $\partial q/\partial\phi$ are not zero. We have in fact

$$\left.\begin{aligned} p&=\omega_1\cos\phi+\omega_2\sin\phi=p\cos 0+q\sin 0,\\ q&=-\omega_1\sin\phi+\omega_2\cos\phi=-p\sin 0+q\cos 0.\end{aligned}\right\} \qquad\text{(7)}$$

Thus

$$\frac{\partial p}{\partial\phi}=q\cos 0=q,\quad \frac{\partial q}{\partial\phi}=-p\cos 0=-p, \qquad\text{(8)}$$

and therefore

$$\frac{\partial T}{\partial\phi}=(A-B)pq. \qquad\text{(9)}$$

5. *Proof of Euler's equations by vectors.* In matters of this sort vector considerations conduce to clearness of conception, and the method of forming equations of motion, so often employed in this work, is really only a simple process of applying vector ideas to give at once the ordinary scalar equations. But it may be desirable to give here a brief vector treatment of the points just discussed.

If **i**, **j**, **k** be unit vectors along the axes OA, OB, OC the resultant angular momentum, considered as a vector **H**, is given by

$$\mathbf{H}=Ap\mathbf{i}+Bq\mathbf{j}+Cr\mathbf{k}. \qquad\text{(1)}$$

Since the axes are principal axes and move with the body $\dot A=\dot B=\dot C=0$, and we have

$$\dot{\mathbf{H}}=A\dot p\mathbf{i}+B\dot q\mathbf{j}+C\dot r\mathbf{k}+Ap\frac{d\mathbf{i}}{dt}+Bq\frac{d\mathbf{j}}{dt}+Cr\frac{d\mathbf{k}}{dt}. \qquad\text{(2)}$$

But **i**, **j**, **k** being unit vectors do not vary in length as they are changed in direction. The direction of $d\mathbf{i}/dt$ is perpendicular to **i**, and obviously its scalar components in the directions of j and **k** are $r$ and $-q$. Similarly we obtain components for $d\mathbf{j}/dt$, $d\mathbf{k}/dt$. Thus we have

$$\frac{d\mathbf{i}}{dt}=\mathbf{j}r-\mathbf{k}q,\quad \frac{d\mathbf{j}}{dt}=\mathbf{k}p-\mathbf{i}r,\quad \frac{d\mathbf{k}}{dt}=\mathbf{i}q-\mathbf{j}p, \qquad\text{.....................(3)}$$

and therefore (1) becomes

$$\dot{\mathbf{H}}=\mathbf{i}\{\mathrm{A}\dot{p}-(\mathrm{B}-\mathrm{C})qr\}+\mathbf{j}\{\mathrm{B}\dot{q}-(\mathrm{C}-\mathrm{A})rp\}+\mathbf{k}\{\mathrm{C}\dot{r}-(\mathrm{A}-\mathrm{B})pq\}. \quad\text{...(4)}$$

Thus, since $\dot{\mathbf{H}}=\mathbf{i}\mathrm{L}+\mathbf{j}\mathrm{M}+\mathbf{k}\mathrm{N}$,

the resultant couple, we get

$$\mathrm{A}\dot{p}-(\mathrm{B}-\mathrm{C})qr=\mathrm{L},\quad \mathrm{B}\dot{q}-(\mathrm{C}-\mathrm{A})rp=\mathrm{M},\quad \mathrm{C}\dot{r}-(\mathrm{A}-\mathrm{B})pq=\mathrm{N}. \quad\text{...(5)}$$

The vector method thus accepts the components of A.M. as they are given without any explicit appearance of $\phi$, and by taking account of the motion of the axes virtually takes account of the variation of the angular momenta with $\phi$.

An example of the necessity for introducing additional terms to take account of the motion of the axes in forming the equations of motion is found in the discussion of the motion of a solid through a liquid, when referred to axes moving with the body. If $u$, $v$, $w$ be the components of linear velocity of the solid, $p$, $q$, $r$ its angular speeds about the axes, and A, B, C, P, Q, R constants, the kinetic energy, under certain conditions as to symmetry, is of the form $\frac{1}{2}(\mathrm{A}u^2+\mathrm{B}v^2+\mathrm{C}w^2+\mathrm{P}p^2+\mathrm{Q}q^2+\mathrm{R}r^2)+\mathrm{K}$, where K depends on the circulation, if any, through the solid. The equations are formed by Lagrange's method with suitable connections [see XIII].

**6.** *Generalised momenta. Hamilton's dynamical equations. Canonical equations.* The equations of the type

$$\frac{\partial \mathrm{T}}{\partial \dot{q}}=p \qquad\text{........................................(1)}$$

give linear equations by which $\dot{q}_1$, $\dot{q}_2$, ... can be calculated in terms of $p_1$, $p_2$, ..., which are called the generalised components of momentum. The values of $\dot{q}_1$, $\dot{q}_2$, ..., thus obtained, can be substituted in the expression for T, which then becomes a quadratic function of the momenta. The equations of motion become transformed into what are known as the Hamiltonian equations of motion, given originally by Sir William Rowan Hamilton.

Consider the function K defined by the equation

$$\mathrm{K}=\Sigma(p\dot{q})-\mathrm{T}, \qquad\text{..................................(2)}$$

where T is supposed to be expressed in terms of the $\dot{q}$s. When T, or K, is expressed in terms of the momenta we shall indicate the fact by the suffix $m$ as in $\mathrm{T}_m$ or $\mathrm{K}_m$. It will be observed that $\Sigma(p\dot{q})$ is not 2T, unless T is a homogeneous quadratic function of the $\dot{q}$s.

We find first $\dfrac{\partial \mathrm{K}_m}{\partial p}$, $\dfrac{\partial \mathrm{K}_m}{\partial q}$, where the notation $K_m$ indicates that, both in $\Sigma(p\dot{q})$ and in T, the $\dot{q}$s have been replaced by functions of the $p$s and $q$s. We get at once

$$\left.\begin{aligned}\frac{\partial K_m}{\partial p} &= \dot{q}+\Sigma\left(p\frac{\partial \dot{q}}{\partial p}\right)-\Sigma\left(\frac{\partial T}{\partial \dot{q}}\frac{\partial \dot{q}}{\partial p}\right)=\dot{q},\\ \frac{\partial K_m}{\partial q} &= \Sigma\left(p\frac{\partial \dot{q}}{\partial q}\right)-\frac{\partial T_s}{\partial q}-\Sigma\left(\frac{\partial T}{\partial \dot{q}}\frac{\partial \dot{q}}{\partial q}\right)=-\frac{\partial T_s}{\partial q}.\end{aligned}\right\} \quad \text{.................. (3)}$$

The results in (3) are of great importance. We get from the second of (3), by substitution in the typical Lagrangian equation (5), 3, the form

$$\frac{dp}{dt}+\frac{\partial K_m}{\partial q}=Q, \quad \text{..............................(4)}$$

and from the first of (3) the companion equation

$$\frac{\partial K}{\partial p}-\dot{q}=0. \quad \text{..................................(5)}$$

If we suppose that Q is derivable from a potential V, a function of the $q$s, so that $-\partial V/\partial q=Q$, and write H for $K_m+V$, we obtain the equations

$$\dot{p}+\frac{\partial H}{\partial q}=0, \quad \dot{q}-\frac{\partial H}{\partial p}=0, \quad \text{..........................(6)}$$

and a similar pair of equations holds for each coordinate $q$. These are the so-called *canonical equations* of dynamics. They were given by Hamilton for the case of $H=T+V$, and are the fundamental differential equations of motion in modern physical astronomy.

We may notice further that, if H be an explicit function of $t$, we have

$$\frac{dH}{dt}=\Sigma\left(\frac{\partial H}{\partial q}\dot{q}\right)+\Sigma\left(\frac{\partial H}{\partial p}\dot{p}\right)+\frac{\partial H}{\partial t}=\frac{\partial H}{\partial t} \quad \text{......................(7)}$$

by (6). Thus H is constant if $t$ does not appear in it explicitly.

But since $H=K+V$, and $K=\Sigma(p\dot{q})-T$, (7) may be written

$$\frac{dH}{dt}=\frac{\partial K}{\partial t}+\frac{\partial V}{\partial t}=-\frac{\partial T}{\partial t}+\frac{\partial V}{\partial t}. \quad \text{.........................(8)}$$

Thus, if H is not an explicit function of $t$, it is a constant, $h$ say, that is $h=K+V$, or there is conservation of $K+V$ $(=T+V)$ in the system. This is also the case if

$$\frac{\partial T}{\partial t}=\frac{\partial V}{\partial t}. \quad \text{...................................(9)}$$

Now, if the time appears explicitly in the kinematical relations of the system, T is no longer a homogeneous quadratic function of the $\dot{q}$s, but consists of the sum of such a function $T_2$, a linear function $T_1$ of the $\dot{q}$s, and a function $T_0$ of the coordinates, and, it may be, of $t$, that is we have

$$K=2T_2+T_1-T=T_2-T_0. \quad \text{..........................(10)}$$

Hence, when $K+V=h$, we have

$$T_2-T_0+V=h,$$

or

$$T-T_1-2T_0+V=h. \quad \text{..............................(11)}$$

If $t$ does not appear in the kinematical equations the $\alpha$, $\beta$, $\gamma$ of (5), 2, are zero, and $T_1$ and $T_0$ disappear from (11). The system is then *conservative*.

It follows from (10) that we may, in forming the Lagrangian equations of motion, regard the homogeneous quadratic function $T_2$ as the kinetic energy, and $V-T_0$ as the potential energy. There will be conservation of $K+V$ if (9) is fulfilled, and conservation of *energy* if $T_1$ and $T_0$ are zero.

When Q is in whole or in part derivable from a function V, we may write L for T − V, and put (5), 3, in the form

$$\frac{d}{dt}\frac{\partial L}{\partial \dot{q}}-\frac{\partial L}{\partial Q}=Q', \quad \text{.......... (12)}$$

where Q′ is the part of Q, if any, not derivable from V. L is called the *kinetic potential*.

7. *Systems which are not holonomous.* If the system is not holonomous,* $a$, $b$, $c$ are not partial differential coefficients of finite functions, which express $x$, $y$, $z$ for each particle in terms of $q_1, q_2, \ldots, q_k$. The process of derivation just used for Lagrangian equations is no longer applicable and another must be sought. In the discussion of this question we shall no longer refer to the Cartesian equations of motion, though we shall suppose that as a rule the kinetic and potential energies have been expressed, in some primary system of reference, in terms of coordinates and velocities, from which it has been necessary to transform to the generalised coordinates and the corresponding velocities.

Let the kinetic energy be primarily expressed by the time-rates of change $\dot{u}, \dot{v}, \dot{w}, \ldots$ of quantities $u, v, w, \ldots$, which fulfil the equations

$$\left.\begin{aligned}\delta u &= a_1\delta q_1+a_2\delta q_2+\ldots+a_i\delta q_i+e_1\delta s_1+e_2\delta s_2+\ldots+e_j\delta s_j,\\ \delta v &= b_1\delta q_1+b_2\delta q_2+\ldots+b_i\delta q_i+f_1\delta s_1+f_2\delta s_2+\ldots+f_j\delta s_j.\\ &\ldots\ldots\ldots\ldots\ldots\ldots\end{aligned}\right\} \quad \text{..........(1)}$$

For a rigid body $\dot{u}, \dot{v}, \ldots$ may be taken as the product of the square roots $M^{\frac{1}{2}}, N^{\frac{1}{2}}, \ldots$ of inertia coefficients M, N, ... by velocity components of the centroid, or the products of angular velocities of the body about given axes by the square roots of the proper inertia coefficients for this case; and so for other systems. Thus, if T be the kinetic energy, we shall have

$$T=\tfrac{1}{2}(\dot{u}^2+\dot{v}^2+\dot{w}^2+\ldots). \quad \text{..........(2)}$$

We thus simplify the equations by avoiding the introduction of mass coefficients, which are easily supplied for any illustrative example.

The parameters are divided in (1) into two sets, the $q$s and the $s$s, for a purpose which will appear later in connection with ignoration of coordinates. By (1)

$$\left.\begin{aligned}\dot{u} &= a_1\dot{q}_1+a_2\dot{q}_2+\ldots+a_i\dot{q}_i+e_1\dot{s}_1+e_2\dot{s}_2+\ldots+e_j\dot{s}_j,\\ \dot{v} &= b_1\dot{q}_1+b_2\dot{q}_2+\ldots+b_i\dot{q}_i+f_1\dot{s}_1+f_2\dot{s}_2+\ldots+f_j\dot{s}_j,\\ &\ldots\ldots\ldots\ldots\ldots\ldots,\end{aligned}\right\} \quad \text{..........(3)}$$

* The following discussion of the dynamics of systems which are not holonomous and of the general dynamics of gyrostatic systems is taken in the main from a paper by the author in *Proc. R.S.E.*, 1909.

if we suppose, as we do for the present, that $t$ does not appear in the kinematical equations. We suppose these values of $\dot{u}$, $\dot{v}$, ... substituted in (2).

Also by the signification of $\dot{u}$, $\dot{v}$, ... we have primary equations of motion, which may be written

$$\ddot{u} = \mathrm{U}, \quad \ddot{v} = \mathrm{V}, \ldots, \qquad \text{(4)}$$

and therefore obtain a series of equations of the form

$$\left.\begin{array}{l} a_1\ddot{u} + b_1\ddot{v} + \ldots = a_1\mathrm{U} + b_1\mathrm{V} + \ldots = \mathrm{Q} \\ a_2\ddot{u} + b_2\ddot{v} + \ldots = a_2\mathrm{U} + b_2\mathrm{V} + \ldots = \mathrm{Q}_2, \\ \ldots\ldots\ldots\ldots\ldots\ldots\ldots\ldots, \end{array}\right\} \qquad \text{(5)}$$

where $\mathrm{Q}_1$, $\mathrm{Q}_2$, ... are generalised "forces" according to the meanings of $\dot{u}$, $\dot{v}$, .... There are $i+j$ such equations, since there are now supposed to be $i+j$ independent parameters $q_1, q_2, \ldots, s_1, s_2, \ldots$.

By (2), (3), and (5) we obtain at once

$$\left.\begin{array}{l} \dfrac{d}{dt}\dfrac{\partial \mathrm{T}}{\partial \dot{q}_1} - (\dot{a}_1\dot{u} + \dot{b}_1\dot{v} + \ldots) = \mathrm{Q}_1, \\ \dfrac{d}{dt}\dfrac{\partial \mathrm{T}}{\partial \dot{q}_2} - (\dot{a}_2\dot{u} + \dot{b}_2\dot{v} + \ldots) = \mathrm{Q}_2, \\ \ldots\ldots\ldots\ldots\ldots\ldots\ldots\ldots. \end{array}\right\} \qquad \text{(6)}$$

These equations are applicable to all cases, whether the system is holonomous or not. When the system is not holonomous the terms in brackets on the left cannot be replaced by $\partial\mathrm{T}/\partial q_1$, $\partial\mathrm{T}/\partial q_2$, ..., for the coefficients $a$, $b$, $c$ are not partial differential coefficients, as they are in holonomous systems.

It is to be noticed with reference to (2) that different modes of breaking up the kinetic energy into a sum of squares are not always equivalent, but may involve different sets of forces. For example, the expression $\frac{1}{2}\mathrm{A}(\dot{\theta}^2 + \sin^2\theta \,.\, \dot{\psi}^2)$ which occurs in the expression for the kinetic energy of a symmetrical top, and which is already a sum of two squares, may also be written in the form $\frac{1}{2}\mathrm{A}(\dot{\theta}\cos\phi + \dot{\psi}\sin\theta\sin\phi)^2 + \frac{1}{2}\mathrm{A}(-\dot{\theta}\sin\phi + \dot{\psi}\sin\theta\cos\phi)^2$. The former is the kinetic energy of the top (apart from kinetic energy of rotation) when the axis of figure is inclined to the vertical at the angle $\theta$, and the axes chosen are axes OD and OE as shown in Fig. 104, while the latter is the same kinetic energy when the axes chosen are OA and OB. The couples are different in the two cases, as may be seen by supposing that the A.M. about the axes of figure is C$n$, and writing down the equations. But we have here two perfectly reconcilable solutions of the same problem, which was to be expected since the system is holonomous.

It is otherwise for systems which are not holonomous. It is possible in fact to specify two distinct cases of motion, which have exactly the same expressions for the kinetic and potential energies, but which have entirely unreconcilable equations of motion. A disk or hoop which is symmetrical about an axis perpendicular to its plane, rolls, without sliding, on a horizontal plane, on an edge (radius $a$) which is the terminating circle of the plane drawn through the centroid perpendicular to the axis. In the other case, the disk or hoop rests with the edge on a horizontal plane without friction, while its centroid is constrained to move along a fixed vertical circle of radius $a$ and centre in the horizontal plane. By supposing the mass of the same in amount in the two cases, but differently distributed, the moments of inertia in the second case can be made such that the potential and kinetic energies in the two cases are identical. In their ordinary form Lagrange's equations are applicable to the second case; they are not applicable to the former case.

In general Lagrange's method is useless for finding certain of the equations of a rolling disk or hoop. It may however be applied to find the $\theta$-equation, where $\theta$ is the inclination of the axis of figure of the hoop or disk to the vertical.

Correct results may always be obtained if the equations of motion are found by first principles. Lagrange's equations do not "fail" any more than Taylor's theorem does; they are inapplicable in cases which are excluded by the conditions involved, tacitly or explicitly, in the proof of the equations.

**8.** *Conditions necessary for usual form of Lagrange's equations.* It is not difficult to show* that if the conditions

$$\left.\begin{aligned}\frac{\partial a_1}{\partial q_2}=\frac{\partial a_2}{\partial q_1},\quad \frac{\partial a_1}{\partial q_3}=\frac{\partial a_3}{\partial q_1},\ \dots,\\ \frac{\partial b_1}{\partial q_2}=\frac{\partial b_2}{\partial q_1},\quad \frac{\partial b_1}{\partial q_3}=\frac{\partial b_3}{\partial q_1},\ \dots,\end{aligned}\right\} \quad\dots\dots\dots(1)$$

are fulfilled. Lagrange's equations are of the usual form typified by

$$\frac{d}{dt}\frac{\partial \mathrm{T}}{\partial \dot{q}_1}-\frac{\partial \mathrm{T}}{\partial q_1}=\mathrm{Q}_1. \quad\dots\dots\dots(2)$$

Equations (1) form one set of the conditions which make equations (3), 7, the derivatives of a set of finite equations; and if all these conditions are fulfilled, every equation of motion takes the usual form (2). The fulfilment of each set involves the validity of a corresponding equation of motion in the usual form.

If one set, say (1), of these conditions holds, we can write (using our present notation)

$$\left.\begin{aligned}\delta u&=\delta \mathrm{F}+\alpha_2\delta q_2+\alpha_3\delta q_3+\dots,\\ \delta v&=\delta \mathrm{G}+\beta_2\delta q_2+\beta_3\delta q_3+\dots,\\ &\dots\dots\dots\dots,\end{aligned}\right\} \quad\dots\dots\dots(3)$$

where $\delta\mathrm{F}$, $\delta\mathrm{G}$, are perfect differentials and $\delta q_1$ does not appear.

If, then, we notice when the fundamental equations are written down that the terms corresponding to any coordinate are expressed in a form corresponding to that shown here for $q_1$, we know that the equation of motion corresponding to that coordinate can be found by the ordinary Lagrangian process.

**9.** *Equations of motion for systems which are not holonomous.* In what follows we shall, for brevity, adopt the notation

$$\left.\begin{aligned}\phi_1\mathrm{T}&=\dot{a}_1\dot{u}+\dot{b}_1\dot{v}+\dots,\\ \phi_2\mathrm{T}&=\dot{a}_2\dot{u}+\dot{b}_2\dot{v}+\dots,\\ &\dots\dots\dots\dots\end{aligned}\right\} \quad\dots\dots\dots(1)$$

If the form of T be modified in any way, for example by the substitution of the values of $\dot{s}_1, \dot{s}_2, \dots$ in terms of the momenta $\partial\mathrm{T}/\partial\dot{s}_1, \partial\mathrm{T}/\partial\dot{s}_2, \dots$, and the other velocities $\dot{q}_1, \dot{q}_2, \dots$, then if T′ be the modified form of T, $\phi_1\mathrm{T}', \phi_2\mathrm{T}', \dots$ will be understood to denote the operations indicated in (1), but performed with the new values which are then given to the coefficients of $\dot{q}_1, \dot{q}_2, \dots, \dot{q}_k$.

It follows from what has been stated that, as has already been pointed out in 8, the operations indicated in (1) cannot be performed without reference to the fundamental equations

* See the paper cited in 8 above.

from which that expression has been derived. For example, two terms in T might be $\frac{1}{2}A\dot{\theta}^2+\frac{1}{2}B\dot{\psi}^2$. These might be derived either from $\dot{u}=A^{\frac{1}{2}}\dot{\theta}$, $\dot{v}=B^{\frac{1}{2}}\dot{\psi}$, or from

$$\dot{u}=A^{\frac{1}{2}}\sin\theta\,.\,\dot{\theta}+B^{\frac{1}{2}}\cos\theta\,.\,\dot{\psi},\quad \dot{v}=A^{\frac{1}{2}}\cos\theta\,.\,\dot{\theta}-B^{\frac{1}{2}}\sin\theta\,.\,\dot{\psi}.$$

The former mode of derivation would satisfy the conditions of integrability so far as these terms are concerned, the second would not. It is possible, in fact, to specify two distinct cases of motion which have precisely the same expressions for the kinetic and potential energies, but which have not the same equations of motion. An example is given towards the end of 7 above.

**10.** *Ignoration of coordinates.* Let now the form of T be modified by the substitutions indicated above. Our object is to inquire what modification is required in the process of "ignoration of coordinates" by the non-integrability of the relations between the generalised coordinates and the functions of these coordinates and their velocities from which the kinetic energy is derived. We shall suppose therefore that the coordinates $s_1, s_2, \ldots$, are absent from the kinetic energy, and from the function V of the coordinates from which the forces are derived, if that function exists. Writing them for a moment $r_1=\partial T/\partial\dot{s}_1$, $r_2=\partial T/\partial\dot{s}_2, \ldots$, we see that if the fundamental relations were integrable $r_1, r_2, \ldots$, would be constants, since then we should have

$$\frac{\partial T}{\partial s_1}=0,\quad \frac{\partial T}{\partial s_2}=0,\ \ldots,$$

on the supposition that besides $\partial V/\partial s_1$, $\partial V/\partial s_2, \ldots$, which are all zero, no generalised forces corresponding to $s_1, s_2, \ldots$, exist. The equations of motion are now however

$$\left.\begin{array}{l}\dfrac{d}{dt}\dfrac{\partial T}{\partial\dot{s}_1}-\chi_1 T=0,\\ \dfrac{d}{dt}\dfrac{\partial T}{\partial\dot{s}_2}-\chi_2 T=0,\\ \ldots\ldots\ldots\ldots\ldots\ldots,\end{array}\right\}\quad\ldots\ldots(1)$$

where

$$\left.\begin{array}{l}\chi_1 T=\dot{e}_1\dot{u}+\dot{f}_1\dot{v}+\ldots,\\ \ldots\ldots\ldots\ldots\ldots\ldots,\end{array}\right\}\quad\ldots\ldots(2)$$

so that $\chi_1, \chi_2, \ldots$, are the operators for the $s$ coordinates that $\phi_1, \phi_2, \ldots$, are for the $q$ coordinates.

The conditions for constancy of the momenta $\partial T/\partial\dot{s}_1$, $\partial T/\partial\dot{s}_2, \ldots$, are therefore now

$$\chi_1 T=0,\quad \chi_2 T=0.\quad\ldots\ldots(2')$$

These conditions are fulfilled by (2) when $\dot{e}_1, \dot{f}_1, \ldots$, are zero, which is the case in various problems of the motions of tops and gyrostats, where none of the coefficients $e_1, f_1, \ldots, e_2, f_2, \ldots$, contains the time or any of the coordinates $q_1, q_2, \ldots$. We shall not assume, unless it is so stated, that $e_1, f_1, \ldots, e_2, f_2, \ldots$, are absolute constants.

We shall assume however that $\dfrac{\partial T}{\partial\dot{s}_1}=c_1,\ \dfrac{\partial T}{\partial\dot{s}_2}=c_2,\ \ldots\ldots(3)$

where $c_1, c_2, \ldots$, are constants. These conditions are fulfilled in a large number of problems regarding rotating flywheels in which the coordinates $q_1, q_2, \ldots$, determining the positions of the axes of rotation have no influence on the momenta $\partial T/\partial\dot{s}_1, \ldots$. When the system is holonomous the constancy of these momenta is secured by the fact that the differential equations (2′) become $\partial T/\partial s_1=0, \ldots$. Equations (3) extended are

$$\left.\begin{array}{l}(e_1a_1+f_1b_1+\ldots)\dot{q}_1+(e_1a_2+f_1b_2+\ldots)\dot{q}_2+\ldots+(e_1^2+f_1^2+\ldots)\dot{s}_1+(e_1e_2+f_1f_2+\ldots)\dot{s}_2+\ldots=c_1,\\ (e_2a_1+f_2b_1+\ldots)\dot{q}_1+(e_2a_2+f_2b_2+\ldots)\dot{q}_2+\ldots+(e_1e_2+f_1f_2+\ldots)\dot{s}_1+(e_2^2+f_2^2+\ldots)\dot{s}_2+\ldots=c_2,\\ \ldots\ldots\ldots\ldots\ldots\ldots\ldots\ldots\ldots\ldots\ldots\ldots\ldots\ldots\ldots\ldots\ldots\ldots;\end{array}\right\}(4)$$

The coefficients of $\dot{q}_1, \dot{q}_2, \ldots, \dot{s}_1, \dot{s}_2, \ldots$, are those of the products $\dot{q}_1\dot{s}_1, \dot{q}_2\dot{s}_1, \ldots, \dot{s}_1\dot{s}_1, \dot{s}_1\dot{s}_2, \ldots$, in the first line, and of $\dot{q}_1\dot{s}_2, \dot{q}_2\dot{s}_2, \ldots, \dot{s}_1\dot{s}_2, \dot{s}_2\dot{s}_2, \ldots$, in the second, and so on. Denoting these coefficients as in the scheme

$$(q_1, s_1),\quad (q_2, s_2), \ldots, (s_1, s_1),\quad (s_1, s_2), \ldots,$$
$$(q_1, s_2),\quad (q_2, s_2), \ldots, (s_1, s_2),\quad (s_2, s_2), \ldots,$$

we can write equations (4) as follows:

$$\left.\begin{array}{l}(s_1, s_1)\dot{s}_1+(s_2, s_1)\dot{s}_2+\ldots=c_1-(q_1, s_1)\dot{q}_1-(q_2, s_1)\dot{q}_2-\ldots, \\ (s_1, s_2)\dot{s}_1+(s_2, s_2)\dot{s}_2+\ldots=c_2-(q_1, s_2)\dot{q}_1-(q_2, s_2)\dot{q}_2-\ldots, \\ \ldots\ldots\ldots\ldots\ldots\ldots\ldots\ldots\ldots\ldots\ldots\ldots\end{array}\right\} \quad\ldots\ldots(5)$$

From these $\dot{s}_1, \dot{s}_2, \ldots$, can be determined in terms of $c_1, c_2, \ldots$, and $\dot{q}_1, \dot{q}_2, \ldots$. These values then substituted in the expression for T convert it into a function T′ of $\dot{q}_1, \dot{q}_2, \ldots$, $c_1, c_2, \ldots$, so that all trace of the variables $s_1, s_2, \ldots$, is now removed. We have to inquire what form the equations of motion take when this substitution is made, or, as it is usually put, *when the coordinates $s_1, s_2, \ldots$, are ignored.* First we form expressions for $\dot{s}_1, \dot{s}_2, \ldots$. Let $(c_1, c_1), (c_1, c_2), \ldots$, denote the ratios of the consecutive first minors of the determinant of equations (5) to that determinant, and put

$$K=\tfrac{1}{2}\{(c_1, c_1)c_1^2+2(c_1, c_2)c_1c_2+\ldots\}. \quad\ldots\ldots(6)$$

Then

$$\left.\begin{array}{l}\dot{s}_1=\dfrac{\partial K}{\partial c_1}-(\dot{q}_1A_1+\dot{q}_2B_1+\ldots), \\ \dot{s}_2=\dfrac{\partial K}{\partial c_2}-(\dot{q}_1A_2+\dot{q}_2B_2+\ldots), \\ \ldots\ldots\ldots\ldots\ldots\ldots,\end{array}\right\} \quad\ldots\ldots(7)$$

where

$$\left.\begin{array}{ll}A_1=(c_1, c_1)(q_1, s_1)+(c_1, c_2)(q_1, s_2)+\ldots, & A_2=(c_2, c_1)(q_1, s_1)+(c_2, c_2)(q_1, s_2)+\ldots, \\ B_1=(c_1, c_1)(q_2, s_1)+(c_1, c_2)(q_2, s_2)+\ldots, & B_2=(c_2, c_1)(q_2, s_1)+(c_2, c_2)(q_2, s_2)+\ldots, \\ \ldots\ldots\ldots\ldots\ldots\ldots & \end{array}\right\} \ldots(8)$$

**11.** *Ignoration of coordinates. Formation of equations of motion.* In the paper referred to in 7 it is shown that if T′ be defined as above,

$$\frac{d}{dt}\frac{\partial T'}{\partial \dot{q}_1}+\Sigma\left(c\frac{dA}{dt}\right)=\frac{d}{dt}\frac{\partial T}{\partial \dot{q}_1},\quad \frac{d}{dt}\frac{\partial T'}{\partial \dot{q}_2}+\Sigma\left(c\frac{dB}{dt}\right)=\frac{d}{dt}\frac{\partial T}{\partial \dot{q}_2}, \ldots,$$

and that

$$\phi_1 T'=\phi_1 T-\Sigma\left(c\frac{dA}{dt}\right), \ldots.$$

It follows therefore that when the coordinates $s_1, s_2, \ldots$, are ignored in the case of a system which is not holonomous the equations of motion have the form

$$\left.\begin{array}{l}\dfrac{d}{dt}\dfrac{\partial T'}{\partial \dot{q}_1}-\phi_1 T'=Q_1, \\ \dfrac{d}{dt}\dfrac{\partial T'}{\partial \dot{q}_2}-\phi_2 T'=Q_2, \\ \ldots\ldots\ldots\ldots\ldots\ldots\end{array}\right\} \quad\ldots\ldots(1)$$

Hence we have the very simple rule: modify the expression for T by substituting for $\dot{s}_1, \dot{s}_2, \ldots$, their values from (7), 11, and then proceed as if no velocities of coordinates $s_1, s_2, \ldots$, had ever entered into the expression for the kinetic energy. The equations of motion obtained are of course applicable also to holonomous systems.

To verify the results obtained, we write the first of (1) in the form

$$\frac{d}{dt}\frac{\partial T'}{\partial \dot{q}_1}-\frac{\partial T'}{\partial q_1}-\left(\phi_1 T'-\frac{\partial T}{\partial q_1}\right)=Q_1, \quad\ldots\ldots(2)$$

and consider what it becomes when the system is made holonomous. We have

$$-\phi_1 T'=-\phi_1 T+\Sigma\left(c\frac{dA}{dt}\right)=-(\dot{a}_1\dot{u}+\dot{b}_1\dot{v}+\ldots)+\Sigma\left(c\frac{dA}{dt}\right), \quad\ldots\ldots(3)$$

and since we have also now

$$\dot{u}=\{a_1-\Sigma(e\mathsf{A})\}\dot{q}_1+\{a_2-\Sigma(e\mathsf{B})\}\dot{q}_2+\dots,$$
$$\dot{v}=\{b_1-\Sigma(f\mathsf{A})\}\dot{q}_1+\{b_2-\Sigma(f\mathsf{B})\}\dot{q}_2+\dots,$$
$$\dots\dots\dots\dots\dots\dots\dots\dots\dots\dots,$$

we obtain

$$\frac{\partial \mathrm{T}}{\partial q_1}=\dot{u}\left[\left\{\frac{\partial a_1}{\partial q_1}-\frac{\partial}{\partial q_1}\Sigma(e\mathsf{A})\right\}\dot{q}_1+\left\{\frac{\partial a_2}{\partial q_1}-\frac{\partial}{\partial q_1}\Sigma(e)\mathsf{B}\right\}\dot{q}_2+\dots\right]$$
$$+\dot{v}\left[\left\{\frac{\partial b_1}{\partial q_1}-\frac{\partial}{\partial q_1}\Sigma(f\mathsf{A})\right\}\dot{q}_1+\left\{\frac{\partial b_2}{\partial q_1}-\frac{\partial}{\partial q_1}\Sigma(f)\mathsf{B}\right\}\dot{q}_2+\dots\right]$$
$$+\dots\dots\dots\dots\dots\dots\dots\dots\dots\dots \qquad (4)$$

If now $\partial a_2/\partial q_1=\partial a_1/\partial q_2, \dots \partial b_2/\partial q_1=\partial b_1/\partial q_2, \dots$, (4) becomes

$$\frac{\partial \mathrm{T}}{\partial q_1}=\dot{a}_1\dot{u}+\dot{b}_1\dot{v}+\dots-\dot{u}\left\{\dot{q}_1\frac{\partial}{\partial q_1}\Sigma(e\mathsf{A})+\dot{q}_2\frac{\partial}{\partial q_1}\Sigma(e\mathsf{B})+\dots\right\}$$
$$-\dot{v}\left\{\dot{q}_1\frac{\partial}{\partial q_1}\Sigma(f\mathsf{A})+\dot{q}_2\frac{\partial}{\partial q_1}\Sigma(f\mathsf{B})+\dots\right\}$$
$$-\dots\dots\dots\dots\dots\dots\dots\dots\dots\dots, \qquad (5)$$

by (2′), 11. Hence finally

$$\frac{d}{dt}\frac{\partial \mathrm{T}'}{\partial \dot{q}_1}-\frac{\partial \mathrm{T}'}{\partial q_1}+\dot{q}_2\Sigma\left\{c\left(\frac{\partial \mathsf{A}}{\partial q_2}-\frac{\partial \mathsf{B}}{\partial q_1}\right)\right\}+\dots=\mathrm{Q}_1, \qquad (6)$$

and similarly for the other equations. The terms in $\dot{q}_1, \dot{q}_2, \dots$ are called gyrostatic terms. The coefficient of $\dot{q}_1$ in the first equation, of $\dot{q}_2$ in the second, and so on, is zero.

Equation (6) is thus of the form

$$\frac{d}{dt}\frac{\partial \mathrm{T}'}{\partial \dot{q}_1}-\frac{\partial \mathrm{T}'}{\partial q_1}+\mathrm{L}_1=\mathrm{Q}_1, \qquad (7)$$

and we have as many such equations as there are coordinates $q_1, q_2, \dots$. If we multiply the first equation by $\dot{q}_1$, the second by $\dot{q}_2, \dots$, and add, we get as the sum of the quantities on the right

$$\mathrm{Q}_1\dot{q}_1+\mathrm{Q}_2\dot{q}_2+\dots.$$

All the terms $\mathrm{L}_1, \mathrm{L}_2, \dots$, the gyrostatic terms, have disappeared. These terms disclose themselves by certain dynamical phenomena, due to motions and connections, which an ordinary inspection of the body and even some dynamical experiments may fail to detect.

**12.** *Routh's rule for ignoration of coordinates in a holonomous system. Gyrostatic terms.* This simple modification of the expression for the kinetic energy by which the equations of motion are obtained when certain coordinates are ignored, may be compared with the modification given by Routh for the case of holonomous systems (*Stability of Motion*, p. 60). If T and T′ have the meanings assigned above, we have now

$$\frac{\partial \mathrm{T}'}{\partial q_1}=\frac{\partial \mathrm{T}}{\partial q_1}+\Sigma\left(c\frac{\partial \dot{s}}{\partial q_1}\right),$$

and obtain

$$\left.\begin{array}{l}\dfrac{d}{dt}\dfrac{\partial \mathrm{T}'}{\partial \dot{q}_1}-\dfrac{\partial \mathrm{T}'}{\partial q_1}-\Sigma\left\{c\left(\dfrac{d}{dt}\dfrac{\partial \dot{s}}{\partial \dot{q}_1}-\dfrac{\partial \dot{s}}{\partial q_1}\right)\right\}=\dfrac{d}{dt}\dfrac{\partial \mathrm{T}}{\partial \dot{q}_1}-\dfrac{\partial \mathrm{T}}{\partial q_1}=\mathrm{Q}_1,\\ \dots\dots\dots\dots\dots\dots\dots\dots\dots\dots\end{array}\right\} \qquad (1)$$

But, by the result stated at the beginning of 11,

$$\left.\begin{array}{l}\Sigma\left(c\dfrac{d}{dt}\dfrac{\partial \dot{s}}{\partial \dot{q}_1}\right)=-\Sigma\left(c\dfrac{\partial \mathsf{A}}{dt}\right),\quad \Sigma\left(c\dfrac{\partial \dot{s}}{\partial q_1}\right)=-\dot{q}_1\Sigma\left(c\dfrac{\partial \mathsf{A}}{\partial q_1}\right)-\dot{q}_2\Sigma\left(c\dfrac{\partial \mathsf{B}}{\partial q_1}\right)-\dots,\\ \dots\dots\dots\dots\dots\dots\dots\dots\dots\dots,\end{array}\right\} \qquad (2)$$

so that (1) becomes

$$\left.\begin{array}{l}\dfrac{d}{dt}\dfrac{\partial \mathrm{T}'}{\partial \dot{q}_1}-\dfrac{\partial \mathrm{T}'}{\partial q_1}+\dot{q}_1\Sigma\left\{c\left(\dfrac{\partial \mathsf{A}}{\partial q_1}-\dfrac{\partial \mathsf{A}}{\partial q_1}\right)\right\}+\dot{q}_2\Sigma\left\{c\left(\dfrac{\partial \mathsf{A}}{\partial q_2}-\dfrac{\partial \mathsf{B}}{\partial q_1}\right)\right\}+\dots+\dfrac{\partial \mathrm{K}}{\partial q_1}=\mathrm{Q}_1,\\ \dots\dots\dots\dots\dots\dots\dots\dots\dots\dots.\end{array}\right\} \qquad (3)$$

Equations (1) show that if we modify the expression T to T′ by substituting in it the values of $\dot{s}_1, \dot{s}_2, \ldots$ given by (7), 11, and then write

$$T'' = T' - c_1\dot{s}_1 - c_2\dot{s}_2 - \ldots, \qquad (4)$$

we can use T″ to obtain the equations of motion for the coordinates $q_1, q_2, \ldots, q_k$ for a holonomous system by the ordinary process. Equations (3) show that the so-called gyrostatic terms flow from the added expression $-c_1\dot{s}_1 - c_2\dot{s}_2 - \ldots$. In equations (1), 11, these added terms are dispensed with, and the equations are applicable to all kinds of systems.

Now in (4) regard T′ for a moment as still a function of $\dot{q}_1, \dot{q}_2, \ldots, \dot{q}_k, \dot{s}_1, \dot{s}_2, \ldots, \dot{s}_j$, that is in fact T, while T″ on the left represents the right-hand side of (4) as modified by the substitution specified. We get by direct differentiation,

$$\frac{\partial T''}{\partial c_1} = \Sigma \frac{\partial T}{\partial \dot{s}} \frac{\partial \dot{s}}{\partial c_1} - \dot{s}_1 - \Sigma\left(c \frac{\partial \dot{s}}{\partial c_1}\right) = -\dot{s}_1. \qquad (5)$$

Moreover T′, as modified, and T″ involve $\dot{q}_1, \dot{q}_2, \ldots$ in the same way. Therefore

$$\frac{\partial T''}{\partial \dot{q}} = \frac{\partial T'}{\partial \dot{q}}. \qquad (6)$$

The modified T″ consists of three groups of terms, (1) a homogeneous quadratic function of $\dot{q}_1, \dot{q}_2, \ldots$, which we shall denote by T, (2) a function bilinear in $\dot{q}_1, \dot{q}_2, \ldots, c_1, c_2, \ldots$, and (3) a homogeneous quadratic function, K, of $c_1, c_2, \ldots$. But from (4) we get

$$T' = T'' - c_1 \frac{\partial T''}{\partial c_1} - c_2 \frac{\partial T''}{\partial c_2} - \ldots \qquad (7)$$

On the right the bilinear terms appear twice over, once in T″ with one sign, and again in the products $c\,\partial T''/\partial c$, with the opposite sign. Thus the kinetic energy T′ is the sum of the two homogeneous quadratic functions T and K.

**13.** *Cyclic systems. Kinetic potential.* Gyrostatic terms are characteristic of what have been called *cyclic systems*. These have the peculiarity that certain coordinates, of the type $s$, do not appear in the expression of the kinetic energy, which depends only on quantities of the type $q$, the $\dot{q}$, and the $\dot{s}$. Hence coordinates of the type $s$ are sometimes referred to as "speed coordinates." Also, as stated in 10, no forces of the type S exist. Thus the equations of motion for the $s$ coordinates are of the type $d(\partial T/\partial \dot{s})/dt = 0$, that is we have

$$\frac{dc_1}{dt} = 0, \quad \frac{dc_2}{dt} = 0. \qquad (1)$$

The $q$-equations of motion are at once obtained from Routh's modified value, T″, of the kinetic energy given by

$$T'' = T' - c_1\dot{s}_1 - c_2\dot{s}_2 - \ldots, \qquad (2)$$

in which, on the right, $\dot{s}_1, \dot{s}_2, \ldots$ are supposed to have been replaced by their values in terms of quantities of the type $q$, $\dot{q}$, and $c$. These equations contain gyrostatic terms. When the Qs are derivable, in whole or in part, from a potential V we may include these forces in the forms on the left of the equations by using, instead of T′ and T″, T′ − V and T″ − V, or as they are usually symbolised, L′ and L″, the so-called *kinetic potentials* [see 6 above]. Then the complete equations are of the form

$$\frac{d}{dt}\frac{\partial L''}{\partial \dot{q}} - \frac{\partial L''}{\partial q} = Q_1, \qquad (3)$$

where $Q_1$ is the part of Q, if any, not derivable from a potential function. Here also, of course, the terms derived from the chain of products $-c_1\dot{s}_1 - c_2\dot{s}_2 - \dots$ are gyrostatic.

In the equations of motion thus obtained no $\dot{s}$ (and of course no $s$) appears. Thus, by the process of modifying the kinetic energy adopted, it is possible, in the case of holonomous systems, to "ignore" the $s$ coordinates and their velocities. But this "ignoration of coordinates," as it is called, is only possible when, as explained, arrangements are made to enable the $s$ motions to have their proper influence on the system.

It ought to be mentioned that M. Appell has invented a dynamical method in generalised coordinates which is applicable to all systems whether holonomous or not. He constructs a function S, which has been called the "kinetic energy of the accelerations," since it is obtained from the expression

$$\tfrac{1}{2}\Sigma\{m(\ddot{x}^2+\ddot{y}^2+\ddot{z}^2)\},$$

by substituting for $\ddot{x}_1, \ddot{x}_2, \dots$, their values from the equations [see (5′), 2] of the form

$$\ddot{x} = a_1\ddot{q}_1 + a_2\ddot{q}_2 + \dots, \dots.$$

The partial derivatives of S, $\partial S/\partial\ddot{q}_1, \partial S/\ddot{q}_2, \dots$, equated to the values of the generalised forces $Q_1, Q_2, \dots$, then give the equations of motion (see Appell, *Mécanique Rationelle*, t. 2, p. 374).

**14.** *Reversibility of the motion of a system.* The motions of systems which are not gyrostatic are in a certain sense reversible. If we suppose that, starting from a certain instant, the time element is reversed, no change takes place in the equations of motion. For there are no linear terms in the kinetic energy, and so no term in an equation of motion is altered in value. If then the forces for each configuration be the same as before, we may suppose time to flow on, and the velocities to be reversed at the instant in question. The system will pass through the configurations of the direct motions in the reverse order, and where there was positive acceleration for any coordinate there will now be retardation for the backward passage, and *vice versa.*

If there be gyrostatic terms, this *total* reversal will not be possible for reversal of the $\dot{q}$s only; the $\dot{s}$s must all be reversed as well. Reversal as regards the $q$-motions is however possible. The motion of a top is a good example. The axis moves between two limiting inclinations $\theta$ to the vertical. As the top moves from the smaller limiting value of $\theta$ to the larger, it passes through a succession of values of $\theta$, and the backward passage is, in a sense, an image of the forward passage, in which the values of $\theta$ are assumed in the reverse order, and the $\dot{\theta}$s are the same for the same $\theta$s but reversed. The azimuthal motions however are not reversed; the axis swings round in the same direction in rising as in falling. For the same forces the azimuthal motion cannot be reversed except by reversal of the spin.

The kinetic energy of the cyclic motions, which are left when every $\dot{q}$ is made zero, may be expressed either as a homogeneous quadratic function T′ of the velocities typified by $\dot{s}$, or as a homogeneous quadratic function K of the $c$s. As we have seen, we have for any $q$-coordinate

$$-\frac{\partial T'}{\partial q}=\frac{\partial K}{\partial q}.$$

As before, we have relative equilibrium of the system when for variation of each coordinate $q$, $V-T'$ or $V+K$ is stationary, that is all the partial differential coefficients

$$\frac{\partial}{\partial q}(V-T') \quad \text{or} \quad \frac{\partial}{\partial q}(V+K)$$

are zero.

**15.** *Stability of the motion of a cyclic system.* In the second case K is equivalent to an addition to the potential energy V. It can be proved, just as it is proved that static equilibrium is stable when V has a minimum value, that the relative equilibrium—the state of motion for which the $\dot{q}$s are all zero—is stable when $V-T'$, or $V+K$, has a minimum value, that is the variation $\delta(V+K)$ caused by any small disturbance from steady motion must be positive.

The best example is that of a top spinning with its axis inclined at an angle $\theta$ to the upward vertical, and supported at a point on the axis. The kinetic energy is

$$T=\tfrac{1}{2}A\dot{\theta}^2+\tfrac{1}{2}A\sin^2\theta\,.\,\dot{\psi}^2+\tfrac{1}{2}Cn^2, \quad\text{.................................(1)}$$

where $n=\dot{\phi}+\dot{\psi}\cos\theta$, the constant angular speed of the spin. The potential energy is $Mgh\cos\theta$. Here $Cn$, the A.M. about the spin-axis, has a constant value $c$, and the A.M., G, about the vertical through the point of support, given by

$$G=Cn\cos\theta+A\sin^2\theta\,.\,\dot{\psi}, \quad\text{.....................................(2)}$$

is also constant. Thus we have

$$V+K=Mgh\cos\theta+\tfrac{1}{2}\frac{(G-c\cos\theta)^2}{A\sin^2\theta}+\tfrac{1}{2}\frac{c^2}{C}. \quad\text{..........................(3)}$$

Now for relative equilibrium with $\theta$ constant the expression on the right must be a minimum. Differentiating with respect to $\theta$, and equating the derivative to zero, we get

$$A\dot{\psi}^2\cos\theta-Cn\dot{\psi}+Mgh=0, \quad\text{.....................................(4)}$$

the condition for steady motion when $\theta$ is not zero.

Differentiating again, we find for the condition that $V+K$ should be a minimum, the inequality

$$\frac{c(c-G\cos\theta)+G(G-c\cos\theta)}{A\sin^2\theta}>4Mgh\cos\theta. \quad\text{............................(5)}$$

If we suppose the top upright, so that $\theta=0$, we must take $c=G$, since we have then each of these quantities equal to $Cn$. Supposing then $\theta$ very small, we have for the inequality, very approximately,

$$2c^2\frac{1-\cos\theta}{\sin^2\theta}>4AMgh\cos\theta, \quad\text{....................................(6)}$$

or in the limit when $\theta$ is evanescent

$$c^2>4AMgh, \quad\text{..........................................(7)}$$

a result obtained otherwise in 12, V, above.

**16.** *Visible and concealed coordinates. Nature of potential energy.* The association of K with V in the criterion for the stability of the steady motion of a cyclic system, and its treatment as equivalent to so much potential energy, suggest that all energy may be kinetic, that in fact energy of configuration, or potential energy, of ordinary bodies, is kinetic energy of motions of the system depending on coordinates which are concealed from us, and which we cannot control. In the dynamics of the motion of solids through a fluid the configuration of the solids is specified by the $q$-coordinates, and by means of these and their velocities the energy of the system is expressed, and the equations of motion of the solids obtained by the Lagrangian method. In this process the coordinates, infinite in number, which determine the positions of the particles of the fluid, are ignored, and the corresponding components of momentum are taken as zero, except in the case of circulation in cyclic regions, such as channels formed by perforations in the solids, when momenta must be introduced which are exactly analogous to those of the flywheels of a gyrostatic system (see 14, XIII).

Thus the total energy of a gravitational system is capable of expression by the visible coordinates and velocities of molar matter, but the potential energy is no doubt really the kinetic energy of unseen and individually uncontrollable particles, among which the bodies observed move just as solids move in a fluid.

**17.** *Case of a group of constant velocities.* If, instead of the momenta of the $s$-coordinates, the velocities of that group are constrained to remain constant, the dualism of such systems, as illustrated by the Lagrangian and Hamiltonian equations of motion, shows that there arise here again terms of gyrostatic form. Let us suppose that the kinetic energy is of the form

$$\mathrm{T}_1 = \mathrm{T} + \mathrm{T}' + (a_{11}\dot{q}_1 + a_{21}\dot{q}_2 + \ldots)\dot{s}_1 + (a_{12}\dot{q}_1 + a_{22}\dot{q}_2 + \ldots)\dot{s}_2 + \ldots, \quad \ldots\ldots\ldots(1)$$

where T and T′ are homogeneous quadratic functions of the $\dot{q}$s and the $\dot{s}$s respectively. Also let the kinetic energy be independent of the $s$ coordinates; then the coefficients, $a_{11}, a_{21}, \ldots, a_{12}, \ldots a_{22}, \ldots, \ldots$ are functions of the $q$s only. The ordinary Lagrangian equations are of the form

$$\frac{d}{dt}\frac{\partial \mathrm{T}}{\partial \dot{q}_f} - \frac{\partial \mathrm{T}}{\partial q_f} - \frac{\partial \mathrm{T}'}{\partial q_f}$$

$$+\left(\frac{\partial a_{f1}}{\partial q_2}\dot{q}_1 + \frac{\partial a_{f1}}{\partial q_2}\dot{q}_2 + \ldots\right)\dot{s}_1 + \left(\frac{\partial a_{f2}}{\partial q_1}\dot{q}_1 + \frac{\partial a_{f2}}{\partial q_2}\dot{q}_2 + \ldots\right)\dot{s}_2 + \ldots$$

$$-\left(\frac{\partial a_{11}}{\partial q_f}\dot{q}_1 + \frac{\partial a_{21}}{\partial q_f}\dot{q}_2 + \ldots\right)\dot{s}_1 - \left(\frac{\partial a_{12}}{\partial q_f}\dot{q}_1 + \frac{\partial a_{22}}{\partial q_f}\dot{q}_2 + \ldots\right)\dot{s}_2 - \ldots = \mathrm{Q}_f. \quad \ldots\ldots(2)$$

We can write this as

$$\frac{d}{dt}\frac{\partial T}{\partial \dot{q}_f}-\frac{\partial T}{\partial q_f}-\frac{\partial T'}{\partial q_f}$$
$$+\left\{\left(\frac{\partial \alpha_{f1}}{\partial q_1}-\frac{\partial \alpha_{11}}{\partial q_f}\right)\dot{s}_1+\left(\frac{\partial \alpha_{f2}}{\partial q_1}-\frac{\partial \alpha_{12}}{\partial q_f}\right)\dot{s}_2+\dots\right\}\dot{q}_1$$
$$+\left\{\left(\frac{\partial \alpha_{f1}}{\partial q_2}-\frac{\partial \alpha_{21}}{\partial q_f}\right)\dot{s}_1+\left(\frac{\partial \alpha_{f2}}{\partial q_2}-\frac{\partial \alpha_{22}}{\partial q_f}\right)\dot{s}_2+\dots\right\}\dot{q}_2$$
$$+\dots\dots\dots = Q_f. \quad \dots\dots(2')$$

The reader will see that if, for example, we put first $f=1$, then $f=2$, the coefficients of $\dot{q}_2$ in the first case and of $\dot{q}_1$ in the second have the same numerical value but are opposite in sign.

An equation of motion can be written down for each coordinate. These equations contain no terms in $\ddot{s}_1, \ddot{s}_2, \dots$, and $\partial T'/\partial s_1, \partial T'/\partial s_2, \dots$ are all zero. Thus

$$\frac{dT'}{dt}=\frac{\partial T'}{\partial q_1}\dot{q}_1+\frac{\partial T'}{\partial q_2}\dot{q}_2+\dots. \quad \dots\dots(3)$$

Hence, if we multiply the first $q$-equation of motion by $\dot{q}_1$, the second by $\dot{q}_2$, and so on, and add the products, we get

$$\frac{d(T-T')}{dt}=Q_1\dot{q}_1+Q_2\dot{q}_2+\dots+Q_i\dot{q}_i. \quad \dots\dots(4)$$

If the forces are derivable from a potential V we get

$$\frac{d(T-T'+V)}{dt}=0. \quad \dots\dots(5)$$

If for any configuration all the $\dot{q}$s are zero T is zero, and we have from the equations of motion, the conditions, $i$ in number,

$$-\frac{\partial T'}{\partial q_f}=Q_f; \quad \dots\dots(6)$$

or in the case when $Q_f=\partial V/\partial q_f$, that is for a conservative system,

$$\frac{\partial(V-T')}{\partial q_f}=0, \quad \dots\dots(6')$$

that is $V-T'$ does not vary with any coordinates, or in other words $V-T'$ is stationary.

We might express the kinetic energy of the $s$ motions as a homogeneous quadratic function K of the $s$ momenta. We have then $\partial T'/\partial q_f=-\partial K/\partial q_f$, and (6′) becomes

$$\frac{\partial(V+K)}{\partial q_f}=0. \quad \dots\dots(7)$$

**18.** *Relative potential energy. Stability of relative equilibrium.* The expression $V-T'$ may be interpreted as relative potential energy. A simple example is that of a conical pendulum, inclined at an angle $\theta$ to the vertical, and in steady motion. The velocity $\dot{\theta}$ is zero, but the azimuthal angular speed $\dot{\phi}$ is not. The coordinate $\phi$ does not appear in the expression for T′, which is $\frac{1}{2}ml^2\sin^2\theta\,.\,\dot\phi^2$, where $l$ is the length of the supporting thread. The potential energy is $-mgl\cos\theta$, so that we have

$$V-T'=-ml(g\cos\theta+\tfrac{1}{2}l\sin^2\theta\,.\,\dot\phi^2).$$

This is to be stationary, and so we must have

$$\frac{\partial}{\partial\theta}(V-T')=ml\sin\theta(g-l\cos\theta\,.\,\dot\phi^2)=0.$$

Thus the azimuthal angular speed is given by

$$\dot\phi^2=\frac{g}{l\cos\theta},$$

the well-known condition for steady motion.

It will be seen that we have obtained the condition of *relative* equilibrium, that is for steady motion, by treating V − T′ as potential energy. Here T′ is the kinetic energy corresponding to the so-called centrifugal force.

It is not difficult to show that this relative equilibrium is stable, and it will be found by a second differentiation, that for the steady motion with angular speed $(g/l\cos\theta)^{\frac{1}{2}}$, the value of V − T′ is a minimum.

But now consider the A.M. $c$ of the bob. It is $ml^2\sin^2\theta\,.\,\dot\phi$. Then in terms of $c$ the kinetic energy, which we have denoted by T′, has the value $\frac{1}{2}c^2/ml^2\sin^2\theta$. We now denote this by K, and find the result of considering V + K as stationary. Differentiating, and equating to zero, we obtain exactly the same result as before, that is $\dot\phi^2=g/l\cos\theta$.

When we differentiate again with respect to $\theta$ we find that V + K is a minimum.

**19.** *Illustrations of the general equations of motion.* (1) As a first and very simple example, we take the motion of a particle of mass $m$ in a plane curve. If at time $t$ the radius-vector drawn from a fixed point be of length $r$, and make an angle $\theta$ with an axis of $x$ drawn from the same origin, the coordinates of the particle are

$$x=r\cos\theta,\quad y=r\sin\theta.$$

Hence for the kinetic energy T we have

$$2T=m\{(\dot r\cos\theta-r\dot\theta\sin\theta)^2+(\dot r\sin\theta+r\dot\theta\cos\theta)^2\},\ \ldots\ldots(1)$$

or

$$2T=m(\dot r^2+r^2\dot\theta^2).\ \ldots\ldots(2)$$

In applying (6), 8, to the problem of finding the $r$, $\theta$, equations of motion of the particle, we have to take the first expression for the kinetic energy. We obtain

$$\frac{d}{dt}\frac{\partial T}{\partial\dot r}=m\ddot r.$$

By (6), 8, we have to subtract from this

$$m(\dot r\cos\theta-r\dot\theta\sin\theta)\frac{d}{dt}(\cos\theta)+m(\dot r\sin\theta+r\dot\theta\cos\theta)\frac{d}{dt}(\sin\theta);$$

that is $mr\dot\theta^2$. The same result would, of course, be obtained by calculating $\partial T/\partial r$. Hence the $r$-equation of motion is $\quad m(\ddot r-r\dot\theta^2)=R,\ \ldots\ldots(3)$

where R is the applied force in the outward direction along $r$.

For the $\theta$-equation we have $\quad \dfrac{d}{dt}\dfrac{\partial T}{\partial\dot\theta}=m(r^2\ddot\theta+2r\dot r\dot\theta).$

We have to subtract from this

$$-m(\dot r\cos\theta-r\dot\theta\sin\theta)\frac{d}{dt}(r\sin\theta)+m(\dot r\sin\theta+r\dot\theta\cos\theta)\frac{d}{dt}(r\cos\theta)$$

or zero. Thus the $\theta$-equation of motion is

$$m(r^2\ddot\theta+2r\dot r\dot\theta)=\Theta,\ \ldots\ldots(4)$$

where $\Theta$ is the applied force perpendicular to the radius-vector.

This method, if it had been applied to the value of T in (2), would have failed. T is here a sum of squares referred to a set of axes so specialised that in the formation of T the quantities $\cos\theta$, $\sin\theta$ have taken the special values 1 and 0; and unless we go back to the fundamental expressions, for the velocities along the unspecialised axes $Ox$, $Oy$, it is not apparent how the process is to be carried out.

It will be observed that in (1) we have

$$\frac{\partial}{\partial\theta}(\cos\theta) = -\frac{\partial}{\partial r}(r\sin\theta), \quad \frac{\partial}{\partial\theta}(\sin\theta) = \frac{\partial}{\partial r}(r\cos\theta),$$

so that the integrability conditions are fulfilled. Thus it is possible to proceed in the ordinary way by calculating $\partial T/\partial r$ and subtracting it from $m\ddot{r}$. The function of $r$ involved in (1) and (2) is the same, and so in the latter case the ordinary process remains applicable, though then, *apparently*, the integrability conditions seem unfulfilled. This explains why in many cases, *e.g.* in the next example, when specialised axes are taken, the ordinary method is applicable, while the other, set forth in § 9 above, is not. The latter can only be applied when the values of $\dot{u}$, $\dot{v}$, ... are perfectly general.

**20.** *Illustrations of the general equations of motion.* (2) *Gyrostatic pendulum.* As a second example we take the gyrostatic pendulum problem already discussed above [21, VII]. The pendulum as ordinarily made is a rigid body symmetrical about a longitudinal axis, and containing a fly-wheel with its axis of rotation along the axis of symmetry. The suspension is by a Hooke's joint, or by means of a piece of steel wire so short that it may be taken as untwistable, while yielding equally freely to bending forces in all vertical planes containing the wire (see Fig. 105).

FIG. 105.

Suppose the axis of the flywheel to contain O, and denote this axis by OC. Let $\theta$ be the inclination of OC to the vertical, $\psi$ the angle which the vertical plane through OC makes with a fixed plane through the vertical containing the point of support, $\phi$ the angle which a plane containing OC, and fixed in the wheel, makes with a vertical plane also containing OC. We shall not suppose in the first instance that the pendulum, apart from the flywheel, is symmetrical, but take C as its moment of inertia about the axis of the wheel, which we shall suppose to be a principal axis of moment of inertia, and A and B as the other two principal axes for the point of support. We shall also denote the moment of inertia of the flywheel about its axis by C′, and its moment of inertia about any axis at right angles to this through the point of support by A′.* It is easy to show from Fig. 105 that the angular velocity of the pendulum (apart from the flywheel) about the axis of symmetry is $-\dot{\psi}(1-\cos\theta)$. That of the flywheel about the same axis is $\dot{\phi}+\dot{\psi}\cos\theta$. We refer in what follows to Fig. 106, below.

We suppose now that the principal axis about which the moment of inertia is B is inclined at the instant under consideration at the angle $\phi$ to the vertical plane containing the axis of the flywheel, so that, if without other change of position of that plane the axis of rotation of the wheel were brought to the vertical, this principal axis would lie

* There is here a deviation from the notation of 20, VII above; but it can hardly be avoided unless the case containing the gyrostat is symmetrical. When there is symmetry the deviation consists simply in the interchange of C and C′. In comparing the solution here given in 20, 21 with that of 21, VII, the reader will also observe that the $\omega$ of the latter is the angular speed of an axial plane of the flywheel relative to an axial plane fixed in the turning case, so that $\dot{\phi} = \omega - \dot{\psi}$. For the turning case, or body, $\dot{\phi} = -\dot{\psi}$.

in the plane from which $\psi$ is measured. The pendulum is turning with angular velocity $\theta$ about OD, taken perpendicular to the vertical plane through the axis of the flywheel, and with angular velocity $\dot{\psi}\sin\theta$ about an axis in that plane and at right angles to the flywheel axis. The angular velocities about the axes of A and B (taken as related to the third axis as the usual axes O$x$, O$y$ are to O$z$) are therefore $\dot{\theta}\cos\phi+\dot{\psi}\sin\theta\sin\phi$ and $-\dot{\theta}\sin\phi+\dot{\psi}\sin\theta\cos\phi$ respectively.

The equation for the kinetic energy can now be written down, and is

$$\mathrm{T}=\tfrac{1}{2}\{\mathrm{A}(\dot{\theta}\cos\phi+\dot{\psi}\sin\theta\sin\phi)^2+\mathrm{B}(-\dot{\theta}\sin\phi+\dot{\psi}\sin\theta\cos\phi)^2+\mathrm{C}(1-\cos\theta)^2\dot{\psi}^2\}$$
$$+\tfrac{1}{2}[\mathrm{A}'(\dot{\theta}^2+\dot{\psi}^2\sin^2\theta)+\mathrm{C}'\{\dot{\phi}+\dot{\psi}\cos\theta\}^2]. \qquad (1)$$

From this, by the ordinary process, since the coordinates of any particle of the body are connected with the angles here specified by finite relations, we can obtain the equations of motion; and this, of course, is the simplest mode of proceeding. All the data for forming the equations in this manner will be found worked out in Thomson and Tait, § 330. The following is another discussion of the problem.

**21.** *Discussion of gyrostatic pendulum.* We refer the motion of any particle of mass $m$ in the body to fixed axes coinciding for the instant under consideration with the principal axes A, B, C, and denote the angular velocities about these axes by $p, q, r$. The corresponding angular velocities for the flywheel are $p, q, r'$. If $x, y, z$ be the coordinates with reference to these axes of a particle of mass $m$ in the pendulum apart from the flywheel, and $x', y', z'$, those of a particle of mass $m'$ in the flywheel, we have

$$\dot{x}=qz-ry, \quad \dot{y}=rx-pz, \quad \dot{z}=py-qx,$$
$$\dot{x}'=qz'-r'y', \quad \dot{y}'=r'x'-pz', \quad \dot{z}'=py'-qx'.$$

Hence
$$2\mathrm{T}=\Sigma[m\{(qz-ry)^2+(rx-pz)^2+(py-qx)^2\}]$$
$$+\Sigma[m'\{(qz'-r'y')^2+(r'x'-pz')^2+(py'-qx')^2\}], \qquad (1)$$

where the second line refers to the flywheel and the first to the rest of the pendulum.

From this we obtain

$$\frac{\partial\mathrm{T}}{\partial p}=\Sigma[m\{-(rx-pz)z+(py-qx)y\}]+\Sigma[m'\{-(r'x'-pz')z'+(py'-qx')y'\}].$$

Calculating the total time-rate of variation of this, and taking account of the fact that the axes of reference coincide with the principal axes of the body, we get

$$\frac{d}{dt}\frac{\partial\mathrm{T}}{\partial p}=\Sigma[m\{-\dot{z}(rx-pz)-z(-\dot{p}z+r\dot{x}-p\dot{z})\}]$$
$$+\Sigma[m\{\dot{y}(py-qx)+y(\dot{p}y+p\dot{y}-q\dot{x})\}]$$
$$+\text{similar expressions for the flywheel.} \qquad (2)$$

From this we are to subtract, according to (6), 7, above,

$$\Sigma[m\{-\dot{z}(rx-pz)+\dot{y}(py-qx)\}]+\Sigma[m'\{-\dot{z}'(r'x'-pz')+\dot{y}'(py'-qx')\}],$$

which vanish identically, and we have $\Sigma\{m(\dot{x}z+x\dot{z})\}=(\mathrm{A}-\mathrm{C})q$, $\Sigma\{m(\dot{x}y+x\dot{y})\}=(\mathrm{B}-\mathrm{A})r$. With these identities, and the substitutions $\dot{y}=rx-pz$, $\dot{z}=py-qx$, $\dot{y}'=r'x'-pz'$, $\dot{z}'=py'-qx'$, $\mathrm{B}q=\Sigma\{m(\dot{x}z-\dot{z}x)\}$, $\mathrm{A}'q=\Sigma\{m'(\dot{x}'z'-\dot{z}'x')\}$, (2) becomes, with equation to the moment of external forces round the axis of A,

$$(\mathrm{A}+\mathrm{A}')\dot{p}-(\mathrm{B}-\mathrm{C})qr-(\mathrm{A}'-\mathrm{C}')qr'=\mathrm{L}. \qquad (3)$$

Similarly we obtain another equation

$$(\mathrm{B}+\mathrm{A}')\dot{q}-(\mathrm{C}-\mathrm{A})rp-(\mathrm{C}'-\mathrm{A}')r'p=\mathrm{M}. \qquad (3')$$

These are really Euler's equations of motion for the pendulum, and might of course have been written down at once. We may now pass to the special axes O(D, E, C), which have been so often used above, and which with O(A, B, C), and the vertical OZ, are indicated in Fig. 106. The axis OE moves with the plane ZOC, and we may take EOB as the angle through which the body in its turning about OC has outstripped the plane ZOC. Denoting this angle by $\phi$, we have $\dot{\phi}$ for the angular speed relative to this plane. From the diagram we have the following values of $p$, $q$ (which have already been given in 20),

$$p = \theta \cos\phi + \dot{\psi} \sin\theta \sin\phi, \quad q = -\dot{\theta} \sin\phi + \dot{\psi} \sin\theta \cos\phi,$$

and we have seen above that

$$r = -\dot{\psi}(1 - \cos\theta), \quad r' = \dot{\phi} + \dot{\psi} \cos\theta.$$

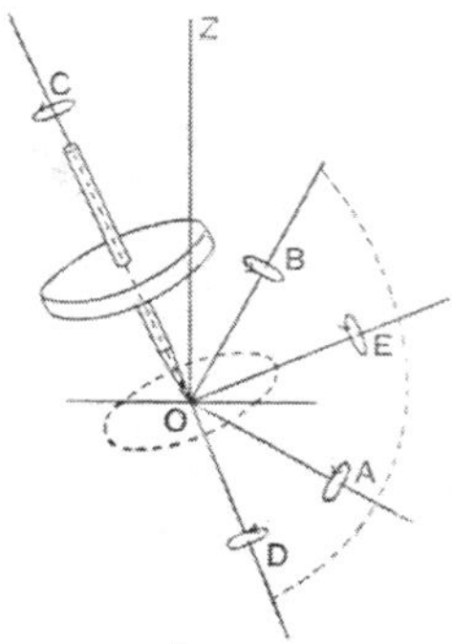

FIG. 106.

Substituting in (3) and (3′), multiplying the first transformed equation by $\cos\phi$, and the second by $\sin\phi$, and subtracting the second product from the first, we get the equation of motion, relative to the axis OD. Multiplying the first transformed equation by $\sin\phi$ the second by $\cos\phi$, and subtracting the second product from the first, we get the equation of motion for OE.

The reader may work all this out and compare with the results of the Lagrangian process. He may then pass at once to the case of symmetry about an axis coincident with that of the flywheel. We may however pass to the case of symmetry by supposing the angle $\phi$ of Fig. 105 equal to zero, and inserting* $\ddot{\theta} + \dot{\psi}\dot{\phi}^2 \sin\theta$ for $\dot{p}$, $\dot{\psi}\sin\theta$ for $q$, $-\dot{\psi}(1-\cos\theta)$ for $r$, $\dot{\phi}+\dot{\psi}\cos\theta$ for $r'$, we get the $\theta$-equation of motion. Similarly the other equations can be obtained.

If the pendulum be symmetrical about the axis of the flywheel, (3) becomes, with the substitutions specified above,

$$A\ddot{\theta} - \{C(1-\cos\theta)\sin\theta + A\sin\theta\cos\theta\}\dot{\psi}^2 + C'\sin\theta \, . \, n\dot{\psi} = L, \quad \ldots\ldots\ldots\ldots\ldots (4)$$

where $n$ denotes the speed, $\dot{\phi} + \dot{\psi}\cos\theta$, of the flywheel about its axis, and A is put for A + A′.

The other equations of motion will be found to be

$$\left.\begin{aligned} \{A\sin^2\theta - C(1-\cos\theta)\cos\theta\}\ddot{\psi} + 2\{(A-C)\cos\theta + C\}\sin\theta \, . \, \dot{\theta}\dot{\psi} - C'n\sin\theta \, . \, \dot{\theta} = 0, \\ \frac{d}{dt}(\dot{\phi} + \dot{\psi}\cos\theta) = 0. \end{aligned}\right\} \ldots (5)$$

In the first of the last-written equations, account of the second has been taken, in putting $n$ for $\dot{\phi} + \dot{\psi}\cos\theta$, and treating $n$ as a constant. When the difference of notation, referred to in the footnote on p. 423, is taken account of, it will be found that the equations of motion here found agree with those obtained from first principles in 21, VII, above.

The angular speed of the flywheel is simply that, $\dot{\phi} + \dot{\psi}\cos\theta$, of a top or gyrostat, about the axis of figure. The expression $\omega - (1-\cos\theta)\dot{\psi}$ has the same meaning, but involves explicitly the turning of the body of the pendulum due to the mode of suspension.

* Referring to Fig. 105, we may introduce the Eulerian angles by supposing axes drawn from the point of suspension O, as there indicated. These axes O(D, E, C) are to be taken as coincident with the axes O(A, B, C) of the text. If $\theta$ be the inclination of OC to the downward vertical, the angular speeds about the axes are, for the body, $\dot{\theta}$, $\dot{\psi}\sin\theta$, $-\dot{\psi}(1-\cos\theta)$, and for the axes themselves $\dot{\theta}$, $\dot{\psi}\sin\theta$, $\dot{\psi}\cos\theta$. The rate of growth of angular speed about OD, regarded for the moment as a fixed axis, is what we have denoted by $\dot{p}$, and this is made up of $\ddot{\theta}$ due to acceleration, together with, by our rule, the two parts $-\dot{\psi}\sin\theta \, . \, \dot{\psi}\cos\theta$, $-\dot{\psi}(1-\cos\theta)\dot{\psi}\sin\theta$, due to the turning of OE about OC, and the turning of OC about OE respectively. Thus, or by differentiating the value of $p$ given in 20, and remembering that, in the rotation of the pendulum body (*but not in that of the flywheel*), $\dot{\phi} = -\dot{\psi}$, we have

$$\dot{p} = \ddot{\theta} - \dot{\psi}^2 \sin\theta\cos\theta - \dot{\psi}^2(1 - \cos\theta \sin\theta) = \ddot{\theta} - \dot{\psi}^2 \sin\theta.$$

Putting now $\partial T/\partial\dot\phi$ or $C'(\dot\phi+\dot\psi\cos\theta)=\kappa$, we have

$$\dot\phi=\frac{\kappa}{C'}-\cos\theta\,.\,\dot\psi,\qquad(6)$$

and therefore, since $\dot\phi$ is the velocity for the ignored coordinate $\phi$, we have from (8), 10, $\mathsf{B}=\cos\theta-1$, $\mathsf{A}=0$, where $\mathsf{B}$ and $\mathsf{A}$ are the functions from which the gyrostatic terms are derived.

Now we can write

$$\begin{aligned}2T&=\dot\theta\frac{\partial T}{\partial\dot\theta}+\dot\psi\frac{\partial T}{\partial\dot\psi}+\dot\phi\kappa\\&=A\dot\theta^2+A\dot\psi^2\sin^2\theta+C\dot\psi^2(1-\cos\theta)^2+C'n\dot\psi\cos\theta+\dot\phi\kappa\\&=A\dot\theta^2+A\dot\psi^2\sin^2\theta+C\dot\psi^2(1-\cos\theta)^2+\frac{\kappa^2}{C'}=2(T_1+K),\qquad(7)\end{aligned}$$

where $K=\frac{1}{2}\kappa^2/C'$.

Thus the kinetic energy is reduced to the sum of two quadratic functions, one, $T_1$, of the velocities $\dot\phi$, $\dot\theta$, the other, K, of the momentum $\kappa$ corresponding to $\dot\phi$.

The equations of motion are reduced to two, which have the form

$$\left.\begin{aligned}A\ddot\theta-\{C(1-\cos\theta)\sin\theta+A\sin\theta\cos\theta\}\dot\psi^2+\kappa\dot\psi\sin\theta+mgh\sin\theta=0,\\\{A\sin^2\theta-C(1-\cos\theta)\cos\theta\}\ddot\psi+2\{(A-C)\cos\theta+C\}\sin\theta\,.\,\dot\theta\dot\psi-\kappa\dot\theta\sin\theta=0.\end{aligned}\right\}\quad(8)$$

The terms $\kappa\dot\psi\sin\theta$, $-\kappa\dot\theta\sin\theta$ are the gyrostatic terms. One is $-\dot\psi\kappa\partial(\cos\theta)/\partial\theta$, the other is $\dot\theta\kappa\partial(\cos\theta)/\partial\theta$.

**22.** *Gyrostatic pendulum vibrating through small range.* These equations can be reduced, when $\theta$ is small throughout the motion, to $x$, $y$ coordinates, and take then a symmetrical form which exhibits better the gyrostatic terms. They are best obtained by transforming the kinetic energy to the new coordinates $x$, $y$, taken in a horizontal plane through the centre of inertia of the pendulum, with the projection of O upon this plane as origin.

We have, approximately, $1-\cos\theta=r^2/2h^2$, where $r\{=(x^2+y^2)^{\frac{1}{2}}\}$ is the distance of the centre of inertia from the new origin, $\dot\theta=(x\dot x+y\dot y)/hr$, $\dot\psi=(x\dot y-y\dot x)/r^2$, so that, neglecting terms of higher order than $r^2/h^2$, we get

$$2T=\frac{A}{h^2}(\dot x^2+\dot y^2)+\frac{\kappa^2}{C'}=T_1+K.\qquad(1)$$

For the calculation of the gyrostatic terms we have

$$\begin{aligned}\dot\phi=\frac{\kappa}{C'}-\dot\psi\cos\theta&=\frac{\kappa}{C'}-\frac{x\dot y-y\dot x}{r^2}\left(1-\frac{r^2}{2h^2}\right)\\&=\frac{\kappa}{C}+\mathsf{A}\dot x+\mathsf{B}\dot y,\qquad(2)\end{aligned}$$

where

$$\mathsf{A}=y\left(\frac{1}{r^2}-\frac{1}{2h^2}\right),\quad \mathsf{B}=-x\left(\frac{1}{r^2}-\frac{1}{2h^2}\right).\qquad(3)$$

Since $\partial T_1/\partial x$, $\partial T_1/\partial y$, $\partial K/\partial x$, $\partial K/\partial y$ are all zero, the equations of motion are

$$\left.\begin{aligned}\frac{d}{dt}\frac{\partial T_1}{\partial\dot x}-\kappa\left(\frac{\partial\mathsf{B}}{\partial x}-\frac{\partial\mathsf{A}}{\partial y}\right)\dot y+\frac{\partial V}{\partial x}=0,\\\frac{d}{dt}\frac{\partial T_1}{\partial\dot y}+\kappa\left(\frac{\partial\mathsf{B}}{\partial x}-\frac{\partial\mathsf{A}}{\partial y}\right)\dot x+\frac{\partial V}{\partial y}=0,\end{aligned}\right\}\qquad(4)$$

or, since $\frac{d}{dt}\frac{\partial T_1}{\partial\dot x}=A\ddot x/h^2$, $\frac{d}{dt}\frac{\partial T_1}{\partial\dot y}=A\ddot y/h^2$, $\frac{\partial V}{\partial x}=mg\frac{x}{h}$, $\frac{\partial V}{\partial y}=mg\frac{y}{h}$, $\frac{\partial\mathsf{A}}{\partial y}-\frac{\partial\mathsf{B}}{\partial x}=-\frac{1}{h^2}$,

$$\left.\begin{aligned}A\ddot x-\kappa\dot y+mghx=0,\\A\ddot y+\kappa\dot x+mghy=0.\end{aligned}\right\}\qquad(5)$$

The analogy of these equations to those for the simplest vibrations of an electron in a magnetic field has already been pointed out (see 9, IX, also 6, XX, below).

**23.** *Motion of a hoop or disk treated by modified Lagrangian equations.* We now sketch as a final example the modified Lagrangian solution of the problem of a hoop or disk, which is interesting as a case in which the conditions of integrability (1), 8, are in part fulfilled. The reader may compare the solution here given with that obtained by first principles in 7, XVIII.

We refer the motion of the centroid to rectangular axes O$x$, O$y$, O$z$ in and perpendicular to the plane in which the hoop rolls, with origin O at the point of contact, and denote the coordinates of the centroid by $x$, $y$, $z$. Let $\psi$ be the angle which the vertical plane through the axis of the hoop makes with the fixed vertical plane containing O$x$, and $\dot\phi$ be the angular velocity of the hoop relatively to the former plane. The angular velocity of the hoop about its axis of figure is thus $\dot\phi+\dot\psi\cos\theta$.

A consideration of the geometry of the problem shows that the following equations hold: (1) If the inclination, $\theta$ say, of the axis of the hoop to the vertical remain unaltered, and the hoop roll through an angle $\delta\chi$,

$$\delta x_1=-y\,\delta\chi,\quad \delta y_1=y\cot\psi\,.\,\delta\chi.$$

(2) Due to the alteration of $\theta$ we have

$$\delta x_2=-a\sin\theta\cos\psi\,.\,\delta\theta,\quad \delta y_2=-a\sin\theta\sin\psi\,.\,\delta\theta,\quad \delta z=a\cos\theta\,\delta\theta.$$

Combining these, we get

$$\left.\begin{aligned}\dot x&=-a(\dot\phi+\dot\psi\cos\theta)\sin\psi-a\sin\theta\cos\psi\,.\,\dot\theta,\\ \dot y&=a(\dot\phi+\dot\psi\cos\theta)\cos\psi-a\sin\theta\sin\psi\,.\,\dot\theta,\\ \dot z&=a\cos\theta\,.\,\dot\theta,\end{aligned}\right\}\quad\text{(1)}$$

which are the kinematical conditions.

The kinetic energy of the motion of the centroid is

$$\begin{aligned}T_c&=\tfrac{1}{2}m(\dot x^2+\dot y^2+\dot z^2)\\ &=\tfrac{1}{2}ma^2[(\sin\theta\cos\psi\,.\,\dot\theta+\cos\theta\sin\psi\,.\,\dot\psi+\sin\psi\,.\,\dot\phi)^2\\ &\qquad+(-\sin\theta\sin\psi\,.\,\dot\theta+\cos\theta\cos\psi\,.\,\dot\psi+\cos\psi\,.\,\dot\phi)^2+\cos^2\theta\,.\,\dot\theta^2].\quad\text{(2)}\end{aligned}$$

This reduces to $$T_c=\tfrac{1}{2}ma^2(\dot\theta^2+\cos^2\theta\,.\,\dot\psi^2+\dot\phi^2+2\cos\theta\,.\,\dot\phi\dot\psi),\quad\text{(2')}$$

but for our present purpose it is necessary to leave it in the expanded form.

Along with this we have the kinetic energy of rotation

$$T_r=\tfrac{1}{2}\{A(\dot\theta^2+\dot\psi^2\sin^2\theta)+C(\dot\phi+\dot\psi\cos\theta)^2\},\quad\text{(3)}$$

where C is the moment of inertia of the body about its axis of symmetry, and A that about any other axis at right angles to the axis of symmetry and passing through the centroid. Then $$T=T_c+T_r.\quad\text{(4)}$$

It will be seen from an examination of the expression for T above that the integrability conditions are fulfilled as between $\theta$ and $\phi$, and $\theta$ and $\psi$. As regards $T_r$ the relative coordinates are integral functions of $\theta$, $\phi$, $\psi$, and the ordinary methods apply. Hence the $\theta$-equation for the hoop or disk can be found in the ordinary way by the equation

$$\frac{d}{dt}\frac{\partial T}{\partial\dot\theta}-\frac{\partial T}{\partial\theta}=-\frac{\partial V}{\partial\theta},\quad\text{(5)}$$

where V $(=mga\sin\theta)$ is the potential energy.

The $\psi$ and $\phi$ equations are however

$$\left.\begin{aligned}&\frac{d}{dt}\frac{\partial T}{\partial\dot\psi}-\frac{\partial T_r}{\partial\psi}+ma\left\{\dot x\frac{d}{dt}(\cos\theta\sin\psi)-\dot y\frac{d}{dt}(\cos\theta\cos\psi)\right\}=0,\\ &\frac{d}{dt}\frac{\partial T}{\partial\dot\phi}+ma\left\{\dot x\frac{d}{dt}(\sin\psi)+\dot y\frac{d}{dt}(\cos\psi)\right\}=0,\end{aligned}\right\}\quad\text{(6)}$$

since $\partial T_r/\partial\phi=0$. The last two equations may be written out in full by the reader. The $\phi$-equation reduces to

$$\frac{d}{dt}\frac{\partial T}{\partial\dot\psi}-ma^2\sin\theta\,.\,\dot\theta\dot\psi=0.$$

The $\theta$-equation (5) worked out has the form

$$(A+ma^2)\ddot{\theta}+(C+ma^2-A)\dot{\psi}^2\sin\theta\cos\theta+(C+ma^2)\dot{\phi}\dot{\psi}\sin\theta=-mga\cos\theta. \quad \text{......(7)}$$

This equation agrees with that obtained by an appeal to first principles in 6, XVIII, above.

**24.** *Hamilton's principal function. Integration of the canonical equations.* We now give a short account of the general method in dynamics invented by Sir William Rowan Hamilton and completed by Jacobi, and apply the method to the discussion of the motion of a top. The reader must refer back to the canonical equations proved in 6. The solution of these equations was reduced by Hamilton to the integration of a certain partial differential equation, (7) below, or of a conjugate differential equation. The method here given of deriving the differential equations is taken from a paper by the author in *Proc. R.S.E.*, Feb. 19, 1912.

Hamilton and Jacobi reduced the solution of dynamical problems to the determination of a function (Hamilton's "Principal Function") S of the $q$s, $t$, and $k$ constants $a_1, a_2, \ldots, a_k$, which depend on the initial configuration or motion of the system. When this function is known, the finite equations of motion are, as we shall prove, the partial derivatives of S with respect to these constants, each equated to another constant, that is

$$-\frac{\partial S}{\partial a_1}=b_1, \quad -\frac{\partial S}{\partial a_2}=b_2, \ldots, -\frac{\partial S}{\partial a_k}=b_k. \quad \text{.................(1)}$$

Any problem is solved when the $q$s for time $t$ are found in terms of $t$, the initial coordinates, and the initial speeds, or a set of $2k$ distinct constants by which the initial configuration and motion can be expressed.

The solution consists therefore in a set of $k$ equations of the form

$$q_i=f_i(a_1, a_2, \ldots, a_k, b_1, b_2, \ldots, b_k, t), \quad \text{.........................(2)}$$

which we shall write in the abbreviated form

$$q_i=f_i(a, b, t). \quad \text{.......................................(3)}$$

From (2) or (3) we obtain $\dot{q}_i=\partial f_i/\partial t$, so that the speeds can also be expressed in terms of $(a, b, t)$. Now from (2) we can find the values of the $b$s in terms of the $q$s, the $a$s, and $t$. These substituted in (3) give equations of the form $\dot{q}_i=\phi_i(q, a, t)$. These values of the speeds, used in (1) of 6, give $k$ equations of the form $p_i=F_i(q, a, t)$. By differentiation of this, and use of the canonical equations, we get for any $p$,

$$\frac{\partial H}{\partial q}+\frac{\partial H}{\partial p_1}\frac{\partial p}{\partial q_1}+\frac{\partial H}{\partial p_2}\frac{\partial p}{\partial q_2}+\cdots+\frac{\partial H}{\partial p_k}\frac{\partial p}{\partial q_k}=-\frac{\partial p}{\partial t}, \quad \text{.................(4)}$$

where H is a function of the $q$s, the $p$s, and $t$, as explained in 6. But differentiation of H as thus expressed gives

$$\frac{\partial H}{\partial q}+\frac{\partial H}{\partial p_1}\frac{\partial p_1}{\partial q}+\frac{\partial H}{\partial p_2}\frac{\partial p_2}{\partial q}+\cdots+\frac{\partial H}{\partial p_k}\frac{\partial p_k}{\partial q}=\left(\frac{\partial H}{\partial q}\right), \quad \text{.................(5)}$$

where $(\partial H/\partial q)$ on the right is the partial derivative of H, when H is made a function of $(q, a, t)$ by substitution from (4) in the value of K as given in (2), 6. Hence (4) and (5) give $k$ equations of the form

$$\Sigma_i\left\{\frac{\partial H}{\partial p_i}\left(\frac{\partial p_i}{\partial q}-\frac{\partial p}{\partial q_i}\right)\right\}=\left(\frac{\partial H}{\partial q}\right)+\frac{\partial p}{\partial t}. \quad \text{(6)}$$

The left-hand side of (6) suggests that each $p$ may be the partial derivative $\partial S/\partial q$ of a function S of the $q$s and $t$, so that all the terms on the left vanish identically.

If, moreover, the value of S cause the right-hand side of (6) to vanish, we have, since $p=\partial S/\partial q$, $\partial(\partial S/\partial t+H)/\partial q=0$. Hence, if we insert $\partial S/\partial q$ for the corresponding $p$ in H expressed in terms of the $q$s, the $p$s, and $t$, and use the abridged notation, we get

$$\frac{\partial S}{\partial t}+H\left(q, \frac{\partial S}{\partial q}, t\right)=0 \quad \text{(7)}$$

for the determination of the function S, which, with $p=\partial S/\partial q$, as suggested above, will fulfil all the conditions involved in the equations of motion. This is Hamilton's partial differential equation.

We have $dS/dt=\partial S/\partial t+\Sigma(\dot q\,\partial S/\partial q)$, so that by (7),

$$\frac{dS}{dt}=\Sigma\left(\dot q\frac{\partial S}{\partial q}\right)-H=\Sigma(p\dot q)-H. \quad \text{(8)}$$

If the right-hand side of this be denoted by L, it will, when $T+V=h$, coincide with L as defined in 3. Thus to a constant we have

$$S=\int L\,dt, \quad \text{(9)}$$

which may be used for the calculation of S.

Equation (8) gives to an additive constant

$$S=\Sigma(pq)-\int\{(\dot p q)+H\}\,dt=\Sigma(pq)-S', \quad \text{(10)}$$

which defines a second function S′. We have $dS'/dt=\Sigma(\dot p q)+H$, and if we suppose S′ to be a function of the $p$s and $t$, we get

$$dS'/dt=\partial S'/\partial t+\Sigma(\dot p\,\partial S'/\partial p).$$

Further, if we put $\partial S'/\partial t=H$, and $q=\partial S'/\partial p$, we shall have

$$\frac{\partial S'}{\partial t}-H\left(p, \frac{\partial S'}{\partial p}, t\right)=0, \quad \text{(11)}$$

which is Hamilton's second differential equation. If, instead of the equations $p_i=F_i(q, a, t)$, we obtain, as we may, equations of the form $q_i=G_i(p\ a, t)$, and proceed [as we did for $\dot p$ and $(\partial H/\partial q)$] to express $\dot q$ and $(\partial H/\partial q)$, on the supposition now that H is a function of $(p, a, t)$, we are led to (11), so that S′ is a function of the $p$s and $t$, by which S may be replaced.

It may be proved that $\partial S/\partial a = -\partial S'/\partial a$, except when $a$ is the value of a constant $p$, taken as one of the constants in S. Obviously $\partial S/\partial t = -\partial S'/\partial t$.

The second partial differential equation (11) may be obtained by a process in which the $p$s and $q$s play a part precisely similar to that played by the $q$s and $p$s in the process by which the first, (7), was arrived at. This fact illustrates the reciprocity of the two equations, which corresponds to the duality of the canonical equations. It will be seen that (7) or (11) is the proper partial differential equation according as H is expressed as a function of the $q$s and $t$, or as a function of the $p$s and $t$.

**25.** *Jacobi's theorem.* Jacobi showed that if a *complete* integral of (7) is known, that is an integral which contains $k$ constants, $a_1, a_2, \ldots, a_k$, besides the additive constant, the integrals of the canonical equations are (1) of 24, with the $k$ equations

$$p_i = \frac{\partial S}{\partial q_i}. \quad \ldots\ldots(1)$$

This can be proved directly by showing that $d(\partial S/\partial a)/dt$ is constant, so that (1), 24, are consistent with the equations of motion. We have

$$\frac{d}{dt}\frac{\partial S}{\partial a} = \frac{\partial}{\partial a}\left\{\frac{\partial S}{\partial t} + \Sigma\left(\dot{q}\frac{\partial S}{\partial q}\right)\right\} - \Sigma\left(\frac{\partial S}{\partial q}\frac{\partial \dot{q}}{\partial a}\right) = \frac{\partial L}{\partial a} - \Sigma\left(\frac{\partial S}{\partial q}\frac{\partial \dot{q}}{\partial a}\right). \quad \ldots\ldots(2)$$

But $L = \Sigma(p\dot{q}) - H$, and therefore

$$\frac{\partial L}{\partial a} = \Sigma\left(\frac{\partial p}{\partial a}\dot{q} + p\frac{\partial \dot{q}}{\partial a} - \frac{\partial H}{\partial p}\frac{\partial p}{\partial a}\right) = \Sigma\left(p\frac{\partial \dot{q}}{\partial a}\right).$$

This, since $\partial S/\partial q = p$, makes the expression on the right of (2) zero, and the theorem is proved.

The $k$ equations (1) of 24 enable the coordinates (the $q$s) to be found in terms of $t$ and the $2k$ arbitrary constants, which express the initial configuration and motion; and these equations [(2) above] are therefore the complete solution of the dynamical problem proposed.

**26.** *Case in which H does not contain t.* If the function does not depend on $t$ as an explicit variable, (7) of 25 becomes

$$\frac{\partial S}{\partial t} + h = 0, \quad \ldots\ldots(1)$$

where $h$ is put for the constant value which, as we have seen, H now possesses. Integrating from $t_0(=0)$ to $t$, we obtain

$$S = -ht + W(q_1, q_2, \ldots, q_k, a_1, a_2, \ldots, a_{k-1}, h). \quad \ldots\ldots(2)$$

The function W is the value of Hamilton's *Characteristic Function*

$$S + Ht - H_0 t_0$$

when H does not contain $t$. It is a complete integral of (1) for each value of $h$; that is, it is possible to choose the constants $a_1, \ldots, a_{k-1}$ so as to give for any chosen value of $h$ arbitrary values of $\partial W/\partial q_1, \ldots, \partial W/\partial q_{k-1}$; and, conversely, if these constants can be so assigned, W is a complete integral of (1). For if we desire that $\partial W/\partial q_1, \ldots, \partial W/\partial q_{k-1}$ shall have certain arbitrary values $(p_1)_0, \ldots, (p_2)_0, \ldots, (p_k)_0$, for $t_0, (q_1)_0, (q_2)_0, \ldots, (q_k)_0$, we put $h=h_0$, the corresponding value of $h$, and then the value of $\partial W/\partial q_k$ is that given by the value $h_0$ of H.

The equations $$\frac{\partial W}{\partial \alpha_1}=b_1, \ldots, \frac{\partial W}{\partial \alpha_{k-1}}=b_{k-1}, \quad \frac{\partial W}{\partial h}=t-t_0 \quad \ldots\ldots(3)$$

give the "path" by the first $k-1$, and the time of passage is given by the last.

The reader may prove that, if coordinates $q_1, q_2, \ldots, q_i$ do not appear in H,

$$S=c_1q_1+c_2q_2+\ldots+c_iq_i+U(t, q_{i+1}, \ldots, q_k),$$

where $c_1, c_2, \ldots, c_i$ are the (constant) components of momentum corresponding to the absent coordinates.

**27.** *A top on a horizontal plane without friction. Jacobi's solution by the Hamilton-Jacobi method.* As an example we take the motion of a top on a horizontal plane without friction. We take $x$, $y$ as the horizontal coordinates of the centroid, C and A as the moments of inertia about the axis of figure and about an axis transverse to the axis of figure and through the centroid, respectively, and the rest of the notation as usual. The horizontal components of velocity are $\dot{x}$, $\dot{y}$, and the vertical velocity (downwards, since $\theta$ is taken with reference to the upward vertical) is $h\sin\theta\,.\,\dot{\theta}$. The kinetic and potential energies are

$$T=\tfrac{1}{2}\{M(\dot{x}^2+\dot{y}^2+h^2\sin^2\theta\,.\,\dot{\theta}^2)+A\dot{\theta}^2+A\sin^2\theta\,.\,\dot{\psi}^2+C(\dot{\phi}+\dot{\psi}\cos\theta)^2\}, \quad V=Mgh\cos\theta, \ldots(1)$$

where of course $h$ is not the energy constant, but the distance of the point of support from the centroid.

The coordinates are thus

$$q_1=x, \quad q_2=y, \quad q_3=\theta, \quad q_4=\phi, \quad q_5=\psi. \quad \ldots\ldots(2)$$

The last two and the first two do not appear in T or V. The momenta corresponding to these are

$$p_1=M\dot{x}, \quad p_2=M\dot{y}, \quad p_3=(Mh^2+A)\sin^2\theta\,.\,\dot{\theta}, \quad p_4=C(\dot{\phi}+\dot{\psi}\cos\theta),$$
$$p_5=A\sin^2\theta\,.\,\dot{\psi}+C\cos\theta(\dot{\phi}+\dot{\psi}\cos\theta). \ldots(3)$$

Solving these last equations for the velocities, and substituting in the value of T+V we get

$$H=\frac{1}{2M}(p_1^2+p_2^2)+\tfrac{1}{2}\left\{\frac{p_3^2}{(Mh^2+A)\sin^2\theta}+\frac{p_4^2}{C}+\frac{(p_5-p_4\cos\theta)^2}{A\sin^2\theta}\right\}+Mgh\cos\theta. \ldots(4)$$

We can now form the canonical equations

$$\dot{p}_1=-\frac{\partial H}{\partial q_1}, \quad \dot{p}_2=-\frac{\partial H}{\partial q_2}, \quad \dot{p}_3=-\frac{\partial H}{\partial q_3}, \quad \dot{p}_4=-\frac{\partial H}{\partial q_4}, \quad \dot{p}_5=-\frac{\partial H}{\partial q_5}.$$

With these of course are the companion equations.

$$\dot{q}_1=\frac{\partial H}{\partial p_1}, \quad \dot{q}_2=\frac{\partial H}{\partial p_2}, \ldots.$$

The first two and the last two of the canonical equations are

$$\dot{p}_1=0, \quad \dot{p}_2=0, \quad \dot{p}_4=0, \quad \dot{p}_5=0, \quad \ldots\ldots(5)$$

since $q_1, q_2, q_4, q_5$ do not appear in H.

Thus we have $$p_1=c_1, \quad p_2=c_2, \quad p_4=c_4, \quad p_5=c_5, \quad \ldots\ldots(6)$$

where $c_1$, $c_2$, $c_4$, $c_5$ are constants. These are first integrals, and they state results we are already familiar with, namely

$$M\dot{x}=c_1,\quad M\dot{y}=c_2,\quad C(\dot{\phi}+\dot{\psi}\cos\theta)=c_4,\quad A\sin^2\theta\dot{\psi}+Cn\cos\theta=c_5. \quad\text{.............(7)}$$

The last two state the constancy of angular speed $\dot{\phi}+\dot{\psi}\cos\theta=n$, and of angular momentum $Cn$, about the axis of figure, and the constancy of A.M. about the vertical through the centroid, while the first two assert that the centroid moves with uniform velocity.

The Hamiltonian differential equation in the present case is

$$\frac{\partial S}{\partial t}+H=0, \quad\text{.............(8)}$$

where in H has been substituted for $p_1$, $p_2$, $p_3$, $p_4$, $p_5$ their values as partial differential coefficients $\partial S/\partial x$, $\partial S/\partial y$, ... of the "principal function" S. Thus, by (4),

$$H=\frac{1}{2}\left\{\left(\frac{\partial S}{\partial x}\right)^2+\left(\frac{\partial S}{\partial y}\right)^2+\frac{1}{A+Mh^2\sin^2\theta}\left(\frac{\partial S}{\partial \theta}\right)^2+\frac{1}{A\sin^2\theta}\left(\frac{\partial S}{\partial \psi}-\frac{\partial S}{\partial \phi}\cos\theta\right)^2+\frac{1}{C}\left(\frac{\partial S}{\partial \phi}\right)^2\right\}$$
$$+Mgh\cos\theta. \quad\text{.............(9)}$$

We notice in the first place that $T+V=h'$, a constant, and that $\theta$ is the only coordinate which appears in (9). Hence, by 26, we assume

$$S=-h't+c_1x+c_2y+c_4\phi+c_5\psi+F(\theta)=-h't+W, \quad\text{.............(10)}$$

where F is a function of $\theta$ to be found. [Here and in what follows $h'$ is put for the energy constant, to prevent confusion.] Now we have $\partial S/\partial\theta=F'(\theta)$, and so get, by (9),

$$\frac{1}{A+Mh^2\sin^2\theta}\left(\frac{\partial S}{\partial\theta}\right)^2=2h'-2Mgh\cos\theta-c_1^2-c_2^2-\frac{1}{C}c_4^2-\frac{1}{A\sin^2\theta}(c_5-c_4\cos\theta)^2. \quad\text{...(11)}$$

Thus we have
$$F(\theta)=\int\{(A+Mh^2\sin^2\theta)f(\theta)\}^{\frac{1}{2}}d\theta, \quad\text{.............(12)}$$

where $f(\theta)$ is given by (11). Hence $p_3=\partial S/\partial\theta=F'(\theta)$, and

$$p_3=\{(A+Mh^2\sin^2\theta)f(\theta)\}^{\frac{1}{2}}. \quad\text{.............(12')}$$

Thus finally we obtain

$$S=-h't+c_1x+c_2y+c_4\phi+c_5\psi+\int\{(A+Mh^2\sin^2\theta)f(\theta)\}^{\frac{1}{2}}d\theta, \quad\text{.............(13)}$$

which containing, as it does, five distinct constants, is the complete integral of (8).

By (3), 26, the integral equations corresponding to the canonical equations are

$$\frac{\partial S}{\partial c_1}=\frac{\partial W}{\partial c_1}=b_1,\quad \frac{\partial S}{\partial c_2}=\frac{\partial W}{\partial c_2}=b_2,\ \dots,\ \frac{\partial S}{\partial h'}=-t+\frac{\partial W}{\partial h'}=-t_0, \quad\text{.............(14)}$$

where $b_1$, $b_2$, $b_4$, $b_5$, $t_0$ are all constants.

If now we write $R=\{(A+Mh^2\sin^2\theta)/f(\theta)\}^{\frac{1}{2}}$, we get, from (14) and (11),

$$\left.\begin{aligned}&x-c_1\int R\,d\theta=b_1,\quad y-c_2\int R\,d\theta=b_2,\quad t-\int R\,d\theta=t_0,\\&\phi-\int R\left\{\frac{c_4}{C}-\frac{\cos\theta}{A\sin^2\theta}(c_5-c_4\cos\theta)\right\}d\theta=b_4,\quad \psi-\int R\frac{c_5-c_4\cos\theta}{A\sin^2\theta}d\theta=b_5.\end{aligned}\right\}\quad\text{......(15)}$$

The first two and the last two of these equations, (15), give the "path" of the top, the remaining equation, the third, the time of passage from an initial to a final configuration. From the third equation we get

$$\int R\,d\theta=t-t_0,$$

so that the first two become

$$x-b_1=c_1(t-t_0),\quad y-b_2=c_2(t-t_0), \quad\text{.............(16)}$$

which again shows that the centroid moves along a straight line with uniform speed.

# CHAPTER XX

## THEORY OF GYROSTATIC DOMINATION

1. *Cycloidal Motion.* An important feature of the second edition of Thomson and Tait's *Natural Philosophy*, published in 1879, is the discussion there given of what the authors called *cycloidal systems*, which in their most general form contain rotating flywheels. The simplest example of cycloidal motion (apart from the motion of a particle under gravity in a cycloidal path) is that of a particle hung by a vertical spiral spring of negligible mass. The equation of motion is

$$\ddot{z}+\frac{g}{l}z=0,$$

where $l$ is the distance through which the spring is drawn out when the particle hangs in statical equilibrium. In this case the displacement $z$ from the equilibrium position is a simple harmonic function of the time, which is also the law fulfilled by the bob of a cycloidal pendulum as regards the displacement of the bob along the cycloidal path from the lowest position. For if $s$ be that displacement the differential equation for motion along the arc, situated with the vertex as the lowest point, is

$$\ddot{s}+\frac{g}{4a}s=0,$$

where $a$ is the radius of the generating circle of the cycloid.

Here the solution of the differential equation is given by the roots of the quadratic

$$\zeta^2+\frac{g}{4a}=0,$$

which are pure imaginaries, so that the motion is simple harmonic in the real period $2\pi(4a/g)^{\frac{1}{2}}$, that is the period is that of a simple pendulum of length equal to the length of the cycloid from the middle point to either cusp.

On the other hand, if a particle were constrained to move under gravity in a cycloid with its cusps turned downward, the equation of motion would be

$$\ddot{s}-\frac{g}{4a}s=0,$$

for the force per unit mass along the path, which is numerically $gs/4a$ in both cases, is, contrary to the former case, now directed away from the middle point. The roots of the quadratic are real, and the finite equation is

$$s = Ce^{\alpha t} + C'e^{-\alpha t},$$

where $\alpha = (g/4a)^{\frac{1}{2}}$. This might perhaps be regarded as simple harmonic motion in an imaginary period $2\pi i(4a/g)^{\frac{1}{2}}$.

Both these examples are typical of what Thomson and Tait called cycloidal motion. In such motion each component of displacement is in the general case of the form

$$Ae^{\lambda t} + Be^{\lambda' t} + \dots,$$

where $\lambda$, $\lambda'$, ... are real or complex quantities. With real values of the coefficients of the equation of motion the complex roots occur in pairs $(\alpha + \beta i,\ \alpha - \beta i)$, and every such pair furnishes a term of the form

$$Ce^{\alpha t}\cos(\beta t - f),$$

where C and $f$ are constants of integration.

2. *Cycloidal systems containing flywheels.* We can only give a very short discussion of cycloidal systems which contain gyrostatic flywheels. A general cycloidal system is one acted on by forces of two kinds, "positional forces" which are proportional to displacements, and "motional forces" which are proportional to velocities, and is such that its kinetic energy, T, is a homogeneous quadratic function of the velocities with constant coefficients. Thus if $q_1, q_2, \dots, q_k$ be the coordinates defining the configuration of such a system, the equations of motion of the system are

$$\left.\begin{aligned} &\frac{d}{dt}\frac{\partial T}{\partial \dot{q}_1} + a_1\dot{q}_1 + a_2\dot{q}_2 + \dots + \alpha_1 q_1 + \alpha_2 q_2 + \dots = 0,\\ &\frac{d}{dt}\frac{\partial T}{\partial \dot{q}_2} + b_1\dot{q}_1 + b_2\dot{q}_2 + \dots + \beta_1 q_1 + \beta_2 q_2 + \dots = 0,\\ &\dots\dots\dots\dots\dots\dots\dots\dots\dots\dots\dots\dots \end{aligned}\right\} \quad\dots\dots\dots(1)$$

If $\alpha_2 = \beta_1$, $\alpha_3 = \gamma_1$, ... the positional forces are said to be conservative; if $a_2 = -b_1$, $a_3 = -c_1$, ... we have gyrostatic terms, which also involve no dissipation of energy. Non-gyrostatic motional forces are not here considered.

If a particular solution of the equations of motion be

$$q_1 = h_1 e^{\lambda t}, \quad q_2 = h_2 e^{\lambda t}, \ \dots,$$

then, if $T_0$ be the same quadratic function of $h_1, h_2, \dots$, that T is of $\dot{q}_1, \dot{q}_2, \dots$, we obtain by substitution in the differential equations

$$\left.\begin{aligned} &\lambda^2\frac{\partial T_0}{\partial h_1} + \lambda(a_1h_1 + a_2h_2 + \dots) + \alpha_1h_1 + \alpha_2h_2 + \dots = 0,\\ &\lambda^2\frac{\partial T_0}{\partial h_2} + \lambda(b_1h_1 + b_2h_2 + \dots) + \beta_1h_1 + \beta_2h_2 + \dots = 0,\\ &\dots\dots\dots\dots\dots\dots\dots\dots\dots\dots\dots\dots,\\ &\dots\dots\dots\dots\dots\dots\dots\dots\dots\dots\dots\dots. \end{aligned}\right\} \quad\dots\dots\dots(2)$$

If now $2T = a_{11}\dot{q}_1^2 + 2a_{12}\dot{q}_1\dot{q}_2 + \ldots + a_{22}\dot{q}_2^2 + 2a_{23}\dot{q}_2\dot{q}_3 + \ldots,$

these equations become

$$\left.\begin{array}{l}(a_{11}\lambda^2 + a_1\lambda + \alpha_1)h_1 + (a_{12}\lambda^2 + a_2\lambda + \alpha_2)h_2 + \ldots = 0,\\ (a_{21}\lambda^2 + b_1\lambda + \beta_1)h_1 + (a_{22}\lambda^2 + b_2\lambda + \beta_2)h_2 + \ldots = 0,\\ \ldots\ldots\ldots\ldots\ldots\ldots\ldots\ldots\ldots\ldots\ldots\ldots,\\ \ldots\ldots\ldots\ldots\ldots\ldots\ldots\ldots\ldots\ldots\ldots\ldots,\end{array}\right\} \quad \ldots\ldots\ldots\ldots(3)$$

with $a_{12} = a_{21}$, $a_{13} = a_{31}$, $a_{23} = a_{32}$, ..., so that we get the determinantal equation

$$\begin{vmatrix} a_{11}\lambda^2 + a_1\lambda + \alpha_1, & a_{12}\lambda^2 + a_2\lambda + \alpha_2, \ldots \\ a_{21}\lambda^2 + b_1\lambda + \beta_1, & a_{22}\lambda^2 + b_2\lambda + \beta_2, \ldots \\ \ldots\ldots\ldots\ldots, & \ldots\ldots\ldots\ldots, \ldots \\ \ldots\ldots\ldots\ldots, & \ldots\ldots\ldots\ldots, \ldots \end{vmatrix} = 0, \quad \ldots\ldots\ldots\ldots(4)$$

which is of degree $2k$ in $\lambda$, and has $2k$ roots on which the possible particular solutions depend. If we use any one of the roots in (3) we shall obtain the ratios of the values of $h_2, h_3, \ldots$ to $h_1$ (or of any $k-1$ of the $h$s to the remaining $h$), that is we obtain a solution with one arbitrary constant $h_1$. Each root thus gives a particular solution with one arbitrary constant, and the general solution is the sum of these particular solutions. By proper choice of the arbitrary constants, the general solution may be made to satisfy any given initial conditions.

3. *Effect of repeated roots of the determinantal equation on stability.* But these conditions presuppose that the roots $\lambda$ are all different. The determinantal equation may however have multiple roots, and in either or both of two ways: (1) For a particular value of $\lambda$ not only the determinant D of (4) may vanish, but all the minor determinants, up to and including those obtained by striking out $m-1$ rows and $m-1$ columns from the main determinant, may vanish. If this is the case, the determinantal equation has $m$ identical roots; but no terms containing powers of $t$ appear in the solution. The only effect of the root of multiplicity $m$ is that $m$ of the coefficients $h_1$, $h_2$, ... are left undetermined. These may be chosen arbitrarily, and the necessary $m$ independent particular solutions obtained. (2) It is possible for the determinantal equation to have a multiple root without evanescence of the minor determinants in the manner specified. The necessary particular solutions can then only be obtained by supposing that $h_1$, $h_2$, $h_3$, ... are not constants, but rational integral functions of the time.

To prove shortly these statements, we go back to (1), 2, and regarding $\lambda$ as $d/dt$, to which for the solution assumed it is equivalent, we write (3) as

$$\left.\begin{array}{l}(a_{11}\lambda^2 + a_1\lambda + \alpha_1)q_1 + (a_{12}\lambda^2 + a_2\lambda + \alpha_2)q_2 + \ldots = 0,\\ (a_{21}\lambda^2 + b_1\lambda + \beta_1)q_1 + (a_{22}\lambda^2 + b_2\lambda + \beta_2)q_2 + \ldots = 0,\\ \ldots\ldots\ldots\ldots\ldots\ldots\ldots\ldots\ldots\ldots\ldots\ldots\end{array}\right\} \quad \ldots\ldots\ldots\ldots(1)$$

These equations may be solved by assuming one of the $q$s to be known by its coefficient $h$, then by omitting one of the equations, $k-1$ equations are made available to give the remaining $k-1$ $q$s in terms of the assumed value. In this way, if $M_1$, $M_2$, ... be first minors of the determinant D, being of course functions of $\lambda$, and V be the function defined in (3) below, we get

$$q_1 = M_1 V, \quad q_2 = M_2 V, \quad q_3 = M_3 V, \ldots \quad (2)$$

and it is clear that

$$DV = 0.$$

It is of no consequence what set of first minors is chosen.

V is a function of $t$, and is called by Routh [*Adv. Rigid Dynamics*, Ch. VI] the type of the solution. It is that function which operated on by the functions $M_1$, $M_2$, ... yields the values of the coordinates. If the roots of the equation $D=0$ be $\lambda', \lambda'', \ldots$, each of the expressions $e^{\lambda' t}, e^{\lambda'' t}, \ldots$, satisfies the equation, and so the general form of V is given by

$$V = L' e^{\lambda' t} + L'' e^{\lambda'' t} + \ldots, \quad (3)$$

where $L', L'', \ldots$ are the arbitrary constants. Should however $m$ roots, $\lambda$, be equal, we shall have, by the usual theory of equality of roots, for V the equation

$$V = (L_0 + L_1 t + \ldots + L_{m-1} t^{m-1}) e^{\lambda t} + \text{etc.}, \quad (4)$$

that is

$$V = \left\{ L_0 + L_1 \frac{\partial}{\partial\lambda} + \ldots + L_{m-1} \left(\frac{\partial}{\partial\lambda}\right)^{m-1} \right\} e^{\lambda t} + \text{etc.} \quad (5)$$

Thus, by (2), we have

$$\left.\begin{aligned} q_1 &= \left\{ L_0 M_1 + L_1 M_1 \frac{\partial}{\partial\lambda} + \ldots + L_{m-1} M_1 \left(\frac{\partial}{\partial\lambda}\right)^{m-1} \right\} e^{\lambda t} + \text{etc.}, \\ q_2 &= \left\{ L_0 M_2 + L_1 M_2 \frac{\partial}{\partial\lambda} + \ldots + L_{m-1} M_2 \left(\frac{\partial}{\partial\lambda}\right)^{m-1} \right\} e^{\lambda t} + \text{etc.}, \\ &\ldots\ldots\ldots\ldots, \end{aligned}\right\} \quad (6)$$

where the etc. denotes terms depending on other roots.

Now we know that if M be a rational integral function of the operator $d/dt$ $(=\lambda)$ and $n$ be any integer,

$$M \frac{\partial^n e^{\lambda t}}{\partial \lambda^n} = \left(\frac{\partial}{\partial\lambda}\right)^n (M e^{\lambda t}),$$

and so obtain

$$\left.\begin{aligned} q_1 &= \left\{ L_0 M_1 e^{\lambda t} + L_1 \frac{\partial}{\partial\lambda}(M_1 e^{\lambda t}) + \ldots + L_{m-1} \left(\frac{\partial}{\partial\lambda}\right)^{m-1} (M_1 e^{\lambda t}) \right\} + \text{etc.}, \\ q_2 &= \ldots\ldots\ldots\ldots, \\ &\ldots\ldots\ldots\ldots. \end{aligned}\right\} \quad (7)$$

From this it follows that the necessary and sufficient condition that there shall be no terms involving powers of $t$ in the solution, is that all the minors of D up to and including the $(m-1)$th should vanish. For the investigation shows that the root $\lambda$ of multiplicity $m$ must, if such terms are to be excluded, be repeated $m-1$ times in every first minor of D.

But if the root is thus repeated it is repeated also $m-2$ times in every second minor, $m-3$ times in every third minor, and so on. These minors appear in consequence of the differentiations in (7) and all vanish in consequence of the repetition of the root.

The general rule as to the stability of a moving system is that the real roots, and the real parts of the complex roots, should be either all negative or all zero. These are the general necessary and sufficient conditions, subject to the following reservations as to equal roots, which we give substantially as stated by Routh (*loc. cit.*). As a rule there are, in the case of equality of roots, terms containing powers of $t$ as factors. If $\lambda$ be positive a term $Lt^s e^{\lambda t}$ renders the system unstable. If $\lambda$ be negative the highest possible numerical value of this term is $|L(s/e\lambda)^s|$, and, if $\lambda$ be small, an initial disturbance may be such as to make the oscillatory disturbance of the $q$s a serious departure from the stable state. But if the initial disturbance does not have this effect, the term ultimately disappears and the motion is stable.

Of course if $\lambda$ is a pure imaginary, $ni$, the term is of the form $t^s \sin nt$ and the motion is unstable.

If there are more than $k$ equal roots, there will be powers of $t$ in the solutions as factors.

**4.** *Motional forces. Dissipation function.* With regard to motional forces it may be noted that (1), 2, give (with $\partial V/\partial q_1$, $\partial V/\partial q_2$, ... for the positional forces, supposing these conservative) by multiplication of the first by $\dot{q}_1$, of the second by $\dot{q}_2$, ... and addition,

$$\frac{d(T+V)}{dt} = -\{a_1\dot{q}_1^2 + (a_2+b_1)\dot{q}_1\dot{q}_2 + (a_3+c_1)\dot{q}_1\dot{q}_3 + \dots + b_2\dot{q}_2^2 + \dots\}. \quad \dots(1)$$

The quadratic function of the velocities shown in brackets on the right of (1) is essentially positive for all natural motional forces, and is the rate of diminution of the energy of the system. It has been called by Lord Rayleigh the "dissipation function" and by Thomson and Tait the "dissipativity." Denoting the dissipation function by F, we get by (1)

$$T+V = E_0 - \int_0^t F\,dt, \quad \dots\dots(2)$$

where $E_0$ is the total energy when $t=0$. When the system is at rest in an equilibrium position T and F are zero, otherwise they are both positive. Hence, as time advances, while the system moves, the time integral of F must continually increase towards infinity unless the system comes asymtotically to rest, or V diminishes towards $-\infty$. If V is positive the system will come more and more nearly to rest in the configuration for $V=0$, and unless V is positive in the zero configuration, the equilibrium is unstable [see Thomson and Tait, § 345[II] *et seq.*].

It is to be observed that gyrostatic motional forces contribute nothing to dissipation, and do not appear in the dissipation function, though they have a very real existence in the equations of motion. If for example $a_2\dot{q}_2$, $b_1\dot{q}_1$ are wholly gyrostatic, $a_2+b_1=0$, and the term $(a_2+b_1)\dot{q}_1\dot{q}_2$ disappears from the dissipation function on the right of (1).

5. ***Motional forces zero. Gyrostatic systems.*** The algebraic conditions are simplified if the motional forces are zero, for then the determinant has $k$ roots, each a value of $\lambda^2$. The $2k$ values of $\lambda$ may be denoted by $\pm\lambda'$, $\pm\lambda''$, ..., and the complete solution of the equations of motion is

$$\left.\begin{aligned} q_1 &= \mathrm{A}'e^{\lambda' t}+\mathrm{B}'e^{-\lambda' t} + \mathrm{A}''e^{\lambda'' t}+\mathrm{B}''e^{-\lambda'' t}+\dots, \\ q_2 &= \frac{h'_2}{h'_1}(\mathrm{A}'e^{\lambda' t}+\mathrm{B}'e^{-\lambda' t})+\frac{h''_2}{h''_1}(\mathrm{A}''e^{\lambda'' t}+\mathrm{B}''e^{-\lambda'' t})+\dots, \\ &\dots\dots\dots\dots\dots\dots \end{aligned}\right\} \quad \dots\dots\dots(1)$$

But now equations (3), 2, become

$$\left.\begin{aligned} (a_{11}\lambda^2+\alpha_1)h_1+(a_{12}\lambda^2+\alpha_2)h_2+\dots=0, \\ (a_{21}\lambda^2+\beta_1)h_1+(a_{22}\lambda^2+\beta_2)h_2+\dots=0, \\ \dots\dots\dots\dots\dots\dots \end{aligned}\right\} \quad \dots\dots\dots(2)$$

To see the effect of equality of roots let $\lambda'^2=\lambda''^2$, and assume that the first minors of D vanish. Then twice over in (2) we can take an arbitrary value of $h_2/h_1$, and in terms of this obtain the other ratios. If these values of the ratios $h_2/h_1$ be denoted by $r'$, $r''$ we have for the arbitrary constants

$$\mathrm{A}'+\mathrm{A}'',\ \mathrm{B}'+\mathrm{B}'',\ r'\mathrm{A}'+r''\mathrm{A}'',\ r'\mathrm{B}'+r''\mathrm{B}'',\ \mathrm{A}''',\ \mathrm{B}''',\ \dots,\ \mathrm{A}^{(k)},\ \mathrm{B}^{(k)},$$

in all $2k$. We have therefore the general solution. In a similar way a triple root, a quadruple root, etc., may be shown to leave the number of particular solutions unaltered.

For the discussion of gyrostatic systems it is desirable to suppose the kinetic energy reduced to a sum of squares of velocities with positive coefficients each equal to unity [as in (2), 7, XIX], and V to a sum of squares of coordinates with positive or negative coefficients as the case may be.* This amounts to supposing that the coordinates are those which would be "normal coordinates" if there were no motional forces. It will conduce to clearness also to adopt Thomson and Tait's notation of 12 for $a_2$, 13 for $a_3$, ..., 21 for $b_1$, 23 for $b_3$, .... As we shall suppose (unless it is otherwise stated) that no motional forces appear that are not gyrostatic, we have $a_1=0$, $b_2=0$, $c_3=0$, .... Hence, since the motional forces are gyrostatic, we have

$$12=-21,\ 13=-31,\ 23=-32,\ \dots \quad \dots\dots\dots(3)$$

$$\mathrm{T}=\tfrac{1}{2}(\dot{q}_1^2+\dot{q}_2^2+\dot{q}_3^2+\dots),\ \mathrm{V}=\tfrac{1}{2}(\varpi_1 q_1^2+\varpi_2 q_2^2+\dots), \quad \dots\dots\dots(4)$$

* For the general theory of vibrations about equilibrium, and the conditions of transformation, see Whittaker, *Analytical Dynamics*, Chapter VII. See also Routh, *Advanced Rigid Dynamics*, Chapter VII.

and the equations of motion become

$$\left.\begin{array}{l}\ddot{q}_1+12\dot{q}_2+13\dot{q}_3+\ldots+\varpi_1 q_1=0,\\ \ddot{q}_2+21\dot{q}_1+23\dot{q}_3+\ldots+\varpi_2 q_2=0,\\ \ddot{q}_3+31\dot{q}_1+32\dot{q}_2+\ldots+\varpi_3 q_3=0,\\ \ldots\ldots\ldots\ldots\ldots\ldots\ldots\ldots\ldots\end{array}\right\} \quad \ldots\ldots\ldots\ldots\ldots\ldots(5)$$

The determinantal equation now becomes

$$\begin{vmatrix}\lambda^2+\varpi_1, & 12\lambda, & 13\lambda, & \ldots\\ 21\lambda, & \lambda^2+\varpi_2, & 23\lambda, & \ldots\\ 31\lambda, & 32\lambda, & \lambda^2+\varpi_3, & \ldots\\ \ldots & \ldots & \ldots & \ldots\end{vmatrix}=0. \quad \ldots\ldots\ldots\ldots\ldots\ldots(6)$$

Since $12=-21$, $13=-31$, ..., this is a skew determinant.

It will be of use in examples of gyrostatic systems to have the expanded determinantal equation for two, three, and four equations of motion. These are

$$\lambda^4+(\varpi_1+\varpi_2+12^2)\lambda^2+\varpi_1\varpi_2=0. \quad \ldots\ldots\ldots\ldots\ldots\ldots(7)$$

$$\lambda^6+(\varpi_1+\varpi_2+\varpi_3+12^2+23^2+31^2)\lambda^4+\{\varpi_1(\varpi_2+31^2)$$
$$+\varpi_2(\varpi_3+12^2)+\varpi_3(\varpi_1+23^2)\}\lambda^2+\varpi_1\varpi_2\varpi_3=0. \quad \ldots\ldots\ldots(8)$$

$$\lambda^8+(\varpi_1+\varpi_2+\varpi_3+\varpi_4+12^2+23^2+34^2+13^2+34^2+41^2)\lambda^6$$
$$+\{\varpi_1\varpi_2+\varpi_2\varpi_3+\varpi_3\varpi_4+\varpi_4\varpi_1+\varpi_4\varpi_2+(\varpi_1+\varpi_2)34^2$$
$$+(\varpi_2+\varpi_3)14^2+(\varpi_3+\varpi_4)12^2+(\varpi_4+\varpi_1)23^2+(\varpi_1+\varpi_3)42^2$$
$$+(\varpi_2+\varpi_4)13^2+12\,.\,34+13\,.\,42+14\,.\,23\}\lambda^4$$
$$+(\varpi_1\varpi_2\varpi_3+\varpi_2\varpi_3\varpi_4+\varpi_3\varpi_4\varpi_1+\varpi_1\varpi_2\varpi_4+\varpi_1\varpi_2 34^2$$
$$+\varpi_2\varpi_3 41^2+\varpi_3\varpi_1 42^2+\varpi_1\varpi_4 23^2+\varpi_4\varpi_2 13^2+\varpi_4\varpi_3 12^2)\lambda^2$$
$$+\varpi_1\varpi_2\varpi_3\varpi_4=0. \quad \ldots\ldots\ldots\ldots\ldots\ldots(9)$$

**6.** *Example: Gyrostatic pendulum, two freedoms.* We take as an example the equations of the gyrostatic pendulum (5), 22, XIX, above. These, as already stated, are precisely similar to the equations for the simplest case of motion of an electrified particle, an electron, in a plane at right angles to the lines of force of a magnetic field. The equations may be written

$$\ddot{x}-k\dot{y}+\varpi x=0, \quad \ddot{y}+k\dot{x}+\varpi y=0, \quad \ldots\ldots\ldots\ldots\ldots\ldots(1)$$

where $k=\kappa/\mathrm{A}$, $\varpi=mgh/\mathrm{A}$.

Here $12=k$, and $\varpi=mgh/\mathrm{A}$, and the determinantal equation is

$$(\lambda^2+\varpi)^2+12^2\lambda^2=0.$$

If the given equations had been of the more general form,

$$\mathrm{A}\ddot{x}-\kappa\dot{y}+\mathrm{E}x=0, \quad \mathrm{B}\ddot{y}+\kappa\dot{x}+\mathrm{F}y=0,$$

we could, by writing $\xi=x(\mathrm{A})^{\frac{1}{2}}$, $\eta=y(\mathrm{B})^{\frac{1}{2}}$, have transformed them to

$$\ddot{\xi}-k\dot{\eta}+\varpi_1\xi=0, \quad \ddot{\eta}+k\dot{\xi}+\varpi_2\eta=0,$$

where $k=\kappa(\mathrm{AB})^{\frac{1}{2}}$, $\varpi_1=\mathrm{E}/\mathrm{A}$, $\varpi_2=\mathrm{F}/\mathrm{B}$.

We shall consider this case a little in detail. The determinantal equation is

$$(\lambda^2+\varpi_1)(\lambda^2+\varpi_2)+k^2\lambda^2=0.$$

This equation involves only even powers of $\lambda$, and so the roots will occur in pairs $\pm(\mu+i\nu)$. For stability it is necessary that every $\mu$ should be zero, otherwise terms $e^{\pm\mu t}{}^{\sin}_{\cos}\nu t$ would appear in the solutions.

There are two possibilities of stability, $\varpi_1$ and $\varpi_2$ may be both positive or both negative. If they are both positive we see at once that both roots of the quadratic $\lambda'^2$, $\lambda''^2$ are negative, so that all four roots $\pm\lambda'$, $\pm\lambda''$ are pure imaginaries, and the solution is oscillatory and stable in the usual sense.

If however $\varpi_1$ and $\varpi_2$ are both negative, another condition must be fulfilled. The product of the roots, $\lambda'^2\lambda''^2$, is positive, and so they have the same sign. Their sum will be negative, and they will be real if

$$k^2 > -(\varpi_1+\varpi_2)+2(\varpi_1\varpi_2)^{\frac{1}{2}},$$

or as we may write it, taking $k$ positive,

$$k > (-\varpi_1)^{\frac{1}{2}}+(-\varpi_2)^{\frac{1}{2}}.$$

These are the conditions of stability. Now the rotational coefficient is in gyrostatic systems usually very large. Taking it so, we see that the sum of the roots is numerically very large in comparison with their product, so that one root must be large and the other small numerically. The approximate numerical value of the greater of the two roots is thus $k^2$, and that of the smaller is therefore $\varpi_1\varpi_2/k^2$. More exactly the roots are

$$\lambda'^2 = -(k^2+\varpi_1+\varpi_2)\left(1-\frac{\varpi_1\varpi_2}{k^4}\right),$$

$$\lambda''^2 = -(k^2+\varpi_1+\varpi_2)\frac{\varpi_1\varpi_2}{k^4}.$$

If no rotational terms existed, and $\varpi_1$, $\varpi_2$ were positive, the system would have two fundamental periods of oscillation, of periods $2\pi/\varpi_1^{\frac{1}{2}}$, $2\pi/\varpi_2^{\frac{1}{2}}$, having $2\pi/(\varpi_1\varpi_2)^{\frac{1}{4}}$ as their geometric mean. It will be seen that the geometric mean of the periods when rotation exists is the same as that for zero rotation.

When $\varpi_1$, $\varpi_2$ are negative the irrotational motions are not oscillatory, and there is instability in both freedoms. Sufficiently rapid rotation converts these *two* instabilities into stability. But if without rotation the motion according to one freedom is stable, and that according to the other is unstable, that is if $\varpi_1$ and $\varpi_2$ have opposite signs, the establishment of rotation still leaves one freedom stable, the other unstable. It is clear that the general theory shows that only an even number of instabilities can be stabilised by rotation.

The different mountings of a gyrostat shown in Figs. 46, 47, above illustrate the different ways in which the freedoms are arranged. These are described with sufficient fulness in 3, 4, 5, VIII.

**7.** *Gyrostatic system with three freedoms. Electric and magnetic analogue.* The case of three freedoms differs only slightly from that just treated. The equations are

$$\left.\begin{aligned}\ddot{q}_1+k_3\dot{q}_2-k_2\dot{q}_3+\varpi_1 q_1=0,\\ \ddot{q}_2+k_1\dot{q}_3-k_3\dot{q}_1+\varpi_2 q_2=0,\\ \ddot{q}_3+k_2\dot{q}_1-k_1\dot{q}_2+\varpi_3 q_3=0.\end{aligned}\right\} \quad\cdots\cdots\cdots(1)$$

The similarity of these to the equations of motion of a body with respect to axes rotating about themselves with angular speeds $-k_1$, $-k_2$, $-k_3$, as described in (2), 10, II, will strike the reader at once. But it is clear that the equations should have this form, for, if we impose equal but opposite rotations on both axes and body, we get exactly the equations just written as those for the now rotating body referred to fixed axes.

It is clear from the remark just made that the turning would be about an axis the direction cosines of which are proportional to $k_1$, $k_2$, $k_3$. This suggests changing to this line through the origin as one coordinate, say that of $z$, and two other axes, of $x$ and $y$, at right angles to themselves and to this direction. Multiplying then the first equation by $k_1$, the second by $k_2$, and the third by $k_3$, and adding, we get

$$k_1\ddot{q}_1+k_2\ddot{q}_2+k_3q_3+k_1\varpi_1q_1+k_2\varpi_2q_2+k_3\varpi_3q_3=0,$$

or, if we suppose the left-hand side divided by $(k_1^2+k_2^2+k_3^2)^{\frac{1}{2}}$ $(=2\omega)$, $z+\mathrm{Z}=0$.

Now let $l$, $m$, $n$ be the direction cosines of the axis of $x$; those of the axis of $y$ are then $(k_3m-k_2n,\ k_1n-k_3l,\ k_2l-k_1m)/2\omega$. Multiplying the first equation by $l$, the second by $m$, and the third by $n$, and adding, we get $x-2\omega\dot{y}+\mathrm{X}=0$.

Similarly, multiplying the equations respectively by the cosines of the axis of $y$, and noting that $l^2+m^2+n^2=1$, $k_1l+k_2m+k_3n=0$, we find $y+2\omega\dot{x}+\mathrm{Y}=0$.

Thus the equations of motion are

$$\dot{x}-2\omega\dot{y}+\mathrm{X}=0,\quad y+2\omega\dot{x}+\mathrm{Y}=0,\quad z+\mathrm{Z}=0. \qquad\text{.......................(2)}$$

These may be regarded as the equations of motion of a particle attached by massless springs to a body revolving uniformly about the axis of $z$ This is Thomson and Tait's interpretation of the case, but other interpretations are possible. For example, they are exactly the equations of motion of an electron in a combined electric and magnetic field in which the magnetic force acts along the axis of $z$. The first two equations give the effect of the magnetic field on the electron in its motion in the electric field, the final terms in all three equations take account of the forces acting against the inertia of the electron, and of the electric forces in the three component directions.

The Zeeman effect in its most elementary form is due to the action of a magnetic field in changing the period of vibrational motion of an electron in a plane (the plane of $x$, $y$) at right angles to the magnetic field intensity. If there were no magnetic field the equations of motion would be, we shall suppose,

$$x+\varpi x=0,\quad \ddot{y}+\varpi y=0,\quad \ddot{z}+\varpi z=0, \qquad\text{.............................(3)}$$

where $\varpi$ is positive. The imposition of the magnetic field in the $z$-direction gives the equations

$$x-2\omega\dot{y}+\varpi x=0,\quad y+2\omega\dot{x}+\varpi y=0,\quad z+\varpi z=0. \qquad\text{.....................(4)}$$

Thus we have $12=2\omega$, $23=31=0$, and the determinantal equation is, by (4),

$$\lambda^6+(3\varpi+4\omega^2)\lambda^4+(3\varpi^2+4\omega^2\varpi)\lambda^2+\varpi^3=0. \qquad\text{........................(5)}$$

The first two equations of (4) are quite independent of the third, and so we have from them

$$\lambda^4+(2\varpi+4\omega^2)\lambda^2+\varpi^2=0. \qquad\text{...................................(6)}$$

Multiplying the last result by $\varpi$, and subtracting the product from (5), we obtain

$$\lambda^6+(2\varpi+4\omega^2)\lambda^4+\varpi^2\lambda^2=0.$$

This gives $\qquad \lambda^2=0,\quad$ and $\quad \lambda^4+(2\varpi+4\omega^2)\lambda^2+\varpi^2=0. \qquad$ ..........................(7)

In this case of three freedoms the motion at right angles to the axis of $z$, taken by itself, is stable whether $\varpi$ be negative or positive, provided the condition stated in 6 is fulfilled. If $\varpi$ be positive, the whole motion is stable; if $\varpi$ is negative, the $z$ motion is not stable.

8. *Gyrostatic domination. Large and small roots of the determinantal equation.* When the gyrostatic terms are due to very rapidly rotating flywheels they "dominate" the motion; for example, in (6) the term $4\omega^2$ is very great. Hence there is a large root of (7) and a small one. Approximately these are given respectively by

$$\lambda^2=-4\omega^2\left(1+\frac{\varpi}{2\omega^2}\right),\quad \lambda^2=-\frac{\varpi^2}{4\omega^2}.$$

The large root is practically independent of the applied forces which enter through $\varpi$; the small root depends very directly on these forces, but is rendered small by the gyrostatic coefficient. Thus we have an example of what has been called "gyrostatic domination."

To a first approximation the large root is $-4\omega^2$, and to this approximation does not depend on the applied forces at all. Hence Thomson and Tait have called the motion corresponding to this root "adynamic." We shall not use this term: one motion is as dynamical as the other, and it may be preferable to refer to one as the rapid motion and to the other as the slow motion. In ordinary applications $2\pi/i\lambda$ is the period of precession. When $\lambda$ is the large root the precession is fast, when $\lambda$ is the small root the precession is slow.

When however $\omega^2$ must be taken as small in comparison with $\varpi$, as for the electron in the magnetic field, the determinantal equations give again (6); but we have now approximately

$$\lambda^2 = -\varpi\left(1 \pm 2\frac{\omega}{\varpi^{\frac{1}{2}}}\right).$$

The period of "precession," being $2\pi/i\lambda$, is in this case $2\pi/(\varpi^{\frac{1}{2}} \pm \omega)$.

If, for example, the electron goes round in a circle, it does so in one or other of these two periods, according to the direction of motion.

We have thus treated again, with rather more detail, the problem of the gyrostatic pendulum when restricted to motion of small amplitude. Up to a certain point the analogy between this pendulum and the electron moving in an electric and magnetic field is complete.

9. *System with four freedoms.* The reader can now write down at once the equations for a system of four freedoms with six gyrostatic links connecting each freedom with the three others. Such a system is fully discussed in Thomson and Tait's *Natural Philosophy*, §§ $345^{\text{xi}}$... $345^{\text{xxi}}$, and we shall here only state the main results, leaving the verification, which is not difficult, to the reader.

(1) If no forces act, that is if $\varpi_1 = \varpi_2 = \varpi_3 = \varpi_4 = 0$, the determinantal equation (8), 5, has two zero roots, and, except as a special case, two others (values of $-\lambda^2$) which are real and positive. The modes of motion corresponding to the zero roots are neutral as regards stability, the other two modes are stable.

(2) If the forces be not zero but very small, the determinantal equation has the two roots just referred to, and two other roots, which replace the zero roots of the case of no forces. These are given (nearly) by the quadratic equation obtained by leaving out from (9), 5, powers of $\lambda$ greater than the fourth, and retaining only the terms involving the six gyrostatic coefficients 12, 13, ..., 34, in the coefficients of $\lambda^4$ and $\lambda^2$.

(3) All four modes of motion are stable, but for this it is not necessary that $\varpi_1$, $\varpi_2$, $\varpi_3$, $\varpi_4$ should be all of one sign. The product must be positive,

so that two must be negative, two positive. If $\varpi_1$, $\varpi_2$ be positive, $\varpi_3$, $\varpi_4$ negative, the following two inequalities are the necessary and sufficient conditions of stability,

$$\left\{\frac{12}{(\varpi_1\varpi_2)^{\frac{1}{2}}} \pm \frac{34}{(\varpi_3\varpi_4)^{\frac{1}{2}}}\right\}^2 > \left\{\frac{13}{(-\varpi_1\varpi_3)^{\frac{1}{2}}} \pm \frac{42}{(-\varpi_4\varpi_2)^{\frac{1}{2}}}\right\}^2 + \left\{\frac{14}{(-\varpi_1\varpi_4)^{\frac{1}{2}}} \pm \frac{23}{(-\varpi_2\varpi_3)^{\frac{1}{2}}}\right\}^2,$$

where the *plus* signs give one inequality, the *minus* signs the other. All the conditions are fulfilled when the rotation is sufficiently rapid and the coefficients have suitable values.

For further particulars and for conclusions as to general systems the reader should consult the *Natural Philosophy*, §§ 345$^{\text{xi}}$ to 345$^{\text{xxviii}}$.

# CHAPTER XXI

## GEOMETRICAL REPRESENTATION OF THE MOTION OF A TOP

1. *Motion of a rigid body under no forces.* A new mode of presenting the theory of the motion of a top was worked out by Darboux in his notes on Despeyrous' *Cours de Mécanique*, on the basis of a theorem due to Jacobi connecting the motion with that of a body under the action of no external forces. We consider therefore first Poinsot's theory of the latter motion.

A rigid body turns about a fixed point, O, under no applied forces; its angular speeds are $p, q, r$ about the principal axes O(A, B, C) drawn through O. The moments of inertia are of course A, B, C. First we suppose that the applied couples are not zero, but have values L, M, N. The equations of motion are three, of the form

$$A\dot{p}-(B-C)qr=L, \qquad \text{.................................(1)}$$

and are applicable even if there is translational motion of the point O.

If now we multiply the first equation by $p$, the second by $q$, and the third by $r$, and add, we obtain

$$Ap\dot{p}+Bq\dot{q}+Cr\dot{r}=Lp+Mq+Nr.$$

But if T be the kinetic energy of the rotational motion this equation is

$$\frac{dT}{dt}=Lp+Mq+Nr, \qquad \text{.................................(2)}$$

that is the time-rate of increase of this kinetic energy is the rate at which the couples do work. Hence, if L, M, N are zero, the value of T is constant.

Again, multiply the equations of motion by $Ap$, $Bq$, $Cr$, respectively; we get, if $H^2=A^2p^2+B^2q^2+C^2r^2$,

$$\frac{dH}{dt}=\frac{1}{H}(LAp+MBq+NCr). \qquad \text{.........................(3)}$$

H is the resultant A.M., and $(Ap, Bq, Cr)/H$ its direction cosines. Thus the equation states that the rate of growth of H at any instant is equal to the total moment of the couples about the localised vector direction of H. If L, M, N are all zero the resultant H remains unchanged both in magnitude and in direction. This direction will be indicated by OH (or by OL) in the diagrams which follow.

That OL remains thus stationary may be seen as follows. Denote the positions of the principal axes at the beginning and end of $dt$ by $O(A_1, B_1, C_1)$ and $O(A_2, B_2, C_2)$. These are fixed positions, and the corresponding angular momenta are $Ap$, $Bq$, $Cr$ and $A(p+\dot{p}\,dt)$, $B(q+\dot{q}\,dt)$, $C(r+\dot{r}\,dt)$, while at the final instant of $dt$ the angular momenta about $O(A_1, B_1, C_1)$ have become

$$Ap+\{A\dot{p}-(B-C)qr\}dt, \quad Bq+\{B\dot{q}-(C-A)rp\}dt, \quad Cr+\{C\dot{r}-(A-B)pq\}dt.$$

But if $L=M=N=0$, these reduce to $Ap$, $Bq$, $Cr$, that is the position of OL with respect to $O(A_1, B_1, C_1)$ has not altered.

2. *Poinsot's representation of the motion of a body under no forces.* If OI be the instantaneous axis (I.A. for brevity) its direction cosines with respect to $O(A, B, C)$ at time $t$ are $(p, q, r)/\omega$, if $\omega^2=p^2+q^2+r^2$. We have then

$$\cos \text{LOI}=\frac{1}{\omega H}(Ap^2+Bq^2+Cr^2).$$

But if T be the kinetic energy this can be written

$$\omega \cos \text{LOI}=\frac{2T}{H}. \quad \text{.................................}(1)$$

The product on the left of (1) is the component of angular velocity about OL, which we have seen is invariable in direction if $L=M=N=0$. Hence when this is the case the body turns with uniform angular speed about OL.

In the general case the directions of OL and OI are both changing, and the body is turning about OI as instantaneous axis. Hence the component, $H \sin\angle \text{LOI}$, of H at right angles to OI is being turned towards the position of the normal to the plane LOI at rate $\omega$, so that the rate of production of A.M. about the normal is $H\omega \sin \text{LOI}$.

It is clear that the motion can give rise to no other rate of change of A.M. Now $(B-C)qr$, $(C-A)rp$, $(A-B)pq$ are the components of rate of growth of A.M. due to the motion, and their resultant must be perpendicular to the plane HOI and be equal to $H\omega \sin \text{LOI}$. Hence

$$H\omega \sin \text{HOI}=\{(B-C)^2q^2r^2+(C-A)^2r^2p^2+(A-B)^2p^2q^2\}^{\frac{1}{2}}. \quad \text{........}(2)$$

We may prove this otherwise. The cosines of OL and OI are

$$(Ap, Bq, Cr)/H, \quad (p, q, r)/\omega.$$

Hence the cosines of a normal to the plane LOI are

$$\{(B-C)qr, \ (C-A)rp, \ (A-B)pq\}/\omega H \sin \text{LOI}.$$

Squaring these cosines and equating the sum of squares to 1, we get again (2).

The quantity on the right of (2) has been called the centrifugal couple, a name which tends to obscure the true meaning of the expression.

We now introduce the method, due to Poinsot,* of discussing the motion of the body by means of the corresponding motion of the momental ellipsoid (M.E.). The conclusions obtained when the motion is restricted to be about a fixed point O will be applicable in other cases, for example to the oscilla-

* *Théorie nouvelle de la rotation des corps*, 1851.

tions and rotations with respect to the centroid, of a quoit or stick thrown into the air, since these motions are not affected by the action of gravity when the body is free in the air.

We prove first that $\omega$ varies as OP, the length of the coincident radius vector of the M.E. If $x$, $y$, $z$ be the coordinates of P with respect to the principal axes, we have

$$\frac{p}{x}=\frac{q}{y}=\frac{r}{z}=\frac{\omega}{\mathrm{OT}},$$

from which we obtain at once

$$\frac{\mathrm{A}p^2+\mathrm{B}q^2+\mathrm{C}r^2}{\mathrm{A}x^2+\mathrm{B}y^2+\mathrm{C}z^2}=\frac{\mathrm{A}^2p^2+\mathrm{B}^2q^2+\mathrm{C}^2r^2}{\mathrm{A}^2x^2+\mathrm{B}^2y^2+\mathrm{C}^2z^2}=\frac{\omega^2}{\mathrm{OP}^2}. \quad\text{.................... (3)}$$

The numerator of the first fraction is 2T, the denominator by the equation of the M.E. [$\mathrm{A}x^2+\mathrm{B}y^2+\mathrm{C}z^2=k^4$, where, taking the mass of the body as unity, we must suppose $k$ a length to keep the dimensions right in formulae which follow] is $k^4$. Hence we have

$$\frac{\omega^2}{\mathrm{OP}^2}=\frac{2\mathrm{T}}{k^4}, \quad\text{.................................... (4)}$$

a constant. Thus $\omega$ varies as OP.

We have seen that $\omega\cos\mathrm{LOI}=2\mathrm{T}/\mathrm{H}$. Hence (4) gives

$$\cos\mathrm{LOI}=\frac{\omega}{\mathrm{OP}^2}\frac{k^4}{\mathrm{H}}. \quad\text{.................................. (5)}$$

If we take the second relation in (3),

$$\frac{\mathrm{A}^2p^2+\mathrm{B}^2q^2+\mathrm{C}^2r^2}{\mathrm{A}^2x^2+\mathrm{B}^2y^2+\mathrm{C}^2z^2}=\frac{\omega^2}{\mathrm{OP}^2}, \quad\text{.............................. (6)}$$

we notice that the perpendicular from the centre O to the tangent plane, which touches the M.E. at the point P in which it is intersected by the I.A.,

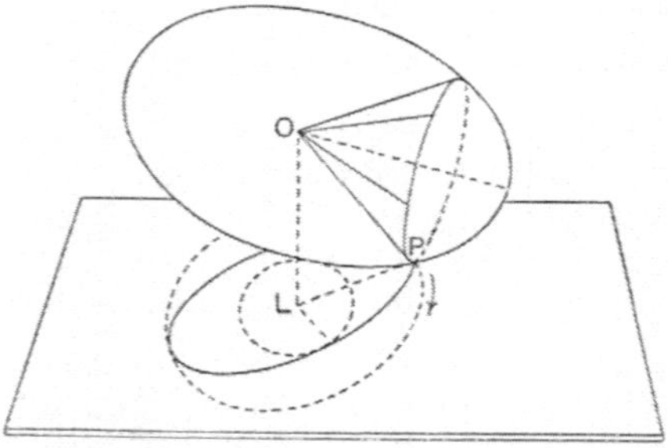

FIG. 107.

has cosines proportional to $\mathrm{A}x$, $\mathrm{B}y$, $\mathrm{C}z$, that is to $\mathrm{A}p$, $\mathrm{B}q$, $\mathrm{C}r$. The perpendicular therefore coincides with the fixed line OH. Its length $\varpi$ is $k^4/(\mathrm{A}^2x^2+\mathrm{B}^2y^2+\mathrm{C}^2z^2)^{\frac{1}{2}}$, that is by (6) $k^4\omega/(\mathrm{H}\,.\,\mathrm{OP})$, $=k^2(2\mathrm{T})^{\frac{1}{2}}/\mathrm{H}$, which is constant. Thus the M.E. always touches a plane perpendicular to the axis of resultant A.M. at the constant distance $k^2(2T)^{\frac{1}{2}}/H$ from the centre of the ellipsoid. A plane through O parallel to the plane just determined is called the *invariable plane* (I.P.). In this plane an impulsive couple, of moment H, would have to be laid to produce the motion from rest. The fixed line OL is called the *invariable line* (I.L.). [See Fig. 107, in which OH coincides with OL and OP is the radius vector of the momental ellipsoid which coincides with OI.]

3. *Measurement of the time of motion. Sylvester's theorem.* The M.E. is turning about OI, which is coincident with OP. At P the M.E. touches a plane parallel to the I.P., and so rolls on that plane. The angular speed $\omega$ resolves into two components, $\omega \cos HOI = 2T/H$ about OL, and $\omega \sin LOI = (\omega^2 H^2 - 4T^2)^{\frac{1}{2}}/H$ about OG, the intersection of the plane LOI with the I.P. If, then, we suppose the I.P. to turn with the M.E. about OL with speed 2T/H, the motion of the ellipsoid relative to that plane will be simply one of rolling about OG with speed $\omega \sin LOI$. The line OG describes a cone in the body, and in space it sweeps over a part of the I.P. about O. The angle, $\chi$ say, turned through by the I.P. about OH is proportional to the time $t$, in fact $t = \chi H/2T$.

A method of determining the time by the motion of an ellipsoid, confocal with the M.E., and rolling on a plane parallel to the I.P. was given by Sylvester. Such an ellipsoid has the equation

$$\frac{A}{1+hA}x^2 + \frac{B}{1+hB}y^2 + \frac{C}{1+hC}z^2 = k'^4. \qquad \text{......................(1)}$$

It touches a plane parallel to the I.P. in a point Q of coordinates

$$x,\ y,\ z = R\{(1+hA)p, \quad (1+hB)q, \quad (1+hC)r\}, \qquad \text{................(2)}$$

where R is the common value of the ratios

$$\frac{x}{(1+hA)p} = \frac{y}{(1+hB)q} = \frac{z}{(1+hC)r}.$$

Substituting for $x$, $y$, $z$ in (1) we get

$$R^2\{Ap^2(1+hA) + Bq^2(1+hB) + Cr^2(1+hC)\} = k'^4,$$

that is

$$R^2 = \frac{k'^4}{2T+hH^2}. \qquad \text{....................................(3)}$$

The perpendicular distance of this plane from O is $k'^2(2T+H^2)^{\frac{1}{2}}/H$. But

$$OQ^2 = x^2+y^2+z^2 = R^2\{p^2(1+hA)^2 + \ldots\},$$

so that

$$OQ^2 = \frac{k'^4(\omega^2 + 4hT + h^2H^2)}{2T+hH^2}. \qquad \text{...........................(4)}$$

The point Q lies in the plane IOL, if L be the point in which the invariable line meets the tangent plane. For the direction cosines of a normal to the plane IOL fulfil the two conditions $lp + mq + nr = 0$, $lAp + mBq + nCr = 0$, and so $l(1+hA)p+m(1+hB)q+n(1+hC)r=0$, or $lx+my+nz=0$, that is the normal is perpendicular to the line OQ.

Now the body turns about OI as I.A., and the M.E. and its confocals move with it. Hence, if $v$ be the speed of the point Q, we have

$$v=\omega \,.\, \mathrm{OQ} \sin \mathrm{QOI}. \quad \text{.............................(5)}$$

The angular speed $\omega'$ of Q about the invariable line OL is therefore given by

$$\omega'=\omega\frac{\sin \mathrm{QOI}}{\sin \mathrm{QOL}}. \quad \text{.................................(6)}$$

It is easy to prove that

$$\mathrm{OQ}\,.\cos \mathrm{QOI}=\frac{\mathrm{R}}{\omega}(\omega^2+2h\mathrm{T}), \quad \mathrm{OQ}\,.\cos \mathrm{QOL}=\frac{\mathrm{R}}{\mathrm{H}}(2\mathrm{T}+h\mathrm{H}^2) \quad \text{.......(7)}$$

From these we find

$$\frac{\sin^2 \mathrm{QOI}}{\sin^2 \mathrm{QOL}}=\frac{\mathrm{H}^2}{\omega^2}\,\frac{\omega^2(\omega^2+4h\mathrm{T}+h^2\mathrm{H}^2)-(\omega^2+2h\mathrm{T})^2}{\mathrm{H}^2(\omega^2+4h\mathrm{T}+h^2\mathrm{H}^2)-(2\mathrm{T}+h\mathrm{H}^2)^2}=\frac{h^2\mathrm{H}^2}{\omega^2},$$

and therefore

$$\omega\frac{\sin \mathrm{QOI}}{\sin \mathrm{QOL}}=h\mathrm{H}. \quad \text{..............................(8)}$$

Thus the point Q turns with constant angular speed $h$H round the I.L., and the time of motion from any epoch is $\theta/h\mathrm{H}$, where $\theta$ is the angle turned through in the interval considered.

Sylvester gave a generalisation of Poinsot's theory, which may be stated as follows: Let a polhode E be given, traced on a surface S (the surface of an ellipsoid or any surface of the second degree), that is let the locus of points traced out on the surface, as it rolls on a fixed plane, be given. Then, if from each point on E we set off, outwards say, along the normal at the point, a given constant distance, the extremities of the lines so drawn is a polhode E, traced on a surface S, confocal with a surface similar and similarly situated to S. We do not enter here into the mathematical discussion, but refer to Sylvester's original paper (*Phil. Trans.*, 156, 1866), or to Darboux's Note XVII, appended to Despeyrous' *Cours de Mécanique*.

Following on the discussion of Poinsot's theory just referred to, Darboux gives in his Note XVIII a representation of the motion of the axis of a top by means of a deformable hyperboloid of one sheet, of which we shall give an account later in this chapter.

4. *Polhode and herpolhode. Body-cone and space-cone.* As the M.E. rolls on the fixed plane, the successive points of contact trace out two loci, one on the ellipsoid, the other on the plane. The former is called the *polhode*, the latter the *herpolhode*. For all points on the polhode the tangent planes to the ellipsoid are at a constant distance $\varpi$ from O, and we have seen above that

$$\varpi=k^2\left(\frac{2\mathrm{T}}{\mathrm{H}^2}\right)^{\frac{1}{2}}. \quad \text{...................................(1)}$$

But we have from the equation of the M.E.

$$Ax^2+By^2+Cz^2=k^4, \quad A^2x^2+B^2y^2+C^2z^2=k^4\frac{H^2}{2T}.$$

Hence we may write

$$A(2AT-H^2)x^2+B(2BT-H^2)y^2+C(2CT-H^2)z^2=0,$$

or

$$A(k^4-\varpi^2A)x^2+B(k^4-\varpi^2B)y^2+C(k^4-\varpi^2C)z^2=0. \quad \text{.........(2)}$$

This is the equation of a cone, fixed relatively to the body, and called the body-cone, which contains the successive lines in the body with which the instantaneous axis coincides as the ellipsoid rolls. Its intersection with the M.E. is the polhode.

It is clear that the body-cone rolls on a cone fixed in space the intersection of which with the plane of contact is the herpolhode. This is called the space-cone. The body-cone does not exist unless $H^2/2T$, that is $k^4/\varpi^2$, lies between the greatest and the least of A, B, C, since otherwise the sum of three quantities of the same sign would be zero. If C be the greatest and A the least moment of inertia, we have

$$\left.\begin{aligned}2AT-H^2&=B(A-B)q^2+C(A-C)r^2,\\ 2CT-H^2&=A(C-A)p^2+B(C-B)q^2,\end{aligned}\right\} \quad \text{.....................(3)}$$

so that $2AT-H^2$ is negative and $2CT-H^2$ positive. Hence

$$C>\frac{H^2}{2T}>A.$$

If A=B the body has an axis of symmetry, the M.E. is of revolution, and the equation of the body-cone is

$$x^2+y^2+\frac{C}{A}\frac{2CT-H^2}{2AT-H^2}z^2=0. \quad \text{..........................(4)}$$

The axis of $z$ is the axis of symmetry of the body, and the polhode is a circle described round O$z$. Euler's equations give, since there are no forces,

$$\dot{p}=\frac{A-C}{A}qr, \quad \dot{q}=\frac{C-A}{A}rp, \quad \dot{r}=0. \quad \text{.......................(5)}$$

Thus $r$ is constant, and $p\dot{p}+q\dot{q}+r\dot{r}=0$, so that $p^2+q^2+r^2$, that is $\omega^2$, is constant. But $\omega\cos\mathrm{HOI}=2T/H$, and therefore $\angle$HOI remains constant as the body moves.

Thus the instantaneous axis OI remains at a constant inclination to the axis of symmetry, and OI is always inclined at the same angle, $\cos^{-1}(2T/\omega H)$, to the invariable line.

Where in what follows we consider only the motion of the M.E. we shall generally suppose that $k=1$, so that then $\varpi=(2T)^{\frac{1}{2}}/H$.

**5.** *Case of axial symmetry: M.E. is of revolution.* With an axis of symmetry the M.E. may be oblate or prolate. In the first case C is the maximum, in the latter the minimum, moment of inertia. We have

$$\cos\mathrm{IO}z=\frac{r}{\omega}, \quad \cos\mathrm{LO}z=\frac{Cr}{H}.$$

If the M.E. is oblate $Cr/C\omega < Cr/H$, and therefore $\angle IOz > \angle LOz$. In this case the space-cone lies within the body-cone, and the concave surface of the latter rolls round the convex surface of the former.

If the M.E. is prolate $Cr/C\omega > Cr/H$, and therefore $\angle IOz < \angle LOz$. The body-cone is now external to the space-cone, and the two convex surfaces are in contact.

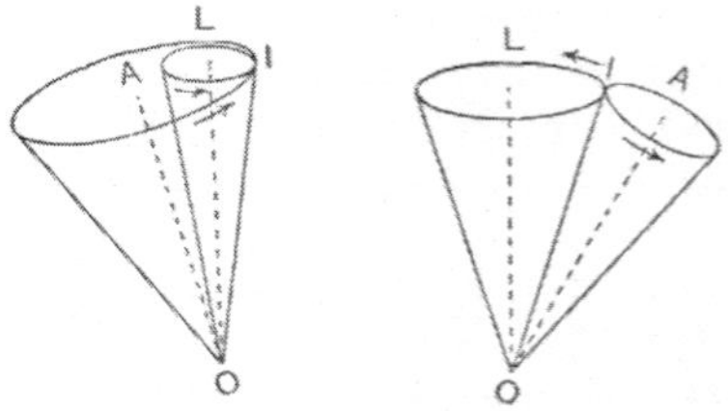

(a) FIG. 108. (b)

The diagram (*a*), (*b*) of Fig. 108 shows the two cases. OA ($Oz$ in the text) indicates the axis of the body-cone, OL that of the space-cone.

The rolling of the ellipsoid (supposed of revolution and prolate) on a fixed plane, parallel to the invariable plane through O, is illustrated in Fig. 107, which gives an idea of the polhode, shown on the ellipsoid, and of the herpolhode as described on the plane.

6. *Cone described in the body by the projection of the instantaneous axis on I.P. Separating polhodes.* Now consider the motion of the line OG, joining O with the projection of P on the invariable plane. Thus GP is parallel to the invariable line OL, and equal in length to $\varpi$, that is $(2T)^{\frac{1}{2}}/H$ (if $k=1$). As stated in 3 above, OG describes a cone in the body. Let G have coordinates $\xi$, $\eta$, $\zeta$, while those of P are $x$, $y$, $z$; the projections of GP on the axes are then $\xi - x$, $\eta - y$, $\zeta - z$. These are proportional to the direction cosines of the normal at P to the M.E., and therefore

$$\frac{\xi - x}{Ax} = \frac{\eta - y}{By} = \frac{\zeta - z}{Cz} = \mu, \text{ say.} \quad \ldots\ldots\ldots\ldots(1)$$

But also OG is perpendicular to the invariable line, so that

$$Ax\xi + By\eta + Cz\zeta = 0. \quad \ldots\ldots\ldots\ldots(2)$$

Equations (1) give

$$\xi = (A\mu + 1)x, \quad \eta = (B\mu + 1)y, \quad \zeta = (C\mu + 1)z. \quad \ldots\ldots\ldots\ldots(3)$$

Multiplying these by $Ax$, $By$, $Cz$, respectively, and adding, we obtain

$$Ax^2 + By^2 + Cz^2 + \mu(A^2x^2 + B^2y^2 + C^2z^2) = 0, \quad \ldots\ldots\ldots\ldots(4)$$

or (since now $k=1$) $1 + \mu/\varpi^2 = 0$, that is $\mu = -\varpi^2$. Thus we obtain from (3)

$$Ax = \frac{A\xi}{1 - \varpi^2 A}, \quad By = \frac{B\eta}{1 - \varpi^2 B}, \quad Cz = \frac{C\zeta}{1 - \varpi^2 C};$$

and therefore, instead of (2),

$$\frac{A\xi^2}{1 - \varpi^2 A} + \frac{B\eta^2}{1 - \varpi^2 B} + \frac{C\zeta^2}{1 - \varpi^2 C} = 0, \quad \ldots\ldots\ldots\ldots(5)$$

the equation of a cone.

Going back now to the equation of the body-cone we see that if C be the greatest moment of inertia and A the least, the equation becomes

$$\left.\begin{aligned} &B(B-A)y^2+C(C-A)z^2=0,\\ \text{or}\quad &A(A-C)x^2+B(B-C)y^2=0,\end{aligned}\right\} \quad\ldots\ldots\ldots\ldots(6)$$

according as $A=1/\varpi^2$ or $C=1/\varpi^2$. Each of these represents a pair of imaginary planes; the former pair meets in the axis of $x$, the latter in the axis of $z$.

The cone degenerates into two real planes if $B=1/\varpi^2$, where B is the intermediate moment of inertia. For then

$$A(A-B)x^2-C(B-C)z^2=0. \quad\ldots\ldots\ldots\ldots(7)$$

These planes intersect in the axis of $y$ and separate the polhodes, which are closed curves surrounding the axes of greatest and least moment of inertia, as shown in Fig. 109. Their intersections with the M.E. are called the *separating polhodes*.

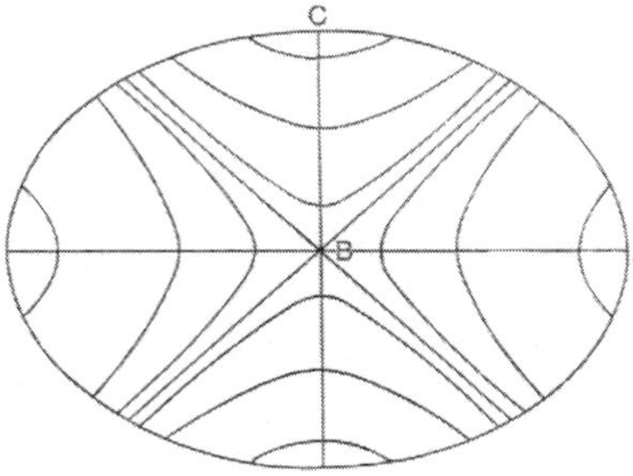


FIG. 109.

7. *Herpolhodes.* The herpolhode is a curve consisting of different parts, corresponding to successive repetitions of the polhode. From the manner of its description by the rolling of the M.E. it must always have its concavity turned towards the point L (Fig. 107), and therefore cannot have a point of inflexion. The distance of P from L at any instant is $(OP^2-\varpi^2)^{\frac{1}{2}}$; and it is evident from the form of the polhode, as displayed by its projections just indicated, that, in each fourth part of a description, OP varies between a maximum and a minimum value. Thus the distance $(OP^2-\varpi^2)^{\frac{1}{2}}$ similarly varies, and so the herpolhode is a curve lying between two circles which have the projection of the centre of the M.E. as their common centre, and touching the outer circle internally and the inner externally as shown in Fig. 107. The herpolhode is not however in general a closed or re-entering curve: unless the angle turned through by LP, from contact with one circle to contact with the other, be commensurable with $2\pi$, the curve will not be repeated.

When $\varpi^2 = 1/B$ the polhode passes through the extremity B of the principal axis of intermediate moment, and is therefore one of the ellipses which form the separating polhodes. When the point B is in contact with the fixed plane $LP = (OP^2 - \varpi^2)^{\frac{1}{2}} = 0$, and so the radius of the inner limiting circle is zero.

Let the motion of the M.E. begin at any point of the polhode distant from the extremity of the axis OB at, say, the maximum value of OP; then the motion consists, as we have seen in 2 and 3, of a spin about the invariable line at angular speed 2T/H, and a turning about the line OG with angular speed $\omega \sin IOL$. If we suppose now the invariable plane to turn about OL with the M.E., the whole motion of the latter is summed up by saying that it is at each instant one of rolling of the M.E. about OG, supposed to be at the same time turning, with the invariable plane, about the invariable line, with angular speed 2T/H, so that OG, fixed in the invariable plane, is at each instant in the plane LOP.

As then the ellipsoid moves and its point of contact with the fixed plane approaches B, the motion becomes more and more nearly that due to the spin alone, and so the herpolhode consists of constantly diminishing arcs of a spiral closing down on a pole P. The spiral is given double by its geometrical definition, but only one half of it is described by the point of contact.

8. *Variation of the radius vector of the herpolhode with time.* Denoting by $\rho$ the distance LP, the radius vector of the herpolhode from L taken as origin, and putting as before $x, y, z$ for the coordinates of P, with reference to the principal axes of the moving ellipsoid, we have

$$x^2+y^2+z^2=\rho^2+\varpi^2, \quad Ax^2+By^2+Cz^2=1, \quad \varpi^2(A^2x^2+B^2y^2+C^2z^2)=1. \quad \text{...(1)}$$

Hence, solving these equations, we get

$$x^2, y^2, z^2 = \{BC(C-B)(\rho^2-\alpha),\ CA(A-C)(\rho^2-\beta),\ AB(B-A)(\rho^2-\gamma)\}\frac{1}{(A-B)(B-C)(C-A)}, \quad \text{......(2)}$$

where
$$\alpha = -\frac{(\varpi^2 B-1)(\varpi^2 C-1)}{\varpi^2 BC}, \quad \beta=\cdots, \quad \gamma=\cdots. \quad \text{................(3)}$$

Connecting $x, y, z$ with $p, q, r$ we have by (4), 2 (with $k=1$),

$$p/x = q/y = r/z = \omega/OP = (2T)^{\frac{1}{2}}.$$

Hence also $\dot{p}/\dot{x} = \dot{q}/\dot{y} = \dot{r}/\dot{z} = (2T)^{\frac{1}{2}}$, and we can write Euler's equations in the form

$$\left.\begin{aligned} A\dot{x}-(2T)^{\frac{1}{2}}(B-C)yz=0, \quad B\dot{y}-(2T)^{\frac{1}{2}}(C-A)zx=0,\\ C\dot{z}-(2T)^{\frac{1}{2}}(A-B)xy=0.\end{aligned}\right\} \quad \text{.........(4)}$$

These give
$$x\dot{x}+y\dot{y}+z\dot{z} = -(2T)^{\frac{1}{2}}\frac{(A-B)(B-C)(C-A)}{ABC}xyz. \quad \text{.............(5)}$$

But since $\varpi^2$ is constant and $x^2+y^2+z^2=\rho^2+\varpi^2$, the quantity on the left of (5) is $\rho\dot{\rho}$. From this and the values of $x, y, z$ given by (2), we get

$$\rho\dot{\rho}=(2\mathrm{T})^{\frac{1}{2}}\{-(\rho^2-\alpha)(\rho^2-\beta)(\rho^2-\gamma)\}^{\frac{1}{2}}, \quad \text{......................(6)}$$

from which $\rho^2$ can be found in terms of $t$. It is to be remembered that here $k$ has been taken as unity, which explains the apparent difference in the "dimensions" of the expressions on the two sides of (6). To restore $k$ we have to multiply the right-hand side by $k^2$.

Now it is proved in (4), 2, that (if $k=1$) $\omega^2=2\mathrm{T}\,.\,\mathrm{OP}^2$, so that $\omega\dot{\omega}=2\mathrm{T}\rho\dot{\rho}$. Hence (6) becomes

$$\omega\dot{\omega}=(2\mathrm{T})^{\frac{3}{2}}\{-(\rho^2-\alpha)(\rho^2-\beta)(\rho^2-\gamma)\}^{\frac{1}{2}}. \quad \text{......................(7)}$$

We may write this also in the form

$$\omega\dot{\omega}=\{-(\omega^2-\omega_\alpha{}^2)(\omega^2-\omega_\beta{}^2)(\omega^2-\omega_\gamma{}^2)\}^{\frac{1}{2}}. \quad \text{......................(8)}$$

For we have $\rho^2=r^2-\varpi^2=\omega^2/2\mathrm{T}-\varpi^2$, so that the equation has the form stated if

$$\omega_\alpha{}^2=2\mathrm{T}(\varpi^2+\alpha), \quad \omega_\beta{}^2=\cdots, \quad \omega_\gamma{}^2=\cdots \quad \text{......................(9)}.$$

The value assigned to $k$ does not affect (8).

**9.** *The differential equation of the herpolhode.* The differential equation of the herpolhode can be found in the following manner. The double rate of description of areas by the projection of the radius vector OP on the plane of $yz$, is, by (4), 8,

$$y\dot{z}-\dot{y}z=(2\mathrm{T})^{\frac{1}{2}}\left(\frac{\mathrm{A}-\mathrm{B}}{\mathrm{C}}y^2-\frac{\mathrm{C}-\mathrm{A}}{\mathrm{B}}z^2\right)x. \quad \text{..................(1)}$$

From this the other components of the double rate of description of areas can be at once written down by symmetry.

The components $y\dot{z}-\dot{y}z$, etc., give a component on any plane parallel to the invariable plane. Let then $l, m, n$ be the direction cosines of the normal to the invariable plane at P. We have $l, m, n=\varpi(\mathrm{A}x, \mathrm{B}y, \mathrm{C}z)$. If $\phi$ be the vectorial angle of the herpolhode corresponding to the radius vector $\rho$,

$$\begin{aligned}\rho^2\dot{\phi}&=l(y\dot{z}-\dot{y}z)+m(z\dot{x}-\dot{z}x)+n(x\dot{y}-\dot{x}y)\\&=\varpi(2\mathrm{T})^{\frac{1}{2}}\left\{\mathrm{A}x^2\left(\frac{\mathrm{A}-\mathrm{B}}{\mathrm{C}}y^2-\frac{\mathrm{C}-\mathrm{A}}{\mathrm{B}}z^2\right)+\mathrm{B}y^2\left(\frac{\mathrm{B}-\mathrm{C}}{\mathrm{A}}z^2-\frac{\mathrm{A}-\mathrm{B}}{\mathrm{C}}x^2\right)\right.\\&\qquad\left.+\mathrm{C}z^2\left(\frac{\mathrm{C}-\mathrm{A}}{\mathrm{B}}x^2-\frac{\mathrm{B}-\mathrm{C}}{\mathrm{A}}y^2\right)\right\}. \quad \text{.......(2)}\end{aligned}$$

But

$$\begin{aligned}\mathrm{A}x^2\left(\frac{\mathrm{A}-\mathrm{B}}{\mathrm{C}}y^2-\frac{\mathrm{C}-\mathrm{A}}{\mathrm{B}}z^2\right)&=\frac{\mathrm{A}x^2}{\mathrm{BC}}\{\mathrm{A}(\mathrm{A}x^2+\mathrm{B}y^2+\mathrm{C}z^2)-(\mathrm{A}^2x^2+\mathrm{B}^2y^2+\mathrm{C}^2z^2)\}\\&=\frac{1}{\varpi^2}\frac{\mathrm{A}x^2}{\mathrm{BC}}(\varpi^2\mathrm{A}-1);\end{aligned}$$

and corresponding values hold for the other terms on the right of (2), and so (2) becomes

$$\rho^2\dot{\phi}=\frac{(2\mathrm{T})^{\frac{1}{2}}}{\varpi}\left(\mathrm{A}x^2\frac{\varpi^2\mathrm{A}-1}{\mathrm{BC}}+\mathrm{B}y^2\frac{\varpi^2\mathrm{B}-1}{\mathrm{CA}}+\mathrm{C}z^2\frac{\varpi^2\mathrm{C}-1}{\mathrm{AB}}\right). \quad \text{..........(3)}$$

By the values of $x^2$, $y^2$, $z^2$ in (2), 8, this reduces to

$$\rho^2\dot{\phi}=\varpi(2T)^{\frac{1}{2}}(\rho^2+E), \quad \text{.......(4)}$$

where $$E=(\varpi^2A-1)(\varpi^2B-1)(\varpi^2C-1)\frac{1}{\varpi^4 ABC}=-(-\alpha\beta\gamma)^{\frac{1}{2}}\frac{1}{\varpi}, \quad \text{.......(5)}$$

by (3), 8.

By the value of $\rho\dot{\rho}$, given in (6), 8,

$$\frac{\dot{\phi}}{\dot{\rho}}=\frac{d\phi}{d\rho}=-\frac{\varpi(\rho^2+E)}{\rho\{-(\rho^2-\alpha)(\rho^2-\beta)(\rho^2-\gamma)\}^{\frac{1}{2}}}. \quad \text{.......(6)}$$

**10.** *Radius of curvature of the herpolhode.* From (4), 9, we might calculate the radius of curvature of the herpolhode for any point, and verify that it can nowhere be infinite, so that the curve cannot have a point of inflexion. The result which is, as suggested above, almost obvious, may also be established by a process due to M. de Saint Germain (*Comptes rendus*, 100, 1885), which is less laborious than a direct calculation of curvature. At a given instant two generators of the space and body cones are in contact. Along these, from the common vertex of the cones, take a length $OP=\omega$, and through P draw a plane cutting both cones at right angles to the coincident generators, and let R, R′ be the radii of curvature of the body-cone and the space-cone, respectively, in this normal plane of section. If the distance travelled in $dt$ by P, taken as fixed on the body-cone, be $ds$, we have

$$\omega\, dt=ds\left(\frac{1}{R}+\frac{1}{R'}\right). \quad \text{.......(1)}$$

If R′ be infinite at P we have $$\omega\, dt=\frac{ds}{R}, \quad \text{or} \quad \omega=\frac{s}{R}. \quad \text{.......(2)}$$

This relation is impossible. To prove this statement take an adjacent generator OQ, so that $PQ=ds$. Then the area of the triangle OPQ is $\frac{1}{2}\omega\, ds$, *if* OP *represents* $\omega$. But the same area may be expressed in terms of the coordinates

$$0,\ 0,\ 0,\quad p_0,\ q_0,\ r_0,\quad p_0+\dot{p}\,dt,\ q_0+\dot{q}\,dt,\ r_0+\dot{r}\,dt$$

of O, P, Q, and the equation $$\omega^2\dot{s}^2=\frac{a^2p^2+b^2q^2+c^2r^2}{A^2B^2C^2}, \quad \text{.......(3)}$$

where $a, b, c=A(2AT-H^2), \ldots, \ldots,$ is obtained.

Changing the origin to P, and taking two other points, M, N, on the normal section, through P, of the body-cone, and distant $ds$ and $ds+d^2s$ from P, we find the distance of the point N from the tangent plane to the cone at P, and hence that the angle between the tangent to the section at M, and the tangent at N is

$$2(ap_0d^2p+bq_0d^2q+cr_0d^2r)\frac{1}{ds\,.\,N},$$

where $N=(a^2p_0{}^2+b^2q_0{}^2+c^2r_0{}^2)^{\frac{1}{2}}$.

But from the fact that PM is perpendicular at once to OP and to the normal to the surface at P, so that the relations

$$p_0dp+q_0dq+r_0dr=0, \quad ap_0dp+bq_0dq+cr_0dr=0$$

hold, and by the equation of the body-cone, which is now

$$ap^2+bq^2+cr^2=0,$$

it can be shown that the value of the angle just found can also be written $ds\,.\,abc\omega^2/N^3$. But the angle is also $ds/R$, and so we have

$$R=\frac{N^3}{abc\omega^2}. \quad \text{.......(4)}$$

But if R′ were infinite we should have by this result $s = N^3/\omega abc$, and by (3) also $s = N/\omega ABC$.

Thus the necessary condition is $N^2 = \dfrac{abc}{ABC}$.

The equation of the surface and the values of 2T and $H^2$ give three equations which enable $N^2$ to be expressed in terms of $p^2$, or in terms of $r^2$. It will then be seen that $N^2$ cannot have the value here stated, and a point of inflexion on the herpolhode is impossible. [This is necessarily true only for the *ordinary* herpolhode here considered.]

It is impossible also that R can be zero: this follows from the equations for 2T and $H^2$. A point of sudden change of direction on the herpolhode would be characterised by $R' = 0$, and this would involve also $R = 0$.

**11.** *A special case of the herpolhode.* In the case referred to above, in which $\varpi^2 = 1/B$, the differential equation (6), 9, reduces to

$$\frac{d\phi}{d\rho} = \frac{1}{\rho B^{\frac{1}{2}}(\beta - \rho^2)^{\frac{1}{2}}}.$$

Substituting $\rho = 1/y$, we get

$$-\frac{d\phi}{dy} = \frac{1}{B^{\frac{1}{2}}} \frac{1}{(\beta y^2 - 1)^{\frac{1}{2}}} \quad \text{and} \quad \frac{d^2 y}{d\phi^2} = B\beta y,$$

which gives

$$\frac{1}{\rho} = \frac{e^{(B\beta)^{\frac{1}{2}}\phi} + e^{-(B\beta)^{\frac{1}{2}}\phi}}{2\beta^{\frac{1}{2}}}.$$

As $\phi$ is supposed to start from an apse of the curve, where $d\phi/d\rho = \infty$, and there $\rho = \beta^{\frac{1}{2}}$, both terms on the right have the same coefficient $1/2\beta^{\frac{1}{2}}$.

**12.** *Stability of the motion of a symmetrical body under no forces.* A result of interest in gyrostatics can now be deduced from Euler's equations of motion for a rigid body turning about a fixed point under the action of no forces. The motion is, in a certain sense, stable when the axis of greatest or least moment of inertia is the instantaneous axis. This is of importance in the case of a body thrown into the air, such as a quoit or an elongated projectile, and left to move under the action of gravity. If gravity alone acts there is no couple on the body, and the translational motion and acceleration may be ignored.

We suppose then that the body is symmetrical about an axis of figure, and that the moment of inertia about that axis is either greater or less than the moment of inertia about any other axis through the centroid. A quoit is an example of the former case, a rifle bullet or a spear spinning about its longitudinal axis is an example of the latter. Spears do not seem to be so thrown, but a juggler, when he throws knives from hand to hand, sometimes at least, spins them in this way.

If the axis of rotation coincide with OC, so that $p = q = 0$, the equations of motion are

$$A\dot{p} = 0, \quad B\dot{q} = 0, \quad C\dot{r} = 0. \qquad \text{(1)}$$

If however the axis of resultant angular velocity deviate slightly from OC, so that the angular speeds are $p$, $q$, $r_0 + r'$ where $p$, $q$, $r'$ are small, we can prove that under certain circumstances $p$, $q$ can never become large.

If products of small quantities be neglected; the equations of motion are now

$$A\dot{p}-(B-C)qr_0=0,\quad B\dot{q}-(C-A)r_0p=0,\quad C\dot{r}'=0. \qquad (2)$$

Differentiating the second equation, and eliminating $\dot{p}$ between the result and the first equation, we obtain

$$\ddot{q}+\frac{(C-A)(C-B)}{AB}r_0^2q=0. \qquad (3)$$

Now $(C-A)(C-B)$ is positive if C is either the greatest or the least of the three principal moments of inertia. If this condition is fulfilled we have

$$\ddot{q}+n^2q=0, \qquad (4)$$

where $n^2\ [=(C-A)(C-B)r_0^2/AB]$ is real and positive. For initial values $q_0$ and $\dot{q}_0$ of $q$ and $\dot{q}$ we get the solution

$$q=q_0\cos nt+\frac{\dot{q}_0}{n}\sin nt.$$

But initially $\dot{q}=(C-A)p_0r_0/B$, so that

$$q=q_0\cos nt+\frac{C-A}{Bn}r_0p_0\sin nt. \qquad (5)$$

Hence, if $p_0$, $q_0$ be small initially, $q$ cannot acquire more than the small value given by (5), and a similar result can be obtained for $p$. The instantaneous axis thus remains in the vicinity of OC.

By substituting for $\dot{q}$ from (5) in the equation $B\dot{q}-(C-A)r_0p=0$, we obtain

$$p=p_0\cos nt+\frac{Bn}{A-C}\frac{q_0}{r_0}\sin nt. \qquad (6)$$

Now we have

$$\tan COI=\frac{(p^2+q^2)^{\frac{1}{2}}}{r_0},\quad \tan COL=\frac{(A^2p^2+B^2q^2)^{\frac{1}{2}}}{Cr_0}. \qquad (7)$$

In the case we have been considering these tangents are both small. According as C is the greatest or the least moment of inertia, the fixed cone lies within or without the moving cone.

13. *Space-cone and body-cone according as* C *is the greatest or the least moment of inertia.* If C be the greatest or the least moment, and $A=B$, the instantaneous axis describes in the body a right cone round OC, the axis of figure, and this cone rolls on a right cone fixed in space. Here

$$n^2=\frac{(C-A)^2r_0^2}{A^2}\quad\text{or}\quad n=\frac{C-A}{A}r_0, \qquad (1)$$

and we have

$$p=p_0\cos nt-q_0\sin nt,\quad q=q_0\cos nt+p_0\sin nt. \qquad (2)$$

These may be written

$$p=R\cos(nt+\epsilon),\quad q=R\sin(nt+\epsilon), \qquad (3)$$

if $R=(p_0^2+q_0^2)^{\frac{1}{2}}$ and $\tan\epsilon=q_0/p_0$. The resultant of $p$ and $q$ is therefore an angular speed about an axis OD, which lies in the plane of the axes OA,

OB, and makes an angle $nt+\epsilon$ with OA. That angle increases at rate $n$, and OD moves round from OA in the direction of rotation about OC if $C > A$, and in the contrary direction if $C < A$.

It is clear from the Poinsot representation of the motion that as the M.E. (now of revolution) moves, the instantaneous axis is always inclined at the same angle to the invariable line OH. For, as we have seen in (1), 2,

$$\cos \mathrm{IOL} = \frac{2\mathrm{T}}{\omega \mathrm{H}},$$

and Euler's equations give

$$p\dot{p} + q\dot{q} + r\dot{r} = \tfrac{1}{2}\frac{d\omega^2}{dt} = 0, \quad \text{.........................} (4)$$

so that $\omega$ is constant.

Equation (7), 12, now becomes

$$\tan \mathrm{COI} = \frac{\mathrm{R}}{r_0}, \quad \tan \mathrm{COL} = \frac{\mathrm{AR}}{\mathrm{C}r_0}. \quad \text{.........................} (5)$$

The angle COL is greater or less than COI according as $C < A$ or $C > A$. In the former case the cones are external to one another, and roll with their convex surfaces in contact, in the latter the fixed cone lies within the moving cone, and the concave surface of the latter rolls on the convex surface of the former. See Fig. 108 above.

**14.** *Illustrations of the stability of a body under no forces.* As stated above, we have a good illustration in a well thrown quoit. A moderate rotation about the axis of figure is given, that rotation remains unchanged during the flight, except for the effect of air resistance, and the direction of the axis changes comparatively slowly, if at all. The action of the air is rendered perfectly regular, and the mark aimed at is more certainly reached. The cones have been already illustrated in Fig. 108. In the present case, if R have any sensible small value, the angles COI and COL are both small, but the latter is smaller than the former.

For a quoit C is not very different from 2A, and so $n$ is approximately equal to $r_0$. Thus the axis of figure OC (OA in Fig. 108) turns round the invariable line with nearly the angular speed $r_0$. The angle COL is small and approximately $\frac{1}{2}\mathrm{R}/r_0$. The larger this angle is the more the quoit has of the usual regular wobbling motion, which is simply the precession of the axis of figure about the invariable line.

With the foregoing case we contrast that in which C is the small moment of inertia, *e.g.* in the case of the knife thrown by the juggler. Here again there is stability for rapid rotation about the axis of figure. But when the body is long and slender $(\mathrm{A}-\mathrm{C})/\mathrm{A}$ is nearly equal to unity, so that again $n$ is nearly equal to $r_0$, but has the opposite sign. Now $\tan \mathrm{COI} = \mathrm{R}/r_0$ and $\tan \mathrm{COL} = \mathrm{AR}/\mathrm{C}r_0$, of which the latter is much the larger, and the cones lie as in Fig. 108 ($b$). In this case the wobble due to the motion of the axis of figure about the invariable line is very much more marked.

**15.** *Illustrations of stability of a top under no forces. Diabolo.* The top called diabolo, which was very popular some years ago, illustrates the principles set forth in the preceding articles. It consists of a kind of spool, constructed as shown in the diagram of frustums of two equal cones, put together with their axes in line and turned in opposite directions. The surface at the junction is slightly rounded out to receive the spinning cord.

This spool is supported with its axis of figure horizontal, or nearly so, on a vertical loop of cord suspended from two handles held by the operator, also as shown in the diagram. The spin is produced by successive strokes each made by raising the right handle (say) quickly and then lowering it, but more slowly. A friction couple about the axis of the spool is applied in the first upward motion. In the succeeding downward motion of this handle, which of course is accompanied by upward motion of the left handle, there is little or no couple applied to the spool; for the motion of the spool at the point of contact is now in the opposite direction to that of the cord, and the relative motion of the spool and cord at the point is a maximum.

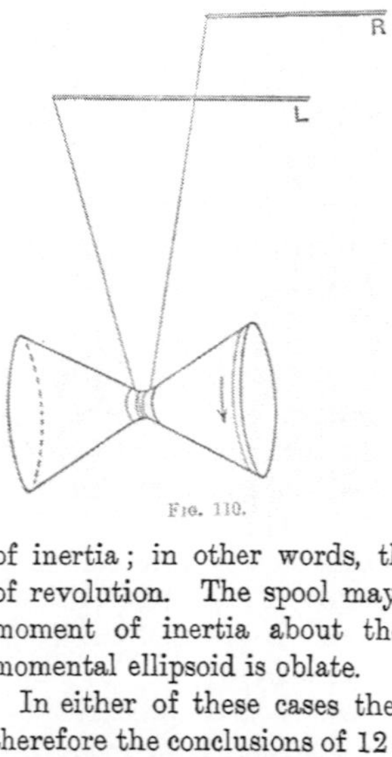

Fig. 110.

As the diabolo is usually constructed its axis of figure is an axis of minimum moment of inertia; in other words, the momental ellipsoid is a prolate ellipsoid of revolution. The spool may however be made so as to have maximum moment of inertia about the axis of symmetry, that is, so that the momental ellipsoid is oblate.

In either of these cases the rotational motion is stable, and illustrates therefore the conclusions of 12 and 13. The spool when spinning fast can be thrown into the air from the string, and preserves its axis unchanged in direction, so that it can easily be caught again on the string by the performer. Should a component of A.M. about an axis transverse to the axis of symmetry exist at any instant, the action of friction will, as explained above, tend to bring the axis of resultant A.M. into coincidence with the axis of figure.

If the spool be made so that the moment of inertia about the axis of figure is equal to that about any transverse axis through the centroid it will be found impossible to get up any spin in the ordinary way. For there is nothing to determine stability, and disturbing couples acting on the body will cause the instantaneous axis to change its position in the body without any tendency to approximate to the axis of A.M.

Two equal cones (or rather a single right circular cone, of equal sheets) made of uniform sheet metal, and united at their vertices with their axes in line, would, if the angle of the cone were $\tan^{-1} 2^{\frac{1}{2}}$, give a spool the momental ellipsoid of which would be a sphere. This ideal construction is difficult; for one thing we cannot unite the two cones if their vertices are sharp. The required result can however be obtained by making the momental ellipsoid oblate, that is with an excess of the moment of inertia C, about the axis of figure, over that A about a transverse axis, and providing it with a tubular hollow along this axis. A rod of wood filling this hollow symmetrically will leave the centroid unchanged in position, and will add to both the moments of inertia, but, unless it is very short, more to A than to C. If the rod is made of proper length the ellipsoid of inertia becomes a sphere. The diabolo, which could be spun without difficulty before, becomes incapable of stable spin after the insertion of the rod.

The equality of the moments of inertia about the principal axes may be tested by hanging the diabolo by a torsion wire, and observing the period (1) with the axis of figure vertical, (2) with that axis horizontal. The periods should not differ so much as 2 per cent.

This experiment is due to Mr. C. V. Boys [*Proc. Phys. Soc.*, Nov. 22, 1907].

If the stick is not inserted in the hollow symmetrically, the centroid will be shifted along the axis, and the moment of inertia about a former transverse axis will be increased by an amount $ma^2$, where $m$ is the mass of the stick and $a$ the distance of its centroid from that of the spool. If the difference is great enough there will be stable spin, together with precession about the vertical transverse axis, with angular speed $mga/\mathrm{C}n$. The vector of A.M. of the spool is to be drawn towards the operator, and turns of course in the precession towards the axis of the couple.

If the operator, after spinning the balanced diabolo with his right hand, draws that hand towards him, a couple about a downward vertical axis is applied, and the end of the spool near the operator turns downward. The contrary turning takes place if the operator draws in his left hand. Equal drawing in of both hands will apply a couple about a horizontal axis pointing towards his left, and the spool, if it has been spun with the right hand, will turn in azimuth in the clock direction as seen from above, that is the outer end of the spool will turn towards the performer's right.

The reader will easily see how these results are modified if the spin is in the opposite direction, that is, has been produced by the left hand.

**16.** *Stability when the body under no forces is unsymmetrical.* Returning to the general case of unequal moments of inertia, with $\mathrm{C} > \mathrm{B} > \mathrm{A}$, we consider the motion starting from approximate coincidence of the axis OI with OB. If the initial values of $p, q, r$ be $p_0, q_0, r_0$, $p_0, r_0$ are very small, and therefore $q$ varies slowly. We have the equations $\mathrm{A}\dot{p} - (\mathrm{B}-\mathrm{C})q_0 r = 0, \quad \mathrm{C}\dot{r} - (\mathrm{A}-\mathrm{B})pq_0 = 0,$ ..........................(1)

since we suppose that so short an interval of time has elapsed that $p$ and $r$ are still small and $q_0$ is little different from its initial value. Eliminating $\dot{p}$, we find

$$AC\ddot{r}-(A-B)(B-C)q_0^2 r=0. \quad \text{......(2)}$$

A – B and B – C are both negative, so that we can write this equation in the form

$$\ddot{r}-n^2 r=0, \quad \text{......(3)}$$

where $n^2=(A-B)(B-C)q_0^2/AC$, and is positive. Hence, if K, L are constants,

$$r=Ke^{nt}+Le^{-nt}. \quad \text{......(4)}$$

When $t=0$, $r_0=K+L$, and $\dot{r}_0=n(K-L)$, so that

$$K=\tfrac{1}{2}\left(r_0+\frac{\dot{r}_0}{n}\right), \quad L=\tfrac{1}{2}\left(r_0-\frac{\dot{r}_0}{n}\right).$$

Thus

$$r=\tfrac{1}{2}r_0(e^{nt}+e^{-nt})+\frac{\dot{r}_0}{2n}(e^{nt}-e^{-nt}). \quad \text{......(5)}$$

After a short interval of time $\tau$, we have

$$r=r_0(1+\tfrac{1}{2}n^2\tau^2)+\dot{r}_0\tau, \quad \text{......(6)}$$

to the second order of small quantities. There is no oscillation of the value of $r$.

Since

$$\dot{r}_0=\frac{A-B}{C}p_0 q_0,$$

we may, by the value of $n^2$, write (5) in the form

$$r=\tfrac{1}{2}r_0(e^{nt}+e^{-nt})+\tfrac{1}{2}\left\{\frac{A(A-B)}{C(B-C)}\right\}^{\frac{1}{2}}p_0(e^{nt}-e^{-nt}). \quad \text{......(7)}$$

From this, by the second of (1), we obtain

$$p=\tfrac{1}{2}p_0(e^{nt}+e^{-nt})+\tfrac{1}{2}\left\{\frac{C(B-C)}{A(A-B)}\right\}^{\frac{1}{2}}r_0(e^{nt}-e^{-nt}). \quad \text{......(8)}$$

Of course it is to be remembered that these integrals cannot be used except for small values of $t$, otherwise it will not be possible to consider $q$ as retaining sufficiently nearly its initial value $q_0$.

17. *Extension of Poinsot's theory to the motion of a top.* We now consider an extension of Poinsot's theory to a top spinning under gravity about a fixed point in the line of its axis of figure OC. If, as we shall suppose, the angular speed $n$ about OC be constant, and a distance OO′ be laid off on OC to represent $n$, the extremity P of the instantaneous axis lies in the plane of OC and the resultant of $p$ and $q$. As the top moves this plane moves also, and a curve is described in it by P, which shows how the instantaneous axis moves in the body. The angular speeds about the principal axes O(A, B, C) are $p, q, r$ $(r=n)$, so that we have $O'P\ (=\rho)=(p^2+q^2)^{\frac{1}{2}}=(\dot{\theta}^2+\dot{\psi}^2\sin^2\theta)^{\frac{1}{2}}$.

Apart from spin the kinetic energy is $\tfrac{1}{2}A(\dot{\theta}^2+\dot{\psi}^2\sin^2\theta)=\tfrac{1}{2}A(p^2+q^2)$. The energy equation is thus

$$\dot{\theta}^2+\dot{\psi}^2\sin^2\theta=\frac{2E}{A}-\frac{2Mgh}{A}\cos\theta, \quad \text{......(1)}$$

where E is the total energy, less the constant energy of rotation. Putting as at 10, V,

$$2E/A=\alpha \quad \text{and} \quad 2Mgh/A=a,$$

we get

$$\dot{\theta}^2+\dot{\psi}^2\sin^2\theta=p^2+q^2=\alpha-az. \quad \text{......(2)}$$

Thus we can write the energy equation in the compact form

$$\rho^2 = a - az \quad \text{or} \quad z = \frac{a-\rho^2}{a}. \qquad \text{.........(3)}$$

As we have seen (*loc. cit.*) the equation of A.M. about the vertical through O can be written

$$\dot\psi(1-z^2) = \beta - bnz, \qquad \text{.........(4)}$$

and elimination of $\dot\psi$ between this equation and (2) leads to the equation

$$\dot z^2 = (a-az)(1-z^2) - (\beta - bnz)^2 = f(z), \qquad \text{.........(5)}$$

which has already been discussed to some extent.

Now from (3) we have $\dot z = -2\rho\dot\rho/a$, and from (5),

$$\rho\dot\rho = -\tfrac{1}{2}a\left\{f\left(\frac{a-\rho^2}{a}\right)\right\}^{\frac{1}{2}}. \qquad \text{.........(6)}$$

To find the vectorial angle ($\chi$, say) corresponding to $\rho$, we calculate $\tan^{-1}(q/p)$. The angular speeds $\dot\theta$, $\dot\psi\sin\theta$, are those about OD, OE (Fig. 4, p. 48), and by (2), 2, IV

$$p = \dot\theta\sin\phi - \dot\psi\sin\theta\cos\phi, \quad q = \dot\theta\cos\phi + \dot\psi\sin\theta\sin\phi.$$

Hence

$$\tan\chi = \frac{\cot\phi + \dfrac{\dot\psi\sin\theta}{\dot\theta}}{1 - \dfrac{\dot\psi\sin\theta}{\dot\theta}\cot\phi}.$$

Now $\sin\theta/\dot\theta = -(1-z^2)/\dot z = -(1-z^2)/\{f(z)\}^{\frac{1}{2}}$ and $\dot\psi = (\beta - bnz)/(1-z^2)$, and therefore

$$\tan\chi = \tan\left\{\tfrac{1}{2}\pi - \phi - \tan^{-1}\frac{\beta - bnz}{\dot z}\right\}. \qquad \text{.........(7)}$$

Thus we may write

$$\chi = \tfrac{1}{2}\pi - \phi - \tan^{-1}\frac{\beta - bnz}{\dot z}. \qquad \text{.........(8)}$$

Differentiating, remembering that

$$\dot z = \{f(z)\}^{\frac{1}{2}} \quad \text{and} \quad \dot\phi = n - \dot\psi\cos\theta = n - (\beta - bnz)z/(1-z^2),$$

we get, after reduction,

$$(a-az)\dot\chi = \tfrac{1}{2}(bna - a\beta) + \tfrac{1}{2}n(b-2)(a-az),$$

that is, since $a - az = \rho^2$,

$$\rho^2\dot\chi = \tfrac{1}{2}n(b-2)\rho^2 + \tfrac{1}{2}(bna - ab). \qquad \text{.........(9)}$$

**18.** *The outer extremity of the* I.A. *for the top lies on a fixed spherical surface.* The locus of the extremity of the instantaneous axis is a sphere fixed in space. For, let $p_1, q_1, r_1$ be the angular speeds of the top referred to fixed axes, of which the vertical OZ is one, and let $r_1$ be the angular speed about OZ. Then we have $p_1^2 + q_1^2 + r_1^2 = p^2 + q^2 + n^2 = a - az + n^2$. But, clearly,

$$r_1 = \dot\psi\sin^2\theta + n\cos\theta = \beta + (1-b)nz. \qquad \text{.........(1)}$$

Eliminating $z$ between this equation and $p_1^2 + q_1^2 + r_1^2 = a - az + n^2$, we obtain

$$(1-b)n(p_1^2 + q_1^2 + r_1^2) + ar_1 - a\beta - (1-b)n(a+n^2) = 0. \qquad \text{.........(2)}$$

But $p_1, q_1, r_1$ are the coordinates of the extremity P of the instantaneous axis, and the equation just found shows that P describes in space a curve on a spherical surface the centre of which is on the vertical at a distance $\frac{1}{2}a/(1-b)n$ below O. The radius of the sphere is

$$\left[a+n^2+\frac{a}{(1-b)n}\left\{\beta+\frac{a}{4(1-b)n}\right\}\right]^{\frac{1}{2}}.$$

If the top be "spherical" $b=C/A=1$, and the equation of the surface reduces to

$$r_1=\beta, \qquad \text{.........(3)}$$

the equation of a plane at distance $\beta$ above the fixed point O.

The motion of the top is given by the rolling of a body-cone on a space-cone the generators of which are the lines joining O to the successive points of the spherical curve just defined. The curve given by the successive positions of P in the body is the polhode, and the successive positions of OP in the body make up the body-cone. This cone rolls on the space-cone and the polhode on the spherical curve.

**19.** *Reduction of the locus of the extremity of the* I.A. *to a plane. Spherical top.* If the top is "spherical" its motion is represented by a motion of the momental ellipsoid in which the polhode rolls on the curve in the plane $r_1=\beta$, to which the spherical curve now reduces. To find this curve let $\rho_1^2=p_1^2+q_1^2$, then, when $b=1$ and $r_1=\beta$,

$$\rho_1^2=\rho^2+n^2-\beta^2. \qquad \text{.........(1)}$$

Also, from Fig. 12, p. 69, we get

$$p_1=-\dot\theta\sin\psi+\dot\phi\sin\theta\cos\psi,\quad q_1=\dot\theta\cos\psi+\dot\phi\sin\theta\sin\psi,\quad r_1=\dot\psi+\dot\phi\cos\theta \quad (2)$$

[and when the top is spherical $\beta=r_1=\dot\phi\cos\theta+\dot\psi$]. Hence

$$\frac{q_1}{p_1}=\frac{\dot\theta\cos\psi+\dot\phi\sin\theta\sin\psi}{-\dot\theta\sin\psi+\dot\phi\sin\theta\cos\psi}=\frac{\cot\psi+\dfrac{\dot\phi\sin\theta}{\dot\theta}}{-1+\dfrac{\dot\phi\sin\theta}{\dot\theta}\cot\psi}=\tan\chi', \text{ say.}$$

Thus

$$\tan\chi'=\tan\left(\psi+\tfrac{1}{2}\pi-\tan^{-1}\frac{\dot\phi\sin\theta}{\dot\theta}\right). \qquad \text{.........(3)}$$

Now, returning to the spherical top, we have, by (1), $\rho_1\dot\rho_1=\rho\dot\rho$, and, since $z=(a-\rho^2)/a$, $f(z)=f\{(a-\rho_1^2+n^2-\beta^2)/a\}$.

Hence

$$\rho_1\dot\rho_1=-\tfrac{1}{2}a\left\{f\left(\frac{a-\rho_1^2+n^2-\beta^2}{a}\right)\right\}^{\frac{1}{2}}. \qquad \text{.........(4)}$$

Expanding the right-hand side to the form

$$f(z)=(a-az)(1-z^2)-(\beta-bnz)^2,$$

and writing $a'$ for $a+n^2-\beta^2$, we get, since $b=1$,

$$\rho_1^2\dot\rho_1^2=\tfrac{1}{4}[(\rho_1^2-n^2+\beta^2)\{a^2-(a'-\rho_1^2)^2\}-(a\beta-na'+n\rho_1^2)^2]. \qquad \text{......(5)}$$

Now let a line OL be drawn from O to represent, in magnitude and *axis*, the resultant A.M. of the top: the velocity of the point L represents the moment of the couple $Mgh\sin\theta$ about OD. The rate of production of A.M.

is therefore represented by the horizontal vector OD, if taken of proper length. [The rate of alteration of angular velocity is given by the motion of the point P in the plane $r_1 = \beta$.] The vector representing the component of $\omega$ at right angles to the vertical is $\rho_1$, and the radial and transversal components of the velocity of P are thus $\dot{\rho}_1$, $\rho_1\dot{\chi}'$. But the rate of change of $\omega$, since there is no change of the angular speed about the axis of figure, has the value $\mathrm{M}gh \sin\theta/\mathrm{A}$, or $\frac{1}{2}a\sin\theta$. Thus

$$\dot{\rho}_1^2 + \rho_1^2\dot{\chi}'^2 = \tfrac{1}{4}a^2\sin^2\theta = \tfrac{1}{4}a^2(1 - z^2). \quad \text{......(6)}$$

Hence, by (4),

$$\rho_1^4\dot{\chi}'^2 = \tfrac{1}{4}a^2\left\{(1 - z^2)\rho_1^2 - f\left(\frac{a' - \rho_1^2}{a}\right)\right\}. \quad \text{......(7)}$$

Here $z = (a' - \rho_1^2)/a$, and so we get after reduction, remembering that for a spherical top $b = 1$,

$$\rho_1^2\dot{\chi}' = \tfrac{1}{2}(an - a'\beta + \beta\rho_1^2). \quad \text{......(8)}$$

We shall call the curve, of which this is the polar equation, the curve $s$.

**20.** *The locus of the extremities of the* I.A. *for the top is a polhode. Jacobi's theorem.* Comparing (5), 19, with (6), 8, we see that $s$ is a herpolhode, in the plane $r_1 = \beta$, which would be described by a properly specified "ellipsoid" of inertia, constructed for an imaginary body turning about O under the action of no forces. In the present case the distance OP, taken along the instantaneous axis, represents $\omega$, the numerical measure of the resultant angular speed of the top; in the case discussed in 8 the multiplier $(2\mathrm{T})^{\frac{1}{2}}$ on the right-hand side of (6) is, in the notation used, $\omega/\mathrm{OP}$. Hence $\omega^2/\mathrm{OP}^2$ corresponds to the multiplier $\frac{1}{4}$ in (5). Thus, reducing to the same scale we see that the angular speed of rolling of the imaginary ellipsoid is half the angular speed of the top.

It is to be observed that what we call here an ellipsoid of inertia may not fulfil the conditions of ordinary ellipsoids of inertia and that, for completeness, it is necessary to generalise the notion of matter so far as to consider inertia as having either sign.

We have now seen that the curve $s$ is described by a body-cone $\mathrm{C}_b$, fixed in the body, rolling on a cone $\mathrm{C}_s$ fixed in space, and that it can also be described by a cone $\mathrm{C}'_b$, fixed in the momental ellipsoid of an imaginary body moving about the point O under the action of no forces, and turning at each instant about the instantaneous axis of the top, with half the top's angular speed. Now let a fourth cone $\mathrm{C}''_b$ roll on $\mathrm{C}_b$ so as to trace out by the relative motion *the polhode in the top*, which lies in the plane perpendicular to the axis of figure at distance $n$ from O. Further let the cones all move so that at each instant each is in contact along one generator of the space-cone.

It is a theorem of Jacobi,* stated however somewhat differently and given by him without proof, that the cones $\mathrm{C}'_b$ and $\mathrm{C}''_b$ are identical.

* Jacobi, *Werke*, Bd. 2, p. 480.

Analytical proofs of this theorem have been given by Halphen * and Darboux, † but the identification can be effected in the following more simple way suggested by M. de Saint Germain.‡

Describe a sphere, centre O, cutting the common generator in a point P, and in $P_1$, $Q_1$, $Q'_1$, $Q''_1$, the four generators which after time $dt$ should be in contact. The arcs $PP_1$, $PQ_1$, $PQ'_1$, $PQ''_1$ must all have the same length $ds$. Let then R be the radius of curvature of the space-cone, $R_1$, $R'_1$, $R''_1$ the radii of curvature of the others, taken in a plane drawn through P at time $t$, at right angles to the coincident generators. We get then

$$\omega=\dot{s}\left(\frac{1}{R}+\frac{1}{R_1}\right),\quad \tfrac{1}{2}\omega=\dot{s}\left(\frac{1}{R}+\frac{1}{R'_1}\right),\quad \tfrac{1}{2}\omega=\dot{s}\left(\frac{1}{R_1}+\frac{1}{R''_1}\right). \quad \text{.........(1)}$$

The first two equations give $\tfrac{1}{2}\omega=\dot{s}\left(\frac{1}{R_1}-\frac{1}{R'_1}\right)$,

so that by the last equation $\quad R'_1=-R''_1. \quad$ ..................................(2)

Since in the kinematic equations (1) the radii of curvature are taken positive when they are turned in opposite directions, that is when convex surface rolls on convex surface, we see that the radii $R'_1$, $R''_1$ in contact at their extremities are turned in the same direction and are equal. This holds for every element of the cones which come into contact, and so the cones are identical. Thus the polhode on $C'_b$, the body-cone for the momental ellipsoid of the imaginary body under no forces, rolling on the space-cone, gives the herpolhode which is the curve $s$, and the rolling motion of the same cone, relative to $C_b$, gives the polhode of the top's motion. A horizontal plane through O is the invariable plane for the absolute motion, while a plane at right angles to the axis OC is the moving (invariable plane) for the relative motion.

**21.** *Case of an unspherical top.* When the top is not spherical the motion can be reduced to that of a spherical top by choosing $n'=Cn/A\ (=bn)$, and using $n'$ as the speed of rotation, when the equations become those for a spherical top. Consider, then, a solid of revolution which has, with respect to the top, an angular speed $(b-1)n$, and its axis of figure always in the same direction as OC. To this body all the results just obtained apply, and to its motion that of the top exactly corresponds. The body-cone for this top is not the same as that of the actual top; for example if $A>C$, we have $n'<n$, and so if OP be the line representing the real angular speed $\omega$ at time $t$, OP′ will be that for the spherical top. The cone with OC′ as representing $n$ will roll on a corresponding space-cone [OC is not necessarily vertical].

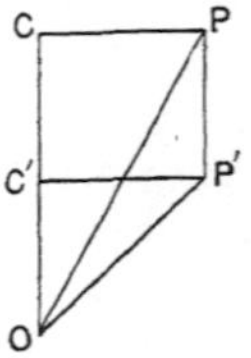

FIG. 111.

* *Journ. de Mathem.*, t. 2, 1886. † *Comptes rendus*, t. 100, p. 1065.

‡ *Résumé de la théorie du mouvement d'un solide autour d'un point fixe.*

**22.** *Passage from one Poinsot movement to another.* If we write $D=H^2/2T$ and $h=2T/H$ the equations of energy and A.M. become

$$Ap^2+Bq^2+Cr^2=Dh^2, \quad A^2p^2+B^2q^2+C^2r^2=D^2h^2, \quad \text{.......................(1)}$$

or, if also we write $Aa=Bb=Cc=Dh$,

$$\frac{p^2}{a}+\frac{q^2}{b}+\frac{r^2}{c}=h, \quad \frac{p^2}{a^2}+\frac{q^2}{b^2}+\frac{r^2}{c^2}=1. \quad \text{..................(2)}$$

The dynamical equations are, in this notation, which is due to Darboux,

$$\dot{p}-a\left(\frac{1}{b}-\frac{1}{c}\right)qr=0, \quad \dot{q}-b\left(\frac{1}{c}-\frac{1}{a}\right)rp=0, \quad \dot{r}-c\left(\frac{1}{a}-\frac{1}{b}\right)pq=0. \quad \text{..............(3)}$$

Consider now for a moment the surface

$$Ax^2+By^2+Cz^2=D\delta^2. \quad \text{........................................(4)}$$

The direction cosines of the normal drawn at the point $x, y, z$ are

$$\varpi\frac{Ax,\ By,\ Cz}{D\delta^2}=\frac{p}{a},\ \frac{q}{b},\ \frac{r}{c}, \quad \text{.................(5)}$$

where $\varpi$ is the length of the perpendicular from the centre on the tangent plane at $x, y, z$.

Instead of the dynamical equations (3) we may use the equations

$$\dot{p}-\alpha qr=0, \quad \dot{q}-\beta rp=0, \quad \dot{r}-\gamma pq=0, \quad \text{...........................(6)}$$

which agree with (3) if

$$\alpha=\frac{a(c-b)}{bc}, \quad \beta=\frac{b(a-c)}{ca}, \quad \gamma=\frac{c(b-a)}{ab}. \quad \text{................ (7)}$$

In this case $\alpha, \beta, \gamma$ satisfy the condition

$$\alpha+\beta+\gamma+\alpha\beta\gamma=0. \quad \text{......................................(8)}$$

If it is satisfied we say that equations (6) represent a Poinsot movement. [Of course $\alpha$, $\beta$ are not to be confused in meaning with the letters $\alpha$, $\beta$ of 16, above.]

From one Poinsot movement another can be derived. For writing

$$p'=\alpha'p, \quad q'=\beta'q, \quad r'=\gamma'r,$$

we obtain for the new coefficients $\alpha_1, \beta_1, \gamma_1$ of equations (6) for the movement $p', q', r'$,

$$\alpha_1=\left(\frac{a}{b}-\frac{a}{c}\right)\frac{\alpha'_1}{\beta'\gamma'}, \quad \beta_1=\left(\frac{b}{c}-\frac{b}{a}\right)\frac{\beta'_1}{\gamma'\alpha'}, \quad \gamma_1=\left(\frac{c}{a}-\frac{c}{b}\right)\frac{\gamma'_1}{\alpha'\beta'}. \quad \text{..................(9)}$$

If this is also a Poinsot movement (8) must be satisfied by the values of $\alpha_1, \beta_1, \gamma_1$ found in (9). The condition for this is found to be

$$a^2(c-b)(\alpha'^2-1)+b^2(a-c)(\beta'^2-1)+c^2(b-a)(\gamma'^2-1)=0. \quad \text{..................(10)}$$

Thus, for a change from one Poinsot movement to another, only a single relation requires to be satisfied by the multipliers $\alpha', \beta', \gamma'$, so that two of these multipliers may be arbitrarily chosen.

The relation (10) is obviously satisfied by $\alpha'=\beta'=\gamma'=-1$. Thus we pass from one Poinsot movement to another by mere reversal of $p, q, r$. If $a', b', c'$ be the constants for the new motion we have

$$\frac{p^2}{a'}+\frac{q^2}{b'}+\frac{r^2}{c'}=h', \quad \frac{p^2}{a'^2}+\frac{q^2}{b'^2}+\frac{r^2}{c'^2}=1. \quad \text{................................(11)}$$

If we multiply the first of (2) by $\lambda$, the second by $\mu$, and identify the sum of the products with the first of (11) we get

$$\frac{1}{a'}=\frac{\lambda}{a}+\frac{\mu}{a^2}, \quad \frac{1}{b'}=\frac{\lambda}{b}+\frac{\mu}{b^2}, \quad \frac{1}{c'}=\frac{\lambda}{c}+\frac{\mu}{c^2}, \quad \lambda h+u=h'. \quad \text{.....................(12)}$$

Similarly, multiplying by $\lambda_1$, $\mu_1$, and identifying the sum of products with the second of (11) we get
$$\frac{1}{a'^2}=\frac{\lambda_1}{a}+\frac{\mu_1}{a^2},\quad \frac{1}{b'^2}=\frac{\lambda_1}{b}+\frac{\mu_1}{b^2},\quad \frac{1}{c'^2}=\frac{\lambda_1}{c}+\frac{\mu_1}{c^2},\quad \lambda_1 h+\mu_1=1. \quad\ldots\ldots (13)$$

From (12) and (13) we have
$$\frac{a^2}{a'}=a\lambda+\mu,\quad \frac{a^2}{a'^2}=a\lambda_1+\mu_1, \quad\ldots\ldots (14)$$
and similar pairs of equations hold for $b$, $b'$ and $c$, $c'$.

Eliminating $a'$ and $a'^2$, we obtain for $a$ the cubic equation
$$\lambda_1 a^3-(\lambda^2-\mu_1)a^2-2\lambda\mu a-\mu^2=0 \quad\ldots\ldots (14')$$
An exactly similar equation holds when $b$, and also when $c$, replaces $a$. Hence $a$, $b$, $c$ are roots of the cubic
$$\lambda_1 x^3-(\lambda^2-\mu_1)x^2-2\lambda\mu x-\mu^2=0. \quad\ldots\ldots (15)$$
Thus we have the five equations
$$\lambda^2-\mu_1=\lambda_1(a+b+c),\quad -2\lambda\mu=\lambda_1(ab+bc+ca),\quad \mu^2=\lambda_1 abc,\quad \lambda h+\mu=h',\quad \lambda_1 h+\mu_1=1, \ldots (16)$$
for the determination of $\lambda$, $\mu$, $\lambda_1$, $\mu_1$, $h'$.

Solving these equations we obtain
$$\left.\begin{aligned}&\lambda=-\frac{2Q}{\Omega},\quad \mu=\frac{2R}{\Omega},\quad \lambda_1=\frac{4R}{\Omega^2},\quad \mu_1=\frac{Q^2-4PR}{\Omega^2},\\ &\text{where}\quad P=a+b+c,\quad Q=bc+ca+ab,\quad R=abc,\quad \Omega^2=Q^2-4R(P-h)=4\mu^2/\lambda_1^2.\end{aligned}\right\} \ldots\ldots (17)$$
From (17) we can now find $a'$, $b'$, $c'$, $h'$. Adopting (since $\alpha$, $\beta$, $\gamma$ are again at our disposal) the notation
$$\left.\begin{aligned}\alpha&=-bc+ca+ab=Q-2\frac{R}{a},\\ \beta&=\ \ bc-ca+ab=Q-2\frac{R}{b},\\ \gamma&=\ \ bc+ca-ab=Q-2\frac{R}{c},\end{aligned}\right\} \ldots\ldots (18)$$
we obtain from (12)
$$a'=\frac{a}{\alpha}\Omega,\quad b'=\frac{b}{\beta}\Omega,\quad c'=\frac{c}{\gamma}\Omega,\quad h'=\frac{Qh-2R}{\Omega} \qquad [\Omega=-2\mu/\lambda_1]. \ldots\ldots (19)$$

The dynamical equations for the motion fulfilling the conditions $p+p'=q+q'=r+r'=0$, are therefore
$$\dot{p}'-a'\left(\frac{1}{b'}-\frac{1}{c'}\right)qr=0,\quad \dot{q}'-b'\left(\frac{1}{c'}-\frac{1}{a'}\right)rp=0,\quad \dot{r}'-c'\left(\frac{1}{a'}-\frac{1}{b'}\right)pq=0. \ldots\ldots (20)$$
Comparing with (3), 22, we obtain the relations $a(1/b-1/c)=-a'(1/b'-1/c')$, etc., or
$$\frac{a}{b}+\frac{a'}{b'}=\frac{a}{c}+\frac{a'}{c'},\quad \frac{b}{c}+\frac{b'}{c'}=\frac{b}{a}+\frac{b'}{a'}. \ldots\ldots (21)$$

**23.** ***Passage back from the second movement to the first.*** The problem is thus completely solved of passing from the first of these associated movements to the second. If $\Omega'$, $\alpha'_1$, $\beta'_1$, $\gamma'_1$ be the corresponding quantities for passage back from the second to the first, we have
$$a=\frac{a'}{\alpha'}\Omega'=a\frac{\Omega\Omega'}{\alpha\alpha'},\quad b=\frac{b'}{\beta'}\Omega'=b\frac{\Omega\Omega'}{\beta\beta'},\quad c=\frac{c'}{\gamma'}\Omega'=c\frac{\Omega\Omega'}{\gamma\gamma'},$$
so that
$$\alpha\alpha'=\beta\beta'=\gamma\gamma'=\Omega\Omega'. \ldots\ldots (1)$$
We shall show presently that $\Omega=\Omega'$, so that
$$\frac{\alpha}{\Omega}=\frac{\Omega}{\alpha'},\quad \frac{\beta}{\Omega}=\frac{\Omega}{\beta'},\quad \frac{\gamma}{\Omega}=\frac{\Omega}{\gamma'}. \ldots\ldots (2)$$

We observe that by the values of P, Q, R [see (17), 22] we obtain from (18), 22,
$$\left.\begin{aligned}&\alpha+\beta+\gamma=Q,\quad \alpha\beta+\beta\gamma+\gamma\alpha=4PR-Q^2,\quad \alpha\beta\gamma=4PQR-Q^3-8R^2,\\ &\alpha^2=Q^2-4R(P-a),\quad \beta^2=Q^2-4R(P-b),\quad \gamma^2=Q^2-4R(P-c).\end{aligned}\right\} \ldots\ldots (3)$$

To prove that $\alpha\alpha' = \Omega\Omega' = \Omega^2$, we have, since $-\alpha = a(P-a) - R/a$,

$$\alpha\alpha' = \left\{a(P-a) - \frac{R}{a}\right\}(-b'c' + c'a' + a'b').$$

But, by (19), 22,

$$a'(b'+c') - b'c' = \left\{\frac{a}{\alpha}\left(\frac{b}{\beta} + \frac{c}{\gamma}\right) - \frac{b}{\beta}\frac{c}{\gamma}\right\}\Omega^2 = (-bc\alpha + ca\beta + ab\gamma)\frac{\Omega^2}{\alpha\beta\gamma}.$$

Inserting from (18), 22, the values of $\alpha$, $\beta$, $\gamma$, in the terms in brackets on the extreme right of the last equation, we find that the expression becomes

$$b^2c^2 - a^2(b-c)^2 = \{bc + a(b-c)\}\{bc - a(b-c)\}$$
$$= \left(Q - \frac{2R}{b}\right)\left(Q - \frac{2R}{c}\right) = -Q^2 + 2Qbc + 4Ra.$$

Thus
$$\alpha\alpha' = \left\{a(P-a) - \frac{R}{a}\right\}(-Q^2 + 2Qbc + 4Ra)\frac{\Omega^2}{\alpha\beta\gamma}. \quad \text{.........................(4)}$$

Multiplying out and reducing, we find that this becomes, by the values of P, Q, R,

$$\alpha\alpha' = (4PQR - Q^3 - 8R^2)\frac{\Omega^2}{\alpha\beta\gamma} = \Omega^2,$$

by (3). Or, the first factor on the right of (4) is $\alpha$, the second $\beta\gamma$.

Again we have, by (19), 22,

$$h' - a' = \frac{Qh - 2R}{\Omega} - \frac{a}{\alpha}\Omega = \frac{\alpha(Qh - 2R) - a\Omega^2}{\alpha\Omega}.$$

It will be found that this reduces to

$$h' - a' = -\frac{\beta\gamma}{\alpha\Omega}(h - a).$$

Thus we obtain the three relations,

$$\alpha^2\frac{h'-a'}{h-a} = \beta^2\frac{h'-b'}{h-b} = \gamma^2\frac{h'-c'}{h-c} = -\frac{\alpha\beta\gamma}{\Omega}. \quad \text{.............................(5)}$$

If $a$, $b$, $c$ are in algebraical order of magnitude it is clear by (2), 22 that $h$ lies between $a$ and $c$. The two quantities $Q^2 - 4R(P-a)$, $Q^2 - 4R(P-c)$, that is $\alpha^2$, $\gamma^2$, are essentially positive, and
$$Q^2 - 4R(P-a) > Q^2 - 4R(P-h) > Q^2 - 4R(P-c).$$

Thus $Q^2 - 4R(P-h)$, or $\Omega^2$, is essentially positive, and the expressions above in which $\Omega$ appears are real.

Again, the relations $\Omega^2 = \alpha\alpha' = \beta\beta' = \gamma\gamma'$ show that $\alpha'$, $\beta'$, $\gamma'$ have the same signs as $\alpha$, $\beta$, $\gamma$ respectively. Thus if the surface corresponding to the first motion is an ellipsoid of inertia, so also is the surface corresponding to the associated motion.

The discussion of two associated movements given above is founded on the Notes of Darboux in Despeyrous' *Cours de Mécanique*. A more general discussion was given by Darboux in the *Journal de Mathématiques* for 1885, but the relations found all reduce to those given above when the top is spherical. As all tops can be reduced to this form by the substitution indicated in 21, the restriction to the case of $p+p' = q+q' = r+r' = 0$, does not deprive the discussion of any needful generality.

**24.** *Body-cone and space-cone for associated movements.* Now consider the two surfaces, S, S′ [(4), 22], which correspond to the two motions. We may suppose for the moment that the two sets of principal axes are coincident. Along the system of axes thus given lay off coordinates $p$, $q$, $r$. This will give a point on a polhode. If the whole sequence of points be thus constructed for the successive positions of the surfaces S, S′ (which it must be remembered move in a manner depending on the body's motion), we shall have a polhode, and the lines joining the fixed point O to the points of the sequence will give a cone C, which can be used as the moving cone for either motion.

In the motion of the surface S the moving cone rolls on a cone A fixed in space, the base of which is a herpolhode H; in the motion of the other surface the moving cone rolls in the opposite direction, with the same speed at the same instant, on another space-cone B, the base of which is a herpolhode H′.

Now at a given instant $t_0$ let a generator $g_m$ of C, considered as *rolling on* A, be in contact with a generator $g_0$ of that cone; at time $t$ it will be in contact with another generator $g$. If at time $t_0$ the moving cone, considered as *rolling on* B, have its generator $g_m$ in contact with a generator $g'_0$ of that cone, at time $t$ it will be in contact with another generator $g'$. If now we suppose $g_0$ and $g'_0$ to be in contact at time $t_0$, and B to roll on A, with angular speed opposite to and twice as great as the angular speed of C on B, the generators $g$ and $g'$ will be in contact at time $t$. This motion of B with respect to A is of importance in the discussion of Jacobi's theorem regarding the motion of a top.

**25.** *The polhode as the intersection of two surfaces of the second degree.* The following proposition is also important in this connection. The intersection of two surfaces of the second degree which are concentric, and have their principal axes along the same straight lines, can be considered, in two different ways, as a polhode.

Let the equations of the surface be

$$\mathrm{A}x^2+\mathrm{B}y^2+\mathrm{C}z^2=\mathrm{D},\quad \mathrm{A}'x^2+\mathrm{B}'y^2+\mathrm{C}'z^2=\mathrm{D}'. \quad \text{.........................(1)}$$

The proposition will be proved if it can be shown that equations (1) can be combined linearly so as to give, first, a resultant identical with

$$\frac{x^2}{a}+\frac{y^2}{b}+\frac{z^2}{c}=h, \ldots \quad \text{...........................................(2)}$$

and, second, a resultant identical with

$$\frac{x^2}{a^2}+\frac{y^2}{b^2}+\frac{z^2}{c^2}=l^2. \quad \text{...............................................(3)}$$

Take the first linear combination of equations (1). Multiply the first of (1) by $\lambda$, the second by $\mu$, and add, and identify the coefficients of $x^2$, $y^2$, $z^2$ in the sum with those in (2). We obtain

$$\lambda\mathrm{A}+\mu\mathrm{A}'-\frac{1}{a}=0,\quad \lambda\mathrm{B}+\mu\mathrm{B}'-\frac{1}{b}=0,\quad \lambda\mathrm{C}+\mu\mathrm{C}'-\frac{1}{c}=0. \quad \text{...................(4)}$$

That these equations may hold simultaneously, the condition is

$$\begin{vmatrix} \mathrm{A}, & \mathrm{A}', & \frac{1}{a} \\ \mathrm{B}, & \mathrm{B}', & \frac{1}{b} \\ \mathrm{C}, & \mathrm{C}', & \frac{1}{c} \end{vmatrix}=0. \quad \text{.......................................(5)}$$

In the same way we obtain from (1) and (3) the other condition

$$\begin{vmatrix} \mathrm{A}, & \mathrm{A}', & \frac{1}{a^2} \\ \mathrm{B}, & \mathrm{B}', & \frac{1}{b^2} \\ \mathrm{C}, & \mathrm{C}', & \frac{1}{c^2} \end{vmatrix}=0. \quad \text{.......................................(6)}$$

Equations (5) and (6) are evidently of the form

$$\frac{\alpha}{a}+\frac{\beta}{b}+\frac{\gamma}{c}=0,\quad \frac{\alpha}{a^2}+\frac{\beta}{b^2}+\frac{\gamma}{c^2}=0, \quad \text{......................................(7)}$$

where $\alpha$, $\beta$, $\gamma$ [not the same quantities as in 22] are functions of A, B, C, A′, B′, C′. If we write $r=a/b$, $s=a/c$, and eliminate first $s$ then $r$ between the two equations (7), we get the quadratics for $r$ and $s$:

$$\left.\begin{aligned}\beta(\beta+\gamma)r^2+2\alpha\beta r+\alpha(\alpha+\gamma)=0,\\ \gamma(\beta+\gamma)s^2+2\alpha\gamma s+\alpha(\alpha+\beta)=0.\end{aligned}\right\} \quad\text{..............(8)}$$

The roots of each of these are real if

$$\alpha\beta\gamma(\alpha+\beta+\gamma)<0, \quad\text{..............(9)}$$

and the roots are equal if

$$\alpha\beta\gamma(\alpha+\beta+\gamma)=0. \quad\text{..............(10)}$$

Thus in general, if it is possible at all, it is possible in two ways for the curve of intersection of a given pair of surfaces to give a polhode, that is for a given curve there are, if (9) is satisfied, two real Poinsot motions. These coincide in the limiting case in which the roots are equal.

Examination shows that if (10) is satisfied the surface (2) is either a plane or a sphere, and the curve cannot in general be a polhode.

If two of the quantities $\alpha$, $\beta$, $\gamma$ have a zero sum, for example, if $\alpha+\beta=0$ the surface (2) has the equation

$$x^2+y^2=ah, \quad\text{..............(11)}$$

that is, it is a cylinder of revolution about the axis of $z$. In the Poinsot motion for this form of the moving surface this cylinder remains in contact along a fixed plane, and the generator in contact revolves with constant angular speed about the perpendicular let fall from the fixed point to the plane. This case corresponds to $r=1$, $s=0$; but the other roots for the same curve are $r=(\beta-\gamma)/(\beta+\gamma)$, $s=2\beta/(\beta+\gamma)$, and there is a corresponding Poinsot motion. The surface (2) in this case has the equation

$$\frac{x^2}{a}+\frac{y^2}{b}+\frac{z^2}{ab}(a+b)=h. \quad\text{..............(12)}$$

It will be observed that in this case we have

$$r+1=s \quad\text{or}\quad bc+ca-ab=0. \quad\text{..............(13)}$$

**26.** *Relation of the curve of intersection of two surfaces to a family of confocal surfaces.* We take next the following proposition: *The curve of intersection of two surfaces of the second degree, which have the same lines as principal axes, is normal to an infinite number of confocal surfaces of the second degree forming one of the three families of an orthogonal system.*

To prove this we observe that if $\alpha$, $\beta$, $\gamma$ have values which fulfil the equations (7), 25, and $a$, $b$, $c$ be three new constants suitably chosen, we obtain, taking the surfaces (1) of 22, as the given surfaces, considering a point $x$, $y$, $z$ of the curve, and eliminating in turn $z^2$ and $x^2$,

$$\frac{x^2}{\alpha}-\frac{y^2}{\beta}=a-b, \quad \frac{y^2}{\beta}-\frac{z^2}{\gamma}=b-c.$$

It is obviously possible to find a quantity $\rho$ such that

$$x^2=\alpha(a+\rho), \quad y^2=\beta(b+\rho), \quad z^2=\gamma(c+\rho). \quad\text{..............(1)}$$

The value of $\rho$ will vary from point to point of the curve. We see at once that the surface of which the equation is

$$\frac{x^2}{a+\rho}+\frac{y^2}{b+\rho}+\frac{z^2}{c+\rho}=\alpha+\beta+\gamma \quad\text{..............(2)}$$

passes through the point $x$, $y$, $z$ of the curve of intersection of (1).

Differentiating with respect to an infinitesimal step along the tangent to the curve we get from (1)

$$\frac{dx}{ds}=-\frac{\alpha}{x}\frac{d\rho}{ds},\quad \frac{dy}{ds}=-\frac{\beta}{y}\frac{d\rho}{ds},\quad \frac{dz}{ds}=-\frac{\gamma}{\rho}\frac{d\rho}{ds}. \quad\text{.........................(3)}$$

But $\alpha=x^2/(a+\rho)$, etc. Hence the values of $dx/ds$, $dy/ds$, $dz/ds$ are proportional to $x/(a+\rho)$, $y/(b+\rho)$, $z/(c+\rho)$. But so are also the cosines of the normal to the surface (2) at the point $x$, $y$, $z$. Hence the curve meets the surface (2) at right angles at that point.

If $\alpha+\beta+\gamma=0$ the surfaces are confocal cones, and the curve lies on a sphere. If this relation is not fulfilled the equation (2) may be written (by making $\alpha+\beta+\gamma=1$) in the form

$$\frac{x^2}{a+\rho}+\frac{y^2}{b+\rho}+\frac{z^2}{c+\rho}=1. \quad\text{......................................(4)}$$

Equations (1) prevent passage of any of the quantities $\alpha(a+\rho)$, $\beta(b+\rho)$, $\gamma(c+\rho)$ through zero from a positive to a negative value: hence, when the signs of $\alpha$, $\beta$, $\gamma$ are fixed, (4) represents only one of the three families of orthogonal surfaces, a member of each of which passes through $x$, $y$, $z$.

It is now clear that since the curve meets every member of the family (4) at right angles, it must be the intersection of two other real surfaces belonging to the other two families of the triply orthogonal system. The curve must therefore, by Dupin's theorem, be a line of curvature on each of these intersecting surfaces. We have therefore the theorem (according to Darboux, *loc. cit.*, due to M. de la Gournerie) that every curve which lies on two concentric surfaces of the second degree, the principal axes of which are along the same lines, can be considered as the intersection of two real confocal surfaces on each of which it is a line of curvature.

The intersecting confocal surfaces will have equations of the form

$$\frac{x^2}{a+\rho_1}+\frac{y^2}{b+\rho_1}+\frac{z^2}{c+\rho_1}=1,\quad \frac{x^2}{a+\rho_2}+\frac{y^2}{b+\rho_2}+\frac{z^2}{c+\rho_2}=1. \quad\text{...................(5)}$$

Now let $\alpha$, $\beta$, $\gamma$ be multipliers such that, *for any point of the path*,

$$x^2=\alpha(a+\rho),\quad y^2=\beta(b+\rho),\quad z^2=\gamma(c+\rho). \quad\text{..........................(6)}$$

Substituting in (5), subtracting one equation from the other, and reducing slightly, we find that $\alpha$, $\beta$, $\gamma$ must satisfy the conditions

$$\left.\begin{aligned}\frac{\alpha a}{(a+\rho_1)(a+\rho_2)}+\frac{\beta b}{(b+\rho_1)(b+\rho_2)}+\frac{\gamma c}{(c+\rho_1)(c+\rho_2)}=0,\\ \frac{\alpha}{(a+\rho_1)(a+\rho_2)}+\frac{\beta}{(b+\rho_1)(b+\rho_2)}+\frac{\gamma}{(c+\rho_1)(c+\rho_2)}=0.\end{aligned}\right\} \quad\text{.....................(7)}$$

Obviously these will be satisfied if

$$\alpha=\frac{(a+\rho_1)(a+\rho_2)}{(a-b)(a-c)},\quad \beta=\frac{(b+\rho_1)(b+\rho_2)}{(b-a)(b-c)},\quad \gamma=\frac{(c+\rho_1)(c+\rho_2)}{(c-a)(c-b)}. \quad\text{.............(8)}$$

**27.** *Determination of the parameters of the confocal surfaces.* Now consider the quadratic equation

$$\frac{\alpha}{a+u}+\frac{\beta}{b+u}+\frac{\gamma}{c+u}=0. \quad\text{......................................(1)}$$

It is easy to verify that this is satisfied by $u=\rho_1$, and by $u=\rho_2$, so that $\rho_1$, $\rho_2$ are the roots of the quadratic. Thus if $\alpha$, $\beta$, $\gamma$ are known $\rho_1$, $\rho_2$ are obtained.

If we multiply the second of (5), 26, by a constant, and add the product to the first equation, we obtain the general equation of the surfaces which pass through the intersection of the two surfaces represented by (5), 25. It may be written

$$\frac{(a-k)x^2}{(a+\rho_1)(a+\rho_2)}+\frac{(b-k)y^2}{(b+\rho_1)(b+\rho_2)}+\frac{(c-k)z^2}{(c+\rho_1)(c+\rho_2)}=1. \quad\text{.....................(2)}$$

If this surface roll with successive points of the curve of intersection in contact with a fixed plane, the distance from the (fixed) centre to the tangent plane must be the same for all such points. We have then to express the condition that this distance, $\varpi$, shall be invariable. From (2) we get

$$\frac{1}{\varpi^2}=\frac{(a-k)^2x^2}{(a+\rho_1)^2(a+\rho_2)^2}+\frac{(b-k)^2y^2}{(b+\rho_1)^2(b+\rho_2)^2}+\frac{(c-k)^2z^2}{(c+\rho_1)^2(c+\rho_2)^2}. \quad\text{...............(3)}$$

Substituting for $x^2$, $y^2$, $z^2$ from (6), 26, we find, by (1), 26, for the condition that the term involving $\rho$ should vanish, which, since the aggregate of terms in $\rho^2$ is zero by (8), 26, is the condition that $\varpi$ should be constant as the surface rolls,

$$\frac{(a-k)^2}{(a+\rho_1)(a+\rho_2)(a-b)(a-c)}+\frac{(b-k)^2}{(b+\rho_1)(b+\rho_2)(b-a)(b-c)}+\frac{(c-k)^2}{(c+\rho_1)(c+\rho_2)(c-a)(c-b)}=0. \quad (4)$$

This can be put in the simpler form

$$\frac{(\rho_1-k)^2}{(a+\rho_1)(b+\rho_1)(c+\rho_1)}=\frac{(\rho_2-k)^2}{(a+\rho_2)(b+\rho_2)(c+\rho_2)}, \quad\text{.......................(5)}$$

which is a quadratic equation for $k$, of which the roots can only be real if the ratio

$$\frac{(a+\rho_1)(b+\rho_1)(c+\rho_1)}{(a+\rho_2)(b+\rho_2)(c+\rho_2)}$$

is positive. Thus since the surfaces to which $\rho_1$, $\rho_2$ apply are two of the three surfaces of the triply orthogonal system which have $x$, $y$, $z$ as a common point, all the three factors in the numerator or denominator of this fraction must be positive and two of the other three negative. Thus the surfaces are an ellipsoid and a hyperboloid of two sheets. The surfaces normal to the curve must therefore be a family of hyperboloids of one sheet. The surfaces (5), 26, intersecting in the curve are imaginary unless the normal surfaces be such hyperboloids.

It is interesting to observe that the two rectilineal generators of the hyperboloid of one sheet which passes through the point $f$, $g$, $h$ in which the generators intersect, and is normal to the polhode curve, are normals at that point to the surfaces (2) on which the curve lies. For the normals to the two surfaces have the equations

$$\left.\begin{aligned}&\frac{x-f}{l}=\frac{y-g}{m}=\frac{z-h}{n},\\ \text{where}\quad & l=\frac{(a-k)f}{(a+\rho_1)(a+\rho_2)},\quad m=\frac{(b-k)g}{(b+\rho_1)(b+\rho_2)},\quad n=\frac{(c-k)h}{(c+\rho_1)(c+\rho_2)}.\end{aligned}\right\} \quad\text{..............(6)}$$

The ratios $l$, $m$, $n$ are proportional to the direction cosines of the normals, and there are really two sets, since $k$ has two values.

That these normals should be generators of the hyperboloid

$$\frac{f^2}{a+\rho}+\frac{g^2}{b+\rho}+\frac{h^2}{c+\rho}=\alpha+\beta+\gamma, \quad\text{................................(7)}$$

the conditions are

$$\left.\begin{aligned}&\frac{fl}{a+\rho}+\frac{gm}{b+\rho}+\frac{hn}{c+\rho}=0,\\ &\frac{l^2}{a+\rho}+\frac{m^2}{b+\rho}+\frac{n^2}{c+\rho}=0.\end{aligned}\right\} \quad\text{................................(8)}$$

Since $f^2=\alpha(a+\rho)$, $g^2=\beta(b+\rho)$, $h^2=\gamma(c+\rho)$, where $\alpha$, $\beta$, $\gamma$ have the values given in (8), 26, the first of (8) becomes

$$\frac{a-k}{(a-b)(a-c)}+\frac{b-k}{(b-a)(b-c)}+\frac{c-k}{(c-a)(c-b)}=0, \quad\text{.......................(9)}$$

which is identically true.

The other condition is simply (4), the condition that $\varpi$ should not vary as the surface rolls on which the polhode is traced.

**28.** *Motion of the axis of a top represented by a deformable hyperboloid.* This theorem of the representation of the normals to the surfaces, on which the polhode is traced, by the pairs of generators of the members of the family of hyperboloids of one sheet, which are normal to the polhode at their points of intersection with the curve, is of great importance for a representation of the motion of the axis of a top. But it is necessary for this representation that the hyperboloid should go through a continuous process of deformation from the extreme case in which the generators are flattened down in one plane, that of the focal ellipse

$$\frac{x^2}{a}+\frac{y^2}{b}=\alpha+\beta+\gamma, \quad \text{.......................................(1)}$$

to any other possible configuration which the generators can take, subject to the condition that the intersections of all pairs of generators of opposite systems remain always the same. This deformation cannot be carried out completely by a model made with rods in the ordinary way, as these must be of finite thickness, and cannot interpenetrate one another.

Sir George Greenhill describes the march of this deformation in the following theorem: A hyperboloid of one sheet is capable of a deformation in which the rectilineal generators remain rectilinear generators, and the lengths of the sides of all quadrilaterals formed by these generators and their unvarying points of crossing remain unaltered. If the centre and the axes remain fixed the successive hyperboloids will be confocals, and the trajectory of every point will be orthogonal to the system of confocal hyperboloids.

To prove this we note that the coordinates $f_1, g_1, h_1, f_2, g_2, h_2$ of the two points $P_1$, $P_2$ on a generating line both fulfil the equation of the surface

$$\frac{x^2}{a+\rho}+\frac{y^2}{b+\rho}+\frac{z^2}{c+\rho}=\alpha+\beta+\gamma, \quad \text{..................................(2)}$$

and since they lie on a tangent plane to the surface, also the equation

$$\frac{x_1x_2}{a+\rho}+\frac{y_1y_2}{b+\rho}+\frac{z_1z_2}{c+\rho}=\alpha+\beta+\gamma. \quad \text{..................................(3)}$$

Similar equations hold for the two points $P'_1$, $P'_2$ of coordinates $f'_1, g'_1, h'_1, f'_2, g'_2, h'_2$, on a generating line of the confocal

$$\frac{\xi^2}{a+\rho+\sigma}+\frac{\eta^2}{b+\rho+\sigma}+\frac{\zeta^2}{c+\rho+\sigma}=\alpha+\beta+\gamma. \quad \text{..........................(4)}$$

If these two points *correspond* to the former pair of points, that is be points which have been carried out from their initial positions to their final positions, along trajectories orthogonal to the successive surfaces, we have

$$\frac{f_1}{f'_1}=\left(\frac{a+\rho}{a+\rho+\sigma}\right)^{\frac{1}{2}}, \quad \frac{g_1}{g'_1}=\left(\frac{b+\rho}{b+\rho+\sigma}\right)^{\frac{1}{2}}, \quad \frac{h_1}{h'_1}=\left(\frac{c+\rho}{c+\rho+\sigma}\right)^{\frac{1}{2}}. \quad \text{..................(5)}$$

The ratios $f_2/f'_2$, $g_2/g'_2$, $h_2/h'_2$ have the same values.

The squares of the distances $\overline{P_1P_2}$, $\overline{P'_1P'_2}$ are

$$(f_1-f_2)^2+(g_1-g_2)^2+(h_1-h_2)^2, \quad (f'_1-f'_2)^2+(g'_1-g'_2)^2+(h'_1-h'_2)^2,$$

and by the preceding relations we have

$$\overline{P'_1P'_2}^2-\overline{P_1P_2}^2=\sigma\left\{\frac{(f_1-f_2)^2}{a+\rho}+\frac{(g_1-g_2)^2}{b+\rho}+\frac{(h_1-h_2)^2}{c+\rho}\right\}. \quad \text{.....................(6)}$$

The expression on the right is zero by equations (2) and (3).

Hence any quadrilateral made up of the portions $P_1P_2$, $P_2P_3$, $P_3P_4$, $P_4P_1$ of crossing generating lines of the two systems remains unaltered in the lengths of its sides as the deformation proceeds, and each point moves along a trajectory orthogonal to the successive surfaces.

**29.** *Forces acting on the body carried by the body-cone.* Now calling the body-cone B and the space-cone A, and going back to the motion of the cone B on A described in 24, in which the relative rotation of B on A has components $2(p, q, r)$, we recall that for the two motions, from which this total motion comes, we have the equations

$$\left.\begin{aligned}\frac{p^2}{a}+\frac{q^2}{b}+\frac{r^2}{c}=h, \quad \frac{p^2}{a^2}+\frac{q^2}{b^2}+\frac{r^2}{c^2}=1,\\ \frac{p'^2}{a'}+\frac{q'^2}{b'}+\frac{r'^2}{c'}=h', \quad \frac{p'^2}{a'^2}+\frac{q'^2}{b'^2}+\frac{r'^2}{c'^2}=1,\end{aligned}\right\} \quad \text{.........................(1)}$$

with $p+p'=q+q'=r+r'=0$. The two motions give herpolhodes in planes, the direction cosines of the normals to which are respectively $p/a$, $q/b$, $r/c$, $p'/a'$, $q'/b'$, $r'/c'$.

We now suppose that in the motion of B relative to A, B carries with it a moving body, and find the forces acting on this body. We shall suppose that the momental ellipsoid of this body is a sphere for all axes passing through O. It has been seen that if this ellipsoid is not a sphere, the motion may, by a substitution, be reduced to that case. If A be the moment of inertia about any diameter of the sphere, the components of the A.M. are $2Ap$, $2Aq$, $2Ar$. We shall suppose that the normal, the direction cosines of which we take as $p/a'$, $q/b'$, $r/c'$, is the axis of the body (or rather we shall call it the axis of the body), and that the perpendicular, the direction cosines of which are $p/a$, $q/b$, $r/c$, coincides with the vertical drawn upward. The axis of the body (the top in Jacobi's theorem) and the vertical are thus parallel to the two generators of the hyperboloid of one sheet, which passes through the point of contact $f$, $g$, $h$ of the herpolhode $s$ [see 19].

The angular speed about the axis of the body is

$$-\left(\frac{p}{a'}2p+\frac{q}{b'}2q+\frac{r}{c'}2r\right)=-2h',$$

while the A.M. about the vertical is

$$A\left(\frac{p}{a}2p+\frac{q}{b}2q+\frac{r}{c}2r\right)=2Ah-A\beta.$$

Clearly these quantities will be constant if there is no couple about the axis of the body, and no couple about the vertical. This will be the case if the body is under the action of gravity, and the centroid is on the axis of figure.

If $\theta$ be the inclination of the axis of the body to the vertical, we have

$$z=\cos\theta=\frac{p^2}{aa'}+\frac{q^2}{bb'}+\frac{r^2}{cc'}. \quad \text{..................................(2)}$$

If the momental ellipsoid of the body is really a sphere the kinetic energy is $2A(p^2+q^2+r^2)$. If the body has been made "spherical" by substitution this will not be the kinetic energy even if $p$, $q$, $r$ be the components of angular velocity given by the substitution, for the substitution consists in equating angular momenta. Calling the kinetic energy T we suppose that we can write

$$T=D'z+E', \quad \text{..........................................(3)}$$

where D′ and E′ are constants. This will be possible if we suppose that $D'=-Mgl$, where M is the mass of the body and $l$ is the distance of the centroid from O. Thus a body under gravity and supported at O will satisfy the conditions referred to above and have the energy equation

$$T+Mglz=E'. \quad \text{..........................................(4)}$$

E′ is the total energy.

If the top be truly spherical, that is if the momental ellipsoid be a sphere, so that the kinetic energy for the angular speeds $2(p, q, r)$ may be written $2A(p^2+q^2+r^2)$, we can show that the equation

$$p^2+q^2+r^2=-\tfrac{1}{2}Dz+\tfrac{1}{2}E, \quad \text{..................................(5)}$$

which is what (4) becomes in this case when we put $-D'=AD$, $E'=AE$, is a linear combination of the first two equations in (1) above, that is of

$$\frac{p^2}{a}+\frac{q^2}{b}+\frac{r^2}{c}=h,\quad \frac{p^2}{a^2}+\frac{q^2}{b^2}+\frac{r^2}{c^2}=1. \quad\text{.............................(6)}$$

We have
$$z=\cos\theta=\frac{p^2}{aa'}+\frac{q^2}{bb'}+\frac{r^2}{cc'}, \quad\text{.................................(6')}$$

and so, instead of (5), we can write

$$p^2\left(1+\frac{D}{2aa'}\right)+q^2\left(1+\frac{D}{2bb'}\right)+r^2\left(1+\frac{D}{2cc'}\right)=\tfrac{1}{2}E. \quad\text{.....................(7)}$$

Multiplying the first of (6) by $\lambda$ and the second by $\mu$, and subtracting the first product from the second, we obtain a linear combination of the two equations which we identify with (7). Thus we have

$$1+\frac{D}{2aa'}=\lambda\frac{h-a}{a^2}+\tfrac{1}{2}\frac{E}{a^2},\quad 1+\frac{D}{2bb'}=\lambda\frac{h-b}{b^2}+\tfrac{1}{2}\frac{E}{b^2},\quad 1+\frac{D}{2cc'}=\lambda\frac{h-c}{c^2}+\tfrac{1}{2}\frac{E}{c^2}. \quad\text{........(8)}$$

With this we have, if, as above, $A\beta$ be the A.M. about the vertical, and $n$ the angular speed about the axis of the top,

$$-h'=\tfrac{1}{2}n,\quad h=\tfrac{1}{2}\beta \quad\text{......................................(9)}$$

[It will be observed that $2D'$ in (3) above has the meaning assigned to $-aA$ in 17, so that when $a$ is used in this sense, it is not to be confounded with $a$ as used in (6). The symbol D as used in (5), (7), and (8) has the meaning $\frac{1}{2}a$ as $a$ is defined in 17.]

**30.** *Determination of the constants of the surfaces.* In (19), 22, above, it has been shown that

$$a'=\frac{a\Omega}{Q-2\dfrac{R}{a}},\ \text{etc.},$$

where $Q=bc+ca+ab$, $R=abc$, $\Omega=-\{Q^2-4R(P-h)\}^{\frac{1}{2}}$. Thus the first of (8), 29, becomes

$$2\Omega a^2+D\left(Q-2\frac{R}{a}\right)=2\lambda(h-a)\Omega+E\Omega,$$

and we have exactly similar equations in $b$ and $c$ from the second and third.

Hence it is clear that if we write this equation, with $x$ instead of $a$, in the form

$$2\Omega x^3+2\Omega\lambda x^2+(DQ-2\Omega\lambda h-\Omega E)x-2DR=0, \quad\text{.......................(1)}$$

we have a cubic equation the roots of which are $a$, $b$, $c$. We have therefore, first (by the last term), $\Omega=D$, and also

$$\lambda=-P,\quad E=2Ph-Q. \quad\text{...................................(2)}$$

It remains to show that the values of $a$, $b$, $c$, thus determined are real. By (19), 22, since $\Omega^2=D^2=Q^2-4R(P-h)$, and $h=\frac{1}{2}\beta$, $h'=-\frac{1}{2}n$,

$$h'=\frac{Qh-2R}{\Omega}=\frac{Q\beta-4R}{2\Omega}=-\tfrac{1}{2}n. \quad\text{...........................(3)}$$

Further, since $\Omega=D$,
$$D^2=Q^2-2R(2P-\beta). \quad\text{...................................(4)}$$

Also, by (2) and (3),
$$P\beta-Q=E,\quad Q\beta-4R=-nD. \quad\text{...........................(5)}$$

From these we obtain P, Q, R as expressed in the equations

$$\left.\begin{aligned}Q&=\frac{2D^2\beta+2nDE-nD\beta^2}{\beta^3-2nD-2\beta E},\\ P&=\frac{2D^2-nD\beta+E\beta^2-2E^2}{\beta^3-2nD-2\beta E},\\ 2R&=\frac{D^2(\beta^2-n^2)}{\beta^3-2nD-2\beta E}.\end{aligned}\right\} \quad\text{...........................(6)}$$

These values, substituted in the cubic (1), give $a$, $b$, $c$ in terms of the known quantities $\beta$, $n$, D, E by the process of solving that equation, which has now the form

$$x^3-Px^2+Qx-R=0. \qquad \text{.........(7)}$$

To settle the question of the reality of the roots of this cubic we notice that we have already obtained the result

$$\dot{z}^2=(\alpha-az)(1-z^2)-(\beta-bnz)^2, \qquad \text{.........(8)}$$

where $\alpha=(2E-Cn^2)/A$, $\beta=G/A$, $a=2Mgl/A$, $b=C/A$. But we are here dealing with a spherical top, so that $b=1$, and (8) becomes

$$\dot{z}^2=\left(\frac{2E}{A}-az\right)(1-z^2)-(\beta^2+n^2-2\beta nz). \qquad \text{.........(9)}$$

But we have, by (6′), 29, $\qquad \dot{z}=2\left(\frac{\dot{p}p}{aa'}+\frac{q\dot{q}}{bb'}+\frac{r\dot{r}}{cc'}\right),$

where of course $a$ is quite different from the $a$ in (9). Hence, by (3), 22, and the values of $a'$, $b'$, $c'$ given in (19), 22, and of $\alpha$, $\beta$, $\gamma$ given in (18) 22, we obtain

$$-\Omega\dot{z}=4\frac{(a-b)(b-c)(c-a)}{abc}pqr. \qquad \text{.........(10)}$$

From equations (6) and (6′), 29, we find

$$\frac{p^2}{a^2}=\frac{2ah-\alpha-\Omega z}{2(a-b)(a-c)},\quad \frac{q^2}{b^2}=\frac{2bh-\beta-\Omega z}{2(b-a)(b-c)},\quad \frac{r^2}{c^2}=\frac{2ch-\gamma-\Omega z}{2(c-a)(c-b)}, \qquad \text{.........(11)}$$

where $\alpha$, $\beta$, $\gamma$ have the values set forth in (18) of 22 above. Thus (8) becomes, by (10),

$$\tfrac{1}{2}\Omega^2\dot{z}^2=(+\alpha-2ah+\Omega z)(+\beta-2bh+\Omega z)(+\gamma-2ch+\Omega z). \qquad \text{.........(12)}$$

This can be shown to reduce to (9). The three roots $u_1$, $u_2$, $u_3$, say of the cubic equation $\dot{z}^2=0$, are thus

$$\frac{1}{\Omega}(-\alpha+2ah,\ \ -\beta+2bh,\ \ -\gamma+2ch).$$

Now, identically, $\qquad +\alpha-2ah=2h^2-2Ph+Q-2\frac{(h-a)(h-b)(h-c)}{h-a}. \qquad \text{.........(13)}$

If, to avoid confusion, we use here B in the sense assigned to $\beta$ in (6) and (8), we have, by (2), $\qquad 2h^2-2Ph+Q=\tfrac{1}{2}B^2-E,\quad 2(h-a)(h-b)(h-c)=\tfrac{1}{4}(B^3-2nD-2BE),$

so that we obtain $\qquad \Omega u_1=\tfrac{1}{2}B^2-E-\frac{B^3-2nD-2BE}{2B-4a}. \qquad \text{.........(14)}$

The other roots are given by exactly the same form of equation, with $b$ and $c$ respectively substituted for $a$. The roots are obviously real.

If we substitute for the variable $z$ on the right-hand side of (12), equated to zero (so that a cubic equation equivalent to $\dot{z}=0$ is obtained), the value of $u$ given by (14), we obtain a cubic in $a$, the three roots of which are the squares $a$, $b$, $c$ of the lengths of the axes of the quadric surface.

**31.** *Determination of the constants for the associated motion.* Darboux remarks that the quantities $a'$, $b'$, $c'$, which can be expressed in terms of $a$, $b$, $c$, might have been determined directly by taking as the basis of the discussion the equations

$$\frac{p^2}{a'}+\frac{q^2}{b'}+\frac{r^2}{c'}=h',\quad \frac{p^2}{a'^2}+\frac{q^2}{b'^2}+\frac{r^2}{c'^2}=1. \qquad \text{.........(1)}$$

If this had been done $-n$ would have been the value of B, and $-$B the value of $n$. Thus if we make the substitutions shown in the scheme

$$\begin{array}{cccc} a,\ b,\ c, & B, & n, & a',\ b',\ c', \\ a',\ b',\ c', & -n, & -B, & a,\ b,\ c, \end{array}$$

we shall obtain the equations determining $a', b', c'$, in terms of $a, b, c$. For example, instead of (14), 30, we should have

$$\Omega u = \tfrac{1}{2}n^2 - \mathrm{E} - \frac{n^3 - 2\mathrm{BD} - 2n\mathrm{E}}{2n + 4a'}. \quad \text{.........................(2)}$$

It was proved in 23 that $\Omega$ is not changed by passage from $a, b, c, h$ to $a', b', c', h'$. In fact we have

$$\Omega^2 = \mathrm{D}^2 = (ab + bc + ca)^2 - 4abc(a + b + c - h)$$
$$= (a'b' + b'c' + c'a')^2 - 4a'b'c'(a' + b' + c' - h').$$

It has already been shown that there is no difficulty in bringing under the scope of this discussion the general case in which the momental ellipsoid of the top is spherical. It has also been shown in 18 that the locus of the extremities of the instantaneous axis for the different positions of the top is a curve lying on a fixed sphere, which when the top is spherical reduces to a fixed horizontal plane, at distance $r_1 = \beta$ from the fixed point.

**32.** *Deformation of the hyperboloid of one sheet as the top moves.* We have now to examine the relation of the motion of the hyperboloid of one sheet to the motion of the top somewhat more particularly. Supposing a confocal surface to start from the focal ellipse, we can write its equation in the form

$$\frac{x^2}{\alpha^2 + \lambda} + \frac{y^2}{\beta^2 + \lambda} + \frac{z^2}{\lambda} = 1. \quad \text{.............................(1)}$$

We suppose $\lambda$ to be positive, so that the equation represents an ellipsoid of which the squares of the same axes are $\alpha^2 + \lambda$, $\beta^2 + \lambda$, $\lambda$. We suppose also in what follows that $\alpha^2 > \beta^2$.

For a hyperboloid of two sheets, passing through the same point $x, y, z$, and confocal with the ellipsoid, we have

$$\frac{x^2}{\alpha^2 + \nu} + \frac{y^2}{\beta^2 + \nu} + \frac{z^2}{\nu} = 1. \quad \text{..............................(2)}$$

Here $\nu$ is negative and $\alpha^2 > -\nu > \beta^2$.

A hyperboloid of one sheet passing through the same point $x, y, z$ and confocal with the other two surfaces is

$$\frac{x^2}{\alpha^2 + \mu} + \frac{y^2}{\beta^2 + \mu} + \frac{z^2}{\mu} = 1, \quad \text{.............................(3)}$$

where $\mu$ is negative and $-\mu < \beta^2$. Thus for the three surfaces we have the sequence of magnitudes

$$\infty > \lambda > 0 > \mu > -\beta^2 > \nu > -\alpha^2.$$

Now, as we have seen in 27, the intersection of the ellipsoid and the hyperboloid of two sheets is a polhode for the top to which the surfaces correspond in the manner explained above. These surfaces remain fixed relatively to one another as the top moves, and we have a new position of the hyperboloid of one sheet for each point on the curve of intersection.

The equation is (3), and the range of values is from 0 to $-\beta^2$. We take $x, y, z$ as the extremity H of the radius vector representing the angular momentum. Through that point pass the two generators of the hyperboloid which are parallel to the vertical and to the axis of the top for its position at the instant.

It is clear that $\lambda, \mu, \nu$ are the roots of the cubic equation

$$\frac{x^2}{a^2+k}+\frac{y^2}{\beta^2+k}+\frac{z^2}{k}=1, \quad \ldots\ldots\ldots\ldots(4)$$

for $k$. This fact enables us to find $x^2, y^2, z^2$ from (1), (2), (3) very readily. We write

$$\frac{x^2}{a^2+k}+\frac{y^2}{\beta^2+k}+\frac{z^2}{k}-1=\frac{(\lambda-k)(\mu-k)(\nu-k)}{(a^2+k)(\beta^2+k)k}. \quad \ldots\ldots\ldots\ldots(5)$$

Multiplying by $a^2+k$, and then putting $k=-a^2$, we get at once $x^2$, and a similar procedure gives $y^2$ and $z^2$. Thus we obtain

$$x^2=\frac{(\lambda+a^2)(\mu+a^2)(\nu+a^2)}{(a^2-\beta^2)a^2}, \quad y^2=\frac{(\lambda+\beta^2)(\mu+\beta^2)(\nu+\beta^2)}{(\beta^2-a^2)\beta^2}, \quad z^2=\frac{\lambda\mu\nu}{a^2\beta^2}. \ldots(6)$$

Adding and reducing, we obtain

$$x^2+y^2+z^2=a^2+\beta^2+\lambda+\mu+\nu=\mathrm{OH}^2. \quad \ldots\ldots\ldots\ldots(7)$$

The only quantity that varies in the passage from point to point of the line of intersection is $\mu$, which is therefore the only variable parameter.

**33.** *Summary of results.* The geometrical discussion contained in 22 ... 32 may be summarised by saying that a polhode on a quadric surface rolling on the horizontal invariable plane through C (Fig. 112), traces out on that plane the herpolhode H ($s$ of 19), while a second rolling quadric related to the former traces out, *by means of the same polhode as before*, the herpolhode H′ on the invariable plane drawn through C′ at right angles to the axis of the top, and therefore carried by the top in its motion.

The polhode is considered first as the intersection of the surfaces

$$\mathrm{A}x^2+\mathrm{B}y^2+\mathrm{C}z^2=\mathrm{D}\delta^2, \quad \mathrm{A}^2x^2+\mathrm{B}^2y^2+\mathrm{C}^2z^2=\mathrm{D}^2\delta^2, \quad \ldots\ldots\ldots\ldots(1)$$

or, as given in (2), 22,

$$\frac{p^2}{a}+\frac{q^2}{b}+\frac{r^2}{c}=h, \quad \frac{p^2}{a^2}+\frac{q^2}{b^2}+\frac{r^2}{c^2}=1. \quad \ldots\ldots\ldots\ldots(2)$$

The passage from one form to the other, and the connection with the notation employed in the earlier part of the chapter on Poinsot motions, may be effected by the equalities

$$x, y, z=m(p, q, r), \quad \mathrm{D}\delta^2=2m^2\mathrm{T}, \quad \mathrm{D}\delta=m\mathrm{H}, \quad \delta^2=m^2h^2, \quad \ldots\ldots(3)$$

where T is the kinetic energy of the motion and H the resultant A.M. about an axis through the fixed point.

The length of the perpendicular from the origin on the tangent plane at the point $x, y, z$ of the rolling surface is easily seen to be $\mathrm{D}\delta^2/\mathrm{D}\delta=\delta$. The first equation in (1) and in (2) is the equation of the rolling surface, the polhode in any of its positions is the line of intersection of (1) and (2). The squares of the semi-axes of the rolling quadric are

$$D\delta^2\left(\frac{1}{A}, \frac{1}{B}, \frac{1}{C}\right) = m^2h(a, b, c),$$

so that
$$D\left(\frac{1}{A}, \frac{1}{B}, \frac{1}{C}\right) = \frac{1}{h}(a, b, c). \quad \ldots\ldots\ldots\ldots\ldots\ldots\ldots\ldots(4)$$

The same polhode exists on a quadric surface coaxial with the former, which rolls on the invariable plane of C and gives the herpolhode in that plane. The two associated quadrics in this case have equations which may be written down by merely accenting all the letters in (1) and (2) except $x^2, y^2, z^2$, $p^2, q^2, r^2$. The relations between the two motions, and the equations for passage from one to the other, have been set forth in 22. Again we have

$$x, y, z = m(p, q, r), \quad D'\delta'^2 = 2m^2T', \quad D'\delta' = mH', \quad \delta'^2 = m^2h'^2, \quad \ldots\ldots(5)$$

with $D'\delta'^2\left(\frac{1}{A'}, \frac{1}{B'}, \frac{1}{C'}\right) = m^2h'(a', b', c')$ or $D'\left(\frac{1}{A'}, \frac{1}{B'}, \frac{1}{C'}\right) = \frac{1}{h'}(a', b', c')$. .....(6)

Through every point of the polhode there passes a hyperboloid of one sheet to which the polhode is perpendicular at the point, and this polhode is coaxial with each of the rolling quadrics. There is thus a family of one sheet hyperboloids the members of which are met orthogonally by the polhode, and from the theorem of triply orthogonal quadric surfaces there meet at any point $x, y, z$ of the polhode three surfaces, the hyperboloid of one sheet already referred to, an ellipsoid, and a hyperboloid of two sheets; these are all confocal, and to the latter surfaces the polhode is a line of curvature. while to each of the family of one sheet hyperboloids, which it meets in its course for any one position in space, it is normal. The ellipsoid and two sheet hyperboloid are definite for each position of the rolling quadric and give the whole curve for that position.

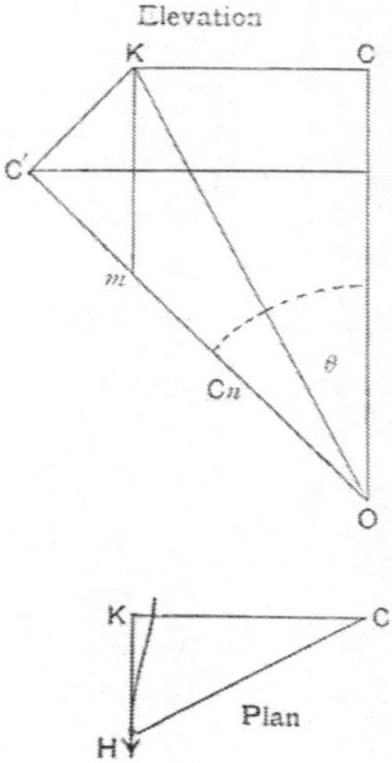

Fig. 112.

34. *Calculation of the motion of the axis. Different forms of the energy equation.* Measuring $\theta$ from the upward vertical drawn from O we have for the A.M., G say, about that vertical the equation

$$Cn\cos\theta + A\dot{\psi}\sin^2\theta = G, \quad \ldots\ldots\ldots\ldots(1)$$

and its value, in the absence of any couple about the vertical OZ, remains invariable. Fig. 112 shows the position of the axis OC of the top and the angle it makes with the vertical OZ. The length OC laid off along the vertical OZ represents the A.M. G, while OC′ along the axis of the top represents $Cn$. In the diagram CK represents $(Cn - A\dot{\psi}\cos\theta)\sin\theta$, and KH represents $A\dot{\theta}$. Thus OH represents the resultant A.M.

The single couple $Mgh\sin\theta$ about an axis through O parallel to KH acts on the body, and this must measure the rate at which H is moving at the instant. This motion is in the direction KH. Thus CH is increasing in length at rate $Mgh\sin\theta\sin KCH = Mgh\sin\theta \,.\, KH/CH$. We have therefore, since $KH = A\dot{\theta}$,

$$CH \,.\, \frac{d}{dt}(CH) = MghA\dot{\theta}\sin\theta, \quad \text{.........................(2)}$$

or, by integration,

$$CH^2 = -aA^2(\cos\theta - E), \quad \text{.............................(3)}$$

where $a = 2Mgh/A$, and E is a constant. [$Mgl$ is used for $Mgh$ in 29.]

Similarly,

$$OH\frac{d}{dt}(OH) = MghA\dot{\theta}\sin\theta, \quad C'H\frac{d}{dt}(C'H) = MghA\dot{\theta}\sin\theta;$$

and therefore

$$OH^2 = -aA^2(\cos\theta - F), \quad C'H^2 = -aA^2(\cos\theta - K), \quad \text{...........(4)}$$

where F and K are constants.

We put $2l$, $2l'$ for $G/A$, $Cn/A$ respectively and $Am^2$ $(=\frac{1}{2}Aa)$ for $Mgh$. From the relations connecting CH, OH, and C′H, we find that the constants E, F, K fulfil the condition

$$E + \frac{G^2}{aA^2} = F = K + \frac{C^2n^2}{aA^2} \quad \text{or} \quad E + 2\frac{l^2}{m^2} = F = K + 2\frac{l'^2}{m^2}. \quad \text{..........(5)}$$

Also we have

$$\dot{\psi} = \frac{G - Cn\cos\theta}{A(1-\cos^2\theta)} = \frac{1}{2A}\left(\frac{G - Cn}{1-\cos\theta} + \frac{G + Cn}{1+\cos\theta}\right) = \frac{l - l'}{1-\cos\theta} + \frac{l + l'}{1+\cos\theta}. \quad \text{...(6)}$$

Again,

$$\dot{\phi} = n - \dot{\psi}\cos\theta = \left(1 - \frac{C}{A}\right)n + \frac{Cn - G\cos\theta}{A\sin^2\theta}$$

$$= \left(\frac{1}{C} - \frac{1}{A}\right)Cn + \frac{1}{2A}\left(\frac{Cn - G}{1-\cos\theta} + \frac{Cn + G}{1+\cos\theta}\right)$$

$$= 2\left(\frac{A}{C} - 1\right)l' + \frac{l' - l}{1-\cos\theta} + \frac{l' + l}{1+\cos\theta}. \quad \text{................(7)}$$

Hence

$$\tfrac{1}{2}(\dot{\phi} + \dot{\psi}) = \left(\frac{A}{C} - 1\right)l' + \frac{l + l'}{1+\cos\theta}, \quad \tfrac{1}{2}(\dot{\phi} - \dot{\psi}) = \left(\frac{A}{C} - 1\right)l' + \frac{l' - l}{1-\cos\theta}. \quad \text{...(8)}$$

Now by Fig. 112 we have

$$KH^2 = OH^2 - OK^2 = OH^2 - C'K^2 - OC'^2$$

$$= 2A^2m^2(F - \cos\theta) - (G^2 - 2CnG\cos\theta + C^2n^2)/\sin^2\theta.$$

Thus, since $\dot{z} = -\sin\theta \,.\, \dot{\theta}$ and $KH = A\dot{\theta}$,

$$\dot{z}^2 = 2m^2(F - z)(1 - z^2) - 4(l^2 - 2ll'z + l'^2)$$

$$= 2m^2(E - z)(1 - z^2) - 4(l' - lz)^2 = 2m^2(K - z)(1 - z^2) - 4(l - l'z)^2. \quad \text{...(9)}$$

Each of these is a form of the energy equation, with account taken of the constancy of A.M. about the upward vertical and about the axis of the top. We can write each equation in the form

$$\dot{z}^2=2m^2Z, \quad \text{.....................................(9')}$$

where $Z=(F-z)(1-z^2)-2\dfrac{l^2-2ll'z+l'^2}{m^2}=(E-z)(1-z^2)-2\left(\dfrac{l'-lz}{m}\right)^2$

$$=(K-z)(1-z^2)-2\left(\frac{l-l'z}{m}\right)^2. \quad \text{...(10)}$$

There is no difficulty in establishing these results analytically; the geometrical investigation given here is due to Greenhill (*R.G.T.*, p. 43 *et seq.*).

**35.** *Calculation of t in terms of z.* The calculation of $t$ in terms of $z$ can, as we have already seen, be carried out by the methods of computation of elliptic integrals. To recapitulate, if $z_1$, $z_2$, $z_3$ be the roots, in order of magnitude, of the equation $z=0$, we have

$$Z=(z-z_1)(z_2-z)(z_3-z), \quad \text{.............................(1)}$$

and the roots are situated as in the sequence

$$\infty>z_1>1>z_2>z>z_3>-1.$$

Each value of $z$ concerned in the motion must lie between $z_2$ and $z_3$. Thus $\dot{z}=m(2Z)^{\frac{1}{2}}$, and so, if $z=z_3$ when $t=0$,

$$mt=\int_{z_3}^{z}\frac{dz}{(2Z)^{\frac{1}{2}}}. \quad \text{.....................................(2)}$$

Thus the motion is periodic in the complete period

$$T=\frac{2}{m}\int_{z_3}^{z_2}\frac{dz}{(2Z)^{\frac{1}{2}}}. \quad \text{.....................................(3)}$$

Also, since $m\,dt=dz/(2Z)^{\frac{1}{2}}$, the azimuthal motion is given by the equation

$$\psi=\frac{l-l'}{m}\int_{z_3}^{z}\frac{dz}{(1-z)(2Z)^{\frac{1}{2}}}+\frac{l+l'}{m}\int_{z_3}^{z}\frac{dz}{(1+z)(2Z)^{\frac{1}{2}}}, \quad \text{.................(4)}$$

that is $\psi$ is the sum of two elliptic integrals of the third kind, the poles of which are respectively the points $z=1$, $z=-1$.

**36.** *Path of the extremity of the vector representing the resultant* A.M. H describes a curve in the horizontal plane through C. If $\rho$ be the radius vector from C as origin, and $\varpi$ the vectorial angle, the speed of H at right angles to CH is $\rho\dot{\varpi}$. But since $Mgh\sin\theta$ is the rate at which H is moving at the instant, and $Mgh=Am^2$, this speed is $Am^2\sin\theta\cos KCH$. But $\cos KCH=CK/CH=(Cn-G\cos\theta)/\rho\sin\theta$. Hence

$$\rho^2\dot{\varpi}=Am^2(Cn-G\cos\theta). \quad \text{.............................(1)}$$

But, by (3), 34, $\rho^2=2A^2m^2(E-z)$, and therefore by the values of $l$, $l'$,

$$\dot{\varpi}=\frac{l'-lz}{E-z}=l+\frac{l'-lE}{E-z}. \qquad (2)$$

Substituting $dz/m\,(2Z)^{\frac{1}{2}}$ for $dt$, and integrating, we obtain

$$\varpi=lt+\frac{l'-lE}{m}\int_{z_3}^{z}\frac{dz}{(E-z)(2Z)^{\frac{1}{2}}} \qquad (3)$$

for the vectorial angle turned through in time $t$ by the radius vector CH, in terms of $lt$, and an elliptic integral of the third kind, the pole of which is the point $z=E$.

Multiplying (2) by $\rho^2=2A^2m^2(E-z)$, we get

$$\rho^2(\dot{\varpi}-l)=2A^2m^2(l'-lE). \qquad (4)$$

Thus the curve $(\rho, \varpi-lt)$ is a central orbit, for $\rho^2(\dot{\varpi}-l)$ is a constant double rate of description of area by the radius vector. The law of force is $k\rho^3+l\rho$.

From Fig. 112 we have

$$\varpi-\psi=\angle\,KCH=\tan^{-1}\frac{KH}{CK}=\tan^{-1}\frac{A\dot{\theta}\sin\theta}{Cn-G\cos\theta},$$

that is

$$\varpi-\psi=\tan^{-1}\frac{-m(2Z)^{\frac{1}{2}}}{2(l'-lz)}. \qquad (5)$$

But also $\angle\,KCH=\cos^{-1}(CK/CH)=\cos^{-1}[(2l'-2lz)/m\{(1-z^2)2(E-z)\}^{\frac{1}{2}}]$, for, since $\dot{\theta}$ vanishes when $\cos\theta=z_3$, we take

$$\angle\,KCH=\sin^{-1}\frac{A\dot{\theta}}{Am\{2(E-z)\}^{\frac{1}{2}}}=\sin^{-1}\frac{-(2Z)^{\frac{1}{2}}}{\{(1-z^2)2(E-z)\}^{\frac{1}{2}}}. \qquad (6)$$

Thus writing $\chi$ for $\angle$KCH, we get

$$\sin\theta(\cos\chi+i\sin\chi)=\sin\theta\,.\,e^{i\chi}=\left\{2\frac{l'-lz}{m}-i(2Z)^{\frac{1}{2}}\right\}\frac{1}{\{2(E-z)\}^{\frac{1}{2}}}. \qquad (7)$$

In (6), 34, (3), and (5) above we have obtained the known result

$$\frac{l-l'}{m}\int_{z_3}^{z}\frac{dz}{(1-z)(2Z)^{\frac{1}{2}}}+\frac{l+l'}{m}\int_{z_3}^{z}\frac{dz}{(1+z)(2Z)^{\frac{1}{2}}}-\frac{l'-lE}{m}\int_{z_3}^{z}\frac{dz}{(E-z)(2Z)^{\frac{1}{2}}}$$
$$=lt-\sin^{-1}\frac{-(2Z)^{\frac{1}{2}}}{(1-z^2)\{2(E-z)\}^{\frac{1}{2}}}, \qquad (8)$$

which expresses the sum of two elliptic integrals of the third kind.

Now from (1), 35, and (10), 34, putting $z=-1$, $z=+1$, $z=E$ in succession, taking account of (4), and defining $\theta_1$ by writing $z_1=\cosh\theta_1$, we obtain

$$\left.\begin{aligned}\left(\frac{l+l'}{m}\right)^2&=\tfrac{1}{2}(1+z_1)(1+z_2)(1+z_3)=4\cosh^2\tfrac{1}{2}\theta_1\cos^2\tfrac{1}{2}\theta_2\cos^2\tfrac{1}{2}\theta_3,\\ \left(\frac{l-l'}{m}\right)^2&=-\tfrac{1}{2}(1-z_1)(1-z_2)(1-z_3)=4\sinh^2\tfrac{1}{2}\theta_1\sin^2\tfrac{1}{2}\theta_2\sin^2\tfrac{1}{2}\theta_3,\end{aligned}\right\} \qquad (9)$$

$$\left(\frac{l'-lE}{m}\right)^2=\tfrac{1}{2}(E-z_1)(E-z_2)(E-z_3). \qquad (10)$$

Thus by (4), 35, we obtain

$$\psi=\sinh\tfrac{1}{2}\theta_1\sin\tfrac{1}{2}\theta_2\sin\tfrac{1}{2}\theta_3\int_{z_3}^{z}\frac{dz}{\sin^2\frac{1}{2}\theta(2Z)^{\frac{1}{2}}}$$

$$+\cosh\tfrac{1}{2}\theta_1\cos\tfrac{1}{2}\theta_2\cos\tfrac{1}{2}\theta_3\int_{z_1}^{z}\frac{dz}{\cos^2\frac{1}{2}\theta(2Z)^{\frac{1}{2}}}.\quad\text{...........(11)}$$

But also, since $\psi=\varpi-\chi$, we have by (3) and (10)

$$\psi=lt+\frac{1}{2^{\frac{1}{2}}}\{(E-z_1)(E-z_2)(E-z_3)\}^{\frac{1}{2}}\int_{z_3}^{z}\frac{dz}{(E-z)(2Z)^{\frac{1}{2}}}$$

$$-\sin^{-1}\frac{-Z^{\frac{1}{2}}}{\{(1-z^2)(E-z)\}^{\frac{1}{2}}}.\quad\text{.................................(12)}$$

The vector CH is $\rho e^{i\omega}$, and so the vector velocity is $(\dot\rho+i\rho\dot\varpi)e^{i\omega}$, where $\omega$ is used for $\varpi$ in the exponentials. The velocity $\dot\rho$ is the rate at which the vector is lengthening, $\rho\dot\varpi$ is the rate at which the extremity H of the vector is moving in the plane at right angles to O*c*.

The couple has magnitude $Mgh\sin\theta$, and direction coinciding with KH. But the angle which KH makes with the initial direction of CK is (see Fig. 112) $\psi+\frac{1}{2}\pi$. Hence as a vector the couple is

$$Mgh\sin\theta\,.\,e^{i(\psi+\frac{1}{2}\pi)}=Am^2\sin\theta\,.\,ie^{i\psi}.$$

Thus $$Am^2\sin\theta\,.\,ie^{i\psi}=\frac{d}{dt}(\rho e^{i\omega})=(\dot\rho+i\rho\dot\varpi)e^{i\omega}.$$

**37**. *Hodograph for the motion of a top.* The point H moves with speed equal to the moment of the couple, and the direction of its motion is at each instant that of the line KH at the instant. If lines equal to the successive values of $Am^2\sin\theta$ are drawn from O, each parallel to the corresponding position of KH, the extremities of these lines will give a curve the tangent to which at each point is the direction of the change of the couple. The speed of a particle imagined to move along the curve, so as to be at each instant at the extremity of the line representing the couple at that instant, will represent the speed of variation of the couple. The curve is the hodograph of the point H. The idea of this hodograph and the vector specification given in 36 above are due to Greenhill [*R.G.T.*, p. 46, and *The Mathematical Theory of the Top, Verhandl. Intern. Math. Kongresses*, 1904. The reader is specially referred to the *R.G.T.* for a fuller discussion. He will notice that there $z$ is given the opposite sign].

# CHAPTER XXII

## ANALOGY BETWEEN A BENT ROD AND THE MOTION OF A TOP. WHIRLING OF SHAFTS, CHAINS, ETC.

1. *Flexure and torsion of a thin bar: analogy to the motion of a top.* It was pointed out by Kirchhoff (*Crelle's Journal*, 56, 1858) that an exact parallel exists between the motion of a top and the combined bending and torsion of a thin elastic bar. We suppose that the bar is held fast at one end, and that at the other a force and a couple, of determinate amount and direction, are applied. Consider a disk of the bar, intercepted between two near cross-sections which, in the straight undeformed state of the bar, were parallel. The substance of the bar beyond the disk on each side applies to the adjacent face of the disk a force and a couple.

Now each force and associated couple can be converted into a force, acting at the centroid of the cross-section, and a couple about a determinate axis. We shall choose for each cross-section three axes, one along the line of centroids of the successive cross-sections, the others at right angles to one another, and parallel to the sides of the cross-section, if that is rectangular. In any case, they are taken so as to be the principal axes of bending moment for the section. As each cross-section, being only slightly distorted, is practically a plane at right angles to the line of centroids where it meets that line, we get thus for each section three axes at right angles to one another, two of which are in the section, and the third at right angles to it. The first two we call transverse axes, the third we may call the axis of the bar at the cross-section considered.

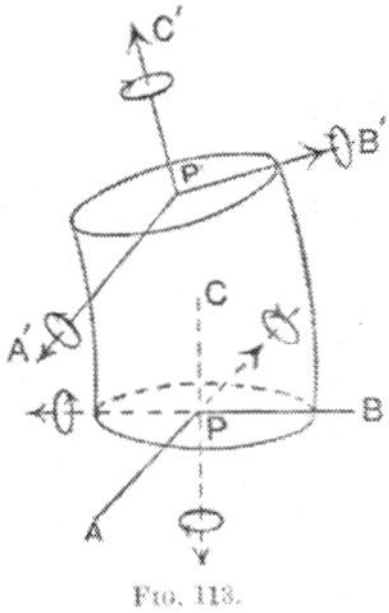

FIG. 113.

The three axes are shown in Fig. 113 for the two sections which form the ends of the disk considered. P(A, B, C) are the axes of $x$, $y$, $z$ for one end face, P′(A′, B′, C′) are those for the other end face. The directions of the latter would be the same as those of the former if the turnings of the second set relative to the first were undone.

2. *Equations of equilibrium.* In strictness the bending of the bar brings into existence distortions of the cross-sections, but these distortions are negligible if the radius of curvature of the bar is great in comparison with every dimension of cross-section. The same kind of result holds as regards the effects of torsion about the axis of the bar, provided the twist is very small. If the twist about the $z$-axis be $\zeta$, that is, if $\zeta$ be the angle turned through per unit distance along $z$, we shall suppose that $1/\zeta$ is also great in comparison with the greatest dimension of cross-section, as when this is the case we may neglect any distortion of the sections from planarity.

We now specify the forces and couples applied to the disk by the substance of the bar on the two sides as follows. Along the axes, and in the opposite sense to the axis in each case, act forces P, Q, R, and about the same lines the couples F, G, H. The system is fully indicated by the dotted lines and arrows. At the opposite face of the disk, along and about the axes P′(A′, B′, C′) act the forces and couples $\mathrm{P}+d\mathrm{P}$, $\mathrm{Q}+d\mathrm{Q}$, $\mathrm{R}+d\mathrm{R}$, $\mathrm{F}+d\mathrm{F}$, $\mathrm{G}+d\mathrm{G}$, $\mathrm{H}+d\mathrm{H}$.

We now suppose that the directions of the axes P′(A′, B′, C′) may have been obtained from those of P(A, B, C) by small turnings $\xi\,ds$, $\eta\,ds$, $\zeta\,ds$ about the directions PA, PB, PC, simultaneously, or by turning first through $\xi\,ds$ about PA held fixed, then through $\eta\,ds$ about PB held in its new position, and finally through $\zeta\,ds$ about PC held in the position to which it has been brought by the two preceding turnings. The superposition of the turnings leads to a result which is independent of the order of imposition if the displacements are small.

We now resolve along the axes P(A, B, C) the forces which act at P′ along the axes P′(A′, B′, C′). P′A′ and P′B′ are now inclined to PC at the respective angles $\frac{1}{2}\pi+\eta\,ds$, $\frac{1}{2}\pi-\xi\,ds$. The force $\mathrm{P}+d\mathrm{P}$ along P′A′ has therefore a component parallel to PC of amount $-(\mathrm{P}+d\mathrm{P})\eta\,ds$, while in the same direction the component of $\mathrm{Q}+d\mathrm{Q}$ is $(\mathrm{Q}+d\mathrm{Q})\xi\,ds$. $\mathrm{R}+d\mathrm{R}$ along P′C′ gives in the limit simply $\mathrm{R}+d\mathrm{R}$ along PC. The force account for the element of the bar, worked out in this way for each of the directions P(A, B, C), is found to give the following balances of force applied by the adjacent matter to the element,

$$d\mathrm{P}-\mathrm{Q}\zeta\,ds+\mathrm{R}\eta\,ds,\quad d\mathrm{Q}-\mathrm{R}\xi\,ds+\mathrm{P}\zeta\,ds,\quad d\mathrm{R}-\mathrm{P}\eta\,ds+\mathrm{Q}\xi\,ds,$$

for these directions respectively. These must be reduced to zero by the application of the external forces $\mathrm{X}\,ds$, $\mathrm{Y}\,ds$, $\mathrm{R}\,ds$. Hence we have the equations of equilibrium,

$$\frac{d\mathrm{P}}{ds}-\mathrm{Q}\zeta+\mathrm{R}\eta+\mathrm{X}=0,\quad \frac{d\mathrm{Q}}{ds}-\mathrm{R}\xi+\mathrm{P}\zeta+\mathrm{Y}=0,\quad \frac{d\mathrm{R}}{ds}-\mathrm{P}\eta+\mathrm{Q}\xi+\mathrm{Z}=0. \quad ...(1)$$

There are also three equations of moments which can be established in a similar way. The forces give moments about axes of $x$ and $y$ of amounts

$-Q\,ds$, $P\,ds$. The equations are (if there be no couples applied to the element from without the system)

$$\frac{dF}{ds}-G\zeta+H\eta-Q=0,\quad \frac{dG}{ds}-H\xi+F\zeta+P=0,\quad \frac{dH}{ds}-F\eta+G\xi=0. \quad \ldots(2)$$

F and G are the couples of bending moment applied about the principal axes, and H the torsion couple. They are equivalent to a single couple $(F^2+G^2)^{\frac{1}{2}}$ about an axis in the plane of $xy$ inclined at the angle $\tan^{-1}(F/G)$ to the axis of $x$. If no external forces or couples are applied, except at the terminals, X, Y, Z are zero in (1) for an internal element; equations (2) stand as they are.

The element is bent through the angles $\xi\,ds$, $\eta\,ds$ in the planes $yz$, $xz$ respectively. Hence $\xi$, $\eta$ are the curvatures of the element in these planes, while $\zeta$ is the twist at the element. We may write $H=C\zeta$, where C depends on the material and on the extent and form of the cross-section. If the cross-section is circular, C is the product of the rigidity modulus $n$ and the "moment of inertia of the cross-section" about the axis of the bar, that is if $r$ be the distance of a ring of the cross-section from the axis, the integral $2\pi\int r^3 dr$ for the cross-section.

Now by the theory of elasticity we have $F=A\xi$, $G=B\eta$, if A, B be the flexural rigidities about the principal axes, that is if

A = Young's modulus × moment of inertia of cross-section about GA,

B = Young's modulus × moment of inertia of cross-section about GB.

Hence equations (2) become

$$A\frac{d\xi}{ds}-(B-C)\eta\zeta=Q,\quad B\frac{d\eta}{ds}-(C-A)\zeta\xi=-P,\quad C\frac{d\zeta}{ds}-(A-B)\xi\eta=0. \quad (3)$$

These are exactly Euler's equations of the motion of a rigid body about a fixed point under applied couples P, Q, R, if $\xi$, $\eta$, $\zeta$ be interpreted as the angular velocities about principal axes through the fixed point, A, B, C as the moments of inertia of the body about the axes, and $ds$ as an element of time.

Thus, passing from element to element, $ds$, along the curve, we have an exact parallelism between the values of $\xi$, $\eta$, $\zeta$ and those of the angular velocities of a rigid body, for which the corresponding quantities are as stated, and which is started under initial conditions which correspond to the terminal conditions of the bar.

We will now suppose P and Q to be zero. We have then from (3) the equations

$$A^2\xi^2+B^2\eta^2+C^2\zeta^2=I^2,\quad A\xi^2+B\eta^2+C\zeta^2=J^2, \ldots\ldots\ldots\ldots\ldots\ldots(4)$$

where I and J are constants. The expression on the left in the second of these equations is twice the potential energy of the combined bending and twisting per unit length, and the fact that J is a constant shows that this, when P and Q are zero, is the same at each cross-section. The two equations in (4) correspond to the equations of angular momentum and kinetic energy in the case of a rigid body turning about a fixed point under no forces.

If we do not suppose that P and Q are zero we may multiply the first of (3) by $\xi$, the second by $\eta$, and the third by $\zeta$, and add the results to the third of (1), and obtain, since $Q\xi - P\eta = R$,

$$\frac{d}{ds}\{R + \tfrac{1}{2}(A\xi^2 + B\eta^2 + C\zeta^2)\} = 0,$$

that is $$R + \tfrac{1}{2}(A\xi^2 + B\eta^2 + C\zeta^2) = \text{const.}, \quad \text{.....................(5)}$$

which is the analogue of the energy equation for the moving body.

3. *Case of bending in one plane.* It is interesting to consider various particular cases of this analogy. We take first the case in which the bending is in one plane, that of $x$, $z$. Then $\xi$ and $\zeta$ are zero. Let UV (Fig. 114) be an element $ds$ of the rod, and a force S in the plane of bending be applied at A. We have, by (2), 2, for the axial force R, the shearing force P, and the couple G, the equations

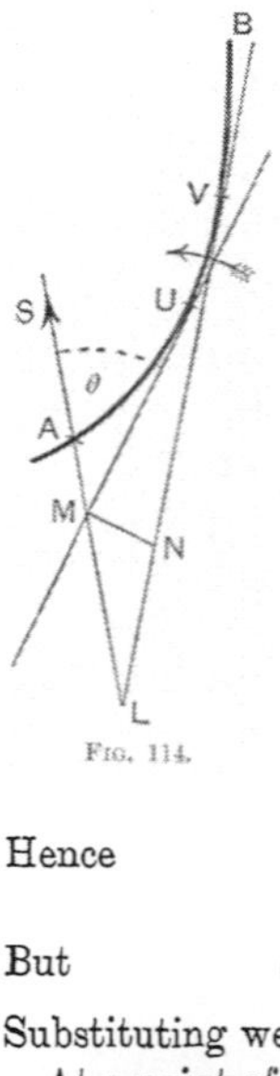

FIG. 114.

$$R = S\cos\theta, \quad P = -S\sin\theta, \quad \frac{dG}{ds} + P = 0. \quad \text{.........(1)}$$

But the elastic reaction couple is $-B\,d\theta/ds = G$. Thus we get, from the values of P and G and the third equation,

$$B\frac{d^2\theta}{ds^2} + S\sin\theta = 0 \quad \text{........................(2)}$$

as the equation of equilibrium.

We may establish this equation from first principles. Clearly we have by the diagram, if $UM = l$,

$$B\frac{d\theta}{ds} + Sl\sin\theta = 0 \quad \text{........................(3)}$$

Hence $$B\frac{d^2\theta}{ds^2} + Sl\cos\theta\frac{d\theta}{ds} + S\sin\theta\frac{dl}{ds} = 0.$$

But $$dl = ds + LN = ds + MN\frac{\cos\theta}{\sin\theta} = ds - l\,d\theta\frac{\cos\theta}{\sin\theta}.$$

Substituting we get again (2).

At a point of inflexion on the curve of the bar $d\theta/ds$ is zero; hence by (3) points of inflexion exist where the line of S intersects the curve, but nowhere else.

Multiplying (2) by $d\theta/ds$, and integrating from an inflexion, where $d\theta/ds = 0$, we get

$$B\left(\frac{d\theta}{ds}\right)^2 = 2S(\cos\theta - \cos\alpha) = 4S(\sin^2\tfrac{1}{2}\alpha - \sin^2\tfrac{1}{2}\theta), \quad \text{......................(4)}$$

where $\alpha$ is the value of $\theta$ at the inflexion. This equation can be integrated by means of elliptic functions. We write $u = s(S/B)^{\frac{1}{2}}$, so that (4) becomes

$$\left(\frac{d\theta}{du}\right)^2 = 4(\sin^2\tfrac{1}{2}\alpha - \sin^2\tfrac{1}{2}\theta). \quad \text{..................................(5)}$$

Now, putting $\sin\frac{1}{2}\theta = \sin\frac{1}{2}\alpha \sin\phi = k\sin\phi$, we get after reduction

$$\left(\frac{d\phi}{du}\right)^2 = 1 - k^2\sin^2\phi. \qquad \text{.........(6)}$$

Thus for the arc from $\phi = \frac{1}{2}\pi$ to $\phi = 0$, that is from the inflexion to the point where the curve is parallel to the force S, we have

$$u_1 = \int_{\frac{1}{2}\pi}^{0} \frac{d\phi}{(1 - k^2\sin^2\phi)^{\frac{1}{2}}} = -\mathrm{K}.$$

Hence also, for the range from $\frac{1}{2}\pi$ to $\phi$, we have

$$u = \int_{\frac{1}{2}\pi}^{0} \frac{d\phi}{(1 - k^2\sin^2\phi)^{\frac{1}{2}}} + \int_{0}^{\phi} \frac{d\phi}{(1 - k^2\sin^2\phi)^{\frac{1}{2}}}, \qquad \text{.........(7)}$$

that is

$$\int_{0}^{\phi} \frac{d\phi}{(1 - k^2\sin^2\phi)^{\frac{1}{2}}} = \mathrm{K} + u. \qquad \text{.........(8)}$$

Thus

$$\sin\tfrac{1}{2}\theta = k\,\mathrm{sn}\,(\mathrm{K}+u), \quad \frac{d\theta}{du} = 2k\,\mathrm{cn}\,(\mathrm{K}+u). \qquad \text{.........(9)}$$

If now we take fixed axes of $x$ and $y$ along and at right angles to the direction of S, we have, by Fig. 114, $dx/ds = \cos\theta$, $dy/ds = -\sin\theta$. Hence, integrating and reducing by the results just obtained, we find

$$\left.\begin{aligned} x &= \left(\frac{\mathrm{B}}{\mathrm{S}}\right)^{\frac{1}{2}} \int \cos\theta\, du = \left(\frac{\mathrm{B}}{\mathrm{S}}\right)^{\frac{1}{2}} \int_{\frac{1}{2}\pi}^{\phi} \frac{1 - 2k^2\sin^2\phi}{(1 - k^2\sin^2\phi)^{\frac{1}{2}}}\, d\phi = \left(\frac{\mathrm{B}}{\mathrm{S}}\right)^{\frac{1}{2}} \{2\mathrm{E}(\phi, k) - 2\mathrm{E} - (\mathrm{K}+u)\}, \\ y &= -\int \sin\theta\, ds = -\left(\frac{\mathrm{B}}{\mathrm{S}}\right)^{\frac{1}{2}} \int_{\frac{1}{2}\pi}^{\phi} 2k\sin\phi\, d\phi = \left(\frac{\mathrm{B}}{\mathrm{S}}\right)^{\frac{1}{2}} 2k\,\mathrm{cn}\,(\mathrm{K}+u). \end{aligned}\right\} \text{...(10)}$$

**4.** ***Bending in one plane represented by pendulum motion.*** Consider then the curve between two points of inflexion, when points of inflexion exist. The kinetic analogue is a pendulum of length B$g$/S vibrating through a finite arc, from rest at a position corresponding to one inflexion to rest at a position corresponding to the other. Of course the forces S at the two points of inflexion are in opposite directions. These forces alone—that is without couples—keep the rod in equilibrium. If the end of the rod remote from A, say B in Fig. 114, is fixed, then B is a place of maximum bending, and corresponds to the middle of the range of the corresponding pendulum.

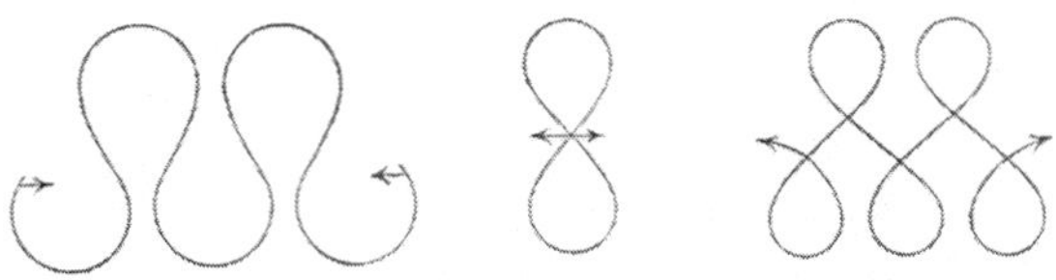

Fig. 115.

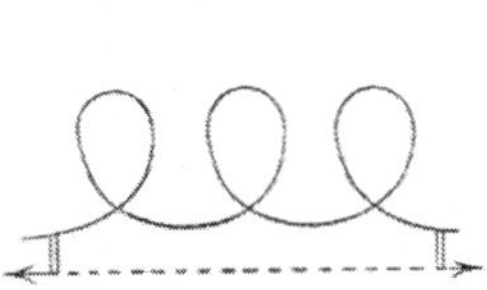

Fig. 116.

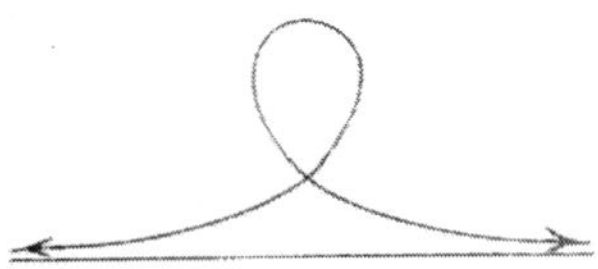

Fig. 117.

Fig. 115 shows forms which have inflexions. Fig. 116 shows a non-inflexional form, and Fig. 117, a form intermediate between the inflexional and non-inflexional forms. The non-inflexional form corresponds to a revolving pendulum, the intermediate form to a pendulum which starts from a position infinitely near the upward vertical, and oscillates through a complete revolution to a point infinitely near the upward vertical on the other side. The infinite time required for the first finite deflection from the upward vertical, or the final finite angle of turning before again reaching it, is shown by the infinite parts of the curve as asymptotic to the line of force.

The discussion of the form of the curve, whether inflexional or non-inflexional, by means of elliptic functions, does not differ from that given in Chap. XV above for the pendulum oscillating through a finite arc or making complete revolutions. A single open loop of the inflexional form of the rod gives the shape of an initially straight uniform rod, when the two ends are pulled towards one another by a cord under tension —a bowstring. The equation of the curve is, as we have seen in 3,

$$\frac{\mathrm{B}}{\rho}+\mathrm{S}y=0, \quad \ldots\ldots(1)$$

where $1/\rho=d\theta/ds$, the curvature at the element considered, and $y=l\sin\theta$. If we put $a^2$ for B/S, and measure $x$ along the bowstring, say from one end, we get from the usual formula for radius of curvature

$$a^2\frac{d^2y}{dx^2}=-y\left\{1+\left(\frac{dy}{dx}\right)^2\right\}^{\frac{3}{2}}, \quad \ldots\ldots(2)$$

where on the right we are to take the positive value of the square root. If $p=dy/dx$, the equation becomes

$$y\,dy=-a^2\frac{p\,dp}{(1+p^2)^{\frac{3}{2}}},$$

and therefore

$$y^2=\mathrm{C}+2a^2\frac{1}{(1+p^2)^{\frac{1}{2}}}, \quad \ldots\ldots(3)$$

where C is a constant. Since $p=0$ when $y=h$, say, we have $\mathrm{C}=h^2-2a^2$, and therefore

$$y^2-h^2+2a^2=\frac{2a^2}{(1+p^2)^{\frac{1}{2}}}. \quad \ldots\ldots(4)$$

From this we obtain

$$\frac{1}{p}=\frac{dx}{dy}=\frac{y^2-h^2+2a^2}{\{4a^4-(y^2-h^2+2a^2)^2\}^{\frac{1}{2}}}; \quad \ldots\ldots(5)$$

and therefore

$$x=\int\frac{(y^2-h^2+2a^2)\,dy}{\{4a^4-(y^2-h^2+2a^2)^2\}^{\frac{1}{2}}}. \quad \ldots\ldots(6)$$

The determination of the length of the curve from one inflexion to the next corresponds to the determination of the period of oscillation of the finite pendulum. Now

$$ds=dy\left(1+\frac{1}{p^2}\right)^{\frac{1}{2}}=dy\{(1+p^2)^{\frac{1}{2}}/p\}.$$

Hence, by (4) and (5),

$$ds=\frac{2a^2dy}{\{4a^4-(y^2-h^2+2a^2)^2\}^{\frac{1}{2}}}=\frac{2a^2dy}{(h^2-y^2)^{\frac{1}{2}}\{4a^2-(h^2-y^2)\}^{\frac{1}{2}}}; \quad \ldots\ldots(7)$$

or if we put $y=h\cos\phi$, and integrate from $y=0$ to $y=h$, that is from $\phi=\frac{1}{2}\pi$ to $\phi=0$, we get, for the whole length of the curve from one inflexion to the next,

$$s=2a\int_0^{\frac{1}{2}\pi}\frac{d\phi}{(1-k^2\sin^2\phi)^{\frac{1}{2}}}=2a\mathrm{K}, \quad \ldots\ldots(8)$$

where K is the complete elliptic integral of the first kind to modulus $k=h/2a$.

For any arc specified by the limits 0 and $\phi$, we have

$$s=2a\int_0^\phi \frac{d\phi}{(1-k^2\sin^2\phi)^{\frac{1}{2}}}=2a\mathrm{F}(\phi, k). \quad\text{.....(9)}$$

If the term $(dy/dx)^2$ be omitted from (2) the differential equation becomes that of a curve of sines, and it is generally assumed that when the slope of the curve is very small a curve of sines will give, for a fixed distance from inflexion to inflexion and a given value of $a$, a sufficiently close approximation for practical purposes to the ordinates of the elastica—the deflections of the bow from the straight. This assumption is quite warranted provided the maximum ordinate of the curve of sines is chosen equal to that of the elastica; it is not so however if the lengths along the two curves from end to end are exactly to agree

The exact length of the elastica is given by (8), and the reader may easily prove that the exact length of the curve of sines,

$$y=h'\sin\frac{x}{a}, \quad\text{.....(10)}$$

extending from $x=0$ to $x=\pi a$, and having $h'$ as maximum ordinate, is

$$s=2(a^2+h'^2)^{\frac{1}{2}}\mathrm{E}\quad[\text{mod. } h'/(a^2+h'^2)^{\frac{1}{2}}], \quad\text{.....(11)}$$

where E is the complete elliptic integral of the second kind. If the $s$ in (9) and the $s$ in (11) have the same value, there is no difficulty in showing by expanding the integrands in K and E and integrating term by term, that approximately, when the deflections are small, $h'=\frac{1}{2}h$. This remark is made by Mr. R. W. Burgess in a paper in the *Physical Review*, March, 1917.

This however does not vitiate the approximation, under the condition specified above, of equality of maximum ordinates, and of distance from end to end along the bowstring. The lengths of the curves will no doubt differ slightly, the curve of sines, which has slightly longer ordinates, will exceed the elastica in length by about $\pi h^2/16a$, which is only a small fraction of $2a$. For slightly deflected elastic rods, for example flexible shafts rendered stable by rotation, it will be possible under proper conditions to neglect the factor $1/\{1+(dy/dx)^2\}^{\frac{3}{2}}$ in the expression for the curvature.

From (2) we have, neglecting the multiplier $(1+p^2)^{\frac{3}{2}}$ on the right,

$$\frac{d^2y}{dx^2}+\frac{1}{a^2}y=0.$$

This gives the curve of sines, $$y=h\sin\left(\frac{x}{a}+\alpha\right),$$

where $y=0$ when $x=0$ and when $x=l$. As we have seen, this agrees with the elastica closely when the deflections are small and the maximum ordinates are made the same. The terminal conditions are fulfilled if $\alpha=n\pi$ and $l/a+\alpha=(n+1)\pi$. Thus $l/a=\pi$. Hence

$$l=\pi a=\pi\left(\frac{\mathrm{B}}{\mathrm{S}}\right)^{\frac{1}{2}} \quad\text{or}\quad \mathrm{S}=\pi^2\frac{\mathrm{B}}{l^2}.$$

Thus the distance $l$ along the chord will diminish if S be made greater than $\pi^2\mathrm{B}/l^2$, and increase if S be brought below this limit. Hence when the bar is straight the value of S required to produce any bending at all is $\pi^2\mathrm{B}/l^2$.

It follows that a bar, acting as a pillar supporting a load S, must not be longer than $\frac{1}{2}\pi(\mathrm{B}/\mathrm{S})^{\frac{1}{2}}$, if the lower end is vertical.

**5.** *A thin bar bent into a helix is analogous to a top in steady motion.* We now consider a thin bar bent into a helix and held in equilibrium by a force S applied at the free end along the axis of the helix, while the other end is held fixed More generally the force system applied at the free end includes the couple G, the axis of which is coincident in direction with S, so that the system is a wrench about the axis of the helix. Let us now suppose that the two principal flexural rigidities of the bar are equal. Since there are no forces except those applied at the ends, the equations of equilibrium are

$$A\frac{d\xi}{ds}-(A-C)\eta\zeta=0, \quad A\frac{d\eta}{ds}-(C-A)\zeta\xi=0, \quad \frac{d\zeta}{ds}=0. \quad \text{.........(1)}$$

The third of these shows that $\zeta$ is the same at all points, just as the axial spin of a top remains constant in the absence of a retarding or accelerating couple.

The two first equations give

$$A(\xi^2+\eta^2)=\text{const.}, \quad \text{.................................(2)}$$

that is, the curvature is the same at all points; in other words the form, being in three dimensions, is a helix.

If $\rho$ be the radius of curvature, we have

$$\frac{1}{\rho}=(\xi^2+\eta^2)^{\frac{1}{2}}. \quad \text{..................................(2')}$$

The bending about the $x$-axis gives convexity at the element $ds$ with respect to the positive direction of the $y$-axis, and similarly, the bending about the $y$-axis, convexity with respect to the $x$-axis. Thus the direction cosines of the normal at a point P are $\eta\rho$, $-\xi\rho$, 1, while those of the binormal are $\xi\rho$, $\eta\rho$, 0. These are the projections of unit length of the binormal on the axes. At a distance $ds$ along the curve, at P′, the values are $\xi\rho+d(\xi\rho)$, $\eta\rho+d(\eta\rho)$. At this latter point the axes have been turned from their directions at P, through the angles $\xi\,ds$, $\eta\,ds$, $\zeta\,ds$ about the axes, P($x$, $y$, $z$) [that is the axes P(A, B, C) of Fig. 113], respectively. The projections of the binormal at P′ on the axes at P are

$$\xi\rho+d(\xi\rho)-\eta\rho\zeta\,ds, \quad \eta\rho+d(\eta\rho)+\xi\rho\zeta\,ds, \quad 0.$$

The direction cosines with reference to the axes at P have therefore changed by

$$d(\xi\rho)-\eta\rho\zeta\,ds, \quad d(\eta\rho)+\xi\rho\zeta\,ds.$$

The angle $d\beta$ through which the binormal has turned is given by

$$d\beta^2=\{d(\xi\rho)-\eta\rho\zeta\,ds\}^2+\{d(\eta\rho)+\xi\rho\zeta\,ds\}^2. \quad \text{.................(3)}$$

But the angle $\phi$ between OD and OA in Fig. 118, p. 491, is the angle between the binormal and the axis PA, and so we have $\xi\rho=\cos\phi$, $\eta\rho=-\sin\phi$. Thus

$$d\beta^2=\left(-\frac{d\phi}{ds}+\zeta\right)^2 ds^2 \quad \text{or} \quad \frac{d\beta}{ds}=-\frac{d\phi}{ds}+\zeta. \quad \text{................(4)}$$

The binormal is carried round the axis of the helix at angular speed $d\psi/ds$, with the plane containing it and the axis of the rod. Thus if $\theta$ be the inclination of the axis of the rod to the axis of the helix, we have $d\beta/ds = \cos\theta \, . \, d\psi/ds$. Hence

$$\zeta = \frac{d\phi}{ds} + \frac{d\psi}{ds}\cos\theta. \quad \text{.................................(5)}$$

This exactly corresponds to the equation

$$n = \dot{\phi} + \dot{\psi}\cos\theta,$$

which holds for a top, supported at O [see Fig. 118] with its axis OC

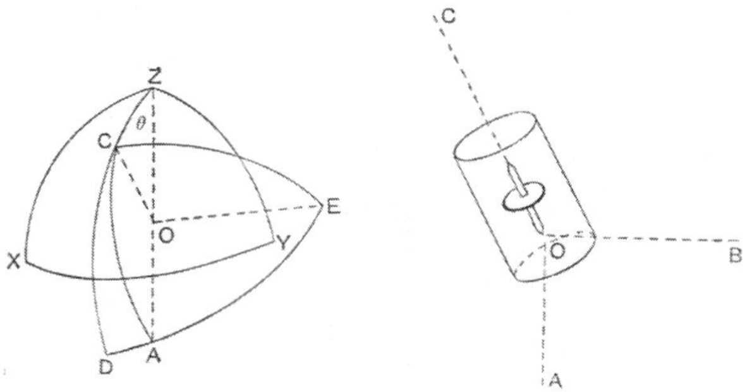


FIG. 118.

inclined at an angle $\theta$ to the vertical OZ, while a plane through the axis turns with angular speed $\dot{\phi}$ with reference to the plane ZOC, and this latter plane turns at the same time round the vertical with speed $\dot{\psi}$. As already stated, the angle ACD in the diagram is $\phi$.

Multiplying the first of (1) by $\eta$, the second by $\xi$, and subtracting the second product from the first, we get

$$\mathrm{A}\left(\eta\frac{d\xi}{ds} - \xi\frac{d\eta}{ds}\right) = (\mathrm{A} - \mathrm{C})\zeta(\xi^2 + \eta^2),$$

that is

$$-\frac{d\phi}{ds} = \frac{\mathrm{C} - \mathrm{A}}{\mathrm{A}}\zeta.$$

Hence

$$-\frac{d\phi}{ds} + \zeta = \frac{\mathrm{A}}{\mathrm{C}}\zeta, \quad \text{.........................................(6)}$$

which, as we have seen, is the rate of turning of the binormal per unit distance along the curve. This rate is constant.

The preceding discussion can be briefly summed up as follows. Take two axes PE, PC in the tangent plane to the curve at the element P (Fig. 113). Then for the helix we have PC turning at constant rate $d\psi/ds$ about the axis of the curve (the "vertical"). This gives a rate of turning $d\psi/ds \, . \cos\theta$ about PC, which obviously is the rate of turning of the binormal, the tortuosity. Besides this there is the turning of an axial plane fixed in the bar, with reference to the plane CPZ, so that $\zeta = d\phi/ds + d\psi/ds \, . \cos\theta$.

Again the rate of production of the analogue of A.M. about the axis OD is

$$A\frac{d^2\theta}{ds^2}+\left(C\zeta-A\frac{d\psi}{ds}\cos\theta\right)\frac{d\psi}{ds}\sin\theta=0.$$

The first term is zero, and so we get, since $\theta$ is not zero,

$$\frac{d\psi}{ds}\cos\theta=\frac{C}{A}\zeta. \quad\ldots\ldots\ldots\ldots\ldots\ldots\ldots\ldots(7)$$

We can put the radius of curvature and the torsion of the helix into other simple forms as follows. Let a narrow strip of paper be wound round a right circular cylinder so that its middle line is a helix. The paper is bent at each point about a generator of the cylinder. This bending can be resolved into two components, one across the band about a line perpendicular to the middle line, and one about the middle. The angle subtended at the axis of the cylinder by the element $ds$, in consequence of the bending about the generating line, is $ds\sin\theta/a$, if $a$ be the radius of the cylinder. The components just specified are therefore $ds\sin^2\theta/a$ and $ds\cos\theta\sin\theta/a$. Hence the curvature and tortuosity are respectively

$$\frac{\sin^2\theta}{a} \quad\text{and}\quad \frac{\sin\theta\cos\theta}{a}.$$

6. *A helix held in equilibrium either by a couple or by axial force.* The helix can be held in equilibrium either by a couple alone or by an axial force alone, applied at the free end. In every cross-section of the bar two couples of elastic reaction exist, one of amount $C\zeta$ about the axis of the bar, the other $A\sin^2\theta/a$, about an axis at right angles to the osculating plane. These have a resultant $(C^2\zeta^2+A^2\sin^4\theta/a^2)^{\frac{1}{2}}=A\sin\theta/a$, the axis of which lies in the tangent plane to the helix and is inclined to the axis of the bar at the angle $\theta$, that is, is parallel to the cylinder axis (the vertical, in the analogous top motion). If in addition a force S acts along the axis of the cylinder we have the wrench specified above. But particular cases are that in which the force is zero and that in which the couple is zero.

When both act, the total moment about the axis of the bar is $G\cos\theta+Sa\sin\theta$. Also the moment about the perpendicular to the osculating plane is $G\sin\theta-Sa\cos\theta$. Thus we have the equations

$$G\cos\theta+Sa\sin\theta=C\zeta,\quad G\sin\theta-Sa\cos\theta=A\frac{\sin^2\theta}{a}.$$

Hence $$S=C\zeta\frac{\sin\theta}{a}-A\frac{\sin^2\theta\cos\theta}{a^2},\quad G=C\zeta\cos\theta+A\frac{\sin^3\theta}{a}. \quad\ldots\ldots\ldots(1)$$

If the force is zero we have

$$\zeta=\frac{A}{C}\frac{\sin\theta\cos\theta}{a},\quad G=A\frac{\sin\theta}{a}=\frac{A}{\rho\sin\theta}. \quad\ldots\ldots\ldots(2)$$

If the couple is zero we have

$$\zeta = -\frac{A}{C}\frac{\sin^3\theta}{\cos\theta}\frac{1}{a}, \quad S = -A\frac{\sin^2\theta}{a^2\cos\theta}. \quad \text{.....................(3)}$$

It follows from the first of (1), that if the step of the helix be very small (that is if $\theta$ is very nearly $\frac{1}{2}\pi$) and $a$ be not small, the axial force depends almost entirely on torsion. An ordinary spring balance works by torsion.

7. *A gyrostat on an overhanging flexible shaft. Equilibrium of the shaft.* So far we have been dealing with a bent and twisted bar in its analogy with the motion of a top. We now suppose the bar, loaded in some definite manner, to be subjected to rotation about a specified axis. To fix the ideas, we take first the practical problem of the motion of a gyrostat mounted at the free end of a flexible shaft. In the case of steady motion the shaft is bent at each instant in one plane. We shall suppose that the slope of the curve from the straight position is everywhere small.

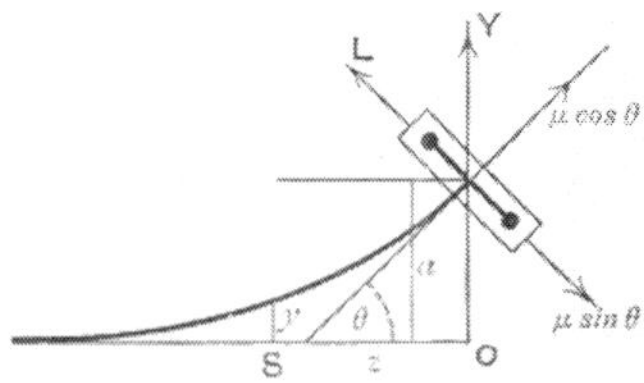

FIG. 119.

The diagram shows the directions of the axes chosen. First we consider the bending. Denoting as before the flexural rigidity by B, supposing that a couple of moment L and a force Y act in the directions shown by the arrows, we have for the cross-section S,

$$B\frac{d^2y}{dz^2} = L + Yz. \quad \text{..................................(1)}$$

The couple L we shall see is due to gyrostatic action. But since the motion is steady and the centroid of the gyrostat moves in a circle of radius $a$, with angular speed $\mu$, the shaft must exert an inward pull on the gyrostat of moment $M\mu^2 a$, where M is the mass of the gyrostat, and therefore the gyrostat exerts on the shaft an outward force of the same amount.

Hence (1) becomes
$$B\frac{d^2y}{dz^2} = L + M\mu^2 az. \quad \text{..............................(2)}$$

Integrating we obtain
$$B\frac{dy}{dz} = Lz + \tfrac{1}{2}M\mu^2 az^2 + c, \quad \text{.............................(3)}$$

where $c$ is a constant. But when $z = l$, $dy/dz = 0$, and so

$$c = -Ll - \tfrac{1}{2}M\mu^2 al^2.$$

Hence when $z = 0$, that is at the attachment of the gyrostat,

$$B\tan\theta = Ll + \tfrac{1}{2}M\mu^2 al^2. \quad \text{..............................(4)}$$

Integrating (3) from $z = 0$ to $z = l$ we get

$$-Ba = \tfrac{1}{2}Ll^2 + \tfrac{1}{6}M\mu^2 al^3 + cl,$$

that is
$$Ba = \tfrac{1}{2}Ll^2 + \tfrac{1}{3}M\mu^2 al^3. \quad \text{...............................(5)}$$

From (4) and (5) we can eliminate L, and obtain

$$\tan\theta = \frac{2a}{l} - \frac{1}{6}\frac{M\mu^2 a l^2}{B}. \quad \text{.............................(6)}$$

8. *Determination of the gyrostatic couple.* We have now to determine the couple L. For this we consider the rate of growth of A.M. about an axis perpendicular to the paper through the centre of the flywheel. If $Cn$ be the A.M. of the flywheel about its axis, C′ the moment of inertia of the case about the same axis, A the moment of inertia of the flywheel, and A′ that of the case, about a transverse axis through the centroid, we have, considering the turnings about the axes indicated,

$$(A + A' - C')\mu^2 \sin\theta\cos\theta - Cn\mu\sin\theta = L. \quad \text{..................(1)}$$

Substituting in (4) and (5) of 7, we get the two equations

$$-B\tan\theta + \{(A + A' - C')\mu^2 \sin\theta\cos\theta - Cn\mu\sin\theta\}l + \tfrac{1}{2}M\mu^2 l^2 a = 0, \quad \text{...(2)}$$

$$\tfrac{1}{2}\{(A + A' - C')\mu^2 \sin\theta\cos\theta - Cn\mu\sin\theta\}l^2 - (B - \tfrac{1}{3}M\mu^2 l^3)a = 0. \quad \text{...(3)}$$

If $\theta$ is very small, as indeed it must be, since we have supposed the curvature to be $d^2y/dz^2$, these equations are

$$[-B + \{(A + A' - C')\mu^2 - Cn\mu\}l]\theta + \tfrac{1}{2}M\mu^2 l^2 a = 0, \quad \text{..............(2')}$$

$$\tfrac{1}{2}\{(A + A' - C')\mu^2 - Cn\mu\}l^2\theta - (B - \tfrac{1}{3}M\mu^2 l^3)a = 0. \quad \text{............(3')}$$

From (2) and (3), or from (2′) and (3′), we see that if the equations are independent they are satisfied only by $\theta = 0$, $a = 0$. If however they are not independent their determinant vanishes, that is we get, from (2′) and (3′)

$$(B - \tfrac{1}{3}M\mu^2 l^3)[-B + \{(A + A' - C')\mu^2 - Cn\mu\}l]$$
$$+ \tfrac{1}{4}M\mu^2 l^4\{(A + A' - C')\mu^2 - Cn\mu\} = 0. \quad \text{...........(4)}$$

We may write this in the form

$$(A + A' - C')\mu^2 - Cn\mu = \frac{B}{l}\left(4 + \frac{36B}{M\mu^2 l^3 - 12B}\right). \quad \text{..................(5)}$$

If, as is usual, $\mu$ be very great, the right-hand side reduces to $4B/l$, and we see that the quartic has two roots not differing much from the roots, one large and one small and of opposite signs, of the quadratic to which (5) approximates.

These results have been obtained on the supposition that the flexible axis turns freely in its bearings. If however it were a piece of untwistable steel wire prevented from turning and held fast at one end in the direction from which $\theta$ is measured, we should have for the terms in C′ the expression $C'\mu^2(1-\cos\theta)\sin\theta$ [see terms in C (not C′) in (4), 21, XIX], which, since the angle $\theta$ is small, would be approximately zero. Thus (5) would become

$$(A + A')\mu^2 - Cn\mu = \frac{B}{l}\left(4 + \frac{36B}{M\mu^2 l^3 - 12B}\right). \quad \text{.................(6)}$$

We have therefore a quartic equation which gives values of $\mu$ consistent with any values of $\theta$ and $\alpha$, which have the ratio

$$\frac{\theta}{\alpha}=\frac{\frac{1}{2}M\mu^2l^2}{B-\{(A+A'-C')\mu^2-Cn\mu\}l}=\frac{2B-\frac{2}{3}M\mu^2l^3}{\{(A+A'-C')\mu^2-Cn\mu\}l^2}. \quad \ldots(7)$$

For a value of the speed given by a root of this quartic the position of the axis of the gyrostat is indeterminate; but if such speeds are avoided the values of $\theta$ and $\alpha$ are zero. The rotating body is then self-centring.

If A, A′, C, C′ are so small as to be negligible, we obtain from (5), by putting the expression on the left equal to 0,

$$M\mu^2l^3=3B,$$

that is a critical speed is given by

$$\mu=\left(\frac{3B}{Ml^3}\right)^{\frac{1}{2}},$$

a result which will be found in accordance with results set forth in the next article. The purely gyrostatic action is in general of only secondary importance.

These results and those which follow are of great importance for the running of rapidly rotating machinery, such as high-speed turbines. A flexible shaft is used, and critical speeds [those which satisfy conditions analogous to (7)] are avoided. The rotating body then revolves stably with perfect balance of centrifugal forces.

**9.** *A rotor carried midway between the two bearings of a flexible shaft.* If the flexible axle is held on two bearings, with the gyrostat or wheel attached midway between them, we may have one or other of two arrangements. In the first the ends, though fixed in position, are free as regards direction, in the second the bearings are made to maintain the directions of the ends in coincidence. If we neglect the inertia of the shaft itself we obtain at once the deflections of the shaft in the two cases from the theory of the bending of a thin rod. If the deflections produced by the turning be $y-y_0$, where $y_0$ is the initial deflection from the straight at the mid-point, the deflecting force be F, and the distance from bearing to bearing be $2l$, we have for steady rotation in the two cases

$$B(y-y_0)=\tfrac{1}{6}Fl^3, \quad B(y-y_0)=\tfrac{1}{24}Fl^3. \quad \ldots\ldots\ldots\ldots(1)$$

But $F=M\omega^2y$, where M is the mass of the flywheel or rotor, and $\omega_1$ is the speed at which the wheel is run. Thus

$$y=\frac{y_0}{1-\frac{1}{6}\frac{Ml^3}{B}\omega^2} \quad \text{or} \quad y=\frac{y_0}{1-\frac{1}{24}\frac{Ml^3}{B}\omega^2}. \quad \ldots\ldots\ldots\ldots(2)$$

These two equations give infinite values of $y$ for $\omega^2=6B/Ml^3$ and $\omega^2=24B/Ml^3$, respectively. Of course the condition for small strains is deviated from long before these speeds are reached, and the equations must not be taken as holding except for small strains. But they indicate speeds that must not be approached if there is to be safety, and, which is of great importance, show that speeds very much higher than these critical speeds $[(6B/Ml^3)^{\frac{1}{2}}, (24B/Ml^3)^{\frac{1}{2}}]$ are consistent with small deflections, which approach zero as the value of $\omega$ is increased beyond limit.

In practice serious deflections of the flexible shaft are avoided by means of stops, and the speed is quickly and safely run up to a high value, at which extreme smoothness of running is obtained. De Laval chose for his turbines $\omega = 7\omega_c$, where $\omega_c$ denotes the critical speed.

Take fixed axes $Ox$, $Oy$ of coordinates in a plane at right angles to the line of bearings and containing the centre of the wheel: the equations of motion of the centroid are

$$\ddot{x} + f(x-\xi) = 0, \quad \ddot{y} + f(y-\eta) = 0, \quad \text{......(3)}$$

where $\xi$, $\eta$ are the coordinates of the point of radial coordinate $y_0$ or $r_0$, which we shall assume is made to revolve about the line of bearings with angular speed $\omega$. For steady turning we have just found

$$F = 6\frac{B}{l^3}(r - r_0) \quad \text{or} \quad F = 24\frac{B}{l^3}(r - r_0). \quad \text{......(4)}$$

We may express both these results by

$$F = M\omega_c^2(r - r_0); \quad \text{......(4')}$$

so that

$$M\omega^2 r = M\omega_c^2(r - r_0). \quad \text{......(5)}$$

From this result we assume that $f(x-\xi) = \omega_c^2(x-\xi)$, $f(y-\eta) = \omega_c^2(y-\eta)$, so that (3) becomes

$$\ddot{x} + \omega_c^2 x = \omega_c^2 \xi, \quad \ddot{y} + \omega_c^2 y = \omega_c^2 \eta. \quad \text{......(6)}$$

But the point $\xi, \eta$, from which the displacements $x-\xi$, $y-\eta$ are measured, goes round in a circle with angular speed $\omega$. Hence we write (6) as

$$\ddot{x} + \omega_c^2 x = r_0\omega_c^2 \cos \omega t, \quad \ddot{y} + \omega_c^2 y = r_0\omega_c^2 \sin \omega t. \quad \text{......(7)}$$

If we put $z = x + iy$, we can write (7) in one equation,

$$\ddot{z} + \omega_c^2 z = r_0\omega_c^2(\cos \omega t + i \sin \omega t). \quad \text{......(8)}$$

For the forced vibration in period $2\pi/\omega$, we assume

$$z = A(\cos \omega t + i \sin \omega t).$$

Substituting in (8), we get $A = r_0 \dfrac{\omega_c^2}{\omega_c^2 - \omega^2}$.

The complete solution is thus

$$z = A \cos \omega_c t + B \sin \omega_c t + r_0 \frac{\omega_c^2}{\omega_c^2 - \omega^2}(\cos \omega t + i \sin \omega t). \quad \text{......(9)}$$

If $A = \alpha + i\alpha'$, $B = \beta + i\beta'$, we have

$$\left.\begin{aligned} x &= \alpha \cos \omega_c t + \beta \sin \omega_c t + r_0 \frac{\omega_c^2}{\omega_c^2 - \omega^2} \cos \omega t, \\ y &= \alpha' \cos \omega_c t + \beta' \sin \omega_c t + r_0 \frac{\omega_c^2}{\omega_c^2 - \omega^2} \sin \omega t. \end{aligned}\right\} \quad \text{......(10)}$$

We see thus that there is vibration in the period $2\pi/\omega_c$ about the circular motion of period $2\pi/\omega$. These vibrations may die out, but they are renewed again with every irregularity of running. This circular motion will not remain of small radius $r_0\omega_c^2/(\omega_c^2-\omega^2)$ if $\omega^2$ approaches $\omega_c^2$ in value. But for large values of $\omega^2$ we have

$$\left.\begin{aligned} x &= a\cos\omega_c t+\beta\sin\omega_c t-r_0\frac{\omega_c^2}{\omega^2-\omega_c^2}\cos\omega t,\\ y &= a'\cos\omega_c t+\beta'\sin\omega_c t-r_0\frac{\omega_c^2}{\omega^2-\omega_c^2}\sin\omega t.\end{aligned}\right\}\quad\text{.................(11)}$$

The forced vibration becomes opposite in phase to the free vibration when the value $\omega^2=\omega_c^2$ is overpassed.

10. *Free period of an oscillatory disturbance is the period of revolution at the critical speed.* It will be observed that it follows from (6) or from (11) that if the flywheel be displaced in any way from the steady revolutionary motion, provided there is no tilting of the wheel, the period of the oscillatory disturbance set up is $2\pi/\omega_c$, that is the period is that of revolution at the critical speed. This result was given by Greenhill for a whirling shaft in a paper entitled "On the strength of shafting when exposed both to torsion and to end thrust," in the *Proceedings of the Institution of Mechanical Engineers*, 1883.* Clearly it is a general result which might be predicted without any calculation. For the rotation in period $2\pi/\omega$ is a forced oscillation which, when performed in the free period $2\pi/\omega_c$, will be continually augmented at the expense of energy derived from the driving motor.

The effects of torsion and end-thrust on the behaviour of shafts are discussed by Greenhill in the paper cited in 10 above. The lateral oscillations of the shafting itself are considered also: in the discussion above the inertia of the shaft has been treated as negligible. If the speed of rotation is continually increased, one critical speed of the shaft after another is passed, one for each mode of lateral vibration. Each mode gives a musical note, and these notes show the frequencies of the modes of lateral vibration of which the shaft as mounted is capable.

With regard to actual dimensions of shafts, the choice will depend on the particular case. Of course shafts of sufficiently great cross-section are required for the transmission of power, which is done by torsion in a propeller shaft, and indeed in all cases in which a motor does work and is driven by pulleys at some distance on the shaft. For gyroscopes weighing from 1 kilogramme to 4 kilogrammes, Dr. J. G. Gray uses steel wire 2 millimetres thick, and about 10 centimetres in length. These shafts are cased round loosely by tubular guards to prevent damage from disturbances at the critical speed.

* See also articles by Sir George Greenhill in *Engineering*, March and April, 1918.

**11.** *Quasi-rigidity of a moving chain. Equations of motion.* The quasi-rigidity conferred on a flexible chain by motion is, like gyrostatic action, of much interest in connection with possible kinetic explanations of the properties of bodies. The equations of motion of a perfectly flexible inextensible chain, of very small links, are easily found by the method so often employed above. Let $mu\,ds$, $mv\,ds$ be the momenta of an element $ds$ along the tangent and the normal at the element, and $\phi$ the angle which the forward direction of the tangent at the element makes at the instant with a fixed direction in the plane. The angular speed with which the tangent is turning round is then $\dot\phi$: we shall suppose this to be positive. The rates of production of momentum along the tangent and normal respectively are $m(\dot u - v\dot\phi)\,ds$, $m(\dot v + u\dot\phi)\,ds$. The forces in these directions are $\mathrm{S}m\,ds + d\mathrm{T}/ds\,.\,ds$, $\mathrm{N}m\,ds$, where T is the traction in the chain and S, N are the tangential and normal applied forces, per unit of mass of the element. Thus we get at once the equations of motion,

$$m(\dot u - v\dot\phi) = m\mathrm{S} + \frac{d\mathrm{T}}{ds},\quad m(\dot v + u\dot\phi) = m\mathrm{N} + \frac{\mathrm{T}}{\mathrm{R}},\quad \text{.................(1)}$$

where R is the radius of curvature.

There are two kinematical equations. Since the chain is inextensible the tangential speed is the same at all points. Consider then the distribution of velocity as it exists at time $t$, and imagine a particle to pass along the curve supposed kept in that position, taking at successive elements the values of $u, v$ at time $t$. The change of $u$ between the two ends of the element $ds$ is $(du/ds - v/\mathrm{R})ds$ [since $ds/\mathrm{R}$ is the angle between the two ends], and this must be zero. Hence we have for a given instant

$$\frac{du}{ds} - \frac{v}{\mathrm{R}} = 0. \quad \text{.................................(2)}$$

Again the rate at which the element $ds$ is turning round is clearly $(v + dv - v)/ds + u/\mathrm{R}$. For the angle $d\phi$ turned through in an element of time $dt$ by the direction of motion of a particle of the chain is $dv/ds\,.\,dt + ds/\mathrm{R}$ $= (dv/ds + u/\mathrm{R})dt$. Hence we have (again at a given instant)

$$\frac{dv}{ds} + \frac{u}{\mathrm{R}} = \dot\phi. \quad \text{.................................(3)}$$

This angular speed $\dot\phi$ is not to be confused with $u/\mathrm{R}$, which is the rate of change of direction due to the curvature of the chain. The two ends of the element are moving at different rates along the respective normals, and the rate of turning $dv/ds$ is due to this. In steady motion however there is not at any element a component of speed along the normal and so $dv/ds = 0$.

If the chain be in steady motion, that is move uniformly without change of figure or change of position of its curve in space (except uniform translation as a whole), then $\dot u = \dot v = 0$, and $v = 0$. Also $\dot\phi = u/\mathrm{R}$. Equations (1) are now

$$m\mathrm{S} + \frac{d\mathrm{T}}{ds} = 0,\quad \frac{mu^2}{\mathrm{R}} - m\mathrm{N} = \frac{\mathrm{T}}{\mathrm{R}}. \quad \text{........................(4)}$$

If the chain be under gravity and the inclination of the normal to the upward vertical be $\theta$, we have $S = g\sin\theta$, $N = -g\cos\theta$. Thus we have

$$mg\sin\theta + \frac{dT}{ds} = 0, \quad m\left(\frac{u^2}{R} + g\cos\theta\right) = \frac{T}{R}. \qquad (5)$$

The chain moves in the equilibrium curve of the chain at rest. The only difference is that T exceeds its value for stationary equilibrium by $mu^2$. If the speed $u$ is very great the value of T is very great in comparison with the applied forces, and therefore these forces have little effect on the figure of the chain.

12. *A bend in a moving chain is not carried along the chain by the motion of the links.* Thus, if a chain [Fig. 120] hanging over a rapidly rotating pulley in a vertical plane be struck by a hammer, the indentation caused by the blow will remain in the same position with respect to surrounding bodies, and be very slowly obliterated by the action of gravity. The moving chain is stiff and bends under the blow like a solid inelastic bar.

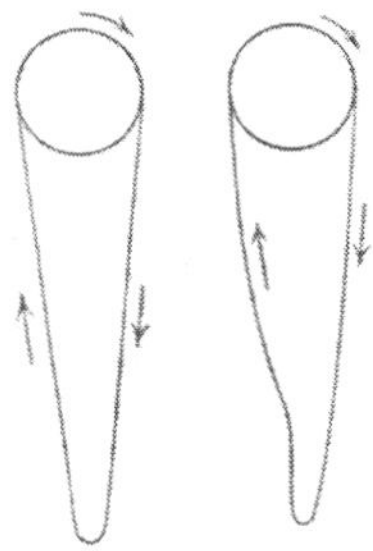

FIG. 120.

The experiment may be made with an ordinary curb chain weighing perhaps half a pound per foot. The pulley may be a foot in diameter, be deeply grooved, and be driven at a high speed by a suitable motor or by a counter-shaft arrangement. The chain may be 10 or 12 feet long, and should be struck by a hammer-head capable of rotating about an axis transverse to the chain when in the act of striking.

Some experiments with lighter chains are interesting as showing rigidity imparted to a flexible chain by motion of its parts. An overhanging vertical pulley carried by a whirling table has a shallow groove round its edge. In this is laid an endless chain only a little longer than the circumference of the pulley. Just under the pulley is arranged a plane sloping slightly downwards in the direction of motion of the top of the pulley. When the chain has been got into rapid motion it is knocked off the pulley by a sharp tap with a pencil delivered sideways, and drops on the inclined plane, along which it runs for several feet like a rigid hoop before it collapses in a flaccid heap.

13. *Theory of a stationary bend on a moving chain, or of a moving bend on an otherwise stationary chain.* To understand how these effects take place, let us suppose that there are no tangential applied forces. The value of T is then $mu^2$. For let the chain move without friction through a tube of any shape under no forces except those applied by the sides of the tube. Thus no force along the chain is applied by the tube, and the

stretching force is not affected by the tube. This force is equal to the pull applied at and in the direction of the ends of the tube where the chain is free. But if $T=mu^2$ there, as it must be by (4), 11, that equation gives $N=0$ everywhere, that is no action is exerted on the chain by the tube. We have

$$u=\left(\frac{T}{m}\right)^{\frac{1}{2}}.$$

Let the diagram represent an open loop on the chain imposed by the shape of the tube, and let the ends of the chain be straight and in line. The chain enters the loop at one end A and emerges at the other B,

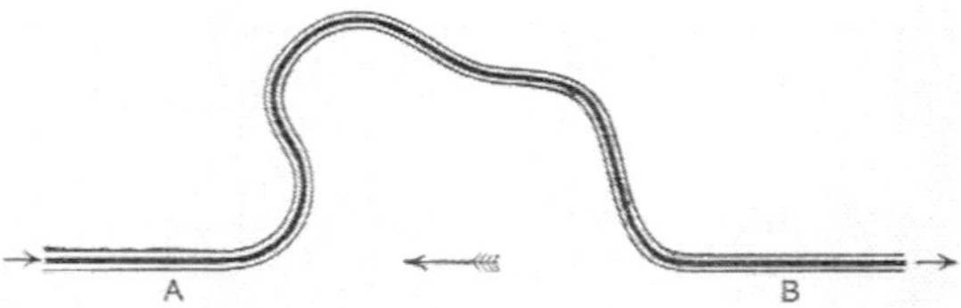


FIG. 121.

with speed $u$ in each case; and as we have seen there is no action between chain and tube. This fact will not be altered if the whole system of chain and tube be moved with speed $u$ in the opposite direction to the motion of the ends of the chain. Thus we have the straight parts of the chain brought to rest, and the loop or bend is made to travel along the chain with speed $u=(T/m)^{\frac{1}{2}}$. Since there is no action of the tube it may be removed.

Thus the speed of propagation of a bend of any shape along a uniform flexible chain or cord under stretching force T is $(T/m)^{\frac{1}{2}}$. This is the result given by the ordinary mathematical theory of the propagation of waves of transverse displacement along cords or chains, where, however, to avoid mathematical difficulties, the transverse displacements are supposed to be small—here no such limit is imposed.

Again consider a flexible tube through which runs a stream of a fluid, such as water, which completely fills the tube. If at rest, the tube will apply, per unit of its length, at any cross-section at which the bend imposed by the tube has radius of curvature R, a force $mu^2/R$ to the fluid. This will arise if there exist a stretching force $T=mu^2$ in the tube applied by external action. The combination of stretched tube and flowing water therefore imitates the quasi-rigidity of the moving chain.

**14.** *A revolving chain under no forces. General case.* We consider also in this connection the problem of a uniform chain revolving with steady angular speed $n$ about an axis $Oz$ while under the action of no forces. Let $x, y, z$ be the coordinates of an element $ds$ of the chain, $m$ the mass per unit length, T the tension at $ds$. The equations of relative equilibrium are

$$\frac{d}{ds}\left(T\frac{dx}{ds}\right)+n^2mx=0, \quad \frac{d}{ds}\left(T\frac{dy}{ds}\right)+n^2my=0, \quad \frac{d}{ds}\left(T\frac{dz}{ds}\right)=0. \quad \text{......(1)}$$

These give at once, since $(dx/ds)^2+(dy/ds)^2+(dz/ds)^2=1$,

$$\frac{dT}{ds}+n^2m\left(x\frac{dx}{ds}+y\frac{dy}{ds}\right)=0,$$

and

$$\frac{dT}{ds}\left(x\frac{dy}{ds}-y\frac{dx}{ds}\right)+T\left(\frac{d^2y}{ds^2}x-\frac{d^2x}{ds^2}y\right)=0.$$

Hence we obtain the three first integrals of the equations (1),

$$T+\tfrac{1}{2}n^2m(x^2+y^2)=H, \quad T\left(x\frac{dy}{ds}-y\frac{dx}{ds}\right)=N, \quad T\frac{dz}{ds}=Z, \quad \text{.........(2)}$$

where H, N, Z are constants.

The second and third of (2) combined give

$$x\frac{dy}{dz}-y\frac{dx}{dz}=\frac{N}{Z}, \quad \text{..................................(3)}$$

the so-called equation of areas. If the radius vector, drawn from the chosen origin to the projection of $ds$ on the plane of $x$, $y$ be $\rho$, and the vectorial angle be $\chi$, we get for (3) the polar equation

$$\rho^2\frac{d\chi}{dz}=\frac{N}{Z}. \quad \text{.....................................(3')}$$

Again we have

$$\tfrac{1}{4}\left(\frac{\partial\rho^2}{\partial z}\right)^2=(x^2+y^2)\left\{\left(\frac{dx}{dz}\right)^2+\left(\frac{dy}{dz}\right)^2\right\}-\left(x\frac{dy}{dz}-y\frac{dx}{dz}\right)^2$$

$$=\rho^2\left\{\left(\frac{ds}{dz}\right)^2-1\right\}-\frac{N^2}{Z^2}.$$

But $ds/dz=T/Z$. Hence it is found that

$$\left(\frac{\partial\rho^2}{\partial z}\right)^2=\left(\frac{1}{2}\frac{n^2m}{Z}\right)^2 S, \quad \text{..............................(4)}$$

if

$$S=4\rho^2\left\{\left(\rho^2-\frac{2H}{n^2m}\right)^2-4\frac{Z^2\rho^2+N^2}{(n^2m\rho)^2}\right\}. \quad \text{.....................(5)}$$

Thus we obtain by integration

$$z=\frac{2Z}{n^2m}\int\frac{d\rho^2}{S^{\frac{1}{2}}}. \quad \text{..................................(6)}$$

Also, eliminating $z$ from (3') by the relation

$$dz=\frac{2Z}{n^2m}\frac{d\rho^2}{S^{\frac{1}{2}}},$$

we get

$$\chi=\frac{2N}{n^2m}\int\frac{d\rho^2}{\rho^2S^{\frac{1}{2}}}. \quad \text{..................................(7)}$$

**15.** *Revolving chain in a plane containing the axis of rotation.* We have thus obtained the equations for a revolving chain in the general case in which the chain lies in a tortuous curve. If however it lies at each instant in a plane containing O$z$, we have an example of a revolving catenary. Each particle moves in a circle about O$z$, and if we project on the plane of $x$, $y$ the paths of two particles which are on levels slightly differing, so that points on the same radius have coordinates $x$, $y$, $x+dz$, $y+dy$, we see that $y/x=dy/dx$. Hence by (2), 14, N$=0$, and

$$S=4\rho^2\left\{\left(\rho^2-\frac{2H}{n^2m}\right)^2-4\left(\frac{Z}{n^2m}\right)^2\right\}, \quad\text{.........(1)}$$

and
$$\left(\frac{d\rho}{dz}\right)^2=\left(\frac{n^2m}{2Z}\right)^2\left\{\left(\rho^2-\frac{2H}{n^2m}\right)^2-4\left(\frac{Z}{n^2m}\right)^2\right\}. \quad\text{.........(2)}$$

If we write
$$y^2=\frac{\rho^2n^2m}{2(H-Z)},\quad k^2=\frac{H-Z}{H+Z}, \quad\text{.........(3)}$$

we obtain from (2), after a little reduction,

$$\frac{nm^{\frac{1}{2}}}{2^{\frac{1}{2}}}\frac{(H+Z)^{\frac{1}{2}}}{Z}dz=\frac{dy}{\{(1-y^2)(1-k^2y^2)\}^{\frac{1}{2}}}.$$

Thus, if $y=\sin\phi$,
$$\rho=\frac{2^{\frac{1}{2}}(H-Z)^{\frac{1}{2}}}{nm^{\frac{1}{2}}}\,\mathrm{sn}\,F(k,\phi). \quad\text{.........(4)}$$

The curve may lie in a plane perpendicular to the axis O$z$, and in this case $dz/ds=0$, so that $Z=0$. Hence, by (7), 14,

$$\chi=\frac{2N}{n^2m}\int\frac{d\rho^2}{\rho^2\left\{4\rho^2\left(\rho^2-\frac{2H}{n^2m}\right)^2-16\left(\frac{N}{n^2m}\right)^2\right\}^{\frac{1}{2}}}. \quad\text{.........(5)}$$

If we write $\zeta$ for $\rho^2$, this becomes

$$\chi=\frac{2N}{n^2m}\int\frac{d\zeta}{\zeta\left\{4\zeta\left(\zeta-\frac{2H}{n^2m}\right)^2-16\left(\frac{N}{n^2m}\right)^2\right\}^{\frac{1}{2}}}, \quad\text{.........(6)}$$

an expression which is capable of being dealt with at once by the Weierstrassian elliptic function analysis.

**16.** *Remarks on a more general problem of the motion of a chain.* The more general problem in which the chain is supposed to be in motion with some uniform tangential speed $v$, in a definite plane path, while at the same time the path as a whole is whirled with constant angular speed $n$ about an axis O$z$ at right angles to the path, is interesting, but has little practical bearing on gyrostatics. Its solution, and the determination of possible paths, involve difficulties of integration which do not appear when the motion is wholly due to the uniform rotation about O$z$. If, for example, it is attempted to deal with the problem, with or without special conditions, by the method of revolving axes, then it is to be noticed that besides the rotational speed $n$ about the axis O$z$, each element has a rotational speed $v/R$, where R is the radius of curvature at the point of the trajectory at which the centre of the element is situated at the instant. But R varies from point to point of the trajectory, and so in the integrals for the determination of $\rho$ and $\chi$, this varying radius, which is introduced in the calculation of the accelerations along and at right angles to the moving tangent, makes the problem almost, if not entirely, intractable. It is however an excellent example of the theory of moving axes to calculate these accelerations. We shall find first the normal acceleration.

If $p$ be the length of the perpendicular from O on the tangent to an element with its centre at a point P fixed relatively to the trajectory, so that with axes O$x$ outward from O and parallel to the normal at P, and O$y$ drawn from O forward parallel to the tangent at P, then $x$ $(=p)$ and $y$ are the coordinates of P. The velocities U, V along fixed directions coinciding with the axes are given by

$$\mathrm{U}=\dot{p}-\left(n+\frac{v}{\mathrm{R}}\right)y, \quad \mathrm{V}=\dot{y}+\left(n+\frac{v}{\mathrm{R}}\right)p.$$

The acceleration towards the centre of curvature is

$$-\dot{\mathrm{U}}+\left(n+\frac{v}{\mathrm{R}}\right)\mathrm{V}=-\ddot{p}+2\left(n+\frac{v}{\mathrm{R}}\right)\dot{y}-v\frac{y\dot{\mathrm{R}}}{\mathrm{R}^2}+\left(n+\frac{v}{\mathrm{R}}\right)^2 p.$$

This is the acceleration of a particle of the chain; the acceleration of a point such as P *fixed relatively to the trajectory* is a different matter, and does not involve the variation of R.

For the reduction of the expression thus found we note that if $ds$ be the length of an element of the path, the time required by a particle to traverse it is $ds/v$. Thus $d/dt=v d/ds$. Hence $\dot{y}=v\,dy/ds$, and this is easily seen to be given by

$$\dot{y}=v\frac{dy}{ds}=v\frac{ds-p\dfrac{ds}{\mathrm{R}}}{ds}=v\left(1-\frac{p}{\mathrm{R}}\right),$$

while

$$\dot{p}=v\frac{dp}{ds}=v\frac{y}{\mathrm{R}},$$

since $dp/y=ds/\mathrm{R}$. Substituting, we find after reduction

$$-\dot{\mathrm{U}}+\left(n+\frac{v}{\mathrm{R}}\right)\mathrm{V}=\frac{v^2}{\mathrm{R}}+n^2p+2nv.$$

Hence we have the equation of motion

$$m\left(\frac{v^2}{\mathrm{R}}+n^2p+2nv\right)=\frac{\mathrm{T}}{\mathrm{R}}. \qquad \text{.......(1)}$$

The same process gives for the acceleration along the tangent the value $-n^2y$. This must be produced by the excess $d\mathrm{T}$ of the tractive force in the chain at the forward end of $ds$ over that at the other end, so that

$$d\mathrm{T}=-mn^2y\,ds. \qquad \text{.......(2)}$$

But, as a diagram will show at once, $\rho\,d\rho=y\,ds$, and we get

$$d\mathrm{T}=-mn^2\rho\,d\rho.$$

Integrating, we obtain

$$\mathrm{T}=-\tfrac{1}{2}mn^2\rho^2+\mathrm{H}, \qquad \text{.......(3)}$$

where H is a constant.

We have seen, (1) above, that

$$\frac{\mathrm{T}}{m}=2nv\mathrm{R}+n^2p\mathrm{R}+v^2,$$

so that

$$\frac{1}{n^2}\left(\frac{\mathrm{T}}{m}-v^2\right)=\frac{2}{n}v\mathrm{R}+p\mathrm{R}.$$

Replacing R by its value $\rho\,d\rho/dp$, and writing $h$ for $2(\mathrm{T}/m-v^2)/n^2$, we obtain, by integration,

$$hnp=2v\rho^2+np\rho^2+\mathrm{C}, \qquad \text{.......(4)}$$

where C is a constant, the value of which depends on the origin chosen for the integration. Let us suppose that the integration is started from a point on the trajectory where the radius vector of length $\rho_1$, say, is normal to the curve, and is therefore equal to the length $p_1$ of the perpendicular from O at that point. Hence we obtain for C the value

$$-(n\rho_1^3+2v\rho_1^2)+hn\rho_1.$$

We have thus the cubic equation

$$n\rho^3+2v\rho^2-hn\rho-n\rho_1^3-2v\rho_1^2-hn\rho_1=0 \quad\text{.........................(5)}$$

for the determination of the locus of the points at which the radius vector is normal to the curve. No generality will be lost by supposing $n$ to be positive. Equation 5 would appear to give three circles which must be touched by the curve. There are three cases of real roots, (1) all three of one sign, (2) two positive and one negative, (3) two negative and one positive. In (1) the direction of $v$ is either with or against that of $\rho_1 n$ at all three circles, in (2) and (3) $v$ and $\rho_1 n$ either agree or are opposed in sign at two circles, while they are opposed or agree in sign at the third, as the case may be. For three real and positive roots the quantities

$$-\frac{2v}{n}, \quad -h, \quad n\rho_1^3+2v\rho_1^2-nh\rho_1$$

must all be positive, that is $v$ and $n$ must have opposite signs (in other words the motion $v$ must be opposed to that due to the rotation $n$), and $h$ must be negative, that is $v^2>T/m$.

Also since for three positive roots $n\rho_1^3+2v\rho_1^2-hn\rho_1$ must be positive, and $\rho_1$ is positive, $n\rho_1^2+2v\rho_1-nh$ must be positive ($=k^2$, say). Thus, for the avoidance of complex values of $\rho$, for all possible values of $k^2$, $v^2/n^2+h$ must be positive, that is

$$\frac{v^2}{n^2}>-h.$$

Going back to (4) and differentiating with respect to $\rho$, we get, after reduction,

$$\frac{1}{\rho}\frac{dp}{d\rho}=\frac{4v+2np}{n(h-\rho^2)}=\frac{1}{R}. \quad\text{..................................(6)}$$

Thus the curvature is infinite if $h=\rho^2$. If $h$ is negative, as it is for three positive roots of the cubic, no point of infinite curvature can occur.

From (4), with the value of the constant of integration found above, we obtain

$$pn+2v=-(\rho_1 n+2v)(\rho_1^2-h)/(\rho^2-h),$$

so that

$$\frac{1}{\rho}\frac{dp}{d\rho}=\frac{1}{R}=-\frac{2}{n}\frac{(\rho_1 n+2v)(\rho_1^2-h)}{(\rho^2-h)^2}. \quad\text{.............................(7)}$$

Hence R is zero, that is the curvature is infinite if $\rho^2=h$. This cannot happen if $h$ is negative.

For three real and positive roots the curve would appear to resemble the horizontal projection of the horizontal projection of the curve of motion of the spherical pendulum as given in Fig. 80 [2, XV, above]. For the curvature at the different circles is given by the insertion of the proper values of $\rho^2$ in the denominator of the expression on the right of (7). In each case it is necessary, in order that the curve should be concave towards O, that $v$ should be negative and $|2v|>\rho_1 n$. It is clear that the curve must be thus concave at the outer circle; and if it were to become convex at a less distance from O R would have to change sign, which the expression on the right of (7) shows is impossible.

There remain the cases (2) and (3). In case (2) $n\rho_1^3+2v\rho_1^2-nh\rho_1$ is negative, in case (3) it is positive.

There are different possible sub-cases, which we cannot discuss here. The following general fact may however be noted. In cases (2) and (3) there will be loops which touch one circle. Points of inflexion must exist if the loops touch the outer circle, but not if they touch the inner circle. An enumeration and description of all the cases would occupy much space.

**17.** *Calculation of the vectorial angle, in terms of the radius vector, for the curve of the chain.* Returning for a moment to (2), 16, we consider the case in which T, and also H, varies as $n^2$. Using now $v$ in the sense of the tangential speed due to the rotation about $Oz$, we have $v=np$, where $p$ is the length of the perpendicular on the curve from O, for the point considered. But now the components of force on an element both involve $n^2$ in the same way, and thus the direction of the element does not depend on $n^2$. From (6), 15, it appears that in this case the vectorial angle also does not depend on $n^2$.

As before, we put $\zeta=\rho^2$, and also write $v^2/n^2=\zeta'$. Thus we have $h+2\zeta'=2\mathrm{H}/mn^2$, and as we shall see presently, $(h+2\zeta'-\zeta_1)\,\zeta_1^{\frac{1}{2}}=4\mathrm{N}^2/(n^2m)^2$. Hence (6), 15, can be written

$$\chi=(h+2\zeta'-\zeta_1)\zeta_1^{\frac{1}{2}}\int\frac{d\zeta}{2\zeta\{\zeta^3-2(h+2\zeta')\zeta^2+(h+2\zeta')^2\zeta-(h+2\zeta'-\zeta_1)^2\zeta_1\}^{\frac{1}{2}}}. \quad (1)$$

With regard to the substitution for $4\mathrm{N}^2/(n^2m)^2$, it is to be noticed that by the second of (2), 1, we have $\mathrm{T}n\zeta_1/v=\mathrm{N}$, or $\mathrm{T}_1\,\zeta_1^{\frac{1}{2}}=\mathrm{N}$, where $\mathrm{T}_1$ is the tension of the chain when $\rho^2=\rho_1^2=\zeta_1$. But by the first of (2), 1, we have $\frac{1}{2}mn^2(h+2\zeta'-\zeta_1)\zeta_1^{\frac{1}{2}}=\mathrm{N}$, which is the substitution.

Writing the cubic expression in the denomination of the integrand in (1) as

$$\mathrm{Z}=(\zeta-\zeta_1)(\zeta-\zeta_2)(\zeta-\zeta_3), \quad \text{.........................(2)}$$

we shall suppose that $\zeta_3>\zeta_2>\zeta_1$, where $\rho_1^2=\zeta_1$. The variable $\zeta$ must lie between $\zeta_3$ and $\zeta_2$, so that according to the supposition made $\zeta_1$, $\zeta_2$ are the squares of the radii of two limiting circles between which the path lies.

**18.** *Integration by elliptic functions.* The integration indicated in (1), 17 can be carried out in the following manner. Writing $z+k$ for $\zeta$ in the expression for 4Z we obtain

$$4\mathrm{Z}=4(z+k)^3-8(h+2\zeta')(z+k)^2+4(h+2\zeta')^2(z+k)-4(h+2\zeta'-\zeta_1)^2\zeta_1. \quad \text{...........(1)}$$

Expanding this, and choosing $k$ so that the coefficient of $z^2$ is zero, we find

$$4\mathrm{Z}=4z^3-3k^2z+k^3-4(h+2\zeta'-\zeta_1)^2\zeta_1, \quad k=\tfrac{2}{3}(h+2\zeta'). \quad \text{.....................(2)}$$

Thus 4Z has been converted to the Weierstrassian form $4z^3-g_2z-g_3$, with

$$g_2=3k^2, \quad g_3=-k^3+4(h+\zeta'-\zeta_1)^2\zeta_1. \quad \text{..............................(3)}$$

Putting $\wp u$ for $z$, and $e_3$ for $z_3$, we have in the usual notation,

$$\omega_3=\int_{e_3}^{\infty}\frac{d\wp}{(4\wp^3-g_2\wp-g_3)^{\frac{1}{2}}}, \qquad u=\int_{\wp}^{\infty}\frac{d\wp}{(4\wp^3-g_2\wp-g_3)^{\frac{1}{2}}}. \quad \text{....................(4)}$$

Thus

$$-du=\frac{d\wp}{(4\wp^3-g_2\wp-g_3)^{\frac{1}{2}}},$$

so that we get for the integral sought,

$$-\tfrac{1}{2}\int\frac{du}{z+k}=-\tfrac{1}{2}\int\frac{du}{\wp u-\wp v}, \quad\text{......(5)}$$

if we write $-k=\wp v$. The limits are the values of $u$ when $z=\wp u$ and $z=e_3$ respectively. Now we have

$$\int\frac{du}{\wp u-\wp v}=\frac{1}{\wp' v}\left\{2u\zeta v+\log\frac{\sigma(u-v)}{\sigma(u+v)}\right\},$$

or taking the integral between the limits stated,

$$-\int\frac{du}{\wp u-\wp v}=\frac{1}{\wp' v}\left\{2(\omega_3-u)\zeta v+\log\frac{\sigma(\omega_3-v)\sigma(u+v)}{\sigma(\omega_3+v)\sigma(u-v)}\right\}. \quad\text{......(6)}$$

The value of $1/\wp' v$ can be calculated by the equation

$$-\wp' v=(4\wp^3 v-g_2\wp v-g_3)^{\frac{1}{2}}. \quad\text{......(7)}$$

These problems are of some interest in the theory of rotation, but they are not direct examples of gyrostatic action. Further particulars will be found in Greenhill, *R.G.T.*, pp. 78, 193.

# CHAPTER XXIII

## EXAMPLES OF GYROSTATIC ACTION AND ROTATIONAL MOTION

**1.** *A gyroscope mounted on the earth.* A gyroscope is mounted on the earth so that the axis of the flywheel is constrained to remain in a plane fixed in the earth. It is required to find the motion.

Referring to Fig. 12, we take OEC as the plane in which the axis is free to move, and O$z$ as the projection on that plane of the earth's axis of rotation. The axis OD is at right angles to the axes OE and OC. If the angle between the axis of rotation of the earth and O$z$ be $\alpha$, and $\omega$ be the earth's angular speed, the components about O$z$ and OD are $\omega \cos \alpha$ and $\omega \sin \alpha$ respectively. We denote the angle CO$z$ by $\chi$. The angular speed about OC due to the rotation of the earth is thus $\omega \cos \chi \cos \alpha$, and that of the wheel of the gyroscope is $\dot{\phi} + \omega \cos \chi \cos \alpha$ $(=n$, say). The angular speed about OE is $\omega \sin \chi \cos \alpha$. It is supposed that the system has no gravitational stability.

The rate of production of A.M. about OD is made up of $\mathrm{A}\ddot{\chi}$ due to acceleration, $\mathrm{C}n\omega \sin \chi \cos \alpha$ due to the turning of OC (with its associated A.M. C$n$) about OE with angular speed $\omega \sin \chi \cos \alpha$, and $-\mathrm{A}\omega^2 \cos^2 \alpha \sin \chi \cos \chi$ due to the turning of OE (with the associated A.M., $\mathrm{A}\omega \sin \chi \cos \alpha$), with angular speed $\omega \cos \chi \cos \alpha$ about OC. Thus we get

$$\mathrm{A}\ddot{\chi} + (\mathrm{C}n\omega - \mathrm{A}\omega^2 \cos \alpha \cos \chi) \cos \alpha \sin \chi = 0.$$

Since $\omega$ is small this gives small oscillations of OC about O$z$ in the period

$$2\pi (\mathrm{A}/\mathrm{C}n\omega \cos \alpha)^{\frac{1}{2}},$$

that is in the period of a simple pendulum of length $g\mathrm{A}/\mathrm{C}n\omega \cos \alpha$.

It will be noticed that if $\alpha = 0$, that is if O$z$ be the direction of the earth's axis, the equilibrium position is that in which the axis OC coincides with the earth's axis. The apparent speed of rotation of the gyrostat is (if $\chi$ is taken an acute angle) $n \pm \omega \cos \chi \cos \alpha$, according as the gyrostat axis points towards the south or the north end of the earth's axis. [Quet, *Liouville's Journal*, 1853.] The discussion (*loc. cit.*) is carried out by means of the general equations, and is very long. The instability which exists when the gyrostat points south is noticed. It will be seen that this is another example of the instability described in 7, VIII.

To find the constraining couple required to keep the axis in the plane fixed in the earth we notice that the rate of production of A.M. about OE due to the motion, and calculated by the method exemplified above, is $(-\mathrm{C}n\omega + \mathrm{A}\omega^2 \cos \alpha \cos \chi) \sin \alpha$. This must be balanced by the couple of constraint.

**2.** *A gyroscope with its axis of spin a generator of a cone fixed in the earth.* In papers by Quet and Bour (*Liouville's Journal*, 1853 and 1863) the previous problem, rendered more general by the condition that the axis of spin is constrained to remain in contact as a generator with the surface of a right circular cone fixed in the earth, is discussed. The following is a brief direct solution from first principles.

Draw a line ON, to represent the earth's axis of rotation, through the vertex of the cone O, and let OC be the position of the spin-axis. Take a section of the cone at right angles to the axis of the surface. Then $\omega$, the earth's angular velocity, can be resolved into two components, one $\omega \cos \beta$ about the axis of the cone (where $\beta$ is the inclination of the cone-axis to the axis of rotation), and the other $\omega \sin \beta$ about an axis through O parallel to C′N, since OC′ and C′N are mutually at right angles.

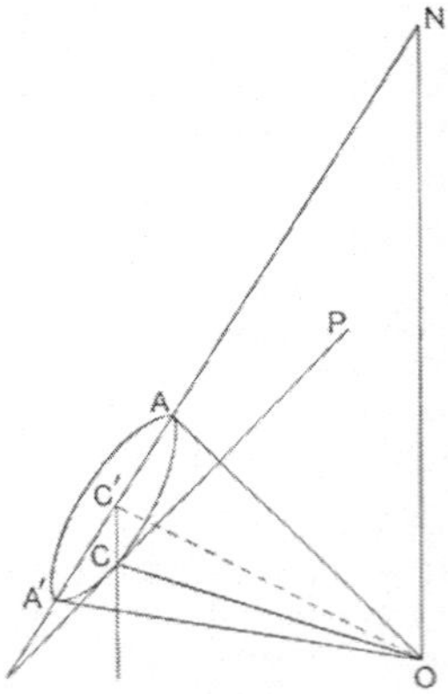

FIG. 122.

The A.M. of the gyrostat about OC can be resolved into two components, $Cn \cos \epsilon$ about the cone-axis OC′, and $Cn \sin \epsilon$ about an axis parallel to C′C. We now resolve the angular velocity $\omega \sin \beta$, which has been found for the axis parallel to C′N, at right angles to the radius C′C of the section of the cone. Denoting the angle CC′A′ by $\vartheta$ we get $\omega \sin \beta \sin \vartheta$. The rate of production of A.M. about an axis parallel to C′C is thus $Cn\omega \sin \beta \sin \vartheta$. The wheel will therefore turn about an instantaneous axis parallel to C′C with angular speed $\dot{\eta}$, and we shall have, measuring $\eta$ round from A′ towards C, tangential to the cone,

$$A\ddot{\eta} - Cn\omega \sin \beta \sin \vartheta = 0.$$

But if $l$ denote the length OC, we have $l\dot{\eta} = l \sin \epsilon . \dot{\vartheta}$, and therefore the equation just written becomes

$$A\ddot{\vartheta} - Cn\omega \frac{\sin \beta}{\sin \epsilon} \sin \vartheta = 0.$$

If we measure $\eta$ and $\vartheta$ round the other way, that is from A to C, we get

$$A\ddot{\vartheta} + Cn\omega \frac{\sin \beta}{\sin \epsilon} \sin \vartheta = 0.$$

The top will therefore oscillate about the generator OA as the equilibrium position. The period of a small oscillation is therefore

$$2\pi \left( \frac{A \sin \epsilon}{Cn\omega \sin \beta} \right)^{\frac{1}{2}}.$$

It will be noticed that if $\beta = 0$, that is if the axis of rotation of the earth coincides with the axis of the cone, we have $\ddot{\vartheta} = 0$, and the spin-axis moves continuously round the cone.

The period found agrees with that in Example 1.

To find the couple constraining the spin-axis to remain on the cone, the reader may calculate the angular speeds and momenta about (1) OC, (2) an axis OE, at right angles to OC in the plane COC′, and (3) an axis OD at right angles to this plane. The time rate of change of the A.M. about the moving axis OD, together with the rates of production of A.M. about OD, due to the motion of the other axes, gives the total rate of production of A.M. which must be balanced by an applied couple, if the spin-axis is to remain on the cone.

**3.** *A gyroscope on gimbal rings. Equation of energy.* A gyroscope is mounted symmetrically so that it can turn about an axis $I_1$ fixed diametrically in a circular ring. That ring turns on an axis $I_2$, at right angles to $I_1$, pivoted similarly on a second ring. The second ring turns about a diametral axis $I_3$, at right angles to $I_2$, and fixed in space.

We denote deviation of the plane of the first ring from planarity with the second by $\theta$, and the angle through which the second ring has been turned about $I_3$ by $\psi$. Thus $\theta$ and $\psi$ correspond to the angles so designated in the theory of the top, as set forth above.

We denote the moment of inertia of the flywheel about its axis of figure by C, and about a transverse axis through its centre by A, and similarly the moments of inertia of the rings by $C_1$, $A_1$, $C_2$, $A_2$.

It is easy to prove that the energy equation is

$$2E=(A+A_1)\dot{\theta}^2+(A\sin^2\theta+A_1\cos^2\theta)\dot{\psi}^2+(C_1\sin^2\theta+A_2)\dot{\psi}^2+Cn^2.$$

Also the A.M. about the axis $I_3$ is

$$\{(A+C_1)\sin^2\theta+A_1\cos^2\theta+A_2\}\dot{\psi}+Cn\cos\theta=G,$$

where G is constant. Thus we have

$$\dot{\psi}=\frac{G-Cn\cos\theta}{(A+C_1)\sin^2\theta+A_1\cos^2\theta+A_2}.$$

Substituting this value of $\dot{\psi}$ in the energy equation, we obtain the relation

$$(A+A_1)\dot{\theta}^2+\frac{(G-Cn\cos\theta)^2}{(A+C_1)\sin^2\theta+A_1\cos^2\theta+A_2}=2E-Cn^2,$$

from which $t$ can be found in terms of $\theta$ by integration.

This result was given by Lottner in 1857 [*Crelle*, 54, 1857]. It was found however by investigating the differential equations of motion for moving axes, and from these finding a first integral. It is instructive to work out for two axes corresponding to OD, OE of I, IV above, and then integrating. OD is at right angles to the vertical plane containing the spin-axis OC of the flywheel, and OE is in that plane and at right angles to OC. The three axes OD, OE, OC form an ordinary system of axes.

4. *Differential equations of motion for example 3.* We can find the differential equations of the last example for the axes OC, OE, OD as follows. A little consideration shows that the components of A.M. about these axes are as follows:

about OC, $Cn+(A_1+A_2)\dot{\psi}\cos\theta$,
about OE, $(A+A_2+C_1)\dot{\psi}\sin\theta$,
about OD, $(A+A_1)\dot{\theta}$.

The angular speeds of the systems of axes are

$$\dot{\psi}\cos\theta \text{ about OC}, \quad \dot{\psi}\sin\theta \text{ about OE}, \quad \dot{\theta} \text{ about OD}.$$

Hence we obtain by the method so often employed above the rates of growth about these axes; and we equate these rates to zero, since there are no applied couples. The equations of motion are therefore respectively

$$\left.\begin{aligned}(A_1+A_2)\ddot{\psi}\cos\theta+(C_1-2A_1)\dot{\theta}\dot{\psi}\sin\theta=0,\\ (A+A_2+C_1)\ddot{\psi}\sin\theta+(2A+C_1)\dot{\theta}\dot{\psi}\cos\theta-Cn\dot{\theta}=0,\\ (A+A_1)\ddot{\theta}-(A-A_1+C_1)\dot{\psi}^2\sin\theta\cos\theta+Cn\dot{\psi}\sin\theta=0.\end{aligned}\right\} \quad \ldots\ldots(1)$$

Multiplying the first of these by $\cos\theta$, and the second by $\sin\theta$, and adding, we get

$$(A\sin^2\theta+A_1\cos^2\theta+C_1\sin^2\theta+A_2)\ddot{\psi}+2(A-A_1+C_1)\dot{\theta}\dot{\psi}\sin\theta\cos\theta-Cn\dot{\theta}\sin\theta=0. \quad \ldots(2)$$

Multiplying this equation by $\dot{\psi}$, and the third of (1) by $\dot{\theta}$ and adding, we find

$$(A+A_1)\dot{\theta}\ddot{\theta}+(A\sin^2\theta+A_1\cos^2\theta+C_1\sin^2\theta+A_2)\dot{\psi}\ddot{\psi}+(A-A_1+C_1)\dot{\theta}\dot{\psi}^2\sin\theta\cos\theta=0. \quad \ldots(3)$$

This gives the integral

$$(A+A_1)\dot{\theta}^2+\{(A+C_1)\sin^2\theta+A_1\cos^2\theta+A_2\}\dot{\psi}^2=2E-Cn^2, \quad \ldots\ldots(4)$$

where E is, as before, the total energy.

Equation (2) can be integrated at once, and gives the constant value G of the A.M. about the vertical.

Thus we obtain again the results already established for this arrangement of gyrostat and rings.

5. *A gyroscope in a spherical case hung by a string.* A gyrostat consists of a heavy symmetrical flywheel mounted in a heavy spherical case, and is suspended from a fixed point by a string of length $l$ fixed to a point in the case. The centres of gravity of the flywheel and case are coincident. The whole revolves in steady motion round the vertical with angular speed $\mu$. Find the steady motion equations.

Let $a$, $b$ be the coordinates of the point of attachment of the string to the case, measured along and at right angles to the axis of rotation, $\alpha$, $\beta$ the inclinations of the string and axis of the top to the vertical, M the whole suspended mass, $Cn$ the A.M. of the flywheel about its axis, and A the moment of inertia of the whole about an axis through the centre of the flywheel at right angles to its axis.

We have, referring to the motion of the centre of gravity of the whole in a circle about the vertical through the point of support, $Mg \tan \alpha$ for the inward pull towards the centre exerted by the cord. But the radius of the circle is $l \sin \alpha + a \sin \beta + b \cos \beta$. Hence equating the two values of the centreward acceleration which we thus have, we get

$$\mu^2(l \sin \alpha + a \sin \beta + b \cos \beta) = g \tan \alpha.$$

The pull in the cord is clearly $Mg \sec \alpha$. Hence the total couple about the centroid applied by the cord is, as a figure will show at once,

$$Mg \sec \alpha \{a \sin (\beta - \alpha) + b \cos (\beta - \alpha)\}.$$

This is equal to the rate of production of A.M. about an axis through the centre of the flywheel at right angles to the vertical plane containing the axis of rotation. But we have many times seen that this is $(Cn - A\mu \cos \beta)\, \mu \sin \beta$. Thus we obtain

$$(Cn - A\mu \cos \beta)\, \mu \sin \beta = Mg \sec \alpha \{a \sin (\beta - \alpha) + b \cos (\beta - \alpha)\},$$

and the motion is completely determined.

6. *A gyrostat suspended by a string.* A gyrostat is suspended from a fixed point by a string of length $a$ fastened to a point P in the axis of rotation, and is in steady motion with the axis horizontal. Prove that if $\alpha$ be the angle which the string makes with the vertical, $n$ the angular speed of the flywheel, $h$ the distance of the point from the centre of gravity of the gyrostat, M the mass of the gyrostat, and C the moment of inertia of the wheel about its axis,

$$\tan \alpha = g \frac{M^2 h^2}{C^2 n^2} (h + a \sin \alpha).$$

The string applies a horizontal force $M\mu^2(h + a \sin \alpha)$, and the gravity couple is $Mgh$. Thus $\mu^2 = M^2 g^3 h^2 / C^2 n^2$, and so the horizontal force applied by the string is

$$M^3 g^2 h^2 (h + a \sin \alpha) / C^2 n^2.$$

The ratio of this to $Mg$ is $\tan \alpha$.

7. *Constraining couple required for rolling of a body-cone on a space-cone.* A uniform solid has a figure of revolution, and moves about a point so that its motion may be represented by the uniform rolling of a cone of semi-vertical angle $\alpha$, and fixed in the body, upon an equal cone fixed in space, the axis of the former being that of revolution. Show that the couple necessary to maintain the motion has moment

$$\tfrac{1}{2} \omega^2 \tan \alpha \{C + (C - A) \cos 2\alpha\},$$

where $\omega$ is the resultant angular speed, and A and C the principal moments of inertia about the axis of revolution and a transverse axis through the point, respectively, and that the couple lies in the plane of the axes of the cones.

The resultant angular speed here is that of the rolling cone about the generator in contact with the fixed cone. Hence the angular speed about the axis of revolution of the body is $\omega \cos \alpha$, and we have $Cn = C\omega \cos \alpha$. Moreover, the angular speed $\mu$ about the axis of the fixed cone is $\omega \sin \alpha / \sin 2\alpha$. The only rate of production of A.M. is that about an axis through the fixed point at right angles to the plane of the axes of the cones. And by the equations used frequently above this rate is

$$(Cn - A\mu \cos \theta)\,\mu \sin \theta = \left(C\omega \cos \alpha - A\omega \frac{\sin \alpha}{\sin 2\alpha} \cos 2\alpha\right) \omega \frac{\sin \alpha}{\sin 2\alpha} \sin 2\alpha$$
$$= \tfrac{1}{2}\omega^2 \tan \alpha \{C + (C - A) \cos 2\alpha\}.$$

8. *Pseudo-elliptic case of the motion of a top.* A top moves about a fixed point in its axis of symmetry. It is given a spin $(AMgh3^{\frac{1}{2}})^{\frac{1}{2}}/C$ about its axis, and then left to itself with its axis of figure at rest inclined at the angle $\cos^{-1}(1/3^{\frac{1}{2}})$ to the downward vertical. Show that the axis will describe the cone

$$\sin^2 \theta \cos 2\psi = \frac{2 \,.\, 2^{\frac{1}{2}}}{3^{\frac{1}{2}} \,.\, 3^{\frac{1}{4}}} (3^{\frac{1}{2}}/2 + \cos \theta)^{\frac{1}{2}},$$

where $\psi$ as usual denotes the azimuthal angle turned through about the vertical [*Math. Tripos*, I, 1894].

By (1), 11 and 5, 13, of V above, with $\beta = bn$, we have, if $z$ refers to the upward vertical,

$$\frac{d\psi}{dz} = \frac{bn}{a^{\frac{1}{2}}} \frac{(z_0 - z)^{\frac{1}{2}}}{(1 - z^2)\left\{1 - z^2 - \dfrac{b^2 n^2}{a}(z_0 - z)\right\}^{\frac{1}{2}}}.$$

But since $b = C/A$, $a = 2Mgh/A$, the data given convert this equation into

$$\frac{d\psi}{dz} = 3^{\frac{1}{4}} \frac{(-z - 1/3^{\frac{1}{2}})^{\frac{1}{2}}}{(1 - z^2)\{(2z + 3^{\frac{1}{2}})(3^{\frac{1}{2}} - z)\}^{\frac{1}{2}}}.$$

It can be verified by differentiation that this gives

$$2\psi = \tan^{-1}[3^{\frac{3}{4}} 2^{-1} (-z - 1/3^{\frac{1}{2}})^{\frac{3}{2}} (-z + 3^{\frac{1}{2}})^{\frac{1}{2}} (2z + 3^{\frac{1}{2}})^{-\frac{1}{2}}],$$

which can be identified with the result stated above.

*Note.*—The value of $\psi$ is obtained here directly in terms of ordinary functions for the case in which $b^2 n^2 z_0 = \frac{1}{2}a$. Another and simpler case is that in which $b^2 n^2 z_0 = a$, so that one of the roots of $1 - z^2 - b^2 n^2 (z_0 - z)/a$ is zero, and the other is $1/z_0$. For a full discussion of these pseudo-elliptic integrals, see Greenhill, *Proc. Lond. Math. Soc.*, 25, 1894, and *R.G.T.*, Chap. V. The case in the problem above appears in § 84 of the *L.M.S.* paper cited.

9. *A symmetrical shell containing a gyrostat and rolling on a horizontal plane.* A symmetrical shell contains a gyrostat the centroid and axis of which coincide with those of the shell. The gyrostat has A.M. K about its axis of spin, M is the mass of shell and gyrostat, A the moment of inertia of the two taken together about a line through the centroid transverse to the axis, C the moment of inertia of the shell about its axis, $a$ the distance, measured parallel to the axis, of the point of contact O of the shell with the supporting plane, from the centroid, $b$ the distance of O from the axis, and $\theta$ the inclination of the axis to the vertical. It is required to prove that if the motion is steady and the centroid moves in a circle of radius $c$ with angular speed $\Omega$ about the centre, the equation of motion is

$$K\Omega + \{C(a \sin \theta + c) - Ab \cos \theta + Mbc(a \cot \theta + b)\}\Omega^2 - Mgb(a - b \cot \theta) = 0.$$

We take axes (1) OD at right angles to the vertical plane containing the axis of symmetry and the point of contact O, (2) OC′ parallel to the axis of symmetry, (3) OE at right angles to OC′ and in the vertical plane.

Let L be the centre of the circle (radius $a$) in which the point of contact O moves on the shell. OE intersects this circle in a diameter. If $n$ be the angular speed of the shell about its axis the speed with which L is moving in space is $bn$ : it is also $(c+a\sin\theta)\,\Omega$. Hence

$$n=\frac{c+a\sin\theta}{b}\,\Omega.$$

The angular speeds of the axes are $\Omega\cos\theta$ about OC′ and $\Omega\sin\theta$ about OE.

The speed of the centroid in the circular path is $c\Omega$. Hence the components of A.M. are as follows:

$$\text{A.M. about OC}'=\mathrm{C}\frac{c+a\sin\theta}{b}\,\Omega+\mathrm{M}bc\Omega+\mathrm{K},$$

$$\text{A.M. about OE}=\mathrm{A}\Omega\sin\theta-\mathrm{M}ac\Omega.$$

The rate of production of A.M. about OD is therefore

$$-(\mathrm{A}\Omega\sin\theta-\mathrm{M}ac\Omega)\Omega\cos\theta+\left\{\left(\mathrm{C}\frac{c+a\sin\theta}{b}+\mathrm{M}bc\right)\Omega+\mathrm{K}\right\}\Omega\sin\theta.$$

The couple applied by gravity has moment $\mathrm{M}g(a\sin\theta-b\cos\theta)$. Equating this to the rate of growth of A.M. and reducing, we obtain the relation stated above.

The problem here discussed is a Cambridge Tripos question made more general by restatement. The case containing the gyrostat was supposed to be a prolate ellipsoid of revolution, but clearly this restriction was not necessary to simplify the solution, and in fact is not represented in the result. The reader will be able, with the help of the equation just indicated above, to write down the complete equation of motion and find the period of a small deviation from steady motion. It is only necessary to add the proper acceleration term.

**10.** *A cylinder rolling on the circular edge of one end.* A homogeneous right circular cylinder, whose altitude is twice the radius of the base, rolls on a rough horizontal with its axis inclined at an angle of 45° to the vertical. If $n$ be the angular velocity about the axis, prove that in steady motion the vertical plane through the axis turns round a fixed vertical line with an angular speed $\mu=30(2)^{\frac{1}{2}}n/31$. Show that the instantaneous axis divides the axis of the cylinder in the ratio 31 : 29. Prove also that the period $2\pi/\lambda$ of the small oscillations about the steady motion is given by

$$\lambda^2+\frac{12(2)^{\frac{1}{2}}}{31}\frac{g}{b}=\frac{1800}{31^2}n^2,$$

where $b$ is the radius of the base.

We notice that the value of $\mu$ is given by the last example if we put $a=b$, $\theta=45°$, $\mathrm{C}=\frac{1}{2}\mathrm{M}b^2$, $\mathrm{A}=\mathrm{M}(\frac{1}{3}b^2+\frac{1}{4}b^2)$. The quantity $a\sin\theta-b\cos\theta$ is zero, and $c=b\,\{n-\mu\,(2)^{\frac{1}{2}}/2\}/\mu$. Thus we find after reduction

$$\frac{5}{2}n-\left(\frac{7}{24}+1\right)2^{\frac{1}{2}}\mu=0,$$

or

$$\mu=\frac{30}{31}2^{\frac{1}{2}}n.$$

The instantaneous axis passes through the point of contact. The angular speed about the vertical contributes to the angular speed $n$ about the generator through that point. The independent angular speeds are thus $30\,(2)^{\frac{1}{2}}n/31$ about the vertical, and $n-\{30(2)^{\frac{1}{2}}n/31\}(2)^{\frac{1}{2}}/2$, or $n/31$, about the generator. Thus we have the two rectangular components, about the vertical and horizontal, respectively,

$$\left\{\frac{30(2)^{\frac{1}{2}}}{31}+\frac{1}{31(2)^{\frac{1}{2}}}\right\}n \quad\text{and}\quad \frac{1}{31}\frac{2^{\frac{1}{2}}}{2}n.$$

Hence, if $\phi$ be the angle which the instantaneous axis makes with the horizontal, we have $\tan\phi=61$. If O be the origin, the equation of the instantaneous axis is $y=61x$, and that of the axis of the cylinder is

$$x+b(2)^{\frac{1}{2}}=y.$$

These lines intersect in the point

$$x=\frac{2^{\frac{1}{2}}}{60}b,\quad y=\frac{61}{60}2^{\frac{1}{2}}b.$$

Hence, since $y=61(2)^{\frac{1}{2}}b/60$, the portion of the axis, intercepted between the point of intersection of the axis of the cylinder produced with the horizontal plane, is $y(2)^{\frac{1}{2}}=61b/30$. But $y(2)^{\frac{1}{2}}-b=31b/30$. The point $x, y$ thus divides the axial length of the cylinder into the two segments of lengths $31b/30$, $29b/30$ respectively.

The last part is left as an exercise in finding the complete equation of motion.

**11.** *A hollow cone revolving about a vertical generator: motion of a sphere on the surface.* A hollow cone rotates with uniform angular speed $\Omega$ about a vertical generator. A sphere is in contact with the diametrically opposite generator and spins about the common normal with angular speed $\omega$, and is prevented by friction from *slipping* down. If $a$ be the radius of the sphere, R the distance of the point of contact from the vertex, $k$ the radius of gyration of the sphere about a diameter, and $\alpha$ the angle of the cone, show that the point of contact will remain at rest on the sphere and on the cone, if

$$\omega=\left(1+\frac{a^2}{k^2}\right)\Omega\sin\alpha-\frac{a\text{R}}{k^2}\Omega\sin\alpha\tan\alpha+\frac{ga}{k^2\Omega}.$$

Consider axes OD, OC′, OG, at right angles to the diametral plane of the cone through the point of contact O, along the generator through that point, and along the diameter of the sphere, respectively. If the angular speed $\omega$ is such that the rate of production of A.M. about OD is equal to the moment $\text{M}ga\cos\alpha$ about that axis, the sphere can rest in relative equilibrium.

The angular speeds of the axes, about an axis through the centroid parallel to the generator in contact, and about OG, respectively, are $\Omega\cos\alpha$ and $\Omega\sin\alpha$. The speed of the centroid is $(\text{R}\sin\alpha-a\cos\alpha)\Omega$. The A.M. about OC′ is therefore

$$\text{M}k^2\Omega\cos\alpha-\text{M}(\text{R}\sin\alpha-a\cos\alpha)a\Omega,$$

and that about OG is $\text{M}k^2\omega$. The rate of production of A.M. about OD, computed by the usual rule, is

$$-\text{M}k^2\Omega^2\sin\alpha\cos\alpha+\text{M}(\text{R}\sin\alpha-a\cos\alpha)a\Omega^2\sin\alpha+\text{M}k^2\omega\Omega\cos\alpha.$$

Equating this to the gravity couple and solving for $\omega$, we get the equation stated above.

If the spin $\omega$ about the axis OG is too great to give the balance here considered, the sphere will roll upwards or downwards along the generator. Consideration of the signs of the quantities concerned shows that if $\Omega$ be in the counter-clock direction about the upward vertical, and $\omega$ be also in the counter-clock direction to an observer looking from beyond G, the sphere will roll upwards if $\omega$ be too great.

**12.** *A horizontal circular disk turning about its axis of figure: motion of a sphere on the surface.* A circular disk capable of motion about a vertical axis through its centre perpendicular to its plane is set in motion with angular speed $\Omega$. A uniform sphere is placed on the disk so as to touch at an eccentric point: prove that, if there is no slipping, the sphere will describe a circle on the disk, and that the disk will revolve with angular speed $7\text{M}k^2\Omega/(7\text{M}k^2+2mr^2)$, where $\text{M}k^2$ is the moment of inertia of the disk about the centre, $m$ the mass of the sphere, and $r$ the radius of the circle traced on the disk by the point of contact.

It is obvious that the sphere will trace out a circle about the centre of the disk. For the friction between it and the disk will set it impulsively in motion in a plane at right angles to the radius through the point of initial contact, and will further constrain it to revolve in a plane at right angles to each successive radius of the disk with which it comes into contact.

Let P be the impulse initially applied to the sphere by the disk. Then we have by the reaction on the disk

$$Pr = Mk^2(\Omega - \omega'), \quad\quad (1)$$

where $\omega'$ is the angular speed of the disk after the impulse has acted.

Also, transferring P to the centre of the sphere, we have

$$mv = P, \quad\quad (2)$$

where $v$ is the speed of the centre of the sphere in the direction of motion of the disk. Hence

$$mvr = Mk^2(\Omega - \omega'). \quad\quad (3)$$

Again, we have for the angular speed $\omega$ of rotation of the sphere

$$aP = \tfrac{2}{5}ma^2\omega, \quad\quad (4)$$

and clearly $v = r\omega' - a\omega$, so that by (3)

$$mr^2\omega' - mra\omega = Mk^2(\Omega - \omega'). \quad\quad (5)$$

Now (4) gives $\omega = P/\tfrac{2}{5}ma$, and by (1) this is $Mk^2(\Omega - \omega')/\tfrac{2}{5}mar$. This value of $\omega$ substituted in (5) gives

$$mr^2\omega' - \tfrac{5}{2}Mk^2(\Omega - \omega') = Mk^2(\Omega - \omega'),$$

or

$$\omega' = \frac{7Mk^2}{7Mk^2 + 2mr^2}\Omega, \quad\quad (6)$$

which was to be proved.

**13.** *A sphere rolling on a vertical plane which turns about a vertical axis.* A sphere rolls without sliding on a vertical plane which revolves with constant angular speed $n$ about a vertical line in itself: to find the motion of the sphere.

Take an origin O on the vertical axis of rotation, an axis $Ox$ horizontal and fixed in the revolving plane, an axis $Oy$ drawn outwards from the plane on the side on which the sphere is situated, and an axis $Oz$ downward. Denote the radius of the sphere by $a$, and the angular speeds about axes through the centre of the sphere parallel to the axes specified by $\omega_x$, $\omega_y$, $\omega_z$. Since there is no sliding, we get by considering as usual the growth of the components of the vector of A.M.,

$$(k^2 + a^2)\dot\omega_x - nk^2\omega_y = ga, \quad \dot\omega_y + n\omega_x = 0, \quad k^2\dot\omega_z + a(vn - \dot u) = 0, \quad\quad (1)$$

with, for the velocities in the instantaneous directions of $Ox$ and $Oz$,

$$u = \dot x - an, \quad a\omega_x = w. \quad\quad (2)$$

The first equation can, by the second of (2), be written

$$(k^2 + a^2)\frac{\dot w}{a} - nk^2\omega_y = ga. \quad\quad (3)$$

Differentiating this and using the second of (1) in the result, we obtain

$$\ddot w + m^2 w = 0, \quad\quad (4)$$

where $m^2 = n^2k^2/(k^2 + a^2)$.

From (3) we find by integration $\quad w = A \sin mt, \quad\quad (5)$

if the time be reckoned from an instant at which the vertical motion of the sphere is zero. Integrating again with respect to the time, we have, since A is a constant,

$$z = \int w\, dt = \frac{A}{m}(1 - \cos mt).$$

Thus the vertical motion is oscillatory, and the sphere rises and falls on the vertical plane in simple harmonic motion.

Substitution of $\dot{w}$ [obtained from (5)] in (3) gives, since $\omega_y = 0$ initially, $A = ga^2/nk(k^2+a^2)^{\frac{1}{2}}$. Thus

$$n^2 z = g\frac{a^2}{k^2}(1 - \cos mt). \qquad \text{.....................(6)}$$

For a uniform sphere $k^2 = \frac{2}{5}a^2$, and therefore in that case

$$n^2 z = \tfrac{5}{2}g\{1 - \cos(\tfrac{2}{7}n^2)^{\frac{1}{2}}t\}. \qquad \text{.....................(7)}$$

The utmost distance this sphere can descend from the position of relative rest is therefore $5g/n^2$.

It will be observed, from (3), that it is the turning about the axis parallel to O$y$, the spin, that is, of the sphere on the turning plane, which prevents continuous motion of the sphere downward. This spin, by the second of (1), grows up at rate $-n\omega_x$. The gravity couple is occupied, in degree varying from zero to the whole value, in turning the axis of spin O$y$ round in azimuth.

Example 11 is another illustration of this effect of spin.

**14.** *The vertical plane of last example, turning also about a normal axis.* The vertical plane of last example revolves with angular speed $n$ as stated, and also spins with angular speed $\mu$ about a normal, intersecting in a point O the vertical axis about which the plane turns. Prove that the position of the centre of the sphere at any time $t$ will be determined by the equations

$$7\ddot{z} + 2\mu(\dot{x} + n^2 x) + 2n^2\dot{z} = 0,$$
$$7\ddot{x} - 2\mu\dot{z} - 5n^2 x = 0,$$

where $x$ and $z$ have the meanings stated in the preceding example [Routh, *Adv. Rigid Dynamics*, 6th edn., p. 208].

The equations of motion of the centre of the sphere are

$$\left.\begin{aligned} u &= \dot{x} - an, \quad v = xn, \quad w = \dot{z}, \\ \dot{x} &= \mu z - \omega_z a, \quad \dot{z} = -\mu x + \omega_x a; \end{aligned}\right\} \qquad \text{.....................(1)}$$

with

$$\left.\begin{aligned} \dot{u} - vn &= \ddot{x} - xn^2 = F, \quad \dot{v} + un = 2\dot{x}n - an^2 = R, \\ \dot{w} &= -\mu\dot{x} + \dot{\omega}_x a = g + F'; \end{aligned}\right\} \qquad \text{.....................(2)}$$

with the rotational equations

$$k^2(\dot{\omega}_x - n\omega_y) = -F'a, \quad \dot{\omega}_y + n\omega_x = 0, \quad k^2\dot{\omega}_z = Fa. \qquad \text{.....................(3)}$$

From the third of (2) and the first of (3), eliminating F′ and differentiating, we get

$$-a\mu\ddot{x} + (k^2 + a^2)\ddot{\omega}_x + k^2 n^2\omega_x = 0. \qquad \text{.....................(4)}$$

But we have $\ddot{w} = -\mu\ddot{x} + \ddot{\omega}_x a$, so that

$$\omega_x = \frac{w + \mu x}{a}, \quad \ddot{\omega}_x = \frac{\ddot{w} + \mu\ddot{x}}{a}.$$

Thus, since $w = \dot{z}$ and $\ddot{w} = \dddot{z}$, we obtain, after a little reduction and putting $k^2 = \frac{2}{5}a^2$, instead of (4),

$$7\dddot{z} + 2n^2\dot{z} + 2\mu(\ddot{x} + n^2\dot{x}) = 0, \qquad \text{.....................(5)}$$

which is the first of the two equations stated above.

We have also, by the second of (2) and the third of (3), with the relation $a\omega_z = \mu z - \dot{x}$,

$$a(\ddot{x} - xn^2) + k^2\frac{\ddot{x} - \mu\dot{z}}{a} = 0,$$

that is

$$7\ddot{x} - 5n^2 x - 2\mu\dot{z} = 0, \qquad \text{.....................(6)}$$

the other equation.

The sphere is set into rotation impulsively at the beginning of the motion. The components of rotation have initial values

$$\omega_x = \frac{5}{7}\frac{\mu x_0}{a}, \quad \omega_z = \frac{5}{7}\frac{\mu z_0}{a} + n. \qquad \text{.....................(7)}$$

For the surface in contact is suddenly started in one case with speed $\mu x_0$, and the centre of oscillation which is at distance $\frac{7}{5}a$ from the point of contact is instantaneously at rest. Similarly the other case is dealt with. Besides the rotation due to the impulse, $\omega_z$ contains also the angular speed $n$. Hence, from the equations

$$\dot{z}_0 = -\mu x_0 + \omega_x a, \quad \dot{x}_0 = \mu z_0 - \omega_z a,$$

we obtain for the initial values of $\dot{x}$, $\dot{z}$,

$$7\dot{z}_0 + 2\mu x_0 = 0, \quad 7\dot{x}_0 + 7an - 2\mu z_0 = 0. \quad \text{.........................(8)}$$

**15.** *A rigid body turning about a principal axis, while another principal axis lies in a plane through the former and a fixed line.* OA, OB, OC are the principal axes of a rigid body which is in motion about a fixed point O. The axis OC has a constant inclination $\theta$ to a line OZ fixed in space, and revolves with uniform angular speed $\mu$ round it, and the axis OA always lies in the plane ZOC. Prove that the constraining couple has its axis along OB, that its moment is $(A-C)\mu^2 \sin\theta\cos\theta$ [Routh, *Adv. Rigid Dynamics*, 6th ed., p. 208].

A diagram for this is furnished by Fig. 4, p. 48 above, if OA is put for OE, OB for OD, and OC is retained as shown. The angular speeds about OA and OC are $\mu\sin\theta$, $\mu\cos\theta$, and the A.M. $A\mu\sin\theta$, $A\mu\cos\theta$, respectively. In consequence of the motion of the axes there is a steady rate $-A\mu\sin\theta \,.\, \mu\cos\theta + C\mu\cos\theta \,.\, \mu\sin\theta = -(A-C)\mu^2\sin\theta\cos\theta$ of growth of A.M. about OB. This must be balanced by a constraining couple L, since the A.M. about OB must remain zero. Hence we have

$$L - (A-C)\mu^2 \sin\theta\cos\theta = 0.$$

**16.** *A heavy flywheel with added eccentric weight carried round in uniform precession.* A heavy flywheel is pivoted at the extremities of a horizontal diameter AOB, and this diameter is carried round a vertical axis through its centre O with uniform angular speed $\mu$. At a point P at distance $a$ from the centre on a diameter at right angles to AB an additional weight $w$ is attached. Find the equation of motion.

Take as axes OA, OP and the axis OC of the wheel drawn from O on the other side of the vertical from OP. If $\theta$ be the angle of inclination of OC to the upward vertical, the angular speeds about OP and OC are $\mu\sin\theta$ and $\mu\cos\theta$, while the A.M. are $A\mu\sin\theta$ and $(C+wa^2)\mu\cos\theta$. The total rate of growth of A.M. about OA is therefore

$$(A+wa^2)\ddot{\theta} + (C+wa^2-A)\mu^2\sin\theta\cos\theta,$$

and the moment of applied force is $wga\sin\theta$. The equation of motion is therefore

$$(A+wa^2)\ddot{\theta} + (C+wa^2-A)\mu^2\sin\theta\cos\theta - wga\sin\theta = 0.$$

The equation of motion for OP is found in like manner. The A.M. about OA is $(A+wa^2)\dot{\theta}$, and the rate of production of A.M. about OP is

$$A\frac{d}{dt}(\mu\sin\theta) - (C+wa^2)\mu\cos\theta \,.\, \dot{\theta} + (A+wa^2)\dot{\theta} \,.\, \mu\cos\theta.$$

Hence, since there is no couple about OP, the equation of motion is

$$A\dot{\mu}\sin\theta - (C-2A)\dot{\theta}\mu\cos\theta = 0.$$

**17.** *A top constrained by two vertical planes parallel to the axis.* A top is set in rapid rotation and is placed on a frictionless horizontal plane with its axis inclined at an angle $\theta_0$ to the vertical, and is constrained by two smooth planes parallel to the angle $\theta_0$, so that its axis must remain in that plane. Prove that the top must fall.

No action of the constraining planes can alter the energy of rotation, or the angular speed $\dot{\theta}$. If $\dot{\theta}=0$ when $\theta=\theta_0$ and A be the moment of inertia of the top about a horizontal axis through its centroid, we have the energy equation

$$\tfrac{1}{2}(A + Mh^2\sin^2\theta)\dot{\theta}^2 = Mgh(\cos\theta_0 - \cos\theta). \quad \text{........................(1)}$$

The left-hand side is positive, so $\theta$ must be greater than $\theta_0$.

This result will be clearer perhaps if the equation of motion for the axis OD is considered. The usual equation is

$$A\theta+(Cn-A\mu\cos\theta)\,\mu\sin\theta=Mgh\sin\theta. \quad \text{.........................(2)}$$

But by the constraining planes $\mu$ is made and kept zero. The equation is therefore

$$A\theta=Mgh\sin\theta, \quad \text{.....................................(3)}$$

so that angular speed $\dot{\theta}$ grows up and the top falls.

The constraining couple which added on the right of 2 gives (3) is thus

$$(Cn-A\mu\cos\theta)\,\mu\sin\theta,$$

and acts about OD. It is equal and opposite to the gyrostatic couple.

[It is important to remember that any constraint must affect the stability of a top or gyrostat. Conclusions, for instance, derived from the behaviour of a top mounted on a tray, as in 7, VII, where a certain diameter of the flywheel is constrained to remain horizontal, cannot be regarded as necessarily holding for a top perfectly free to precess.]

**18.** *Stability of a ring of wire spinning on the top of a sphere.* A ring of wire of radius $c$ rests on the top of a smooth fixed sphere of radius $a$, and is set rotating about the vertical diameter of the sphere with angular speed $n$. Prove that the motion is unstable if $n^2c^4<2g(2a^2-c^2)(a^2-c^2)^{\frac{1}{2}}$ [*Math. Tripos*, 1885].

Since the ring moves on the surface of the sphere it may be regarded as a top turning about the centre of the sphere. It is proved in 17, II, that the top is unstable unless the inequality

$$C^2n^2>4AMgh\cos\theta$$

is fulfilled. In the present case we have

$$\theta=0,\quad C=Mc^2,\quad A=\tfrac{1}{2}Mc^2+M(a^2-c^2),\quad h=(a^2-c^2)^{\frac{1}{2}}.$$

Hence we obtain the result stated.

**19.** *A gyrostat sliding freely between two rods the plane of which revolves.* Two intersecting rods are at right angles to one another. One is placed vertical, the other can turn in a horizontal plane about the lower end of the first. The ends of the axis of a gyrostat slide freely on these rods, and the axis (of length $2a$) is initially inclined at an angle $\theta_0$ to the vertical, when also the horizontal rod is turning with angular speed $\dot{\psi}_0$. If at time $t$ the inclination of the axis to the vertical be $\theta$, and the azimuthal speed $\dot{\psi}$, prove that

$$(Ma^2+A)(\dot{\psi}\sin^2\theta-\dot{\psi}_0\sin^2\theta_0)+Cn(\cos\theta-\cos\theta_0)=0,$$

$$(Ma^2+A)\ddot{\theta}+\{Cn-(Ma^2+A)\dot{\psi}\cos\theta\}\dot{\psi}\sin\theta-Mga\sin\theta=0,$$

where M is the mass of the gyrostat and C and A are its principal moments of inertia for axes through the centroid. [The moment of inertia of the case about the axis of symmetry is neglected, and the gyrostat is supposed to be midway between the ends of the axis.]

The first equation is clearly of constancy of A.M. about the vertical through the point of contact, and requires no proof.

The second equation is that which is at once obtained when the motion is referred to axes through the centroid as explained in Chapter V.

**20.** *A sphere started spinning about an axis parallel to the earth's axis, and constrained to keep the axis of spin in the meridian.* At a point P on the earth's surface a sphere has its centre fixed, and is spinning, in the direction of the earth's rotation and relatively to the earth, about a diameter OC with angular speed $\dot{\phi}$. The axis OC is constrained to move only in the meridian. Show that if the sphere be disturbed the axis OC will oscillate in the meridian in period $2\pi/(\dot{\phi}\omega)^{\frac{1}{2}}$, where $\omega$ is the angular speed of the earth about its axis.

Let a small deviation $\theta$ of the diameter OC from the direction of the earth's axis be imposed, and the sphere be then left to itself. The equation of motion for the axis OD, so often used above, becomes for this case

$$A\ddot{\theta}+(Cn-A\mu\cos\theta)\mu\sin\theta=0,$$

where $\mu=\omega$. But here also C=A, and $\theta$ is small, so that we have

$$\ddot{\theta}+(n-\omega)\omega\theta=0,$$

that is

$$\ddot{\theta}+\dot{\phi}\omega\,.\,\theta=0.$$

The period of oscillation is therefore $2\pi/(\dot{\phi}\omega)^{\frac{1}{2}}$ as stated.

**21.** *A body supported at its centroid and under the action of a constant couple.* A body is supported at its centre of inertia and has an initial velocity of rotation $\Omega$ about an axis at right angles to that of symmetry. A constant couple of moment N is applied about the axis of symmetry; show that the cone described in the body by the instantaneous axis has the equation

$$\tan^{-1}\frac{y}{x}=\frac{\Omega^2 C}{2N}\frac{C-A}{A}\frac{z^2}{x^2+y^2}.$$

The equations of motion are

$$\left.\begin{array}{l}A\dot{\omega}_1-(A-C)\omega_2\omega_3=0,\quad A\dot{\omega}_2-(C-A)\omega_3\omega_1-0,\\ C\dot{\omega}_3=N.\end{array}\right\} \quad\text{.......(1)}$$

The third equation gives $\quad C\omega_3=Nt, \quad$ .......(2)

and the first two combined $\quad \omega_1{}^2+\omega_2{}^2=\Omega^2. \quad$ .......(3)

Also, multiplying the second equation by $i$ and adding to the first, we get

$$A(\dot{\omega}_1+i\dot{\omega}_2)=(C-A)(\omega_1+i\omega_2)i\frac{Nt}{C},$$

that is

$$\frac{d}{dt}\{\log(\omega_1+i\omega_2)\}=\frac{C-A}{A}\frac{iNt}{C}. \quad\text{.......(4)}$$

Hence we obtain by integration

$$\log(\omega_1+i\omega_2)=\frac{C-A}{A}\frac{iNt^2}{2C},$$

or

$$\omega_1+i\omega_2=e^{\frac{C-A}{A}\frac{iNt^2}{2C}}. \quad\text{.......(5)}$$

This gives

$$\tan^{-1}\frac{\omega_2}{\omega_1}=\frac{C-A}{A}\frac{Nt^2}{2C}. \quad\text{.......(6)}$$

To find the equation of the cone described by the instantaneous axis we have (since $x/\omega_1=y/\omega_2=z/\omega_3=(x^2+y^2)^{\frac{1}{2}}/\Omega$), so that $Nt^2/2C=C\Omega^2z^2/2N(x^2+y^2)$.

Hence, by (6),

$$\tan^{-1}\frac{y}{x}=\frac{C-A}{A}\frac{\Omega^2 C}{2N}\frac{z^2}{x^2+y^2},$$

which is the equation of the cone required.

**22.** *A spherical gyrostat contained within a rolling sphere.* A sphere rotates within a spherical concentric light shell of radius $a$, which rolls without slipping on a horizontal plane about an axis through the common centre: find the motion. If $\theta$ be the inclination of the axis to the vertical, $\omega_3$ the velocity of rotation of the sphere about the axis, and $\dot{\psi}$ the angular speed of the vertical plane through the axis, prove that $\dot{\psi}\sin^2\theta+\omega_3\cos\theta$ is constant.

Show that a state of steady motion is possible in which the centre of the sphere describes a circle of radius $r$ with speed $v$, while spinning with angular speed $\omega$, if the axis is inclined to the vertical at an angle $\alpha$ given by the equation

$$k^2\sin\alpha(v\cos\alpha-r\omega)=var, \quad\text{.......(1)}$$

where $k$ is the radius of gyration of the sphere about a diameter [Besant, *Dynamics*, 2nd edn., p. 420].

The reader will observe that this is a particular case of the problem treated in 14, XVI above. The inertia of the containing shell is neglected, and the internal gyrostat is a sphere. The first result to be established is that the A.M. of the internal sphere about the vertical is constant, which obviously is expressed by the equation stated.

The result also as to the radius of the circle in which the centre of the sphere moves in steady motion agrees with (7), 14, XVI, when properly modified. We give here a solution of this simplified problem from first principles.

We obtain by the process set forth in 16, XVI, the complete equation (2) of that article, which, since we may put $K=k^2\omega$, $A=k^2$, we can write

$$(k^2+a^2)\ddot{\theta}+(k^2\omega+a^2n)\dot{\psi}\sin\theta-(k^2+a^2)\dot{\psi}^2\sin\theta\cos\theta=0. \qquad\text{(2)}$$

For steady motion, putting $\mu$ for $\dot{\psi}$, and $\alpha$ for $\theta$, we get from this the relation

$$k^2\omega+a^2n=(k^2+a^2)\mu\cos\alpha. \qquad\text{(3)}$$

But $\mu=v/r$, and we have $\dfrac{v}{r}=\dfrac{an\sin\alpha}{r+a\sin\alpha\cos\alpha}$.

Eliminating $n$ and $\mu$ from the steady motion equation (3), we obtain equation (1)

**23.** *The ordinary problem of a rapidly spinning top.* The angular speed of a top is communicated to it by unwinding rapidly a string from the axis when the inclination of the axis to the vertical is $\theta_0$: prove that if the angular speed is great the inclination $\theta$ at any time $t$, reckoned from the starting of the top, is given by

$$\theta=\theta_0+r\sin\theta_0(1-\cos pt),$$

and the azimuthal angular deflection at time $t$ by

$$\psi=\mu t-r\sin pt,$$

where $r=A\mu^2/Mgh$ and $p=Cn/A$.

Hence show that the axis describes very nearly a right cone round its position in the steady motion, in the same direction as the axis rotates.

Find also the forces applied to the apex (the peg) of the top: also find the arc described in a half-period by the centroid.

The first two paragraphs of this example are fully dealt with in 14, V above, and a more exact discussion is given in 1 ... 3, VI.

The last part is dealt with in substance in (3), 1, V. The forces there denoted by Y and Z come out in the present case as

$$Y=-Mh\{\ddot{\theta}\cos\theta-(\dot{\theta}^2+\mu^2)\sin\theta\}$$
$$=-Mh\left\{rp^2\sin\theta\cos\theta\cos pt-r^2p^2\sin^2\theta\sin^3 pt-\sin\theta\frac{M^2g^2h^2}{C^2n^2}\right\},$$
$$Z=Mg-Mhrp^2\sin^2\theta(\cos pt+r\cos\theta\sin^2 pt).$$

As regards the surface described by the axis of the top, we may add that the angular deviations from the steady motion at time $t$ are $-r\sin\theta\cos pt$ and $-r\sin pt$, the former in the vertical plane through the axis of the top, the latter about the vertical. Thus, if $h$ be the distance of the upper extremity of the axis from the apex, the linear displacement are $-hr\sin\theta\cos pt$ and $-hr\sin\theta\sin pt$ respectively. Squaring these, we get for the resultant $-hr\sin\theta$, which is independent of $t$. The amplitude $r$ is very small, and therefore the axis describes a cone as stated. It is easy to verify that the direction is that specifie

The arc described in any time $dt$ is, if $\alpha$ be the steady motion value of $\theta$,

$$h\sin\alpha(\mu^2-2\mu pr\cos pt+p^2r^2)^{\frac{1}{2}}dt,$$

which, since $r = \mathrm{A}\mu^2/\mathrm{M}gh$, $p = \mathrm{C}n/\mathrm{A}$, reduces to $2h \sin \alpha \sin \frac{1}{2}pt \,.\, dt$. Hence

$$\left.\begin{array}{r}\text{Arc described}\\ \text{as specified}\end{array}\right\} = 2\mu h \sin \alpha \int_0^{\pi/p} \sin \tfrac{1}{2}pt \,.\, dt = 4\frac{\mu h}{p} \sin \alpha.$$

The arc described in time $\pi/p$ in the steady motion is clearly $\pi\mu h \sin \alpha/p$. Hence the vibration increases the distance in the ratio $4/\pi$.

**24.** *The arc described by a point on the axis of a rapidly spinning top.* A symmetrical top is set in motion on a horizontal plane with an angular speed $n$ about its axis of figure, which is initially inclined to the vertical at an angle $\alpha$. Prove that between the greatest approach to and recess from the vertical the centroid describes an arc $h\beta$, where $(p - \cos \alpha) \tan \beta = \sin \alpha$ and $p = \mathrm{C}^2n^2/4\mathrm{AM}gh$.

In the notation explained in 10, V, but with $\alpha'$ and $\beta'$ for the $a$ and $\beta$ there used, we get from the equation of energy

$$\dot{\theta}^2 + \dot{\psi}^2 \sin^2 \theta = \alpha' - a \cos \theta.$$

But if $\dot{\chi}$ be the rate at which the axis of the top is changing direction, we have

$$\dot{\chi} = (\dot{\theta}^2 + \dot{\psi}^2 \sin^2 \theta)^{\frac{1}{2}} = (\alpha' - a \cos \theta)^{\frac{1}{2}}.$$

By the equation of constancy of A.M. about the vertical we eliminate $\dot{\psi}^2 \sin^2 \theta$, and get with $z = \cos \theta$,

$$\dot{\theta} = \frac{1}{\sin \theta}\{(\alpha' - az)(1 - z^2) - (\beta' - bnz)^2\}^{\frac{1}{2}}.$$

Eliminating $dt$ between the two last equations, we find, after a little reduction based on the initial values $\theta = \alpha$, $\dot{\theta} = \dot{\psi} = 0$, which lead to $\alpha' = a \cos \alpha$, $\beta' = bn \cos \alpha$,

$$d\chi = \frac{\sin \theta \, d\theta}{(\sin^2 \theta + 2p \cos \theta - 2p \cos \alpha)^{\frac{1}{2}}},$$

where $2p = b^2n^2/a$. If we write $x^2 = (p - \cos \theta)^2$ and $c^2 = 1 - 2p \cos \alpha + p^2$, we get

$$d\chi = \frac{dx}{(c^2 - x^2)^{\frac{1}{2}}}.$$

Thus

$$\beta = \left[\sin^{-1} \frac{x}{c}\right]_{\theta = \alpha}^{\theta = \cos^{-1}\{p - (p^2 - 2p \cos \alpha + 1)^{\frac{1}{2}}\}},$$

that is

$$\beta = \tfrac{1}{2}\pi - \sin^{-1} \frac{p - \cos \alpha}{(1 - 2p \cos \alpha + p^2)^{\frac{1}{2}}}.$$

This last result gives $\cos \beta = \dfrac{p - \cos \alpha}{(1 - 2p \cos \alpha + p^2)^{\frac{1}{2}}}$,

and therefore

$$\tan \beta = \frac{\sin \alpha}{p - \cos \alpha},$$

or

$$(p - \cos \alpha) \tan \beta = \sin \alpha,$$

the relation stated in the problem.

**25.** *A top supported on a horizontal plane without friction.* If the top be supported on a plane which offers no resistance to the horizontal displacement of the point O, and if no horizontal force act on it, its centroid will continue in its united state of rest or uniform motion. This case of motion may be discussed by means of the equations set forth in 1, V. [See also 27, XIX.]

We take axes GD, GE, GC drawn from G the centroid in the directions so many times specified above. If F be the vertical reaction of the plane on the point O of the top, we have

$$\mathrm{A}\ddot{\theta} + (\mathrm{C}n - \mathrm{A}\dot{\psi} \cos \theta)\dot{\psi} \sin \theta = \mathrm{F}h \sin \theta. \quad \text{.....(1)}$$

The speed of G vertically downwards is $h\dot{\theta} \sin \theta$, and therefore

$$\mathrm{M}\frac{d}{dt}(h\dot{\theta} \sin \theta) = \mathrm{M}g - \mathrm{F},$$

or

$$\mathrm{F} = \mathrm{M}g - \mathrm{M}h(\sin \theta \,.\, \ddot{\theta} + \cos \theta \,.\, \dot{\theta}^2). \quad \text{.....(2)}$$

This value of F substituted in (1) gives

$$(A+Mh^2\sin^2\theta)\ddot{\theta}+Mh^2\sin\theta\cos\theta.\,\dot{\theta}^2+(Cn-A\dot{\psi}\cos\theta)\dot{\psi}\sin\theta=Mgh\sin\theta. \quad\text{.......(3)}$$

The equation of A M. about the vertical has the usual form

$$Cn\cos\theta+A\dot{\psi}\sin^2\theta=G, \quad\text{.......(4)}$$

where G is a constant.

The kinetic energy in the present case is

$$\tfrac{1}{2}\{Mh^2\sin^2\theta.\,\dot{\theta}^2+A(\dot{\theta}^2+\dot{\psi}^2\sin^2\theta)+Cn^2\},$$

and the potential energy is $Mgh\cos\theta$, so that we have

$$(A+Mh^2\sin^2\theta)\dot{\theta}^2+A\dot{\psi}^2\sin^2\theta+Cn^2+2Mgh\cos\theta=2E. \quad\text{.......(5)}$$

$Cn$ is constant, and we can write by (4),

$$\dot{\psi}=(G-Cn\cos\theta)/A\sin^2\theta.$$

Thus the equation of energy can be written

$$(A+Mh^2\sin^2\theta)\dot{\theta}^2+\frac{(G-Cn\cos\theta)^2}{A\sin^2\theta}+Cn^2+2Mgh\cos\theta=2E. \quad\text{.......(6)}$$

The second term on the left of (6) can be put into different forms convenient for calculation [see 4 and 19, XII]. Another form of (6) which is of practical importance is

$$A\{A+Mh^2(1-z^2)\}\dot{z}^2=-(G-Cnz)^2+2A(c-Mghz)(1-z^2), \quad\text{.......(7)}$$

where $z=\cos\theta$ and $2c=2E-Cn^2$, so that $c$ is the whole energy of the top, *minus* the energy of spin.

**26.** *Motion relative to the earth.* Considering the earth as a rotating body with its centroid at rest in space, find the equations of motion of a particle with reference to axes $Ox$, $Oy$, $Oz$ drawn from the centroid parallel to the horizontally southward, the eastward, and the vertically upward directions at a point $P_0$ on the surface.

The coordinates of $P_0$ are thus 0, 0, $a$, where $a$ is the vertical distance of $P_0$ from O, approximately the earth's radius at the point. Let the direction of the gravitational force G, on unit mass at $P_0$, make a small angle $\theta$ with the vertical. The components of G at $P_0$ are $G\cos(\frac{1}{2}\pi+\theta)$ along $Ox$, zero along $Oy$, and $G\cos\theta$ along $Oz$. The angular speeds about the axes are $n\cos\lambda$, 0, $n\sin\lambda$, if $n$ be the angular speed of rotation of the earth, and $\lambda$ the geographical latitude. The coordinates of any other point P with reference to these axes are $x$, $y$, $z$, and the components of force there are $X+G\cos(\frac{1}{2}\pi+\theta)$, Y, $Z-G\cos\theta$, where X, Y, Z are due to the constraint imposed (*e.g.* in the case of a pendulum by the suspension cord). If there are any other forces besides these and the components of gravity, they may be denoted by X′, Y′, Z′. The equations of motion are

$$\begin{aligned}
&\ddot{x}-2n\dot{y}\sin\lambda+n^2\sin\lambda(z\cos\lambda-x\sin\lambda)=X+X'-G\sin\theta,\\
&\ddot{y}+2n\dot{x}\sin\lambda-2n\dot{z}\cos\lambda-n^2y=Y+Y',\\
&\ddot{z}-2n\dot{y}\cos\lambda-n^2\cos\lambda(z\cos\lambda-x\sin\lambda)=Z+Z'-G\cos\theta.
\end{aligned}$$

The reader will note the terms in $\dot{x}$, $\dot{y}$, $\dot{z}$, which are in form gyrostatic terms, arising from the motion of the earth. If the equation of relative energy is formed these terms disappear.

If, as is generally convenient, the axes be taken in the directions specified, but from $P_0$ as origin, it is only necessary to substitute $z-a$ for $z$ in these equations, and to add, on the right, components of force equal and opposite to those required to give the acceleration $n^2a\cos\lambda$ of a particle at P towards the earth's axis of rotation.

**27.** *Theory of Foucault's pendulum experiment.* Apply the equations of last example to a simple pendulum suspended from $P_0$, and executing small oscillations under gravity.

Change the origin to $P_0$, as explained in the last example, and then neglect terms in $n^2x$, $n^2y$, $n^2z$. Then, if F be the pull applied by the thread to the bob,

$$X = -Fx/l, \quad Y = -Fy/l, \quad Z = -Fz/l = F.$$

These are the only applied forces besides those due to gravity.

Verify that the third equation gives $-F = G\cos\theta = g$, nearly, where $g$ is the apparent force of gravity on unit mass.

Verify also that if $\omega$ be written for $n\sin\lambda$ and $G\sin\theta$ be neglected, the first two equations of motion are

$$x - 2\omega\dot{y} + \frac{g}{l}x = 0, \quad y + 2\omega\dot{x} + \frac{g}{l}y = 0.$$

Show that these equations are satisfied to terms involving $\omega^2$ by

$$x = a\cos mt\cos\omega t, \quad y = -a\cos mt\sin\omega t,$$

where $m^2 = g/l$. Hence, when $t=0$, $x=a$, $y=0$, and at time $t$,

$$r = (x^2+y^2)^{\frac{1}{2}} = a\cos mt, \quad \tan^{-1}\frac{y}{x} = -\omega t.$$

The plane of vibration therefore turns round, relatively to the axes $Ox$, $Oy$, in the direction opposed to the earth's rotation, with angular speed $\omega = n\sin\lambda$, an effect which is due to the turning of the earth under the pendulum.

This is the theory of Foucault's celebrated pendulum experiment for demonstrating the earth's rotation experimentally. After some preliminary trials it was carried out on a large scale at the Panthéon in Paris in 1851. The pendulum there consisted of a ball of lead weighing about 28 kilos., carried by a steel wire 67 metres long. Underneath the pendulum, with centre vertically below the point of support, was a circle of wood 6 metres in diameter, divided to fourths of a degree. Round part of this was placed a thin ridge of sand, which was cut through by a spike projecting below the bob, and gave a register of turning of the plane of vibration relative to the earth. A smaller concentric circle enabled the turning to be traced for a longer time, about five or six hours in all.

The period of turning at the latitude of the Panthéon is theoretically 31 h. 47 m. 14·6 s., and the pendulum appears to have shown a period of about 32 hours. The experiment was repeated successfully immediately after in the cathedrals of Rheims and Amiens, and at other places. Extreme care is necessary to make the suspension perfectly symmetrical [see *Travaux Scientifiques de Foucault*, Paris, 1878].

**28.** *Analogy of Foucault's pendulum to a gyrostatic pendulum.* Show that in a complete form the equations of motion of the pendulum (origin at $P_0$) are

$$\ddot{x} - 2\omega\dot{y} + \left(\frac{g}{l} - \omega^2\right)x = 0, \quad \ddot{y} + 2\omega\dot{x} + \left(\frac{g}{l} - \omega^2\right)y = 0,$$

and prove that they are satisfied by $x = a\cos mt$, $y = a\sin mt$, where $m$ is a root of

$$m^2 + 2\omega m - \frac{g}{l} + \omega^2 = 0.$$

Show that the motion of the bob is in a horizontal circle, in one case in the direction of the earth's rotation in period $2\pi/\{(g/l)^{\frac{1}{2}} - \omega\}$, in the other in the opposite direction in period $2\pi/\{(g/l)^{\frac{1}{2}} + \omega\}$.

These periods, it will be seen, exactly correspond to those of the circular vibrations of a gyrostatic pendulum. The rotation of the earth has thus the same effect on the apparent motion of the pendulum as the spin in the bob has on the actual motion. Thus again we have an analogy to the circular motion of an electron in a magnetic field [see 3, VIII and 9, IX above].

The reader may refer to Gray's *Dynamics* for the application of the equations of Example 26 to find the relative motion of a falling body and the deviations of a projectile.

**29.** *Revolving balance showing the earth's rotation.* The beam of a balance can turn about a horizontal axis through its centroid O, and is symmetrical about a longitudinal axis. The whole arrangement is made to turn at uniform angular speed $n$ about the vertical OZ. It is required to find the motion of its axis about the horizontal position, under the influence of the earth's rotation.

We suppose that the moment of inertia of the beam about the axis of symmetry is C, and about any transverse axis through O is A. If $\lambda$ be the latitude of the place, and $\omega$ the earth's angular speed, there are two components of $\omega$, one $\omega \sin \lambda$ about the upward vertical at the place, and the other $\omega \cos \lambda$ about a line drawn northward horizontally from O.

We take three axes of coordinates, one OC along the rod, inclined at an angle $\theta$ to the upward vertical OZ, and two OA, OB at right angles to OC, and turning with the system. Of the latter we take OB as coincident with the horizontal axis about which the beam turns. The reader may make a diagram. OC may be indicated sloping upward and away from the observer. OB is directed somewhat towards his right, and supposed to be horizontal, while OA is drawn so as to suggest that it is at right angles to the plane BOC, and therefore nearly vertical in the plane of the angle $\theta$, when $\theta$ is nearly 90°.

Let the vertical OZ have been drawn, and a horizontal line, OD towards the reader and in the meridian, indicated in the diagram. This line may be taken as drawn northward, so that DOA is an angle $nt$ in azimuth towards the west from north, if $t$ be reckoned from the instant at which OA coincides with OD. But OD is turning about the vertical with angular speed $\omega \sin \lambda$, if $\omega$ be the speed of rotation of the earth and $\lambda$ the latitude of the place. The earth's rotation gives also a component $\omega \cos \lambda$ about OD.

We find the A.M. about each of the axes *in the order* O(B, A, C). The angular speeds are respectively $\dot{\theta} + \omega \cos \lambda \cos nt$, $(n + \omega \sin \lambda) \sin \theta + \omega \cos \lambda \sin nt \cos \theta$,

$$(n + \omega \sin \lambda) \cos \theta - \omega \cos \lambda \sin nt \sin \theta.$$

The angular momenta about O(A, B, C) are the angular speeds multiplied by the corresponding moments of inertia A, A, C. If there is no applied couple the equation of motion for the axis OB is

$$A(\dot{\theta} - n\omega \cos \lambda \sin nt) - \tfrac{1}{2}(C - A)[\{(n + \omega \sin \lambda)^2 - \omega^2 \cos^2 \lambda \sin^2 nt\} \sin 2\theta + 2(n + \omega \sin \lambda)\omega \cos \lambda \sin nt \cos 2\theta] = 0. \quad \text{......(1)}$$

If we suppose, as we may, that $n$ is great in comparison with $\omega$, we may neglect the terms in $\omega^2$, and obtain

$$A(\ddot{\theta} - n\omega \cos \lambda \sin nt) - \tfrac{1}{2}(C - A)\{(n^2 + 2n\omega \sin \lambda) \sin 2\theta + 2n\omega \cos \lambda \sin nt \cos 2\theta\} = 0. \quad \text{...(2)}$$

So long as the beam is nearly horizontal $\sin 2\theta$ is very small and $\cos 2\theta$ is nearly $-1$. Then we have

$$A\ddot{\theta} - (2A - C)n\omega \cos \lambda \sin nt = 0. \quad \text{......(3)}$$

If the beam is long and slender C is negligible in comparison with A, and this equation becomes

$$\ddot{\theta} - 2n\omega \cos \lambda \sin nt = 0. \quad \text{......(4)}$$

Thus, when the beam is nearly horizontal, the angular speed with which it is turning about the axis OA is given by

$$\dot{\theta} = -2\omega \cos \lambda \cos nt. \quad \text{......(5)}$$

There is no constant of integration, since we reckon $t$ from the instant at which $\ddot{\theta}$ is zero.

We may now suppose that the turnings about the different axes are resisted by air-friction couples which are proportional to the angular speeds relative to the air. The angular speed about the vertical is $n + \omega \sin \lambda$, and is unvarying. The angular speed about OA relative to the air is $\dot{\theta}$, hence we assume that a couple of moment $k\dot{\theta}$ resists the variation of $\theta$. The equation of motion for a nearly horizontal position is now

$$A(\ddot{\theta} - 2n\omega \cos \lambda \sin nt) + k\dot{\theta} = 0, \quad \text{......(6)}$$

on the supposition of course that $k\dot{\theta}$ is great in comparison with the quantities neglected in arriving at equation (4), for example in comparison with $An^2 \sin 2\theta$, or with $A\omega^2$. For this, as a little consideration shows, the balance must be made small.

From (6) we get by integration

$$A(\dot{\theta}+2\omega \cos \lambda \cos nt)+k\theta=0, \quad \text{.....(7)}$$

or, neglecting $A\dot{\theta}$,

$$k\theta+2A\omega \cos \lambda \cos nt=0. \quad \text{.....(8)}$$

Thus $\theta$ is the simple harmonic function of the time given by (8).

Assuming the approximations, the whole matter could be put in a very brief common-sense way as follows. The axis OB is in advance of the meridian by the angle $nt$. Consider a perpendicular axis also horizontal, which is $\frac{1}{2}\pi - nt$ behind the meridian. About this the horizontal revolution of the earth's speed of rotation gives an angular speed $\omega \cos \lambda \sin nt$. This turns the nearly vertical axis OA, about which the A.M. is, approximately, $An$, so as to increase the angle between it and the instantaneous position of OB. Hence A.M. about OB is being produced at rate $-An\omega \cos \lambda \sin nt$. [The reader will see that the other turnings produce no effect on the whole.] But if no couple acts this must be balanced by the rate of change of A.M. due to acceleration. This is

$$A(\ddot{\theta}-n\omega \cos \lambda \cos nt),$$

since the angular speed about OA is $\dot{\theta}+\omega \cos \lambda \cos nt$. Hence we get as before, on the hypothesis of no couple,

$$\dot{\theta}-2n\omega \cos \lambda \sin nt=0.$$

The couple $-k\theta$ can now be introduced and the solution completed as above.

We have thus found that under the influence of the resisting action of the air, and the periodic disturbing forces due to the rotation, an oscillatory variation of $\theta$ is produced the period of which is $2\pi/n$ (the period of the imposed rotation), while the amplitude is $2A\omega \cos \lambda/k$, that is twice the result of dividing by $k$ (the damping coefficient) the product of the earth's horizontal component of the earth's angular velocity by the moment of inertia A.

This case of motion is interesting as an example of rotation, and as affording a new method by which it may be possible to observe the rotation of the earth. The method was suggested to its inventor Baron Eötvös, of Budapest, by variations found in records of measurements of gravity at sea, and due undoubtedly to effects of the velocity of the ship, according as it was directed east or west, on the apparent value of gravity.

The apparatus is described in a general way, without details of dimensions and masses, with some particulars of the method of experimenting, in a paper by D. Korda in the *Archives des sciences physiques et naturelles*, of Geneva, for Nov. 15, 1917. The theoretical amplitude quoted, without proof, by M. Korda agrees with that found above.

The question of the accuracy of this result has been raised [see an interesting paper by Mr. C. V. Boys in *Nature* for March 21, 1918, in which attention was first directed to the device], but there can be no doubt that, under the conditions stated, the theory given above is correct. The action of the device is gyrostatic, and is in no way dependent on the variation of apparent gravity with direction of motion.*

It is to be noticed however that a very slight varying couple is neglected, that due to the difference between the apparent gravity on the end moving west and that on the end moving east. But the balance was small and the period of revolution was about a minute, so that the action of this couple is entirely negligible in comparison with the effect taken account of in the equations found above. To bring the centrifugal force effect into play, it would be desirable to make the beam of the balance long—a thin rigid rod tipped with massive spheres would be best—and run it at as high a speed as possible. The theory of such an arrangement is not difficult, and the effect might easily have been included in the discussion given above, if it had been worth while.

* This has also been pointed out by Professor J. B. Dale, *Nature*, June 27, 1918.

# INDEX

*The references are to pages*

*The references are to pages*

*The references are to pages*

*The references are to pages*

*The references are to pages*

*The references are to pages*



GLASGOW: PRINTED AT THE UNIVERSITY PRESS BY ROBERT MACLEHOSE AND CO. LTD.

1059174R0

Printed in Great Britain by
Amazon.co.uk, Ltd.,
Marston Gate.